Marine Policy & Econom

Editor-in-Chief

John H. Steele

Marine Policy Center, Woods Hole Oceanographic Institution, Woods Hole,
Massachusetts, USA

Editors

Steve A. Thorpe

National Oceanography Centre, University of Southampton,
Southampton, UK
and
School of Ocean Sciences, Bangor University, Menai Bridge, Anglesey, UK

Karl K. Turekian

Yale University, Department of Geology and Geophysics, New Haven,
Connecticut, USA

Subject Area Volumes from the Second Edition

Climate & Oceans edited by Karl K. Turekian
Elements of Physical Oceanography edited by Steve A. Thorpe
Marine Biology edited by John H. Steele
Marine Chemistry & Geochemistry edited by Karl K. Turekian
Marine Ecological Processes edited by John H. Steele
Marine Geology & Geophysics edited by Karl K. Turekian
Marine Policy & Economics guest edited by Porter Hoagland, Marine Policy Center,
Woods Hole Oceanographic Institution, Woods Hole, Massachusetts
Measurement Techniques, Sensors & Platforms edited by Steve A. Thorpe
Ocean Currents edited by Steve A. Thorpe
The Coastal Ocean edited by Karl K. Turekian
The Upper Ocean edited by Steve A. Thorpe

MARINE POLICY & ECONOMICS

A DERIVATIVE OF ENCYCLOPEDIA OF OCEAN SCIENCES, 2ND EDITION

Editor

PORTER HOAGLAND

ELSEVIER

AMSTERDAM • BOSTON • HEIDELBERG • LONDON • NEW YORK • OXFORD
PARIS • SAN DIEGO • SAN FRANCISCO • SINGAPORE • SYDNEY • TOKYO
Academic Press is an imprint of Elsevier

ACADEMIC PRESS

Academic Press is an imprint of Elsevier
32 Jamestown Road, London NW1 7BY, UK
30 Corporate Drive, Suite 400, Burlington, MA 01803, USA
525 B Street, Suite 1900, San Diego, CA 92101-4495, USA

Notice
No responsibility is assumed by the publisher for any injury and/or damage to persons or property as a matter of products liability, negligence or otherwise, or from any use or operation of any methods, products, instructions or ideas contained in the material herein, Because of rapid advances in the medical sciences, in particular, independent verification of diagnoses and drug dosages should be made

British Library Cataloguing in Publication Data
A catalogue record for this book is available from the British Library

Library of Congress Cataloging-in-Publication Data
A catalog record for this book is available from the Library of Congress

ISBN: 978-0-08-096481-2

For information on all Academic Press publications
visit our website at www.elsevierdirect.com

Printed and bound by CPI Group (UK) Ltd, Croydon, CR0 4YY

Transferred to Digital Print 2012

Working together to grow
libraries in developing countries
www.elsevier.com | www.bookaid.org | www.sabre.org
ELSEVIER   BOOK AID   Sabre Foundation
              International

CONTENTS

[†]Deceased.

EXPLOITATION OF NATURAL AND CULTURAL RESOURCES

MARINE POLLUTION

[†]Deceased.

NATURAL HAZARDS

CLIMATE CHANGE AND IMPACTS

APPENDICES

INDEX

MARINE POLICY AND ECONOMICS: INTRODUCTION

Marine policy emerged as a recognizable field arguably as early as 1493 with the issuance by Pope Alexander VI of the *Inter Caetera*, a Papal Bull that eventually led to a Castilian monopoly over territorial and trade rights in lands to the west of a line 100 marine leagues west of the Azores. The close connection between science (exploration) and policy was evident even then, as the edict was issued quickly after the return of Columbus from his discoveries in the New World. A century later, Hugo Grotius, the distinguished Dutch jurist, responded to the policy by arguing in *Mare Liberum* (1609) for the international (common) nature of the seas. His purpose at that time was to break the Portugese and Spanish trade monopolies that resulted from *Inter Caetera*.

Grotius had based his argument on precedent found in Roman law. During the second century, Roman law asserted that the sea, mainly the Mediterranean, was common to all. Of modern day relevance, one reason for this policy was to provide the legal authority to control acts of piracy, which was seen as a risk, and therefore a cost, to merchant shipping. Roman law has as its legacy the modern doctrine of the "public trust," which assigns a priority for certain public uses of the coastal oceans, primarily navigation and fishing. Its legacy persists as well in areas well beyond national jurisdiction, known as the High Seas and the Area (the deep seabed). Although considerable attention now is being drawn to the problems of the over-harvest of high seas fisheries, international agreement to rationalize this use has been difficult to achieve.

The idea of the sea as a commons developed well before concerns over the exhaustibility of marine resources. By the 18th Century, Coastal States had begun to exert territorial jurisdiction both for military reasons and to regulate trade. The notion of a three-mile territorial sea was said to reflect the distance of a cannon shot from a coastal fortress, suggesting that the expanded jurisdiction could come only with the ability to enforce the claim. As marine resources came into increasing use, the need for fair and efficient institutions for deciding upon resource allocations and resolving conflicts became necessary. The application of theories and methods in economics and the other social sciences helps to outline policy approaches to this end. Modern marine policy reflects laws and policies the purpose of which is to regulate the use of ocean resources, encouraging both development and conservation.

Science and technological innovations have played a central role in the development of marine policy, as new resources are discovered and the dynamic physical characteristics of the ocean are revealed. Many classes of marine resources now exist, comprising fisheries; marine aquaculture; offshore hydrocarbons; hard minerals; wind, wave, current, and thermal energy recovery; biotechnology and pharmaceuticals; and submerged cultural resources. Science has helped characterize the transport and fate of pollutants, predict the occurrence of marine natural hazards, and perceive the signal of climate change in the chemistry and biota of the oceans. Scientific revelations, and the ongoing monitoring of ocean parameters, can be expected to lead to new institutions in marine policy. In this way, developments in marine policy both lag scientific developments and set constraints on human uses, including further scientific investigations and technological applications. Understanding this interplay so that policy effectiveness can be improved is a central problem in the field.

This volume samples many of the current issues in marine policy. It has an international flavor, but many of the articles draw upon national examples too. It begins with an overview of the academic field of marine policy and a general characterization of institutions, including international organizations with the responsibility for ocean science and management, the evolving law of the sea, and the conceptual basis for and implementation of large-scale area-based management ideas: large marine ecosystems, ocean zoning, and coastal zone management.

Areas of topical importance are addressed next. These include significant attention to theories and practice in the management of commercial fisheries. Commercial fisheries represent the classic example of the "tragedy of the commons" (where the commons is understood to be "open-access" or an area where regulation is lacking or feckless). Advances in both natural and social sciences hold the promise of reducing uncertainty in the distributions, age structures, and sizes of fish stocks and the application of methods of property rights to conserve and optimize yields. Where uncertainties persist, conservation measures such as marine protected areas (MPAs) may comprise a precautionary approach to fisheries management. Increasing amounts of the world's supply of seafoods are now supplied by marine aquaculture, and expansion of activity in this industry holds both the promise of relieving pressure on wild stocks and the threat of pollution.

As human uses become more widely distributed and more invasive, policies have emerged to protect species of special interest, including mammals, reptiles, and birds. Some of these species, especially the great whales, are recovering yet from the depredations of commercial harvests (occurring still in Scandinavia, Japan, and by indigenous peoples). These species and their habitats are thought to have a non-market value, in economic parlance, and many are now the focus of emerging marine eco-tourism industries worldwide. Protected species must swim in polluted oceans and bear the consequences of anthropogenic noise, and all face the looming prospect of a warmer, more acidic sea as the consequence of climate change.

The human uses of the ocean are manifold, and this volume includes chapters on uses both current and prospective. Marine pharmaceuticals, new biotechnologies, and alternative forms of renewable energy have captured the attention of many recently, but most of these uses may still be of the future. More mundane, but economically productive, uses include shipping, seaports, offshore oil and gas development, sand and gravel recovery, and the exploration of historic shipwrecks.

All human uses of the coasts and oceans have the potential for side-effects, termed social costs by economists. The almost ubiquitous absence of property rights in the ocean means that these costs go un-controlled, unless governments enact and enforce environmental laws to control polluting activities. His-torically, and even today, the ocean has provided considerable benefits as the receiver of human wastes, including everything from sewage sludge to radioactive waste to nitrogen. The sheer volume of wastes and the increasing uses of the oceans imply that social costs exist and are growing. Approaches to pollution control may include controls on quantities (output limits, technological mandates), prices (taxes), or com-binations of the two.

Storms and tsunamis are some of the most feared natural phenomena. Increasingly, human populations are migrating to the coasts, putting more and more people in harm's way. In combination with predictions of rising sea levels due to climate change, coastal properties and structures are at risk, and public health is threatened by flooding. Policy responses are fundamentally either to protect life and property or to retreat from the coast. Much attention has been directed recently at removing public assistance — such as subsidized flood insurance or government funding for beach nourishment — that may encourage humans to continue to be threatened by these hazards.

Finally, this volume includes several chapters on climate change and its impacts. After providing a broad overview of the phenomenon and the carbon dioxide cycle, chapters focus on direct and indirect effects on marine resources, including plankton, fisheries, and protected species. The economics of sea level rise are reviewed, and two large-scale proposed technological solutions to the problem of atmospheric carbon are suggested: sequestration via direct injection into the ocean and iron fertilization. Both solutions involve much uncertainty and potentially high costs, but these and others may need to be considered seriously as part of a multi-pronged approach to the problem of climate change.

This volume represents the work of a large number of natural and social scientists, government officials, and industry practitioners who have much to contribute to the debates over appropriate marine policy and the fair and efficient use and conservation of coastal and ocean resources. The wide diversity of talents and interests represented by these authors is a testament to the import and reach of the field of marine policy. Given the human propensity to live on the coast, and the technological developments that inexorably reduce the costs of travel, work, and leisure on and under the oceans, we can expect marine policy to continue to increase in both relevance and utility in the future.

Porter Hoagland
Editor

MARINE INSTITUTIONS AND POLICY

MARINE POLICY OVERVIEW

P. Hoagland, Woods Hole Oceanographic Institution, Woods Hole, MA, USA
P. C. Ticco, Massachusetts Maritime Academy, Buzzards Bay, MA, USA

Introduction

Marine policy is an academic field in which approaches from social science disciplines are applied to problems arising out of the human use of the oceans. Usually, human actions affecting ocean resources take place within an institutional context: laws establish a system of enforceable property rights, and goods and services are exchanged through markets. Most marine policy problems involve institutional imperfections or 'failures.' Governance failures include ill-defined property rights, the incomplete integration of the actions of public agencies operating under separate authorities, and wasteful 'rent seeking' on the part of stakeholders. Market imperfections include oil spills, nutrient runoffs leading to eutrophication in coastal seas, and over-exploitation of commercial fish stocks, among others. Even in the absence of technically defined institutional failures, problems may arise when decisions allocating marine resources are perceived to be unfair.

Most marine policy issues are subsets of broader policy areas. Some examples are presented in **Table 1**. Marine policy can be distinguished from these more general policy areas because legal property rights in the ocean often differ from those found on land. One reason for this difference is the relatively high cost of monitoring and enforcing private property rights in a remote and sometimes hostile environment. Other reasons include the fugitive nature of biological resources and the ease with which nutrients and pollutants are dispersed by currents and other physical processes.

The existence of these characteristics argues for collective action (i.e., the exercise of public authority) as a means of optimizing human uses and managing conflicts among users. The nature of collective action covers a spectrum from a centralized system of government 'command and control' to the implementation of decentralized 'market-based approaches.' The goal of marine policy analysis is to identify alternative courses of action for addressing a problem of ocean resource use and to inform public and private decision makers about the likely consequences. Consequences include physical, ecological, economic, and distributional (equity) effects.

Table 1 Some examples of overlaps of marine policy with general public policy areas

General policy area	Marine policy focuses
Environment	Ocean and climate change; macronutrient fluxes; eutrophication and hypoxia; treated and untreated sewage effluent; oil and hazardous material spills; industrial chemical and heavy metal effluents; thermal effluents from power plants
Natural resources	Commercial and recreational fisheries management; ocean minerals exploration and management; aquaculture regulation; conservation of protected species (mammals, birds, reptiles, fish, corals); ecosystem management; marine protected areas; conservation of biological diversity
Energy	Offshore oil and gas development; tidal power; ocean thermal energy conversion
Land use	Coastal zone management; planning; zoning uses; barrier beach protection
Waste management	Solid waste disposal; sewage sludge disposal; marine debris; nuclear waste disposal; incineration at sea
Transportation	Shipping and ports; underwater cables and pipelines; safety of life at sea; aids to navigation; international rights of passage; salvage; admiralty law
Defense	Zoned training and testing areas; atomic free zones; acoustic pollution; rights of passage for military vessels
Foreign policy	Legal geography; piracy; international trade in protected species; refugees; high seas fisheries; transboundary pollution
Emergency management	Weather prediction; hurricanes; coastal flood insurance; tsunamis; harmful algal blooms; search and rescue
Science policy	Funding for oceanographic research; technology transfer; basic versus applied research; large-scale science programs

In any particular situation, the universe of policy alternatives is constrained by the environmental characteristics of the ocean, the range of feasible technological responses, financial resources, and, sometimes, institutional frameworks and processes.

History of the Field

The emergence of marine policy as a distinct field of research dates back only about 40 years, coincident with rapid increases in ocean uses, the maturation of oceanography as a scientific field of study, and the rise of environmentalism. A number of journals specializing in public policy topics concerning the oceans, estuaries, and the coastal zone began publication in the early 1970s. Among these journals are: *Coastal Management, Marine Policy, Marine Policy Reports, Marine Resource Economics, Maritime Law and Commerce, Maritime Policy and Management, Ocean Development and International Law,* and *Ocean and Shoreline Management.* More recent additions to this list include: the *International Journal of Marine and Coastal Law* and the *Ocean and Coastal Law Journal.* Many marine policy problems predate this period, such as those relating to national security, international boundary determinations, resource exploitation, and shipping. In earlier periods, however, marine policy was not easily distinguishable from other, more general policy areas.

The negotiations on the third United Nations Convention on the Law of the Sea (UNCLOS) (1970–1982) may have spurred the development of the field, as many academic institutions in the West established programs in marine policy in the early 1970 s. For example, the Marine Policy Center (then the Marine Policy and Ocean Management program)

was established in 1972 at the Woods Hole Oceanographic Institution by Paul Fye, who was then director of the Institution. At that time, one main purpose of the Center was to follow and analyze the potential international regulation of marine scientific research, a focus of debate at the UNCLOS negotiations. Two institutions, the Law of the Sea Institute (1966 to present) and the Center for Ocean Law and Policy at the University of Virginia (1976 to present), have published on a continuous basis the proceedings of their annual meetings on international law of the sea issues. Since 1978, the International Ocean Institute, located jointly at the University of Malta and Dalhousie University, has published an annual *Ocean Yearbook* that features scholarly articles on marine policy topics, compiles descriptive statistics of ocean uses and legal geography, and summarizes the activities of marine policy research centers worldwide.

Social Science Disciplines

Marine policy is often described as a multidisciplinary field. Although academic degrees are issued in the United States and Europe in the field of marine policy or 'marine affairs,' progress in understanding marine policy problems typically occurs within the confines of more traditional social science disciplines. Alternative points of view may arise from the application of methods from different disciplines to a specific policy problem.

The social sciences are divided into a number of well-established disciplines. Some of these disciplines are listed in **Table 2**, along with examples of recent research topics to which they have been applied. Notably, considerable overlap may exist in the

Table 2 Social science disciplines and some examples of research foci

Discipline	Some example research foci
Cultural anthropology	Analysis of the effects of fisheries management on fishing communities; underwater archaeology research
Economics	Development of bioeconomic models of fisheries; estimating the net benefits of fisheries regulation; valuation of the nonmarket benefits of coastal and marine recreation; measurement of damages from marine pollution; evaluation of the net benefits of alternative policy instruments for controlling marine pollution
Geography	Mapping and analysis of demographic, resource, and economic data using geographic information systems
History	History of oceanography as a science; characterization of laws, social norms, and customs from earlier societies
Law	Analysis of legal institutions governing the use of marine resources; interpretation of common and statutory law with respect to ocean resource use
Philosophy	Identification and interpretation of the principles of environmental ethics as they apply to marine resource uses and conservation
Planning	Forecasting coastal and marine resource uses; demographic trends in the coastal zone; zoning the marine environment; marine protected areas; control of land use in the coastal zone
Political science	Analyzing common property institutions; characterizing the effectiveness of international environmental institutions; international regime formation
Sociology	Effects of fisheries management on fishing communities; importance of institutions in control of resource use

disciplinary coverage of certain topics, such as fisheries management.

Ocean Resources and Uses

The uses of the ocean for transportation, as a source of protein, and as a sink for wastes are among its oldest. In ancient times, the supply of ocean space and fish were thought to be virtually without limit. Modern humans have demonstrated that some uses of the ocean can preclude other uses, underscoring the existence of limits to the supply of space and resources, and giving rise to the potential for conflicts across uses. Since its modern development in the 1930s, oceanographic research has made significant strides in characterizing the distribution of ocean resources, although substantial uncertainties persist.

A broad range of ocean uses can be mapped into a small set of ocean resources. These resources include ocean space, living resources and their habitats, nonliving resources, and energy. **Table 3** lists the most prominent uses of the ocean along with a summary of typical marine policy issues that arise as a consequence of institutional imperfections.

Institutional Frameworks

Marine resources, their utilization, and ocean space are all managed through a myriad of legal instruments. These instruments exist at all levels of governance, including those policies directed at local or subnational concerns, and those designed to address issues of national, regional, or global importance. A seventeenth century *laissez-faire* concept of 'freedom of the seas' was based upon the premise that the ocean was infinite, its resources inexhaustible, its degradation impossible. These assumptions have proved to be both unrealistic and detrimental. It is now widely acknowledged that complete freedom of the seas would lead to resource waste and exploitation, economic inefficiency, and increased conflict among users.

Enclosure of Ocean Space

From a pragmatic perspective, the management of ocean space involves methods of enclosure. Theoretically, the enclosure of ocean space can be derived from both national and international management regimes. In practice, it has been accomplished through the seaward extension of national jurisdictions by establishing zones of authority and use (e.g., the territorial sea and exclusive economic zone; *see* Law of the Sea). The primary thrust has been toward the expansion of sovereignty over ocean space previously considered open-access. Although large-scale ocean enclosures have led to reductions in international conflicts over resource use within the proscribed enclosure, such conflicts continue to persist among domestic users and over resources (e.g., straddling fish stocks) that transgress enclosure boundaries.

Global Institutions

International cooperation to address marine and coastal concerns has been codified through several formal commitments. In international affairs, this institutionalization usually takes the form of a treaty or customary practice, although certain important intergovernmental organizations also exist. On the

Table 3 Ocean uses and some leading policy issues

Use	Some leading policy issues
Commercial fishing	Overharvesting due to inappropriate management measures; overcapacity due to government subsidization; shifts to fishing lower trophic levels; impacts on habitat, species diversity, ecological functions, protected species; loss of gear; human safety risks
Recreational fishing	Overharvesting due to inappropriate management measures
Aquaculture	Macronutrient pollution; spread of disease; escaped fish; interactions with protected species; loss of gear
Shipping	Cabotage laws; cartelization; infrastructure investments, including harbor dredging; piracy; oil and hazardous material spills; marine debris; transport of invasive species; interactions with protected species; acoustic pollution; safety of life at sea
Channel dredging	Disposal of contaminated material; government subsidization
Ocean dumping	Radioactive waste disposal; chemical waste disposal; transport of pollutants from disposal sites
Minerals	Oil spills; benthic disturbances; habitat impacts; acoustic pollution; commercial and recreational fishery impacts
Recreation	Loss of ecosystems and habitat to other uses; impacts of global climate change; impacts of recreation on protected species, coral reefs; recreational boating safety
Defense	Weapons tests; acoustic tests; runoff of pollutants from military sites; oil and hazardous waste spills; marine debris
Coastal development	Erosion; industrial runoff; habitat loss; limits to public access
Agriculture	Macronutrient and pesticide runoffs; hypoxia; hypothesized links to harmful algal blooms

global level, both broadly based and issue-specific treaties that affect a majority of national interests have been developed. **Table 4** describes some prominent examples of these agreements.

The proclivity of most States to cooperate in world affairs also extends to regional arrangements. Many coastal and ocean resources transcend political boundaries and thus do not conform to jurisdictional constraints. Therefore, several regional agreements address concerns that extend beyond national jurisdictions to the interests of neighboring states. **Table 5** provides several illustrations of regional institutional governance.

National Institutions

Virtually all coastal nations have enacted domestic marine policies and laws to legitimize their claims to ocean resources and space. Despite the inefficiencies of fragmented policy administrations and a general lack of public input and future planning, the resulting governance regimes have brought order to the management of various ocean uses. These legislative actions have often been taken as a reaction to real or perceived threats to the health of the ocean or the overexploitation of resources. Often, these laws are designed to work in conjunction with regional and international treaties, but sometimes they do not. The US Marine Mammal Protection Act is one example of a national institution that aims to conserve specific marine resources but which has come into sharp conflict with international trade law.

Institutional Integration

In general, most laws governing the use of ocean space and resources are sectoral and issue-specific. Examples include legislation pertaining to fisheries management, offshore oil and gas development, and coastal mineral extraction. The primary concern is that as ocean space, particularly coastal waters, becomes the subject of increasingly intense and diversified uses, the activities of one user group will frequently affect the interests of others. A goal of institutional integration is to discover ways in which all uses can be optimized or, at least, coexist without rancor. The hope is that integration can reduce conflicts between uses. The integration of marine policies, sometimes through the implementation of so-called multiple-use management regimes, works to eliminate the inefficiency of single-sector regulatory schemes and is believed to mirror more closely the dynamic complexity of the ocean system. For example, some marine protected areas exhibiting high degrees of marine

Table 4 Prominent global agreements and organizations for ocean management

Year	Institution	Description
1992	UNCED (United Nations Conference on Environment and Development)	International 'soft law' that helped to set the context for several international agreements targeting the interdependence of global environmental protection, sustainable development, and social equity. Most prominent for ocean management was Chapter 17 of Agenda 21 that stresses both the importance of oceans and coasts in the global life support system and the positive opportunity for sustainable development that ocean and coastal areas represent
1982	UNCLOS (United Nations Convention on the Law of the Sea)	An overarching framework convention that provides both a foundation for global ocean law, and a means for individual States to direct specific coastal and marine activities
1973	MARPOL (International Convention for the Prevention of Pollution from Ships)	The first comprehensive global convention that prevents or limits the type and amount of vessel-source pollution including oil, garbage, noxious liquid substances, sewage and plastics.
1972	London Convention	Established the first global standards to govern the dumping of wastes into the ocean, including specific mandates as to what materials may be legally dumped through a permit system.
1971	Ramsar Convention	Requires national initiatives by each signatory to conserve wetlands as regulators of water regimes and as habitats of distinctive ecosystems of global importance
1958	IMO (International Maritime Organization)	Facilitates international cooperation on matters of safety and environmental protection in maritime navigation and shipping. Its principal environmental responsibilities are to prevent marine oil pollution, provide remedies when prevention fails, and to assist the development of jurisdictional powers to prescribe and enforce pollution control standards through intergovernmental cooperation
1946	IWC (International Whaling Commission)	Regulates, but does not preclude, the global sustainable taking of whales through a system of quotas designed to prevent their overexploitation and possible extinction. Various management procedures and moratoria (including stout opposition to the moratoria by some commercial whaling nations) have provided an institutional framework but not a cessation of stock depletion

Table 5 Important regional institutions for coastal and ocean management

Institution	Description
UNEP (United Nations Environment Program) Oceans and Coastal Areas Program	Designed to address coastal and marine environmental problems (e.g., marine pollution, fisheries conservation and development, species protection) and socioeconomic issues such as tourism common to those nations that share a communal body of water. At present, over 100 hundred States participate in twelve regional oceans and coastal area programs
Large Marine Ecosystems (LME)	This concept has been proposed but it does not presently constitute a legal institution. An LME is a large region of ocean space, generally over 200 000 km^2 (77 000 square miles) and situated typically within exclusive economic zones, that have unique bathymetry, hydrology, and productivity and encompass a regional functional ecological unit. Managing this comprehensive ecosystem for both the protection of biological diversity and sustainable uses requires broad regional cooperation between States
Man and the Biosphere Program	The United Nations Educational, Scientific, and Cultural Organization's (UNESCO) Man and the Biosphere Program is an international program of concerted scientific cooperation among countries directed towards finding practical solutions to environmental problems. A major function is the establishment of protected areas (including several marine and coastal reserves) of ecological significance
Cartagena Convention	The Cartegena Convention addresses the myriad of environmental concerns (including marine oil spills) associated with the cultural, economic and political differences exhibited throughout the wider Caribbean. As a supplement, the protocol establishes protected areas to conserve and maintain species and ecosystems, and promotes the sustainable management and use of flora and fauna to prevent their endangerment
ICCAT (International Convention for the Conservation of Atlantic Tunas)	Primary goal is the conservation of tuna-like fishes and billfishes throughout the Atlantic Ocean and adjacent seas. Member nations must conduct most research, carry out analyses, and enforce ICCAT recommendations for their own nationals
Antarctic Treaty System	Composed of the 1972 Convention on the Conservation of Antarctic Seals; the 1980 Convention on the Conservation of Antarctic Marine Living Resources (CCAMLR); the 1988 Convention on the Regulation of Antarctic Mineral Resource Activities (CRAMRA); and the 1991 Protocol on Environmental Protection; the Antarctic Treaty System (ATS) seeks to bring institutional order to the activities of those States claiming sovereignty over Antarctic territory, or those interested in resource exploitation
The Great Lakes Program	A comprehensive management regime for the protection and management of the Great Lakes established through the cooperation of the federal governments of the United States and Canada, eight US states, the Canadian province of Ontario, and many local and regional organizations. Targeted issues include nonpoint source pollution, water levels, navigation, recreational activities, and fishing.

biological diversity are zoned also for human uses such as tourism within a multiple-use management system.

The primary purposes of integration are to ensure that links among issues are not neglected in the creation and implementation of public policies and to internalize the external costs that normally accompany the misuse of open-access resources. Integration also emphasizes responsiveness to the legitimate needs of current users while exercising stewardship responsibility on behalf of future generations. Unfortunately, a number of obstacles to the integration of marine policies remain, including incomplete scientific information, boundary disputes, lack of political will, fractionalization of government efforts, and the existence of short-term ocean management programs that may not be optimal for solving persistent problems.

Integrated Coastal Zone Management

A prime illustration of the movement toward policy integration is found in the management of the coastal zone and its resources. Integrated coastal zone management (ICZM) is a process that attempts to resolve coastal conflicts, promote the sustainability of resources, and enhance economic benefits to coastal communities. Despite some reservations as to the practicality of the concept, ICZM is designed to overcome the traditional sectoral approach to managing coastal uses by accommodating all sectors within the context of a larger planning scheme. Management tools including zoning, special area planning, land acquisition and mitigation, easements, and coastal permitting are employed to implement an ICZM program. Evolving ICZM efforts are ongoing in such diverse nations as the United Kingdom, Thailand, South Korea, and Tanzania.

Analytical Approaches

Approaches to the analysis of marine policy issues are diverse, ranging from highly quantitative models to qualitative and descriptive techniques. Whether mathematical or descriptive, these approaches are unified by the presentation of policy options and the comparison, using disciplinary criteria, of alternative courses of action. Economic and political science models tend to be more quantitative, whereas models from other social science disciplines tend to be less mathematically oriented. All social science applications to marine policy problems may test hypotheses, employing rigorous statistical methods for the analysis of empirical data. These methods include the standard regression techniques as well as modern nonparametric, time series, and limited dependent variable techniques. Different analytical approaches in marine policy can be complementary, and they are commonly informed by oceanographic research findings and theory.

Economic Analysis

The economic theory focusing on the management of marine resources provides the most common example of the application of a quantitative approach. Neoclassical economics emphasizes the selection of a course of action that optimizes the welfare of society through the supply and consumption of goods and services. In the marine environment, natural resources, such as fish, marine mammals, coral reefs, or entire ecosystems, represent these goods and services, and the dynamics of the ecosystem, including its response to human exploitations, provides a natural constraint to welfare optimization.

A basic model, developed in the 1950s, seeks to maximize welfare in the form of producer surplus (profits, broadly defined) in a fishing fleet of identical vessels from the harvest of a single fish stock. Numerous extensions of the basic model include the addition of other ecologically related species, the incorporation of uncertainty, the investigation of nonlinear dynamics, the consideration of a non-uniform distribution of fish stocks, the analysis of consumer surplus, the introduction of competing fleets or nations (game theory), and so forth. The model has become an important tool in the analysis of the economic and biological effects of the implementation of conservation and management measures in a fishery, such as marine reserves or individual transferable quotas. Given significant uncertainty and lags in the response of the marine ecosystem to human perturbations, fisheries economists and scientists now think in terms of managing a fishery adaptively by observing how the system responds to variations in the level of fishing pressure.

The economic optimization model has been utilized most commonly in the analysis of fishery management questions. Other, related applications include those in the areas of marine pollution, the environmental risks of offshore oil and gas development, shipping infrastructure, ocean dumping, and marine aquaculture.

Economic analysis is also directed at estimating the willingness of members of society to pay for goods and services that are not traded on established markets. Several approaches have been used to value these so-called 'nonmarket' commodities, some of which have generated considerable controversy. Nonmarket goods for which demand has been estimated include beach visits, water quality, marine mammals, marine protected areas, and coral reefs, among others. The purpose of estimating nonmarket values is to allow a comparison in common units of the economic values of market and nonmarket commodities when deciding on the net benefits of alternative courses of action.

Organizational Studies

Social organization and cultural norms are institutional forms that may shape the feasible set of policy alternatives for any particular marine policy issue. Researchers in disciplines such as geography, sociology, history, and cultural anthropology, among others, focus their research efforts on broad- and fine-scale characterizations and mappings of social organization. Their studies include understanding the development of resource-based communities and enclaves and the ways in which coastal and marine resources are used and conserved. Through induction, empirical studies lead to theories of the natural emergence of organizational principles for the management of marine resources, including collective choice arrangements, enclosures, property right definition and enforcement, and modes of conflict resolution. One such theory that appears in the fisheries context involves the concept of co-management, through which management responsibilities and functions are shared, according to specified rules, between the owners of the resource, or their agents, and those who are involved in its exploitation.

Legal Studies

During the last 30 years, the body of law governing the human uses of the ocean has expanded and diversified at a rapid pace. At both national and international levels, virtually all uses of the sea are now regulated in some fashion. Ocean law is a dynamic institution that responds to changing ecological parameters, economic conditions, and technological and scientific advances. Legal analysts track the changing nature of the law, interpret the way in which legal institutions affect the allocation of marine resources, and characterize the actual and potential impacts of these institutions on human behavior.

One could easily argue that the courts, legislative bodies, and executive agencies with responsibility for ocean management rely on legal analysis to a much larger extent than other types of marine policy analysis. Methods of legal analysis can be characterized generally as descriptive and interpretive, relying upon: the practice of nations; the content of treaties, statutes, and rules; the interpretations of courts; and uncodified societal norms. Legal analysis may be further characterized as subjective, in the nature of advocating a particular policy to benefit the interests of one or more agencies or stakeholders.

Institutional Effectiveness

In the field of international political relations and in domestic policy reviews, analysts attempt to understand the extent to which an institution is effective at attaining agreed-upon goals. For example, the degree to which an institution, such as an international agreement to control land-based marine pollution, is effective at improving the quality of the marine environment would be based upon observed changes in environmental quality measures over time. In contrast with economic analysis, studies of institutional effectiveness put forth no normative standards, such as the optimization of social welfare. The goals are determined by the participants (stakeholders, national legislatures, legations) who establish the institution. If the goal is attained, then, holding constant other motivations, such as political power, changes in economic conditions, or external influences, it is assumed that the institution has been effective in motivating its participants to take action.

Lesson-Drawing

Another useful analytical approach is known as 'lesson-drawing.' As a form of comparative political analysis, lesson-drawing focuses on the set of circumstances through which marine policies observed to be effective in one jurisdiction are potentially transferable to another. Confronted with a common problem or consistent behaviors, policy makers may be able to learn from how their counterparts elsewhere respond, and conclude that the implementation of policies in other places may be of use in their own circumstances. Lesson-drawing is particularly useful in nations that share some commonalities such as resource availability or cultural norms. The methodology involves an initial search for similar contexts and policies in other jurisdictions, the

development of a conceptual model of the application of the policy, a comparison of practices across jurisdictions, and a prediction or forecast of success after the lesson has been drawn and the policy approach adopted.

Notably, the search for and discovery of lessons does not imply that there must be a common application. Realistically, one cannot expect that policies can be successfully transferred without considering the idiosyncratic characteristics of jurisdictions that may allow the policy to be effective in one place but not in another. For example, in the case of preserving marine biodiversity by zoning, there is no generic type of marine protected area that is capable of meeting every situation. The nature of a reserve, its design, and its regulatory framework all depend on the primary objectives it seeks to achieve. These identified objectives will influence the size, shape, and other design constraints of the protected area, and its implementation.

Future Prospects

Marine policy will continue to grow in importance as human populations place increasing pressure on coastal space, ocean resources, and marine ecosystems. These pressures, driven by such forces as population growth, human migration to coastal areas, and expanding demand for both living and nonliving resources, will disrupt ecosystems, lead to genetic losses, and exacerbate user conflicts. As many of these problems involve institutional failures, in the future, historical customs and institutions will need to be re-examined. Solutions involving the establishment of new (or clarification of existing) property rights and their enforcement, utilizing technologies that lower the costs of monitoring and enforcing such rights, will undoubtedly come to the fore.

Policy choices affecting the allocation of ocean resources lead to questions of effectiveness, or the ability of institutions to meet agreed-upon goals. Despite the steady advance of marine science and technology, policy makers must face choices across options with a high degree of uncertainty. In the face of uncertainty, policy analyses can be neither comprehensive nor fully conclusive, leading policy makers to turn increasingly toward precautionary approaches. Substantial alterations to the current institutional framework supporting coastal and ocean activities are necessitated by the shift to a precautionary approach, including a movement away from sectoral management and toward the greater integration of policies.

See also

Further Reading

Anderson LG (1986) The Economics of Fisheries Management. Baltimore: Johns Hopkins University Press.

Armstrong JM and Ryner PC (1981) Ocean Management: A New Perspective. Ann Arbor, MI: Ann Arbor Science.

Broadus JM and Vartonov RV (eds.) (1994) The Oceans and Environmental Security. Washington: Island Press.

Caldwell LK (1996) International Environmental Policy. Durham, NC: Duke University Press.

Calvert P (1993) An Introduction to Comparative Politics. Hertfordshire, UK: Harvester Wheatsheaf.

Cicin-Sain B and Knecht RW (1998) Integrated Coastal and Ocean Management. Washington: Island Press.

Clark CW (1985) Bioeconomic Modeling and Fisheries Management. New York: Wiley-Interscience.

Eckert RD (1979) The Enclosure of Ocean Resources: Economics and the Law of the Sea. Stanford, CA: Hoover Institution Press, Stanford University.

Ellen E (1989) Piracy at Sea. Paris: ICC International Maritime Bureau.

Farrow RS (1990) Managing the Outer Continental Shelf Lands: Oceans of Controversy. New York: Taylor Francis.

Freeman AM (1996) The benefits of water quality improvements for marine recreation: a review of the empirical evidence. Marine Resource Economics 10: 385–406.

Friedheim RL (1993) Negotiating the New Ocean Regime. Columbia, SC: University of South Carolina Press.

Gould RA (2000) Archaeology and the Social History of Ships. Cambridge: Cambridge University Press.

Hsü KJ and Thiede J (eds.) (1992) Use and Misuse of the Seafloor. New York: John Wiley.

International Maritime (1991) The London Dumping Convention: The First Decade and Beyond. London: International Maritime Organization.

Ketchum BH (1972) The Water's Edge: Critical Problems of the Coastal Zone. Cambridge, MA: MIT Press.

Mahan AT (1890) The Influence of Sea Power Upon History: 1660–1783. Newport, RI: US Naval War College.

Mangone GJ (1988) Marine Policy for America, 2nd edn. New York: Taylor Francis

Mead WJ, Moseidjord A, Muraoka DD, and Sorenson PE (1985) Offshore Lands: Oil and Gas Leasing and Conservation on the Outer Continental Shelf. San Francisco: Pacific Institute for Public Policy Research.

Miles EL (ed.) (1989) Management of World Fisheries: Implications of Extended Coastal State Jurisdiction. Seattle: University of Washington Press.

Mitchell RB (1994) International Oil Pollution at Sea: Environmental Policy and Treaty Compliance. Cambridge, MA: MIT Press.

Norse EA (ed.) (1993) Global Marine Biological Diversity. Washington: Island Press.

Ostrom E (1990) Governing the Commons: The Evolution of Institutions for Collective Action. New York: Cambridge University Press.

Richardson JQ (1985) Managing the Ocean: Resources, Research, Law. Mt Airy, MD: Lomond Publications.

Rose R (1991) What is Lesson-Drawing? Journal of Public Policy, 3–30.

Sherman K, Alexander LM, and Gold BD (eds.) (1990) Large Marine Ecosystems: Patterns, Processes, and Yields. Washington: American Association for the Advancement of Science.

Underdal A (1980) Integrated Marine Policy – What? Why? How? Marine Policy 4(3): 159–169.

Vallega A (1992) Sea Management: A Theoretical Approach. New York: Elsevier Applied Science.

Wenk E Jr (1972) The Politics of the Ocean. Seattle, WA: University of Washington Press.

INTERNATIONAL ORGANIZATIONS

M. R. Reeve, National Science Foundation, Arlington VA, USA

Introduction

This article is limited to a description of the most important organizations related to Ocean Sciences, rather than exhaustively cataloging perhaps hundreds of entities which may cross one or more national borders. I have classified the major international organizations into governmental (or more accurately intergovernmental) and nongovernmental organizations, and within those groupings, global and regional organizations. With the proliferation of the world wide web, much information can be readily accessed from each organization's web page, some of which I have edited and utilized.

Intergovernmental Organizations (Global)

The overarching global intergovernmental organization is the United Nations (UN). Within it, the Intergovernmental Oceanographic Commission (IOC) has the main responsibility for the coastal and deep oceans. Organizationally, the IOC is a component commission of the United Nations Educational, Scientific and Cultural Organization (UNESCO). The World Meteorological Organization (WMO), a UN specialized agency, has strong interactions with the ocean interests, by virtue of the coupling of weather and climate with the circulation of the oceans and its other properties. The Fisheries and Agricultural Organization of the UN also has strong ties to the oceans through its work in marine fisheries.

Intergovernmental Oceanographic Commission (IOC)

The work of the IOC, founded in 1960, has focused on promoting marine scientific investigations and related ocean services, with a view to learning more about the nature and resources of the oceans. The IOC focuses on four major themes: (1) facilitation of international oceanographic research programs; (2) establishment and coordination of an operational global ocean observing system; (3) education and training programs and technical assistance; and (4) ensuring that ocean data and information are made widely available.

The IOC is currently composed of 126 Member States, an Assembly, an Executive Council and a Secretariat. The Secretariat is based in Paris, France. Additionally the IOC has a number of Subsidiary Bodies. Each Member State has one seat in the Assembly, which meets once every two years. The Assembly is the principal organ of the Commission which makes all decisions to accomplish the objectives of the IOC. The Secretariat is the executive arm of the organization. It is headed by an Executive Secretary who is elected by the Assembly and appointed by the Director-General of UNESCO. Countries contribute dues to support the work of the IOC.

Scientific/technical subsidiary bodies of IOC include Ocean Science In Relation To Living Resources, Ocean Science In Relation to Non-Living Resources, Ocean Mapping (OM), Marine Pollution Research and Monitoring, Integrated Global Ocean Services System, Global Ocean Observing System, and International Oceanographic Data And Information Exchange. There are various subprograms attached to most of these including the Global Coral Reef Monitoring Network.

IOC also has regional subsidiary bodies with responsibilities for carrying out region-specific programs voted by the Assembly. These are the Subcommission for the Caribbean and Adjacent Regions, Regional Committee for the Southern Ocean, Regional Committee for the Western Pacific, Regional Committee for the Cooperative Investigation in the North and Central Western Indian Ocean, Regional Committee for the Central Indian Ocean, Regional Committee for the Central Eastern Atlantic and Regional Committee for the Black Sea.

The work of the Secretariat is accomplished by a small permanent staff and a larger number of scientists usually funded by institutions or agencies in their own countries, who have interest in a specific aspect of the activities of the IOC. The funds provided to do this are in addition to country dues. This mechanism to focus staff support on such areas of interest is good, and helps to counterbalance the opposite tendency of the General Assembly, meeting every two years, to direct the General Secretary to undertake new activities. Because such activities must usually be funded out of the existing budget, the tendency inevitably dilutes existing activities, sometimes to a subcritical level.

World Meteorological Organization (WMO)

The World Climate Program (WCP) is the main activity of WMO related to the oceans. Established in 1979, the WCP includes the World Climate Research Program (WCRP), of which the Global Climate Observing System, encompassing all components of the climate system, atmosphere, biosphere, cryosphere, and oceans, is a component. WMO and its WCRP are treated fully elsewhere and only a brief summary is provided here, relating to oceans.

The World Meteorological Convention, by which the World Meteorological Organization was created, was adopted in 1947, and in 1951 was established as a specialized agency of the United Nations. There at least 185 member countries and territories. The World Meteorological Congress, which is the supreme body of WMO, meets every four years. In order to assess available information on the science, impacts and the cross-cutting economic and other issues related to climate change, in particular possible global warming induced by human activities, WMO and the United Nations Environment Program (UNEP) established the Intergovernmental Panel on Climate Change in 1988.

In the early years of their development, WCRP incorporated the Tropical Ocean – Global Atmosphere Study initiated by the Scientific Committee on Oceanic Research (SCOR) (see below), and the World Ocean Circulation Experiment (WOCE). The former was successfully completed and the latter is in its later stages of analysis and synthesis, the major field programs including the WOCE Hydrographic Survey, having been completed. Both programs have been very successful, to the point that an offspring is beginning to be implemented – the Climate Variability and Predictability Study. This 15-year program to observe the atmosphere and oceans will incorporate new technologies developed as a result of the predecessor programs, and rapidly increasing sophistication of computer hardware and software to develop new coupled models. Past climates will also be reconstructed.

Related to this effort is the Global Ocean Data Assimilation Experiment which is planned for the first half of this decade. This program was a product of the joint IOC/WCRP Ocean Observations Panel for Climate.

In 1999 the governing bodies of WMO and IOC agreed to set up a Joint Technical Commission for Oceanography and Marine Meteorology to act as a coordinating mechanism for the full range of WMO and IOC existing and future operational marine program activities, including the coordinating and managing the implementation of an operational Global Ocean Observing System.

The World Bank

The Global Environmental Facility (GEF) of the World Bank was launched in 1991 as an experimental facility and evolved 'to serve the environmental interests of people in all parts of the world'. By 2000, more than 36 nations pledged $2.75 billion in support of GEF's mission to protect the global environment and promote sustainable development. This has been complemented by $5 billion in co-financing from GEF partners, which include the UN Development Program and the UN Environment Program, as well as host countries. GEF funds projects in four areas: biodiversity, climate change, international waters, and ozone. Up to 1999 the GEF had allocated over $155 million to international waters initiatives. The term 'international' refers to fresh as well as ocean waters, and the projects seek to reverse the degradation of bodies of water controlled by a mosaic of regional and international water agreements.

A list of currently funded projects is available from the GEF web site. Examples include Black Sea Environmental Management, Gulf of Guinea Large Marine Ecosystem and various coral reef rehabilitation and projection projects. The GEF Secretariat is located within the World Bank in Washington DC, USA.

Intergovernmental Organizations (Regional)

Besides regional activities of global intergovernmental organizations referred to above, there are two main regional intergovernmental organizations. The first, the International Council for the Exploration of the Sea (ICES) (for the North Atlantic), and the second is the North Pacific Marine Science Organization.

International Council for the Exploration of the Sea (ICES)

Oceanographic investigations form an integral part of the ICES program of multidisciplinary work aimed at understanding the features and dynamics of water masses and their ecological processes. In many instances emphasis is placed on the influence of changes in hydrography (e.g., temperature and salinity) and current flow on the distribution, abundance, and population dynamics of finfish and shellfish stocks. These investigations are also relevant to marine pollution studies because physical oceanographic conditions affect the distribution and transport of contaminants in the marine environment. ICES promotes the development and calibration of oceanographic equipment and the maintenance of appropriate standards of quality and comparability of oceanographic data.

ICES is the oldest intergovernmental organization in the world concerned with marine and fisheries science. Since its establishment in Copenhagen in 1902, ICES has been a leading scientific forum for the exchange of information and ideas on the sea and its living resources, and for the promotion and coordination of marine research by scientists within its member countries. Each year, ICES holds more than 100 meetings of its various working groups, study groups, workshops, and committees. These activities culminate each September when ICES holds its Annual Science Conference, which attracts 500–1000 government, academic, and other participants. Proceedings of these meetings, and other related activities are published by ICES.

Membership has increased from the original eight countries in 1902 to the present 19 countries which come from both sides of the Atlantic and include Canada, the USA and all European coastal states except the Mediterranean countries from Italy eastward. Each country has a vote in the governance, through two delegates, all of whom come together annually at the 'statutory meeting' held at the time of the science conference. The Council elects a President at three year intervals from its members, as well as a small executive group (the Bureau) to conduct business intercessionally.

The ICES Secretariat in Denmark maintains three databanks, the Oceanographic databank, the Fisheries databank, and the Environmental (marine contaminants) databank. Since the 1970s, a major area of ICES work as an intergovernmental marine science organization has been to provide information and advice to member country governments and international regulatory commissions (including the European Commission) for the protection of the marine environment and for fisheries conservation. This advice is peer-reviewed by the Advisory Committee on Fishery Management and the Advisory Committee on the Marine Environment before being passed on.

The structure and operation of ICES has continuously evolved, to meet current needs of the advice-seeking member nations and the oceanographic community. In earlier years it focused mostly on the relationship of oceanography to fisheries, but over the past 15 years, the need has arisen increasingly for advice over the broad range of environmental issues from marine contaminants to effects of fishing activities on the environment, particularly the seafloor ecosystems. The annual meeting, which was traditionally a business occasion where standing committees reviewed the activities of working groups and passed recommendations on to the Council, has evolved into a major north Atlantic science meeting on oceanography and its application to regional societal problems.

ICES has tried to capture its past century of achievements through special lectures, a History Symposium and a soon to be published written history volume. Its future is mapped through the development of its first strategic plan, and ongoing organizational evolution.

North Pacific Marine Science Organization (PICES)

PICES held its first Annual Meeting in October 1992, in Victoria, British Columbia. From the beginning, the PICES approach has been multidisciplinary, with standing committees concerned with biological oceanography, fishery science, physical oceanography and climate, and marine environmental quality. There has been growing interaction among these specialties, with joint scientific sessions, interdisciplinary symposia, and a broad study of climate change and carrying capacity (the CCCC program) in the region. Most recently, PICES has taken the lead in joining forces with other international organizations to organize an intersessional Conference under the title of El Niño and Beyond (March 2000). Although PICES is an infant compared with its prototype, ICES, it has already become a major focus for international cooperation in marine science in the northern North Pacific.

PICES is an intergovernmental scientific organization. Its present members are Canada, People's Republic of China, Japan, Republic of Korea, Russian Federation, and the USA. The purposes of the organization are: to promote and coordinate marine research in the northern North Pacific and adjacent seas especially northward of 30°N; advance scientific knowledge about the ocean environment, global weather and climate change, living resources and their ecosystems, and the impacts of human activities; and promote the collection and rapid exchange of scientific information on these issues.

PICES annual meetings, symposia and workshops provide fora at which marine scientists interested in the North Pacific can exchange latest results, data, and ideas and plan joint research. These meetings have been effective in stimulating and accumulating the interest of the scientific community in member countries to coordinate marine science on the basin scale. PICES has been successful at bringing oceanographers from a number of Pacific Rim countries and disciplines (physical, chemical, and biological) together talking and working with fisheries scientists to understand and eventually forecasts temporal variability of ocean ecosystems, and their key species. It has also encouraged and facilitated collaborative marine scientific research in the North Pacific. International collaboration in this region is essential since the open North Pacific is too large for any one country to study adequately on its own, and oceanic circulation and biological species do not recognize international boundaries. Comparisons of similar species and ocean environments on the eastern and western sides of the Pacific will be much more revealing about the large-scale processes affecting ecosystem and fish dynamics than isolated studies in national waters alone.

PICES was established, in large measure, by the tireless efforts of Warren Wooster, of the University of Washington, USA. He had long been active in ICES, including as its President, and had become convinced that there was a great need for a similar organization focused on the North Pacific.

Other Regional Commissions

Throughout the world, governments have formed regional organizations to protect the environment and regulate activities. Two European examples of these are the Helsinki Commission, otherwise known as the Baltic Marine Environment Protection Commission, and OSPAR Commission for the Protection of the Marine environment of the north-east Atlantic. Their members comprise the countries bordering these specific marine environments. There are also several such commissions organized to protect and regulate specific fisheries and/or fishery regions (e.g., the North Atlantic Salmon Commission), listings and explanations of which go beyond the scope of this article.

Nongovernmental Organizations (Global)

International Council for Science (ICSU)

Formerly known as the International Council of Scientific Unions, ICSU is a nongovernmental organization, founded in 1931 to bring together natural scientists in international scientific endeavor. It comprises 95 multidisciplinary National Scientific Members (scientific research councils or science academies) and 25 international, single-discipline Scientific Unions to provide a wide spectrum of

scientific expertise enabling members to address major international, interdisciplinary issues which none could handle alone. ICSU also has 28 Scientific Associates.

The Council seeks to break the barriers of specialization by initiating and coordinating major international interdisciplinary programs and by creating interdisciplinary bodies which undertake activities and research programs of interest to several members. It acts as a focus for the exchange of ideas and information and the development of standards. Hundreds of congresses, symposia and other scientific meetings are organized each year around the world, and a wide range of newsletters, handbooks, and journals is published.

The principal source of finance for the ICSU is the contributions it receives from its members. Other sources of income are grants and contracts from UN bodies, foundations, and agencies, which are used solely to support the scientific activities of the ICSU Unions and interdisciplinary bodies. ICSU has a three-tier system of governance. They are the General Assembly (the highest organ), the Executive Board, and the Officers. These are assisted by a Secretariat responsible for the day-to-day work of the Council.

Interdisciplinary ICSU bodies are created by the General Assembly as the need for these arises in cooperative projects. Two of these, the International Geosphere-Biosphere Program (IGBP) and SCOR, are currently of particular interest to the ocean sciences community.

IGBP, planning for the International Geosphere-Biosphere Program: A Study of Global Change, was begun in 1986. The IGBP is a research program with the objective to describe and understand the interactive physical, chemical, and biological processes that regulate the total Earth system, the unique environment that it provides for life, the changes that are occurring in this system, and the manner in which they are influenced by human actions. The program is focused on acquiring basic scientific knowledge about the interactive processes of biology and chemistry of the earth as they relate to global change. Priority is placed on those areas in each of the fields involved that deal with key interactions and significant changes on timescales of decades to centuries, that most affect the biosphere, that are most susceptible to human perturbations, and that will most likely lead to a practical, predictive capability.

SCOR, the Scientific Committee on Oceanic Research, established in 1957, is the oldest of ICSU's interdisciplinary bodies. The recognition that the scientific problems of the oceans required a truly interdisciplinary approach was embodied in plans for the International Geophysical Year. Accordingly, SCOR's first major effort was to plan a coordinated, international attack on the least-studied ocean basin of all, the Indian Ocean. The International Indian Ocean Experiment of the early 1960s was the result.

For the next 30 years, the reputation of SCOR was largely based on the successes of its scientific working groups. These small international groups of not more than ten members are established in response to proposals from national committees for SCOR, other scientific organizations, or previous working groups. In general, they are designed to address fairly narrowly defined topics (often new, 'hot' topics in the field) which can benefit from international attention.

Although SCOR does not have the resources to fund research directly, many of its scientific groups have organized international meetings and produced important publications in the scientific literature. Others have proposed and planned large international collaborative efforts such as the Joint Global Ocean Flux Study and Global Ocean Ecosystem Dynamics. Scientists from the 39 SCOR member countries participate in its working groups and steering committees for the larger programs. SCOR often works in association with intergovernmental organizations such as IOC and ICES. The work of SCOR falls into two major categories. The first of these is the traditional mechanism of the SCOR working group, addressing topics which range from the ecology of sea ice to the role of wave breaking on upper ocean dynamics and from coastal modeling to the biogeochemistry of iron in sea water and the responses of coral reefs to global change. For longer-term, complex activities, such as the planning and implementation of large-scale programs SCOR establishes scientific committees.

Over the past 15 years there has been a growing awareness of the influence of the ocean in large-scale climate patterns and in moderating global change. By the early 1980s the promise of increased computing capabilities and new satellite instruments for remote sensing of the global ocean permitted oceanographers to conceive of large-scale, internationally planned and implemented experiments of the sort never before possible.

The first two of these were the World Ocean Circulation Experiment (WOCE) and the Tropical Ocean–Global Atmosphere Study (TOGA, 1985–1995). Both grew out of SCOR's former Committee on Climatic Changes and the Ocean which was also cosponsored by the IOC. A few years ago WOCE was incorporated into the World Climate Research program; its field program is now complete and WOCE is now embarking upon the critical phase of analysis, interpretation, modeling and synthesis.

Since the late 1980s SCOR has played a major role in fostering the development of two newer global change programs, both of which now form part of the IGBP effort. These are the Joint Global Ocean Flux Study and Global Ocean Ecosystem Dynamics.

SCOR consists of its 'members' – the national committees for oceanic research of its 39 member countries, each of which is represented by three individual oceanographers. The biennial general meetings elect an executive committee, which also includes *ex officio* members from allied disciplinary organizations, namely, the International Association for Physical Sciences of the Ocean, the International Association for Biological Oceanography, and the International Association for Meteorological and Atmospheric Sciences.

The Ocean Drilling Program (ODP)

The ODP is an international partnership of scientists and research institutions organized to explore the evolution and structure of the earth. It uses drilling and data from drill holes to improve fundamental understanding of the role of physical, chemical, and biological processes in the geological history, structure, and evolution of the oceanic portion of the earth's crust. The ODP provides researchers around the world access to a vast repository of geological and environmental information recorded far below the ocean surface in seafloor sediments and rocks.

The National Science Foundation (US Federal Government) supports approximately 60% of the total international effort. Other partners (Germany, France, Japan, the UK, the Australia/Canada/Chinese Taipei/Korea consortium, European Science Foundation consortium, and the People's Republic of China), comprising 20 other countries, provide 40% of the program costs.

Currently, specific studies include documenting the history of volcanic plumes in the western Pacific, examining the formation of mineral deposits near west Pacific island arcs, instrumentation of boreholes to study seismicity of the north-west Pacific, and recovery of gas hydrate deposits to examine their formation along the Oregon margin. Support will also be provided for new scientific and operational developments to extend capabilities for deep biosphere investigations for ocean biocomplexity studies.

Other Nongovernmental Organizations (Global and Regional)

Within this category there are many scientific unions and societies, as well as other organizations, which are either explicitly ocean-oriented or have ocean components. Several, as noted above are related to ICSU. Others include the American Society of Limnology and Oceanography and the American Geophysical Union (both primarily north American, but with a substantial international membership), the European Geophysical Society, the Oceanography and Challenger Societies, and many others.

Acknowledgements

I wish to gratefully acknowledge the assistance of my colleague Kandace Binkley in the preparation of this article.

Memorial

I dedicate this article as a small memorial for my friend and colleague Dr. George Grice, whose untimely death occurred in March 2001. A former Associate Director of the Wood's Hole Oceanographic Institution and Deputy Director of the Northeast Fisheries Science Center in Wood's Hole, George was involved with international organizations all his professional life, particularly ICES and IOC. He was serving the latter institution at the time of his death as Senior Science Advisor to the Executive Secretary, Dr. Patricio Bernal.

Links to International Organizations

- American Society of Limnology and Oceanography (ASLO) – http://aslo.org

- American Geophysical Union (AGU) – http://www.agu.org
- Challenger Society for Marine Science – http://www.soc.soton.ac.uk/OTHERS/CSMS
- European Geophysical Society (EGS) – http://www.mpae.gwdg.de/EGS/EGS.html
- Global Ocean Ecosystem Dynamics (GLOBEC) – http://www1.npm.ac.uk/globec
- International Association for Meteorology and Atmospheric Sciences (IAMAS) – http://iamas.org
- International Association for the Physical Sciences of the Oceans (IAPSO) – http://www.olympus.net/IAPSO

- International Council for the Exploration of the Sea (ICES) – http://www.ices.dk
- International Council for Science (ICSU) – http://www.icsu.org
- International Geosphere-Biosphere Program (IGBP) – http://www.igbp.kva.se
- Intergovernmental Oceanographic Commission (IOC) – http://ioc.unesco.org/iocweb
- Joint Global Ocean Flux (JGOFS) – http://ads.smr.uib.no/jgofs/jgofs.htm
- Baltic Marine Environment Protection Commission (HELCOM) – http://www.helcom.fi/oldhc.html
- North Pacific Marine Science Organization (PICES) – http://pices.ios.bc.ca
- Ocean Drilling Program (ODP) – http://www-odp.tamu.edu
- OSPAR Commission for the Protection of the Marine Environment of the North-East Atlantic – http://www.ospar.org
- Scientific Committee on Antarctic Research (SCAR) – http://www.scar.org
- Scientific Committee on Oceanic Research (SCOR) – http://www.jhu.edu/~scor
- The Oceanography Society (TOS) - http://tos.org
- Tropical Ocean – Global Atmosphere Coupled Ocean/Atmosphere Response Experiment (TOGA) – http://trmm.gsfc.nasa.gov/trmm_office/field_campaigns/toga_coare/toga_coare.html
- United Nations (UN) – http://www.un.org
- United Nations Environment Program (UNEP) – http://www.unep.ch/index.html
- United Nations Educational, Scientific and Cultural Organization (UNESCO) – http://www.unesco.org
- World Bank, Global Environmental Facility (GEF) – http://www.gefweb.org
- World Meteorological Organization (WMO) – http://www.wmo.ch
- World Ocean Circulation Experiment (WOCE) – http://www.soc.soton.ac.uk/OTHERS/woceipo/ipo.html

LAW OF THE SEA

P. Hoagland, J. Jacoby and M. E. Schumacher,
Woods Hole Oceanographic Institution, Woods Hole,
MA, USA

Introduction

The law of the sea is a body of public international law governing the geographic jurisdictions of coastal States and the rights and duties among States in the use and conservation of the ocean environment and its natural resources. The law of the sea is commonly associated with an international treaty, the Convention on the Law of the Sea (UNCLOS), negotiated under the auspices of the United Nations, which was signed in 1982 by 117 States and entered into force in 1994. At present 133 States have signed and ratified UNCLOS; Canada, Israel, Turkey, USA, and Venezuela are the most prominent among those that have not ratified. This treaty both codified customary international law and established new law and institutions for the ocean. UNCLOS is best understood as a framework providing a basic foundation for the international law of the oceans intended to be extended and elaborated upon through more specific international agreements and the evolving customs of States. These extensions have begun to emerge already, making the law of the sea at once broader, more complex, and more detailed than UNCLOS *per se*.

The law of the sea can be distinguished from two closely related bodies of law: maritime and admiralty. Maritime law is the private law relating to ships and the commercial business of shipping. Admiralty law, often used synonymously with maritime law, applies to the private law of navigation and shipping, in inland waters as well as on the ocean. The latter may also refer more parochially to the legal jurisdiction of specialized Admiralty courts. There may be important overlaps between the public international law of the sea and private maritime law, as may occur through the application of rules for vessel passage through a jurisdiction or the enforcement of domestic law in the ocean.

The historical development of the law of the sea is sometimes traced back to a Papal Bull of 1493, which divided the world's oceans between Portugal and Spain, thereby solidifying Spain's claim to Columbus' discovery of the New World. In the early seventeenth century, an important 'debate' took place between the Dutch jurist Hugo Grotius, who, in 1608, argued on the basis of natural law for freedom of the seas, and the English academic, John Selden, who argued in 1635 for the establishment of sovereign rights over areas of the ocean. In modern times, both regimes persist, although scientific and technological advances have combined to reduce that portion of the seas that is not subject to the authority of coastal States, and international rules have been developed to regulate many types of activities that occur beyond the reach of national jurisdictions.

This article outlines the public international law of the sea, focusing mainly on UNCLOS. Important extensions of the UNCLOS framework are highlighted. The development of the law of the sea can be conceptualized as a tree with UNCLOS as its trunk. Its roots are historical customs, some centuries old, and agreements that emerged mostly after World War II. Its branches are customs, agreements, and soft law that is only now beginning to take shape. Six topical areas are covered: underlying principles, jurisdictions, fishery resources, mineral resources, marine science and technology, environmental protection, and dispute settlement.

Underlying Principles

UNCLOS and its related agreements articulate certain distinctive, but closely related, principles of international environmental law. One of these, concerning sovereignty over resources, can be considered a general principle of customary international law. Others, including precautionary action, the common heritage of mankind, the duty to conserve the environment, sustainable development, and international cooperation, are just now emerging. These latter are philosophical concepts helping to shape the law of the sea that may one day achieve the status of general principles.

Sovereignty over Resources

One of the most widely accepted norms of international environmental law is found in Principle 21 of the Stockholm Declaration of 1972. Its objective is to strike a balance between a State's sovereignty and its responsibility to ensure that its activities and the activities of its citizens do not cause environmental harm to other States or to areas beyond national jurisdiction. The UNCLOS rendering of Principle 21 reads:

> States have the sovereign right to exploit their natural resources pursuant to their environmental policies and in accordance with their duty to protect and preserve the marine environment.

Further,

> States shall take measures necessary to ensure that activities under their jurisdiction or control are so conducted as not to cause damage by pollution to other States and their environment, and that pollution arising from incidents or activities under their jurisdiction or control does not spread beyond the areas where they exercise sovereign rights...

This principle applies to the actions of the citizens of a State within its territorial sea and exclusive economic zone, as well as on ships flying its flag, wherever they may steam.

Precautionary Action

First applied to the marine environment in 1987 after the development of UNCLOS, the principle of precautionary action refines and strengthens Principle 21. During the last decade, it was incorporated increasingly into international agreements and soft law, such as the 1992 Rio Declaration and its accompanying Report on the United Nations Conference on Environment and Development (popularly known as Agenda 21). The articulation of the principle has been inconsistent, leading to varying interpretations in different contexts. In the context of marine pollution, a fair, but general, reading of the principle is that the release of substances thought to be potentially harmful should be regulated (or prohibited) prior to the establishment, according to scientific methods, of a causal link between the release and environmental damage. The principle implies a shift in the burden of proof from the pollutee or regulator, who previously had to prove that the release of a substance was harmful, to the polluter, who now must prove that it is not. The principle has an analogous interpretation in the fisheries context.

Common Heritage

Five principal elements characterize the common heritage of mankind: (1) common space areas are owned by no one but are managed by everyone; (2) universal popular interests have priority over national interests; (3) the economic benefits of natural resources exploited from the commons must be shared among all States; (4) the use of the commons must be limited to peaceful purposes; and (5) scientific research is permissible as long as there is no

threat to the environment. The principle is stated in connection with the Area (see next section on Jurisdictions), which, along with its resources, is defined explicitly in UNCLOS as the common heritage of mankind. Some commentators have argued, however, that the exploitation of the resources of the Area is still a high seas freedom, not subject to the common heritage principle. This latter interpretation may be particularly relevant for States that have not ratified UNCLOS.

Environmental Conservation

UNCLOS specifies that:

> All States have the duty to take, or to cooperate with other States in taking, such measures for their respective nationals as may be necessary for the conservation of the living resources of the high seas.

Thus, a State whose nationals fish on the high seas is obliged to adopt conservation laws for its own citizens. The principle of obligatory environmental conservation under UNCLOS has influenced subsequent environmental agreements, including the 1985 ASEAN agreement through which its parties contracted to take measures to safeguard ecological processes, and soft law, including Chapter 17 (concerning the Oceans) of Agenda 21.

Sustainable Development

Sustainable development is another principle that emerged after the development of UNCLOS. It was articulated most clearly in the Rio Declaration, and it appears (referred to as sustainable use) in the 1992 Convention on Biological Diversity. As a general guiding principle, it implies economic or resource development in a way and at a rate such that the needs of both present and future generations can be met. Early conceptions of this principle appeared in UNCLOS, particularly with respect to the sustainable yield in fisheries, and it can be seen as closely related to the principles of environmental conservation and precautionary action.

International Cooperation

In addition to the general obligation of members of the United Nations to cooperate in good faith with the organization and among themselves, UNCLOS expresses a particular need to cooperate to conserve the seas. The convention calls for international cooperation in the conservation and management of living and nonliving resources, the use of scientific study for the benefit of mankind, the peaceful settlement of all sea-related disputes, regulation of pollution, technology transfer to developing nations, and enforcement of all the provisions of the Convention. International cooperation is facilitated by international organizations, thus the Convention provides mechanisms to aid dialogue among member States. Some examples include: the International Sea Bed Authority, the International Tribunal on the Law of the Sea, and the Commission on the Limits of the Continental Shelf.

Jurisdictions

The world's oceans are divided into six basic zones in which the types and degrees of State jurisdiction vary. These zones are: the territorial sea, the contiguous zone, the exclusive economic zone (EEZ), the continental shelf, the high seas, and the Area. The seaward limits of the territorial sea, contiguous zone, and EEZ are defined in terms of distance from a baseline, which is essentially the waterline at low tide. The construction of baselines may follow any of several methods; in theory, the baseline might shift with changes in coastal geomorphology. The drawing of straight baselines is permitted across deeply indented coastlines or to connect islands along the coast of a State. (Islands, differentiated from mere rocks, must be capable of sustaining human habitation or an economic life of their own.) Baselines may not extend more than 24 nautical miles across the mouth of a bay.

Territorial Sea

The territorial sea extends to a limit of 12 nautical miles from the baseline of a coastal State. Within this zone, the coastal State exercises full sovereignty over the air space above the sea and over the seabed and subsoil. A coastal State may legislate on matters concerning the safety of navigation, the preservation of the environment, and the prevention, reduction, and control of pollution without any obligation to make these rules compliant with international standards. Resource use within the territorial sea is strictly reserved to the coastal State.

All States have the right of innocent passage through the territorial sea of another state, although there is no right of innocent air space passage. Innocent passage is considered moving through the territorial sea in a way that is not prejudicial to the security of the coastal State, including any stopping and anchoring necessary to ordinary navigation. Innocent passage implies two important limits to the power of coastal State jurisdiction in the territorial sea: (1) the obligation not to hamper, deny, or impair the right of innocent passage; and (2) the recognition

of innocent passage even in the case of vessel-source pollution as long as the pollution is not willful and serious. With notice, innocent passage may be suspended in specified areas of the territorial sea for security reasons.

Even warships are to be accorded innocent passage (submarines must remain on the surface); however, in practice, many States require prior authorization for warships entering their territorial sea, and the law is unsettled here. Following the decision of the International Court of Justice in an infamous case in which Albania failed to notify Great Britain of the presence of underwater mines in the Corfu Channel, the coastal State must notify other States of its knowledge of navigational hazards. Regimes exist also for transit passage through international straits and archipelagic sea lanes passage in designated sea lanes through archipelagos, such as the Philippines.

Contiguous Zone

The contiguous zone is a region adjacent to the territorial sea in which the coastal State may exercise control to prevent and punish infringement of its customs, fiscal, immigration, or sanitary laws. It may not exceed a distance of 24 nautical miles from the baseline. The coastal State may take action only with respect to offenses committed within its territory or territorial sea – not to those occurring within the contiguous zone or beyond. Although not sanctioned by UNCLOS, States such as India, Pakistan, and Yemen have asserted security jurisdiction in their contiguous zones. Such practices are becoming more widely accepted as customary international law.

Exclusive Economic Zone (EEZ)

The EEZ is an area beyond and adjacent to a coastal State's territorial sea to a limit of 200 nautical miles from the baseline. Within this zone, the coastal State may exercise sovereign rights over exploration, exploitation, conservation, and management of natural resources and other economic activities, such as the production of wind or tidal power. All States, whether coastal or land-locked, enjoy the right of navigation and overflight and the laying of submarine cables and pipelines within any EEZ. The coastal State alone, however, has the right to construct and operate artificial islands and other structural installations with accompanying 500 meter safety zones. Within the EEZ, the coastal State is primarily responsible for the conservation of living resources. The coastal State has the right to regulate both marine scientific research and pollution in the EEZ. It also has legislative and enforcement competence within its EEZ to deal with the dumping of waste from vessels and pollution from seabed activities.

The practice of claiming an EEZ is one example of how UNCLOS has given rise to customary international law. The United States, for example, is not a party to UNCLOS but claims an EEZ that extends up to 200 nautical miles from its baseline. Canada has even adapted UNCLOS provisions to meet its needs for an exclusive fishing zone.

Continental Shelf

The continental shelf is geologically defined as the submerged prolongation of the land mass of the coastal State, consisting of the seabed and subsoil of the shelf, slope, and rise. It does not include the deep ocean floor. The significance of the continental shelf is that it may contain valuable minerals and shellfish. UNCLOS addresses the issue of jurisdiction over these resources by allocating sovereign rights to the coastal State for exploration and exploitation.

The shelf has been defined as extending either to the edge of the continental margin or to 200 nautical miles from the baseline, whichever is further. Unlike the case of an EEZ, coastal States do not have to proclaim a continental shelf, but they must define its limits. Where the physical limits of the continental shelf extend beyond 200 nautical miles, the coastal State must delineate it, according to one of several formulas, using straight lines that do not exceed 60 nautical miles in length. A Commission on the Limits of the Continental Shelf makes recommendations to coastal States on matters related to the establishment of outer limits of the continental shelf where they extend beyond 200 nautical miles.

High Seas

UNCLOS defines the high seas to be:

> All parts of the sea that are not included in the EEZ, the territorial sea, the internal waters of a State, or in the archipelagic waters of an archipelagic State.

On the high seas, all States enjoy freedoms of navigation, overflight, fishing, scientific research, the laying of submarine cables and pipelines, and the construction of artificial islands and installations. Because the high seas are open to all States, no State may attempt to subject any part of them to its sovereignty.

Jurisdiction over ships on the high seas is reserved for the flag State. There must be a genuine link between the State and the ship that flies its flag, and States must fix their own conditions for granting nationality to ships and for registration. Warships and government vessels are accorded complete immunity from the jurisdiction of any State other than their flag State. High seas fishing States have a duty to take conservation measures for their own nationals either alone or in cooperation with other nations. In instances of piracy, unauthorized broadcasting, slave trading, illicit drug trafficking, or statelessness, nonflag States may exercise enforcement jurisdiction.

The Area

The Area is defined as 'the sea-bed and ocean floor and subsoil thereof, beyond the limits of national jurisdiction'. The Area has significance because of the occurrence of mineral resources, such as polymetallic nodules. Like the rest of the high seas, the Area and its resources are considered to be the common heritage of mankind. Each State must ensure that the activities of its own nationals are controlled, with the understanding that damage to the Area may entail State liability.

An International Seabed Authority, established in Jamaica, regulates all activities in the Area, from marine scientific research to resource exploration and development. The Authority also has the right to conduct scientific research and to enter into research contracts. Finally, it enjoys the right to make rules and regulations preventing pollution to the marine environment and protecting natural resources. All installations are subject to these rules and regulations.

Boundary Determinations

Several territorial and continental shelf boundaries were decided prior to the signing of UNCLOS, but there are many international boundaries that still must be drawn. In 1984, in the Gulf of Maine Case, the International Court of Justice decided the first combined EEZ and continental shelf boundary, between the United States and Canada. There appear to be no hard and fast rules for boundary determinations. Rationales for claims have ranged from historic uses to economic significance, leading the Court to decide most cases on the basis of equitable principles and relevant circumstances.

Fishery Resources

The last half-century bore witness to significant growth in worldwide yields of marine fish stocks, starting at around 20 million metric tons in 1950 and peaking at 93 million metric tons in 1997. Although each fishery has its own unique characteristics, fisheries scientists now believe that, at the global level, aggregate yields have approached a natural limit. There are well-known examples of fisheries that have been exploited at inefficiently high rates, leading in some cases to severe stock depletion (e.g., north–west Atlantic cod). Evidence continues to mount of a shift from the exploitation of species at high trophic levels to those at lower levels, revealing a natural constraint to further expansion of wild harvests. Any increases in the production of seafood from the ocean and its value are likely to require both the implementation of more effective management measures that seek to optimize economic yields and the continuing development of husbandry (aquaculture).

In the face of production trends and constraints in wild harvest fisheries, there is a critical need for the implementation of management measures that lead to sustainable yields. Although this need has been recognized for decades, it has rarely been achieved because of the difficulties of allocating shares of harvests across different groups in the face of limits to understanding the dynamics of intertwined ecological and environmental systems. As a practical matter, the international law of the sea relating to fisheries conservation provides only a crude framework within which to work. Domestic and regional institutions implement specific management measures within the broader context of this framework.

Regional Fishery Management

Regional institutions were established as early as a century ago primarily for the purposes of conducting scientific research on fisheries and ecosystems that would lead, it was hoped, to recommendations for management (namely, the International Council for Exploration of the Seas in 1902). In the period since the end of World War II, these regional institutions proliferated. Today, more than 30 regional fishery bodies exist worldwide, most of which now strive to couple fisheries science to the active management of stocks that straddle the fisheries jurisdictions of multiple States or stocks that are located in part beyond any national jurisdiction (the so-called high-seas and highly migratory stocks). Where stocks are actively managed, national quotas tend to be the instrument of choice, although enforcement problems are rife. Even with this institutional presence, at any time, dozens of fisheries conflicts are occurring between the nationals of different States. In the extreme, fishery conflicts have been known to escalate to the level of military intervention.

The impetus for extending national territories that led to the basic jurisdictional zones codified in UNCLOS was driven by the perceived value of marine resources adjacent to coastal States. For fishery resources in the developed world, this value increased as demand expanded and technological

innovations reduced costs. In 1958, an international Convention on Fishing and Conservation of the Living Resources of the High Seas was signed, providing the basic framework that remains little changed to this day: local management coupled with the encouragement to cooperate internationally where nationals from different States prosecute the same fishery. According to the 1958 Convention, States were permitted to implement conservation and management measures for their own nationals fishing 'high seas' stocks adjacent to their coasts and were urged to cooperate with other States fishing there.

Fishery Conservation Zones

One shortcoming of the 1958 Convention was that the geographic boundary defining the high seas was left undefined. This problem was rectified by UNCLOS, which permitted States to claim an EEZ within which they could exercise 'sovereign rights' over the exploitation of their natural resources, including fisheries. Several conditions were placed on this exercise of sovereign rights, but, in practice, these conditions are not seen as limiting. For example, States may determine the total allowable catch and are to manage EEZ fisheries at levels that can produce a maximum sustainable yield. However, management for maximum sustainable yield may be qualified at the State's discretion by economic, environmental, ecological, or distributional reasons. These qualifications could be used as arguments for setting allowable catch, and thereby fishing effort, at levels either above or below those that might maximize sustainable yield. If a coastal State does not have the capacity, as measured by itself, to harvest its allowable catch, then, by agreement, it shall give other States, including landlocked and geographically disadvantaged States, access to any surplus. (This provision does not apply to sedentary shellfish stocks anywhere on the continental shelf.) In practice, the discretion accorded a State in determining fishing capacity and allowable catch implies that any surplus could be defined away easily. However, some States, notably Pacific Island States, have used these provisions to rent out their EEZ fisheries to the fleets of major distant water fishing nations, such as Japan.

Coastal States within whose internal waters and EEZs anadromous fish (e.g., salmons) originate or catadromous fish (e.g., eels) spend the greater part of their life cycle are responsible for management of these species. Unless otherwise agreed to on a regional or an international basis, such species are to be fished inside the EEZ. Highly migratory species (e.g., tunas, billfishes, sharks, cetaceans) are to be managed through regional or international organizations to ensure conservation and to promote optimum utilization. Importantly, marine mammal conservation may be regulated more strictly within a coastal State's EEZ than provided for by international rules.

Straddling and High Seas Stocks

UNCLOS also provides a framework for straddling and high seas stocks. This framework has been elaborated further in a 1995 international Agreement on Straddling Fish Stocks and Highly Migratory Fish Stocks. Problems remain, however, including the specification of multiple and potentially mutually exclusive management objectives (e.g., maximize yield and minimize by-catch). Again, regional bodies are asked to undertake the tough job of operational management. States are encouraged to join existing or to establish new regional management institutions. However, where such bodies already exist, the basis for incorporating new entrants into decision making and for allocating to them a limited quota remain unclear.

Mineral Resources

Ocean mineral resources, particularly offshore oil and natural gas, contribute significantly to worldwide supply. Offshore deposits now provide almost 10% of oil and 20% of natural gas production worldwide. Hard mineral deposits are much less important, although in some areas their production is meaningful to local economies. Tin has been produced for decades by dredging high-grade deposits located in the nearshore waters of Thailand and Indonesia. Diamonds are now profitably recovered off the coast of Namibia. Sulfur and salt are mined in conjunction with offshore oil production. Sand and gravel and calcium carbonate for use as a construction aggregate and to forestall beach erosion are dredged in many parts of the world. Other minerals on the continental shelves include phosphorite deposits and heavy mineral sands. Interest in the exploration of these latter occurrences continues, but these resources cannot yet be classified as economic reserves.

Certain types of deep ocean mineral deposits are plentiful, including polymetallic nodules, ferromanganese crusts, and polymetallic sulfides. Much political effort was expended to establish in UNCLOS an international legal regime governing the exploitation of these classes of minerals. Although deep ocean resources are thought to be vast, the cost of recovery and processing, including the

major risks of operating on the high seas, cannot now or in the foreseeable future justify their commercial exploitation.

Continental Shelf Minerals

Because of the costs of operating in the offshore environment, much of the production of ocean minerals takes place in shallow, near-shore waters. Deep-water facilities, which at present are operational only for oil and natural gas, such as those in the North Sea, require very large or high-grade deposits to generate viable scale economies. Where production takes place within the territorial sea, the legal regime is well developed, differing little from domestic rules onshore. Consequently, the most significant legal provisions in the international law of the sea relating to mineral resources concern the establishment of a regime for the continental shelf.

Production from seabed pools of oil and natural gas began at the turn of the century off the coast of California and in the Gulf of Mexico. But it was not until after World War II that an international legal regime governing the disposition of the resources of the continental shelf began to take shape. In 1945, US President Harry Truman issued a Proclamation asserting US jurisdiction and control over the Continental Shelf seabed and the natural resources of the subsoil. No seaward limit to the shelf was specified, although it was suggested that the shelf could be considered to extend to a depth of 100 fathoms ($\cong 183$ meters). The Truman Proclamation (and its companion proclamation concerning fishery resources) helped set off a series of jurisdictional claims of varying geographic and legal coverages in Latin America, the Middle East, and elsewhere. In 1958, a Convention on the Continental Shelf entered into force, defining the continental shelf as an area adjacent to a State's coast – but beyond its territorial sea – to a depth of 200 m. The adjacent coastal State could exercise sovereign rights over the exploration and exploitation of the natural resources of its continental shelf. Importantly, this jurisdiction could be extended 'to where the depth of the superjacent waters admits of the exploitation' of the natural resources. In this sense, jurisdiction could be expected to 'creep' with technological advance and changes in market conditions.

With respect to ocean mineral development, the activity surrounding the legal regime for the deep seabed arguably has drawn attention away from a more important part of UNCLOS: the royalty provisions concerning the development of the continental shelf. UNCLOS provides that nonliving resource production occurring on that portion of the continental shelf extending beyond 200 nautical miles is subject to financial payments or contributions in kind to the International Seabed Authority, which is to share them equitably among the parties to UNCLOS. Payments begin at 1% of the value or volume of production in the sixth year of production. The payment increases at 1% a year until it reaches 7% in the 12th year, where it remains fixed.

Deep Seabed Minerals

As the Continental Shelf Convention was being finalized, economic geologists and mining engineers began to examine more closely the potential for exploiting the vast deposits of polymetallic nodules occurring on the deep seabed. Polymetallic nodules are composed of a number of metals, including iron, manganese, nickel, copper, and cobalt. Recent economic analyses focus on nodules mainly as a nickel ore, with cobalt, copper, and, in some scenarios, manganese to be produced as by-products. Early analyses, conducted in the late 1950s and early 1960s, suggested that the nodule resource was commercially exploitable, while noting that there was no legal mechanism for allocating rights to areas thought to be so far offshore as to be beyond national jurisdiction.

In 1967, the Maltese Ambassador to the United Nations, Arvid Pardo, called for an international agreement to prevent the national appropriation of the deep seabed, to establish the seabed and its resources as a common heritage of mankind, and to employ any resource rents for the development of poor nations. Although these basic principles were eventually incorporated into UNCLOS, their acceptance by the international community, especially by the developed West and the Soviet bloc, was not immediate. By 1970, however, the administration of US President Richard Nixon proposed a common heritage mining regime for an International Seabed Area, located beyond the 200 m isobath, that laid the basis for the UNCLOS negotiations.

When UNCLOS was ready for signature in 1982, the deep seabed regime had become so extraordinarily complex and restrictive as to be unpalatable to some of the western market-oriented States. The common heritage principle was to be the centerpiece of the postcolonial new international economic order, through which the development of the world's poorer States would be boosted by mandatory technology transfer and the promise of financial payments flowing from mineral royalties. This conception was made to appear realistic in light of predictions of world resource limits, such as those made by groups like the Club of Rome, and short-term upward trends in metal commodity prices. Those States concerned about the effects on their own mineral sectors from seabed mine production were appeased in part with the promise of production limits. The final treaty provided for a parallel system of mining. Each pioneer investor (either a State or an industrial consortium sponsored by a State) would stake a mining claim and offer an additional claim of equivalent expected value to the International Seabed Authority's Enterprise. The Enterprise would mine the parallel claims using technology transferred to it by the industry.

Although the parallel system was a US proposal, in 1982 the incoming administration of US President Ronald Reagan would have nothing to do with the deep seabed mining provisions. The United States, Germany, and Great Britain, all with industrial interests in deep seabed mining, refused to sign the Convention, arguing that the nonseabed provisions reflected customary international law. Other developed States with seabed mining interests, including Japan, France, Canada, The Netherlands, Australia, and the Soviet Union, signed the Convention but delayed ratification. In lieu of the UNCLOS regime, a reciprocating States regime was organized by the West, permitting claims to the deep seabed to be staked and recognized among the participants to that agreement. The combination of the alternative regime, a steep decline in commodity prices in the 1980s, and delayed ratifications resulted in an agreement in 1994 to modify the deep seabed mining regime. Among other provisions, the revised UNCLOS deep seabed mining regime eliminated production controls and mandatory technology transfers, reduced license fees, and put the claims of miners registered under the reciprocating States regime on an equal footing with pioneer investors registered under the Convention.

Marine Science and Technology

UNCLOS was the first international agreement to establish a regime for the conduct of marine scientific research in the ocean. The regime recognizes the right of a coastal State to control access to ocean areas under its authority for the study of the physical characteristics of the ocean and its natural resources. Although the regime has been characterized by some in the scientific community as unnecessarily burdensome and too discretionary, and although problems in obtaining permission for scientific research commonly arise, the regime has proven to be workable. Seeking permission to conduct marine scientific research in the EEZ or on the continental shelf of another coastal State requires careful advance planning and, frequently, close cooperation with the scientific community in the coastal State.

The Convention recognizes that any State or competent international organization has the legal right to conduct marine scientific research. This right is conditioned only on the rights and duties of other States. Marine scientific research must be conducted for peaceful purposes, using appropriate scientific methods, and in such a way so as not to interfere unjustifiably with other legitimate uses of the ocean.

Consent Regime

Within its territorial sea, each coastal State has the right to regulate, authorize, or conduct marine scientific research, as a specific exercise of its sovereignty there. The conduct of marine scientific research in the territorial sea of a coastal State requires its express consent. Within its EEZ and on its continental shelf, each coastal State has the right to regulate, authorize, or conduct marine scientific research, as a specific exercise of its jurisdiction there. The conduct of marine scientific research in the EEZ or on the continental shelf of a coastal State requires its consent. A coastal State may not exercise its discretion to withold consent for research on the continental shelf beyond 200 nautical miles unless such research is proposed in areas that have been specifically designated by the coastal State for exploration or exploitation. All States have the right to conduct marine scientific research in the water column beyond a coastal State's EEZ and in the Area.

States seeking consent to conduct marine scientific research must provide detailed information about a proposed research project at least six months in advance. Although it is free to do so, a coastal State is under no obligation to grant its consent for scientific research in its territorial sea. Conversely, under normal circumstances, a coastal State is to grant consent for EEZ or continental shelf research, and it must establish rules so that requests for research are not delayed or denied unreasonably. However, coastal States are given considerable discretion to withhold their consent for EEZ and Continental Shelf research. Notably, consent may be withheld if a scientific research project is of significance for resource exploration or exploitation, involves drilling or the use of explosives or harmful substances, involves the construction, operation, or use of artificial islands; if the request for consent contains inaccurate information about the nature and objectives of the project, or if the requesting State has outstanding obligations from a prior research project. If consent is granted, the coastal State has the right to participate or to be represented in the research project and must be given access to all data and samples, assessments of data, and preliminary and final project results. Unless the coastal State acts to withhold consent within four months of the request or requires supplementary information, or unless outstanding obligations on the part of the requesting State exist, the consent of the coastal State is deemed to have been implied, and the research project may proceed without an affirmative grant of consent.

Technology Transfer

As an element of the new international economic order, language encouraging marine technology transfer was incorporated into UNCLOS to accelerate the social and economic development of the developing States. However, the technology transfer provisions are mainly hortatory, promoting international cooperation and suggesting options for program development. Importantly, there is no obligation to transfer technology other than on fair and reasonable terms and conditions, respecting the rights and duties of holders, suppliers, and recipients of marine technologies.

Environmental Protection

UNCLOS was designed, in part, to serve as the unifying framework for international law on marine environmental protection, which it does primarily by clarifying the rights and duties of States in this regard. It provides general goals and a few recommendations for combating all forms of marine pollution and environmental degradation but no specific pollution-control standards or required actions. To the extent that specific standards and requirements exist, they are set forth in other multilateral agreements that address either a particular form or source of pollution or a particular area of ocean space.

In addition to creating new legal instruments elaborating rules pursuant to the general goals of UNCLOS, States are called upon to cooperate in notifying other countries of imminent threats of pollution, eliminating the effects of such pollution and minimizing the damage, developing contingency response plans, undertaking research programs, exchanging data, establishing appropriate scientific criteria for pollution-control rules and standards, and implementing and further developing international law relating to responsibility and liability for damage assessment and compensation.

Although UNCLOS requires States to take measures against pollution from any source, it places particular emphasis on certain categories of pollutant substances and sources. States must take measures designed to minimize to the fullest extent possible: (1) releases of toxic, harmful, or noxious substances, especially those that are persistent, from land-based sources, from or through the atmosphere, or by dumping; (2) pollution from vessels; (3) pollution from installations and devices used in exploration or exploitation of the natural resources of the seabed and subsoil; and (4) pollution from other installations and devices operating in the marine environment. Among the pollutant sources enumerated in UNCLOS, only vessel discharges and dumping by ships and aircraft are currently subject to detailed standards and regulations at the global level.

Vessel Discharges

The main instrument addressing operational discharges by vessels is the 1973 International Convention for the Prevention of Pollution from Ships, as modified by the 1978 Protocol thereto. Known as MARPOL 73/78, this treaty system includes five annexes containing regulations for the prevention of pollution by oil, by noxious liquid substances in bulk, by harmful substances carried at sea in packaged forms or in freight containers, portable tanks, or road and rail wagons, by sewage from ships, and by garbage from ships. The annexes covering oil and noxious liquid substances in bulk are mandatory for all contracting parties, but the others are optional.

Ocean Dumping

Pollution by dumping includes the deliberate disposal at sea of wastes or other matter from vessels, aircraft, platforms, or other artificial structures, as well as the deliberate disposal of the vessels, aircraft, or structures themselves. Dumping is regulated at the global level under the 1972 Convention on the Prevention of Marine Pollution by Dumping of Wastes and Other Matter, also known as the London Convention. The Convention prohibits the dumping of certain hazardous materials and limits the dumping of other wastes or matter by requiring prior permits, including special permits for some materials according to criteria relating to the nature of the material, the characteristics of the dumping site, and the method of disposal. An important category of wastes not covered by the London Convention are those derived from the exploration and exploitation of seabed mineral resources.

Under UNCLOS, such wastes remain subject to regulation by individual States for activities conducted in areas under their jurisdiction, while wastes resulting from activities in the Area beyond national jurisdiction are subject to regulation by the International Seabed Authority. UNCLOS also mitigates a more general shortcoming of the London Convention – the fact that it has only 78 Contracting Parties representing just 68% of world merchant-marine tonnage. The main benefits of UNCLOS in this regard are that it clarifies the rights of coastal States to prohibit dumping in waters under their jurisdiction and requires all of its Contracting Parties to enact domestic measures that are at least as stringent as the London Convention requirements.

Movement of Hazardous Wastes

Another global agreement, of relevance to both vessel-source pollution and dumping, is the 1989 Convention on the Control of Transboundary Movements of Hazardous Wastes and Their Disposal (Basel Convention). Under the Basel Convention, transboundary movements of hazardous or other wastes can take place only upon prior written notification by the exporting State to the States of import and transit, and each shipment of waste must be accompanied by a detailed movement document.

Land-Based Marine Pollution

Land-based marine pollution (LBMP), although it accounts for an estimated 80% of all contaminants entering the sea, is regulated only at the national level throughout most of the world, with the exception of six regional seas where multilateral agreements are in force. Adoption of a global treaty on LBMP was the object of intensive diplomatic effort from the mid-1980s until 1995, when 109 States adopted instead the nonbinding Global Programme of Action for the Protection of the Marine Environment Against Land-Based Activities. Among the main factors discouraging adoption of a binding global convention have been the largely disappointing results of the regional agreements and the fact that the causes and effects of LBMP operate primarily at regional or smaller geographic scales. The main arguments in favor of a global convention have centered on the pervasiveness and seriousness of the problem in virtually all regions of the world and the inability of developing countries and regions to address it effectively in the absence of a legal mechanism that provides for the transfer of relevant technologies and other forms of assistance from the developed world.

Airborne Marine Pollutants

No multilateral agreements are in force whose primary purpose is the regulation of airborne marine pollutants, but such pollutants are included within the general scope of several regional agreements that address a broad range of marine pollution sources. Of these, only the agreements covering the Baltic, North-East Atlantic, and Mediterranean include any specific regulatory measures. In addition, the 1979 Geneva Convention on Long Range Transboundary Air Pollution provides for detailed regulation of emissions of numerous airborne pollutants by participating Northern Hemisphere countries. Although it does not target marine pollution directly, the Geneva Convention presumably provides indirect benefits to the marine environment.

Persistent Organic Pollutants

Potentially among the most significant international instruments for controlling marine pollutants that are both land-based and airborne is a draft global convention slated for adoption in late 2000. Commonly known as the POPs Treaty, the agreement will regulate the production, sale, and use of initially one dozen persistent organic pollutants (POPs), most of them pesticides, whose characteristics include the tendencies to bioaccumulate in the marine food chain and to undergo long-range oceanic and atmospheric transport.

Habitat and Ecosystem Protection

UNCLOS calls for States' pollution-control measures to include measures to protect habitats and ecosystems, but it does not make an explicit call for cooperation in this regard or for ecosystem-based management of marine resources. UNCLOS thus leaves large marine ecosystems, which typically straddle two or more jurisdictional zones, subject to potentially conflicting management approaches and enforcement standards. Protection of marine *habitats* is provided under two major international treaties – the 1975 Convention on Wetlands of International Importance Especially as Waterfowl Habitat (Ramsar Convention) and the 1992 Convention on Biological Diversity – and under several Regional Seas protocols and other regional agreements. Protection of marine ecosystems is far less well developed in international law, no doubt in large part because ecosystem science and management are themselves comparatively new and undeveloped fields. This circumstance may also account for what some legal scholars consider to be an incoherent approach to ecosystem protection in UNCLOS.

The lack of clarity as to the locus of authority to enforce ecosystem protections is uncharacteristic of UNCLOS, which otherwise exhibits an overriding concern with jurisdictional clarity in the balance it

strikes between the competing interests of international navigation and the environmental protection concerns of coastal States. In general, UNCLOS limits the authority of States to enforce national and international environmental regulations where such authority conflicts with other principles established under the various legal regimes relating to different categories of ocean space. For example, coastal State authority to enforce national laws is subordinated to the right of innocent passage in the territorial sea; and on the high seas, only the flag State of an offending vessel has authority to enforce international environmental regulations, in deference to the principle of freedom of navigation. Because of such provisions, in the view of some environmentalists, UNCLOS does not provide the basis for full and effective protection of the marine environment, even if its entire agenda of elaborating agreements is eventually completed.

Dispute Settlement

Following the UN Charter, which requires that all States settle their international disputes by peaceful means and without endangering international security, UNCLOS provides a binding framework for the peaceful settlement of sea-related disputes. The Convention stipulates that if States cannot resolve their disagreements peacefully on their own, they are to submit them to one of the following international bodies of their choice: (1) the International Tribunal of the Law of the Sea; (2) an arbitral tribunal constituted in accordance with Annex VII of the Convention; (3) a tribunal set up in accordance with Annex VIII; or (4) the International Court of Justice.

International Tribunal of the Law of the Sea

The Tribunal applies the provisions of UNCLOS and other rules of international law in deciding disputes. Its decisions are final and must be complied with by parties to the dispute. The decisions have binding force only among the parties and with respect to their particular dispute.

The jurisdiction of the Tribunal comprises all disputes between parties to the Convention and the agreement relating to the implementation of the deep seabed mining provisions. The Tribunal is called upon to settle three types of claims: (1) claims that application of the International Seabed Authority's rules and procedures are in conflict with obligations of the parties; (2) claims concerning excess jurisdiction or misuse of power; and (3) claims for damages to be paid for failure to comply with conventional or contractual obligations. The Tribunal also has jurisdiction over disputes concerning the Area through a special Seabed Disputes Chamber and can conduct judicial reviews of the International Seabed Authority. It cannot, however, substitute its own decision or measure for that of the Authority or annul any underlying rule, regulation, or procedure established by it.

In 1999, a dispute between Saint Vincent and the Grenadines and Guinea was one of the first to be settled by the Tribunal. The M/V *Saiga*, a vessel flying the flag of Saint Vincent and the Grenadines, had been pursued and arrested by the Guinean Navy in international waters south of Guinea's EEZ because illegal bunkering was alleged to have taken place within Guinea's EEZ. The Tribunal was charged with making a judgment on whether Guinea could apply its customs laws in an area beyond its territorial sea. Although the ship's master was eventually found guilty on several counts, the Tribunal found that Guinea's application of customs laws in its EEZ was contrary to UNCLOS.

Annex VII Arbitration

When parties to a dispute do not select a specific type of arbitration under Article 287, an arbitral tribunal under Annex VII is automatically formed. Arbitral tribunals formed under Annex VII are five-member tribunals. Each party appoints one member and the remaining three must be approved by both parties and must be nationals of third-party States. Decisions are made by a majority vote of its members.

Annex VIII Arbitration

Arbitration under Annex VIII entails the establishment of four lists of experts from which arbitral tribunals may be constituted to hear special cases. Each party to the convention may nominate two experts in each of the fields. The lists are then established and maintained by four different international institutions: (1) the Food and Agriculture Organization for fisheries; (2) the UN Environment Programme for marine environmental protection; (3) the International Maritime Organization for navigation and ocean dumping; and (4) the Intergovernmental Oceanographic Commission for marine scientific research. Five-member tribunals are set up by these institutions to perform fact-finding and settle disputes.

International Court of Justice

The International Court of Justice (ICJ) is an independent forum for dispute settlement that was established under the UN Charter and whose authority is recognized by UNCLOS. Some disputes regarding the law of the sea have already been brought before the ICJ. An important difference of arbitration under the ICJ is that once States accept the court's jurisdiction, under the Statute of the International Court of Justice, acceptance of the final decision in any case cannot be withdrawn once proceedings are underway. Another difference is that parties cannot select the members of the court who will be hearing the case. Although arbitration before the ICJ depends on the willingness of both parties to agree to and to participate in the arbitration, the Court is powerful enough to exert influence over parties to a dispute who refuse arbitration. For example, in the 1974 Fisheries Jurisdiction case, although Iceland did not appear before the court, the fact that the case made it to the level of international arbitration put considerable pressure on Reykjavik to comply with applicable rules of international law.

Future Prospects

The law of the sea will continue to evolve as the rising worldwide population places greater pressures on the natural resources and ecological systems of the coastal ocean. Most of these pressures, without question, will be situated in the territorial seas and exclusive economic zones. For example, the continuing and growing releases of macronutrients, such as nitrogen, from agricultural operations in all States may have far-reaching and cumulative impacts on coastal environments. To the extent that marine science can unveil the complex physical and ecological links among national marine jurisdictions, the relevance and import of the international law of the sea will grow. Further, a consequence of the scientific portrayal of coupled ocean–atmosphere systems will draw the law of the sea more tightly into the fold of international environmental law, integrating the broader field, and thereby rendering the law of the sea less distinguishable as a selfstanding body of law.

See also

Fishery Management. International Organizations. Manganese Nodules. Mariculture, Economic and Social Impacts. Maritime Archaeology.

Further Reading

Center for Earth Science Information Network (2000) www.ciesin.org.

Charney JI and Alexander LM (1991) *International Maritime Boundaries*. Boston: Martinus Nijhoff.

Churchill RR and Lowe AV (1988) *The Law of the Sea.* Manchester: Manchester University Press.

D'Amato A and Engel K (1996) *International Environmental Law Anthology.* Cincinnati: Anderson.

Food and Agriculture Organization (2000) www.fao.org.

Group of Experts on the Scientific Aspects of Marine Pollution (1990) *The State of the Marine Environment.* UNEP Regional Seas Reports and Studies No. 115. Nairobi: United Nations Environment Programme.

International Maritime Organization (2000) Summary of status of conventions as at 31 July 2000. www.imo.org/imo/convent/summary.htm.

Joyner CC (2000) The international ocean regime at the new millenium: a survey of the contemporary legal order. *Ocean and Coastal Management* 43: 163–203.

Schoenbaum TJ (1994) *Admiralty and Maritime Law,* 2nd end, Vol. 1. St Paul: West Publishing.

United Nations Law of the Sea (2000) www.un.org/Depts/los.htm.

United Nations Secretary General (1989) Law of the sea: protection and preservation of the marine environment. *Report of the Secretary General* A/44/461.

LARGE MARINE ECOSYSTEMS

K. Sherman, Narragansett Laboratory, Narragansett, RI, USA

Introduction

Coastal waters around the margins of the ocean basins are in a degraded condition. With the exception of Antarctica, they are being degraded from habitat alteration, eutrophication, toxic pollution, aerosol contaminants, emerging diseases, and overfishing. It has also been recently argued by Pauly and his colleagues that the average levels of global primary productivity are limiting the carrying capacity of coastal ocean waters for supporting traditional fish and fisheries and that any further large-scale increases in yields from unmanaged fisheries are likely to be at the lower trophic levels in the marine food web and likely to disrupt marine ecosystem structure.

Large Marine Ecosystems

Approximately 95% of the world's annual fish catches are produced within the geographic boundaries of 50 large marine ecosystems (LMEs) (**Figure 1A**). The LMEs are regions of ocean space encompassing coastal areas from river basins and estuaries out to the seaward boundary of continental shelves, and the outer margins of coastal currents. They are relatively large regions, on the order of $200\,000\,km^2$ or greater, characterized by distinct bathymetry, hydrography, productivity, and trophically dependent populations. The close linkage between global ocean areas of highest primary productivity and the locations of the large marine ecosystems is shown in **Figure 1B**. Primary productivity at the base of marine food webs is a critical factor in the determination of fishery yields. Since the 1960s through the 1990s, significant changes have occurred within the LMEs, attributed in part to the affects of excessive fishing effort on the structure of food webs in LMEs.

Food Webs and Large Marine Ecosystems

Since 1984, a series of LME conferences, workshops, and symposia have been held during the annual meeting of the American Association for the Advancement of Science (AAAS). In the subsequent intervening 15 years, 33 case studies of LMEs were prepared, peer-reviewed, and published (see Further Reading). From the perspective of actual and potential fish yields of the LMEs an 'ECOPATH'-type trophic model, based on the use of a static system of linear equations for different species in the food web, has been developed by Polovina, Pauly and Christensen (eqn [1]).

$$P_i = \mathrm{Ex}_i + \sum B_i(Q/B_i)(DC_{ji}) + B_i(P/B) - (IEE_i) \quad [1]$$

P_i is the production during any normal period (usually one year) of group i; Ex_i represents the exports (fishery catches and emigration) of i; $\sum_i$ represents the summation over all predators of I; B_j and B_i are the biomasses of the predator J and group I, respectively; Q/B_j is the relative food consumption of j; DC_{ji} the fraction that i constitutes of the diet of B_i is the biomass of i and $(I - EE_i)$ is the other mortality of I, that is the fraction of i's production that is not consumed within or exported from the system under consideration. A practical consideration of food web dynamics in LMEs is the effect that changes in the structure of marine food webs could have on the long-term sustainability of fish species biomass yields.

Biomass Yields and Food Webs

South China Sea Large Marine Ecosystem

An example of the use of fisheries yield data in constructing estimates of combined prey consumption by trophic levels is depicted in **Figure 2** for shallow waters of the South China Sea (SCS) LME. The trophic transfers up the food web from phytoplankton to apex predators is shown in **Figure 3** for open-ocean areas of the SCS. The differences in fish/fish predation is approximately 50% of the fish production in the shallow-water subsystem and increases to 95% in the open-ocean subsystem.

Application of the ECOPATH model to the SCS LME by Pauly and Christensen produced an initial outcome of an additional 5.8 Mt annually. This is a rate that is nearly double the average annual catch reported for the SCS up through 1993, indicating some flexibility for increasing catches from the ecosystem, but not fully realizing its potential because of technical difficulties in fishing methodologies.

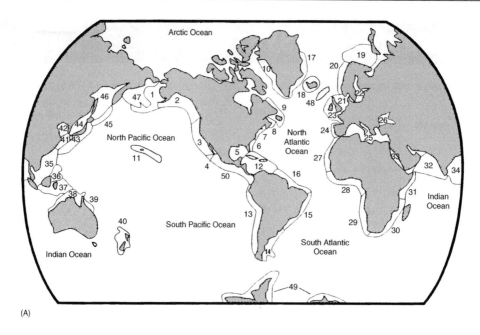

(A)

1. Eastern Bering Sea	11. Insular Pacific-Hawaiian	21. North Sea	31. Somali Coastal Current	41. East China Sea
2. Gulf of Alaska	12. Caribbean Sea	22. Baltic Sea	32. Arabian Sea	42. Yellow Sea
3. California Current	13. Humboldt Current	23. Celtic-Biscay Shelf	33. Red Sea	43. Kuroshio Current
4. Gulf of California	14. Patagonian Shelf	24. Iberian Coastal	34. Bay of Bengal	44. Sea of Japan
5. Gulf of Mexico	15. Brazil Current	25. Mediterranean Sea	35. South China Sea	45. Oyashio Current
6. South-east US Continental Shelf	16. North-east Brazil Shelf	26. Black Sea	36. Sulu-Celebes Seas	46. Sea of Okhotsk
7. North-east US Continental Shelf	17. East Greenland Shelf	27. Canary Current	37. Indonesian Seas	47. West Bering Sea
8. Scotian Shelf	18. Iceland Shelf	28. Gulf of Guinea	38. Northern Australian Shelf	48. Faroe Plateau
9. Newfoundland Shelf	19. Barents Sea	29. Benguela Current	39. Great Barrier Reef	49. Antarctic
10. West Greenland Shelf	20. Norwegian Shelf	30. Agulhas Current	40. New Zealand Shelf	50. Pacific Central American Coastal

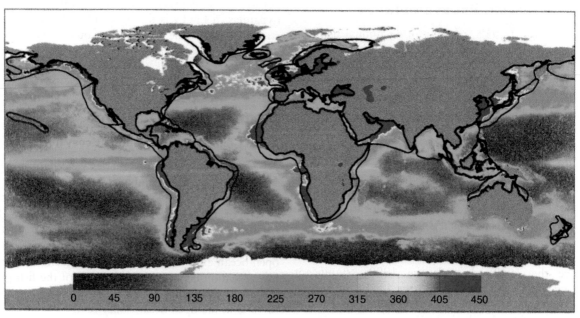

(B)

Figure 1 Boundaries of 50 large marine ecosystems (and) (B) SeaWiFS chlorophyll and outlines of LME boundaries.

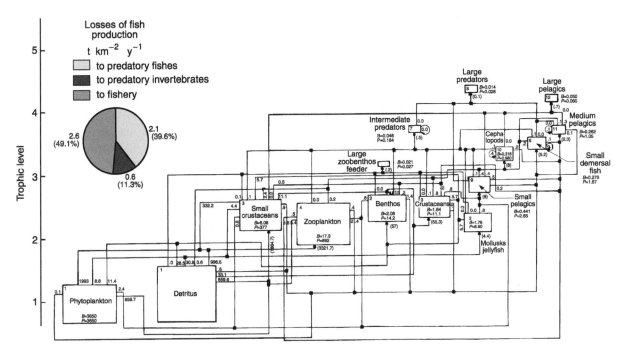

Figure 2   South China Sea shallow-water food web based on the ECOPATH model. (From Pauly and Christensen (1993).)

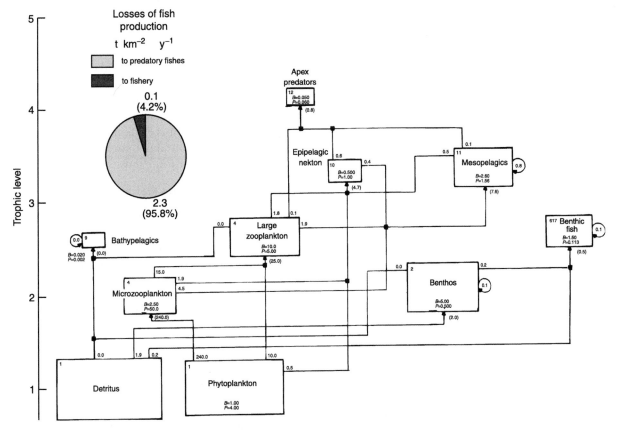

Figure 3   South China Sea open-ocean food web. (From Pauly and Christensen (1993).)

East China Sea Large Marine Ecosystem

Evidence for the negative effects of fishing down the food chain can be found in the report by Chen and Shen for the East China Sea (ECS) LME. For a 30-year period of the early 1960s to the early 1990s, little change was reported in the productivity and community composition of the plankton at the lower end of the food chain of the ECS. However, during the same period major changes were reported for a shift in biomass yields among the 'old traditional' bottom species (yellow croaker) and new species dominated by shrimp, crab, and small pelagic fish species. It appears that the annual catch increase from 0.9 Mt in the 1960s to 5.8 Mt in the early 1990s exceeded the sustainable level of yield for several species. The greatest increases in biomass yield during this period has been in a category designated as 'Other Species.' The species in this category are near the base of the food web. They are relatively small, pelagic, and fast growing, and are not used for human consumption but are used for feeding 'cultured fish or poultry' (**Figure 4**). Collectively, the catches of 'Other Species' provide additional evidence of the effects of 'fishing down the food web.'

Yellow Sea Large Marine Ecosystem

A projection of the Yellow Sea food web is given in **Figure 5**. The decline in the east Asian LMEs of demersal species and what appears to be 'trophic-forcing' down the food web hypothesized by Pauly and Christensen are apparent in the changes that have occurred over 30 years in the Yellow Sea LME (YS LME). The catch statistics indicate a rapid decline of most bottom fish and large pelagic fish from the YS LME from the 1960s through the early 1990s. Recent acoustic survey results indicate that the Japanese

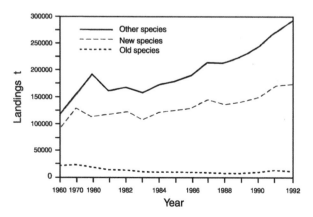

Figure 4 East China Sea fisheries yield 1960s to early 1990s, showing increased annual catches of 'Other Species' used mostly for fish and poultry food. (From Chen and Shen (1999).)

anchovy population in the YS LME has significantly increased from an annual catch level of 1000 Mt in the 1960s to an estimated biomass of 4 Mt in the 1990s.

Overfishing has led to major structural changes in the fish community of the YS LME. In the 1950s and 1960s bottom fish were the major target species in China's fisheries. Small yellow croaker was the dominant preferred demersal market species in the late 1950s, constituting about 40% of research vessel trawl catches. By 1986, pelagic fish dominated the catches ($\sim 50\%$) of research vessel surveys suggesting that they may have replaced depleted demersal stocks and are effectively utilizing surplus zooplankton production no longer utilized by the depleted large pelagics and early life-history stages of depleted fish species.

Large Marine Ecosystem Regime Shifts, Food Webs, and Biomass Yields

In the eastern Pacific, large-scale oceanographic regime shifts have been a major cause of changes in food web structure and biomass yields of LMEs.

Gulf of Alaska Large Marine Ecosystem

Evidence of the food web effects from oceanographic forcing was reported for the Gulf of Alaska LME (GA LME). An increase in biomass of zooplankton, approaching a doubling level between two periods 1956–62 and 1980–89 has been linked to favorable oceanographic conditions leading to increases in primary and secondary productivity and subsequent increases in abundance levels of pelagic fish and squid in the GA LME; it is estimated by Brodeur and Ware that total salmon abundance in the GA LME was nearly doubled in the 1980s.

California Current Large Marine Ecosystem

In contrast to the 1980–89 Gulf of Alaska increases in biomass of the zooplankton and fish biomass components of the GA LME, a declining level of zooplankton has been reported for the California Current LME (CC LME) of approximately 70% over a 45-year monitoring period. The cause according to Roemmich and McGowan appears to be an increase in water column stratification due to long-term warming. The clearest food web relationship reported related to the zooplankton biomass reduction was a decrease in the abundance of pelagic sea-birds.

US North-east Shelf Large Marine Ecosystem

The US North-east Shelf LME is an ecosystem with more structured coherence in the lower food web

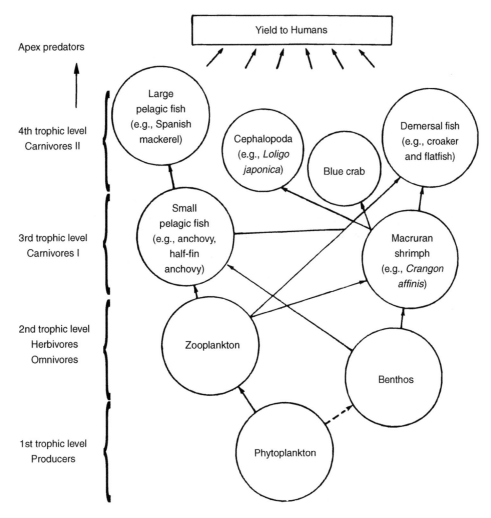

Figure 5 A simplified version of the Yellow Sea food web and trophic structure based on the main resources populations in 1985–1986. (From Tang 1993).)

than in the Gulf of Alaska or California Current systems. Following a decade of overfishing beginning in the mid-1960s, the demersal fish stocks, principally haddock, cod, and yellowtail flounder, declined to historic low levels of spawning biomass. In addition, the herring and mackerel spawning stock levels were reduced in the mid-1970s. By the mid-1980s, the demersal fish biomass had declined to less than 50% of levels in the early 1960s.

Following the 1975 extension of jurisdiction by the United States to 200 miles of the continental shelf, the rebuilding of the spawning stock biomass (SSB) of herring and mackerel commenced. Beginning in 1982 there was a sharp reduction in fishing effort from foreign vessels excluded from the newly designated US Exclusive Economic Zone (EEZ). Within four years the mackerel population recovered from just under 0.5 Mt to 1 Mt in 1986 and an estimated 2 Mt by 1994. Herring recovery was also initiated in the absence of any significant fishing

effort from 1982 to 1990 when increases in SSB went from less than 0.2 Mt to 1 Mt. An unprecedented 3.5 Mt level of herring SSB was reached by 1994.

The NOAA-NMFS time-series of zooplankton collected from across the entire North-east Shelf ecosystem from 1977 to 1999 is indicative of an internally coherent structure of the zooplankton component of the North-east Shelf ecosystem. During the mid to late 1990s and the unprecedented abundance levels of SSB of herring and mackerel, the zooplankton component of the ecosystem showed no evidence of significant changes in biomass levels with annual values close to the long-term annual median of 30 ml/100 m^3 for the North-east Shelf ecosystem. In keeping with the robust character of the zooplankton component is the initiation of spawning stock recovery subsequent to reductions in fishing effort for cod, haddock, and yellowtail flounder. Accompanying the recovery of spawning stock biomass is the production of a strong year-class of haddock in 1998 and a strong year-class of yellowtail flounder in 1997. The initial increases in skate and spiny dogfish populations following the declines in cod, haddock, and flounder stocks have been significantly reduced by targeted fisheries on these species. The reductions in abundance of these predators coupled with the robust character at the lower parts of the North-east Shelf food web enhance probability for recovery of the depleted cod, haddock, and yellowtail flounder stocks.

Large Marine Ecosystem Food Web Dynamics and Biomass Yields

Two major sources of long-term changes in biomass near the top of the food web—fishes and pelagic birds—have been observed and reported in the literature. In the case studies of the Yellow Sea and East China Sea, the multidecadal shift in fish community structure resulting from overfishing appeared to promote the production of small pelagic fish species, indicative of 'fishing down the food web' as hypothesized by Pauly and Christensen, as the abundance levels of predator species decline through overfishing. For the South China Sea, estimates from a Pauly and Christensen ECOPATH model suggests that the mean annual biomass yield of fish was not fully utilized. It appears from the case study that a significant percentage of an additional 5 Mt could be fished if managed in a sustainable manner. In the eastern Pacific the results of oceanographic regime shifts had direct impact in increasing zooplankton and fish biomass in the Gulf of Alaska LME, whereas a multidecadal warming trend in the California Current LME lowered productivity at the base of the food web and resulted in a decrease in pelagic bird biomass. The importance of fish and fisheries to the structure of marine food webs is also an important cause of variability in biomass yields. A clear demonstration of this relationship is found in the application of the ECOPATH model to four continental shelf ecosystems, where it was shown that fish preying on other fish was a principal source of fish biomass loss. The level of predation ranged from 3 to 35 times the loss to commercial fisheries.

Fish are keystone components of food webs in marine ecosystems. The worldwide effort to catch fish using highly effective advanced electronics to locate them, and efficient trawling, gill-netting, and longline capture methodologies, has had an impact on the structure of marine food webs. From case studies examined, evidence indicates that the fishing effort of countries bordering on LMEs has resulted in changes in the structure of marine food webs,

ranging from significant abundance shifts in the fish component of the ecosystem from overfished demersal stocks to smaller faster-growing pelagic fish and invertebrate species (herrings, anchovies, squids) as fisheries are refocused to species down the food web, predation pressure increases on the plankton component of the ecosystem.

The economic benefits to be derived from the trend in focusing fisheries down the food web to low-priced small pelagic species used, in part, for poultry, mariculture, and hog food are less than from earnings derived from higher-priced groundfish species, raising serious questions regarding objectives of ecosystem-based management integrity of ecosystems and sustainability of fishery resources. These are questions to be addressed in the new millennium with respect to the implementation of management practices. As in the case of the US North-east Shelf LME, overfished species can recover with the application of aggressive management practices, when supported with knowledge that the integrity of the lower parts of the food web remain substantially unchanged during the recovery period. However, under conditions of recent large-scale oceanographic regime shifts in the Pacific, evidence indicates that the biodiversity and biomass yields of the north-east sector of the Pacific in the Gulf of Alaska LME were significantly enhanced from increased productivity through the food web from the base to the zooplankton and on to a doubling of the fish biomass yields close to the top level of the food web. In contrast, in the California Current ecosystem the apparent heating and deepening of the thermocline effectively reduced phytoplankton and zooplankton production over a 40-year period, suggesting that in upwelling regions prediction of oceanographic events effecting food web dynamics require increased commitment to long-term monitoring and assessment practices if forecasts on effects of regime shifts on biomass yields are to be improved.

If ecosystem-based management is to be effective, it will be desirable to refine ECOPATH-type models for estimating the carrying capacity of LMEs in relation to sustainability levels for fishing selected species. It was assumed in the early 1980s by Skud, based on the historic record, that herring and mackerel stocks inhabiting the US North-east Shelf ecosystem could not be supported at high biomass levels simultaneously by the carrying capacity of the ecosystem. However, subsequent events have demonstrated the carrying capacity of the ecosystem is now of sufficient robustness to support an unprecedented almost 5.5 Mt of spawning biomass of both species combined. In addition, the ecosystem in its present state apparently has the carrying capacity to support the growing spawning biomass of recovering haddock and flounder stocks. Evidence of the production of strong year-classes for both species supported by high average levels of primary production of $350 \, g \, Cm^2 \, y^{-1}$, a robust level of zooplankton biomass, relatively high levels of epibenthic macrofauna, and apparent absence of any large-scale oceanographic regime shift suggests that integrity of the ecosystem food web will enhance the return of the fish component of the ecosystem to the more balanced demersal–pelagic community structure inhabiting the shelf prior to the massive overfishing perturbation of the 1960s to the 1980s.

Prospectus: Food Webs and Large Marine Ecosystem Management

It is clear from the LME studies examined that time-series measurements of physical oceanographic conditions that are coupled with appropriate indicators of food web integrity (e.g., phytoplankton, chlorophyll primary productivity, zooplankton, fish demography) are essential components of a marine science program designed to support the newly emergent concept of ecosystem-based management.

It is important to consider the dynamic state of LMEs and their food webs in considering management protocols, recognizing that they will need to be considered from an adaptive perspective. To assist economically developing countries in taking positive steps toward achieving improved understanding of food web dynamics and their role in contributing to longer-term sustainability of fish biomass yields, reducing and controlling coastal pollution and habitat degradation, and improving oceanographic and resource forecasting systems, the Global Environment Facility (GEF) and its $2 billion trust fund has been opened to universal participation that builds on partnerships with several UN agencies (e.g., World Bank, UNDP, UNEP, UNIDO). The GEF, located within the World Bank, is an organization established to provide financial support to post-Rio Conference actions by developing nations for improving global environmental conditions in accordance with GEF operational guidelines.

See also

Dynamics of Exploited Marine Fish Populations. Ecosystem Effects of Fishing. Fisheries and Climate. Fisheries Overview. International Organizations. Marine Fishery Resources, Global State of.

Further Reading

Chen Y and Shen X (1999) Changes in the biomass of the East China Sea ecosystem. In: Sherman K and Tang Q (eds.) *Large Marine Ecosystems of the Pacific Rim: Assessment, Sustainability, and Management*, pp. 221–239. Malden MA: Blackwell Science.

Kumpf H, Stiedinger K, and Sherman K (1999) *The Gulf of Mexico Large Marine Ecosystem: Assessment, Sustainability, and Management*. Malden, MA: Blackwell Science.

Pauly D and Christensen V (1993) Stratified models of large marine ecosystems: a general approach and an application to the South China Sea. In: Sherman K, Alexander LM, and Gold BD (eds.) *Large Marine Ecosystems: Stress, Mitigation, and Sustainability*, pp. 148–174. Washington DC: AAAS.

Sherman K and Alexander LM (eds.) (1986) *Variability and Management of Large Marine Ecosystems, AAAS Selected Symposium 99*. Boulder, CO: Westview Press.

Sherman K and Alexander LM (eds.) (1989) *Biomass Yields and Geography of Large Marine Ecosystems, AAAS Selected Symposium 111*. Boulder, CO: Westview Press.

Sherman K, Alexander LM, and Gold BD (eds.) (1990) *Large Marine Ecosystems: Patterns, Processes, and Yields, AAAS Symposium*. Washington, DC: AAAS.

Sherman K, Alexander LM, and Gold BD (eds.) (1991) *Food Chains, Yields, Models, and Management of Large Marine Ecosystems*. AAAS Symposium Boulder, CO: Westview Press.

Sherman K, Alexander LM, and Gold BD (eds.) (1992) *Large Marine Ecosystems: Stress, Mitigation, and Sustainability*. Washington, DC: AAAS.

Sherman K, Jaworski NA, and Smayda TJ (eds.) (1996) *The Northeast Shelf Ecosystem: Assessment, Sustainability, and Management*. Cambridge, MA: Blackwell Science.

Sherman K, Okemwa EN, and Ntiba MJ (eds.) (1998) *Large Marine Ecosystems of the Indian Ocean: Assessment, Sustainability, and Management*. Malden, MA: Blackwell Science.

Sherman K and Tang Q (eds.) (1999) *Large Marine Ecosystems of the Pacific Rim: Assessment, Sustainability, and Management*. Malden, MA: Blackwell Science.

Tang Q (1993) Effects of long-term physical and biological perturbations on the contemporary biomass yields of the Yellow Sea ecosystem. In: Sherman K, Alexander LM, and Gold BD (eds.) *Large Marine Ecosystems: Stress, Mitigation, and Sustainability*, Washington DC, AAAS Press, pp. 79–93.

OCEAN ZONING

M. Macleod, World Wildlife Fund, Washington, DC, USA

M. Lynch, University of California Santa Barbara, Santa Barbara, CA, USA

P. Hoagland, Woods Hole Oceanographic Institution, Woods Hole, MA, USA

Introduction

During the last half-century, as the extent and variety of marine resource uses have expanded with population growth, technological advances, and changing human preferences, ocean space in many areas has become limited and thus more valuable. A multitude of ocean uses compete for available space, including shipping, naval operations, commercial fishing, energy development, waste disposal, marine recreation, oceanography, and environmental conservation. This phenomenon of increased demand for a resource that is in limited supply is most pronounced in coastal marine environments, including bays, estuaries, and near-shore waters. Generally speaking, ocean space becomes more available as one moves further offshore, but the cost of accessing that space becomes prohibitively expensive for all but a few high-valued uses, such as shipping, offshore hydrocarbon production, fishing, and oceanography.

The intense competition for ocean space among human users in near-shore waters leads inevitably to costly disputes, accidental or purposeful destruction of property, and, on occasion, armed conflict. In most cases, the private ownership of ocean space is unauthorized, rendering traditional property markets unavailable for mitigating disputes over its use. Ocean space is typically allocated through a variety of mechanisms of collective action, such as government programs and policies.

Various agencies at the municipal, state, and national levels employ a wide range of management measures in attempts to regulate the use of ocean space and the exploitation of ocean resources. The academic field of marine policy is organized around analyzing the effectiveness and fairness of alternate management measures for resolving such disputes. The absence of restrictions on uses or the adoption of ineffective management measures can be not only wasteful but may lead to the imposition of social costs, including those associated with the degradation of the marine environment.

Ocean zoning is a policy instrument intended to help resolve disputes over the use of ocean space. It involves the designation of areas of the ocean for one or more specific purposes. According to current practice, ocean zoning is a type of command-and-control or nonmarket policy instrument. Except in special cases, such as the sale at auction of Outer Continental Shelf (OCS) oil and gas leases, the implementation of ocean zoning does not usually involve the pricing of resources or ocean areas, *per se*, in order to create incentives for users to undertake socially preferred activities.

The essential characteristics of an ocean zone are a boundary and a set of rules that identify the types and regulate the extent of activities that can occur inside the boundary. Ocean zones range in size from expansive exclusive economic zones (EEZs) extending 200 nautical mile (nmi) out from the shorelines of all coastal nations to relatively much more diminutive 'areas to be avoided' around a deep-water port (**Figure 1** and **Table 1**). The former may involve millions of square nautical miles whereas the latter may involve less than $10\,\text{nmi}^2$. In special circumstances, ocean zones may convey property rights or other legal interests in the ocean or the seabed from the public to private industry or to nongovernmental organizations (NGOs). Such rights may entitle their holders to exclude other human users from the relevant zone.

Ocean zones may vary in the degree to which uses are restricted. The Great Barrier Reef Marine Park in Australia, for instance, includes some ocean zones where essentially all human uses are prohibited, other than certain scientific studies. At the other extreme, zones designated as essential fish habitats (EFHs) in the United States are merely a means of classification, aiming to inform fisheries policy but not restricting human uses *per se*.

Depending upon the nature of the activities involved, an ocean zone may be designated exclusively for a specific use, or it may permit multiple uses. In some circumstances, zones can be nested, such as the location of oil and gas lease tracts within a regional 5-year planning area, which is itself located within the EEZ or on the OCS. Further, ocean zones may be of limited duration, such as temporary fishery closures, or they may be designed to exist in perpetuity, such as marine sanctuaries.

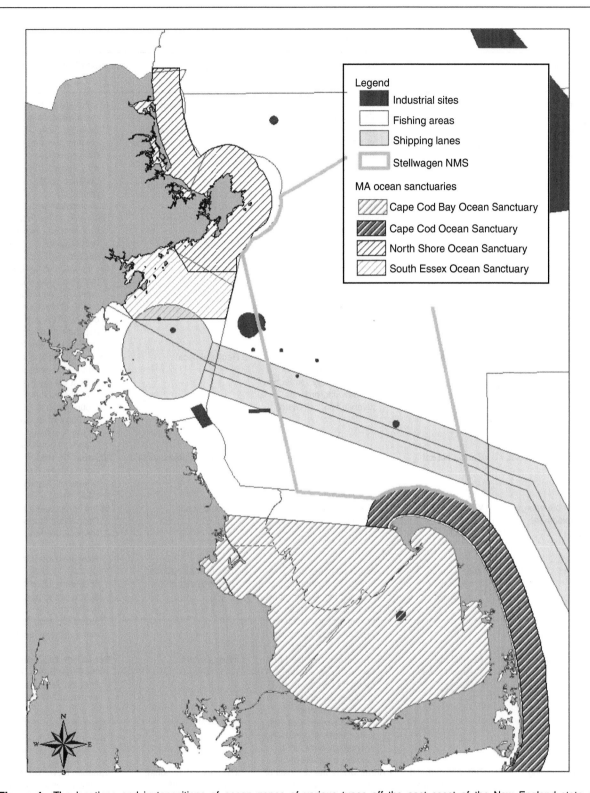

Figure 1 The locations and juxtapositions of ocean zones of various types off the east coast of the New England state of Massachusetts (MA) in the western Gulf of Maine.

Most ocean zones are fixed geographically. Some, however, have been designed to move with a resource, such as the buffers surrounding large whales that are the object of whale-watching ecotours. The ocean has been zoned for national defense purposes, vessel traffic separation, fishery conservation, offshore oil and gas development, marine protection, ocean dumping, scientific research, among many other purposes.

Table 1 Some examples of spatial management schemes

Human use	Ocean zone example	Description	Planning level	Example of spatial scale
General	EEZ and OCS	Zones in which a coastal nation has sovereign rights over the exploitation of living and nonliving natural resources	National, following customary international law and the codified Law of the Sea	Extending 1112 km from the territorial sea baseline (water column and seabed); the continental shelf may exceed this distance under certain geological circumstances
General	Submerged lands	Individual coastal states in the USA have ownership of the submerged lands	Local (state level)	Extending 17 km from the territorial sea baseline, with regional exceptions
Mineral exploration and production	OCS oil and gas leases	Geographic units allocated for hydrocarbon exploration, development, and production	Regional	On the OCS of the USA, the typical lease tract is 31 km^2, as modified by the characteristics of a hydrocarbon deposit or industry pooling agreements
Commercial fisheries	Fishery reserves	Reservation or closure of areas for purposes of stock recovery; protection of spawning grounds, juvenile fish, or habitat	Regional, local	Variable (both spatially and temporally); the largest fishery reserve is a closure to groundfish trawling of 950 000 km^2 of waters off the Aleutian Islands, Alaska, USA
Protection of species	Critical habitat	Protection of endangered or threatened species habitat under the US Endangered Species Act	National, regional, local	Variable; 95 182 km^2 for North Pacific right whales off Alaska, USA
Renewable energy development	Wind park	Areas designated for the siting of wind turbines	National, local	65 km^2 proposed for the Cape Wind facility in Nantucket Sound, off the coast of Massachusetts, USA
Shipping	Vessel traffic separation schemes	Designated lanes for inbound/outbound traffic approaching harbors and ports, crossing areas, and vessel types; associated precautionary areas and areas to be avoided	International, national, regional, local	Varies significantly depending upon the nature of the port and harbor and the roadstead; may be many miles in length
Marine conservation	Particularly sensitive sea areas	Ship operators advised to take extra care within marked boundaries. IMO member states may develop additional regulatory measures (e.g., vessel traffic schemes, no-anchoring zones, and discharge regulations)	International	Ten worldwide, of which the Great Barrier Reef in Australia is the largest at 343 000 km^2
Ocean dumping	Dredged material disposal sites	Areas where the disposal of materials dredged from harbors and channels is permitted	International, national, regional, local	Variable: New York/New Jersey disposal site is 54 km^2, located 3.5 m off NJ and 7.7 m off NY coast
Marine aquaculture	Shellfish leases	Nonpermanent lease of areas for purposes of shellfish (clam, oyster, mussel) cultivation	Local	Fractions of km^2 (variable)

Comprehensive Ocean Zoning

The term 'ocean zoning' also has been used recently to refer to the application of land-based zoning principles to the ocean and in this context has become part of the lexicon of ocean management. This vision of ocean zoning, which often is a key implementation tool for so-called 'ocean planning' or 'comprehensive ocean management' schemes, may involve the designation of areas for particular uses before specific projects are proposed. Under some ocean planning schemes, the entire body of water in question is zoned; only certain regions are zoned in others; and many other ocean areas remain loosely regulated or largely open access. Potentially competing uses may be separated while compatible uses may be clustered, based on economic, ecological, or infrastructural criteria.

Proponents of ocean zoning see the expansion of the concept as a moving away from what they perceive to be an *ad hoc*, fragmented, and uncoordinated regulatory environment and toward a more effective method for setting down a sensible mosaic of ocean uses. Detractors argue that such schemes could be unfair to historic users of the ocean and would be unable to address uncertain and changing future use patterns. A key element of any debate about ocean zoning concerns the ease with which zoning decisions can be modified as environmental conditions change or as human preferences shift. Opponents also claim that ocean zoning could prove to be an inefficient and expensive form of regulation, disproportionately subject to the influence of special interest groups and plagued by enforcement difficulties.

The examples in **Table 1** reflect the wide range of scales of implementation, spatial and temporal extents, and regulatory details that characterize spatial ocean management. EEZs or continental shelves comprise broad jurisdictions, establishing the authority of nations to regulate activities within their limits. Other more parochial zones, such as fisheries closures and shipping lanes, are examples of historically unregulated transient ocean uses; zoning for these uses perpetuates the public character of the ocean. Still others, such as offshore oil and gas lease tracts on the US OCS, wind farms off the coast of Denmark, or aquaculture development areas in Hawaiian territorial waters, are examples of zones for nontransient ocean uses in which private industry can occupy ocean areas or exploit ocean resources.

Fisheries closures are unique among the examples in **Table 1** in that they may be temporary, instituted in response to a narrowly defined management need. When the northwest Atlantic cod fishery in the Gulf of Maine collapsed in the early 1990s, a series of emergency closures were established in order to allow stocks to recover from severe levels of recruitment overfishing. Such emergency closures can last from several months to several years. Spatial fisheries closures also can be utilized as part of a long-lasting management plan, either in the form of permanent no-take reserves or seasonal closures of breeding, feeding, or hatching grounds. Indeed, one set of groundfish reserves along the western boundary of the Gulf of Maine involves a wave of sequential closing and opening of areas during the fishing season to protect spawning grounds and juvenile fish.

Fisheries closures further illustrate the varying degrees and methods by which uses can be restricted in an area. Nautical charts of Georges Bank, a highly productive fishing area off the coast of New England, include both strictly enforced do-not-enter closures alongside areas that are closed only to certain types of boats or gear. Closures also might apply to only a portion of the water column (e.g., from 1000 to 2000 m) or, when selective gear can be deployed, to particular species.

The Challenges of Ocean Management

The uses of many ocean resources provide quintessential examples of Professor Hardin's tragedy of open access, by which individual resource users, acting in their own self-interest, undermine collective long-term sustainability. Unfortunately, most extant public policies render traditional solutions, such as privatization and mutual agreements, at best challenging and at worst illegal. A thoughtful literature on the theory and practice of commons management has recognized certain resource and user characteristics, the existence of which increase the likelihood that unregulated ocean areas may become a true commons, to be managed collectively for the public good. These characteristics comprise a small zone, within the bounds of which a moderate level of resource use is undertaken, involving insignificant external effects. Further, resource users should share common cultural beliefs and social norms, exhibit an instinctive understanding of resource dynamics, and face nominal costs of monitoring resource stocks and flows. Regrettably, in only a few circumstances do these characteristics describe the ocean and its modern users.

The reality of ocean management is characterized by complexity, inchoate science, and steep costs of monitoring and enforcement. The ocean is a fluid environment, and its many living resources exhibit wide-ranging and incompletely understood migratory patterns. Resource users may not fully understand the cause-and-effect relationships that relate their actions to the responses of either a targeted resource or the larger ecosystem. In many places, the science of marine resource dynamics is incomplete or even unstudied, rendering monitoring and assessment, even when technologically and economically possible, uncertain and inconclusive. Further, ocean habitats are dynamic, sometimes chaotically so, implying that the sampled environmental characteristics used to designate a zone at one point in time could become immaterial at a later date. These unique characteristics of the ocean may preclude the use of spatial management schemes in some contexts or may require more careful (and costly) ongoing attention to the dimensional aspects of the ocean, such as current flows, temperature and salinity fluxes, plankton distributions, migration routes, species interactions, among many other variables.

Appropriate Scales for Ocean Zoning

The physical and ecological features of the marine environment, in combination with unique institutional regimes, make the choice and application of management scales more problematic than on land. Comprehensive ocean planning involving zoning has received increased attention of late from academics and policymakers because it offers the hope that disputes over ocean uses may be mitigated by physically separating potentially competing uses. This approach assumes that there are appropriate spatial scales for particular uses, and that these scales, which may differ drastically across uses or for different resources, can be implemented successfully in practice.

One of the most prevalent uses of ocean zoning has been the designation of various types of marine protected areas (MPAs). According to a definition adopted by the International Union for the Conservation of Nature: "[an MPA is] an area of intertidal or subtidal terrain, together with its overlying water and associated flora, fauna, historical and cultural features, which has been reserved by law or other effective means to protect part or all of the enclosed environment." Importantly, many MPAs are implemented with multiple objectives, so that human uses are allowed to continue at some level, while natural features and processes are conserved.

The most restrictive of MPAs are the 'no-take' fishery reserves, within which no fishing or other extractive uses are permitted. Fishery reserves are implemented commonly for one of two purposes: (1) fisheries enhancement, for which effectiveness is not always apparent; or (2) biodiversity conservation. While reserves have a mixed record with regards to fisheries enhancement, they have repeatedly been shown to be an effective means of preserving biodiversity within their bounds.

Most marine ecologists agree that the success of spatial conservation depends strongly upon reserve size and location. There is a growing awareness that optimal design depends upon many factors, including regional oceanographic characteristics, biological variables, such as patterns of larval dispersal, and the management goals of a particular reserve. Yet there remain many unanswered questions about the design of marine reserves and whether reserves should be linked together into a regional network. The optimal scale for reserves can vary by several orders of magnitude, and the need for and design of reserve networks further factors into considerations of scale.

Linear programming approaches have received much recent attention as methods for determining the optimal spatial distribution of a group of protected or regulated areas within a region. These approaches involve the use of methods from the field of operations research to minimize the total area or boundary length of an MPA subject to user-defined conservation constraints. The effectiveness of such spatial modeling approaches depends critically upon the quality of the underlying data and the relationship of the data to the conservation goals. For example, one such application involves the use of mapped geological characteristics of the seafloor as a means for identifying which groundfish spawning grounds should be protected in a fishery. The effectiveness of this application obviously depends on the correlation between the relevant seafloor characteristics and fish-spawning behavior.

Programming approaches also have been used to try to minimize the economic impact of the displacement of fishermen from protected areas. Of note, these programs tend to model a static situation, characterized by unchanging environmental parameters and historical patterns of fishing. Most applications to date have been limited by the lack of complementary models that predict the dynamic behavioral responses of the displaced fishermen. Notwithstanding these problems, the design of spatial progams for optimizing ocean management schemes continues to be a very active area of interdisciplinary research. In the future, other models and programs, some using emergent geographic information system capabilities, may be drawn upon to help design comprehensive, multiple-use ocean-zoning schemes.

A scientific consensus apparently is coalescing around the idea that biological connectivity over distance may be best preserved through networks of MPAs. Thus, networks arguably provide greater conservation benefits than isolated individual reserves. Consequently, the implementation of ocean zoning within regional planning units may make it easier for natural resource managers to coordinate resource protection, enhance fishery yields, and conserve biodiversity on the scale of ecosystems. Because of the multiplicity of jurisdictions and the large numbers of stakeholders, however, attempts at ocean zoning at a regional level often face daunting political obstacles.

Economic and Distributional Implications of Ocean Zoning

Zoning occurs as a consequence of the public perception of inefficient or unfair distributions of existing human uses of the ocean. For example, the overexploitation that occurs in an open-access fishery may lead to the dissipation of economic rents associated with targeted fish stocks. Changes in ecological states that accompany overexploitation also may adversely affect commercial fishermen targeting other fisheries, recreational fishermen, and ecotourists who like to watch whales. The establishment of MPAs, fishery closures, EFHs, and other forms of zoning is one reaction to the wasteful use of the ocean as an open-access fishery.

The delimitation of a zone is inherently an economic decision that reveals societal preferences for particular forms of the production of goods and services from the ocean. When a particular use or combination of uses of ocean space is perceived to be more appropriate than a historical pattern of uses, a zone may be created to modify the 'production technology'. Thus a wind farm might be sited in an area that fishermen have historically used to trawl for fish. The term 'ocean use' is defined broadly here to include the production of goods and services that trade in existing markets (fish, transportation services, offshore oil and natural gas recovery, and sand and gravel mining) as well as the enjoyment of those that have no markets (natural habitat, biological diversity, esthetic views, and disposal of dredge spoils).

Among the social sciences, only neoclassical economics provides a normative framework for judging whether or not the production technology represented by the allowed uses and nonuses of an ocean zone is socially optimal (i.e., economically efficient). Following that framework, ocean zones ought to be established to allow those combinations of uses that maximize net present value as measured in monetary terms. In order for a combination of ocean uses to be considered economically efficient, the benefits of that combination of uses must exceed the opportunity costs of all other potentially excluded uses, either alone or in feasible combinations. Thus, in principle, there is an economically optimal mix of uses in an ocean area.

This simple characterization of a decision rule for deciding upon efficient zoning in the ocean belies many complexities. Foremost among these is the problem of estimating the value of patterns of alternative uses and nonuses, such as habitat preservation. The net economic surpluses from alternative uses or compatible combinations of uses must be estimated, forecast into the future, and discounted into present values. These then would be compared with similar estimates of potentially excluded uses. Where nonuses are involved, specialized methodologies must be applied to estimate nonmarket or 'passive use' values. All valuation approaches also may involve uncertainty and risk aversion, which can make the decision process more complex.

To circumvent these issues, some jurisdictions invoke legal decision rules, such as the so-called 'public trust doctrine', that establish priorities for certain uses over others. For example, navigation and fishing, which tend to be transitory uses of the ocean, usually are regarded as priority public trust uses applied at the individual state level in the United States in decisions over the allocation of uses in coastal waters and tidelands. Although the public trust doctrine may simplify decision making, and it may appear to be fair to some user groups, it does not necessarily imply that ocean-use decisions are economically efficient.

The economic criterion tends to ignore the distributional effects of zoning decisions. For example, if an area of the ocean is zoned as a deep-water port, commercial fishermen may be displaced from the area. It may be economically efficient to establish such a zone, particularly when it displaces an open-access fishery in which resource rents have been dissipated, but the displaced fishermen often will complain about the inequitable result, particularly if they have been fishing in the zoned location for many years or even for generations. In the absence of well-defined property rights in ocean space or legal interests in ocean uses, it can be difficult for historical users to demand compensation for the loss of access to a newly zoned area. Notably, where such rights exist, users have a stronger argument for compensation. For example, fishermen in New Zealand who own transferable quota rights for wild fisheries have argued for compensation for areas of the fishery that have been zoned for marine aquaculture.

Institutional Approaches to Zoning

The allocation of ocean space for specific uses through zoning can be implemented by executive fiat, legislative decree, or mutual agreement arrived at among a set of stakeholders acting independent of government. In some cases, ocean zones may be established through a joint effort of government agencies and stakeholders, a method of collective decision-making referred to by some commentators as 'co-management'. Good examples exist of many of these means of implementation, and analyzing their pros and cons is an area of active social science research. Recently, ocean zoning by centralized management institutions, which exercise authority by fiat or decree, has fallen into disfavor. Co-management or stakeholder agreements have received increasing attention as socially preferred forms of collective decision making, leading to gains in fairness and possibly even efficiency in the form of lower administrative costs.

Although designation by executive fiat is one of the oldest forms of ocean zoning, for example, the establishment in 1935 by US President Franklin D. Roosevelt of the Fort Jefferson National Monument in the Dry Tortugas, it has been used less frequently in developed nations in recent decades. An exception concerns the recent designation in June 2006 of the Paphānaumokuākea Marine National Monument by US President George W. Bush. This $137\,792\,km^2$ MPA, the largest MPA in the world at the time of this writing, is located in a remote region of small islands and atolls in the northwestern Hawaiian Islands. The national monument was established to phase out commercial fishing, ban seabed resource extraction, prohibit ocean dumping, preserve access for native Hawaiian cultural activities, enhance visitation around Midway Island, and provide for education and scientific research. The designation by executive fiat was justified in part on the need to forestall a co-management process that was perceived to be contentious and needlessly slow.

The Canada Oceans Act of 1997 provides an example of a co-management approach to ocean zoning. The act defines 'integrated management' as a collaboration that brings together all interested parties to incorporate social, cultural, environmental, and economic values in developing and implementing spatially explicit 'ocean-use plans'. Under the auspices of the act, a pilot planning project has been organized for the integrated management of $325\,000\,km^2$ of eastern Canada's Scotia Shelf. Human uses of the Scotian Shelf comprise commercial fishing, oil and natural gas exploration and production, national defense, laying of submarine cables, scientific research, and recreation and tourism. Triggered by an oil and gas development proposal in a sensitive underwater canyon habitat, the Eastern Scotian Shelf Integrated Management (ESSIM) ocean-use plan has been touted as an ecosystem-based approach that focuses on defining management goals that will attempt to balance commercial uses with environmental protection of the area. The Canadian Department of Fisheries and Oceans (DFO) has been developing the plan with industry, conservation organizations, academics, First Nations, communities, and all levels of government through a multitiered stakeholder process.

Conclusions

As the variety and scale of human uses of the coastal ocean increase, ocean space becomes more scarce and therefore more valuable in an economic sense. Ocean zoning is a nonmarket policy instrument for allocating scarce ocean space in order to mitigate real and potential conflicts over its use. The imposition of restrictions on particular uses in certain areas of the ocean is inevitably controversial, leading to the need for the selection of political institutions for implementing ocean zoning. Recently, co-management-type institutions, involving agreements leading to zoning between government and stakeholders, have been in vogue, displacing zoning allocations by executive fiat or legislative decree. If economic efficiency were to become an overriding policy goal, the development of a market in ocean space would be an appropriate institution. Such an institution might involve the identification of certain areas that would be set beyond the reach of the market for conservation purposes. Increasingly, we expect to see the growing use of ocean zoning to carry out the goals of comprehensive, regional ocean management. As zoning is implemented, as legal interests begin to attach to the uses of zoned ocean space, and as technologies reduce the costs of establishing and enforcing these interests, markets for allocating ocean space will almost certainly emerge.

See also

Law of the Sea. Marine Policy Overview. Marine Protected Areas.

Further Reading

Crowder LB, Osherenko G, Young OR, *et al.* (2006) Resolving mismatches in US ocean governance. *Science* 313: 617–618.

Day JC (2002) Zoning – lessons from the Great Barrier Reef Marine Park. *Ocean and Coastal Management* 45: 139–156.

Dietz T, Ostrom E, and Stern PC (2003) The struggle to govern the commons. *Science* 302: 1907–1912.

Eckert RD (1979) *The Enclosure of Ocean Resources: Economics and the Law of the Sea.* Stanford, CA: Hoover Institution Press, Stanford University.

Halpern BS (2003) The impact of marine reserves: Do reserves work and does reserve size matter? *Ecological Applications* 13: S117–S137.

Hanna SS, Folke C, and Mäler K-G (eds.) (1996) *Rights to Nature: Ecological, Economic, Cultural, and Political Principles of Institutions for the Environment.* Washington, DC: Island Press.

Hardin G (1968) The tragedy of the commons. *Science* 162: 1243–1248.

Johnston DM (1993) Vulnerable coastal and marine areas: A framework for the planning of environmental security zones in the ocean. *Ocean Development and International Law* 24: 63–79.

Libecap GD (1989) *Contracting for Property Rights.* New York: Cambridge University Press.

McGrath K (2003) The feasibility of using zoning to reduce conflicts in the exclusive economic zone. *Buffalo Environmental Law Journal* 11: 183–220.

Russ GR and Zeller DC (2003) From mare liberum to mare reservum. *Marine Policy* 27: 75–78.

Relevant Websites

http://www.gbrmpa.gov.au
- Great Barrier Reef Marine Park Authority.
http://www.imo.org
- International Maritime Organization.
http://mpa.gov
- Marine Protected Areas of the United States.
http://www.un.org
- United Nations, Office of Legal Affairs, Division for Ocean Affairs and the Law of the Sea.
http://www.oceancommission.gov
- US Commission on Ocean Policy.
http://www.mms.gov
- US Department of the Interior, Minerals Management Service, Offshore Minerals Management.

COASTAL ZONE MANAGEMENT

D. R. Godschalk, University of North Carolina,
Chapel Hill, NC, USA

Introduction

Home to over half the world's population, coastal area environments and economies make critical contributions to the wealth and well-being of maritime countries. Market values stem from fisheries, tourism and recreation, marine transportation and ports, energy and minerals, and real estate development. Nonmarket values include life support and climate control through evaporation and carbon absorption, provision of productive marine and estuarine habitats, and enjoyment of nature.

Cumulative effects of human activities and coastal engineering threaten the sustainability of coastal resources. We are depleting coastal fisheries, degrading coastal water quality, draining coastal wetlands, blocking natural beach and barrier island movements, and spoiling recreational areas. Population growth is putting pressure on fragile ecosystems and increasing the number of people exposed to natural hazards, such as hurricanes, typhoons, and tsunamis. Maintaining sustainability demands active intervention in the form of coastal management.

Managing the conservation and development of coastal areas is, however, a challenging enterprise. Coastal land and water resources follow natural system boundaries rather than governmental jurisdiction boundaries, fragmenting management authority and responsibility. Coastal scientists and coastal managers operate in different spheres, fragmenting knowledge creation and dissemination. Coastal resource programs tend to focus on single sectors, such as water quality or land use, fragmenting efforts to manage holistic ecosystems. The field of coastal management has evolved to deal with the challenges of managing fragmented transboundary coastal resources.

Evolution of Coastal Zone Management

Coastal zone management developed to protect coastal resources from threats to their sustainability and to overcome the ineffectiveness of single function management approaches. Unlike ocean management issues, such as freedom of navigation and conservation of migratory species, coastal management traditionally focused on issues related to the land–sea interface, such as shoreline erosion, wetland protection, siting of coastal development, and public access. Important conceptual landmarks in the field's development are the intergovernmental framework of the 1972 US Coastal Zone Management Act, the 1987 Brundtland report proposing sustainable development, and the 1992 Earth Summit recommendation for initiation of integrated coastal management (ICM).

Coastal Zone Management

In 1972, the United States enacted the Coastal Zone Management Act to create a formal framework for collaborative planning by federal, state, and local governments. To receive federal funding incentives, states were required to set coastal zone boundaries, define permissible land and water uses, and designate areas of particular concern, such as hazard areas. An additional incentive was the promise that federal government actions would be consistent with the state's approved coastal zone management program. In practice, state programs tended to focus on land-use planning and regulation. Many other developed countries followed suit, establishing their own coastal zone management programs.

North Carolina offers an example of a state program developed under the Coastal Zone Management Act. Its 1974 Coastal Area Management Act established a coastal resource management program for the 20 coastal counties influenced by tidal waters. Policy is made by the Coastal Resources Commission, whose members are appointed by the governor. The act is administered by the state Division of Coastal Management, which issues coastal development permits, manages coastal reserves, and provides financial and technical assistance for local government planning. Required local land-use plans must meet approved standards, include a post-storm reconstruction policy, and identify watershed boundaries, but localities have considerable implementation flexibility. Four Areas of Environmental Concern are designated: estuarine and ocean, ocean hazard, public water supplies, and natural and cultural resources areas. All major development projects within an Area of Environmental Concern must receive a state permit. Small structures must be set back beyond the 30-year line of estimated annual beach erosion and large structures must be set back beyond the 60-year line. Shore-hardening structures are banned. While the North Carolina program has enjoyed considerable success, in practice many coastal localities continue to put economic development ahead of environmental protection.

The New Paradigm of Sustainable Development

In 1987, the UN World Commission on Environment and Development published its report *Our Common Future*, referred to as the Brundtland report in recognition of its chairman, Norwegian Prime Minister Gro Harland Brundtland. The report called for global sustainability, a concept which is now the dominant paradigm in coastal management. Sustainability requires economic development which meets the needs of the present generation to be achieved without compromising the ability of future generations to meet their own needs. Sustainability demands balance among the elements of a triple bottom line, sometimes referred to as 'the three e's': ecological sustainability, social equity, and economic enterprise. The triple bottom line has become a touchstone for accountability reporting of business as well as government actions. The 2002 UN World Summit on Sustainable Development revalidated the concept as a collective responsibility at local, national, regional, and global levels.

The main ideas of sustainable development can be stated in terms of questions to be asked of every environment and development decision:

- How does it improve the quality of human life?
- How does it affect the environment and natural resources?
- Are its benefits distributed equitably?

The Rise of Integrated Coastal Management

The 1992 United Nations Conference on Environment and Development in Rio de Janeiro – the Earth Summit – recommended principles to guide actions on environment, development, and social issues. It also approved Agenda 21, an action plan for sustainable development. While both the principles and actions are nonbinding, they represent a major shift toward understanding that sustainability must address the interdependence of environment and development in both developed countries (North) and developing countries (South). As shown in **Figure 1**, global environmental problems, such as greenhouse gases, generated in the North threaten the ability of the South to develop, while poverty and overpopulation in the South lead to local environmental stresses, such as air and water pollution.

Interdependence generates the need for integration between environment and development in sectors

Figure 1 Interdependence of environment and development. Source: Cicin-Sain B and Knecht RW (1998) *Integrated Coastal and Ocean Management: Concepts and Practices*, p. 83. Washington, DC: Island Press.

and nations. The ocean and coastal issues chapter of Agenda 21 calls for ocean and coastal management to be integrated in content and precautionary and anticipatory in ambit. The precautionary principle holds that lack of full scientific certainty shall not be used as a reason for postponing cost-effective measures to prevent environmental degradation. In other words, a conservative regulatory and management approach should be taken, in the absence of convincing evidence to the contrary.

The Agenda 21 section on integrated management and sustainable development of coastal areas calls on each coastal state to establish coordination mechanisms at both the local and national levels. It suggests undertaking coastal- and marine-use plans, environmental impact assessment and monitoring, contingency planning for both human-induced and natural disasters, improvement of coastal human settlements, conservation and restoration of critical habitats, and integration of sectoral programs such as fishing and tourism into a coordination framework. It also calls for national guidelines for ICM and actions to maintain biodiversity and productivity of marine species and habitats. The section highlights the need for information on coastal and marine systems and natural science and social science variables, along with education and training and capacity building.

The European Commission defines integrated coastal zone management (ICZM) as "a dynamic, multidisciplinary and iterative process to promote sustainable management of coastal zones. It covers the full cycle of information collection, planning (in its broadest sense), decision making, management and monitoring of implementation. ICZM uses the informed participation and cooperation of all stakeholders to assess the societal goals in a given coastal area, and to take actions towards meeting these objectives. ICZM seeks, over the long-term, to balance environmental, economic, social, cultural and recreational objectives, all within the limits set by natural dynamics. Integrated in ICZM refers to the integration of objectives and also to the integration of the many instruments needed to meet these objectives. It means integration of all relevant policy areas, sectors, and levels of administration. It means integration of the terrestrial and marine components of the target territory, in both time and space."

Coastal managers worldwide face many of the same problems: environmental degradation, marine pollution, fishery depletion, and loss of marine habitat. However, each nation's coastal management programs differ due to unique geography, development issues, and political system. Still, similar packages of tools and techniques are seen in many countries. The packages include combinations of regulations, national policies, planning, diagnostic studies of natural and socioeconomic systems and government capacity, incentives provision, and consensus building and participation to respond to conflicts. The ICM influence often is less evident in those developed countries with coastal programs already in place, than in less-developed countries where external funding assistance is needed to generate new coastal management institutions and to sustain traditional coastal livelihoods.

How well does ICM work? Evaluation efforts typically take the form of best practice stories, assessments of process outputs, or descriptions of levels of implementation. Program success indicators based on management evaluations indicate that successful programs tend to be comprehensive (based on ecosystem boundaries, such as watersheds or river basins), participatory (involving stakeholders), cooperative (networking among organizations), contingent (allowing for uncertainty and change), precautionary (acting conservatively to preserve the environment in the absence of conclusive evidence), long-term (with time horizons beyond project deadlines), focused (aimed at perceived issues or problems), incremental (taking small steps toward objectives, as in managed retreat from the shore), and adaptive (using learning to redirect program efforts). When more experience has been acquired, it may be possible to evaluate program outcomes or long-term, large-scale successes.

What types of ICM actions have been taken? Practices may take the form of innovative programs or institutions, as well as innovative applications to particular problems or content areas. Examples of both types of successful ICM best practices are illustrated in **Table 1**.

Regional Integrated Coastal Management Initiatives

With the maturing of ICM, its scope has expanded to include regional or transnational projects and programs. Inherent linkages among terrestrial and ocean processes have led to a number of regional ocean initiatives; 16 regional action plans have been formally adopted as of 2001 under the UN's Regional Seas Program. Each plan focuses on the unique problems of its region. For example, the East Asian Seas region has tackled problems of coastal aquaculture, fisheries exploitation, coral reef restoration, and resource extraction, using an ecoregion approach. Nations in the Western Indian Ocean region, where 70% of the African continent earn their living from natural resources, focus on simultaneously alleviating poverty and conserving natural resources.

Table 1 Successful ICM practices

Long-range planning and marine zoning for Australia's Great Barrier Reef Marine Park, where use and protection zones were mapped and participatory visioning was used to prepare a 25-year long-range plan

Bringing ocean and coastal management together in the Republic of Korea's new Ministry of Maritime Affairs and Fisheries, which integrated coastal zone management at the national level with navigation, ports, and fisheries programs

Involving publics in operation of the special area management zones in Ecuador, through education, training, and outreach

Determining the value of coastal ocean utilization in the pilot ICM program in Xiamen, China, through valuing the use of ocean space and charging users for mariculture and anchorage space

Incorporating tradition management practices in American Samoa's villages, which are responsible for monitoring and enforcing ICM measures on their land and waters

Controlling coastal erosion in Sri Lanka through establishment of a coastal zone development permit system, prohibition of coral mining except for research, and recruiting coastal communities into the development of management plans

Control of nonpoint marine pollution in the Chesapeake Bay, where a partnership among three US states (Virginia, Maryland, and Pennsylvania) and the federal government applies science to restore America's largest and most productive estuary

Protecting coastal resources in Turkey though designating specially protected areas to address rampant tourism development and its impacts on coastal waters and fisheries

Community-based coral reef protection in Phuket Island, Thailand, with a bottom-up program of education and outreach to protect the main tourist attraction, 60% of which has been seriously damaged or degraded

Excerpted from Cicin-Sain B and Knecht RW (1998) *Integrated Coastal and Ocean Management: Concepts and Practices*, pp. 297–299. Washington, DC: Island Press.

These developing regions promote sustainable livelihoods in order to meet both conservation and development goals.

The example of the European Union illustrates the evolution of a regional program. They conducted a 1996 demonstration with 35 projects in member countries and six thematic analyses of key ICM factors, including legislation, information, EU policy, territorial and sectoral cooperation, technical solutions, and participation. Their ICM strategy, approved in 2000, recommended that member states commit to a common vision for the future of their coastal zones, based on durable economic opportunities, functioning social and cultural systems in local communities, adequate open space, and maintenance of ecosystem integrity. It set out principles for a long-term holistic perspective, adaptive management, local specificity, working with natural processes, participatory planning, involvement of relevant administrative bodies, and use of a combination of instruments.

The EU strategy was able to draw on the experience gained from the 1975 Mediterranean Action Plan (MAP), one of the regional seas programs of the United Nations Environment Program (UNEP). Originally focused on marine pollution control, the MAP widened to deal with the land-based sources of 80% of pollution sources. It created coastal area management programs and, following Agenda 21, shifted them to an integrated basis. The Mediterranean region faces rapid coastal urbanization, growing tourist activities, high water consumption, concentrated pollution hot spots, biodiversity losses, and soil erosion. Lessons learned from the MAP experience emphasize the need for building an evaluation and monitoring mechanism at the beginning. Programs start out oriented to specific issues and problems, but must become more comprehensive at later stages in order to deal with complex linkages and provide integrated solutions. Strong political commitment at all levels to the preparation and initiation of initiatives, along with participation of stakeholders and end-users from program design through implementation, are of utmost importance. While ICM has provided a sturdy conceptual framework, full integration remains an elusive goal, as discovered in evaluation studies.

Ten MAP projects implemented during 1987–2001 were evaluated. The projects were located in three entire national coastal areas (Albania, Syria, and Israel), two areas of large parts of the national coast (Malta and Lebanon), two semi-enclosed bays polluted by urban and industrial expansion (Kastela Bay in Croatia and Izmir Bay in Turkey), two islands (Rhodes and Malta), a polluted industrialized urban coast (Sfax, Tunisia), and a relatively virgin coastal area under threat of uncontrolled development (Fuka-Matrouh, Egypt). The evaluation found that many changes occurred in the region during the multiyear implementation period, requiring the ICM projects to adapt to the new conditions, institutional structures, and coastal area priorities. Initial pilot projects focused on data collection, studies of pollution and ecosystem protection, and capacity building. First- and second-cycle projects focused on sectoral studies, resource management, implementation and its basic tools (e.g., environmental impact assessment and geographic information system or GIS), and introducing carrying capacity assessment. Third-cycle projects focused on sustainable development, applying ICM, and newer tools (strategic environmental assessment, systemic sustainability analysis, sustainability indicators, resource valuation, and socioeconomic evaluation).

The evaluation reported that the projects had an overall significant impact on national systems in the

Mediterranean region, with a more significant impact on the developing countries. The impacts directly influenced the practices, capacities, and awareness of the national and local authorities, institutions, teams, and stakeholders, as well as the resident coastal populations. The result was a strengthened awareness of the importance of coastal areas and the need for sustainable development, as well as an improvement in the ICM concepts and practices. Three of the 10 projects were also evaluated by a World Bank/Mediterranean Environmental Technical Assistance team in 1997. They found that, despite weaknesses in preparation and financial support, the overall impact of the projects was good, especially in terms of increased institutional capacity, impact on decision makers, and catalytic role of the projects. Performance problems were weak sectoral integration, weak participation and absence of domestic financial support, and uncertain project follow-up.

Assessments of ICM bring out both its strengths and weaknesses. Its strengths are based on the rationality of an integrated attack on interdependent coastal problems aimed at balancing ecological, economic, and equity values. No single government agency or program is equipped to deal with the complex interactions of oceans and coasts, yet it is precisely these interactions that pose such serious challenges to global sustainability. Its weaknesses stem from its need to reform and reinvent long-standing institutions, political turfs, and development practices. The times necessary to achieve such ambitious reforms far exceed the times of the usual 4–6-year ICM project. In absence of a crisis event to catalyze, and a huge infusion of funds to pay for, the necessary institutional change, progress proceeds slowly and long-term coordination takes a back seat to short-term demands. These same weaknesses plague efforts to manage all transboundary resources: air, water, and land.

Future Directions

All indications are that the need for ICM will continue to grow along with continued integration efforts. Serious global environmental problems are on the rise and more and more people are flocking to coastal areas where they will be exposed to potential catastrophes. The natural response by scientists and planners, governments and nongovernmental organizations, and other concerned stakeholders will be to call for further integration of coastal and ocean programs, institutions, and strategies. Increasing support for sustainable development in the form of

green buildings and smart growth by professional organizations, private businesses, and governmental bodies should provide a positive foundation for strengthening ICM around the world. Still the questions remain: Can we expect the emergence of sufficient political will to reform decades of territorial and institutional balkanization? Can we overcome the perception that ICM actions are an obstacle to economic development of the coast?

Indicators of a Looming Coastal Crisis

One possible catalyst for a worldwide surge in ICM is the impacts of the dangerous combination of climate change, natural hazards, resource depletion, poverty, and coastal population growth. Scientific studies have made it increasingly difficult for politicians to deny the reality of global warming. The transformation of the Earth's atmosphere through carbon dioxide emissions is driving up global temperature, which is expected to cause a string of disasters, including hurricanes, droughts, glacial melting, sea level rise, and ocean acidification. Since the turn of the century, the world experienced two of history's most catastrophic coastal disasters – the 2004 Indian Ocean tsunami with a death toll of 230 000 people triggered by an earthquake off the coast of Indonesia and the 2005 hurricane Katrina which devastated New Orleans and coastal Mississippi causing 1836 deaths and $81.2 billion in property damage. Yet people, rich and poor alike, continue to move to the ocean's edge, fueling the growth of vulnerable settlements and megacities. An estimated 40% of world cities of 1–10 million population and 75% of cities over 10 million population were located on the coast in 1995 and the number is growing. Clearly, continuation of these trends is a recipe for future disasters unless coastal settlement patterns resilient to natural hazards can be ensured.

Increasing Integration Efforts

One example of a major new ICM initiative is the 2004 report of the US Commission on Ocean Policy, *An Ocean Blueprint for the 21st Century*. The report acknowledges the magnitude of the crisis but insists that all is not lost. It envisions a future in which the coasts are attractive places to live, work, and play with clean water, public access, strong economies, safe harbors, adequate roads and services, and special protection for sensitive habitats and threatened species. Future management boundaries coincide with ecosystem regions, managers balance competing considerations and proceed with caution, and ocean governance is effective, participatory, and well coordinated. The report recommends creation of a National Ocean Council within the Executive Office of the President to coordinate federal policy, encourages groups of states to form networked regional ocean councils, and calls for federal laws and programs to be consolidated and modified to improve sustainability, including: amending the Coastal Zone Management Act to add measurable goals and performance measures and to implement watershed-based coastal zone boundaries; changing federal funding and infrastructure programs to discourage growth in fragile or hazard-prone coastal areas; and revising the National Flood Insurance Program to reduce incentives to develop in high-hazard areas. Response to the report has been generally positive, but it remains to be seen as to whether the recommendations will be implemented.

Possible Futures

The future of ICZM is closely tied to the future directions of global development. The Millennium Ecosystem Assessment has attempted to imagine possible future scenarios of global development pathways between 2000 and 2050. The scenarios represent different combinations of governance and economic development (global vs. regional) and of ecosystem management (reactive vs. proactive). They focus on ecosystem services and the effects of ecosystems on human well-being. The four scenarios are:

- Global Orchestration (socially conscious globalization, with emphasis on equity, economic growth, and public goods, and reactive toward ecosystems);
- Order from Strength (regionalized, with emphasis on security and economic growth, and reactive toward ecosystems);
- Adapting Mosaic (regionalized, with proactive management of ecosystems, local adaptation, and flexible governance); and
- TechnoGarden (globalized, with emphasis on using technology to achieve environmental outcomes and proactive management of ecosystems).

None of the scenarios represents a best or worst path, although ecosystems do better in the Techno-Garden scenario. However, they illuminate the potential consequences of choices in institutional design and management which could inform future decisions about coastal zone management.

Whether the catalysts of environmental threats, integrated management reforms, and scenarios of potential ecosystem outcomes can turn the tide toward fully integrated and effective coastal management is an open question. What is clear is that a strong worldwide movement is actively pursuing this goal.

See also

Coastal Topography, Human Impact on. Economics of Sea Level Rise. Tsunami.

Further Reading

Barbiere J and Li H (2001) *Third Millennium Special Issue on Megacities. Ocean and Coastal Management* 44: 283–449.

Beatley T, Brower DJ, and Schwab AK (2002) *An Introduction to Coastal Zone Management*, 2nd edn. Washington, DC: Island Press.

Belfiore S (ed.) (2002) *From the 1992 Earth Summit to the 2002 World Summit on Sustainable Development: Continuing Challenges and New Opportunities for Capacity Building in Ocean and Coastal Management. Special Issue on Capacity Building Ocean and Coastal Management* 45: 541–718.

Burbridge P and Humphrey S (eds.) (2003) *Special Issue: The European Demonstration Programme on Integrated Coastal Zone Management. Coastal Management* 31: 121–212.

Cicin-Sain B and Knecht RW (1998) *Integrated Coastal and Ocean Management: Concepts and Practices.* Washington, DC: Island Press.

Cicin-Sain B, Pavlin I, and Belfiore S (eds.) (2002) *Sustainable Coastal Management: A Transatlantic and Euro-Mediterranean Perspective.* Dordrecht: Kluwer.

Francis J, Tobey J, and Torell E (eds.) (2006) *Special Issue: Balancing Development and Conservation Needs in the Western Indian Ocean Region. Ocean and Coastal Management* 49: 789–888.

Glavovic B (2006) Coastal sustainability – an elusive pursuit: Reflections on South Africa's coastal policy experience. *Coastal Management* 34: 111–132.

Heilerman S (ed.) (2006) *IOC Manuals and Guides 46, ICAM Dossier 2: A Handbook for Measuring the Progress and Outcomes of Integrated Coastal and Ocean Management.* Paris: UNESCO.

Kay R and Alder J (2005) *Coastal Planning and Management*, 2nd edn. London: Taylor and Francis.

Mageau C (ed.) (2003) *Special Issue: The Role of Indicators in Integrated Coastal Management. Ocean and Coastal Management* 46: 221–390.

Sain B and Knecht RW (1998) *Integrated Coastal and Ocean Management: Concepts and Practices.* Washington, DC: Island Press.

Stojanovic T, Ballinger RC, and Lalwani CS (2004) Successful integrated coastal management: Measuring it with research and contributing to wise practice. *Ocean and Coastal Management* 47: 273–298.

Tobey J and Lowry K (eds.) (2002) *Learning from the practice of integrated coastal management. Coastal Management* 30: 283–345.

US Commission on Ocean Policy (2004) *An Ocean Blueprint for the 21st Century.* Final Report. Washington, DC: US Commission on Ocean Policy.

Vallega A (2002) The regional approach to the ocean, the ocean regions, and ocean regionalization – a post-modern dilemma. *Ocean and Coastal Management* 45: 721–760.

Williams E, Mcglashan D, and Finn J (2006) Assessing socioeconomic costs and benefits of ICZM in the European Union. *Coastal Management* 34: 65–86.

Relevant Websites

http://web.worldbank.org
 – Coastal and Marine Management, World Bank.
http://www.unesco.org
 – Coastal Regions and Small Islands (CSI) Portal, UNESCO.
http://www.coastalguide.org
 – EUCC: Coastal Guide.
http://ec.europa.eu
 – EUROPA Coastal Zone Policy.
http://www.keysheets.org
 – Integrated Coastal Management Keysheet.
http://www.coastalmanagement.com
 – Integrated Coastal Management Websites.
http://www.icriforum.org
 – International Coral Reef Initiative.

http://ioc3.unesco.org
 – IOC Marine Sciences and Observations for Integrated Coastal Management.
http://www.millenniumassessment.org
 – Millennium Ecosystem Assessment.
http://www.netcoast.nl
 – NetCoast: A Guide to Integrated Coastal Management.
http://www.oneocean.org
 – OneOcean: Coastal Resources and Fisheries Management of the Philippines.
http://www.pemsea.org
 – PEMSEA Integrated Coastal Management.
http://www.unep.org
 – Regional Seas Programme, United Nations Environment Programme.
http://www.deh.gov.au
 – The Contribution of Science to Integrated Coastal Management.
http://www.globaloceans.org
 – The Global Forum on Oceans, Coasts, and Islands.
http://www.csiwisepractices.org
 – Wise Coastal Practices for Sustainable Human Development Forum.
http://www.ngdc.noaa.gov
 – World Data Center for Marine Geology and Geophysics: National Geophysical Data Center.
http://earthtrends.wri.org
 – World Resources Institute: Earth Trends.

WILD-HARVEST FISHERIES

FISHERIES OVERVIEW

M. J. Fogarty, Northeast Fisheries Science Center, National Marine Fisheries Service, Woods Hole, MA, USA
J. S. Collie, University of Rhode Island, Narragansett, RI, USA

Introduction

The long-standing importance of fishing as a human enterprise can be traced through the diversity of harvesting implements found in ancient archeological sites, artistic depictions throughout prehistory and antiquity, and the recorded history of many civilizations. Sustenance from the sea has been essential throughout human history and has formed the basis for trade and commerce in coastal cultures for millennia. The remains of fish and shellfish in extensive middens throughout the world attest to the prominence of these resources in the diets of early coastal peoples. In Northern Europe, the fortunes of the Hanseatic League during the medieval period were linked to trade in fishery resources, demonstrating a dominant role of fisheries in the trade of nations that extends over many centuries. Today, the critical importance of food resources from the sea has been further highlighted by increased demand related to the burgeoning human population and the recognized benefits of seafood as a high-quality source of protein. In turn, this importance is reflected in the diversity of fishery-related topics included in this encyclopedia. Here, we introduce the topics to be covered in the individual sections on fishing and fisheries resources and provide further background information to set the stage for these contributions.

The fishery resources of the oceans were long thought to be boundless and the high fertility of fishes was thought to render them impervious to human depredation. Thomas Henry Huxley, the preeminent Victorian naturalist, wrote in 1884 "... the cod fishery, the herring fishery, the pilchard fishery, the mackerel fishery, and probably all the great sea-fisheries, are inexhaustible; that is to say that nothing we do seriously affects the number of fish ... given our present mode of fishing. And any attempt to regulate these fisheries consequently ... seems to be useless." (quoted in Smith, 1994). Although Huxley qualified his remarks and limited them to the harvesting methods of his day and to ocean fisheries, the paradigm of inexhaustibility was broadly accepted, shaping the attitudes of fishers, scientists, managers, and politicians and complicating efforts to establish effective restrictions on fishing activities. Experimental studies of the effects of fishing on marine populations were conducted in Scotland as early as 1886. In a decade-long study, bays open and closed to fishing were compared, demonstrating that harvesting did result in declines in the abundance and average size of exploited fishes. However, the results were deemed controversial and the debate concerning the impact of fishing on marine populations continued through the middle of the twentieth century. By the second half of the last century with the development of large-scale industrial fisheries and distant-water fleets, it was abundantly clear that humans have the capacity to outstrip the production capacity of exploited marine populations, resulting in resource depletion and loss in yield.

In contrast to the long history of human harvest of the oceans, scientific endeavors in support of resource management (and the understanding of underlying basic ecological principles) are comparatively recent. Indeed, the term ecology (*oecologie*) itself was not coined until 1866, while written records of large-scale marine fisheries predate this landmark by several centuries. In many instances, fish and shellfish populations had already been substantially altered by fishing prior to the development of a scientific framework within which to evaluate these changes and true baseline conditions can only be inferred.

Because of the broad spatial and temporal scales over which fisheries operate, institutions dedicated to monitoring fisheries and resource species and providing scientific advice in support of management have been established. In Western countries, many of these institutions were formed in the latter nineteenth and early twentieth centuries. For example, the US Fish Commission was established in 1871 in response to concerns over declines in coastal fishery resources at that time. The Fishery Board of Scotland was established in 1883 and the International Council for Exploration of the Sea followed in 1902. These institutions and others such as the Marine Biological Association of the United Kingdom, established in 1884, approached the problem of understanding fluctuations of exploited fish and shellfish species from a broad scientific perspective, including consideration of the physical and biological environments of the organisms. Spencer Fullerton Baird, first US Commissioner of Fisheries, wrote that studies of fish "... would not be complete without a thorough knowledge of their associates in the sea, especially of such as prey upon them or constitute their food" (Baird, 1872). Baird further noted with respect to the importance of understanding the physical setting in the ocean that "... the temperature taken at different depths, its varying transparency, density, chemical composition, percentage of saline matter, its surface- and undercurrents, and other features of its physical condition ... throw more or less light on the agencies which exercise and influence upon the presence or absence of particular fishes." This broad multidisciplinary perspective remains an important component of fisheries investigations. Recently, the importance of an ecosystem perspective has been reemphasized and efforts to understand the potential impact of climate change on fishery resources have assumed high priority.

Attempts to estimate the production potential of the seas, based on energy flow in marine ecosystems, initially indicated that the coastal ocean could sustain yields of approximately 100 million tons of fish and shellfish on a global basis. Worldwide marine landings in 2004 were 86 million tons (*see* Marine Fishery Resources, Global State of). Globally, 3% of the stocks for which information is available are classified as underexploited, 20% are moderately exploited, 52% are considered fully exploited, 17% are overexploited, 7% are depleted, and 1% are listed as recovering. It is clear that we are at or near the limits to production for many exploited populations and have exceeded sustainable levels for many others. Improved management does hold the promise of increasing potential production in overexploited and depleted populations; carefully controlled increases in exploitation of currently underutilized species also may result in some increase in yield. As noted above, the pressures on fishery resources on a global basis are directly related to increases in human population size and the resulting increased demand for protein from the sea. Further, the recently emphasized health benefits of seafood consumption have resulted in increases in per capita consumption of fish in many Western countries that traditionally had comparatively low consumption levels.

Fishing and Fishery Resources

The articles concerning fishery resources in this encyclopedia document the broad spectrum of species taken and modes of capture in fisheries worldwide. The descriptions of fisheries for those species groups summarized herein are linked to overviews of their biological and ecological features elsewhere in the encyclopedia. Reviews are provided for major species groups supporting important fisheries. These reviews cover small-bodied fishes inhabiting the open water column (small pelagic fishes), larger-bodied pelagic fishes, bottom-dwelling organisms (demersal species), salmon, and shellfish (including crustaceans and mollusks). Regions (and associated species) requiring special consideration such as the vulnerable Antarctic marine ecosystem(s), coral reefs, and deep-water habitats are accorded separate treatment. In addition, articles dealing with a number of overarching issues are included to provide a broader context for understanding the importance of fisheries to society and their impacts on natural systems. An overview of fishing methods and techniques employed throughout the world is provided as are review articles on the global status of marine fishery resources, factors controlling the dynamics of exploited marine species, harvesting multiple species assemblages, and the ecosystem effects of fishing. Key considerations in the management of marine resources are documented and the intersection between human interests and motivations and resource management are explored.

The biological and ecological characteristics of the species sought in different fisheries and their behavioral patterns play a dominant role in harvesting methods, vulnerability to exploitation, and overall yields. The general strategies involved in fish capture include the use of entangling gears, trapping, filtering with nets, and hooking or spearing (*see* Fishing Methods and Fishing Fleets). Based on their experience and that of others gained over time, fishermen use detailed information on distribution, seasonal movement patterns, and other aspects of behavior of different species in the capture process. Recently, advances in electronic equipment, ranging from sophisticated hydroacoustic fishfinders to satellite navigation systems, have allowed fishermen to refine their understanding of these characteristics, greatly increasing the efficiency of fishing operations and the consequent impact on fishery resources.

Small-bodied fish such as herring, mackerel, and sardines that characteristically form large schools are often taken by surrounding nets or purse seines in large quantities. These pelagic species typically inhabit mid- to near-surface water depths. Schooling behavior can increase the detectability of these species and therefore their vulnerability to capture, a problem which has resulted in overharvesting of small pelagic species in many areas. Similar considerations hold for larger-bodied pelagic fishes such as the tunas. The harvesting pressure on these larger species is fueled by their high unit value; management is complicated by their extensive movements and migrations, necessitating international management protocols and agreements (*see* Fishery Management).

A diverse assemblage of fish typically inhabiting near-bottom or bottom waters support fisheries of long-standing importance such as the cod fisheries throughout the North Atlantic, halibut fisheries in both the Atlantic and Pacific, and many others (*see* Demersal Species Fisheries).

The exploited demersal species exhibit a wide range of body sizes and life history characteristics that influence their response to exploitation. These species are often captured with nets dragged over the seabed, although traps, entangling nets, and other devices are also used. Such fishing gear often captures many species not specifically targeted by the fishery and may also disrupt the bottom habitat and associated species. As a result, substantial concern has been expressed over both the direct and indirect effects of bottom fishing practices with respect to alterations of food web structure and disturbance to critical habitat (*see* Ecosystem Effects of Fishing).

Species such as Atlantic and Pacific salmon spawn in fresh water but spend a substantial part of their life cycle in the marine environment. These anadromous species are impacted not only by harvesting but also by land- and freshwater-use practices affecting the part of the life cycle occurring in fresh water. Damming of rivers, deforestation, pollution, and overharvesting have all resulted in declines in these species in some areas. Salmon exhibit strong homing instincts and typically return to their natal river system to breed. The concentration of fish as they return to river systems and in the rivers and lakes themselves makes these species particularly vulnerable to capture. Pacific salmon differ considerably from most other fish species that support important fisheries in that they die after spawning only once; it is essential that a sufficient number of adults escape the capture process to replenish the population. Artificial enhancement through hatchery programs has been very widely employed for salmon stocks in an attempt to maintain viable populations although concerns have been raised about effects on the genetic structure of natural stocks and the possibility of transmission of diseases from hatchery stocks.

Fisheries for shellfishes, including those for lobsters, crabs, and shrimp and for clams, snails, and squids (as well as other cephalopods) are among the most lucrative in the world. Mollusks also are taken for uses other than food, such as the ornamental trade in shells. Most shellfish live immediately on or near the seabed and are harvested by traps (lobsters, crabs, some cephalopods such as octopus, and some snails such as whelks), towed nets (e.g., shrimps and squid), and dredges (oysters, clams, etc.) among other devices. The ready availability of some types of shellfish in intertidal habitats has resulted in a very long history of exploitation by coastal peoples. Important commercial and recreational fisheries continue for these shellfish, which are often harvested with very simple implements. The high unit value of shellfish makes aquaculture economically feasible to supplement harvesting of natural populations.

Fisheries prosecuted in some habitats and environments require special considerations. For example, in the waters off Antarctica, the need to protect the potentially vulnerable food web, encompassing endangered and threatened marine mammal populations, has led to the development of a unique ecosystem-based approach to management. Harvesting of krill populations, a preferred prey of a number of whale, seal, and seabird populations, is regulated with specific recognition of the need to avoid disruption of the food web and impacts on these predators. In coral reef systems, the very high diversity of species found and the sensitivity of the habitat to disruption has highlighted the need for an ecosystem approach to management in these systems. Growing concern over destructive fishing practices using toxins and explosives has emphasized the need for effective management in these areas. In deep-water habitats, many resident species exhibit life history characteristics such as slow growth and delayed maturation that make them particularly vulnerable to exploitation. The vulnerability of deep-sea habitats to disturbance by fishing gear is also a dominant concern in these environments.

Issues in Fishery Management

Fishery management necessarily entails consideration of resource conservation, the economic implications of alternative management strategies, and the social context within which management decisions are effected (*see* Fishery Management). The relative weights assigned to these diverse considerations can vary substantially in different settings, resulting in very different management decisions and outcomes. The setting of conservation standards is tied directly to understanding of basic life history characteristics such as the rate of reproduction at low population levels, growth characteristics, and factors affecting the survivorship from the early life stages to the age or size of vulnerability to the fishery (*see* Dynamics of Exploited Marine Fish Populations and Fisheries and Climate). The choice of particular harvesting strategies and levels holds both economic and social implications. Fishery management is ultimately a political process and decisions concerning allocation of fishery resources often engender intense debates (*see* Fishery Management, Human Dimension). These debates are often set within the context of differing perspectives on fishing rights and privileges. In many societies, fishing is viewed as a basic right open to all citizens and fishery resources are often viewed as a form of common property.

Formal designation of fishery resources as *res nullia* (things owned by no one) can be traced to Roman law where ownership was conferred by the process of capture. Traditions of open access to fishery resources in many Western countries persist and remain a principal factor in the global escalation of fishing pressure. This legacy has led to excess capacity and overcapitalization of world fishing fleets, resulting in conflicts between conservation requirements and the social and short-term economic impacts of implementing rational and effective management. Garrett Hardin's (1968) influential statement of the "Tragedy of the commons" – a resource owned by no one is cared for by no one – has been applied to fishery resources and further honed to reflect considerations of the importance of well-defined property rights and attendant responsibilities in natural resources management. Various forms of dedicated access privileges have been implemented in fisheries around the world to reduce overcapitalization and to vest fishermen in the long-term sustainability of fisheries (*see* Fisheries Economics).

Biological reference points provide the basis for specifying objectives for fishery management in many of the major fisheries throughout the world. Limit reference points define the boundaries of a situation that could cause serious harm to a stock, while target reference points are used to determine harvest control rules that are risk-averse and have a low probability of causing serious harm. Limits are conceived as reference levels that should have a low probability of being exceeded and are designed to prevent stock declines through recruitment overfishing. Targets are reference levels providing management goals but which may not necessarily be met under all conditions. Although originally conceived as target reference points, the fishing mortality rate resulting in maximum sustainable yield and the corresponding level of equilibrium biomass are now commonly employed as limit reference points. Yield and spawning stock biomass per recruit analyses have been used to provide both limit and target reference points (*see* Dynamics of Exploited Marine Fish Populations). It is possible to construct a two-dimensional representation of the exploitation status of a stock in relation to the estimated levels of fishing mortality and population size (**Figure 1**). When

coupled with information on threshold levels of fishing mortality and population size used to define limit reference points, decision rules can be defined to assess appropriate courses of action. In instances where the limit fishing mortality reference point is exceeded, 'overfishing' is said to occur; when the stock declines below the limit biomass reference point, the stock is 'overfished' and management action is required. It is further possible to specify target exploitation levels in this context.

Although biological reference points have been widely applied on a global basis and are often required under fisheries legislation, corresponding economic reference points exist and deserve special consideration. For example, the concept of maximum economic yield has served as a cornerstone of resource economic theory (*see* Fisheries Economics). Resource economists have long recognized that in an unregulated open-access fishery, fishing effort increases to a bioeconomic equilibrium at which profits are completely dissipated. Developing ways to understand and create appropriate incentive structures to ensure appropriate and efficient economic utilization of fishery resources is critically important.

Tools available to fishery managers to control the activities of harvesters include constraints on the overall amount of fishing activity (measured as the number of vessels allowed permits, the number of days vessels can spend at sea, etc.), the total amount of the catch allowed in a specified time period and regulated by various forms of quota systems, the types of fishing gear that can be used and their characteristics (e.g., net mesh size), and closures of fishing grounds (including seasonal and year-round closures) (*see* Fishery Management). Often, several of these

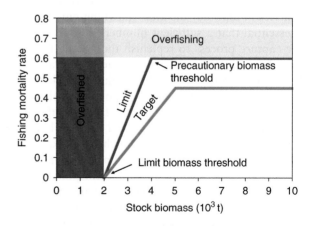

Figure 1 Control diagram for evaluating the status of a marine fish stocks in relation to biological reference points. The green line defines target levels of stock biomass and fishing mortality; the red line defines limits to fishing mortality and biomass thresholds.

tools will be used in combination to meet specified management objectives. As we move toward a paradigm of Ecosystem Approaches to Fishery Management (EAFM), these basic tools will remain the essential elements in tactical fishery management. However, because a broader suite of objectives will be embodied in EAFM, the mix of management tools applied in a given setting will undoubtedly differ relative to single-species management strategies.

Failures in fishery management can often be traced to conflicting goals and objectives in the conservation, economic, and social dimensions. For example, the needs for conservation can be compromised by desires to maintain full employment opportunities in the fishing industry if this leads to political pressure to permit high harvest levels. In the longer term, actions taken to ensure the sustainability of fishery systems also ensure the viability of the industries dependent on these resources. However, the short-term impacts (or perceived impacts) of fisheries regulations on fishers and fishing communities are often dominant considerations in whether particular regulations will be put in place (see Fishery Management, Human Dimension).

Emerging Issues in Fisheries and Fisheries Science

The need for a more holistic view of human impacts in the marine environment is increasingly recognized. Harvesting has both direct and indirect effects on marine ecosystems. The former include removal of biomass and potential impacts on habitat and non-target species (see Ecosystem Effects of Fishing). The latter includes alteration in trophic structure through species-selective harvesting patterns changing the relative balance of predators and their prey. Multispecies considerations in fishery management account for interactions among harvested species and the need to consider factors such as the food and energetic requirements of protected resource species. These interactions must be further viewed in the context of external forcing such as climate change and variability as noted in Fisheries and Climate (see also below). Collectively, these factors can result in shifts in productivity states that must be accounted for in management. Further, species interactions require that we explicitly deal with trade-offs in management (e.g., between predators and their prey).

The global escalation in fishing pressure and recognition of the potential environmental impacts of fishing activities have led to an increased interest in and emphasis on an ecosystem approach to fisheries management. An ecosystem approach to fisheries management seeks to ensure the preservation of ecosystem composition, structure, and function based on an understanding of ecological interactions and processes required for ecosystem integrity. As noted above, ecosystem principles guide the management of marine resources off Antarctica, and broader application of an ecosystem approach to fisheries management approaches is under development in many other regions.

The interest in an ecosystem approach to fisheries management has led to a reevaluation of management tools and an increased emphasis on the use of strategies such as the development of marine protected areas in which harvesting and other extractive activities are strongly controlled. The use of closed areas as a fishery management tool has an extensive history and the current focus on marine reserves can be viewed as an extension of these long-standing approaches to meet broader conservation objectives. Areas in which all extractive practices are banned may reduce overall exploitation rates, protect sensitive habitats and associated biological communities, and preserve ecosystem structure and function. The use of no-take marine reserves has been also advocated as a hedge against uncertainty in our understanding of ecosystem structure and function and in our ability to control harvest rates. While there is substantial evidence of increases in biomass, mean size, and biodiversity within reserves, the effects on adjacent areas through spillover effects are less well documented, although information is accruing. The importance of reserves as source areas for adjacent sites open to harvesting and other activities can be critical for the resilience of spatially linked populations. Used in concert with other measures that restrain fishing activities in areas open to fishing, no-take marine protected areas can be a highly effective component of an ecosystem approach to fisheries management strategies.

The prospect of global climate change and its implications for terrestrial and marine systems is one of the most pressing issues facing us today. The potential impacts of global climate change in the marine environment are receiving increased attention as higher-resolution forecasting models are developed and oceanographic features are more fully represented in general circulation models. The projected climate impacts with respect to changes in temperature, precipitation, and wind fields hold important implications for oceanic current systems, mesoscale features such as frontal zones and eddies, stratification, and thermal structure. In turn, these impacts will affect marine organisms dependent on the physical geography of the sea for dispersal in currents, the size and location of feeding and spawning grounds, and basic biological considerations such as temperature effects on metabolism. Shifts in distribution patterns of fishery resource species, changes in vital rates such as survivorship and growth, and alterations in the structure of marine communities can be anticipated. Persistent changes in environmental conditions can affect the production characteristics of different systems and the production potential of individual species (*see* Fisheries and Climate). In particular, a shift in environmental states can interact synergistically with fishing pressure to destabilize an exploited population. Exploitation rates that are sustainable under a favorable environmental regime may not remain so if a shift to less favorable conditions occurs (*see* Dynamics of Exploited Marine Fish Populations). Large-scale research programs such as the Global Ocean Ecosystem Dynamics (GLOBEC) Program have now been implemented throughout the world to assess the potential effects of global climate change on marine ecosystems, including impacts on resource species.

Collectively, the problems of overexploitation, habitat loss and degradation, alteration of ecosystem structure, and environmental change caused by human activities point to the need to consider humans fully as part of the ecosystem and not somehow apart, and to manage accordingly. Notwithstanding the depleted status of many world fisheries, several important fisheries have recovered from overexploitation in response to management regulations. Wisely managed, fisheries can continue to meet important human needs for food resources from the sea while meeting our obligations to future generations.

See also

Demersal Species Fisheries. Dynamics of Exploited Marine Fish Populations. Ecosystem Effects of Fishing. Fisheries and Climate. Fisheries Economics. Fishery Management. Fishery Management, Human Dimension. Fishing Methods and Fishing Fleets. Marine Fishery Resources, Global State of.

Further Reading

Baird SF (1872) Report on the conditions of the sea fisheries, 1871. *Report of the US Fish Commission 1.* Washington, DC: US Fish Commission.

Cushing DH (1988) *The Provident Sea.* Cambridge, UK: Cambridge University Press.

FAO (2006) *The State of World Fisheries and Aquaculture.* Rome: Food and Agriculture Organization of the United Nations. http://www.fao.org/sof/sofia/index_en.htm (accessed Mar. 2008).

Fogarty MJ, Bohnsack J, and Dayton P (2000) Marine reserves and resource management. In: Sheppard C (ed.) *Seas at the Millennium: An Environmental Evaluation,* ch. 134, pp. 283–300. Amsterdam: Elsevier.

Hardin G (1968) The tragedy of the commons. *Science* 162: 1243–1247.

Jennings S, Kaiser MJ, and Reynolds JD (2001) *Marine Fisheries Ecology.* Oxford, UK: Blackwell Science.

Kingsland SE (1994) *Modeling Nature.* Chicago, IL: University of Chicago Press.

Pauly D (1996) One hundred million tons of fish, and fisheries research. *Fisheries Research* 25: 25–38.

Quinn TJ, II and Deriso RB (1999) *Quantitative Fish Dynamics.* Oxford, UK: Oxford University Press.

Ryther J (1969) Photosynthesis and fish production in the sea. *Science* 166: 72–76.

Sahrhage D and Lundbeck J (1992) *A History of Fishing.* Berlin: Springer.

Smith TD (1994) *Scaling Fisheries: The Science of Measuring the Effects of Fishing, 1855–1955.* Cambridge, UK: Cambridge University Press.

MARINE FISHERY RESOURCES, GLOBAL STATE OF

J. Csirke and S. M. Garcia, Food and
Agriculture Organization of the United Nations, Rome,
Italy

Introduction

The Fisheries Department of the Food and Agriculture Organization of the United Nations (FAO) monitors the state of world marine fishery resources and produces a major review every 6–8 years, with shorter updates presented every 2 years to the FAO Committee on Fisheries (COFI) as part of a more general report *The State of Fisheries and Aquaculture* (*SOFIA*). The latest major FAO review of the state of world marine fishery resources was issued in 2005 with a shorter update in *SOFIA 2006* and a further updating focusing on fishery resources that can be found partly or entirely on the high seas also published in 2006. This article draws significantly from sections of the above FAO publications and uses catch information available from 1950 to 2004 (the last year for which global catch statistics are available).

With a view to offering a comprehensive description of the global state of world fish stocks, the short analysis provided below considers successively: (1) the relation between 2004 and historical production levels; (2) the state of stocks, globally and by regions according to information from the FAO reports above; and (3) the trends in state of stocks since 1974, globally and by region.

Relative Production Levels

The catch data available for the 19 FAO statistical areas (**Table 1**) or regions of the world's oceans indicate that four of them are at or very close to their maximum historical level of production: the Eastern Indian Ocean and the Western Central Pacific Oceans reached maximum production in 2004 while the Western Indian Ocean and the NE Pacific produces 95% and 90% of the maximum, produced in 2003 and 1987, respectively. All other regions are presently producing less than 90% of their historical maximum, for various reasons. This may result, at least in part, from natural oscillations in productivity caused by ocean climate variability or by the fact that short-term exceptionally high initial catches

(a phenomenon known as 'overshooting') can be obtained when a new fishery on a previously unexploited stock develops too fast. However, the lower catch values in recent years may also be indicative of an increase in the number of resources being overfished or depleted.

Global Levels of Exploitation

The recent FAO reviews report on 584 stock or species groups (stock items) being monitored. For 441 (or 76%) of them, there is some more-or-less recent information allowing some estimates of the state of exploitation. These 'stock' items are classified as underexploited (U), moderately exploited (M), fully exploited (F), overexploited (O), depleted (D), or recovering (R), depending on how far they are from 'full exploitation' in terms of biomass and fishing pressure. 'Full exploitation' is used by FAO as loosely equivalent to the level corresponding to maximum sustainable yield (MSY) or maximum long-term average yield (MLTAY).

1. U and M stocks could yield higher catches under increased fishing pressure, but this does not imply any recommendation to increase fishing pressure.
2. F stocks are considered as being exploited close to their MSY or MLTAY and could be slightly under or above this level because of natural variability or uncertainties in the data and in stock assessments. These stocks are usually in need of (and in some cases already have) effective control on fishing capacity in order to remain as fully exploited, and avoid falling in the following category.
3. O or D stocks are exploited beyond MSY or MLTAY levels and have their production levels reduced; they are in need of effective strategies for capacity reduction and stock rebuilding.
4. R stocks are usually at very low abundance levels compared to historical levels. Directed fishing pressure may have been reduced by management or because of profitability being lost, but may nevertheless still be under excessive fishing pressure. In some cases, their indirect exploitation as bycatch in another fishery might be enough to keep them in a depressed state despite reduced direct fishing pressure.

According to information available in 2004 (**Figure 1**), 3% of the world fish stocks (all included)

Table 1 Ratio between recent (2004) total catch and maximum reached catch, by FAO statistical areas

FAO area (code and name)	Year when maximum catch was reached	Total catch in 2004 (10^3)	Ratio of 2004 over maximum catch (%)
18 – Arctic Sea	1968	0	0
21 – Atlantic Northwest	1968	2353	52
27 – Atlantic Northwest	1976	9952	77
31 – Atlantic, Western Central	1984	1652	66
34 – Atlantic, Eastern Central	1990	3392	82
37 – Mediterranean and Black Sea	1988	1528	77
41 – Atlantic, Southwest	1997	1745	66
47 – Atlantic, Southwest	1978	1726	53
51 – Indian Ocean, Western	2003	4147	95
57 – Indian Ocean, Western	2004	5625	100
61 – Pacific, Northwest	1998	21 558	87
67 – Pacific, Northwest	1987	3050	90
71 – Pacific, Western Central	2004	11 011	100
77 – Pacific, Eastern Central	2002	1701	87
81 – Pacific, Southwest	1992	736	80
87 – Pacific, Southwest	1994	15 451	76
48 – Atlantic, Antarctic	1987	125	25
58 – Indian Ocean, Antarctic	1972	8	4
88 – Pacific, Antarctic	1983	3	28

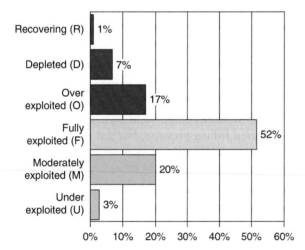

Figure 1 State of world fish stocks in 2004. Reproduced with permission from FAO (2005) Review of the State of World Marine Fishery Resources. *FAO Fisheries Technical Paper, 457*, 235pp, figure A2.1. Rome: FAO.

appeared to be underexploited, 20% moderately exploited, 52% fully exploited, 17% overfished, 7% depleted, and 1% recovering.

On the one hand, this indicates that 25% of the world stocks (O + D + R) for which some data are available are below the level of abundance corresponding to MSY or are exploited with a fishing capacity well above this level. They require management to rebuild them at least to the level corresponding to MSY as provided by the 1992 UN Convention on the Law of the Sea (UNCLOS). As 52% of the stocks appear to be exploited around

MSY and most, if not all, also require that capacity control measures be applied to avoid the negative effects of overcapacity, it appears that 77% (F + O + D + R) of the world stocks for which data are available require that strict capacity and effort control be applied in order to be stabilized or be rebuilt around the MSY biomass levels, and possibly beyond. Some of the fisheries concerned may already be under such management schemes.

On the other hand, **Figure 1** also indicates that 23% (U + M) of the world stocks for which some data are available are above the level of abundance corresponding to MSY, or are exploited with a fishing capacity below this level. Considering again that 52% of the stocks are exploited around MSY, this means that 75% of the stocks (U + M + F) are at or above MSY level of abundance, or are exploited with a fishing capacity at or below this level, and should be therefore considered as compliant with UNCLOS basic requirements.

These two visions of the global situation of fishery stocks indicate that the 'glass is half full or half empty' and both are equally correct depending on which angle one takes. From the 'state of stocks' point of view, it is comforting to see that 75% of the world resources are still in a state which could produce the MSY, as provided by UNCLOS. From the management point of view, it should certainly be noted that 77% of the resources require stringent management and control of fishing capacity. As mentioned above, some of these (mainly in a few developed countries) are already under some form of capacity management. Many, however, would

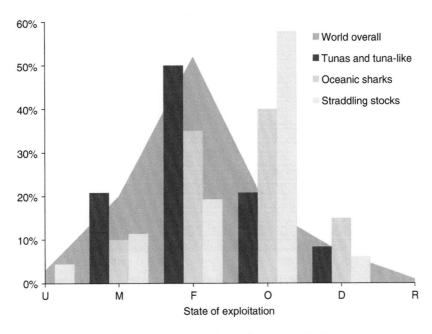

Figure 2 State of exploitation of world highly migratory tuna and tuna-like species, highly migratory oceanic sharks, and straddling stocks (including high seas fish stocks). Reproduced with permission from Maguire J-J, Sissenwine M, Csirke J, Grainger R, and Garcia S (2006) The State of World Highly Migratory, Straddling and Other High Seas Fishery Resources and Associated Species. *FAO Fisheries Technical Paper*, 495, 84pp, figure 59. Rome: FAO.

require urgent action to stabilize or improve the situation. For 25% of them, energetic action is required for rebuilding.

The situation appears more critical in the case of fish stocks that occur partly or entirely, and can be fished in the high seas (**Figure 2**).

While the state of exploitation of highly migratory tuna and tuna-like species is very similar to that of the world overall, the state of exploitation of highly migratory oceanic sharks appears to be more problematic, with more than half of the stocks listed as overexploited or depleted. The state of straddling stock (including high seas stocks) is even more problematic with nearly two-thirds being classified as overexploited or depleted.

State of Stocks by Region

When the available information is examined by regions (**Figure 3**), the percentage of stocks exploited at or beyond levels of exploitation corresponding to MSY (F + O + D + R) and needing fishing capacity control ranges from 43% (for the Eastern Central Pacific) to 100% (in the Western Central Atlantic). Overall, in most regions, 70% of the stocks at least are already fully exploited or overexploited. The percentage of stocks exploited at or below levels corresponding to MSY (U + M + F) ranges from 48% (in the Southeast Atlantic) to 100% (in the

Eastern Central Pacific). An indication of how weak (or strong) management and development performance can be given by the proportion of stocks that are exploited beyond the MSY level of exploitation (O + D + R), that in the latest reviews were ranging from 0% (for the Eastern Central Pacific) to 52% (for the Southeast Atlantic).

Global Trends

The trends in the proportion of stocks in the various states of exploitation as taken from the various FAO reviews since 1974 (**Figure 4**) shows that the percentage of stocks maintained at MSY level, or fully exploited (F) has slightly increased since 1995, reversing the previous decreasing trend since 1974. The underexploited and moderately exploited stocks (U + M), offering some potential for expansion, continue to decrease steadily, while the proportion of stocks exploited beyond MSY levels (O + D + R) increased steadily until 1995, but has apparently leveled off and remained more or less stable at around 25% since. The number of 'stocks' for which information is available has also increased during the same period, from 120 to 454.

Discussion

The perspective view of the state of world stocks obtained from the series of FAO fishery resources reviews

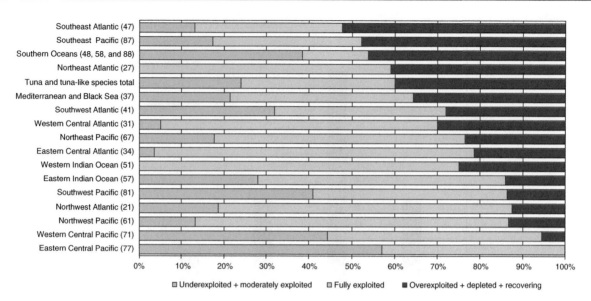

Figure 3 Percentage by FAO fishing areas of stocks exploited at or beyond MSY levels (F + O + D + R) and below MSY levels (U + M). Reproduced with permission from FAO (2005) Review of the State of World Marine Fishery Resources. *FAO Fisheries Technical Paper*, *457*, 235pp, figure A2.2. Rome: FAO.

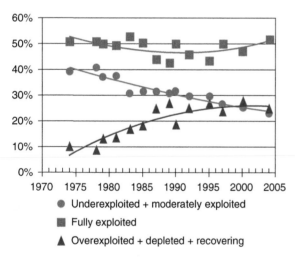

Figure 4 Global trends in the state of exploitation of world stocks since 1974. Reproduced with permission from FAO (2005) Review of the State of World Marine Fishery Resources. *FAO Fisheries Technical Paper*, *457*, 235pp, figure A2.3. Rome: FAO.

indicates clearly a number of trends. Globally, between 1974 and 1995, there was a steady increase in the proportion of stocks classified as 'exploited beyond the MSY limit', that is, overfished, depleted, or recovering (after overexploitation and depletion). These conclusions are in line with earlier findings summarized by Garcia, de Leiva, and Grainger in **Figure 5**. Since the findings by Garcia and Grainger in 1996 were based on a sample of the world stocks, severely constrained by availability of information to FAO, the conclusions

are considered with some caution. A key question is: To what extent does the information available to FAO reflect reality? There are many more stocks in the world than those referred to by FAO. In addition, some of the elements of the world resources referred to by FAO as 'stocks' are indeed conglomerates of stocks (and often of species). One should therefore ask what validity a statement made for the conglomerate has for individual stocks (*stricto sensu*). There is no simple reply to this question and no research has been undertaken in this respect.

However, while recognizing that the global trends observed reflect trends in the monitored stocks, it is also noted that the observations generally coincide with reports from studies conducted at a 'lower' level, usually based on more insight and detailed data. For instance, an analysis on Cuban fisheries using the same approach as used by Garcia and Grainger for the whole world leads to surprisingly similar conclusions, using less coarse aggregations with an even longer time series.

There is of course the possibility that stocks become 'noticed' and appear in the FAO information base as 'new' stocks only when they start getting into trouble and scientists having accumulated enough data start dealing with them, generating reports that FAO can access. This could explain the increase in the percentage of stocks exploited beyond MSY since 1974. This assumption, however, does not hold for at least two reasons.

1. The number of 'stock items' identified by FAO but for which there is not enough information has also

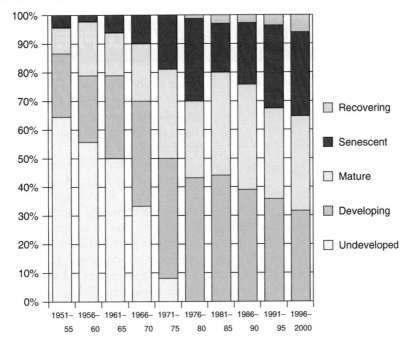

Figure 5 Stage of development of the 200 major marine fishery resources: 1950–2000. Reproduced with permission from FAO (2005) Review of the state of world marine fishery resources. *FAO Fisheries Technical Paper, 457*, 235pp, figure A2.4. Rome: FAO.

increased significantly with time, from seven in 1974 to 149 in 1999, clearly showing that new entries in the system are not limited to well-developed fisheries or fish stocks in trouble.

2. From the 1980s, based on the recognition of the uncertainties behind identification of the MSY level, and recognizing also the declines due to decadal natural fluctuations, scientists have become more and more reluctant to definitely classify stocks as 'overfished'. The apparent 'plateauing' of the proportion of stocks with excessive exploitation in the world oceans may in part be due to this new trend.

While the trend analyses of world fisheries landings (**Figure 5**) tend to suggest that the proportion of senescent and recovery fisheries (taken as grossly corresponding to those being exploited beyond MSY) is still on the increase, the resources analyses summarized in **Figure 4** indicate that there is a 'flattening' in the proportion of fish stocks that are overexploited (or exploited beyond MSY). Even if the latter may be indicative that the 'deteriorioration' process leading to the overexploitation of marine fishery resources has slowed down and that eventually the past trend can be reversed, there is no evidence of clear improvements in the state of exploitation of world stocks, and with 25% the proportion of stocks overexploited or depleted is still high.

See also

Demersal Species Fisheries. Dynamics of Exploited Marine Fish Populations. Ecosystem Effects of Fishing. Fisheries Overview. Fishery Management. Mariculture Overview.

Further Reading

Baisre JA (2000) Chronicles of Cuban Marine Fisheries (1935–1995): Trend Analysis and Fisheries Potential. *FAO Fisheries Technical Paper, 394*. Rome: FAO.

Csirke J Vasconcellos M (2005) Fisheries and long-term climate variability. In: Review of the State of World Marine Fishery Resources. *FAO Fisheries Technical Paper, 457*, pp. 201–211. Rome: FAO.

FAO (2005) Review of the State of World Marine Fishery Resources. *FAO Fisheries Technical Paper, 457*, 235pp. Rome: FAO.

FAO (2007) *The State of World Fisheries and Aquaculture 2006*, 162pp. Rome: FAO.

Garcia SM and De Leiva Moreno I (2000) Trends in world fisheries and their resources: 1974–1999. *The State of World Fisheries and Aquaculture 2000*. Rome: FAO.

Garcia SM and Grainger R (1996) Chronicles of Marine Fishery Landings (1950–1994): Trend Analysis and Fisheries Potential. *FAO Fisheries Technical Paper, 359*. Rome: FAO.

Maguire J-J, Sissenwine M, Csirke J, Grainger R, and Garcia S (2006) The State of World Highly Migratory, Straddling and Other High Seas Fishery Resources and Associated Species. *FAO Fisheries Technical Paper, 495*, 84pp. Rome: FAO.

FISHING METHODS AND FISHING FLEETS

R. Fonteyne, Agricultural Research Centre, Ghent, Oostende, Belgium

Introduction

Since the early days of mankind fishing has been an important source of food supply. The means to collect fish and other aquatic animals evolved from very simple tools to the present, often sophisticated, fishing methods. Nevertheless, many of the ancient fishing gears are still in use today, in one form or another. Even if their contribution to the total world catch is negligible, they are often very important for the economy of local communities. The efficiency of a limited number of fishing methods, such as trawling and purse seining, has become very high. Associated with this efficiency and the increased growth of the world fishing fleet are the many problems that fisheries have to face at present. Many aquatic resources, long regarded as unlimited, are now subject to overfishing and the fishing industry is confronted with criticism on the negative environmental impact of some major fishing methods.

This article deals with the technical aspects of fishing in general. It gives a review of the different fishing methods introduced by some considerations about the basic principles of fishing. In a second section the composition and evolution of the world fishing fleet is dealt with. Finally some of the most stringent problems directly associated with fishing operations are briefly discussed.

Basic Principles of Fishing

Fundamentally fishing is based on a limited number of basic principles. The fishing process can be split up into three fundamental subprocesses:

1. attracting or guiding the fish to the fishing gear;
2. fish capture, consisting of collecting and retaining the fish;
3. removing the fish from the water.

To attract or guide the fish to the fishing gear the fish behavior is manipulated, influenced, or controlled. The main mechanisms are frightening and luring. Both aim at directing the fish towards the fishing gear or into an area where the capture can more easily take place. The stimuli to obtain these reactions are diverse. Fish can be frightened by sounds and moving objects through the water. Light, natural and artificial baits, and also electricity are used to attract fish.

The efficiency of fishing operations depends on a thorough knowledge of the natural behavior of the target species. Knowledge of the distribution in time and space is needed to decide where and when the fishing gear has to be deployed. Aggregation of fish at spawning times allows catching large quantities in relatively short times. The schooling behavior of many pelagic species led to the development of large surrounding nets. Pelagic fishes show a strong preference for water layers in a specific temperature range, which is often related to the availability of food. The performance of static gears such as entangling gears and certain traps depend on the encounter of fish and fishing gear and hence on the movement pattern of fish. The latter is often determined by factors such as water flow, foraging, and antipredator behavior. Appropriate stimuli are particularly important in active gears like seines and trawls. The rigging of these gears is adjusted to make optimal use of the response of fish to external stimuli, e.g., to herd the fish towards the entrance of the net.

To realize fish capture four essential mechanisms are used:

1. tangling or enmeshing;
2. trapping;
3. filtering;
4. hooking/spearing.

These mechanisms are easily recognizable in the different fishing methods described below.

Classification of Fishing Methods

A classification of fishing gears permits an overview and better understanding of the numerous fishing gears in use around the world. In such a classification, fishing gears using the same basic principles and techniques to catch fish are brought together in a limited number of larger groups of fishing methods. Fishing gears can also be classified according to their aim (commercial versus recreational fishing), their way of operation (passive versus active), or their scale of application (artisanal versus industrial), but these groups will eventually fall apart into subgroups based on the catching principle. Each main group in the classification contains a number of gear types. They differ in construction and/or in the way they are operated. Further subdivisions can be made when different species, fishing grounds, water depth, etc. require specific adaptations of the basic gear.

A classification of fishing gears is also required for management and statistical purposes. The classification presented here is based on the FAO classification of fishing gears.

Surrounding Nets

Surrounding nets are used for the capture of pelagic fish. The fish are caught by surrounding them both from the sides and from below thus preventing them escaping in any direction. Often lights are used to concentrate the fish in the capture area. Two types of surrounding nets can be distinguished.

Surrounding nets with purse lines or purse seines A purse seine consists of a large wall of netting, up to several thousand meters long and several hundred meters deep, which is set around the fish school (**Figure 1A**). Numerous floats on the floatline keep the net at the surface. A so-called purse line runs through rings attached to the lower side of the net. By hauling the purse line at the end of the encircling procedure the net is closed underneath, preventing the fish from escaping by diving.

When the purse line, the rings and most of the netting is taken onboard the fish is bailed out or pumped on board the purse seiner. The part of the netting keeping the catch, called the bunt, is stronger and has the smallest mesh size.

The nets are operated by one or by two boats. Single boat operation, with or without an auxiliary skiff, is most usual. Purse seining is probably the most important fishing method for catching pelagic fish.

Surrounding nets without purse lines The most representative net in this group is the lampara net (**Figure 1B**). The net is shaped like a dustpan and is provided with two lateral wings. As for the purse seine the net bag or bunt has a smaller mesh size. The typical protruding bottom is obtained by using a weighted groundrope, which is much shorter than the floated upper line. When the wings are simultaneously hauled the lower part of the net closes preventing the fish from escaping underneath.

Seine Nets

Seining is one of the oldest fishing methods known, and is employed at least since the third millenium BC. Seine nets consist of a wall of netting with very long wings to which long towing ropes are connected. More or less in the middle of the net there is a section capable of holding the catch, called the bunt or bag. This can simply be a section of netting with smaller meshes and more slack or a real bag. The gear is set around an area supposed to contain the fish and towed to a predetermined location by means of the two towing ropes, which also have the important function of herding the fish. The net can be hauled to the shore (beach seines) or a vessel (boat seines). Seine nets are also widely used in freshwater fisheries.

Beach seines Beach seines are operated from land in shallow waters (**Figure 2A**). The net is set by a boat or even just a fisherman, enclosing the area by the first towing rope, followed by the first wing, the bunt or bag, the second wing and finally the second

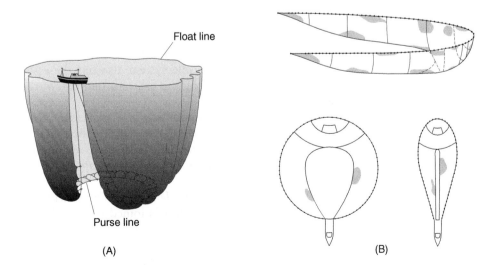

Figure 1 Surrounding nets. (A) Purse seine; (B) lampara net. (Adapted with permission from Nédélec and Prado, 1990.)

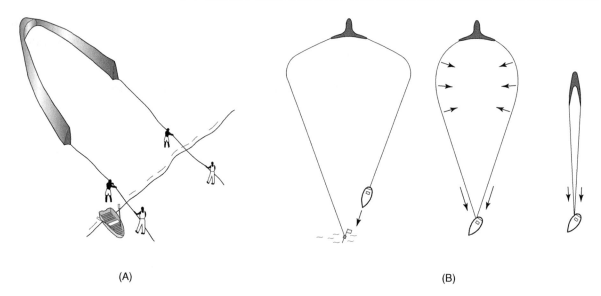

Figure 2   Seine nets. (A) Beach seine; (B) Danish seine. (Adapted with permission from Nédélec and Prado, 1990.)

towing rope, the end of which normally remains ashore. The net, the bottom and the water surface enclose the fish. The first towing rope is led to the shore again where both ropes are simultaneously hauled.

Boat seines   Boat seining in sea fisheries is a much more recent development in which the seine net is operated from a boat. A typical example is the Danish seine or 'snurrevaad' (**Figure 2B**). The seine is set out by a vessel from an anchored buoy to enclose the area to be fished. At the end, the vessel is back at the anchored buoy to start the hauling operation. The two very long seine ropes drive the fish into the net when they are slowly hauled. While hauling, the seiner maintains a fixed position. A variant of the Danish seine is the Scottish seine where the vessel slowly moves forward while hauling the gear.

Trawl Nets

Trawl nets are funnel-shaped nets with two wings of varying length to extend the net opening horizontally and ending in a codend where the catch accumulates. Trawls are towed on the bottom or in the midwater by one or two boats. In certain demersal fisheries two or more nets are towed by one vessel. Modern trawls are among the most-efficient fishing gears.

Bottom trawls   Bottom trawls are operated on the seabed. The vertical net opening varies according to the occurrence in the water column of target species. Consequently a distinction is sometimes made between low opening bottom trawls aiming at the capture of demersal species and high opening bottom trawls suitable for the capture of semidemersal or pelagic species.

Beam trawls   The net of a beam trawl is kept open horizontally by a beam supported at both ends by two trawl heads (**Figure 3A**). In modern fisheries the beam is made of steel and is up to 12 m long. Relatively light beam trawls are used to catch shrimps. Flatfish beam trawls are equipped with an array of tickler chains to disturb the flatfish from the bottom. Generally two beam trawls are towed at the same time by means of two derrick booms (double rig operation) but a single rig operation, mostly over the stern of the vessel, is also possible.

Bottom otter trawls   In these trawls the horizontal opening of the net is obtained by two otter boards, which are pushed sideward by the water pressure (**Figure 3B**). The otter boards are connected to the net by means of wires called bridles. The otter board/bridle system also has the essential function of herding the fish into the path of the trawl.

Bottom pair trawls   Two vessels towing the trawl simultaneously assure the horizontal opening of the net (**Figure 3C**). Since there are no otter boards, the warps are connected directly to the bridles.

Otter multiple trawls   Otter twin trawls consist of two otter trawls joined in a single rig (**Figure 3D**). The inner wings are attached to a sledge or weight and the outer wing of each trawl is connected to an otter board. The twin trawl principle is also used to tow three or even four nets at the same time. The

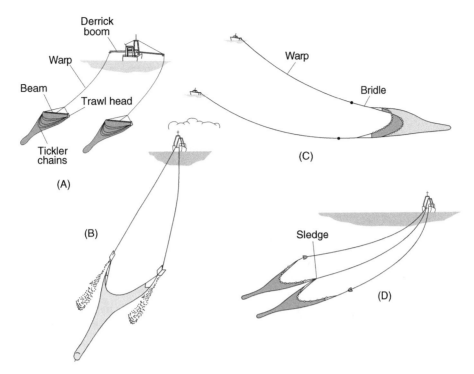

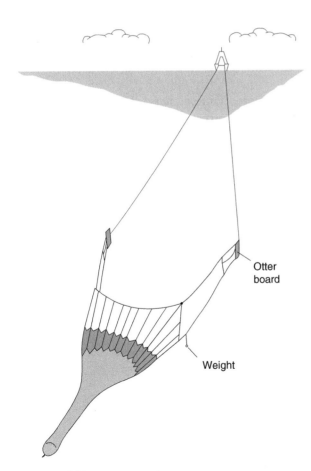

Figure 3 Bottom trawls. (A) Double rig beam trawl; (B) bottom otter trawl; (C) bottom pair trawl; (D) otter twin trawl.

advantage of multiple trawls is that the same area can be fished with a gear with a lower hydrodynamic resistance than a single net with the same horizontal opening, as the total amount of netting is less in the multiple trawl.

Midwater trawls Midwater or pelagic trawls can be operated at any level in the midwater, including surface water. Midwater trawls are usually much larger than bottom trawls in order to be able to enclose large schools of pelagic fish. To reduce the resistance of these large trawls, very large meshes or ropes are used in the net mouth.

Midwater otter trawls In these the horizontal net opening is controlled by two hydrodynamically shaped otter boards. The vertical opening is assured by weights in front of the lower wings (**Figure 4**).

Again, the combination of otter boards, bridles and large meshes or ropes in the front part of the net herd the fish to the center of the trawl. The depth of the trawl is controlled by changing the length of the warps or the towing speed.

Midwater pair trawls In these the net is kept open by two boats towing the trawl simultaneously.

Dredges

Dredges are designed to collect mollusks (mussels, oysters, scallops, clams, etc.) along the bottom.

Figure 4 Midwater otter trawl.

Figure 5 Boat dredges. (Adapted with permission from Nédélec and Prado, 1990.)

A bag net attached to a fixed frame retains the mollusks but releases water, sand and mud.

Boat dredges These are dredges of varying shape, size, and construction, in general consisting of a metal frame to which a bag net, completely or partly constructed of iron rings, is attached. Often the underside of the frame is provided with a scraper or teeth and the upper side with a depressor to keep the dredge close to or into the seabed (**Figure 5**).

Boat dredges are operated in different ways. A vessel can tow one, two, or more dredges, sometimes by means of beams, or vessel and dredge can be towed towards an anchor by means of a winch. To increase the efficiency, especially when the shells are deep in the sediment, high-pressure water jets may be used to dig them out. To handle large catches suction pumps or conveyer belts may be used. These gear systems are usually regarded and classified as 'harvesting gears.'

Hand dredges These dredges are smaller and lighter and are operated by hand in shallow waters.

Lift Nets

In lift nets the gear consists of horizontal or bag-shaped netting with the opening facing upwards, which is kept open by a frame or other stretching device. The net is lowered onto the bottom or at the required depth and the fish above are filtered from the water when the net is lifted. Before lifting, the fish are often first concentrated above the net by the use of light or bait. The nets can be operated by hand or mechanically, from the shore or from a boat.

Portable lift nets These are small, hand-operated lift nets, often used to catch crustaceans.

Boat-operated lift nets Larger lift nets, including the so-called bag nets and blanket nets, are operated from one or more boats (**Figure 6A**). Blanket nets may be very large and the sheet of netting is kept stretched by several vessels or by large, sometimes fixed, constructions.

Shore-operated lift nets These gears are often operated from stationary constructions along the shore (**Figure 6B**).

Falling Gear

The principle here is to cover the fish with the gear. Falling gears are usually used in shallow water.

Cast nets These nets are cast over the fish from the shore or a boat. When hauling the net closes from underneath and encloses the fish (**Figure 7**).

Cover pots, lantern nets These simple hand-operated gears are put over the animal that is taken out through the opening at the top. Used by a single person in shallow water.

Gillnets and Entangling Gear

These passive gears consist of very fine netting kept vertically in the water column by a ballasted

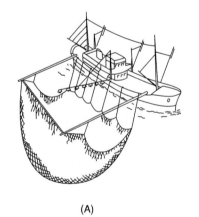

(A)

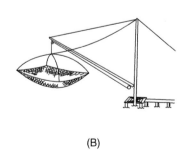

(B)

Figure 6 Lift nets. (A) Boat-operated lift net; (B) shore-operated lift net. (Adapted with permission from Nédélec and Prado, 1990.)

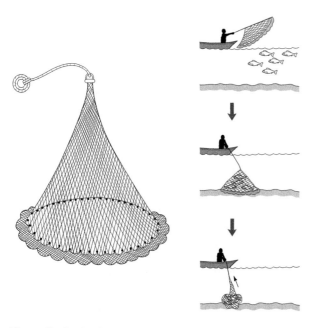

Figure 7 Cast net.

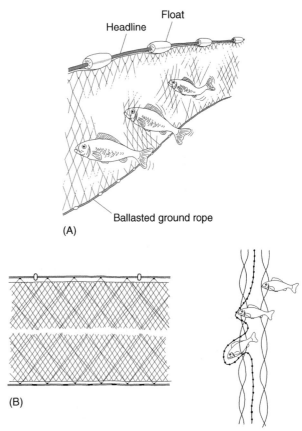

Figure 8 Gillnets and entangling gear. (A) Gillnet; (B) trammel net. (Adapted with permission from Nédélec and Prado, 1990.)

groundrope and a buoyant headline. The fish are gilled, entangled, or enmeshed in the netting which may be single (gillnets) (**Figure 8A**) or triple (trammel nets) walled (**Figure 8B**).

Sometimes different types of nets are combined. Usually a number of nets are connected to each other thus forming a so-called fleet. Depending on the design and rigging the nets can be used on the seabed, in the midwater, or at the surface.

Set gillnets (anchored) These nets are fixed to or near the bottom, usually by means of anchors.

Drifting gillnets These gears are kept near to the surface by means of floats and can freely drift with the currents. Most often they are connected to the drifting boat.

Encircling gillnets These are shallow-water gear operated at the surface. The encircled fish are driven to the nets, by noise or other means, and gill or entangle themselves in the netting.

Fixed gillnets In these the netting is attached to stakes in coastal water and the fish are collected at low tide.

Trammel nets These bottom-set nets consist of three walls of netting (**Figure 8B**). The two outer walls have a larger mesh size than the loosely hung inner wall. The fish can pass through the meshes in the outer walls but are entangled when trying to pass through the inner wall.

Combined gillnets – trammel nets The lower part of these bottom-set nets is a trammel net to catch bottom fish; the upper part is a gillnet to catch semidemersal or pelagic fish.

Traps

These gears are constructions in which the fish can freely enter but are unable to get out again. These gears are very variable in size and operation.

Stationary uncovered pound nets These are usually large netting constructions, anchored or fixed to stakes and open at the surface. They are generally divided into different chambers, closed at the bottom by netting, and provided with leaders (**Figure 9**).

Pots Pots are made from different materials and have one or more entrances (**Figure 10A**). They are usually set on the bottom, often in rows. They are designed to catch fish or crustaceans and are often baited.

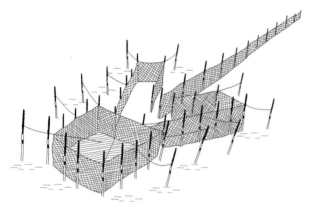

Figure 9 Stationary uncovered pound net. (Adapted with permission from Nédélec and Prado, 1990.)

Fyke nets   These nets consist of a cylindrical or conical-shaped body usually mounted on rings and provided with leaders to guide the fish to the entrance (**Figure 10B**). Fyke nets are used in shallow waters where they are fixed on the bottom by anchors or stakes.

Stow nets   These nets are used in waters with strong currents, e.g., rivers (**Figure 10C**). The conical or pyramidal nets are fixed by means of anchors or stakes and can also be deployed from a boat. The net entrance is usually held open by a frame.

Barriers, fences, weirs, corrals, etc   These gears are made from a variety of materials and are set up in

tidal waters (**Figure 10D**). The fish are collected at low tide.

Aerial traps   Jumping and gliding or flying fish can be collected on the surface in horizontal nets (veranda nets) and also in boxes, boats, rafts etc. Sometimes the fish are frightened to jump out of the water.

Hooks and Lines

The fish are attracted by natural or artificial bait attached to a hook at the end of a line, on which they are caught. Hooks can also catch the fish by ripping them when they come near by. A well-known example is the jigging lines for squid. Hooks and lines are used in both commercial and sport fishing.

Handlines and pole-lines (hand operated)   Handlines may be used with or without a pole or rod, or using reels, mostly in deep waters. This fishing method also includes hand-operated jigging lines.

Handlines. and pole-lines (mechanized)   Mechanized handlines are operated with powered reels or drums (**Figure 11A**). Tuna are sometimes caught with mechanized pole-lines, i.e., the pole movement is automated.

Set longlines   These are bottom longlines, baited or not, set on or near the bottom (**Figure 11B**). They

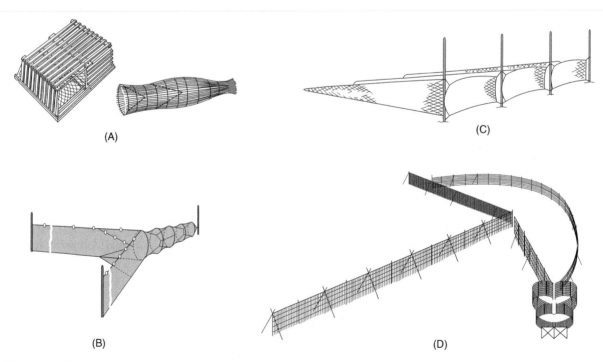

(A)   (C)

(B)   (D)

Figure 10   Traps. (A) Pots; (B) fyke net; (C) stow net; (D) corral. (Adapted with permission from Nédélec and Prado, 1990.)

consist of a main line with snoods attached to it at regular intervals. Longlines may be several hundred or even a thousand meters long.

Drifting longlines The very long lines are kept near to the surface or at a certain depth by means of floats and can drift freely with the currents. Drifting longlines can also be set vertically, hanging from a float at the surface.

Trolling lines Trolling lines are lines with natural or artificial bait that are towed near the surface or at a certain depth. The towing vessel is usually provided with outriggers to tow several lines at the same time (**Figure 11C**).

Grappling and Wounding Gear

This group contains gear for killing, wounding or grappling fish or mollusks. Examples are harpoons, spears, arrows, tongs, clamps, etc.

Harvesting Gear

Modern and very efficient gears are used to extract fish or mollusks directly from the sea by pumping or forced sifting.

Pumps With these the fish, usually attracted by light, are directly pumped on board.

Mechanized dredges Mechanized dredges use strong water jets to dig out mollusks. The mollusks are then transferred to the boat carrying the dredge by a conveyor belt-type device or by suction (**Figure 12**).

Miscellaneous

A great variety of fishing methods have no place under the gear categories described so far. The most important of these are:

- gathering by hand or with simple tools;
- scoop nets and landing nets;
- poisons and explosives;
- trained animals;
- electrical fishing.

Fishing Fleets

In 1995 the FAO estimated the number of vessels in the world fishing fleet at 3.8 million. Only one-third of these vessels is decked and the remaining are generally less than 10 m in length. Also only one-third of the undecked vessels is equipped with an engine. The number of undecked fishing vessels and the total tonnage of decked fishing vessels increased in the 1970s and 1980s (**Figures 13** and **14**). This increase was mainly due to the growth in Asia. Since the end of the 1980s the expansion rate of both decked and undecked vessels has slowed down.

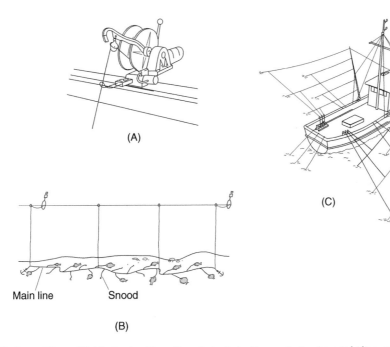

Figure 11 Hooks and lines. (A) Mechanized handlines (adapted with permission from Nédélec, 1982); (B) set longline; (C) trolling line. (Adapted with permission from FAO, 1985.)

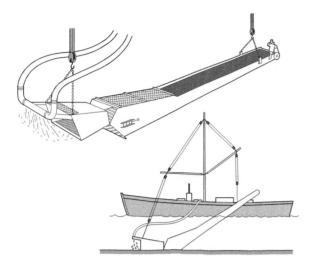

Figure 12 Mechanized dredge. (Adapted with permission from Nédélec and Prado, 1990.)

Only 1% of the decked vessels is larger than 100 gross tonnage (GT), equivalent to about 24 m. This category of vessels is capable of fishing beyond the 200 nautical mile exclusive economic zone (EEZ) limits, although the majority operate on the continental shelf within the EEZ of their own flag state. An analysis of vessels of 100 gross registered tonnage (GRT) or more in the databank of Lloyd's Maritime Information Services (MLIS) showed that the world industrial fishing fleet grew until 1991, with nearly 26 000 vessels, and has declined since to 22 700 in 1997. The Lloyds database, however, contains only 1% of the Chinese fleet. The Chinese fleet has continued to increase, from about 3000 vessels in 1985 to about 15 000 in 1995. With regard to the vessels over 100 tonnes, China accounts for around 40% of the world's fishing fleet. The Chinese vessels, however, are of low horsepower and hence have a lower fishing capacity. The general decline indicated by the Lloyds database is more or less compensated by the growth of the Chinese fleet. As a result the world fishing capacity remained approximately steady over the 1990s. The number of vessels in many of the main fishing nations of the world (Japan, former USSR, USA, Spain) shows a declining trend. A decrease is also noticed in most European countries as a result of the decommissioning policies of the EU. The fishing fleets of Latin American countries and developing countries are still growing.

Since 1988, the annual number of newly built vessels has steadily decreased from 1000 to around 200 in 1997. This is likely to be due to the overcapacity of the world fleet in the early 1990s and to the imposition of licensing systems by many developing countries. As a result of the low renewal rate the world fishing fleet comprises more and more old vessels. Taking into account the building rates, the predicted scrapping rates and a loss factor, it can be estimated that by 2008 the Lloyds database will contain no more than 14 000 vessels. With regard to changes in national fishing fleets, reflagging has become a more important factor than building or scrapping. The trend is for a flow of vessels from

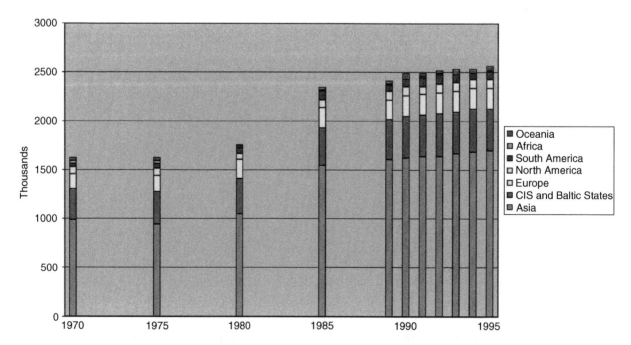

Figure 13 Number of undecked vessels by continent. (Adapted with permission from FAO, 1999.)

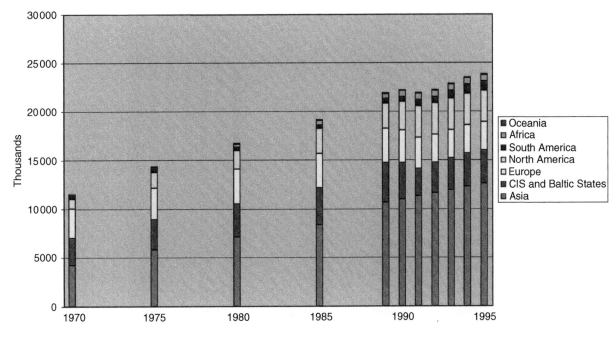

Figure 14 Total tonnage of decked vessels by continent. (Adapted with permission from FAO, 1999.)

developed to underdeveloped countries. The general idea that both the average vessel size and the horse-power are increasing is not supported by the data analysis. In fact, both show only slight changes over the period 1970–1997. It should be borne in mind that these findings are valid for the global fleet but may be different for some national fleets.

Gillnets and lines are the most frequently used fishing gears (**Figure 15**). Trawlers, however, are the largest and most powerful fishing vessels, accounting for about 40% of the total gross tonnage (**Figure 16**).

Problems Related to Fishing Gears and Fishing Fleets

Efficiency

The efficiency of a fishing method depends not only on the fishing gear itself but also on many elements which make up the complete fishing system. Among these are the fishing vessel, navigation and fish detection, handling and monitoring of the gear, schooling and training of the fishermen. As a result, the efficiency of the different fishing methods varies enormously. Artisanal, mostly passive gears, have a low input of modern technology and consequently have a rather low efficiency, as is shown by the labor productivity of fishermen given in **Table 1**.

Active fishing methods, like trawling and purse seining, make use of the best available technical means to increase their efficiency. The use of synthetic netting materials and the mechanization of the

gear handling on board makes it possible to construct and use larger fishing gears. Mechanical power for the propulsion of the fishing vessels together with precise navigation and electronic fish-finding devices have drastically improved the efficiency of modern fishing operations.

Excess Fishing Capacity

For years excessive fishing capacity in many fisheries or even at the global level has been a matter of increasing concern. Excessive fishing capacity is thought to be largely responsible for the degradation of marine resources and has a large impact on the socioeconomic performances of the fleets and industries concerned. The FAO estimates an excess of fishing capacity of at least 30% on the principal high value, mostly demersal, species. These species represent about 70% of the landings but exclude many pelagic and low-value species. When aggregating all resources, on a global scale, there is no overcapacity in relation to maximum sustainable yield. This situation clearly indicates a full or overexploitation of some stocks while other resources, generally of lower-value pelagic species, still offer opportunities for increased landings. Due to the limited mobility of the fleet, the excess in capacity cannot be easily transferred from overexploited to underexploited resources.

By-catch and Discards

In general the catch of a fishing gear does not solely consist of the target species. With the target species

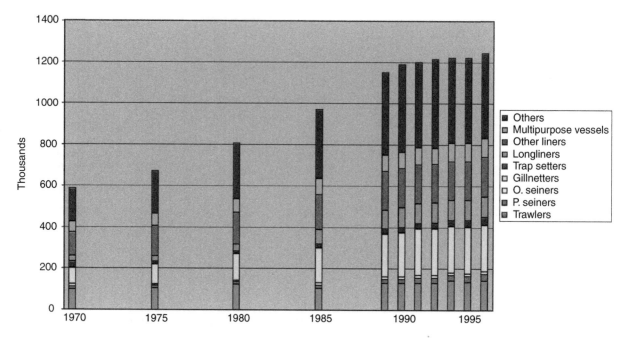

Figure 15 Number of decked vessels by type. (Adapted with permission from FAO, 1999.)

the catch will usually contain a so-called by-catch which may consist of other, not primarily sought, marketable species and also animals that will not be landed but returned at sea or discarded. Discarding has become one of the major problems of modern fisheries. The FAO Technical Consultation on Reduction of Wastage in Fisheries in 1996 estimated the global discards of fish in commercial fisheries at 20 million tonnes. There are several reasons why a

fisher can decide to discard the catch or part of it. It is often illegal to land undersized animals and for management reasons it may not be permissible to land species for which the quota has been reached. Some by-catch species have a low commercial value and animals of a certain size may not be attractive enough for the fishing industry and are discarded for economic reasons. As the survival of most species returned to the sea is low, discarding practices

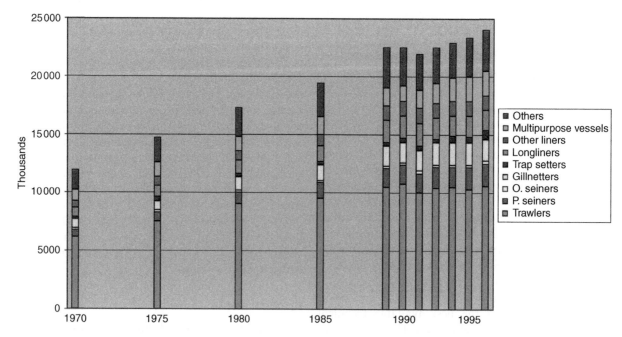

Figure 16 Tonnage of decked vessels by type. (Adapted with permission from FAO, 1999.)

Table 1 Labor productivity of fishermen

Annual catch per fisherman (tonnes)	Gear types
1	Traps, pole and hooked lines and nets from rowing boats
10	Inshore longlines, entangling nets, and trawls from small vessels
100	High seas trawls from large vessels
400	Purse-seines from super-seiners

Adapted, with permission, from Fridman (1986).

contribute significantly to the problem of overfishing. In recent years the catch of nonfish species such as marine mammals, turtles, and seabirds has gained special attention, not least due to pressure from environmental groups.

Fishing Gear and Fishery Management

To protect the resources and to reduce by-catches and discard practices, many fishing gears are subject to a number of technical measures. Most of these measures aim at improving the selectivity of the gear or at a reduction of the fishing effort. Two types of selectivity should be envisaged, namely, length selectivity and species selectivity. The implementation of minimum mesh sizes to prevent catches of immature fish or other marine organisms is a worldwide practice to regulate the length selection of fishing gears constructed of netting. In other types of gears, such as traps, length selectivity is controlled by laying down the space between bars or lathes. The selectivity of hooks is determined by the size and shape of the hooks.

The species selectivity of fishing gears can be improved by technical modifications. Most adaptations have been made to trawls since they are often used in mixed-species fisheries. The techniques used involve separator panels, separator grids or grates and square mesh panels. These techniques take advantage of differences in behavior between species to the fishing gear. In a separator trawl a netting panel divides a bottom trawl in two horizontal sections. Benthic species that stay close to the bottom are caught in the lower section, whereas species that tend to swim upwards are caught in the upper section. Each section can be provided with a cod end with its own appropriate mesh size. Inclined netting is used in various configurations to separate different species but mainly shrimps from finfish. The shrimps can pass through the small meshes in the panel but the larger fish are guided along the panel through an escape opening in the top panel of the trawl.

Separator grids are based on the same principle and are widely used to separate crustaceans from finfish. The netting panel is replaced by a grid or grid system inserted in the aft part of the trawl. Turtle excluder devices are successfully used in US shrimp fisheries to minimize the by-catch of turtles. Comparable gear modifications are applied to give escape to cetaceans caught in midwater and bottom trawls. For similar reasons, traps are often equipped with escape vents.

The ability of square meshes to remain open under tension is exploited to give roundfish better escape opportunities. Conventional diamond-shaped meshes, on the contrary tend to close when the drag increases. As a consequence the selectivity of, for example, cod ends will decrease. Square mesh windows inserted in the cod end or in the top panel of a trawl present a constant mesh opening and offer juvenile fish an escape route. In mixed fisheries, some species will take advantage of the square mesh window while species with a different body form will not.

Management options like time and/or area fishing restrictions and effort reductions are important measures that contribute to solving the by-catch and discard problem. The number of gears employed, e.g., traps, may be limited. In some fisheries the maximum vessel size is restricted. Vessel-size and gear-size restrictions may be limited to certain areas and for well-determined periods, for example, on nursery grounds.

Individual transferable quotas or other forms of management in which property rights are well defined strengthen individual responsibility.

Ecosystem Effects of Fishing

The impact of fisheries on the marine ecosystem and habitats is an issue of growing concern. Especially since the 1980s scientists have paid increasing attention to environmental effects of fishing other than the decrease in population size of the target species. It is now recognized that fishing can cause alteration of the seabed and may adversely affect the seabed communities. As a consequence of discarding practices large-scale changes in the abundance and distribution of scavenging sea birds have been observed in some regions. Discarding also affects the food supply of many marine animals and may even be beneficial for the productivity of commercially important species.

Although the effects of fishing on the marine environment should not be minimized, it should be recognized that the effects of fishing are added to the impact of other anthropogenic activities and natural changes. Moreover, one should realize that it is extremely difficult to distinguish between different causes of the complex changes in the marine ecosystem.

At present a number of measures to reduce adverse environmental effects of fishing are under discussion. Generally effort reduction is regarded as the most effective measure to protect marine ecosystems. Ecologically valuable and vulnerable habitats can be protected by creating permanent closed areas. Technically the impact of fishing gears may be reduced by gear substitution and gear modification. A milestone in the issue of environmental considerations, including the by-catch and discard problem, was the adoption in 1995 by the FAO Conference of the Code of Conduct for Responsible Fisheries.

See also

Demersal Species Fisheries. Marine Fishery Resources, Global State of.

Further Reading

Anon (1992) *Multilingual Dictionary of Fishing Gear*, 2nd edn. Oxford: Fishing News Books: Luxembourg: Office for Official Publications of the European Communities.

FAO (1987) *Catalogue of Small-Scale Fishing Gear* 2nd edn. Farnham, Surrey: Fishing News Books.

FAO (1995) *Code of Conduct for Responsible Fisheries*. Rome: FAO.

FAO (1999) *The State of the World Fisheries and Aquaculture 1998*. Rome: FAO.

Fridman AL (1986) *Calculations for Fishing Gear Designs*. FAO Fishing Manuals. Farnham, Surrey: Fishing News Books.

George J-P and Nédélec C (1991) *Dictionnaire des Engins de Pêche*. Index en six langues. Rennes, France: IFREMER, Editions Ouest-France.

Hall SJ (1999) *The Effects of Fishing on Marine Ecosystems and Communities*. Oxford: Blackwell Science.

Jennings S and Kaiser MJ (1998) The effects of fishing on the marine ecosystems. *Advances in Marine Biology* 34: 201–352.

Kaiser MJ and de Groot SJ (eds.) (2000) *Effects of Fishing on Non-Target Species and Habitats*. Oxford: Blackwell Science.

Nédélec C and Prado J (1990) *Definition and Classification of Fishing Gear Categories*. FAO Technical Paper No. 222, Revision 1. Rome: FAO.

von Brandt A (1984) *Fish Catching Methods of the World*, 3rd edn. Farnham, Surrey: Fishing News Books.

Wardle CS and Hollington CE (eds.) (1993) *Fish Behaviour in Relation to Fishing Operations*. ICES Marine Science Symposia Vol. 196.

DYNAMICS OF EXPLOITED MARINE FISH POPULATIONS

M. J. Fogarty, Woods Hole, MA, USA

Introduction

The development of sustainable harvesting strategies for exploited marine species addresses a critically important societal problem. Understanding the sources of variability in exploited marine species and separating the effects of human impacts through harvesting from natural variability is essential in devising effective management approaches. Population dynamics is the study of the continuously changing abundance of plants and animals in space and time. For exploited marine species, population dynamics studies provide the foundation for evaluation of their resilience to exploitation, the determination of optimal harvesting strategies, and the specification of the probable outcomes of alternative management actions.

In the following, a 'population' is defined as a group of interacting and interbreeding individuals of the same species. A closed population is one in which immigration and emigration are negligible; an open population is one in which dispersal processes do affect abundance levels. A 'stock' is a management unit defined on the basis of fishery and/or distinct biological characteristics. 'Recruitment' is the number of individuals surviving to the age or size of vulnerability to the fishery. A 'cohort' is defined as the individuals born in a specified time interval; a cohort born in a particular year is also referred to as a 'year class'.

Other articles in this encyclopedia focus on population dynamics in the context of multispecies assemblages, climate-related factors, and the ecosystem effects of fishing. Here, the primary emphasis will be on the effects of harvesting at the population level with the recognition that a full understanding of human impacts on exploited marine species can only be attained with a more holistic view incorporating the ecosystem perspective. The conceptual basis for the development of models of exploited populations and the information requirements for these models and an example of an application of these tools in a fishery assessment and management setting is provided.

Production of Marine Populations

The relative balance between increases in biomass due to recruitment and individual growth and losses from mortality due to natural causes and fishing defines the dynamics of exploited populations (**Figure 1**). The change in biomass of a population over time due to variation in these factors is called net production. For an open population, factors affecting immigration and emigration must also be considered in any evaluation of population change. The integration of information on these fundamental biological and ecological processes is a central focus of population dynamics studies. Prediction of the effects of alternative management regulations or changes in the production characteristics of an exploited population depends on a synthesis of fundamental biological and ecological processes in the form of mathematical models of varying degrees of complexity.

Recruitment is often the most variable component of production in many marine populations. Recruitment varies in response to changing environmental conditions as fluctuations in food supply, the activity of predators, and physical conditions affect the growth and survival of the early life stages. The relationship between the reproducing adult population in a given season or year and the resulting recruitment is of particular importance. For a

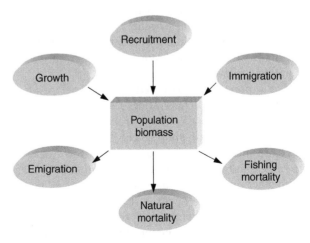

Figure 1 Components of production and dispersal affecting the biomass of exploited marine populations.

renewable resource, it is of course essential that the replenishment of the population through reproduction should not be adversely affected by exploitation.

With respect to the other elements of production, the natural mortality component reflects the effects of biological factors such as predation and disease as well as adverse physical conditions. The increase in size and weight as an individual grows is an important contributor to change in biomass of a cohort over time. Finally, dispersal among populations or subpopulations through immigration and emigration can have important consequences for the persistence of a population and its resilience to exploitation. For example, a harvested population that receives members from an adjacent unexploited subpopulation in effect receives a subsidy that can contribute to its persistence under exploitation.

Sustainability

Understanding how marine species will respond to exploitation and the appropriate levels of fishing pressure to ensure continued harvest is an essential component of population dynamics studies. More specifically, the extraction of a yield that is optimal in some defined biological or economic sense is a broadly accepted goal in resource management. In order for a long-term sustainable harvest to be possible, the population must have some capacity to compensate for reductions in population biomass through increased recruitment and growth and/or decreased mortality at one or more life history stages. Mechanisms that underlie such compensatory responses include cannibalism and competition for critical resources. If the population exhibits some form of density dependence in critical processes affecting vital rates, different equilibrium levels of population biomass will exist, corresponding to different levels of fishing pressure. Fluctuations in the physical and biological environment and their effects on the population will result in variation about the equilibrium level.

For a population governed by density-dependent feedback processes, harvesting can reduce intraspecific competition or other interactions by reducing density and overall abundance. This results in an increase in the overall productivity of the stock. In the unexploited case, the stock is dominated by larger, older individuals; harvesting shifts the population state to one with a higher proportion of younger, faster-growing (and therefore more productive) individuals. The production generated in this way is called surplus production and, in principle, this production can be taken as yield.

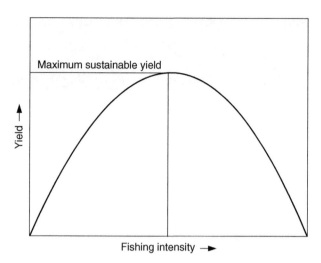

Figure 2 Relationship between yield (surplus production) and fishing intensity (measured as fishing mortality or fishing effort) under a simple conceptual model incorporating density-dependent feedback processes.

The relationship between surplus production and biomass for a species governed by a simple form of compensatory dynamic is dome-shaped, with a peak at some intermediate level of population biomass. In this simple conceptual model, production is expected to be zero when the biomass is at its highest level (the unexploited state) because of intraspecific interactions such as competition or cannibalism. The production–biomass relationship can be readily translated to one linking yield (surplus production) to fishing intensity. Again, a domed-shaped relationship is expected (**Figure 2**). At relatively low levels of fishing pressure, the yield is lower than the maximum because some correspondingly low fraction of the population is removed. Conversely, at higher levels of fishing pressure, the productivity of the stock is reduced by removing too high a fraction of the population. The point where yield is the highest is the maximum sustainable yield. All of the points on the curve except where yield is zero are considered to be sustainable yield levels. However, as the population is reduced to low levels, its viability is jeopardized, particularly under variable environmental conditions. It is therefore important not only to specify sustainability as a broad goal of fishery management but to consider optimum harvesting policies that also minimize risk to the population.

Assessing Population Status

The development of management strategies involves several components including the determination of the current biomass relative to 'desired' levels and projections of how alternative management actions will affect the population in the future. The reconstruction of past and present population levels typically involves a synthesis of information derived from fishery-dependent and fishery-independent sources. Fishery-dependent information includes factors such as the catch removed from the population and either marketed or discarded at sea, the age and/or size composition of the catch, and the amount of fishing effort and its spatial distribution. Fishery-independent information includes studies to determine the relative abundance of fish though the use of scientific surveys aimed at estimating population levels at different life stages. For example, surveys are often conducted to determine abundance of juvenile and adult fish using modified fishing gears, and others to determine the distribution and abundance of the eggs and larvae of marine species. A careful attempt is made to standardize the methods and gear used over time to ensure that the changes measured from one survey to the next reflect true changes in relative abundance. Other special studies such as mark and recapture experiments are also employed to determine population size and mortality rates.

If accurate information on catch levels is available, coupled with information on the size or age composition of the catch, estimates can be made of both population size and mortality rates. The number in the catch removed in a specified time interval, if accurately known, provides an initial minimum estimate of the population size in the sea because at least that many individuals had to have been present to account for the catch. If the fraction of the population removed by harvesting and the fraction dying due to natural sources such as disease and predation can be determined, it is possible to derive an estimate of the actual population size. Knowing the age or size composition of the catch is critical in determining the overall mortality rates. By tracking the changes in the numbers of a cohort over time as it progresses through the fishery, it is possible to estimate the survival rates from one age class to the next and therefore to generate estimates of the population size-at-age. If size but not age composition of the catch is known but we do know the growth rates and the time required to grow from one size class to the next, we can also determine the population size and survival rates.

In some cases, the catch adjusted by the amount of fishing effort to obtain that catch can be used as an index of relative abundance. The utility of this index depends on the accuracy of the catch statistics and on the validity of the measure of fishing effort. The latter can be complicated by the fact that several different types of gear may be employed in a single fishery and it will necessary to standardize among the gear types. Similarly, fishing vessels of different sizes have differing fishing power characteristics that require adjustment factors. Finally, technological developments such as advanced navigation and mapping systems, satellite imagery of oceanographic conditions and increasingly sophisticated echo-sounders used to locate concentrations of target species result in continual increases in the realized fishing power of vessels. Most importantly, it is essential to recognize that fishers are not striving to attain unbiased estimates of overall population size but rather are attempting to maximize their catch rates using all of the experience and tools at their disposal. This must be considered in any attempt to use catch-per-unit-effort as an index of abundance.

Scientific surveys are an attempt to provide an independent check on information derived from the fishery itself. Such surveys have proven to be invaluable tools in determining the status of marine populations and the communities and ecosystems within which they are embedded. Typically, fishery-independent surveys are employed to collect information not only on economically important species but all species that can be adequately sampled by the survey gear, thus providing a broader perspective on changes in the system. In addition, key oceanographic and meteorological measurements are routinely made to index changes in the physical environment.

The information noted above is integrated into mathematical models that describe, for example, the decay of a cohort over time under losses due to fishing and other sources. The information from fishery-independent sources can be used to calibrate models operating on fishery-derived data sources in an integrated analysis. In turn, the estimates of population size derived from these models and estimation procedures can be used in models designed to assess alternative harvesting strategies.

An example of the application of this overall research approach is provided below for the Icelandic cod population. Estimates of cod recruitment at age 3 years based on applications of models to catch-at-age information from the fishery and fishery-independent survey information are provided in **Figure 3**. Estimates of the number of recruits over a 50-year period show fluctuations about a relatively stable level. Estimates of the adult population size for this period are also available and show overall declines during this period as exploitation increased. The relationship between adult biomass and the resulting recruitment for Icelandic cod is provided in **Figure 4** along with the predicted fit from a simple model for this relationship. Note that there are substantial deviations from the deterministic curve,

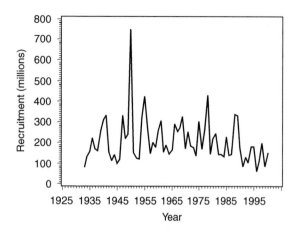

Figure 3 Estimates of recruitment at age 3 years (millions) for the Icelandic cod population over a 50-year period based on stock assessments conducted under the auspices of the International Council for Exploration of the Sea.

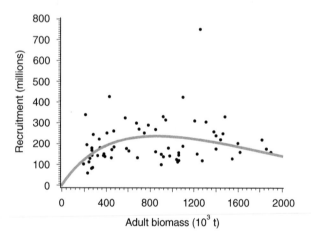

Figure 4 Relationship between recruitment and adult (spawning) biomass of Icelandic cod based on stock assessments conducted under the auspices of the International Council for Exploration of the Sea.

reflecting the effects of variable environmental conditions (and variation attributable to estimation errors). Despite this variability, it is evident that this relationship is nonlinear — recruitment does not increase continuously with increasing adult biomass but rather levels off and even declines at quite high levels of spawning population size. This reflects a form of compensatory response in the adult–recruit relationship. For Atlantic cod populations in general, it is known that cannibalism can be an important regulatory factor that limits recruitment at higher population levels. The model used to develop the predicted relationship between adult biomass and recruitment in **Figure 4** is in fact one that incorporates the effect of predation by adults on their progeny. Recall that some form of compensatory

capacity is essential if a sustainable harvest is to be possible. An illustration of how this information can be combined with other key aspects of the production of cod to predict yield at different levels of fishing pressure is described later.

Management Targets and Limits to Exploitation

Effective management requires a clear specification of goals and objectives to be achieved. Without an appropriate and pre-agreed target for management, the many conflicting and entrenched interests in the fishery management arena cannot be reconciled. The development of biological and economic reference points has played an integral role in fishery management. In the following, the emphasis will be on biologically based targets for management and on the determination of the resilience of the population to exploitation on the basis of models of the dynamics of exploited populations. One important reference point has already been encountered — the maximum sustainable yield (MSY). The corresponding level of fishing pressure resulting in MSY is also of direct interest. Although the concept of MSY has endured a somewhat controversial history, it remains a cornerstone in many national and international fishery policy statements. For example, in the United States, the Magnuson–Stevens Fishery Management Act of 1996 defines the optimum yield as 'equal to maximum sustainable yield as reduced by economic, social, or other factors.' This statement includes an important change in the specification of optimum yield relative to the original legislation introduced in 1976 in which optimum yield was defined as 'equal to maximum sustainable yield as modified by economic, social, or other factors.' In the more recent version, MSY and the corresponding fishing mortality rate is taken as a limit to exploitation.

Other important biological reference points for management are based on consideration of the effect of harvesting on a cohort once it enters or is recruited to the fishery. By tracking the growth and the fishing and natural mortality over the lifespan of the cohort, it is possible to determine the effects of harvesting on yield and the adult biomass as the level of fishing pressure is changed and/or as we modify the age or size of vulnerability to the fishery. The fishing mortality rate at which yield is maximized (denoted F_{max}) is one such reference point (assuming a maximum does in fact exist). An alternative reference point that has been widely applied is defined by the point on the yield-per-recruit curve where the rate of change in yield is one-tenth of the rate at the origin (denoted $F_{0.1}$). An advantage of this reference point is that it always exists (unlike F_{max}); further, in cases where F_{max} does exist, $F_{0.1}$ is always lower and therefore leads to more conservative management. An example for Icelandic cod is provided in **Figure 5**; in this case, $F_{max} \sim 0.35$ and $F_{0.1} \sim 0.2$.

The advantage of this overall approach is that it does not require specific information on the incoming recruitment. Rather, it is possible to express the yield on a per-unit-recruitment basis. A disadvantage of this approach if it is taken alone is that it does not provide direct information on how fishing pressure might affect the replenishment of the population through recruitment.

To fully evaluate the effects of fishing on the population, it is possible to combine information from the yield and adult biomass per recruit analysis with the data on recruitment as a function of adult population size to generate a complete life-cycle representation. Fishing reduces the adult biomass and it is possible to estimate the predicted recruitment for each level of fishing as a function of the spawning population (see **Figure 4** for the case of Icelandic cod). Multiplying the yield per recruit at each of these fishing rates by the predicted recruitment then gives the total yield. This process is illustrated for the Icelandic cod example in **Figure 6** where the predicted equilibrium yield is shown as a function of fishing mortality. Superimposed on this curve are the observed (nonequilibrium) yields for Icelandic cod against the estimated fishing mortality rates. The actual catch data reflect variability in the stock–recruitment relationship due to factors not

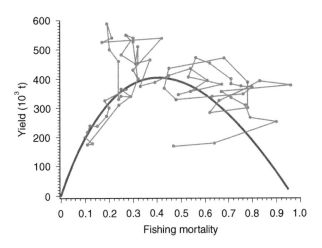

Figure 6 Predicted equilibrium relationship between yield (surplus production) and fishing mortality (solid line) based on an age-structured model for Icelandic cod incorporating information on the adult–recruitment relationship (see **Figure 4**), and yield and adult biomass per recruit analyses (see **Figure 5**). Observed (nonequilibrium) yield and estimated fishing mortality rates for Icelandic cod are shown (dots).

included in the model. The maximum yield is predicted to occur at moderate levels of fishing mortality ($F \sim 0.35$) and the limiting level of fishing mortality is at $F \sim 1.0$. Although estimated fishing mortality rates for Icelandic cod decreased at the very end of the available time series, it is clear that Icelandic cod has been overexploited and that the effects of excessive fishing have adversely affected yields. The limiting level of fishing mortality beyond which the probability of stock collapse is high can be shown to be directly related to the rate of recruitment at low spawning stock sizes. Species characterized by a high rate of recruitment at low adult population size are more resilient to exploitation than those with a lower rate.

Shifting Environmental States

The discussion has concentrated so far on conditions under which the physical and biological environments affecting the population are relatively stable and do not undergo trends or shifts in state. However, persistent shifts in environmental conditions do occur on decadal timescales with important implications for harvested species. It is useful to distinguish environmental variations in physical factors such as temperature and salinity or biological factors such as prey or predator concentrations that occur on relatively short timescales (seasonal to interannual) from those that occur on longer timescales. The distinction between high-frequency variation on seasonal or annual timescales and low-frequency

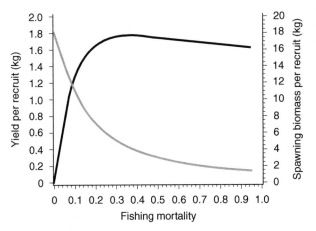

Figure 5 Yield per recruit (peaked curve) and adult (spawning) biomass per recruit (decaying exponential curve) as a function of fishing mortality for Icelandic cod based on stock assessments conducted under the auspices of the International Council for Exploration of the Sea.

variation on decadal timescales has important implications for overall levels of productivity of a population and its resilience to exploitation. With low-frequency variation, the potential interaction between changes in the environment and harvesting is of particular concern because persistent shifts in productivity of the population require change in the biological reference points. Exploitation regimes that are sustainable under one set of environmental conditions may not be under another, lower-productivity pattern. An illustration of this in the context of a simple production model is shown in **Figure 7**, where a change in the basic productivity level of the population under changing environmental conditions is reflected in a change not only in the overall yield levels attainable but also in the level of fishing pressure at which yield is maximized and at which a stock collapse is predicted. Under the lower-productivity regime depicted, the stock collapses at a fishing pressure that is sustainable (although suboptimally) under higher productivity levels. It is clear that a dynamic concept of maximum sustainable yield and other reference points is required that does account for changing conditions in the biological and physical environments experienced by the population. The integration of population dynamics studies with broader ecological investigations and physical oceanographic research is essential if we are to improve understanding of the effects of harvesting on exploited marine species.

Uncertainty, Risk, and the Precautionary Approach

Sustained monitoring of the abundance, demographic characteristics, and productivity of widely distributed marine populations entails special challenges. The precision with which it is possible to measure changes in these key variables depends critically on factors such as funding levels and infrastructure available for both fishery-dependent and fishery-independent programs, and on intrinsic characteristics of the populations themselves such as their degree of heterogeneity in space and time. Some of the principal sources of uncertainty in fishery assessments can be attributed to (a) variability (error) in estimates of population size and demographic characteristics, (b) natural variation in production rates and processes, particularly in recruitment, and (c) lack of complete information on broader ecosystem characteristics that affect the species targeted by harvesting and on the direct and indirect effects of harvesting on the ecosystem. The intrinsic variability of marine populations, communities, and ecosystems

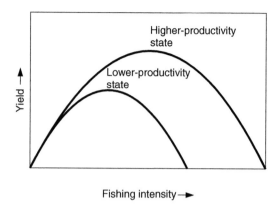

Figure 7 Yield (surplus production) as a function of fishing intensity under two environmental regimes in which the intrinsic rate of increase of the population is affected by changing productivity.

contributes substantially to these components of uncertainty.

It is now commonplace to frame issues concerning human health in terms of risk. Considerations of diet, exposure to chemicals, lifestyle choices, etc. affect the probability that an individual will contract certain diseases. Similarly, it is increasingly common in fisheries stock assessments to describe the risk to the population under alternative management scenarios. Although many specifications of risk are possible, an easily understood definition is that risk is the probability that the population will decline and remain below some specified level. Uncertainty in estimates of key parameters and quantities contributes to risk because of the possibility that errors in estimation or in basic model structures may result in overestimates of population size and productivity, inadvertently resulting in overfishing of the resource.

The recognition that many populations of exploited marine species are now fully exploited or overexploited has led to an important reevaluation of management policies. In particular, the need for a more precautionary approach to management has been recognized and integrated into a number of national and international management policy statements. Under the precautionary approach, more conservative management is required in situations where higher levels of uncertainty concerning the stock status and production characteristics exist. The burden of proof that harvesting activities were detrimental to marine populations, communities, and ecosystems has historically (and implicitly) rested with scientists and managers. An important element of the precautionary approach is the recognition that a shift in the burden of proof is required to show that the users of a resource are not adversely impacting the productivity of a population or ecosystem.

Summary

Questions relating to the stability and resilience of marine species under exploitation involve fundamental ecological considerations such as the role of density-dependent processes in population regulation, the importance of interspecific interactions such as predation and competition, and the role of the physical environment in the production dynamics of the system. Studies of the dynamics of exploited species incorporating these considerations provide an essential framework for quantitative consideration of the effects of alternative harvesting policies on these populations. The importance of fisheries in an economic and social context has led to intensive efforts on a global basis to understand the dynamics of exploited marine species on broad spatial and temporal scales. These studies have become increasingly important as it has become clear that we are near or have exceeded the apparent limits to fishery production from marine systems. Widespread problems such as overcapitalization and excess capacity of fishing fleets and the prevalence open-access fisheries, however, remain substantial impediments to effective management. The extensive information base available for exploited marine species and recent advances in understanding of ocean ecosystem dynamics can provide a strong foundation for improvements in resource management.

See also

Ecosystem Effects of Fishing. Fishery Management. Fishery Management, Human Dimension. Marine Fishery Resources, Global State of.

Further Reading

Gulland JA (ed.) (1977) *Fish Population Dynamics*. New York: Wiley Interscience.

Gulland JA (1983) *Fish Stock Assessment*.

Gulland JA (ed.) (1988) *Fish Population Dynamics*, 2nd edn. New York: Wiley Interscience.

Hilborn R and Walters CJ (1992) *Quantitative Fisheries Stock Assessment*. New York: Chapman and Hall.

Quinn TJ and Deriso R *Quantitative Fish Dynamics*. New York: Oxford University Press.

Rothschild BJ (1986) *Dynamics of Marine Fish Populations*. Cambridge, MA: Harvard University Press.

ECOSYSTEM EFFECTS OF FISHING

S. J. Hall, Flinders University, Adelaide, SA, Australia

Introduction

In comparison with conventional fisheries biology, which examines the population dynamics of target stocks, there have been relatively few research programs that consider the wider implications of fishing activity and its effects on ecosystems. With growing recognition of the need to conduct and manage our activities within a wider, more environmentally sensitive framework, however, the effects of fishing on ecosystems is increasingly being debated by scientists and policy makers around the world. As with many other activities such as waste disposal, chemical usage or energy policies, scientists and politicians are being asked whether they fully understand the ecological consequences of fishing activity.

The scale of biomass removals and its spatial extent make fishing activity a strong candidate for effecting large-scale change to marine systems. Coarse global scale analyses provide a picture of our fish harvesting activities as being comparable to terrestrial agriculture, when expressed as a proportion of the earth's productive capacity. It has been estimated that 8% of global aquatic primary production was necessary to support the world's fish catches in the early 1980s, including a 27 million tonne estimate of discards (see below). Perhaps the most appropriate comparison is with terrestrial systems, where almost 40% of primary productivity is used directly or indirectly by humans. Although 8% for marine systems may seem a rather moderate figure in the light of terrestrial demands, if one looks on a regional basis, the requirements for upwelling and shelf systems, where we obtain most fisheries resources, are comparable to the terrestrial situation, ranging from 24 to 35% (**Table 1**). Bearing in mind that the coastal seas are rather less accessible to humans than the land, these values for fisheries seem considerable, leading many to agree that current levels of fishing – and certainly any increases – are likely to result in substantial changes in the ecosystems involved. It is generally accepted that the majority of the world's fish stocks are fully or overexploited.

When considering ecosystem effects it is useful to distinguish between the direct and indirect effects of fishing. Direct effects can be summarized as follows:

1. fishing mortality on species populations, either by catching them (and landing or throwing them back), by killing them during the fishing process without actually retaining them in the gear or by exposing or damaging them and making them vulnerable to scavengers and other predators;
2. increasing the food available to other species in the system by discarding unwanted fish, fish offal and benthos;
3. disturbing and/or destroying habitats by the action of some fishing gears.

In contrast, indirect effects concern the knock-on consequences that follow from these direct effects, for example, the changes in the abundances of predators, prey and competitors of fished species that might occur due to the reductions in the abundance of target species caused by fishing, or by the provision of food through discarding of unwanted catch.

Table 1 Global estimates of primary production and the proportion of primary production required to sustain global fish catches in various classes of marine system

Ecosystem type	Area $(10^6\,km^2)$	Primary production $(g\,C\,m^{-2}y^{-1})$	Catch $(g\,m^{-2}y^{-1})$	Discards $(g\,m^{-2}y^{-1})$	Mean % of primary production	95% CI
Open ocean	332.0	103	0.01	0.002	1.8	1.3–2.7
Upwellings	0.8	973	22.2	3.36	25.1	17.8–47.9
Tropical shelves	8.6	310	2.2	0.671	24.2	16.1–48.8
Nontropical shelves	18.4	310	1.6	0.706	35.3	19.2–85.5
Coastal reef systems	2.0	890	8.0	2.51	8.3	5.4–19.8

Reproduced from Hall (1999).

By-catch and Discards

In many areas of the world a wide variety of fishing gears are used, each focusing on one or a few species. Unfortunately, this focus does not mean that non-target species, sexes or size-classes are excluded from catches. Target catch is usually defined as 'the catch of a species or species assemblage that is primarily sought in a fishery' – nontarget catch, or by-catch as it is usually called, is the converse. By-catch can then be further classified as incidental catch, which is not targeted but has commercial value and is likely to be retained if fishing regulations allow it and discard catch, which has no commercial value and is returned to the sea.

The problem of by-catch and discarding is probably one of the most important facing the global fishing industry today. The threat to species populations, the wastefulness of the activity and the difficulties undocumented discarding poses for fish stock assessment are all major issues. A recent published estimate of the annual total discards was approximately 27 million tonnes, based on a target catch of 77 million tonnes. This figure, however, did not include by-catch from recreational fisheries, which could add substantially to the total removals, and the estimate is subject to considerable uncertainty. **Figure 1** shows how these discard figures break down on a regional basis. Just over one-third of the total discards occur in the Northwest Pacific,

arising from fisheries for crabs, mackerels, Alaskan pollock, cod and shrimp, the latter accounting for about 45% of the total. The second ranked region is the Northeast Atlantic where large whitefish fisheries for haddock, whiting, cod, pout, plaice and other flatfish are the primary sources. Somewhat surprisingly, capelin is also a rather important contributor to the total, primarily because capelin are discarded due to size, condition and other market-related factors. The third place in world rankings is the West Central Pacific, arising largely through the action of shrimp fisheries. These fisheries, prosecuted mainly off the Thai, Indonesian and Philippine coasts, accounted for 50% of the total by-catch for the region, although fisheries for scad, crab and tuna are also substantial contributors. Interestingly, the South East Pacific ranks fourth, not because the fisheries in the area have high discard ratios (on the contrary, the ratios for the major anchoveta and pilchard fisheries are only 1–3%), but simply due to the enormous size of the total catch. For the remaining tropical regions, by-catch is again dominated by the actions of shrimp fisheries, although some crab fisheries are also significant.

One characteristic difference between temperate and tropical fishery discards is worthy of note. In the tropics, where shrimp fisheries dominate the statistics, discards mainly comprise small-bodied species which mature at under 20 cm and weigh less than 100 g. In contrast, for the temperate and subarctic

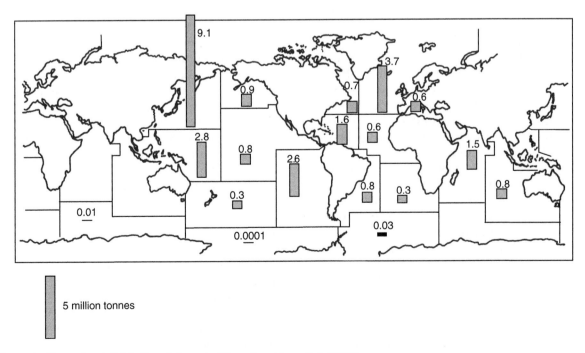

5 million tonnes

Figure 1   The regional distribution of discards. (Reproduced from Hall, 1999.)

regions discards are generally dominated by sublegal and legal sizes of commercially important, larger-bodied species. Thus, in the temperate zone discarding is not only an ecological issue, it is also a fisheries management issue in the strictest sense. Fish are being discarded which, if left alone, would form part of the future commercial catch.

A cause of particular concern is the incidental catch of larger vertebrate fauna such as turtles, elasmobranchs and marine mammals. Catch rates for these taxa are generally highest in gillnet fisheries, which increased dramatically in the 1970s and 1980s, particularly for salmonids, squid and tuna. In some fisheries, the numbers caught can be very substantial. In the high seas longline, purse seine and driftnet fisheries for tuna and billfish, for example, migratory sharks form a large component of the catch, with some 84 000 tonnes estimated to have been caught from the central and south Pacific in 1989.

As with the other nonteleost taxa for which by-catch effects are a concern, the life-history characteristics of sharks make them particularly vulnerable to fishing pressure. Slow growth, late age at maturity, low fecundity and natural mortality, and a close stock recruitment relationship all conspire against these taxa. Such life-history attributes have also led to marked alterations in the absolute and relative abundance of ray species in the North and Irish Sea, which are subject to by-catch mortality from trawl fisheries. In the Irish Sea for example, the 'common skate' (*Raja batis*) is now rarely caught.

For some species (e.g., some species of albatross and turtle species) levels of by-catch are so great that populations are under threat. But even if the mortality rates are not this great (or the data are inadequate) there is a legitimate animal welfare perspective which argues for strenuous efforts to limit by-catch mortalities regardless of population effects. Few people like the idea of turtles or dolphins being needlessly drowned in fishing nets, regardless of whether they will become locally or globally extinct if they continue to be caught.

Although declines in populations as a result of by-catch are the most obvious effect, there are also examples where populations have increased because of the increase in food supply resulting from discarding. The most notable among these are seabirds in the North Sea, where in one year it was estimated that approximately 55 000 tonnes of offal, 206 000 tonnes of roundfish, 38 000 tonnes of flatfish, 2000 tonnes of elasmobranchs and 9000 tonnes of benthic invertebrates were consumed by seabirds. There is good evidence that populations of scavenging seabirds in the North Sea are substantially larger than they would be without the extra food provided by discards.

Solving the By-catch and Discard Problem

There is no universally applicable solution for mitigating by-catch and discard problems. Each fishery has to be examined separately (often with independent observers on fishing vessels) and the relative merits of alternative approaches assessed. One obvious route to reducing unwanted catch, however, is to increase the selectivity of the fishing method in some way. In trawl fisheries, in particular, technical advances, combined with a greater understanding of the behavior of fish in nets has led to the development of new methods to increase selectivity. These methods adopt one of two strategies. The first is to exploit behavioral differences between the various fished species, using devices such as separator trawls, modified ground gear (i.e., the parts of the net that touch the seabed) or modifications to the sweep ropes and bridles that attach to the trawl doors. For example, separator trawls in the Barents Sea have been shown successfully to segregate cod and plaice into a lower net compartment from haddock, which are caught in an upper compartment. In Alaska this approach has been used to allow 40% of bottom-associated halibut to escape while retaining 94% of cod, the target species.

The second approach is to exploit the different sizes of species. In many fisheries it is the capture of undersized fish that is the main problem and regulation of minimum permissible mesh size is of course a cornerstone of most fisheries management regimes. Such a measure can often, however, be improved upon. For example, the inclusion of square mesh panels in front of the codend can often allow a greater number of escapees, because the meshes do not close up when the codend becomes full. In addition, recent work that alters the visual stimulus that the net provides by using different colored netting in different parts has been shown to improve the efficiency of such panels considerably. At the other end of the scale excluding large sharks, rays or turtles from the catch can be achieved by fitting solid grids of various kinds. In some fisheries such devices are now mandatory (e.g., turtle exclusion devices in some prawn fisheries), but there is often resistance from fishermen because they can be difficult to handle and catches of target species can fall. For non-trawl fisheries, examples of technical solutions can also often be found. For example, new methods of laying long-lines have been developed to avoid incidental bird capture and dolphin escapement procedures that are now used in high seas purse seine fleets.

Although many technical approaches have met with considerable success in different parts of the world it is important to recognize that technical fixes are only part of the solution – the system in which they have to operate must also be considered. The regulations that govern fisheries and the vagaries of the marketplace often create a complex web of incentives and disincentives that drive the discarding practices of fishermen. The situation can be especially complicated in multispecies fisheries.

The Effects of Trawling and Dredging on the Seabed

Disturbance of benthic communities by mobile fishing gears is the second major cause for concern over possible ecosystem effects of fishing, threatening nontarget benthic species and perhaps also the longer-term viability of some fisheries themselves if essential fish habitat is being destroyed. With continuing efforts to find unexploited fish resources, hitherto untouched areas are now becoming accessible as new technologies such as chain mats, which protect the belly of the net, are developed. In Australia, for example, new fisheries are developing in deeper water down to depths of 1200 m.

A prerequisite for a rational assessment of fishing effects on benthos is an understanding of the distribution, frequency and temporal consistency of bottom trawling. On a global basis, recent estimates obtained using Food and Agriculture Organization catch data from fishing nations suggest that the continental shelves of 75% of the countries of the world which border the sea were exposed to trawling in 1996. It would appear, therefore, that few parts of the world's continental shelf escape trawling, although it should be borne in mind that in many fisheries trawl effort is highly aggregated. Although we have an appreciation of average conditions, these are derived from a mosaic of patches, some heavily trawled along preferred tows, others avoided by fishermen because they are unprofitable or might damage the gear. Unfortunately, lack of data on the spatial distribution of fishing effort prevents estimates of disturbance at the fine spatial resolution required to obtain a true appreciation of the scale of trawl impacts. Nevertheless, there is little doubt that substantial areas of the world's continental shelf have been altered by trawling activity.

For the most part the responses of benthic communities to trawling and dredging is consistent with the generalized model of how ecologists expect communities to respond, with losses of erect and sessile epifauna, increased dominance by smaller faster-growing species and general reductions in species diversity and evenness. This agreement with the general model is comforting, but we have also learnt that not all communities are equally affected.

For example, it is much more difficult to detect effects in areas where sediments are highly mobile and experience high rates of natural disturbance, whereas boulder or pebble habitats, those supporting rich epifaunal communities that stabilize sediments, reef forming taxa or fauna in habitats experiencing low rates of natural disturbance, seem particularly vulnerable. However, despite the body of experimental data that has examined the impacts of trawling on benthic communities, it is often not possible to deduce the original composition of the fauna in places where experiments have been conducted because data gathered prior to the era of intensive bottom-fishing are sparse. This is an important caveat because recent analyses of the few existing historical datasets suggest that larger bodied organisms (both fish and benthos) were more prevalent prior to intensive bottom trawling. Moreover, in general, epifaunal organisms are less prevalent in areas subjected to intensive bottom fishing. Communities dominated by sponges, for example, may take more than a decade to recover, although growth data are notably lacking. Such slow recovery contrasts sharply with habitats such as sand that are restored by physical forces such as tidal currents and wave action.

Habitat Modification

An important consequence of trawling and dredging is the reduction in habitat complexity (architecture) that accompanies the removal of sessile epifauna. There is compelling evidence from one tropical system, for example, that loss of structural epibenthos can have important effects on the resident fish community, leading to a shift from a high value community dominated by Lethrinids and Lutjanids to a lower value one dominated by Saurids and Nemipterids. Similar arguments have also been made for temperate systems where structurally rich habitats may support a greater diversity of fish species. Importantly, such effects may not be restricted to the large biotic or abiotic structure provided by large sponges or coral reefs. One could quite imagine, for example, that juveniles of demersal fish on continental shelves might benefit from a high abundance of relatively small physical features (sponges, empty shells, small rocks, etc.) but that over time trawling will gradually lower the physical relief of the habitat with deleterious consequences for some fish species. Such effects may account for notable increases in the dominance of flatfish in both tropical and temperate systems. Our current understanding of the functional role of many of the larger-bodied long-lived species (e.g., as habitat features, bioturbators, etc.) is limited and needs to be addressed to predict the outcome of permitting chronic fishing disturbance in areas where these animals occur.

Although fishing-induced habitat modification is probably most widely caused by mobile gears, it is important to recognize that other fishing methods can also be highly destructive. For coral reef fisheries, dynamite fishing and the use of poisons represent major threats in some parts of the world.

Perhaps the only effective approach for mitigating the effects of trawling in vulnerable benthic habitats is to establish marine protected areas in which the activity is prohibited. Given the widespread distribution of trawling, it is not surprising that the establishment of marine protected areas is a key goal for many sectors of the marine conservation movement, although it should be borne in mind that it is not only trawling effects that can be mitigated by the approach. A key driver for the establishment of marine protected areas has come from The World Conservation Union (IUCN) and others who have called for a global representative system of marine protected areas and for national governments to also set up their own systems. A number of nations have already taken such steps, including Australia, Canada and the USA, with other nations likely to follow suit in the future.

Species Interactions

Even species that are not directly exploited by a fishery are likely to be affected by the removal of a substantial proportion of their prey, predator or competitor biomass and there are certainly strong indications that interactions with exploited species should be strong enough to lead to population effects elsewhere. For example, an analysis of the energy budgets for six major marine ecosystems found that the major source of mortality for fish is predation by other fish. Predatory interactions may, therefore, be important regulators for marine populations and removing large numbers of target species may lead to knock-on effects. Unfortunately, however, gathering the data necessary to demonstrate such controls is a major task that has rarely been achieved. Without studies directed specifically at the processes underlying the population dynamics of specific groups of species, it is difficult to evaluate the true importance of the effects of fisheries acting through species interactions in marine systems. Despite this caveat, some general effects appear to be emerging.

Removing Predators

For communities occupying hard substrata, there is good evidence that some fisheries have reduced predator abundances and that this has led to marked changes lower in the food web. Both temperate hard substrates and coral reefs provide good examples where reductions in predator numbers have led to change in the abundance of prey species that compete for space (e.g., mussels or algae), or in prey that themselves graze on sessile species. Such changes have led in turn to further cascading changes in community composition. For example, in some coral reef systems, removal of predatory fish has led to increases in sea urchin abundance and consequent reductions in coral cover.

Examples of strong predator control are much less easy to find in pelagic systems than they are in hard substratum communities. This perhaps suggests that predator control is less important in the pelagos. Alternatively, the lack of evidence may simply reflect our weak powers of observation; it is much harder to get data that would support the predator control theory in the pelagic than it is on a rocky shore.

Removing Prey

Fluctuations in the abundance of prey resources can affect a predator's growth and breeding success. Thus, if prey population collapses are sustained over the longer term due to fishing this will translate into a population decline for the predator. Examples of such effects can be found, particularly for bird species, but also for other taxa such as seals. Since many people have strong emotional attachments to such taxa, there is often intense interest when breeding failures or population declines occur. In the search for a culprit fishing activity is often readily offered as an explanation for the prey decline, or at least as an important contributory factor. In assessing the effect of prey removal, however, one must consider whether the fishery and predator compete for the same portion of the population, either in terms of spatial location or stage in the life cycle. For example, if the predator eats juveniles whose abundance is uncorrelated with the abundance of the fishable stock, the potential for interactions is greatly reduced. Such a feature seems rather common and probably needs to be examined closely in cases where a fishery effect is implicated. Nevertheless, there can be little doubt that unrestrained exploitation increases the likelihood of fisheries collapses and this is turn will take its toll on predator populations.

Removing Competitors

Unequivocal demonstrations of competition in most marine systems are rare. Perhaps the only exception to this is for communities occupying hard substrates where competition for space has been demonstrated and can be important in determining community responses to predators (see above). For other systems (e.g., the pelagic or soft-sediment benthos), we can only offer opinions, based on our assessment of the importance of other factors (e.g., predation, low quality food, environmental conditions). One system where fishing activity has been generally accepted to have an impact through competitive effects is the Southern Ocean, where massive reductions in whale populations by past fishing activity has led to apparent increases in the population size, reproduction or growth of taxa such as seals and penguins. A recent assessment, however, has even cast doubt on this interpretation, concluding that there is little evidence that populations have responded to an increase in available resources resulting from a decline in competitor densities.

Species Replacements

Despite the difficulties of clearly identifying the ecological mechanism responsible for the changes, there are some examples where fishing is heavily implicated in large-scale shifts in the species composition of the system and apparent replacement of one group by another. The response of the fish assemblage in the Georges Bank/Gulf of Maine area is, perhaps, the clearest example (**Figure 2**). During the 1980s the principal groundfish species, flounders and other finfish, declined markedly in abundance after modest increases in the late 1970s. It seems almost certain that the subsequent decline was a direct result of overexploitation by the fishery. In contrast, the elasmobranchs (skates and spiny dogfish) continued to increase during the 1980s. It would appear, therefore, that the elasmobranchs have responded opportunistically to the decline in the other species in the system, perhaps by being able to exploit food resources that were no longer removed by target species. Other possible examples of species replacements are the apparent increase in cephalopod species in the Gulf of Thailand, which coincided with the increase in trawl fishing activity and reduction in the abundance of demersal fish, and the increase in flatfish species that seems to have occurred in the North Sea and elsewhere.

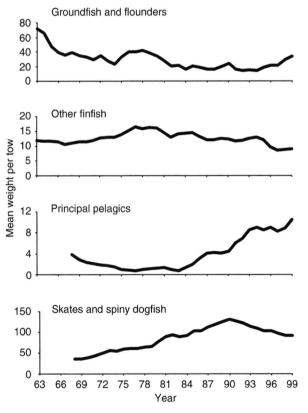

Figure 2 Trends in the relative contribution to total biomass (numbers) made by major taxonomic fish groups. Data from National Oceanographic and Atmospheric Administration, National Marine Fisheries Service, Woods Hole, MA, USA (personal communication).

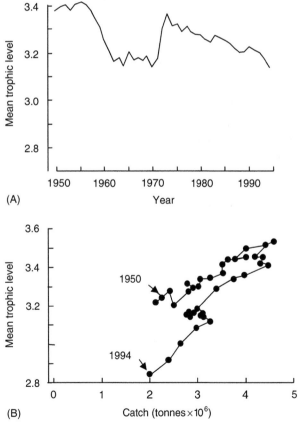

Figure 3 (A) Global trends in mean trophic level of fisheries landings from 1950 to 1994. (B) Plot of mean trophic level versus catch for the north-west Atlantic. (Reproduced from Hall, 1999.)

Conclusion

A final perspective on the system-level effects of fisheries come from an examination of changes over the last 45 years in the average trophic level at which landed fish were feeding (**Figure 3A**). This analysis indicates that there has been a decline in mean trophic level from about 3.3 in the early 1950s to 3.1 in 1994. Very large landings of Peruvian anchoveta, which feeds at a low trophic level, account for the marked dip in the time series in the 1960s and early 1970s. When this fishery crashed in 1972–73 the mean trophic level of global landings rose again. For particular regions, where fisheries have been most developed there have been generally consistent declines in trophic level over the last two decades.

Plots of mean trophic level against catches give a more revealing insight into the system-level dynamics of fisheries (**Figure 3B**). Contrary to expectations from simple trophic pyramid arguments, highest catches are not associated with the lowest trophic levels. This is important because it has been suggested in the past that fishing at lower trophic levels will give greater yields because energy losses from transfers up the food chain will be less. It appears, however, that the global trend towards fishing down to the lower trophic levels yields lower catches and generally lower value species – features indicative of fisheries regimes that are badly in need of restoration. Care needs to be taken when interpreting data such as these, particularly because catches of fish at different trophic levels are influenced by a number of factors including the demand for and marketability of taxa and the level of fishing mortality relative to optimum levels. Declines in catches at the end of the time series, for example, may well reflect depleted stocks of fish at all trophic levels. Nevertheless, these analyses are clear warning signs that global fisheries are operating at levels that are certainly inefficient and probably beyond those that are prudent if we wish to prevent continuing change in the trophic structure of marine ecosystems.

See also

Dynamics of Exploited Marine Fish Populations. Fisheries Overview. Large Marine Ecosystems. Sea Turtles.

Further Reading

Alverson DL, Freeberg MH, Murawski SA, and Pope JG (1994) A global assessment of bycatch and discards. *FAO Fisheries Technical Paper 339*, 233pp, Rome.

FAO (1996) Precautionary approach to fisheries. Part 1: Guidelines on the precautionary approach to capture fisheries and species introductions. *FAO Fisheries Technical Paper 350/1*, Rome.

FAO (1997) Review of the state of the world fishery resources: marine fisheries. *FAO Fisheries Circular no. 920*. Rome.

Hall SJ (1999) *The Effects of Fishing on Marine Ecosystems and Communities*. Oxford: Blackwell Science.

Jennings S and Kaiser MJ (1998) The effects of fishing on marine ecosystems. *Advances in Marine Biology* 34: 201–352.

Kaiser MJ and deGroot SJ (2000) *The Effects of Fishing on Non-target Species and Habitats: Biological, Conservation and Socio-economic Issues*. Oxford: Blackwell Science.

Pauly D and Christensen V (1995) Primary production required to sustain global fisheries. *Nature* 374: 255–257.

Pauly D, Christensen V, Dalsgaard J, Forese R, and Torres F (1998) Fishing down marine food webs. *Science* 279: 860–863.

DEMERSAL SPECIES FISHERIES

K. Brander, International Council for the Exploration of the Sea (ICES), Copenhagen, Denmark

Introduction

Demersal fisheries use a wide variety of fishing methods to catch fish and shellfish on or close to the sea bed. Demersal fisheries are defined by the type of fishing activity, the gear used and the varieties of fish and shellfish which are caught. Catches from demersal fisheries make up a large proportion of the marine harvest used for human consumption and are the most valuable component of fisheries on continental shelves throughout the world.

Demersal fisheries have been a major source of human nutrition and commerce for thousands of years. Models of papyrus pair trawlers were found in Egyptian graves dating back 3000 years. The intensity of fishing activity throughout the world, including demersal fisheries, has increased rapidly over the past century, with more fishing vessels, greater engine power, better fishing gear and improved navigational and fish finding aids. Many demersal fisheries are now overexploited and all are in need of careful assessment and management if they are to provide a sustainable harvest.

Demersal fisheries are often contrasted with pelagic fisheries, which use different methods to catch fish in midwater and close to the water surface. Demersal species are also contrasted with pelagic species (see relevant sections), but the distinction between them is not always clear. Demersal species frequently occur in mid-water and pelagic species occur close to the seabed, so that 'demersal' species are frequently caught in 'pelagic' fisheries and 'pelagic' species in demersal fisheries. For example Atlantic cod (*Gadus morhua*), a typical 'demersal' species, occurs close to the seabed, but also throughout the water column and in some areas is caught equally in 'demersal' and 'pelagic' fishing gear. Atlantic herring (*Clupea harengus*), a typical pelagic species, is frequently caught on the seabed, when it forms large spawning concentrations as it lays its eggs on gravel banks.

Total marine production rose steadily from less than 20 million tonnes (Mt) in 1950 to around 100 Mt during the late 1990s (**Figure 1**). The demersal fish catch rose from just over 5 Mt in 1950 to around 20 Mt by the early 1970s and has since fluctuated around that level (**Figure 2**). The proportion of demersal fish in this total has therefore declined over the period 1970–1998.

The products of demersal fisheries are mainly used for human consumption. The species caught tend to be relatively large and of high value compared with typical pelagic species, but there are exceptions to such generalizations. For example the industrial (fishmeal) fisheries of the North Sea, which take over

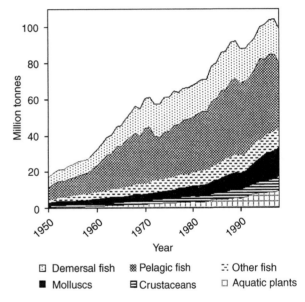

Figure 1 Total marine landings.

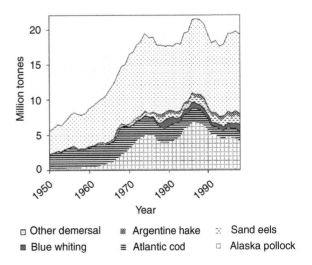

Figure 2 Total demersal fish landings.

half of the total fish catch, are principally based on low-value demersal species which live in or close to the seabed (sand eels, small gadoids).

Demersal fisheries are also often known as groundfish fisheries, but the terms are not exact equivalents because 'groundfish' excludes shellfish, which can properly be considered a part of the demersal catch. Shellfish such as shrimps and lobsters constitute the most valuable component of the demersal trawl catch in some areas.

Principal Species Caught in Demersal Fisheries

For statistical, population dynamics and fisheries management purposes the catch of each species (or group of species) is recorded separately by FAO (UN Food and Agriculture Organization). The FAO definitions of species, categories and areas are used here. FAO groups the major commercial species into a number of categories, of which, the flatfish (flounders, halibuts, soles) and the shrimps and prawns are entirely demersal. The gadiforms (cods, hakes, haddocks) include some species which are entirely demersal (haddock) and others which are not (blue whiting). The lobsters are demersal, but most species are caught in special fisheries using traps. An exception is the Norway lobster, which is caught in directed trawl fisheries or as a by-catch, as well as being caught in traps.

The five demersal marine fish with the highest average catches over the decade 1989–98 are Alaska (walleye) pollock, Atlantic cod, sand eels, blue whiting and Argentine hake. These species all spend a considerable proportion of their time in mid-water. Sand eels spend most of their lives on or in the seabed, but might not be regarded as a typical demersal species, being small, relatively short-lived and of low value. Sand eels and eight of the other 20 top 'species' in fact consist of more than one biological species, which are not identified separately in the FAO classification (**Table 1**).

The vast majority of demersal fisheries take place on the continental shelves, at depths of less than 200 m. Fisheries for deep-sea species, down to several thousand meters, have only been undertaken for the past few decades, as the technology to do so developed and it became profitable to exploit other species.

The species caught in demersal fisheries are often contrasted with pelagic species in textbooks and described as large, long-lived, high-value fish species, with relatively slow growth rates, low variability in recruitment and low mortality. There are so many exceptions to such generalizations that they are likely to be misleading. For example tuna and salmon are large, high-value pelagic species. Three of the main pelagic species in the North Atlantic (herring, mackerel and horse mackerel) are longer lived, have lower mortality rates and lower variability of recruitment than most demersal stocks in that area (**Table 2**).

Table 1 Total world catch of 20 top demersal fish species (averaged from 1989–1998)

Common name	Scientific name	Tonnes
Alaska pollock	*Theragra chalcogramma*	3 182 645
Atlantic cod	*Gadus morhua*	2 317 261
Sand eels	*Ammodytes* spp.	1 003 343
Blue whiting	*Micromesistius poutassou*	628 918
Argentine hake	*Merluccius hubbsi*	526 573
Croakers, drums nei	Sciaenidae	492 528
Pacific cod	*Gadus macrocephalus*	425 467
Saithe (Pollock)	*Pollachius virens*	385 227
Sharks, rays, skates, etc. nei	Elasmobranchii	337 819
Atlantic redfishes nei	*Sebastes* spp.	318 383
Norway pout	*Trisopterus esmarkii*	299 145
Flatfishes nei	Pleuronectiformes	289 551
Haddock	*Melanogrammus aeglefinus*	273 459
Cape hakes	*Merluccius capensis, M. paradox*	266 854
Blue grenadier	*Macruronus novaezelandiae*	255 421
Sea catfishes nei	Ariidae	242 815
Atka mackerel	*Pleurogrammus azonus*	237 843
Filefishes	*Cantherhines* (= *Navodon*) spp.	234 446
Patagonian grenadier	*Macruronus magellanicus*	230 221
South Pacific hake	*Merluccius gayi*	197 911
Threadfin breams nei	*Nemipterus* spp.	186 201

Table 2 The intensity of fishing is expressed as the average (1988–1997) probability of being caught during the next year. The interannual variability in number of young fish is expressed as the coefficient of variation of recruitment. Species shown are some of the principal demersal and pelagic fish caught in the north-east Atlantic

	Probability of being caught during next year	Coefficient of variation of recruitment
Demersal species		
Cod	33–64%	38–65%
Haddock	19–52%	70–151%
Hake	27%	33%
Plaice	32–48%	35–56%
Saithe	29–42%	45–56%
Sole	28–37%	15–94%
Whiting	47–56%	41–61%
Pelagic species		
Herring	12–40%	56–63%
Horse mackerel	16%	40%
Mackerel	21%	41%

Fishing Gears and Fishing Operations

A very wide range of fishing gear is used in demersal fisheries (*see also* Fishing Methods and Fishing Fleets), the main ones being bottom trawls of different kinds, which are dragged along the seabed behind a trawler. Other methods include seine nets, trammel nets, gill nets, set nets, baited lines and longlines, temporary or permanent traps and barriers.

Some fisheries and fishermen concentrate exclusively on demersal fishing operations, but many alternate seasonally, or even within a single day's fishing activity, between different methods. Fishing vessels may be designed specifically for demersal or pelagic fishing or may be multipurpose.

Effects of Demersal Fisheries on the Species They Exploit

Most types of demersal fishing operation are non-selective in the sense that they catch a variety of different sizes and species, many of which are of no commercial value and are discarded. Stones, sponges, corals and other epibenthic organisms are frequently caught by bottom trawls and the action of the fishing gear also disturbs the seabed and the benthic community on and within it. Thus in addition to the intended catch, there is unintended disruption or destruction of marine life (*see also* Ecosystem Effects of Fishing)

The fact that demersal fishing methods are non-selective has important consequences when trying to limit their impact on marine life. There are direct impacts, when organisms are killed or disturbed by fishing, and indirect impacts, when the prey or predators of an organism are removed or its habitat is changed.

The resilience or vulnerability of marine organisms to demersal fishing depends on their life history. In areas where intensive demersal fisheries have been operating for decades to centuries the more vulnerable species will have declined a long time ago, often before there were adequate records of their occurrence. For example, demersal fisheries caused a decline in the population of common skate (*Raia batis*) in the north-east Atlantic and barndoor skate (*Raia laevis*) in the north-west Atlantic, to the point where they are locally extinct in areas where they were previously common. These large species of elasmobranch have life histories which, in some respects, resemble marine mammals more than they do teleost fish. They do not mature until 11 years old, and lay only a small number of eggs each year. They are vulnerable to most kinds of demersal fishery, including trawls, seines, lines, and shrimp fisheries in shallow water.

The selective (evolutionary) pressure exerted by fisheries favors the survival of species which are resilient and abundant. It is difficult to protect species with vulnerable life histories from demersal fisheries and they may be an inevitable casualty of fishing. Some gear modifications, such as separator panels may help and it may be possible to create refuges for vulnerable species through the use of large-scale marine protected areas. Until recently fisheries management ignored such vulnerable species and concentrated on the assessment and management of a few major commercial species.

In areas with intensive demersal fisheries the probability that commercial-sized fish will be caught within one year is often greater than 50% and the fisheries therefore have a very great effect on the level and variability in abundance (**Table 2**). The effect of fishing explains much of the change in abundance of commercial species which has been observed during the few decades for which information is available and the effects of the environment, which are more difficult to estimate, are regarded as introducing 'noise', particularly in the survival of young fish. As the length of the observational time series increases and information about the effects of the environment on fish accumulates, it is becoming possible to turn more of the 'noise' into signal. It is no longer credible or sensible to ignore environmental effects when evaluating fluctuations in demersal fisheries, but a considerable scientific effort is still needed in order to include such information effectively.

Effects of the Environment on Demersal Fisheries

The term 'environment' is used to include all the physical, chemical and biological factors external to the fish, which influence it. Temperature is one of the main environmental factors affecting marine species. Because fish and shellfish are ectotherms, the temperature of the water surrounding them (ambient temperature) governs the rates of their molecular, physiological and behavioral processes. The relationship between temperature and many of these rates processes (growth, reproductive output,

mortality) is domed, with an optimum temperature, which is species and size specific (**Figure 3**). The effects of variability in temperature are therefore most easily detected at the extremes and apply to populations and fisheries as well as to processes within a single organism. Temperature change may cause particular species to become more or less abundant in the demersal fisheries of an area, without necessarily affecting the aggregate total yield.

The cod (*Gadus morhua*) at Greenland is at the cold limit of its thermal range and provides a good example of the effects of the environment on a demersal fishery; the changes in the fishery for it during the twentieth century are mainly a consequence of changes in temperature (**Figure 4**). Cod were present only around the southern tip of Greenland until 1917, when a prolonged period of warming resulted in the poleward expansion of the range by about 1000 km during the 1920s and 1930s. Many other boreal marine species also extended their range at the same time and subsequently retreated during the late 1960s, when colder conditions returned.

Changes in wind also affect demersal fish in many different ways. Increased wind speed causes mixing of the water column which alters plankton production. The probability of encounter between fish larvae and their prey is altered as turbulence increases. Changes in wind speed and direction affect the transport of water masses and hence of the planktonic stages of fish (eggs and larvae). For example, in some years a large proportion of the fish larvae on Georges Bank are transported into the Mid-Atlantic Bight instead of remaining on the Bank. In some areas, such as the Baltic, the salinity and oxygen levels are very dependent on inflow of oceanic water, which is largely wind driven. Salinity

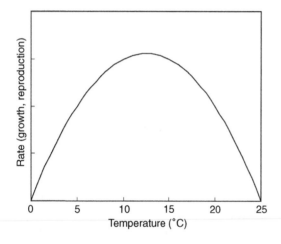

Figure 3 The relationship between temperature and many rate processes (growth, reproductive output, mortality) is domed. The optimum temperature is species and size specific.

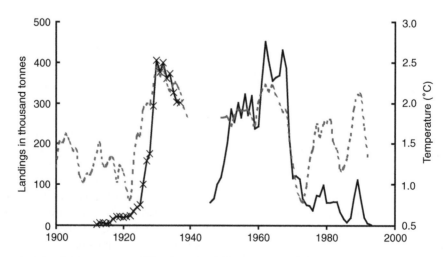

Figure 4 Cod catch and water temperature at West Greenland. Temperature is the running five-year mean of upper layer (0–40 m) values. ×, local catch × 20; ——— international catch; – – –, temperature.

and oxygen in turn affect the survival of cod eggs and larvae, with major consequences for the biomass of cod in the area. These environmental effects on the early life stages of demersal fish affect their survival and hence the numbers which recruit to the adult population.

World Catches from Demersal Fisheries and the Limits

Demersal fisheries occur mainly on the continental shelves (i.e., at depths less than 200 m). This is because shelf areas are much more productive than the open oceans, but also because it is easier to fish at shallower depths, nearer to the coast. The average catch of demersal fish per unit area on northern hemisphere temperate shelves is twice as high as on southern hemisphere temperate shelves and more than five times higher than on tropical shelves (**Figure 5**). The difference is probably due to nutrient supply. The effects of differences in productive capacity of the biological system on potential yield from demersal fisheries are dealt with elsewhere (*see also* Ecosystem Effects of Fishing).

Demersal fisheries provide the bulk of fish and shellfish for direct human consumption. The steady increase in the world catch of demersal fish species ended in the early 1970s and has fluctuated around 20 Mt since then (**Figure 2**). Many of the fisheries are overexploited and yields from them are declining. In a few cases it would seem that the decline has been arrested and the goal of managing for a sustainable harvest may be closer.

A recent analysis classified the top 200 marine fish species, accounting for 77% of world marine fish production, into four groups – undeveloped, developing, mature and declining (senescent). The proportional change in these groups over the second half of the twentieth century (**Figure 6**) shows how fishing has intensified, so that by 1994 35% of the fish stocks were in the declining phase, compared with 25% mature and 40% developing. Other analyses reach similar conclusions – that roughly two-thirds of marine fish stocks are fully exploited or overexploited and that effective management is needed to stabilize current catch levels.

Fish farming (aquaculture) is regarded as one of the principal means of increasing world fish production, but one should recall that a considerable proportion of the diet of farmed fish is supplied by demersal fisheries on species such as sand eel and Norway pout. Fishmeal is also used to feed terrestrial farmed animals.

Management of Demersal Fisheries

The purposes of managing demersal fisheries can be categorized as biological, economic and social. Biological goals used to be set in terms of maximum sustainable yield of a few main species, but a broader and more cautious approach is now being introduced, which includes consideration of the ecosystem within which these species are produced and which takes account of the uncertainty in our assessment of the consequences of our activities. The formulation of biological goals is evolving, but even the most basic, such as avoiding extinction of species, are not being

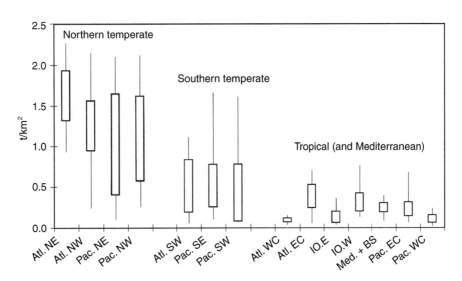

Figure 5 Demersal fish landings per unit area of continental shelf <200 m deep for the main temperate and tropical areas. The boxes show the spread between the upper and lower quartiles of annual landings and the whiskers show the highest and lowest annual landings 1950–1998. Atl, Atlantic; Pac, Pacific; IO, Indian Ocean; Med, Mediterranean; BS, Black Sea.

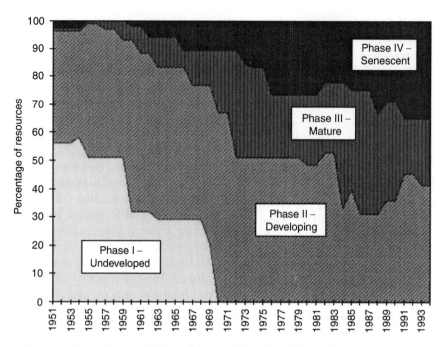

Figure 6 Temporal change in the level of exploitation of the top 200 marine fish species, showing the progression from mainly undeveloped or developing in 1951 to mainly fully exploited or overexploited in 1994.

achieved in many cases. At a global level it is evident that economic goals are not being achieved, because the capital and operating costs of marine fisheries are about 1.8 times higher than the gross revenue. There are innumerable examples of adverse social impacts of changes in fisheries, often caused by the effects of larger, industrial fishing operations on the quality of life and standard of living of small-scale fishing communities. Clearly there is scope for improvement in fisheries management.

From this rather pessimistic analysis of where fisheries management has got us to date, it follows that a description of existing management regimes is a record of current practice rather than a record of successful practice.

Biological management of demersal fisheries has developed mainly from a single species 'yield-per-recruit' model of fish stocks. The output (yield) is controlled by adjusting the mortality and size of fish that are caught. Instruments for limiting fishing mortality are catch quotas (TAC, total allowable catch) and limits on fishing effort. Since it is much easier to define and measure catch than effort, the former is more widely used. In many shared, international fisheries, such as those governed by the European Union, the annual allocation of catch quotas is the main instrument of fisheries management. This requires costly annual assessment of many fish stocks, which must be added to the operating costs when looking at the economic balance for a fishery. Because annual assessments are costly, only

the most important species, which tend to be less vulnerable, are assessed. For example the US National Marine Fisheries Service estimate that the status of 64% of the stocks in their area of responsibility is unknown.

The instruments for limiting the size of fish caught are mesh sizes, minimum landing sizes and various kinds of escape panels in the fishing gear. These instruments can be quite effective, particularly where catches are dominated by a single species. In multispecies fisheries, which catch species with different growth patterns, they are less effective because the optimal mesh size for one species is not optimal for all.

There are two classes of economic instruments for fisheries management: (1) property rights and (2) corrective taxes and subsidies. The former demands less detailed information than the latter and may also lead to greater stakeholder participation in the management process, because it fosters a sense of ownership.

The management of demersal fisheries will always be a complex problem, because the marine ecosystem is complicated and subject to change as the global environment changes. Management will always be based on incomplete information and understanding and imperfect management tools. A critical step towards better management would be to monitor performance in relation to the target objectives and provide feedback in order to improve the system.

One of the changes which has taken place over the past few years is the adoption of the precautionary approach. This seeks to evaluate the quality of the evidence, so that a cautious strategy is adopted when the evidence is weak. Whereas in the past such balance of evidence arguments were sometimes applied in order to avoid taking management action unless the evidence was strong (in order to avoid possible unnecessary disruption to the fishing industry), the presumption now is that in case of doubt it is the fish stocks rather than the short-term interests of the fishing industry which should be protected. This is a very significant change in attitude, which gives some grounds for optimism in the continuing struggle to achieve sustainable fisheries and healthy ecosystems.

Demersal Fisheries by Region

The demersal fisheries of the world vary greatly in their history, fishing methods, principal species and management regimes and it is not possible to review all of these here. Instead one heavily exploited area with a long history (New England) and one less heavily exploited area with a short history (the south-west Atlantic) will be described.

New England Demersal Fisheries

The groundfish resources of New England have been exploited for over 400 years and made an enormous contribution to the economic and cultural development of the USA since the time of the first European settlements. Until the early twentieth century, large fleets of schooners sailed from New England ports to fish, mainly for cod, from Cape Cod to the Grand Banks. The first steam-powered otter trawlers started to operate in 1906 and the introduction of better handling, preservation and distribution changed the market for fish and the species composition of the catch. Haddock became the principal target and their landings increased to over 100 000 t by the late 1920s. The advent of steam trawling raised concern about discarding and about the damage to the seabed and to benthic organisms.

From the early 1960s fishing fleets from European and Asian countries began to take an increasing share of the groundfish resources off New England. The total groundfish landings rose from 200 000 t to 760 000 t between 1960 and 1965. This resulted in a steep decline in groundfish abundance and in 1970 a quota management scheme was introduced under the International Commission for Northwest Atlantic Fisheries (ICNAF). Extended jurisdiction ended the activities of distant water fleets, but was quickly followed by an expansion of the US fleet, so that although the period 1974–78 saw an increase in groundfish abundance, the decline subsequently

continued. Most groundfish species remain at low levels and, even with management measures intended to rebuild the stocks, are likely to take more than a decade to recover.

The changes in abundance of 'traditional' groundfish stocks (cod, haddock, redfish, winter flounder, yellowtail flounder) have to some extent been offset by increases in other species, including some elasmobranchs (sharks and rays). However, prolonged high levels of fishing have resulted in severe declines in the less resilient species of both elasmobranchs (e.g. barndoor skate) and teleosts (halibut, redfish). This is covered more fully elsewhere (*see* Ecosystem Effects of Fishing).

The second half of the twentieth century saw major changes in the species composition of the US demersal fisheries. By the last decade of the century catches of the two top fish species during the decade 1950–59, redfish and haddock, had declined to 0.6% and 2.3% of their previous levels, respectively. Shellfish had become the main element of the demersal catch (**Table 3**).

Demersal Fisheries of the South-West Atlantic (FAO Statistical Area 41)

The catch from demersal fisheries in the SW Atlantic increased steadily from under 90 000 tonnes in 1950 to over 1.2 Mt in 1998 (**Figure 7**). The demersal catch consists mainly of fish species, of which Argentine hake has been predominant throughout the 50-year record. The catch of shrimps, prawns, lobsters and crabs is almost 100 000 t per year and they have a relatively high market value.

Thirty countries have taken part in demersal fisheries in this area, the principal ones being Argentina, Brazil, and Uruguay, with a substantial and continuing component of East European effort. A three year 'pulse' of trawling by the USSR fleet resulted in a catch over 500 000 t of Argentine hake and over 100 000 t of demersal percomorphs in 1967. The fisheries in this area are mostly industrialized and long range. As the stocks on the continental shelf have become fully exploited, the fisheries have extended into deeper water, where they take pink cusk eel and Patagonian toothfish. The main coastal demersal species are whitemouth croaker, Argentine croaker and weakfishes.

The Argentine hake fishery extends over most of the Patagonian shelf. It is now fully exploited and possibly even overexploited. Southern blue whiting and Patagonian grenadier are also close to full exploitation. Hake and other demersal species are regulated by annual TAC and minimum mesh size regulations.

Table 3 Average US catch from the north-west Atlantic region

Species	Scientific name	Average catch (t) 1950–59	Average catch (t) 1989–98
Atlantic redfishes nei	*Sebastes* spp	77 923	518
Haddock	*Melanogrammus aeglefinus*	64 489	1487
Silver hake	*Merluccius bilinearis*	47 770	16 469
Atlantic cod	*Gadus morhua*	18 748	24 112
Scup	*Stenotomus chrysops*	17 904	3846
Saithe (Pollock)	*Pollachius virens*	11 127	6059
Yellowtail flounder	*Limanda ferruginea*	9103	5085
Winter flounder	*Pseudopleuronectes americanus*	7849	5700
Dogfish sharks nei	Squalidae	502	17 655
Raja rays nei	*Raja* spp.	73	10 304
American angler	*Lophius americanus*	41	19 136
American sea scallop	*Placopecten magellanicus*	77 650	81 320
Northern quahog (Hard clam)	*Mercenaria mercenaria*	52 038	25 607
Atlantic surf clam	*Spisula solidissima*	43 923	160 795
Blue crab	*Callinectes sapidus*	28 285	57 955
American lobster	*Homarus americanus*	12 325	29 829
Sand gaper	*Mya arenaria*	12 324	7739
Ocean quahog	*Arctica islandica*	1139	179 312

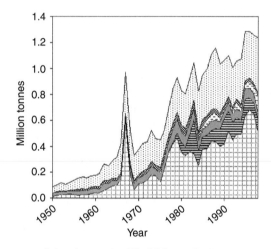

Figure 7 Total demersal fish landings from the south-west Atlantic (FAO Statistical area 41).

Legend: Other demersal; Weakfishes nei; Patagonian grenadier; Whitemouth croaker; Southern blue whiting; Argentine hake

then the catches from some stocks have declined, due to overfishing, while other previously underexploited stocks have increased their yields. The limits of biological production have probably been reached in many areas and careful management is needed in order to maintain the fisheries and to protect the ecosystems which support them.

See also

Ecosystem Effects of Fishing. Fishing Methods and Fishing Fleets.

Conclusions

Demersal fisheries have been a major source of protein for people all over the world for thousands of years. World catches increased rapidly during the first three-quarters of the twentieth century. Since

Further Reading

Cochrane KL (2000) Reconciling sustainability, economic efficiency and equity in fisheries: the one that got away? *Fish and Fisheries* 1: 3–21.

Cushing DH (1996) *Towards a Science of Recruitment in Fish Populations.* Ecology Institute, D-21385 Oldendorf/Luhe, Germany.

Gulland JA (1988) *Fish Population Dynamics: The Implications for Management,* 2nd edn Chichester: John Wiley.

Kurlansky M (1997) *Cod. A Biography of the Fish That Changed the World.* London: Jonathan Cape.

FISHERY MANAGEMENT

T. P. Smith and M. P. Sissenwine, Northeast
Fisheries Science Center, Woods Hole, MA, USA

The world's fisheries are significant from many per-spectives: biological, economic, cultural, and polit-ical. Fisheries may be measured in terms of biological yield; economic returns and contributions to eco-nomic value, income and jobs; production of food; cultural dependence; recreation; relationship to the ecosystem and the environment; and domestic and international trade.

Total fisheries production is reviewed elsewhere in this volume (*see* Marine Fishery Resources, Global State of). The focus of this article is not on the fish-eries themselves or the production from them, but rather the institutions of fisheries management. A broad definition of fisheries management will be used – a set of rules that govern who can fish and how fishing is conducted. Fishery management cuts across all the perspectives mentioned above and provides the bridge to human governance of fishery harvesting and processing.

An understanding of fisheries management is use-ful in interpreting trends in production, and changes in fleets, revenue, jobs, and income. More import-antly, however it is largely the actions of fishery managers that will determine the dynamics and likely future of the world's fisheries.

Introduction

Fisheries management is based on a number of goals. In general, managers seek to maximize long-term production from the fishery. Foremost among the formal concepts of management is the principle of maximum sustainable yield (MSY). MSY is derived from the fact that increasing application of effort will result in increasing catch (yield) up to a point at which additional effort will lead to decreased stock size and, subsequently, reductions in total yield.

It's not only important to maximize total sustain-able production, however. Managers should also seek to maximize the total value of fisheries. This brings into consideration economic returns, costs, and profitability; social and cultural considerations such as jobs, income and preserving a way of life; and the value of the resource as food, the focus of recreational activity, etc. Maximizing the value of the

fishery, however measured, is known as managing for optimum yield (OY). Again management systems around the world tend to manage for Optimum Yield, either directly or indirectly.

Beyond these concepts of maximization are goals associated with fairness and equity such as ensuring equal value to all users, preserving a historic fishery, preventing concentration of the fleet, and so forth.

It is clear that these goals and objectives can be in conflict and that much of fisheries management is devoted to preventing or, at least, minimizing such conflicts.

Why do we need fisheries management or, said another way, why do fishery management insti-tutions exist? A century ago it was believed that fishery resources, particularly offshore marine fish-eries, were inexhaustible – mankind's catch was small relative to the level of existing stocks and the stocks themselves were capable of production far in excess of these needs.

It is clear today that the world's fishery resources are not only exhaustible but also that, for many fisheries, current levels of fishing pressure are not sustainable. Stated more formally, for many of the world's fishery populations, demand at the current cost of production (taking into account the use of the best available technology) exceeds the rate of re-newal of the fish population, thus resulting in over-fishing (unsustainable fishing). As an example of this, the FAO considered the state of nearly 400 stocks as of 1996 and found that 73% were fully exploited or overexploited; 23% were overfished, depleted, or recovering (**Figure 1**).

Advances in technology, adoption of this technol-ogy, and the generally increasing demand for fishery products worldwide (mostly due to population growth but also due to gains in the standard of living and the perceived benefits of consuming fishery products) have contributed to this trend. Yet, fishery managers have been slow to recognize these changes and, even when recognized, management institutions have been slow to react. As a result the majority of the world's major stocks are fully exploited or overexploited and fisheries management is often perceived as having failed.

Management Systems–Institutional Arrangements

Fishery management institutions have some limited sphere of influence. For example, a fishery

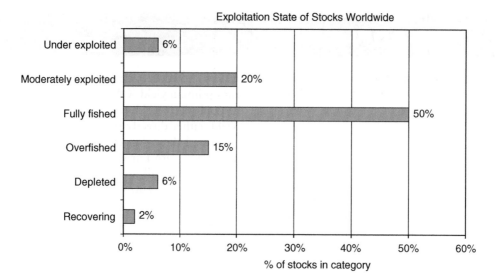

Figure 1 Exploitation state for the world's stocks, 1996. (Adapted from Garcia and DeLeiva Moreno, 1999.)

management plan may manage a state or provincial fishery, a plan may deal with a region's fishery or fisheries, have national scope, and so forth. Moreover, there are many institutional arrangements that have jurisdiction beyond national boundaries, for example, the Northwest Atlantic Fishery Organization (NAFO), or the International Commission for the Conservation of Tuna (ICCAT).

Fishery management is directed toward maximizing the benefits of the production unit (fish stock) that is being managed. Since stock boundaries may transcend national boundaries, many new geopolitical complications arise. Thus it is important to design an institution which promotes compromise among diverse human interests and values. This can be difficult not only conceptually, but practically as well if different management bodies in different countries use different approaches, timing, systems, etc.

Since ultimately the goal is to manage a stock appropriately, it is possible to delegate some management elements to a more local scale, while insuring that the collective impact on the fishery resource is sustainable.

Management institutions-local, regional, national, or international-require supporting infrastructure. This includes research facilities and scientists to determine the current state of the managed stock or stocks, to assess how current management is interacting with these stocks and to provide advice on how future management will impact the sustainability of the managed species. The science that supports management is necessarily multidisciplinary in nature and includes biology, stock dynamics, oceanography and ecosystem considerations, economics, sociology, and institutional behavior. Beyond the science and its delivery to managers there

must be in place data reporting and collections systems, and enforcement systems. Overlying all these systems is the management authority itself, consisting of one or more committees or panels responsible for making the decisions about the particulars of a management system.

If these roles are partitioned into science, reporting/monitoring, enforcement, and decision making, it can be seen that very different skills are appropriate to the various parts of the system. More importantly, stakeholders have different roles and responsibilities. Scientists need to be able to explain in a clear and concise manner the interaction between the fishery and the stock and the likely consequences of various management alternatives. Fishers need to report data in a timely and accurate manner, and data managers need to design data collection instruments that are effective and easy to comply with. Enforcement agencies need to have a dialog with managers to ensure that management plans are capable of being enforced and enforcement has to be applied uniformly, fairly, and consistently across all managed parties. In short, all the different players in the system need to articulate their concerns and the management authority needs to design a mechanism that allows a wide-ranging dialog while still providing for efficient decision making. This is a difficult task, made more problematic by some of the fundamental management difficulties discussed below.

Management institutions can be either formal, such as those established by law, or informal, such as nonlegally binding arrangements. The latter were common in villages or communities that influenced fishing practices of their members, helping to conserve fishery resources within their sphere of influence. Today, formal fishery management arrangements established in law are the norm. Such arrangements have become necessary because of the increases in efficiency and demand, and the increasing mobility of the population as traditional village or community level influences have broken down. It is now recognized that traditional informal management arrangements were insufficient to conserve fishery resources throughout their range. Thus there is a wide variety of national legislation and an elaborate international framework for managing fisheries.

The basis for international management authority flows from the United Nations Convention for the Law of the Sea (1982) (UNCLOS), which codified existing institutions and provided governance structure with respect to science, environmental control, and fishing and other commercial activities. UNCLOS extends jurisdiction to 200 miles, but also includes responsibilities for sustainable use of the resources under the control of each nation. To date some 132 states have become party to the convention. Notably signatories do not include the USA and Peru.

Beyond the fundamental agreement embodied in UNCLOS there have appeared more recent agreements on straddling stocks and highly migratory fish stocks. Essentially these agreements provide common ground for dealing with the conservation of high seas stocks and for regional or subregional management authority for stocks which are transboundary in distribution. In addition, another agreement prohibits nations from allowing vessels to register in their country (known as flying a flag of convenience) in order to avoid enforcement of fishery management regulations of the country in which they fish.

In 1995 the Code of Conduct for Responsible Fishing was agreed to by the Committee on Fisheries of the Food and Agriculture Organization of the United Nations. This nonbinding Code of Conduct establishes norms for fishery management.

From a national perspective, extended jurisdiction to 200 miles for most countries (also flowing from the Law of the Sea) forms the cornerstone of fishery management authority. Within this authority are many approaches. For example, in the USA, legislation now called the Magnuson–Stevens Fishery Conservation Management Act (MSFCMA) was introduced in 1976. The MSFCMA provides for eight regional fishery management councils, representation from each of the region's states on the council, and a detailed protocol for public decision making. Councils develop Fishery Management Plans (FMPs) to be implemented by the US government so long as the plans achieve optimum yield and do not violate specified national standards. (The 10 national standards include maintaining optimum yield while preventing overfishing, scientific standards, management by stock unit,

nondiscrimination among different states, efficiency, recognition of variation and contingencies, importance to fishing communities, minimization of bycatch, and the promotion of safety at sea.) The Act specifies the goal as obtaining optimum yield defined as MSY reduced by ecological, economic, and social factors. Although the MSFCMA has been amended several times since its inception, the principles of regional management, public debate, and decision making (technically advisory to the US Department of Commerce) utilizing technical committees (scientists), advisory panels (industry participants), and a decision-making body (the Council itself) have not changed.

Greater involvement of stakeholders in the fishery management process, is becoming common. For example, Canada has a Fishery Resource Conservation Committee to advise on fishery management and the Australian Fisheries Management Authority forms groups of stakeholders to prepare fishery management plans.

The systems described, by and large, are in place in developed countries. In developing countries, biological complexity and a general lack of scientific and governmental infrastructure make fisheries management problematic. One solution may be to build on traditional community or village rights (which have been dismantled in many places) to develop more realistic options, perhaps along the lines of a participatory management approach.

Management Systems–Controls

Since fishing involves the application of some effort to land a certain amount of fish, harvests can be controlled by either controlling effort or landings (or both). Effort controls may be direct, for example, limiting the total days spent fishing or may be indirect, controlling the amount of inputs used to produce a day of fishing. Thus, managers may limit net size, horsepower, hooks fished, and so forth. Input controls are based on the principle of regulated inefficiency. That is, if the manager restricts the technology that can be applied to catch fish, then the total harvest can be restricted so that it does not exceed a sustainable amount. Input controls are very commonly used in an attempt to overcome increases in total demand and improvements in technology which lead to increases in fishing power.

Note that input controls are usually gear specific. There may be minimum mesh size limits for trawls and gillnets, escape vent size limits for pots and traps, hook limits for longline or set gear, and so forth.

Unfortunately this general approach results in raising the total cost of fishing beyond what could be efficiently supplied. Additional inefficiencies arise when fishers invest in more unregulated inputs, seeking a competitive advantage (known as 'capital stuffing'). Direct effort controls, say by actively limiting days at sea, avoid the inefficiency of regulating individual inputs, but, unfortunately do not avoid the more general capital stuffing problem, as fishers are free to increase per day inputs in any number of dimensions. (Generally, however, such systems do not allow replacement of qualified vessels with new boats that are significantly larger, or have greatly increased horsepower.)

Output Controls

The other fundamental management approach is to control output. Generally output controls take the form of limits on the landings of a species of fish, where the limits are commonly called quotas or Total Allowable Catches (TAC). Quotas generally apply to the annual output from a fishery, but it is not uncommon to have the annual quota further divided into seasonal quotas, region-specific quotas, or both. The ITQ management systems mentioned above also use overall output controls, and, in addition, limit the output of each individual fisher to that allocation implied by their quota share.

One difficulty in output control systems is distinguishing between catch and landings, i.e. accounting for discards. In principle, output limits are aligned with total mortality targets and thus relate to total catch (landings plus discards). In practice, especially if discards are poorly estimated or not estimated at all, only landings are counted and therefore total mortality may be underestimated.

Time and Area Closures

Beyond restrictions on the inputs to catching fish or the total amount of fish caught, managers may also restrict when and where fishing can occur. This class of input management is known as time/area closure and is useful in protecting a stock in a particular place and time, for example, when the fish are aggregated for spawning. Alternatively, prohibiting fishing in certain areas for extended periods (perhaps year round) can provide a refuge for a core population of the regulated stock.

Size Limits

It is not uncommon for managers to place restrictions on the minimum size of the animals taken in the fishery, and less commonly, the maximum size as well. This form of output control recognizes that the fishery is not completely selective and may capture fish that are too small (or too large when maximum size limits are in place). Minimum size limits attempt to eliminate mortality on juvenile fish as it is the survival and growth of these fish that provide for a sustainable fishery. Similarly, maximum size limits are intended to take advantage of the greater reproductive potential of larger more mature animals.

These size limits can lead to discarding. In practice, managers try to match gear restrictions with size limits so as to minimize waste. For example, in a trawl fishery, the minimum mesh size is usually set to allow some portion (e.g. 50%) of the fish below the minimum size to pass through the mesh. Here there is a tradeoff between the capture efficiency of the net and the desire to protect smaller fish. Similarly, managers can regulate hook size to enhance the probability that only larger fish will be captured, specify minimum mesh sizes in gillnets, etc.

To allow for enforcement of regulations on size limits, such limits are usually couched in terms of possession limits, for example, no possession of Atlantic cod that are less than 19″ (48 cm) in length.

Prohibited Species Catch Limits

A special case of possession limit is the situation where managers do not want any individuals of a particular species captured. For example, in most of the world's fisheries it is illegal to capture large marine mammals such as whales or porpoises. If such an animal is caught the fisher is obligated to return it to the sea as quickly as possible with a minimum of harm. Generally speaking, such protections are applied to all marine mammals including seals, sea lions, and the like, and in some cases sea birds as well.

It also is possible that the possession of a species that may be targeted by other fisheries (perhaps under a different management jurisdiction) is not allowed. For example, in the bottom-fish fisheries of the North Pacific targeting primarily pollock, cod, and several species of flatfish, the possession of Pacific halibut, several species of crab, herring, and salmon is not allowed.

Performance Issues

Input versus Output Controls

On a worldwide basis most fisheries are not formally managed. For those that do utilize governance systems, management is, in general, based on input or output controls (or some combination of the two). There is a tendency for management to evolve from input to output controls and, perhaps, subsequently to harvest rights-based systems. (see discussion below).

Initially, managers tend to favor input controls, and in developing countries, management systems that do exist tend to take this form. Systems based on controlling input may be followed by an increasing reliance on output controls, because such systems do not explicitly impose inefficiencies, and more generally, can be more directly related to a sustainable level of fishing mortality and catch. Additionally, output limits such as quotas can easily be allocated between nations (for international fisheries where TAC management first became common), user groups, or individuals. While output controls do not explicitly impose inefficiencies, "quotas" do not necessarily solve the problem of the inefficiency in fisheries resulting from the race for the fish.

Output controls that result in shutdown of the fishery can lead to closures early in the season should the fishery harvest its allocation rapidly. Closure after a short season has considerable negative consequences including loss of jobs, income, social disruption, and the like. Managers can mitigate these effects by assigning quotas to seasons or areas, imposing trip limits, and so forth, but these measures tend to reduce the efficiency of the individual vessel.

Input controls, particularly direct controls on days fishing, avoid these problems by formally controlling fishing effort. Unfortunately, capital stuffing may be encouraged as boats attempt to increase their fishing power per unit of effort. Because of this, systems using input controls usually define a unit of fishing effort (e.g. a 50' vessel with 1000 hp). It is therefore necessary that managers understand the link between a unit of effort, say a day at sea, and a day's catch as, ultimately, it is total fishing mortality (i.e. total output) that must be controlled. To the extent that fishers can add technology, labor, etc. to maximize production on a day at sea, management expectations on the conservation benefits of direct input controls may be too optimistic.

Progress toward attainment of input control limits are more difficult to monitor, requiring either a self-reporting or electronic monitoring system to determine if a vessel is actively fishing. A related difficulty in a system based entirely on input controls is that there may be no controls on output whatsoever. This means that if pre-season expectations on total mortality are not met then adjustments must be made in the next season.

In any case, unless fishing capacity or fishing power is somehow rationalized prior to and independent of the output or input controls, neither system will work effectively as fishers will either rapidly exhaust quotas or available days at sea and contribute to a protracted shutdown of the fishery. If these periods of inactivity are too long operations will not be able to cover their fixed costs and will go out of business.

Allocation

To understand the effectiveness (or lack thereof) of management it is necessary to consider the fundamental principle in fisheries management: allocation. That is, beyond the particulars of the type of controls used to conserve the fishery resource, fisheries management is essentially a method to allocate a scarce resource (the fish) to a number of competing users. This means that the fisheries management process can be characterized as a debate about who are the winners and losers in terms of the opportunity to benefit from the extraction of the fishery resource. The allocation issue itself causes more and more layers of management to be added. For example, quotas might be established for particular areas, for particular gear groups (trawlers, longliners, etc.), for residents of a particular state or province, for vessels of different sizes (e.g. small trawlers, large trawlers) and so forth, resulting in a very Balkanized management system.

More to the point of trying to manage effectively, the allocation debate can easily overshadow the discussion on 'doing the right thing'. Thus, discussion on input versus output controls, open versus closed access versus access granted by harvest rights, may not occur as managers become pre-occupied with dividing up a limited pie among competing users instead of debating how to make the pie bigger.

Competitive Allocation versus Rights-based Allocation

An overarching issue in fisheries management is the choice between rights-based allocations versus competitive allocations. Essentially, open access and limited access systems are competitive allocation systems and individual fishery quotas are a form of rights-based allocation.

Under rights-based management system, the holders of the rights have an incentive to conserve stocks and to provide the appropriate level of inputs for a given catch. And since fishers hold the rights to a certain portion of the catch and can land that amount at a time and place convenient to their operations, fishers operating under rights based systems tend to utilize the most efficient gear, time at sea, and so forth.

One such management system based on harvest rights is the Individual Fishery Quota system (IFQ) or Individual Transferable Quota system (ITQ). Management by individual quota has, for some nations, become the management system of choice (e.g., Iceland and New Zealand).

Rights-based systems may allocate output shares (IFQs) and those shares may be transferable among qualified participants (ITQs). Quotas may be freely transferable, or limited to certain quota holders (e.g. ITQ holders in a certain area). The total amount of shares held may be limited. Entitlements may be made to individuals, vessels, companies, communities, regions, etc. and transfers may be limited to like ownership categories. Allocations may be absolute (e.g. 10 metric tons of species A) or relative (e.g. one-tenth of 1% of the overall quota for species A).

In principle, management based on Individual Transferable Effort (e.g. days at sea) could also be utilized, although we are not aware of any systems that allow effort trading.

The principle of rights-based management can be simply stated. If the harvester has some assigned (and protected) rights to harvest a certain portion of a stock he will have every incentive to utilize inputs most efficiently, choose a vessel which is most appropriate for the way he wishes to prosecute a fishery, fish at a time and place that is both convenient and maximizes opportunities for maximizing catch or profit per unit of effort.

Granting such rights is controversial, however. Concerns include who is granted the initial privilege to fish and how does one deal with the 'windfall profit' provided to those who qualify (the quota holder's share can have significant economic value). Another concern is concentration in the fishery, since efficient operators can buy or lease rights from others and increase their scale of operations. Concentration can lead to a large boat fishery or a fishery with absentee owners. Finally, there is the issue of 'high grading' where fishers may discard less valuable animals (e.g. smaller fish) for which they hold quota so as to land the most revenue per pound of quota held.

Effectiveness and Enforcement

Over time, given all the difficulties mentioned above, the management system can become very complex. Thus, it is not unusual to have a fundamental effort control system such as regulating days at sea, overlain with input restrictions, time/area closures, and the like. This can lead to very complex regulations that are poorly understood by fishers. More to the point, fishers dislike controls which limit their fishing opportunity and thus may not comply with regulations.

Given this, it becomes clear that a significant issue is the enforcement of regulations. Some measures are easy to enforce (e.g. a closed area) and some regulations are not (e.g. discard limits or reporting discards). This leads to two kinds of difficulties: the effectiveness of the regulations may be much less than intended or assumed, and enforcement agents may not be able to successfully prosecute violators.

Problems and Issues

Access

The issue of access, that is open versus limited versus rights-based access, as outlined above is the fundamental issue in management. Entry to fisheries has traditionally been open; newcomers, provided they had the necessary capital, could enter a fishery whenever they wished. Open access, however, leads to the race for the fish, and to the dissipation of rent (the returns beyond normal business profits that arise due to production from the renewable fishery resources). The 'race for fish' results from each individual fisher believing that another day at sea, more or better gear, electronics or other inputs will result in more revenue. This is, of course, true up to a point. However, as participation and effort increases, the actions of that fisher begin to negatively impact other fishers. These negative interactions increase external costs (called externalities by economists) via crowding, interference with the fishing operations of others or, more fundamentally, through depletion of the stock that others are trying to fish.

Failure to account for the externalities of fisheries production has been characterized as the 'tragedy of the commons', using the analogy of overgrazed public greens.

Given these well understood problems with open access, most developed fishing nations have begun to limit fisheries access to a set of individuals with an established history in the fishery (called limited access or limited entry). This barrier to new entry can, in principle, reduce stock externalities. In practice, however, many of the problems discussed above are still prevalent simply because the number of initial qualifiers exceeds what would be appropriate in a rationalized fishery (a fishery where available effort matches available supply). Further exacerbating the problem, qualified entrants have every incentive to increase their individual fishing power while ignoring the contribution of other fishers. Thus limited access, by itself, can be ineffective in matching fishing power or effort to sustainable levels of fishing.

Ratchet Effect

Another reason for the general lack of success in managing fisheries, especially overexploited fisheries, is what has been characterized as the 'ratchet effect'. General scientific principles as well as guidance from the United Nations and the laws governing many national fisheries lead to a prescription of managing to produce the long-term equilibrium yield, usually the maximum sustainable yield (MSY). Thus, scientists provide advice on management goals that are based on long-term productivity considerations. In reality, fishery populations fluctuate in response to environmental conditions and interactions with other fish populations.

Dynamic, rather than stable, population levels can lead to a situation where managers and fishers tend to add capacity when stocks are stable or increasing, all the while attempting to reach a level of production equal to the maximum yield. However, should a downturn in the stock occur, managers are unable to quickly reduce fishing power, because of economic and political pressures to maintain jobs, income, and lifestyle. This feedback loop is very unattractive in that fishing pressure increases in good times but does not contract in bad times.

A similar problem occurs because there is pressure on managers to continue overfishing in the face of scientific uncertainty. This problem occurs because managers seek scientific consensus before taking action. In reality, the level of a fish population is determined via a complex set of interactions among the environment, other fish populations, and the fishery itself. Thus necessary information may be lacking, or timely information on a stock that is very dynamic may not be yet available. Again, the bias is uni-directional. In the face of uncertainty, managers tend to favor the most optimistic forecast and discount the least favorable.

Discarding

Fisheries management institutions themselves can contribute to the 'management failure' when inefficient management rules lead to considerable biological and economic waste. One dimension of waste is fishery discarding and incidental catch or bycatch. Bycatch results from fishing operations that are not completely selective, that is, catch species other than those targeted. Often these fish are discarded, either because regulations require it, or because the product is not marketable. An example of regulatory discarding is the tossing overboard of fish below a minimum size limit.

Discarding of target species can also occur when the size or condition of the animal makes it less attractive to the market and the fisher discards so as to land the maximum value of product given restrictions on time fishing, hold space, or landing limits.

A similar problem is the discard resulting from regulatory landing limits or trip limits. Here, to limit overall fishing mortality or the mortality on a certain segment of the stock, managers restrict the landing of a species to a certain weight (or number) of fish per trip. Since fishery catches are often mixed and since some fish are more valuable than others, a vessel may discard large amounts of a nontargeted species or, alternatively over-harvest the target species, discarding the less valuable or smaller fish.

The waste apparent from discarding in a single species context can be greatly magnified when more than one species is being managed by trip limits. This is because the individual fishery trip limits now interact and the most constraining trip limit becomes the rule that controls total fisheries catch. Thus, considerable potential catch may be foregone. From the opposite perspective, waste may increase, as the mix of species taken in a fishery would only match the proportions implied by the set of all the various species' trip limits by chance (or if the trip limits were 'perfectly' estimated).

One further complication to the discarding problem is that adequate reporting systems may not be in place. Discards may not be reported, catch may be underestimated, and assessments based on a determination of total fishery mortality (landings and discards) may be inaccurate.

Ecosystem Management

The difficulties facing managers due to some of the single-species issues discussed above become greatly complicated when managers attempt to control the harvest of several interacting stocks. Recently there has been a focus on what is called the ecosystem effects of fishing. Here, managers are being asked to not only manage single stocks and groups of stocks interacting in a multispecies fishery, but also to manage the entire ecosystem in which the fish occur; the fish, other nonfish populations, and the fishery habitat. Such a perspective is partly due to concerns about potentially environmentally destructive fishing practices and partly due to renewed interest in a holistic or ecosystem perspective for managing interrelated species. Unfortunately, managing for the ecosystem is much more complex than managing for single species maximum yield; data requirements are more demanding, scientific uncertainty is increased; and our ability to understand all the important biological and economic interactions is limited.

Towards the Future

The problems discussed above and the linkages to decision making are shown in **Figure 2**. Unfortunately, the difficulties outlined are fairly endemic to fishery management and fairly intractable as well.

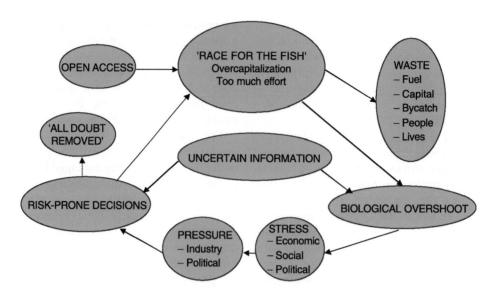

Figure 2 A diagrammatic view of the fisheries management problem, showing the feedback mechanisms, which can lead to management failure. (Adapted with permission from Sissenwine and Rosenberg, 1993.)

Given this, the appropriate role for fisheries management is to put in place governance systems that reduce or eliminate discard and waste, promote efficiency, rationalize effort, and recognize management and scientific limitations.

Most important to effective management is the notion of controlled access and beyond this, rights-based access. Without mechanisms for efficiently controlling and allocating overall effort fishery management goals cannot be met. In this regard, incentive based systems such as ITQs or IFQs harvest systems, or less formal, but equally effective, community-based or regional control systems, offer the most promise.

The Precautionary Approach

An emerging paradigm in fisheries management known as the precautionary approach is becoming the basis for fishery conservation systems worldwide. The precautionary approach is explicitly embodied in the Straddling Stocks and Highly Migratory Stocks agreement and the Code of Conduct mentioned above and is based on (but different from) the Precautionary Principle. The Principle implies an extreme reversal of the burden of proof in that it attempts to prevent irreversible damage to the environment by implementing strict conservation measures, even in the absence of evidence that environmental degradation is human caused. The precautionary approach also reverses the burden of proof but is designed to provide practical guidance for managers in the areas of fisheries management, fisheries research, and fisheries technology.

Formally, the definition offered by Restrepo *et al.* (1998) is:

> In fisheries, the Precautionary Approach is about applying judicious and responsible fisheries management practices, based on sound scientific research and analysis, proactively (to avoid or reverse overexploitation) rather than reactively (once all doubt has been removed and the resources are severely overexploited), to ensure the sustainability of fishery resources and associated ecosystems for the benefit of future as well as current generations.

Adoption of the precautionary approach as the overarching yardstick in developing management systems is important and has led to major changes in many of the developed nations' fisheries management strategies. For example, in the USA, the precautionary approach is the philosophy behind a number of important revisions to the MSFCMA which provided for management limits, risk-averse targets, specification of uncertainty, and the like.

In fact, the precautionary approach offers a road map for the future; a management perspective that can accommodate all the difficulties highlighted in this article. Providing for effective fisheries management on a worldwide (or even national) context is a daunting task but, with the adoption of an explicit risk-averse philosophy and participatory management practices and the rationalization of overall effort there is hope for success, rather than failure, in future management systems.

See also

Demersal Species Fisheries. Marine Fishery Resources, Global State of.

Further Reading

Anderson LG (1986) *The Economics of Fisheries Management*, 2nd edn. Baltimore, MD: Johns Hopkins University Press.

Garcia and DeLeiva Moreno 2000. Trends in World Fisheries and their resources: 1974-1999, in FAO, the State of World Fisheries and Aquaculture 2000 (in press).

Hancock DA, Smith DC, Grant A, and Beumer JP (eds.) (1996) *Developing and Sustaining World Fisheries Resources. The State of Science and Management. 2nd World Fisheries Congress.* Collingwood, Australia: CSIRO Publishing.

Hardin G (1968) The Tragedy of the Commons. *Science* 162: 1243–1248.

Huxley TH (1884) Inaugural address. *Fisheries Exhibition Literature* 4: 1–22.

Kruse G, Eggers DM, Marasco RJ, Pautzke C and Quinn TJ II (eds.) (1993) *Proceedings of the International Symposium on Management Strategies for Exploited Populations.* University of Alaska, Fairbanks: Alaska Sea Grant College Program Report No. 93–102.

Lackey RT and Nielsen LA (eds.) (1980) *Fisheries Management.* Oxford: Blackwell Scientific Publications.

Ludwig D, Hilborn R, and Walters C (1993) Uncertainty, resource exploitation, and conservation: lessons from history. *Science* 260: 17–36.

Miles E. (ed.) 1989 *Management of World Fisheries: Implications of Extended Coastal State Jurisdiction.* Proceedings of a Workshop organized by the World Fisheries. Project, Institute for Marine Studies University of Washington. Seattle: University of Washington Press.

National Research Council/Committee to Review Individual Fishing Quotas, Ocean Studies Board, Commission on Geosciences, Environment and Resources (1999) *Sharing the Fish. Toward a National Policy on Individual Fishing Quotas.* Washington, DC: National Academy Press.

Pikitch EK, Huppert DD, and Sissenwine MP (eds.) (1997) *Global Trends: Fisheries Management.* American Fisheries Society Symposium 20, Seattle, WA, 14–16 June 1994. Bethesda, MD: AFS.

Restrepo VR, Mace PM and Serchuk FM (1998) The Precautionary Approach: A New Paradigm, or Business as Usual? In: *Our Living Oceans.* Report on the status of US living marine resources. US Department of Commerce NOAA Technical Memo. NMFS-F/SPO-41.

Sissenwine MP and Rosenberg A (1993) Marine fisheries at a critical juncture. *Fisheries* 18: 6–10.

Sutinen JG and Hanson LC (eds) (1986) *Rethinking Fisheries Management.* Proceedings from the Tenth Annual Conference, June 1–4, 1986, Center for Ocean Management Studies, University of Rhode Island, Kingston, RI.

Waugh G (1984) *Fisheries Management. Theoretical Developments and Contemporary Applications.* Boulder, CO: Westview Press.

FISHERIES ECONOMICS

U. R. Sumaila and G. R. Munro, University of British Columbia, Vancouver, BC, Canada

Introduction

Natural resources can be classified into nonrenewable and renewable resources. Minerals and hydrocarbons are examples of nonrenewable natural resources, while forests and fishery resources are examples of renewable resources. The basic distinction between the two categories is whether or not the resource is capable of growth.

Fish, being renewable resources, portray the following characteristics: (1) 'utilization' of a unit of the fish resource implies its destruction, that is, the unit is completely and irrevocably lost; and (2) the fish stock can be augmented again to enable a continuing availability through time.

Thus fish, as for other renewable natural resources, have the special feature that new stocks can be created by a process of self-generation. This regeneration occurs at a natural or biological rate, often related to the amount of original stock remaining unutilized. The essence of fisheries economics stems from these characteristics of fish stocks, coupled with the fact that the rate of biomass adjustment of a fish stock is a function of that stock. Essentially, the central problem of natural resource economics at large and fisheries economics in particular is intertemporal allocation. That is, natural resource economists are mainly concerned with the question of how much of the stock should be designated for consumption today and how much should be left in place for the future.

World ocean fisheries are commonly divided between aquaculture, or fish farming, and capture fisheries, the hunting of fish in the wild. While not denying the growing importance of aquaculture, this contribution will be confined to the economics of capture fishery resources. The economics of aquaculture fisheries, and their links to capture fisheries, demand a contribution unto themselves.

World ocean capture fisheries yield annual catches in the order of 80 million t, which, in turn, have a first sale value of approximately US$80 billion. While these fisheries are but a source of modest employment in the developed world, they continue to be an important source of employment in the developing world. It is estimated that world ocean capture fisheries provide employment – direct and indirect – for upward of 200 million men and women.

There is a clear evidence that the output of world ocean capture fisheries has reached, or is reaching, an upper limit because of the finite capacity of the world's oceans to support such resources (*see* Marine Fishery Resources, Global State of). Scope for increased catch must come through improved resource management. The fact that world ocean capture fisheries may be reaching an upper limit should be no great cause for alarm. What is a cause for alarm is the fact that there is a growing evidence of overexploitation of capture fishery resources, and that the overexploitation has intensified over the past several decades.

Sustainable catch of a capture fishery resource essentially involves skimming off the growth of the resource, or what fishery scientists call the 'surplus production'. If the fish biomass is equal to zero, then its growth, by definition, is equal to zero. If the biomass has achieved its natural equilibrium level (no catch), the growth of the resource will also, by definition, be equal to zero. At some point between these two extremes, the growth of the resource ('surplus production') will be at a maximum, which fishery scientists refer to as maximum sustained yield (catch), it (see **Figure 1**). While the use of MSY, as a resource management criterion, has been subject to many criticisms, it is still used widely as a reference point.

Using MSY as a reference point, the Food and Agriculture Organization of the United Nations (FAO) periodically assesses the capture fishery resources of the world. If the resource is being exploited at, or near, the MSY level, it is deemed to be 'fully exploited'. If the resource has been driven down below the MSY level, it is deemed by the FAO to have been 'overexploited'. On the other hand, if the resource is above the MSY level, it is deemed to be less than fully exploited. On this basis, the FAO estimates that, in the first half of the 1950s, under 5% of the world's capture fishery resources fell into the 'overexploited' category. By the early 1970s, the percentage in the 'overexploited' category had doubled to 10%. As of 2004, the figure stood at 25%, while another >50% fell into the 'fully exploited' category. **Figure 2** illustrates this development vividly. While 'fully exploited' does not mean 'overexploited', there is the fear that many 'fully

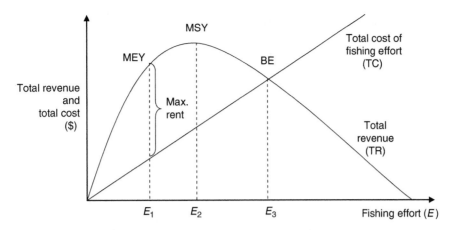

Figure 1 Gordon–Schaefer bioeconomic model (Gordon, 1954). MEY is the maximum economic yield, where economic rent (or profit) is maximized. MSY is the maximum sustainable yield. BE is the bionomic equilibrium, where all rent is dissipated.

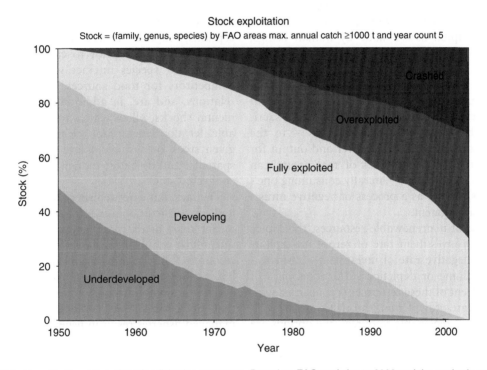

Figure 2 Global trend in the status of marine fisheries resources. Based on FAO statistics to 2003 and the methods and definitions in Froese and Pauly (2003).

exploited' stocks are prime candidates for future inclusion in the 'overexploited' category.

How then did we get into such a situation, and what is to be done about it? It will be argued that economics lies at the heart of each of these questions. This is a view now coming to be held by marine biologists, as well as economists. Having said this, let us immediately follow by stating that economics alone cannot deal with the fisheries management issues at hand. Anyone involved with the questions of capture fisheries management is compelled to acknowledge that the issue is inescapably multidisciplinary in nature.

Fisheries economists have no choice but to work closely with marine biologists, legal experts, and sociologists, to mention but a few.

The Basic Economics of Capture Fishery Resources

One can pin the beginning of fisheries economics to the 1950s, when Gordon (1954) and Scott (1955) published their seminal papers. Of course, the seeds of the discipline were sowed as early as 1911. The models developed by Gordon and Scott, which still form the core of the basic economic models of fishing, were essentially static (snapshots) in nature. Further developments of the static approaches were pursued in the 1960s with contributions such as that of Turvey. Basic versions of dynamic fisheries economics models began to be developed in the 1960s and 1970s. Following the pioneering works were a flurry of papers that extended these initial contributions in various directions (stochastic fisheries models, migratory fish stock models, shared stocks fisheries models, etc.).

Central to the economics of natural resources, including fishery resources, is the concept of 'real' (as opposed to financial) capital. Real capital is any asset that is capable of yielding a stream of economic benefits to society over time. Natural resources fall within this definition of 'real' capital. They differ from capital assets made by human beings, only in that they come to us as endowments from nature.

The process of increasing society's stock of 'real' capital, human-made or natural, is a process of positive investment, which involves making a sacrifice today by saving – consuming less – in the promise of greater productive power and output for society in the future. One's stock of 'real' capital can also be reduced, of course – literally consuming one's capital. This, we see as a process of negative investment, or disinvestment.

In the case of nonrenewable resources, the choice is between an investment rate of zero – no exploitation and a negative rate of investment – disinvestment (i.e., mining or depletion). The economics of the management of these resources is concerned with the optimal rate of depletion, through time.

In the case of renewable resources, the options are considerably broader. First, positive investment in such resources is achievable by utilizing less than the growth of the resource. An investment rate of zero in the resource does not involve refraining from exploiting the resource, but rather utilizing the resource on a sustainable basis.

The economics of capture fishery resource management is thus best seen as a problem in 'real' asset management through time. In an ideal world, society's portfolio of capture fishery resources would be managed to maximize the net economic benefits (broadly defined) from the fishery resources to society, not just for today, but through time. The economics of fisheries management draws upon the economist's theories of capital and investment. In this contribution, we shall do no more than discuss, in general terms, the basic results from the capital theoretic models of the fishery.

Figure 1 describes the basic theory of fisheries economics, which stipulates that fishing cost in open-access fisheries, assumed to be proportional to fishing effort, will continue to increase even though revenues per unit of effort are declining, and that ultimately revenues will decline until they equal costs. The point at which total revenue equals total cost is commonly regarded as the bionomic equilibrium (BE), where both industry profits and resource rents have been completely dissipated.

The Complications: Dealing with Complex Capture Fisheries Problems

Mobility, Visibility, and Uncertainty

Capture fishery resources have always been notoriously difficult to manage effectively. With few exceptions, capture fishery resources are mobile, and are not visible prior to capture. Furthermore, the relevant species interact with one another, as competitors for food sources, or in predator–prey relations, and are, in addition, subject to environmental shocks, all of which are not readily observable, let alone measurable or predictable. Finally, a given stock of a single species may be spread out spatially, and be better thought of as a set of interconnected stocks.

The fact that capture fishery resources are mobile, and are difficult to observe prior to capture, has meant that, in the past, at least, it has been very difficult to establish effective property rights to these resources. This, in turn, has meant that the resources have a long history of being exploited on an open-access, or 'common pool', basis. It has been shown that open-access, and 'common pool' nonjointly managed fisheries result in fishers being faced with a set of economic incentives that are perverse from the point of view of society.

No rational would-be investor will undertake an investment, unless the expected net economic returns from the investment, discounted at an appropriate rate of interest, exceed the current cost of that investment. The sum total of these discounted net economic returns is referred to as the present value of the returns. In open-access and 'common pool' nonjointly managed fisheries, no individual fisher can count on a positive return on an investment in the resource. If a fisher refrains from catching the fish in order to build up the resource, he/she may do nothing more than increase the catch of his/her competitors.

It can be shown that, in such fisheries, fishers will act as if they are applying a rate of discount (interest) to future returns from the fishery equal to infinity.

Tomorrow's returns from the fishery count essentially for nothing. This, in turn, means that the rational fisher is given the incentive to treat the resource as a nonrenewable resource, namely, as a resource to be mined.

What we might term the social rate of discount (interest) never, even in extreme circumstances, approaches infinity. Thus, one can say with assurance that open access or 'common pool' nonjointly managed fisheries will lead to overexploitation, in the sense that the disinvestment in the resources is excessive from society's point of view.

Fisheries economics, to a large extent, is involved in demonstrating the negative economic, and ecological effects (e.g., excessive resource disinvestment) of open-access or 'common pool' nonjointly managed fisheries. It also develops and analyzes approaches for tackling these effects.

Ecosystem-based Fisheries Management

It is agreed that fisheries management, and therefore economic analysis, should properly not be done on a single-species basis, but rather on an ecosystem basis, in which species interactions are explicitly recognized. In general, an ecosystem is a geographically specified system of organisms, including humans, the environment and the processes that control its dynamics. Similarly, an ecosystem-based approach to the management of marine resources (EAM) is geographically specified, it considers multiple external influences, and strives to balance diverse societal objectives. EAM requires that the connections between people and the marine ecosystem be recognized, including the short-term and long-term implications of human activities along with the processes, components, functions, and carrying capacity of ecosystems.

The fact that ecosystems and the EAM are geographically specified implies that for ecosystems that are shared by two or more countries, policies that are transboundary in nature are required to manage them successfully. This is because in such cases, the ecosystems do not respect national borders. Many of the world's 64 large marine ecosystems are shared by two or more countries. For instance, to apply EAM to the management of the Gulf of Guinea Large Marine Ecosystem effectively, policies need to be crafted and adopted by the 16 countries bordering the ecosystem. In terms of policy, getting countries with diverse societal objectives to agree on and implement joint EAMs is certainly a challenge, which must be addressed if EAM is to gain universal applicability. Two ways by which a country's societal objective regarding the use of marine ecosystem

resources can enter economic analysis are (1) how the country values market relative to nonmarket values from the ecosystem, and (2) the discount rate applied to discount flows of net benefits over time from the ecosystem.

The economic analysis of a full ecosystem approach to management is daunting at best. Producing analytical results is next to impossible in ecosystem-economic models. As a result, most fisheries economic models continue to be based on single-species biological models. Current economic-ecosystem models have mostly resorted to simulation and numerical models. The economic issues raised by EAM relate to the economics of internationally shared fishery resources, to which we turn in the next section.

Internationally Shared Fishery Resources

The establishment of the exclusive economic zone (EEZ) regime in the early 1980s brought with it the shared fish stock problem. Two broad, nonmutually exclusive, categories of shared stocks have been identified as transboundary stocks (EEZ to EEZ) and straddling fish stocks (both within the EEZ and the adjacent high seas).

Under the 1982 United Nations (UN) Convention, coastal states sharing a transboundary resource are admonished to enter into negotiations with respect to cooperative management of the resources (UN, 1982, Article 63(1)). Importantly, however, they are not required to reach an agreement. If the relevant coastal states negotiate in good faith, but are unable to reach an agreement, then each coastal state is to manage its share of the resource (i.e., that part occurring within its EEZ), in accordance with the relevant rights and duties laid down by the 1982 UN Convention. We refer to this as the default option.

With the default option in mind, economists find two issues that needed attention:

1. the consequences, if any, of the relevant coastal states adopting the default option, and not cooperating in the management of the resource;
2. the conditions that must prevail, if a cooperative resource management regime is to be stable over the long run.

Two aspects of the problem need to be noted. The first is that the state property rights to the fishery resources are straightforward and clear. The relevant coastal states are to be seen as joint owners of the resources. The second is that, with few exceptions, there will be a strategic interaction between, or among, the coastal states sharing the resource in the following sense. Consider two coastal states, I and II, sharing a transboundary resource. The fishing activities of I can be expected to have an impact upon the fishing opportunities available to II, and vice versa. II (I) will have no choice but to take into account the likely catch plans of I (II) – hence the strategic interaction.

In attempting to analyze issues (1) and (2), economists are compelled to recognize such strategic interactions. The economics of the management of transboundary fish stocks is, as a consequence, a blend of the standard fisheries economics applied to domestic fisheries (i.e., fisheries confined to a single EEZ), and the theory of strategic interaction (or interactive decision theory), more commonly known as the theory of games. Economists studying other shared resources, for example, water, the atmosphere, also find themselves compelled to incorporate game theory into their analysis.

There are two broad classes of games, noncooperative (competitive) and cooperative. We draw upon the theory of noncooperative games to analyze issue (1), the default option of noncooperative management. The key conclusion arising from noncooperative game theory is that the 'players' (coastal states sharing the fishery resources) will be driven inexorably to adopt strategies that they know perfectly well will produce decidedly undesirable results. This outcome is referred to as a 'prisoner's dilemma' outcome, after a famous noncooperative game developed to illustrate the point.

With respect to cooperative management, the analysis draws upon the theory of cooperative games. It is assumed that each 'player' is motivated by self-interest, and is prepared to consider cooperating, only because it believes that it will be better off, than by playing competitively. The chief problem in cooperatively managed fisheries is that of ensuring the stability of the cooperative management regime through time.

The effective economic management of transboundary fish stocks, while a demanding problem, is simple in comparison with the management of straddling fish stocks. The management of these resources, of seemingly minor importance in 1982, became a crisis during the following decade, and compelled the UN to mount another international conference – the UN Conference on Straddling Fish Stocks 1993–95.

The root of the problem lay in the High Seas part of the 1982 UN Convention (UN, 1982, Part VII). Straddling stocks are exploited both by coastal states and distant water fishing states (DWFs), the latter operating in the high seas. The drafters of Part VII attempted to balance the Freedom of the Seas, pertaining to the fisheries, with the rights and needs of the coastal states. The resulting articles are models of opaqueness, which left the rights and duties of (1) coastal states, and (2) DWFs, with respect to the high-seas segments of straddling stocks, unclear. The lack of clarity virtually assured that the stocks would be managed noncooperatively. The model of noncooperative management of transboundary stocks proved to be applicable, without modification, to straddling stocks, and to have a high degree of predictive power. It should be noted that the economics of cooperative management of these resources differs substantially from the economics of management of transboundary stocks.

Concrete examples of where game theory may have informed the creation of joint management arrangements are between (1) Norway and Russia in the management of cod in the Barents Sea; (2) the United States and Canada in the management of Pacific salmon; (3) Angola, Namibia, and South Africa in the management of the fishery resources in the Benguela Large Marine Ecosystem; and (4) New Zealand and Australia in the management of the South Tasman Rise Trawl Fishery targeting orange roughy.

Fisheries Subsidies, Overcapacity, and Overexploitation

Economists have long demonstrated that fishery subsidies greatly impact the sustainability of fishery resources. Subsidies that reduce the cost of fisheries operations and those that enhance revenues make fishing enterprises more profitable than they would be otherwise. Such subsidies result in fishery resources being overexploited, as they contribute directly or indirectly to the buildup of excessive fishing capacity, thereby undermining the sustainability of marine living resources and the livelihoods that depend on them. **Figure 3** demonstrates how subsidies that lower cost from TC_1 to TC_2, will also lower the bionomic equilibrium from BE_1 to BE_2, thus encouraging the growth of fishing effort from E_3 to E_4.

Not all subsidies, it must be conceded, are harmful. Government expenditures on fisheries management, research, enforcement, and enhancement are classified by the OECD as fisheries subsidies. These subsidies can be seen as being, at worst, neutral in their impact upon resource management. One subsidy that is still being viewed by many fisheries scholars and managers as having a positive, conservationist impact is buyback subsidies. Some economists argue that buyback subsidies, by leading to the removal of economically wasteful and resource-

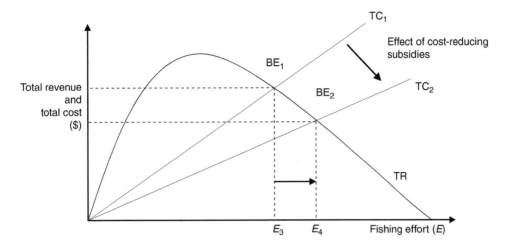

Figure 3 Schematic representation of how subsidies induce overfishing.

threatening excess fleet capacity, make a positive contribution to resource management and conservation. Others, however, argue against the usefulness of buyback subsidies, arguing that these subsidies are basically ineffective, as capital removed by them simply 'seeps' back into the fishery. It has also been demonstrated that even if capital seepage does not occur after a buyback scheme, this class of subsidies could still be ineffective if fishers know beforehand that a buyback scheme will be implemented in the future because they may accumulate more fishing effort than if they did not.

Economics of Illegal Fishing

Illegal fishing is conducted by vessels of countries that are part of a fisheries organization but which operate in violation of its rules, or operate in a country's waters without permission, or on the high seas without showing a flag or other markings. Since those engaging in illegal fishing are mainly economic agents that seek to maximize their profits from the activity, a good understanding of the economics of illegal fishing is important in order to design appropriate measures. Becker's (1968) contribution on the economics of criminal activity forms the theoretic basis for almost all of the contributions in the economic literature on illegal fishing. These contributions argue that criminals behave essentially like other individuals in that they attempt to maximize utility subject to a budget constraint. These models focused on the probability and severity of sanctions as the key determinants of compliance. It has also been shown in the literature that moral and social considerations play a crucial role in determining whether an individual engages in illegal activity or not.

Economics of Marine Protected Areas

Marine protected areas (MPAs) are, as the name implies, designated areas in the ocean within which human activities are regulated more stringently than elsewhere in the marine environment. The protection afforded by MPAs can vary widely, from minimal protection to full protection, that is, no-take reserves. Such areas are carved out to maintain, at least to some extent, the natural environment of the designated area for ecological, economic, cultural, social, recreational, and other reasons.

MPAs and other forms of spatial management are part of the toolkit of fisheries managers. Both theoretical and applied economic analyses are common in the MPA literature. The former focus broadly on obtaining insights into the dynamics and performance of MPAs. On the other hand, applied economic research is oriented toward analyzing specific case studies, both to provide an understanding of particular MPA situations, and to help build a body of case studies that can facilitate a meta-analysis of MPAs.

One of the principal arguments made for MPAs is their use as an 'insurance policy'. It is argued that even if other management measures fail in a fishery, an MPA will ensure a certain minimal level of ecosystem and fish stock health. Thus, many analyses explore the economics of MPAs as insurance policies, and the impact of MPAs within an environment that is prone to structural uncertainty.

Even though MPAs inherently involve spatial considerations, many studies of MPA economics – particularly those with a primarily theoretical focus – are based on aggregated spatially homogeneous models. There has been a recognition of the need for developing spatially heterogeneous models, and progress is being made in this direction, providing needed insights on such issues as the economically optimal placement of an MPA within a certain region, and the desired 'design' of MPAs to optimize economic benefits.

Economics of 'Rights-Based' Management

We will consider three approaches to 'rights-based' or what has come to be known recently as 'dedicated access privileges' (DAPs) management: individual transferable quotas (ITQs), fishers' cooperatives, and community-based fisheries management, sometimes referred to as territorial use rights fisheries (TURFS). One can, in addition, find blends of the three. We argue that there is significant convergence among the three. The most widely used of the three is individual quotas (IQs), usually in the form of transferable quotas (ITQs). The IQ scheme was originally designed to deal with the consequences of ineffectively controlled 'regulated open access' – the race for the fish. Under the scheme, the resource managers continue to set the total allowable catch (TAC), as before. Individual fishers, vessel owners, companies, etc., are then assigned individual shares of the TAC.

Having individual TAC shares, the fishers' incentive to engage in a 'race for the fish' and to overinvest in vessel capacity should be reduced. Obviously, the scheme will not work, if monitoring and enforcement are ineffective.

Of the three approaches, the most studied is the ITQ. Many studies of this approach from around the world show that economic efficiency does indeed improve with the implementation of ITQ schemes. But the role of ITQs with regards to the conservation of the resources and the ecosystems, and ensuring equity in the use of these resources, remains controversial.

Concluding Remarks

In this article, the basic theory of fisheries economics and a selection of topical and central issues in this area are presented. An extensive reading list is provided so that interested readers can dig deeper into the issues discussed in this short piece. The reading list also provides an opportunity to explore other important and topical issues being addressed by fisheries economics, which, unfortunately, could not be covered in this article.

Acknowledgments

The authors wish to express their gratitude for the generous support provided by the *Sea Around Us* Project of the Fisheries Centre, University of British Columbia, which is, in turn, sponsored by the Pew Charitable Trust of Philadelphia, USA. Sumaila also acknowledges the EC Incofish Project Contract 003739 for its support.

See also

Fisheries Overview. Fishery Management. Fishery Management, Human Dimension. Fishing Methods and Fishing Fleets. Marine Fishery Resources, Global State of.

Further Reading

Berman M and Sumaila UR (2006) Discounting, amenity values and marine ecosystem restoration. *Marine Resource Economics* 21(2): 211–219.

Bjørndal T and Munro G (1998) The economics of fisheries management: A survey. In: Tietenberg T and Folmer H (eds.) *The International Yearbook of Environmental and Resource Economics 1998/1999*, pp. 152–188. Cheltham: Edward Elgar.

Clark C (1976) *Mathematical Bioeconomics: The Optimal Management of Renewable Resources*. New York: Wiley.

Clark CW, Clarke FH, and Munro GR (1979) The optimal exploitation of renewable resource stocks: Problems of irreversible investment. *Econometrica* 47: 25–49.

Clark C and Munro G (1975) The economics of fishing and modern capital theory: A simplified approach. *Journal of Environmental Economics and Management* 5: 198–205.

Clark C, Munro G, and Sumaila UR (2005) Subsidies, buybacks and sustainable fisheries. *Journal of Environmental Economics and Management* 50: 47–58.

Copes P (1986) A critical review of the individual quota as a device in fisheries management. *Land Economics* 62: 278–291.

Gordon HS (1954) The economic theory of common property resource: The fishery. *Journal of Political Economy* 62: 124–143.

Hannesson R (1993) *Bioeconomic Analysis of Fisheries*, 131pp. Oxford: Fishing News Books and the FAO.

Lauck T, Clark C, Mangel M, and Munro G (1998) Implementing the precautionary principle in fisheries management through marine reserves. *Ecological Applications* 8: S72–S78.

Quirk JP and Smith VL (1970) Dynamic economic models of fishing. In: Scott AD (ed.) *Economics of Fisheries Management: A Symposium*. Vancouver, BC: Institute of Animal Resource Ecology, University of British Columbia.

Scott AD (1955) The fishery: The objectives of sole ownership. *Journal of Political Economy* 63: 727–756.

Smith VL (1968) Economics of production from natural resources. *American Economic Review* 58: 409–431.

Sumaila UR, Marsden D, Watson R, and Pauly D (2007) Global ex-vessel fish price database: Construction and applications. *Journal of Bioeconomics* 9: 39–51.

Sutinen JG and Kuperan K (1999) A socioeconomic theory of regulatory compliance in fisheries. *International Journal of Social Economics* 26: 174–193.

Turvey R (1964) Optimization and suboptimization in fishery regulation. *American Economic Review* 54: 64–76.

Warming J (1911) Om grundrente av fiskegrunde. *Nationalokonomisk Tidskrift* 499–511.

Wilen J (2006) *TURFs and ITQs: Coordination vs. Decentralized Decision Making.* Contribution prepared for the Workshop on Advances in Rights-Based Fisheries Management, Reykjavik, 28–30 August.

Relevant Websites

http://www.seaaroundus.org
– Sea Around Us Project.

FISHERY MANAGEMENT, HUMAN DIMENSION

D. C. Wilson, Institute for Fisheries Management and Coastal Community Development, Hirtshals, Denmark
B. J. McCay, Rutgers University, New Brunswick, NJ, USA

Introduction

The human dimension is central, not peripheral, to fisheries management. In capture fisheries, the behavior of people can be managed, but not the behavior of fish. Consequently, being able to monitor human behavior and to enforce regulations is an important 'human dimension' of fisheries. Moreover, in all fisheries, management decisions affect individuals and social and cultural groups in different ways. Management decisions have social impacts and come about through political processes. Those processes and impacts are mediated by other aspects of human dimensions that come into play in fisheries management: cultural values and identity (of fishers, managers, scientists, consumers, and society at large); risk perception and behavior; and local, regional, and global demographic, economic, and political forces. We focus on legitimacy, a key aspect of politics. We show that the legitimacy of fisheries management institutions, and hence their success in achieving sustainable fisheries, depends on economic rationality, the use of science in decision-making, the fairness of the processes and decisions that come from it, and how various groups participate in the process.

Fish are an extremely important source of food and income. Worldwide, fish are the largest source of animal protein, even though they rank well behind terrestrial animals in Western countries. In addition, fishing is often an essential source of subsistence and income for people without other means of livelihood. This critical resource is under heavy pressure from increased exploitation and from environmental changes that reduce productivity. The Food and Agriculture Organization of the United Nations (FAO), the leading international agency dealing with fisheries management, believes that 69% of the known fish stocks need management urgently, and that a reduction of 30% is needed in global fishing effort.

Overfishing means removing fish from the water at a higher rate than that which would produce the greatest overall production of fish over time. If any large group of people is allowed to fish without restrictions, the result is likely to be decline in the productivity of the fish stock. The reason is simple: If there are no rules, and one person decides to leave a fish in the water to reproduce or grow bigger, someone else could catch that fish the next day. Neither the first person as an individual, nor the common good, benefits from the first person's restraint. The only one who benefits is the second person who catches the fish. In this situation, no one will voluntarily restrict his or her own fishing. It would be foolish.

Fisheries management is the process that creates and enforces the rules that are needed to prevent overfishing and help overfished stocks rebound. However, it is not about managing fish unless aquaculture is involved. In the case of capture fisheries, the focus of this article, fisheries management is entirely about managing the people who fish. Capture fisheries take many forms. Gigantic factory trawlers catch tonnes of pollock in the Bering Sea and then fillet and freeze the fish on board. This starts the fish on a path through the vast, global chain of processed foods. In the end they may be sold in a supermarket as part of a food product with nothing like 'pollock' appearing on the label. At the other extreme, millions of African and other farmers living near oceans, lakes, swamps, and rivers have small boats that they take out fishing when other tasks permit. These fish feed their families and are sold to small traders for markets in nearby villages and cities.

The rules involved in fisheries management – whether at the level of local fishing communities or of regional and national governments – are about how fishing is done. To understand the human dimensions of fisheries management the most important thing is that in practice the rules will almost always have an allocative effect; that is, the rules will mean that person A is going to get to catch more fish than person B. This means that fisheries management is, in the final analysis, a political process that involves many competing interests, and it has social consequences, in terms of its effects on different groups and kinds of people and communities.

The political process and social consequences become even more complex when the high level of uncertainty about fish stocks is recognized. It is far more difficult to count or even reliably estimate the size and behavior of creatures that live in lakes, rivers, and oceans than of those living on land. Uncertainty is an important aspect of both scientific and folk, or experience-based, ways of trying to understand fish populations and how they are affected by human activities.

Human Dimensions and Management Rules

Fisheries management has a long history. It includes controls on the techniques used in fishing, on fishing effort (the intensity of fishing), on the timing and location of fishing activity, and on the size and amount of fish (or shellfish) that can be taken. 'Traditional' or 'folk' management refers to practices, beliefs, and rules that arise from the experience, knowledge, and sociocultural systems of groups engaged in fishing. They may be formally recognized and enforced by local governments, or they may be informally recognized and reliant on social pressure for enforcement. One traditional mechanism involves changes in technique or fishing pressure in the face of the declining productivity of overexploited areas. Sometimes cultural practices that relate to other activities also protect fisheries. There have been many religious taboos, for example, against fishing in particular areas or for particular species. Some societies have developed forms of fishing etiquette that can reduce or spread fishing effort. A common traditional management mechanism is restriction of rights to fish in particular marine areas. This can be as simple as denying access to outsiders or can be a complex system of valid claims to particular areas. Another is to require the taking of turns at access to a valued fishing spot. Many of these traditional management techniques are still in effect today. Many others were intentionally destroyed by colonial authorities or have broken down under pressure from commercial fish markets. Others are being created all the time, particularly to deal with conflicts over fishing grounds.

Traditional management usually deals with when, where, and how to fish rather than with controls on how much fish is taken. To some extent that is true for 'modern' management too. Modern management means regulation by a government authority that relies heavily on scientific analyses and formal police and judicial powers. Managers often find controls on techniques or closed seasons or areas attractive because they have fewer implications for allocation: it is easier to tell all the fishers that they must not use a specific gear or fishing ground than it is to try to

determine and divide up a quota of fish. Restrictions on techniques, seasons, and areas do not, however, have a good track record for protecting a fish stock's ability to reproduce itself given high demand for fish and other factors influencing how many people fish and how hard. Nor do they completely avoid social and political or allocative questions: A classic way of protecting one group's interest against that of another group is to seek restriction of the techniques, fishing times, or fishing grounds of the other group.

Modern fisheries managers often try to regulate the effects of fishing on fish stocks more directly, by imposing limits on the size or other biological features of fish and on the amount of fish caught. Controls on the amount of fish are based on 'quotas,' which are a mechanism for distributing a 'total allowable catch' (TAC). The TAC is, in turn, ideally based on a scientific stock assessment model using data from fishery catches supplemented by fishery-independent survey and other data. Quota-based management has problems too. The information required for reliable stock assessments and predictions of the consequences of different quota levels is often scarce, heavily biased, or nonexistent, which makes it difficult to come up with consensus about TACs. Quota-based management is not feasible for many anadromous fishes, such as salmon, and it can be difficult to justify for many subtropical and tropical fishes. Where many different species are caught at one time, managing with separate quotas for each species can be extremely difficult. This situation can lead to high rates of discarding and 'high-grading,' in which fishers who are only allowed to land a certain amount of fish throw less-valuable fish overboard. (High-grading can be a problem in relatively 'clean' fisheries targeted at a single species, too, if there are market-based incentives to discard less-valuable sex classes or sizes.) But quota systems have brought back depleted fisheries. Fishers are often able to respond to management measures by changing techniques and finding ways to catch as much fish as before. Quotas can be an effective way to keep them from doing this.

Discussions of human dimensions in the past have focused on the harvesters and fishing communities involved in commercial and subsistence fishing. The kinds of people and social values involved are broader and various, and conflicts have escalated. For example, many people are marine conservationists, whose focus is protecting fish, birds, and marine mammals. Institutional measures they seek, for example, in reducing by-catch and creating large 'marine protected areas,' may cause economic and social distress for commercial fishing communities. A growing number of people around the world are recreationists, who compete with commercial and subsistence fishers for fish and fishing space and who, in some areas, are well enough organized to force the end of commercial fishing. Aquaculturists, boaters, shipping companies, the military, researchers, industries, and sewerage authorities are among the many others with distinct and often competing uses of and ways of valuing marine ecosystems.

The references cited for further reading develop these and related areas. We will narrow our focus to issues of surveillance, enforcement, and legitimacy.

Cooperation with Fisheries Management

Regardless of the strengths and weaknesses of different management measures, the key is to ensure cooperation with them. The cooperation of people with any rule, or a set of related rules that social scientists refer to as an 'institution,' involves three dimensions: enforcement, surveillance, and legitimacy. Enforcement consists of the probability that one will be caught violating an institution, and the severity of the sanctions that follow from being caught. Surveillance is clearly necessary for enforcement. Just as importantly, people need to know about the behavior of other people in order to decide their own behavior. Seeing others, and being seen by others, as conforming to an institution is an independent and essential part of maintaining that institution. Legitimacy is the social and cultural acceptance of an institution. It involves the degree to which people assume that behavior will follow the institution and the degree to which the institution shapes people's understanding of situations. When an institution is legitimate, people will refer to it to justify or explain behavior.

Enforcement

Enforcement is a continuous challenge in fisheries management. Fishing can be a very difficult activity to monitor, and the fines imposed for breaking regulations are often much less valuable than illegally caught fish. Sometimes, especially in Third World countries, the main reason a management measure is chosen is to facilitate enforcement. A good example is found in Zambia, which has extensive, commercial freshwater fisheries. In Zambia a single closed season is imposed on all fishing in all parts of the country regardless of species. This blanket ban makes enforcement feasible. The main reason is that the police on the highways can confiscate any fish they find during the closed season. This curtails the commercial fish trade, which, in turn, reduces the fishing pressure.

International fisheries pose major challenges to enforcement. The main enforcement focus of the United Nations is flag state responsibility. The problem is 'reflagging'. Often when one country changes its fisheries regulations to become more stringent, boats from that country will move to another country with less stringent regulations. This practice has been used to avoid many kinds of maritime regulations and taxes for years. It has made international fisheries agreements ineffective because vessels in states that acceded to international fisheries agreements evade restrictions by reflagging in nonmember states. The World Congress on Fisheries Management and Development in 1984 adopted the principle of flag state responsibility for the behavior of all its vessels.

Sanctions for illegal fishing vary greatly among nations, creating difficulties for regional and international fisheries management. Efforts to harmonize fisheries enforcement are being made in various world regions. These include agreement by nations to impose sanctions on their own fishers when they violate other nation's rules. The threat of blacklisting of repeated violators across large areas of ocean has worked well in the Western Pacific, for example, where regional enforcement provisions authorize the hot pursuit of fishing vessels into foreign jurisdictions.

Surveillance

At the local level, people engaged in fishing can often see at least some part of what the others in the area are doing, which may support local-level management. However, in many of the world's commercial fisheries this is difficult because fishing may take place far from land and can involve vessels from many different ports and nations. Gathering information, or surveillance, depends on national governments and international agreements. One of the most important provisions of the Reflagging Agreement is that it requires flag states to maintain records of their vessels operating in international waters and to report such information to FAO. These records are very extensive and important. They include information about the technical and economic characteristics of the vessels that are gathered during the licensing procedure. At FAO, this information is linked with the vessel's fishing history. This database, which is still in the early stages of development, is a critical tool for understanding both what regulations are necessary and how they can be enforced.

One of the most important characteristics of a fisheries management measure is whether compliance can be monitored from land or whether it involves only behavior at sea. The first kind is, of course, much easier to monitor. Some measures, such as the banning of certain gears, can be fairly easily monitored in port. In most Western countries, the amount and size of fish that a vessel sells can often be monitored through a system of licensed fish dealers.

Most at-sea monitoring still depends on physical inspections from surface craft that are time consuming, expensive, and difficult to do effectively. Another option is to place observers, who can be government officers or private contractors, on fishing vessels. Areal surveillance is also used in fisheries management. The newest trend is toward satellite-based vessel monitoring systems (VMS) that are both more comprehensive and less expensive. Current plans call for these systems to be linked to the FAO database on fishing vessels.

All of these types of surveillance are feasible only on vessels that are large enough that the costs of observation are only a small portion of the value of the fish the vessel lands. For most fishing vessels in the world that is not the case, nor will it be in the foreseeable future. Small-scale, inshore fisheries will continue to rely on the cooperation of fishers both to comply with management measures and to aid in monitoring those who do not.

Legitimacy

Four characteristics of fisheries management are particularly important in determining how acceptable fisheries management will be and thus the probability of compliance with it. These are its economic rationality, its basis in science, its fairness to the various user groups, and who participates in making management decisions.

Economic rationality Almost any management measure introduces economic irrationality because it restricts the ability of a fisher to use his assets in the most efficient way to produce new wealth. This irrationality can produce senseless consequences, such as derby fisheries. The halibut fishery in the North Pacific waters of the United States, for example, had a fishing fleet of 2900 vessels in 1981, which grew to about 4400 vessels in 1991. The fishing season, meanwhile, was reduced from 120 days to 48 hours. Four thousand boats all trying to catch halibut in two days is an enormous waste of fishing assets that results only in a very low price for the fish. In the eyes of many fishers, the worst example of a perverse incentive structure is regulatory discarding. When fishers catch fish that they cannot legally sell, they must throw them back.

With many fishing methods, however, the fish are already dead or dying when they are sorted and the regulation simply causes dead fish to be thrown back in the water.

In the eyes of many people the solution to these examples of management-driven economic irrationality is the Individual Transferable Quota (ITQ) system. This technique creates exclusive and tradable rights in fishing, usually as a percentage of the total allowable catch of a certain stock. A similar technique can be based on exclusive and transferable rights to use certain fishing gears, such as lobster traps. In both cases, markets are relied upon to adjust investments to the actual status of the resource and to correct for distortions caused by other management techniques. For example, in the North Pacific halibut case mentioned above, ITQs have made it easier for some to continue in this fishery and for others to leave, but with something to sell when they leave. They have expanded the season, which in turn adds fresh halibut, rather than just frozen halibut, to the market. In Florida, individual trap certificates have helped reduce the number of traps used in the spiny lobster fishery. However, there is major resistance to this form of fisheries management because of the displacement of labor and other factors when market forces result in major downsizing. Consequently, in recent years attention has also been given to the potentials for community-based property rights, where rights and responsibilities are clearly defined. See the section on Participation below for further discussion.

Science The second critical area for the legitimacy of management is its basis in science. Indeed, it was this role that gave birth to fisheries science. In the late nineteenth century, politicians looking for ways to resolve disputes between fishers brought biologists together and asked them to study fisheries. This was the genesis of today's national fisheries research organizations, such as the American Fisheries Society formed in 1870, as well as the first international society for fisheries science, the International Council for the Exploration of the Sea (ICES) formed in 1902.

Fisheries scientists are confident in their basic approach to management: the principle of regulating fishing mortality to ensure future recruitment and growth of populations. Despite high levels of uncertainty and inadequacy of data, most cases of serious decline in fish stocks have not been the result of inadequate scientific advice as much as of the failure of others to follow the advice. Fisheries science is a form of what sociologists of science call 'mandated science'. It is a science that is trying to respond to political and legal as well as scientific questions. When government managers draw on science, they are looking for clear distinctions about what is at issue, precise decision rules, and efficiencies in presentation and procedure, all difficult to achieve given the practice and requirements of science. Moreover, science applied to policy often produces conflicting results, makes moral and political dilemmas more explicit, and is often accused of corruption. Competing user groups will use both science and gaps in science to define the issues in terms of their own social objectives.

One crucial factor that makes using fisheries science as the basis of policy particularly problematic is that fishers also know a deal about the fisheries resource, but from a different perspective from that of the scientists. This perspective is often referred to as 'traditional ecological knowledge,' or just 'local knowledge.' Many people concerned with management feel that local knowledge should be utilized by management. This is a very difficult goal. Fishers tend to view the resource in much smaller temporal and spatial scales than it is conceived of by managers. They often see fisheries as systems in which small perturbations may have substantial future consequences and are likely to emphasize the importance of habitat over population dynamics. These viewpoints can be incongruous with management, because managers are responsible for large areas and often need to simplify ecological complexity to a point where decisions can be identified and made. However, in many circumstances the knowledge of fishers is virtually the only source of information about a particular fish population or spawning area. It can yield insights and information that help scientists develop improved methods of data collection and analysis. In addition, the willingness of fishers to articulate and share their knowledge can be an extremely important expression of their desire to collaborate with scientists in the development of tools for marine conservation.

Fairness As mentioned above, fisheries management is an inherently political process because any given management decision affects different people and groups in different ways. Every management measure not only seeks to protect fish from overexploitation but also allocates them among potential users. The legitimacy of a system may depend on how fair people see that allocation to be.

A common principle used to decide what is and is not fair in fisheries management is 'historical participation'. People who have been fishing a stock more heavily or for a longer time are said to have a greater 'right' to fish that stock in the future. Except

in the case of the treaty rights of indigenous peoples, historical participation often derives more from a sense among fishers about what is fair than it does from any actual legal claim. In 1998, for example, the Iceland Supreme Court rejected aspects of an individual transferable quota system, which was based on historical participation, as unconstitutional because relying on historical participation discriminated illegally, given Icelanders' constitutional right to work.

Fisheries management begins with the idea that we are being unfair to those in the future if we fail to conserve fish. Because of this issue of intergenerational fairness, many people involved in fisheries have come to believe in the 'precautionary' approach, which says that when knowledge is uncertain we must err on the conservative side. We should be risk-averse. This approach, however, has its own fairness issue because the costs of caution are borne by people fishing today while the benefits will often go to others. Indeed, those who lose out are often the more economically vulnerable people who cannot afford to wait until the fish stock recovers.

Participation The participation of fishermen and fishing communities in resource management has been widely accepted as a desirable policy goal. 'Participation' is used to describe many different activities, but its basic meaning is that people who are concerned with, make a living on, or are otherwise dependent on a fish resource are involved in enhancing the resource and/or preventing its misuse. Participation takes many forms, from top-down processes where the government tells the fishers what to do and the fishers participate by complying, to systems where fishers organize and run their own management schemes. A wide range of advisory systems can be found in between, including co-management, or the active cooperation of resource users and government agencies in management. Various forms of cooperation between governments and the fishing industry have become commonplace in the West since the 1970s. Regarding Third World countries, FAO published a local management manual in 1985 and other donor agencies such as the World Bank are encouraging user group participation in fisheries programs that they support.

Researchers have documented local and co-management schemes for over 20 years. They have found that participation by fishermen and other people from fishing communities aids management in several ways. One is to facilitate access to information that fishermen have about the fishery, how it is fished, and what they will do in response to specific management measures. Participation also increases transparency in decision making, creates greater accountability for officials, and increases the use of and respect for community perspectives. The resulting management systems have greater legitimacy, a more open flow of information, and are more flexible in their response to changes in the fishery. Rights-based management at the community rather than individual level is thus an attractive alternative to either open access at one extreme, or individual transferable quotas on the other.

Local and co-management schemes around the world share certain problems. One is difficulty maintaining local autonomy, especially when the fish stocks involved go beyond local boundaries. This usually means some extra-local government involvement to handle conflicts among local communities and protect fish stocks. Such involvement may threaten local systems of fisheries governance in a variety of ways. For instance, external government agencies can be treated as resources that increase factionalism within the community, weakening its capacity for local management. This almost always must be recognized and addressed. Another challenge is reconciling scientists' and local knowledge about the resource system in a way that is acceptable to all stakeholders and therefore elicits cooperation, while still maintaining the values of objective science. A third problem is having fair representation of different interests in management decision making. Effective co-management requires a democratic decision-making system. At all but the smallest scales it is very difficult to have all interests represented in ways seen as fair without sacrificing other conditions required for effective co-management, such as open communication.

Conclusion

Fisheries management is often seen as a solution to 'tragedies of the commons,' where the lack of exclusive property rights means that the fish stocks are likely to be overfished and capital and labor are used wastefully. Government must intervene. Intervention is unlikely to be successful, however, if the knowledge used is poor, if the economic and social impacts create major political problems for government, and if people are unwilling to comply with the rules. Our discussion of legitimacy highlights the importance of these issues and underscores the value of transparent and participatory decision-making processes to fisheries management.

See also

Demersal Species Fisheries. Marine Fishery Resources, Global State of.

Further Reading

Apostle R, Barrett G, Holm P, *et al.* (1999) *Community, State, and Market on the North Atlantic Rim; Challenges to Modernity in the Fisheries.* Toronto: University of Toronto Press.

Bromley DW (1992) The commons, common property, and environmental policy. *Environmental and Resource Economics* 2: 1–17.

Dyer CL and McGoodwin JR (eds.) (1994) *Folk Management in the World's Fisheries; Lessons for Modern Fisheries Management.* Niwot, CO: University Press of Colorado.

Johannes RE (1981) *Words of the Lagoon: Fishing and Marine Lore in the Palau District of Micronesia.* Berkeley, CA: University of California Press.

Kooiman J, van Vliet M, and Jentoft S (eds.) (1999) *Creative Governance; Opportunities for Fisheries in Europe.* Aldershot, UK: Ashgate Publishing Ltd.

McCay BJ and Acheson JM (eds.) (1987) *The Question of the Commons: The Culture and Ecology of Communal Resources.* Tucson, AZ: University of Arizona Press.

McGoodwin JR (1990) *Crisis in the World's Fisheries.* Stanford, CA: Stanford University Press.

Newell D and Ommer R (eds.) (1999) *Fishing Places, Fishing People; Traditions and Issues in Canadian Small-Scale Fisheries.* Toronto: University of Toronto Press.

Pinkerton E (ed.) (1989) *Co-operative Management of Local Fisheries.* Vancouver: University of British Columbia Press.

Symes D (ed.) (1998) *Property Rights and Regulatory Systems in Fisheries.* Oxford: Fishing News Books.

Young O (1994) *International Governance: Protecting the Environment in a Stateless Society.* Ithaca, NY: Cornell University Press.

MARINE PROTECTED AREAS

P. Hoagland, Woods Hole Oceanographic Institution, Woods Hole, MA, USA

U. R. Sumaila, University of British Columbia, Vancouver, BC, Canada

S. Farrow, Carnegie Mellon University, Pittsburgh, PA, USA

Introduction

Marine protected areas (MPAs) are a regulatory tool for conserving the natural or cultural resources of the ocean and for managing human uses through zoning. MPAs may also be referred to as marine parks, sanctuaries, reserves, or closures; the latter two terms are used most commonly in the context of fisheries management.

Definition

At a conceptual level, zoning in the ocean involves the spatial segregation of a marine area in which certain uses are regulated or prohibited. This general definition might apply to any marine area in which a set of human uses are given preference over others. For example, by law the US President may set aside hydrocarbon deposits on the US outer Continental Shelf as 'petroleum reserves.' However, the typical use of the term 'protected' implies that a primary focus of an MPA is on the conservation of either individual species and their habitats or ecological systems and functions through the regulation of 'extractive' or potentially polluting commercial uses, such as fishery harvests, waste disposal, and mineral development, among others.

MPAs are frequently considered to be a fishery management measure, but they may be used for other purposes as well. For instance, in 1975, the first US national marine sanctuary was created around the wreck of the *U.S.S. Monitor*, a civil war vessel, located off the coast of North Carolina. The sanctuary was established to prevent commercial 'treasure' salvage and looting of the shipwreck, to regulate recreational diving, and to promote archaeological studies. In the discussion below, we focus on the use of MPAs in the field of fishery management because this use represents one of the most relevant and interesting examples.

Size

Although there is no discernible size limitation, the issue of geographic scale may be another defining characteristic of MPAs. On the tidelands of US coastal states, for example, the 'public trust doctrine' gives preference in the common law to transitory public uses, typically navigation, fishing, and hunting, over permanent private uses, such as constructing a dock. Yet the tidelands, which are quite extensive, are not referred to as an MPA. Some fishery closures can be quite large, and we would classify these as one type of MPA. The Great Barrier Reef Marine Park in Australia is the largest MPA in the world, measuring 344 million km^2. Most of the world's existing MPAs are much smaller, however, and focused on unique ocean features or sites, such as coral reefs or underwater banks. The World Bank estimates the median size of a sample of about one thousand of the world's MPAs to be 15 840 km^2 (**Figure 1**).

Number

Worldwide, MPAs have become a popular form of ocean management, and their use has expanded exponentially since they were first introduced in the late nineteenth century (**Figure 2**). The trend in the establishment of MPAs follows on the heels of a more general trend in the regulation of ocean uses, as an MPA represents merely a form of governance distinguishable geographically by type or severity of regulation. Regulation of the ocean has become necessary as human uses of the ocean have increased in scale and variety and as conflicts among mutually exclusive uses and users have arisen.

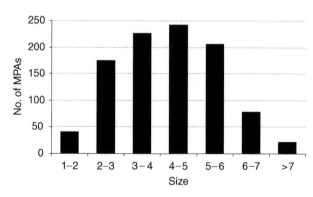

Figure 1 Worldwide size distribution of marine protected areas ($n = 991$). Sizes are grouped by km^2 to the powers of ten.

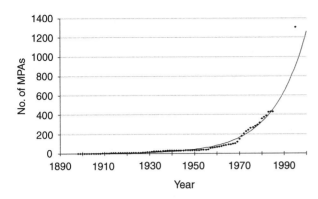

Figure 2 Cumulative worldwide growth in the number of marine protected areas and estimated logarithmic trend 1898–1995.

With the expansion in the establishment of MPAs, marine scientists and policy experts have begun to take a closer look at the likely benefits and opportunity costs associated with zoning the ocean, and several recent studies have emerged. In particular, as ecological models of the marine environment become more realistic, marine policy analysts can begin to make more sophisticated examinations of how to choose among competing human uses, given the constraints presented by the natural system.

Management Objectives

The extent to which an MPA may be considered an effective management tool depends on its management objectives. Here, objectives are classified under the following general categories: Biological (ecological); economic; and distributive (equity). Often, the establishment of an MPA involves objectives from more than one of these categories. To make the discussion in the next three sections more focused, we ignore the complexities of the subject, and return to them in a later section.

Biology

Consider a single species fishery as a starting point, and assume that a biological objective is to increase stock size or biomass. Restrictions on fishing in certain areas are expected to lead to positive 'refuge' and 'stock' effects. The refuge effect is a static concept implying that some portion of the target stock cannot be harvested because it remains within the MPA. As a consequence, the entire stock is not exploited to the same degree as it would be in the unregulated case. The stock effect is a dynamic concept implying that fish within the MPA will grow and reproduce and that either their larva will drift out beyond the boundary of the MPA and eventually recruit to the fishery or new recruits (or possibly older, larger fish) will 'diffuse' across the boundary into the fishery. Where the behavior patterns of fish stocks are well understood, the careful placement of an MPA may be effective from a biological standpoint. One excellent example is the establishment of an MPA around a spawning aggregation in tropical fisheries.

Economics

The economic implications of an MPA depend critically upon the nature of the institutional framework for managing the fishery. Suppose that a fishery supplies only a small part of a large market and that, initially, it is unregulated. The first assumption implies that seafood consumers are not much affected by changes in the supply of fish from the fishery of concern. Assuming that an equilibrium is reached where harvests balance stock growth, theory suggests, and empirical investigations confirm, that the economic value of the fishery is near zero. In an unregulated fishery, fish are an unpriced factor in the production of seafood, and this implicit subsidy encourages too much fishing effort and, consequently, excessive exploitation. In the jargon of economics, 'resource rents' are dissipated. Depending on the scale of the variable costs of fishing, yields may fall below levels considered to be the maximum sustainable.

Now suppose that an MPA is established. The refuge effect implies that the exploitable stock is smaller for any given level of fishing effort. In the absence of any complementary regulation, fishermen will exit the fishery until an open-access equilibrium sets up for the residual exploitable stock. As before, rents are dissipated at this new equilibrium, and no economic value is created through the establishment of the MPA. Over time, the stock effect might lead to an expansion of the exploitable biomass. Again, the existence of economic rents associated with an expanding biomass will attract fishermen until rents dissipate.

It is conceivable that the exploitable biomass could expand to a level exceeding that in the fishery prior to the establishment of the MPA. This might happen where increasing returns exist in the production of eggs as female fish grow older and larger. A common example is the red snapper (*Lutjanus campechanus*), a reef fish native to the Gulf of Mexico. It has been claimed that a 10 kg red snapper produces in a single spawn more than 200 times the eggs of a fish weighing only 1 kg. Only in cases in which the stock effect more than compensates for the refuge effect and surpluses accrue to consumers due to the absence of close seafood substitutes, can a case be made that the establishment of an MPA in an otherwise un-regulated fishery is valuable in an economic sense. And this result is due solely to the expansion of value to the consumer, not to the fisherman.

The establishment of an MPA might be complemented with other forms of regulation. Assuming that the costs of administering fishery regulations are minor, resource rents can be realized through the implementation of management measures in conjunction with an MPA, such as taxes on either fish landings or fishing effort or the introduction of an individual tradeable quota system. However, in theory, the implementation of these alternative management measures by themselves can lead to the capture of resource rents, implying no need for an accompanying MPA. Recent research suggests, however, that, where the stock effect overcomes the refuge effect, the establishment of an MPA can lead to increases in economic value in an otherwise optimally managed fishery.

Distribution of Economic Impacts

The third general category of objectives concerns the distribution of economic benefits and costs across human users. In attempting to achieve either biological or economic objectives, the effects of the establishment of an MPA on individual fishermen are not considered explicitly. For example, an economic decision rule would argue for the creation of an MPA as long as economic benefits exceed economic costs, assuming all relevant sources of benefits and costs are accounted for, without regard for the identities of the recipients of the surplus. Moreover, even if the creation of an MPA results in net benefits, the historical pattern of the distribution of gains may be shifted. One example is the creation of an MPA in the vicinity of a fishing port, forcing fishermen from that port to travel longer distances to fish.

In some circumstances, such as a small fishery in a developed economy, the distributional effects may be minor, as fishermen are able to switch at low cost to other stocks or to other occupations. On the other hand, the distributive effects of an MPA may be more serious for a community that is heavily reliant on a stock for income or as a source of protein. In such cases, an objective of fairness to users may necessitate foregoing potential biological or economic gains through, say, the relocation or reduction in size of an MPA. The political economy of the management regime may dictate such a result, if users are capable of influencing the adoption of an MPA through voting, negotiation, or other means.

In circumstances where some form of regulation must be imposed, it is possible that, on the basis of equity, MPAs may be the preferred choice of fishermen, relative to alternative measures. The reason for this preference is that the establishment of an MPA does not single fishermen out on the basis of gear type or other distinguishing characteristics. Further, it may be difficult to discern *ex ante* which specific fishermen eventually will bear the costs or be forced to exit.

Complexities

There are a number of important issues that increase the complexity of the simple scenarios described above. A few of these issues are touched on here, and the interested reader should refer to the reading list for further detail.

Dynamic Responses

In the discussion above, we have ignored the potential for lags in the response of the system, including the behavior of both fish stocks and fishermen, to the implementation of an MPA. Importantly, it may take more than one fishing season for the stock effect to contribute significantly to recruitment. Further, the refuge effect does not always result in the immediate exit of fishermen from the fishery. When few opportunities exist for redeploying boats and hands elsewhere, fishermen may continue to fish in the short run, as long as they can cover their variable costs (wages, fuel, ice, etc.). In certain circumstances, fishermen might rationally delay exit, expecting the stock effect to lead to a future expansion of the fishery. If fishermen delay exit, the expected stock rebuilding may be prolonged. When environmental conditions and ecological interactions are added in, it is not hard to imagine a scenario in which an MPA appears to have no effect, at least in the short run. The lack of results may lead to political action to remove the MPA.

Both fish stocks and fishing effort may be distributed nonuniformly across the fishery. This spatial distribution can be affected through the establishment of an MPA. As a consequence, location becomes an important consideration when planning an MPA. For example, recent models of plaice fisheries show that a properly located MPA can protect undersized fish when fishing effort becomes redistributed around the borders of the MPA.

Ecological Relationships

MPAs have also been established to protect aggregations of species or components of ecosystems. Even where the management of a single species is of primary concern, a characterization of the biological relationships between the species of focus and other species in the ecosystem is crucial to understanding the biological, economic, and distributional impacts of the establishment of an MPA. Where fishing technologies are nonselective, MPAs may prove beneficial in reducing the by-catch of nontarget species and minimizing the impacts of trawl gear on seafloor habitat.

Biological Diversity

Recent developments in international and domestic law have emphasized the conservation of biological diversity as a biological objective, and MPAs have been suggested as one means of achieving that objective. Although the conservation of biological diversity is an appealing concept at a superficial level, basic definitional questions persist. For example, does biological diversity refer to species richness (i.e., the number of species) or to some other measure, such as the average genetic distance among a set of species? Assuming that an appropriate measure can be agreed upon, economic research has focused on the problem of maximizing a chosen diversity measure subject to limits on financial resources. When coupled with information on species distributions and ecological relationships, this research may be useful in optimizing locations and scaling the size of MPAs.

Insurance and Precaution

The ocean is an uncertain environment. Substantial gaps exist in our understanding of ecological relationships among species, the linkages between environmental conditions and ecosystem states, especially given uncertainties about long-term environmental changes, and the impacts of fishing activity on habitat quality and on ecological relationships. For reasons of tractability, bioeconomic models of fisheries are often based on equilibrium assumptions, when it is not clear that, even if their existence is plausible, steady states can ever materialize. In the context of this uncertainty, MPAs have been touted as a hedge for insuring against stock depletion or collapse.

Although it seems reasonable to conclude that MPAs might be useful as a hedge against uncertainty, we should heed the message of economic theory that, in the long run, some MPAs may not remedy the problem of rent dissipation, especially if they are used as the only means by which to manage fisheries. Furthermore, the presence of ineluctable uncertainty raises the question of the extent of the practical contribution that fisheries scientists and marine ecologists can make to specifying the size and location of MPAs. This issue has led some observers to suggest that 'picking' MPAs is akin to picking securities in the stock market. They conclude that, in the long term, it may be sensible to randomly select a portfolio of MPAs that cover some agreed-upon percentage of the geography in a particular ocean region. Making estimates of the proportion of ocean area to be included in an MPA may also be problematic, as models suggest a wide range, 20–90% of the relevant area.

Irreversibilities

Human uses of the ocean can result in ecological impacts that are costly or impossible to reverse. Examples include the extinction of marine fish or protected species, such as mammals or reptiles, and biomass 'flips,' in which the collapse of commercially important stock groupings are replaced by others. Concerns about these irreversibilities reflect the notion that there may be preferred states for marine ecosystems. Changed ecosystems could result in smaller potential economic surpluses and a different set of options for the use of the system in the future. The latter may include 'nonmarket' damages when protected species or unique ecosystems, such as coral reefs or underwater banks, are affected adversely.

In the presence of uncertainty about human uses or ecosystem states, it may be worthwhile to delay decisions to proceed with human uses, such as fishing, that result in irreversible effects, where the development of new information reduces the uncertainty. The existence of this 'quasi-option value' may be a formal justification for taking the so-called 'precautionary approach' to fisheries management. The precautionary approach, which has now become embodied in international soft law, argues for the maintenance of commercial fish stocks at relatively high levels because, when accounting for uncertainty, the expected losses due to over-exploitation exceed those due to underexploitation. Some analysts have pointed to MPAs as an essential element of a precautionary approach. The value of MPAs in this context may be most apparent when they are employed as a control in a scientific experiment designed to test hypotheses about the impacts of fishing. The partial closure of the US portion of Georges Bank to sea scallop dredging, for example, provided valuable information on the ability of that stock to rebuild in a discrete area.

The designation of an MPA can be conceptualized as a kind of 'administrative' irreversibility, where it may be difficult to modify the designation through political processes. To many observers and interests, this kind of policy inflexibility may be the whole point to designating an MPA. Nevertheless, as environmental conditions change, ecosystems adjust, and it is sensible to have in place a management tool that can also be adjusted. The boundaries of the Canadian 'Endeavor Hot Vents' MPA off the coast of Vancouver, which has been proposed at the site of a deep-sea hydrothermal system, is designed to be adjusted as vents turn on and off and their associated microbial and faunal assemblages appear and disappear.

Administrative Costs

MPAs have been promoted as a management tool that is less costly than alternative measures. Recent research suggests that management costs may decline with the size of an MPA as fixed costs of monitoring and enforcement are spread over larger areas (**Figure 3**). The degree to which MPAs are less costly to manage may depend, however, on the form of management. If MPAs are complemented with other management measures, it may be difficult to argue that the entire management regime is less costly.

Many MPAs have been criticized as being 'paper parks' because monitoring and enforcement are minimal. In such cases, the apparent 'savings' in administrative costs relative to other management measures are illusory. Although some users may be dissuaded from breaking the rules inside the boundaries of an MPA, others weigh the product of the probability of apprehension and the penalty, concluding from this calculation that it is rational to ignore the rules. Even in well-monitored and enforced areas, poaching occurs, as enforcement actions in fishery closures in the US Gulf of Maine demonstrate on a regular basis. Limits on government budgets may imply that some portions of very large MPAs are paper parks.

Summary

MPAs clearly hold promise as a rational way of managing ocean resources, but this promise should not be overstated. In particular, MPAs should not be seen as a panacea to all the problems of fisheries management. Indeed, the best way to see MPAs is probably as part of a collection of management tools and measures. As the marine counterpart to systems of national and international parks, they are conceptually easy to understand and naturally appealing to the public. Yet MPAs differ in important ways from land parks because of their relative inaccessibility, the fugitive nature of fish stocks and the physical transport of pollutants and plankton, the legal characteristics of property rights in the ocean, and the costs of monitoring human activities. As we learn more about the ocean and the workings of its environmental and ecological systems, and as demand for the special characteristics of these systems expands with growing coastal populations, we can expect the use of MPAs to grow as well.

See also

Dynamics of Exploited Marine Fish Populations. Ecosystem Effects of Fishing. Fishery Management. Fishery Management, Human Dimension.

Further Reading

Arnason R (2000) *Marine protected areas: Is there an economic justification. Proceedings, Economics of Marine Protected Areas.* Vancouver: Fisheries Centre, University of British Colombia.

Bohnsack JA (1996) Maintenance and recovery of reef fishery productivity. In: Polunin VC and Roberts CM (eds.) *Reef Fisheries*, pp. 283–313. London: Chapman and Hall.

Crosby M, Geenen KS, and Bohne R (2000) *Alternative Access Managment Strategies for Marine and Coastal Protected Areas: a Reference Manual for Their Development and Assessment.* Washington: Man and the Biosphere Program, US Department of State.

Farrow S (1996) Marine protected areas: Emerging economics. *Marine Policy* 20(6): 439–446.

Great Barrier Reef Marine Park Authority, World Bank, and World Conservation Union (1995) *A Global Representative System of Marine Protected Areas, Vols I–IV.* Washington: Environment Department, The World Bank.

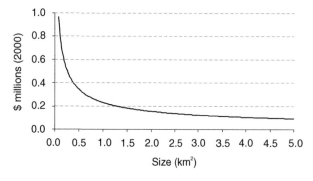

Figure 3 An estimate for small marine protected areas (MPA) of the relationship between size and the average costs of establishing and managing an MPA. The relationship demonstrates economies of scale. Costs include the acquisition of coastal land, demolition of existing structures, development, and operating costs (capitalized at 5%). Average costs are estimated from data pertaining to size alternatives for the proposed Salt River Bay MPA in St Croix, US Virgin Islands.

Gubbay S (1996) *Marine Protected Areas: Principles and Techniques for Management.* New York: Chapman Hall.

Guénette S, Lauck T, and Clark C (1998) Marine reserves: from Beverton and Holt to the present. *Review of Fish Biology and Fisheries* 8: 251–272.

Hoagland P, Broadus JM, and Kaoru Y (1995) *A Methodological Review of Net Benefit Evaluation for Marine Reserves.* Environment Dept. Paper No. 027. Washington: Environment Department, The World Bank.

Holland DS and Brazee RJ (1996) Marine reserves for fisheries management. *Marine Resource Economics* 11: 157–171.

Ocean Studies Board (2000) *Marine Protected Areas: Tools for Sustaining Ocean Ecosystems.* Washington: National Academy Press.

Roberts CM and Polunin NVC (1993) Marine reserves: simple solutions to managing complex fisheries? *Ambio* 22: 363–368.

Rowley RJ (1994) Marine reserves in fisheries management. *Aquatic Conservation: Marine and Freshwater Ecosystems* 4: 233–254.

Sanchirico JN (2000) *Marine Protected Areas as Fishery Policy: an Analysis of No-Take Zones.* Discussion Paper 00-23. Washington: Resources for the Future.

Sumaila UR (1998) Protected marine reserves as fisheries management tools: a bioeconomic analysis. *Fisheries Research* 37: 287–296.

Sumaila UR, Guénette S, Alder J, and Chuenpagdee R (2000) Addressing the ecosystem effects of fishing using marine protected areas. *ICES Journal of Marine Science* 57(3): 752–760.

ARTIFICIAL REEFS

W. Seaman and W. J. Lindberg, University of Florida, Gainesville, FL, USA

Introduction

Artificial reefs are intentionally placed benthic structures built of natural or man-made materials, which are designed to protect, enhance, or restore components of marine ecosystems. Their ecological structure and function, vertical relief, and irregular surfaces vary according to location, construction, and degree to which they mimic natural habitats, such as coral reefs. For ages, humans have taken advantage of the behavior of some organisms to seek shelter at submerged objects by introducing structures into shallow waters, where biological communities could form and fishes could be harvested. Contemporary purposes for artificial reefs include increasing the efficiency of artisanal, commercial, and recreational fisheries, producing new biomass in fisheries and aquaculture, boosting underwater recreation and ecotourism opportunities, preserving and renewing coastal habitats and biodiversity, and advancing research. This article reviews the evolution, spread, and increased scale of such practices globally, discusses their scientific basis, and describes trends concerning artificial reef planning, evaluation, and their appropriate application in natural resource management.

Utilization of Reefs in the World's Oceans

The near-shore ocean ecosystems of all inhabited continents contain artificial reefs. Their scales and purposes vary widely, ranging, for example, from placement in local waters of small structures by individuals in artisanal fishing communities, to deploying complex systems of heavy modules in distant areas by organizations concerned with commercial seafood production, to pilot studies using experimentally based structures for restoration of seagrass, kelp, and coral ecosystems. The origins of artificial reef technology are traced to places as diverse as Japan and Greece. Modern deployment of reefs has the longest history and has been most widespread in Eastern Asia, Australia, Southern Europe, and North America. In recent years, the technology has been introduced or more extensively established in scores of nations worldwide, including in Central and South America and the Indo-Pacific basin. No organization maintains a database for artificial reef development globally, so that assessments of reef-related activity must be derived from the records of economic development, fishery, and environment organizations, as well as from scientific journal articles which serve as a proxy for gauging national efforts.

For centuries structures of natural materials have been used to support artisanal and subsistence fishing in coastal communities, particularly in tropical areas (**Figure 1**). In India, traditionally tree branches have been weighted and sunken. Brush parks made of

Figure 1 In some areas of long-standing artisanal fisheries, structures such as the *pesquero* (left), a bundle of mangrove tree branches, are used for benthic structure, as in Cuba. More recently, small fabricated modules, such as this example from India (right), are used to complement or replace natural materials.

branches stuck into the substrates have been used in estuaries of several African countries, Sri Lanka, and Mexico. In the Caribbean basin, casitas of wood logs provide benthic shelters from which spiny lobsters (*Panulirus argus*) are harvested. Indigenous knowledge of local fisherfolk is important in sustainably managing these reef systems, even when in their initial stages they may function principally to aggregate larger mobile fishes, such as around bundles of mangrove branches, *pesqueros*, in Cuba.

In recent decades, some artisanal fishing communities have deployed newer designs of artificial reefs, and at larger scales. In part, this has been in response to damage of habitat and fisheries by coastal land-use practices and more intense fishing practices, such as trawling. In India, for example, a national effort has deployed reef modules of steel plates, to increase the time fishing cooperatives actually devote to harvest as a means of promoting social and economic well-being. The coastal waters of Thailand received concrete modules as part of a plan to balance small- and large-scale fisheries, whereby areas of 14–22 km^2 received between 2400 and 3300 units.

The most extensive deployment of artificial reefs, for any purpose, is in Japan. There, fishing enhancement is the goal of a national plan begun in 1952 and since greatly expanded. By 1989, 9% of the coastal shelf (<200 m depth) had been affected by reef development. Early emphasis was on engineering and construction of reefs to withstand rigorous high seas environmental conditions. Industrial manufacturers have used materials such as steel, fiberglass, and concrete in structures that are among the largest units in the world, attaining heights of 9 m, widths of 27 m, and volumes of 3600 m^3 (**Figure 2**). Research at sea has been augmented by laboratory studies in which scale models of reefs have been analyzed for effects including deflection of

bottom water currents to create nutrient upwellings for enhancing primary production. Observations of fish behavior led to a classification of the affinity of fishes to associate with physical structure. The Japanese aims have been to develop nursery, reproduction, and fishing grounds to supply seafood including abalone, clams, sea urchins, and finfishes. This partially was in response to the closure of distant waters to fishing by other nations.

In contrast to Japan (and many other nations more recently involved), the United States, while also an early entrant in this field, built thousands of artificial reefs to enhance recreational fishing through numerous, independent small-scale efforts organized by local interest groups or governmental organizations at the state level. Until recently, materials of opportunity predominated, including heavier and more durable materials such as concrete rubble from bridge and building demolition sites. Designed structures are becoming more common, especially as reefs gain acceptance in habitat restoration. One state, Florida, has about half of the nation's total number of reefs, and in one four-county area annual expenditures by nonresidents and visitors on fishing and diving at dozens of artificial reefs were US $1.7 billion (2001 values). Recreational fishing has also been the focus of widespread reef building in Australia, and lesser efforts in areas such as Northern Ireland and Brazil.

Other regions more recently deploying artificial reefs to enhance or simply maintain fisheries to supply human foodstuffs include southern Europe and Southeast Asia. From inception, their approach has been to design, experimentally test, and document reefs. As early as the 1970s, 8 m^3, 13 t concrete blocks (and other designs) were deployed in the Adriatic Sea of Italy, forming reefs of volumes up to 13 000 m^3 and coverage up to 2.4 ha. They serve multiple, fishery-related objectives that are sought in many nations: shelter for juvenile and adult organisms; reproduction sites; capture nutrients in shellfish biomass; protect fish spawning and nursery areas from illegal trawling; and protect artisanal set fishing gear from illegal trawling damage. Higher catches and profits for small-scale fisheries were the result. This is especially true in Spain, where heavy concrete and steel rod reef designs have been used effectively to protect seagrass beds from illegal trawling (**Figure 3**).

Artificial reefs are being used as part of a larger strategy to implement marine ranching programs. In Korea, for example, the decline of distant water fisheries led to a transition from capture-based to culture-based fisheries. From 1971 to 1999, artificial reef placements have affected 143 000 ha of submerged

Figure 2 The largest reef modules have been fabricated and deployed in Japan.

Figure 3 Concrete reef modules with projecting steel beams such as railroad tracks are deployed in protected marine areas of Spain to discourage illegal trawling, so as to promote restoration of seagrass habitats and fish populations. Photograph Courtesy of Juan Goutayer.

Figure 4 This larval enclosure tent is for settlement of the coral, *Montastraea faveolata*, at the site of a ship grounding on Molasses Reef, Florida Keys National Marine Sanctuary. Coral are mass-spawned, larvae cultured through the planktonic phase, and then introduced into the tent. In the laboratory, larvae may settle onto rubble pieces which subsequently are cemented to the reef. Photograph by D. Paul Brown.

coastal areas, out of the total 307 000 ha planned. At certain reef sites, hatchery-reared juvenile invertebrates and fishes have been released, and enhanced fishery yield reported, as part of long-term experiments. Artificial reefs are also used exclusively for aquaculture, such as in settings where excess nutrients stimulate primary production that is transferred into biomass of harvestable species, such as mussels (*Mytilus galloprovincialis*) and oysters (*Ostrea edulis* and *Crassostrea gigas*) in Italy.

Restoration of marine ecosystems using artificial reefs has focused on plant and coral communities. On the Pacific coast of the United States, for example, a project to mitigate for electric powerplant impacts is developing a 61 ha artificial reef for kelp (*Macrocystis pyrifera*) colonization, survival, and growth. Along the coast of the Northwest Mediterranean Sea, artificial reefs are deployed in protected areas to preserve and promote colonization of seagrasses (*Posidonia*). In numerous tropical areas, artificial reefs have been used in repairs to damaged coral reefs (**Figure 4**), or to hasten replacement of dead or removed coral. In the Philippines, for example, hollow concrete cubes are deployed as sites for coral colonization, sometimes being located in marine reserves.

Increasing popularity of artificial reefs to promote tourism is seen in the development of diving opportunities, both for people in submersible vehicles and for scuba divers. In places such as Mexico, the Bahamas, Monaco, and Hawaii, submarine operators provide trips to artificial reefs. On a larger scale, obsolete ships have been sunk to create recreational dive destinations in areas as diverse as British Columbia, Canada, Oman, New Zealand, Mauritius, and Israel. Purposes include generation of economic revenues in

local communities and diversion of diver pressure from natural reefs.

As strictly works of engineering, structures such as rock jetties for shore protection also serve as *de facto* reefs with biological communities. Submerged breakwaters have been used to create waves for recreational surfboarding.

Progress toward Scientific Understanding

Advances in research on artificial reefs include methodologies, subjects and rigor of approaches. Methodological advances include improved underwater biological census techniques, and the application of remote sensing. Investigations have become more ecological process-oriented in explaining biological phenomena such as sheltering, reproduction, recruitment, and feeding, better able to explain physical and hydrodynamic behavior of reefs, and have led to specifying reef designs consistent with the ecology of organisms. Newer approaches include an expansion of hypothesis-driven and experimental research, scaling up of pilot designs into larger reefs, ecological modeling and forecasting, and interdisciplinary studies with various combinations of physical, social, life, and mathematical sciences. Comparative studies of natural and artificial habitats have advanced understanding of both systems. The earlier history of research on artificial reefs may be observed in the subjects, methods, and findings reported at eight international conferences held from

1974 to 2005. Subsequent to a peak number of peer-reviewed articles contained in a proceedings from the 1991 conference (84), fewer resulted from the latest conferences (1999 (56) and 2005 (20)). Scientists studying artificial reefs increasingly are reporting in journals devoted to fish biology, fisheries, and marine ecology.

Colonization of artificial reefs has the oldest and most extensive record of scientific inquiry. The appearance of large fishes at some reef sites almost immediately after placement of structure is the most visible form of colonization (and also a factor in one of the most controversial artificial reef issues, discussed below). However, microbes, plants, and invertebrates also colonize reefs, and patterns of succession occur that converge toward naturally occurring assemblages of species (**Figure 5**). Augmenting earlier studies describing diversity and abundance of artificial reef flora and fauna, later research addressed the influence of environmental variables upon reef ecology, interactions of individual species with the artificial reef structure and biota, and long-term life history studies of selected species. Comparison of patch reefs of stacked concrete blocks with translocated coral reefs in the Bahamas (**Figure 6**) determined that the reefs had similar fish species composition, while the natural reefs supported more individuals and species, owing to their greater structural complexity (i.e., variety of hole sizes) and associated forage base. Meanwhile, long-term studies of the spiny lobster experimentally established that artificial reef structures, with dark recessed crevices approximately one body-length deep and multiple entrances, indeed augmented recruitment in a situation of habitat limitation in the Florida Keys, USA. While these situations allowed observation by scuba diving, in deeper waters hydroacoustic and video-recording instruments have been used to document fish species, such as at petroleum production platforms in the North Sea and Gulf of Mexico.

Prominent among the scientific issues regarding the efficacy of artificial reefs for fisheries production and management is the 'attraction-production question': Do artificial reefs merely attract fishes thereby improving fishing efficiency, or do they contribute to biological production so as to enhance fisheries stocks? While this popularized dichotomy is an oversimplification, and an expectation of one all-encompassing answer is unrealistic, this question has had great heuristic value for the advance of reef science and responsible reef development. The origins of this issue derive from early observations and assumptions about artificial reefs. High catch rates and densities of fish at artificial reefs were popularly

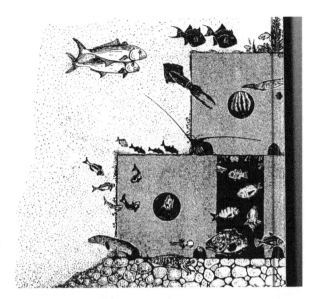

Figure 5 Aspects of attachment surfaces for marine plants and shelter and feeding sites for invertebrates and fishes provided by artificial reefs. Drawing courtesy of S. Riggio.

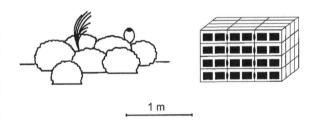

1 m

Figure 6 Experimental modules such as the one pictured on the right from the Bahamas exemplify a global trend toward greater manipulation of artificial reefs as tools for testing of hypotheses concerning ecological processes in the ocean, and comparison with systems such as coral reefs, pictured left. Drawing courtesy of American Fisheries Society.

taken as proof that artificial reefs benefited fisheries stocks, based on assumptions that hard substrata were limiting and reef fish standing stocks were dependent on food webs that have hard substrata as their foundations. However, a few fisheries scientists challenged this reasoning in the 1980s. They noted that before heavy exploitation the existing natural habitat supported abundant reef fishes, presumably at or near carrying capacity. Fishing then reduced populations to some lower level, yet the amount of natural habitat remained the same, still capable of supporting higher population numbers. The fisheries scientists reasoned that with fish stocks substantially below carrying capacity, the amount of hard-bottom habitat could not be the factor limiting population size, so the addition of artificial reefs would not benefit fish stocks. Thus, observed high densities of fish and high catch rates at artificial reefs were considered an artifact of fish behavioral preferences, which simply concentrated them at reef sites and intensified fishing mortality, what conservationists today call an ecological trap.

The fisheries scientists' formal argument was more thorough and complex than the popularized version, as they recognized a continuum from solely attraction, with little if any direct biological benefit, to potentially high levels of added biological production from each unit of additional reef structure. Species fit along that continuum according to their life history characteristics and use of reef habitat ecologically. Thus small, highly site-attached fishes (e.g., blennies and gobies) that derive all resource needs (i.e., food, shelter, and mates) directly from the occupied reef structure, and that complete all but the planktonic phase of their life cycle in one place, would have the greatest net production potential from artificial reef development. Conversely, large, transient reef-visiting fishes (e.g., jacks and mackerels) were expected to have the least production potential from artificial reef development. In between these extremes were the majority of exploited reef fishes for which one would expect some combination of attraction and production affected by reef ecological setting and overall stock abundance. (Implicit in the argument by the fisheries scientists was the assumption that fishing pressure would be intensely focused at the artificial reefs, which has more to do with the management of human activities than the intrinsic ecological characteristics of the reefs themselves.)

Differing professional positions concerning attraction-production still remain, and the debate continues to stimulate research questions and artificial reef management issues. In part, this is because the complexity of reef ecological functions and the effects of spatially explicit fishing mortality are beyond simple answers to the seemingly dichotomous question that has been popularized among lay audiences. And in part, this is because researchers necessarily address this issue from their own disciplinary perspectives and study systems, generally not integrating the multiple levels of biological organization or spatial-temporal domains over which this question can be legitimately addressed. Pertinent studies can focus on life history stages of populations, communities, or ecosystems and encompass individual reefs, broad landscapes, or geographic ranges of fisheries stocks, while covering seasonal, interannual, or long-term time frames. Practical solutions to the fisheries management implications will require more specific attention to the desired species and assemblages, and the processes that sustain them.

Now a pivotal question is how the manipulation of habitat affects demographic rates and exploitation rates across spatial and temporal scales. An example of a way to integrate large sets of ecological data to analyze different scenarios of artificial reef development is through ecosystem simulations such as for Hong Kong where fishing levels in and out of protected areas have been predicted as part of a fishery recovery program.

Advances in Planning, Design, and Construction

The increase of artificial reefs in the world's coastal seas, as measured by the growing number, cost, size, intricacy, and footprint of structures deployed, has prompted an emphasis on their planning. This is to promote more efficient and cost-effective structures, enable work at larger scales and with more precision, satisfy regulatory requirements, reduce conflicts with other natural and human aspects of ecosystems, and minimize the prospects of unintended consequences from improperly constructed reefs. The scientific basis for reefs has been strengthened by research efforts in dozens of countries, coupled with the practical experiences of numerous individuals and groups that have worked independently to build, manage, and, increasingly, evaluate reefs.

Independently in different geographic areas, common approaches and some published guidelines for developing artificial reefs have emerged. The earliest handbooks focused on physical and oceanographic conditions as first encountered by the Coastal Fishing Ground Enhancement and Development Project of Japan. (Translations of certain documents into English in the 1980s by the United States National Marine Fisheries Service disseminated information.) The United States produced a national plan in 1984 (in revision in 2006) which spurred promulgation of plans among the coastal states. In some nations, planning is done through a coordinated reef program run by a federal or provincial marine or fishery resources organization, as in Italy or China. Finally, where reef deployment is directed to selected localities or areas, but not nationwide, master plans or other guidelines for siting, materials, and other aspects have been produced, such as for the Aegean Sea of Turkey and northern Taiwan.

With a legitimate intent assured, the initial component of reef planning is to frame measurable objectives, define expectations for success, and forecast interactions of that reef with the ecosystem. Subsequently, a valid design and site plan for the reef and procedures for its construction must be developed and meet regulatory requirements. The context for management of the reef must be defined. For example, an artificial reef program in Hong Kong was based on an extensive consultation with community interests, leading to stakeholder support for regulatory practices in the local fisheries and protection of marine reserve areas containing some of the reefs. Finally, protocols for evaluation of reef performance and management of assessment data must be established, with communication of results to all interested parties.

Construction is a three-phase process: fabrication, transportation, and placement. Representative costs of larger reefs include US \$16 000 000 for a 56 ha reef in California, USA, and between US \$38 and \$57 million annually during 1995–99 for a national program in Korea. The ecological and economic impacts of poorly designed artificial reefs are exemplified by an effort in the United States – projected to cost US \$5 000 000 – to remove 2 000 000 automobile tires that had been deployed in the 1970s, but which were drifting onto adjacent live coral reefs and beaches.

The design phase of reef development has changed dramatically due to growing emphasis on making the structural attributes of intentional, fabricated reef materials conform and contribute to the biological life history requirements of organisms that are particularly desired as part of the reef assemblage. Reef design should be dictated by: (1) a set of measurable and justified objectives for the reef, (2) the ecology of species of concern associated with the reef, and (3) predicted and understandable environmental and socioeconomic consequences of introduction of the reef into the aquatic ecosystem. This requires a multidisciplinary approach using expertise of biologists, engineers, economists, planners, sociologists, and others.

From a physical standpoint, key aspects of design and construction include stability of a reef site, durability of the reef configuration, and potential adverse impacts in the environment. Also, the use of physical processes to enhance reef performance is a desirable aspect. To build a quantitative understanding of the environment into which a reef is to be placed, large-scale oceanographic processes and local conditions must be determined by site surveys early in reef planning, including water circulation driven by tides, wind and baroclinic/density fields, locally generated wind waves, swells propagated from distance, sediment/substrate composition, distribution and transport, and depth. These factors are then coupled with the attributes of the reef material, such as weight, density, dimensions, and strength, in order to forecast reef physical performance.

From an ecological standpoint, abiotic and biotic influences considered in design include geographic location, type and quality of substrate surrounding the reef site, isolation, depth, currents, seasonality, temperature regime, salinity, turbidity, nutrients, and productivity. Substrate attributes affecting reef ecology include its composition and surface texture, shape, height, profile, surface area, volume and hole size, which taken together contribute to the structural complexity of the reef. Spacing of individual and groups of reef structure is important.

A plethora of designs for artificial reefs exist (**Figure 7**). For a mitigation reef aiming to create new kelp beds off California, USA, biologists and engineers concluded that the most effective design should place boulders and concrete rubble in low-relief piles (<1 m height), at depths of 12–14.5 m on a sand layer of 30–50 cm overlaying hard substrate; success criteria include (1) support of four adult plants per $100 \, m^2$ and (2) invertebrate and fish populations similar to natural reefs. As part of a system of artificial reefs in Portugal, experiments quantified production of sessile invertebrates (upon which fishes could feed) according to location on different facets of cubic settlement structures placed on reefs. For fishes, meanwhile, one of five designs used in Korea is the box reef, a $3 \times 3 \times 3$ m concrete cube (**Figure 8**), targeted to two species: small, dark spaces in the lower two-thirds of the reef are provided for rockfish (*Sebastes schlegeli*), while the upper third is more open to satisfy behavioral preferences of porgy (*Pagrus major*). Repeatedly, authorities cite complexity of structure as a primary factor in design.

As large individual and sets of reefs are planned, more organizations are using pilot projects to determine physical, biological, economic, and even

Figure 7 Artificial reefs of concrete modules are used worldwide, with designs intended to meet ecological requirements of designated species and habitats.

Figure 8 The box reef design used in Korea resulted from a collaboration of engineers and biologists. Modules have been placed in research plots within Tongyong Marine Reserve for use in experiments with stocking juvenile fishes. Photograph by Kim Chang-gil.

political feasibility. The 42 000 t Loch Linne Artificial Reef in Scotland was started in 2001 as a platform for scientific investigation of the performance of different structures, ultimately to establish fisheries for target species, specifically lobster (*Homarus gammarus*). This effort also is representative of a trend toward increased predeployment research – in this case a 4-year study of seabed, water-column, and biological parameters – intended to enable better forecasting of reef impacts and measurement of results.

A special case of working with reef materials concerns the deployment of obsolete ships and petroleum/gas production platforms, due to the need to handle, prepare, and place them in environmentally compatible ways. In Canada, for example, decommissioned naval vessels require extensive removal of electrical wiring and other components to eliminate release of pollutants into the sea, in conformance with strict federal rules. In the northern Gulf of Mexico (where over 4000 platforms provide a considerable area of hard surface for sessile organism attachment), eastern Pacific, North Sea, and Adriatic Sea offshore platforms either act as *de facto* reefs or in a limited number of cases are being toppled in place or transported to new locations to serve as dedicated reefs, which costs less than removal to land.

Integration of Reefs in Ecological and Human Systems

Increasingly, artificial reef technology is being applied globally in fisheries and ecosystem management. Early disappointments and healthy skepticism

have led to more realistic expectations bolstered by two decades of scientific advances. National plans (e.g., Japan, Korea, and United States), regional programs (e.g., Hong Kong, Singapore, and Turkey), and large-scale pilot projects (e.g., Loch Linne, Scotland; San Onofre, USA) concerning artificial reefs are better defining their role in ecosystem and fishery management. International scientific bodies such as the North Pacific Marine Sciences Organization (PICES) have addressed the relevance of artificial reefs to core fisheries issues, including stock enhancement, fishing regulations, and conservation. Integrated coastal management in the Philippines, India, Spain, and elsewhere now includes artificial reefs in multifaceted responses to issues of habitat destruction, fishery decline, and socioeconomic development. Finally, private consultants and businesses have found markets for their services and products, with one patented design being deployed in over 40 countries, and ecological engineers have recognized reefs as constructed ecosystems for use in restoration ecology.

The evaluation of reef performance is fostering an increased acceptance and utilization of artificial reef technology by resource management organizations. Quantitative evaluation is increasingly driven by agency concerns for demonstrating positive returns on their investments, and the increasing scale of artificial reef projects. Evaluations vary in intensity and complexity, ranging from descriptive studies of short duration (e.g., pre- and postdeployment) to extensive studies of ecological processes and dynamics, to the synthesis of complex databases through quantitative modeling. One response of the scientific community was through formation of the European Artificial Reef Research Network, which promulgated priorities and protocols for research across its membership.

A significant trend is for artificial reef projects to be planned and evaluated in an adaptive management framework, in which expectations are more explicitly stated, the projects implemented and rigorously evaluated, and then adjustments made to the management practices based on findings from the evaluations. When applied consistently, this cycle will continue to evolve the application of artificial reef technologies toward an ever-increasing standard of practical effectiveness.

See also

Coastal Zone Management. Fisheries Overview. Fishery Management, Human Dimension. Fishing Methods and Fishing Fleets. Large Marine Ecosystems. Mariculture Diseases and Health.

Mariculture, Economic and Social Impacts. Mariculture of Aquarium Fishes. Mariculture of Mediterranean Species. Mariculture Overview.

Further Reading

American Fisheries Society (1997) *Special Issue on Artificial Reef Management. Fisheries* 22: 4–36.

International Council for the Exploration of the Sea (2002) *ICES Journal of Marine Science* 59(supplement S1-5363).

Jensen AC, Collins KJ, and Lockwood APM (eds.) (2000) *Artificial Reefs in European Seas*. Dordrecht: Kluwer.

Lindberg WJ, Frazer TK, Portier KM, *et al.* (2006) Density-dependent habitat selection and performance by a large mobile reef fish. *Ecological Applications* 16(2): 731–746.

Love MS, Schroeder DM, and Nishimoto MM (2003) The ecological role of oil and gas production platforms and natural outcrops on fishes in southern and central California: A synthesis of information. OCS Study MMS 2003-032. Seattle. WA: US Department of the Interior, US Geological Survey, Biological Resources Division.

Seaman W (ed.) (2000) *Artificial Reef Evaluation*. Boca Raton, FL: CRC Press.

Svane I and Petersen JK (2001) On the problems of epibioses, fouling and artificial reefs: A review. *Marine Ecology* 22: 169–188.

CONSERVATION OF PROTECTED SPECIES

CONSERVATION OF PROTECTED SPECIES

MARINE MAMMALS, HISTORY OF EXPLOITATION

R. R. Reeves, Okapi Wildlife Associates, Quebec, Canada

Introduction

Products obtained from marine mammals – defined to include the cetaceans (whales, dolphins, and porpoises), pinnipeds (seals, sea lions, and walrus), sirenians (manatees, dugong, and sea cow), sea otter, and polar bear – have contributed in many ways to human survival and development. Maritime communities, from the tropics to the poles, have depended on these animals for food, oil, leather, ivory, bone, baleen, and other materials. Some marine mammal products have had strategic value to nations. For example, for several centuries, streets and homes in much of the western world were illuminated with sperm oil candles and whale oil lanterns. Delicate machinery and precision instruments were lubricated with the head oil of toothed whales. Whale oil was an important source of glycerine during World War I and a key ingredient in margarine during and after World War II.

Other uses of marine mammal products have been more frivolous. Seal penises are sold as aphrodisiacs; narwhal (*Monodon monoceros*) and walrus (*Odobenus rosmarus*) tusks and polar bear (*Ursus*

Figure 1 A male narwhal with a 2 m tusk killed in the eastern Canadian Arctic, 1975. The tusk ivory of narwhals and walruses continues to provide an important incentive for hunting them, although both species are also valued as food by native people. Most narwhal tusks are sold and exported, intact, as novelties or trophies. Photo by RR Reeves.

maritimus) hides are displayed as 'trophies' in homes and offices (**Figure 1**). Spermaceti and ambergris, both obtained from sperm whales (*Physeter macrocephalus*), were highly valued by the perfume and cosmetics industries. Baleen used to be a stiffener for ladies' hoop skirts and undergarments. And, of course, the pelts of fur seals and sea otters have always been in great demand in luxury fur markets.

In general, the history of marine mammal exploitation is marked by overuse and abuse, with most wild populations having been severely overhunted. Some species and populations were extirpated or brought to the brink of extinction. Many others have been reduced and fragmented as a result of too much exploitation. It was not until well into the twentieth century that any serious restrictions were imposed on the sealing and whaling industries for the sake of conservation.

Cetaceans

Small Cetaceans (Dolphins, Porpoises, and the Smaller Toothed Whales)

Harpoon hunting of small cetaceans has occurred virtually all around the world, but mainly in coastal and shelf waters (**Figure 2**). It continues most notably in Japan, where 15 000–20 000 Dall's porpoises (*Phocoenoides dalli*) and at least several hundred dolphins, short-finned pilot whales (*Globicephala macrorhynchus*), and false killer whales (*Pseudorca crassidens*) are taken annually with hand harpoons, and about 150 additional pilot whales and Baird's beaked whales (*Berardius bairdii*) are taken each year with mounted harpoon guns. The meat of small cetaceans is highly valued in Japan. Eskimos in Greenland, Canada, and Alaska (USA) continue their long tradition of hunting white whales (*Delphinapterus leucas*) and narwhals. Although they formerly used kayaks, hand harpoons, and lances, today most of the hunting involves outboard-powered boats and high-powered rifles. Only in north-western Greenland are the traditional hunting techniques still used to any extent. Altogether, several thousand white whales and narwhals are taken each year. In addition, Greenlanders kill close to 2000 harbor porpoises (*Phocoena phocoena*) with rifles (**Figure 3**). The skin of small cetaceans is a delicacy in the Arctic. When saved, the meat and viscera are either eaten by people or fed to dogs.

A large commercial hunt for short-beaked common dolphins (*Delphinus delphis*), bottlenose

Figure 2 Whale hunters from Barrouallie, St Vincent, Lesser Antilles, at sea in pursuit of short-finned pilot whales (foreground) in the 1960s. The harpoon mounted on the bow of the sailboat was fired with a shotgun. Killed whales were cut into manageable pieces alongside the boat, and these pieces were brought on board to be taken ashore. On the beach, the pieces were cut in strips, hung on bamboo racks to dry, and sold to buyers from Kingstown. In 1968, the average pilot whale was worth about $40 US. Photo by David K Caldwell.

Figure 3 Meat from harbor porpoises shot in West Greenland is sold at an open-air market in Nuuk. Harbor porpoises contribute to a diverse array of wild foods consumed by the Greenlandic people, including fish, reindeer, seal, and whale. Photo by Steve Leatherwood, September 1987.

dolphins (*Tursiops truncatus*), and harbor porpoises was conducted in the Black Sea, using rifles and purse seine nets, from the nineteenth century into the late twentieth century. In the 1930s, nearly 150 000 dolphins and porpoises were taken in a single year. Although dolphin hunting was banned in the Soviet Union in 1966 and in Turkey in 1983, large kills were still being made in the Turkish sector of the Black Sea as recently as 1991. Oil and animal feed ('fish meal') were the main products, but the hunting was also prosecuted as a means of predator control. Fishermen viewed the cetaceans as serious competitors.

Some small cetaceans, particularly the pilot whales (*Globicephala* spp.), false killer whale, and melon-headed whale (*Peponocephala electra*), strand (i.e., come ashore) *en masse* in numbers ranging from tens to hundreds. This phenomenon remains unexplained but is known to occur naturally. Early coastal people would have welcomed mass strandings, as they represented windfalls of food and other useful products. It is not difficult to imagine their making the leap to a drive fishery, in which groups of animals were 'herded' toward shore, forced into shallow waters, and killed with lances or long knives. The first pilot whale drives in the Faroe Islands apparently took place at least four centuries ago, and similar drive fisheries existed elsewhere in the North Atlantic. In the Faroes, the whales have been used principally as food for humans, but in the other areas oil was a major incentive. In the post-World War II Newfoundland drive fishery, most of the catch (which reached nearly 10 000 pilot whales in 1956) was used to feed ranch mink. Drive hunting of cetaceans in the North Atlantic continues only in the Faroes, where hundreds, and in some years well over a thousand, long-finned pilot whales (*Globicephala melas*) and Atlantic white-sided dolphins (*Lagenorhynchus acutus*) are taken, and in West Greenland, where white whales and occasionally pilot whales are driven.

Drive fisheries for small cetaceans have also developed in the Solomon Islands and Japan. The Solomons example represents one of the more bizarre forms of exploitation of marine mammals. There, fishermen in dugout canoes fan out across a wide expanse of ocean to search for schools of dolphins and small whales. Large stones are struck together underwater to produce aversive sounds and scare the animals in the desired direction. Eventually, the school is guided into an enclosed harbor where the animals are quickly dispatched. Although some of the meat is cooked and eaten, the primary purpose of the hunt is to obtain 'porpoise teeth.' Porpoise-tooth necklaces must be given to a woman's parents as 'bride price,' an essential item in marriage transactions.

Dolphin drive fisheries have existed in Japan since the late fourteenth century. Initially, sail-assisted rowing boats were used, but motor vessels were introduced in the 1920s, allowing the hunters to cover much larger areas in their search for schools of small cetaceans. In recent years, high-speed motor boats have been used to find and drive ashore striped dolphins (*Stenella coeruleoalba*), pantropical spotted dolphins (*Stenella attenuata*), bottlenose dolphins, Risso's dolphins (*Grampus griseus*), pilot whales, and a number of other species.

Long seine nets were used to catch bottlenose dolphins along the Atlantic coast of the United States starting in the late eighteenth century and continuing at Cape Hatteras, North Carolina, until the late 1920s. A line of nets was set parallel to the shore, and when a school of dolphins moved into the area between the net line and the shore, the fishermen used nets to shut off escape, then swept the dolphins onto shore. Oil was the main prize, but a supple, durable shoe leather was also made from the hides. Commercial whalers and traders in the Arctic used large seine nets to trap schools of white whales, beginning as early as the 1750s in Hudson Bay and continuing in some areas (e.g., Svalbard) until as recently as 1960. Hides, oil, and dog food were the main products of these commercial netting operations.

Large Cetaceans (Baleen Whales and the Sperm Whale)

People in the Arctic were hunting bowhead whales (*Balaena mysticetus*) as long ago as the middle of the first millennium AD, and western Europeans were taking right whales (*Eubalaena glacialis*) by the beginning of the second. The technology and culture of subsistence whaling spread eastward within the Arctic and Subarctic from the Bering Strait region. Commercial whaling originated with the Basques, who had begun hunting right whales in the Bay of Biscay by the eleventh century. Initially, small open boats were launched from shore when a whale was sighted. However, the spread of whaling was relentless as Dutch, German, Danish, and British entrepreneurs vied to dominate the rich whaling grounds in the cold latitudes of the North Atlantic. In the 1760s, with the invention of a means to boil blubber on board the ship, it became possible to make extended offshore voyages, often lasting several years. The whaling fleets from New England, Great Britain, and France grew to dominate the industry. From the late eighteenth century to the early 1900s, commercial whaling ships penetrated all of the world's oceans except the Antarctic. The sperm whale bore the brunt of this activity (**Figure 4**). More than 225 000 were killed by American whalers alone from 1804 to 1876. During the peak years from the early 1830s to mid-century, over 100 000 barrels of sperm oil were delivered annually by more than 700 vessels working out of American ports. The nineteenth-century whalers often hunted blackfish (their name for pilot whales) while searching for sperm whales. They also made special voyages in pursuit of right, humpback (*Megaptera novaeangliae*), gray (*Eschrichtius robustus*), and bowhead whales. Only the fast-swimming finner whales – the blue, fin, sei, Bryde's, and minke (*Balaenoptera musculus*, *B. physalus*, *B. borealis*, *B. edeni/brydei*, and *B. acutorostrata/bonaerensis*, respectively) – were beyond their capabilities to capture.

Modern whaling, characterized by engine-driven catcher vessels and deck-mounted harpoon cannons firing explosive grenades, began in Norway in the 1860s. These inventions made possible the routine capture of any species, including the elusive finners. They also led to exploitation of the richest whaling ground on the planet, the Antarctic. In the first three-quarters of the twentieth century, factory ships from several nations, including Norway, Great Britain, Germany, Japan, the United States, and the Soviet Union, operated in the Antarctic. At its pre-War peak in 1937–38, the modern industry's 356 catcher boats, associated with 35 shore stations and as many floating factories, killed nearly 55 000 whales, 84% of them in the Antarctic. Having exhausted the stocks of right, bowhead, gray, and humpback whales in other areas, the industry rapidly proceeded along the same path in the Antarctic, reducing the

Figure 4 A small sperm whale killed by artisanal whalers at St Vincent, Lesser Antilles, during the 1960s. These whalers hunt for a variety of small and medium-sized cetaceans; sperm whales are taken only occasionally. The meat and oil are used locally. Photo by David K. Caldwell.

largest species first and then turning its attention to the next largest.

Commercial whaling declined in the 1970s as a result of conservationist pressure and depletion of the whale stocks. The last whaling stations in the United States and Canada were closed in 1972, and the last station in Australia ceased operations following the 1978 season. By the end of the 1970s, only Japan, the Soviet Union, Norway, and Iceland were still engaged in commercial whaling. With the decision by the International Whaling Commission (IWC) in 1982 to implement a global moratorium on commercial whaling, Japan and the Soviet Union made their final large-scale factory-ship expeditions to the Antarctic in 1986/87, and Japan stopped its coastal hunt for sperm whales and Bryde's whales in 1988. Iceland closed its whaling station in 1990, and shortly thereafter withdrew its membership in the IWC. Contrary to the widespread belief that commercial whaling had ended, however, Norway and Japan continued their hunting of minke whales through the 1990s and into the 2000s. By formally objecting to the IWC moratorium, Norway reserved its right to carry on whaling. Thus, Norwegian whalers have continued to kill more than 500 minke whales each year in the North Atlantic. Using a provision in the whaling treaty that allows member states to issue permits to hunt protected species for scientific research, Japan has continued taking more than 400 Antarctic minke whales and 100 North Pacific minke whales annually. The main incentive for continued commercial whaling is the demand for whale meat and blubber, particularly in Japan. Norway is eager to re-open the international trade in whale products so that stockpiles of blubber can be exported to Japan.

Aboriginal hunters in Russia, the United States (Alaska), and Canada kill several tens of bowheads and 100–200 gray whales every year. This hunting is primarily for human food. However, from the 1960s to early 1990s, gray whales taken by a modern catcher boat and delivered to native settlements in north-eastern Russia were used partly to feed foxes on fur farms. In recent years, native people in Washington State (USA), British Columbia (Canada), and Tonga (a South Pacific island nation) have expressed interest in re-establishing their own hunts for large cetaceans in order to reinvigorate their cultures. In the spring of 1999, the Makah Indian tribe in Washington took their first gray whale in more than 50 years.

Pinnipeds

Sealing began in the Stone Age, when people attacked hauled-out animals with clubs. Later methods included the use of traps, nets, harpoons thrown from skin boats, and gaff-like instruments for killing pups on ice or beaches. The introduction of firearms transformed the hunting of pinnipeds and caused an alarming increase in the proportion of animals that were killed but not retrieved, especially in those hunts where the animals were shot in deep water before first being harpooned. This problem of sinking loss also applies to many of the cetacean hunts mentioned above.

In addition to their meat and fat, the pelts of some seals, especially the fur seals and phocids, are of value in the garment industry. Markets for oil and sealskins fueled commercial hunting on a massive scale from the late eighteenth century through the middle of the twentieth. The ivory tusks and tough, flexible hides of walruses made these animals exceptionally valuable to both subsistence and commercial hunters. Thousands of walruses are still killed every year by the native people of north-eastern Russia, Alaska, north-eastern Canada, and Greenland. The killing is accomplished with high-powered rifles, and in some areas harpoons are still used to secure the animal. Walrus meat and blubber are eaten by people or fed to dogs, and the tusks are either used for carving or sold as curios.

Native hunters in the circumpolar north also kill more than a hundred thousand seals each year, mainly ringed seals (*Pusa hispida*) but also bearded (*Erignathus barbatus*), ribbon (*Histriophoca fasciata*), harp (*Pagophilus groenlandicus*), hooded (*Cystophora cristata*), and spotted seals (*Phoca largha*) (**Figure 5**). Seal meat and fat remain important in the diet of many northern communities, and the skins are still used locally to make clothing, dog traces, and hunting lines. There is also a limited commercial export market for high-quality sealskins and a strong demand in Oriental communities for pinniped penises and bacula. The sale of these items, along with walrus and narwhal ivory, white whale and narwhal skin (maktak), and polar bear hides and gall bladders, has helped offset the economic losses in some native hunting communities caused by the decline in international sealskin markets (**Figure 6**).

The scale of commercial sealing, like that of commercial whaling, has declined considerably since the 1960s. It continues, however, in several parts of the North and South Atlantic. After a period of drastically reduced killing in the 1980s, the Canadian commercial hunt for harp and hooded seals has been expanded, at least in part as a result of governmental support. An estimated 350 000 harp seals were taken by hunters in eastern Canada and West Greenland in 1998. A few tens of thousands of molting pups are clubbed to death on the sea ice, but

(A)

Figure 6 Pelt of a hooded seal stretched to dry on the side of a house in Upernavik, Northwest Greenland, June 1987. Photo by Steve Leatherwood.

(B)

Figure 5 Ringed seal killed by a Greenlander off Northwest Greenland, June 1988. Photo by Steve Leatherwood.

the vast majority of the killing is accomplished by shooting. Norwegian and Russian ships continue to visit the harp and hooded seal grounds in the Greenland Sea ('West Ice') and Barents Sea ('East Ice'), taking several tens of thousands of seals annually. Also in recent years, thousands of South African and South American fur seals (*Arctocephalus pusillus* and *A. australis*, respectively) have been taken in south-western Africa and Uruguay, respectively. These hunts are centuries old, having been driven initially by markets for skins and oil, and more recently by the Oriental demand for reproductive parts. Much of the hunting for pinnipeds is motivated by the desire of fishermen to see their populations reduced. Seals and sea lions are often held responsible for damaged fishing gear, the removal of fish from nets and lines, and the spread of parasitic worms which infect groundfish.

Sirenians

Sirenians have been hunted mainly for meat and blubber, which are highly prized as food. Steller's sea cow (*Hydrodamalis gigas*), a North Pacific endemic and the only modern cold-water sirenian, was hunted to extinction within about 25 years after its discovery by commercial sea otter and fur seal hunters in 1741. Sea cows were easy to catch and provided the ship crews with sustenance as they carried on the hunt for furs and oil from other marine mammals.

Manatee hides were traditionally used by people in South and Central America and in West Africa to make shields, whips, and plasters for dressing wounds. For a time, these hides were also in great demand for making glue and heavy-duty leather products (e.g., machinery belts, hoses, and gaskets). The hides of more than 19 000 manatees were exported for this purpose from Manaus, Brazil, between 1938 and 1942. For a much longer time, from the 1780s to the late 1950s, the commercial exploitation of manatees in South America was driven by the market for mixira, fried manatee meat preserved in its own fat. Although no large-scale commercial hunt takes place today, local people continue to kill manatees for food. It is impossible to make a reasonable guess at how many manatees are killed by villagers in West Africa and South and Central America, but the total in recent years has probably been in the thousands (all three species, *Trichechus manatus*, *T. inunguis*, and *T. senegalensis*, combined). Manatees are captured in many different ways, apart from simply stalking them in quiet dugout canoes and striking them with a lance or harpoon. These involve such things as stationary hunting blinds; drop traps armed with heavy, pointed wooden posts; and fence traps or nets placed strategically in the intertidal zone or in constricted channels.

Dugongs (*Dugong dugon*), like manatees, have long been a prized food source for seafaring people. Hunting continues throughout much of their extensive Indo-Pacific range, even in areas where the species is almost extinct. Dugong hunters in some areas have used underwater explosives to kill their prey. In Torres Strait between Australia and New Guinea, portable platforms are set up on seagrass beds, and the hunter waits there overnight for opportunities to spear unsuspecting dugongs as they graze.

Sea Otter

The sea otter (*Enhydra lutris*) has one of the most luxuriant and thus desirable pelts of any mammal species. As a result, it was eagerly hunted by aboriginal people all round the rim of the North Pacific Ocean. Also, beginning soon after Vitus Bering discovered the Commander Islands in 1741, Russian, and later American and Japanese, expeditions were mounted for the explicit purpose of obtaining sea otter furs, which commanded high prices in the Oriental market. No statistics were kept, but at least half a million sea otters were taken (or received in trade) by commercial hunters between 1740 and 1911, when the species was given legal protection. The hunters sometimes used anchored nets to catch the otters, but more often they lanced them from small boats. Once rifles became available, these were used in preference to lances. In California, sea otters were sometimes shot by men standing on shore, and in Washington, shooting towers were erected at the surfline and Indians were employed to swim out and retrieve the carcasses. Alaskan natives are still allowed to hunt sea otters as long as the furs are used locally to make clothing or authentic handicraft items. The reported annual kill during the mid to late 1990s ranged from 600 to 1200.

Polar Bear

Eskimos traditionally hunted polar bears with dog teams and hand lances. The meat was eaten and the hides used for clothing and bedding. White explorers, whalers, sealers, and traders in the Arctic often killed polar bears with high-powered rifles. They also provided a commercial outlet for hides obtained by the Eskimos. In modern times, the Eskimos hunt polar bears with rifles and search the ice in snowmobiles rather than dogsleds. Norwegian trappers and weather station crews on Svalbard formerly used poison, foot snares, and set guns to kill polar bears. The set gun consisted of a wooden box resting on poles about 75 cm above ground level, with a rifle or shotgun mounted inside. A string connected the gun's trigger to bait placed in front of the box. When the bear took the bait, the trigger was pulled and the gun fired. Sport hunters have taken thousands of polar bears as trophies, particularly in Alaska where guided hunting with aircraft began in the late 1940s and continued until 1972. At least several hundred polar bears are still killed each year, most of them by Eskimos for meat and the cash value of their hides and gall bladders. Hunting permits issued to native communities in Canada are often sold to sport hunters, on the understanding that a local guide will be hired to accompany the hunter and that only the head and hide will be exported.

Live-capture and other Forms of Exploitation

Although the numbers of marine mammals removed from the wild for captive display and research have been small in comparison to the numbers killed for meat, oil, skins, etc., the high commercial value of some species establishes this as an important form of exploitation (**Figure 7**). More than 1500 bottlenose dolphins were live-captured in the United States, Mexico, and the Bahamas between 1938 and 1980. Close to 70 killer whales (*Orcinus orca*) were removed from inshore waters of Washington State (USA) and British Columbia (Canada) and transported to oceanaria between 1962 and 1977, and about 50 were exported from Iceland in the 1970s and 1980s (**Figure 8**). Live killer whales and bottlenose dolphins are presently worth about $1 million and $50 000, respectively. Captive-bred animals and 'strandlings' (animals that come ashore and require rehabilitation) have increasingly been used to stock oceanaria, but this trend applies mainly to North America and involves primarily bottlenose dolphins, killer whales, California sea lions (*Zalophus californianus*), and harbor seals (*Phoca vitulina*). Dolphin, whale, and sea lion displays are becoming more popular in Asia and South America, and new facilities on those continents create a continuing demand for wild-caught animals, especially dolphins (**Figure 9**). Most polar bears and walruses brought into captivity have been young ones, orphaned when their mothers were killed by hunters. In Florida, manatees are often brought into captivity after being

(A)

(B)

Figure 7 Commerson's dolphins are endemic to the coastal waters of southern South America and certain subantarctic islands. There is some demand for them in North American, European, and Japanese oceanaria. (A) Here, a hoop net at the end of a pole is used in an attempt to capture a dolphin from a bow-riding group off the coast of Chile, February 1984. (B) A Commerson's dolphin (foreground) shares an oceanarium tank with a white whale (beluga) at a zoo in Duisberg, Germany. Photos by Steve Leatherwood.

Figure 8 Killer whales are the most valuable marine mammals in the oceanarium trade. Recent success at captive breeding and rearing has relieved some of the pressure on wild populations to stock display facilities. The movie "Free Willy" inspired a campaign to return the whale "Keiko" back to its natal waters near Iceland. Photo by Steve Leatherwood.

Figure 9 Irrawaddy dolphins have a limited coastal and freshwater distribution in southeast Asia and northern Australia. They are fairly popular in Asian oceanaria, and live captures add to the stress on populations caused by incidental mortality in gillnets. This animal was recently on display at a facility in Thailand. Photo by Steve Leatherwood.

Figure 10 Gray whales attract many tourists each year to the nearly pristine waters of Laguna San Ignacio, Baja California, Mexico. Whale-watching in the lagoon is closely regulated by Mexican authorities. The recognized economic value of nature tourism was partly responsible for the government's decision in 2000 to reject a proposal for a large evaporative salt factory on the shores of San Ignacio. Photo by Steve Leatherwood.

injured or orphaned as a result of boat strikes. Nearly 100 sea otters were taken from Alaskan waters for public display between 1976 and 1988.

It should be mentioned that marine mammals are also 'exploited' as the objects of tourism. Whale-, dolphin-, seal-, and sea otter-watching supports an extensive network of tour operations around the world (**Figure 10**). Commercial fishermen 'exploit' pelagic dolphins in the tropical Pacific Ocean by using them to locate schools of tuna, and this can result in large numbers of dolphins being killed by accident.

FurtherReading

Bonner WN (1982) *Seals and Man: A Study of Interactions*. Seattle: University of Washington Press.

Bräutigam A and Thomsen J (1994) Harvest and international trade in seals and their products. In: Reijnders PJH Brasseur S, *et al.* (eds.) *Seals, Fur Seals, Sea Lions, and Walrus: Status Survey and Conservation Action Plan*. Gland, Switzerland: International Union for Conservation of Nature and Natural Resources.

Busch BC (1985) *The War Against the Seals: A History of the North American Seal Fishery*. Kingston, Ontario: McGill-Queen's University Press.

Dawbin WH (1966) Porpoises and porpoise hunting in Malaita. *Australian Natural History* 15: 207–211.

Domning DP (1982) Commercial exploitation of manatees *Trichechus* in Brazil c. 1785–1973. *Biological Conservation* 22: 101–126.

Ellis R (1991) *Men and Whales*. New York: Alfred A. Knopf.

International Whaling Commission. Annual report and special issues. Available from IWC, The Red House, 135 Station Road, Impington, Cambridge, CB4 9NP, UK. (From April 1999, these reports appear as Supplements or Special Issues of The Journal of Cetacean Research and Management, published by the IWC.)

McCartney AP (ed.) (1995) *Hunting the Largest Animals: Native Whaling in the Western Arctic and Subarctic*. Edmonton: Canadian Circumpolar Institute, University of Alberta.

Mitchell E (1975) *Porpoise, Dolphin and Small Whale Fisheries of the World*. Morges, Switzerland: International Union for Conservation of Nature and Natural Resources.

Taylor VJ and Dunstone N (eds.) (1996) *The Exploitation of Mammal Populations*. London: Chapman and Hall.

Tønnessen JN and Johnsen AO (1982) *The History of Modern Whaling*. Berkeley: University of California Press.

Twiss JR Jr and Reeves RR (eds.) (1999) *Conservation and Management of Marine Mammals*. Washington, DC: Smithsonian Institution Press.

MARINE MAMMALS AND OCEAN NOISE

D. Wartzok, Florida International University, Miami, FL, USA

Introduction

The oceans are much more transparent to acoustic energy than to electromagnetic energy in the frequency ranges at which animal sensory systems operate. Consequently it is not surprising that both human and animal underwater communication systems rely on acoustic transmission. Some baleen whales are presumed to hear as low as 10 Hz and some river dolphins as high as 200 kHz. Therefore virtually all human activity on or in the oceans results in the incidental addition of sound in the hearing range of one or more marine mammal species. The intentional and unintentional introduction of acoustic energy into the oceans by human activities constitutes ocean noise so far as marine mammals are concerned. The issue of marine mammals and sound has been the subject of four National Research Council reports and several special issues of journals.

Human-generated sound is not the only sound that constitutes noise for marine mammals. Natural sources of sound include those generated by the interaction of wind with the sea surface, waves breaking on shorelines, precipitation, thunder, earthquakes, and ice cracking. Biological sources of sound include snapping shrimp, fish choruses, and the vocalizations of other marine mammals. Marine mammals have evolved in this cacophony of sounds and have developed mechanisms to compensate for background noise, so they can successfully perform acoustically mediated prey detection, predator avoidance, navigation, and intraspecific communication. However, when human-generated sound exceeds the evolved adaptive capacity though intensity, duration, pervasiveness, particular signal characteristics, or as an additive factor to natural sounds, marine mammals can be negatively affected.

While there is a general consensus that anthropogenic ocean sound has increased over the past decades, there are few instances in which ambient noise measurements have been made at the same locations over a span of decades. A few reports suggest that at certain locations in the Northern Hemisphere ambient noise in the 20–80-Hz band has increased approximately 12 dB over the past four decades. Each 3-dB increase represents a doubling of the ambient noise (see XXX for definition of dB). Shipping noise is the predominant contributor to ambient noise in the 20–80-Hz range. Shipping is expected to increase in gross tonnage and speed of transport, both of which will lead to further increases in ocean noise unless quiet ship technologies are incorporated in the design phase of these new vessels.

Although naval sonar and seismic exploration activities also transmit much energy into the ocean each year, these sources are at higher frequencies (sonar) and thus do not propagate as far, or are not omnidirectional (sonar and seismic), as is shipping. Thus they make less contribution to the global ambient.

It has been difficult to demonstrate specific effects of noise on marine mammals and even more difficult to partition the noise contribution to the suite of natural and anthropogenic factors potentially affecting marine mammals. Multiyear habitat abandonment has been demonstrated for killer whales (*Orcinus orca*) in a British Columbia archipelago where acoustic harassment devices were employed to protect aquaculture facilities and for gray whales (*Eschrichtius robustus*) where a Baja California lagoon was abandoned during the period of industrial activities. In both of these cases, the habitat was reoccupied with the cessation of the acoustic harassment. The other clear example of the effect of anthropogenic sound on marine mammals is the stranding of primarily beaked whales (*Ziphiidae*) in response to mid-frequency naval sonar.

Sound has not been considered a significant factor in any of the documented major declines of marine mammal populations, for example, Steller sea lions (*Eumetopias jubatus*), Aleutian Islands sea otters (*Enhydra lutris*), Alaskan harbor seals (*Phoca vitulina richardsi*), and fur seals (*Callorhinus ursinus*). However, it is important to note that these all involved species much easier to monitor than cetaceans. It has been estimated that the probability of sighting a beaked whale directly on the trackline of the survey vessel is only 23% under ideal conditions and is closer to 2% under typical conditions. Thus population level effects of sound on beaked whales would be difficult to detect if they were occurring. Because of these uncertainties and the difficulty of obtaining definitive data, the National Research Council (2005) concluded "On the one hand, sound may represent only a second-order effect on the conservation of marine mammal populations; on the other hand, what we have observed so far may be only the first early warnings or 'tip of the iceberg' with respect to sound and marine mammals."

Zones of Influence

Ocean noise affects marine mammals in different ways depending on the characteristics of the sound and its intensity. The influence of particular characteristics of the sound is not well understood. For example, the stranding of beaked whales in the presence of naval mid-frequency sonar appears to depend on characteristics of the sound rather than the received intensity, although exactly what those characteristics are and how they result in stranding are unknown. **Figure 1** is a diagram illustrating the range of possible effects of sound on marine mammals based on received intensity of the sound. In most cases, the limit of auditory detectability will be the threshold for any response. Humans can show a range of physiological reactions to nonaudible, low-frequency noise. Whether marine mammals have similar responses is unknown. If a sound is above the background level and above the auditory threshold at a particular frequency, then the animal can detect that sound. Beluga whales detect the return of their echolocation signal when it is only 1 dB above the background and gray whales react to playbacks of predator killer whale vocalizations when the signal is 0 dB above the ambient.

Behavioral Response

Whether the animal responds to the sound and what response the animal has to the sound depend on a suite of factors, including age, gender, season, location, behavioral state, prior exposure, and individual variation. For example, beluga whales are more sensitive to ship noise when they are confined to leads, meters to hundreds of meters wide open water channels between ice sheets, than when not so confined. Migrating gray whales changed their orientation when exposed to a Low Frequency Active sonar that was in their migratory path but ignored the same signal source at even higher received sound levels when it was located seaward of their migratory path. Right whales (*Eubalaena glacialis*) and fin whales (*Balaenoptera physalus*) are more tolerant of stationary noise sources than those moving toward them, whereas bottlenose dolphins (*Tursiops truncates*) show aversive behaviors in response to speedboats and jet skis even when they are not approaching. Humpback whales (*Megaptera novaeangliae*) respond at lower received levels to stimuli with sudden onset than they do to continuous sound sources.

One of the strongest recorded reactions of a marine mammal to a sound is the response of beluga

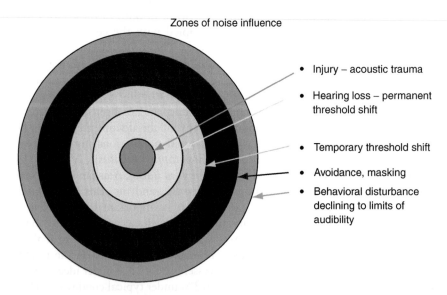

Zones of noise influence

- Injury – acoustic trauma
- Hearing loss – permanent threshold shift
- Temporary threshold shift
- Avoidance, masking
- Behavioral disturbance declining to limits of audibility

Figure 1 Close to an intense source, sound may be loud enough to cause death or serious injury. Somewhat farther away, an animal might have less serious injury, such as hearing loss. Temporary threshold shifts occur at greater distances. Animals may avoid exposures at even greater distances or they may not move from the area but still be affected through masking of important auditory cues from the environment. They may show barely observable behavioral disturbance at distances comparable with the limit of audibility. The different distances for the different effects define different areas for each zone. Reproduced from National Research Council (2005) *Marine Mammal Populations and Ocean Noise: Determining When Noise Causes Biologically Significant Effects.* Washington, DC: The National Academies Press, with permission from The National Academies Press.

whales in the High Arctic to the first icebreaker of the season. Belugas have been recorded fleeing for more than 80 km from the icebreakers and exhibiting a range of other behavioral changes at received sound levels between 94 and 105 dB re 1 µPa, yet 1 or 2 days later showing no reaction to icebreaker sound of 120 dB re 1 µPa. Even greater habituation is shown by belugas in Bristol Bay, Alaska, where they continue to move into the Kvichak River to feed on salmon smolt even when purposely harassed by motorboats. On the other hand, an elephant seal (*Mirounga angustirostris*) showed sensitization rather than habituation in response to repeated presentations of broadband pulsed signals that were similar to predatory killer whale echolocation clicks.

Masking

As the received sound level rises above the limits of detectability and possible behavioral response, the sound has the potential to interfere with the natural uses of sound by marine mammals in that frequency range. The noise-induced reduction of acoustic information with respect to conspecifics, predators, prey, and other environmental clues is termed masking. Mammals have dealt with masking throughout their evolutionary history and have developed a number of ways to maintain normal functions in the presence of masking sounds.

Noise is only effective in masking a signal if it is within a certain critical band around the signal's frequency. The actual degree of masking depends upon the amount of noise energy within this critical frequency band. Although critical bands have been measured in only a few marine mammals, at frequencies above 1 kHz, the critical bands for cetaceans are narrower than they are for most other mammals.

In addition to narrower critical bands, marine mammals also reduce masking by directional hearing. Unless the masking sound is directly on axis with the sound of interest, directional hearing provides a significant gain in the signal-to-noise ratio. Odontocetes have good directional hearing above 1 kHz. The directivity index (DI) is a measure of the ability of a receiver to reduce the effects of omnidirectional noise and is expressed as the decibel level above the signal the omnidirectional noise must be increased in order to mask the signal. For bottlenose dolphins, the DI increases from 10.4 dB at 30 kHz to 20.6 dB at 120 kHz.

Animals adapt to masking in a number of ways. For example, a beluga whale that was moved to a location with higher levels of continuous background noise increased both the average level and frequency of its vocalizations. A beluga that was required to echolocate on an object placed in front of a source of noise reduced masking by reflecting its sonar signals off the water surface to ensonify to the object. The strongest echoes from the object returned along a path that was off axis from the noise. This animal's ready application of such complex behavior suggests the existence of many sophisticated strategies to reduce masking effects.

Masking compensation responses in noncaptive situations include: beluga whales increasing call repetition and shifting to higher peak frequencies in response to boat traffic; gray whales increasing the amplitude of their vocalizations, changing the timing of vocalizations, and using more frequency-modulated signals in noisy environments; humpback whales increasing the duration of their songs by 29% when exposed to low-frequency active sonar; and killer whales increasing call duration over time as the number of whale-watching boats increased.

No hearing thresholds have been measured in baleen whales. Their vocalizations emphasize frequencies in the same range as that of the increased background noise due to shipping. This raises the possibility of significant masking of important signals, particularly conspecific communication signals for these whales.

The adaptations used to enhance signaling in the presence of background noise such as increased amplitude, longer duration, and repetition are not available in the situations where the animals are passively listening for acoustic cues such as waves breaking on distant shores or noises produced by schools of fish prey. These cues can be only a few decibels above the normal ambient and are easily masked by increased noise levels.

The passive sonar equation can be used to show that as noise increases by a set amount, the range at which a signal can be detected decreases by a constant proportion, termed the range reduction factor (RRF). For example, a 6-dB increase in noise has an RRF of 2 for transmission loss by spherical spreading and an RRF of 4 for transmission loss by cylindrical spreading. In many situations spatial distribution is important (e.g., breeding humpback whales), and in these situations the reduction in area monitored may be a more realistic measure of the effect of increased noise. The area will be reduced proportional to RFF^2.

Temporary Threshold Shift

When animals are exposed to more intense sound levels, they experience a temporary loss of hearing sensitivity, or temporary threshold shift (TTS). The amount of the TTS depends on the energy of the stimulus. A reduction in sensitivity in response to an intense or prolonged stimulus is the usual response of a sensory system and within normal ranges is reversible. TTS has been studied extensively in humans and laboratory animals and results from a combination of anatomic and metabolic alterations occasioned by the intense sound.

There have been few studies of TTS in marine mammals. Among the odontocetes, only the bottlenose dolphin and the beluga whale have been tested and among the pinnipeds only harbor seals (*Phoca vitulina vitulina*), California sea lions (*Zalophus californianus*), and elephant seals. For some species, only a single animal has been tested. Given the small number of animals tested and recognizing the substantial variation in TTS in more extensively studied laboratory species, any conclusions regarding TTS in marine mammals are tentative.

A just measurable TTS of about 6 dB has been produced with stimuli ranging from an impulse of 1-ms duration at 226-dB peak-to-peak re 1 μPa at 1-m to 30-min exposure to octave bandwidth noise of 179-dB re 1 μPa. A doubling of exposure time has approximately the same effect as a reduction in intensity of 3 dB. This relationship between exposure time and sound intensity shows that the various stimuli producing a just measurable TTS all delivered approximately the same amount of energy to the auditory system. For a 1 s exposure, this TTS threshold is 195 dB re 1 μPa.

When the time course of recovery from TTS was tested immediately following the cessation of the fatiguing stimulus, recovery was within 4–40 min depending on experimental conditions and magnitude of the TTS. Even TTS of 23 dB usually recovered within 30 min of exposure. When the time course of recovery was not measured immediately after cessation, there was full recovery when the animals were next tested 24 h later.

Permanent Threshold Shift

As the name implies, an animal does not recover hearing sensitivity when it experiences a permanent threshold shift (PTS). PTSs are both functionally and anatomically different from TTSs. PTS results in losses of the cells that convert sound energy into neural signals. No experiments have been done on marine mammals that have resulted in a PTS. The TTS measured so far is well below the level that would be expected to grade into PTS. For terrestrial animals approximately 40-dB TTS is required for a PTS. As with TTS, PTS is less likely to occur if the noise bursts are shorter and the intervening periods between intense noise are longer.

Nonauditory Effects

Direct nonauditory effects of sound have not been demonstrated in marine mammals, but some have been observed in other animals or humans or *in vitro*. Humans exposed to intense sounds can experience dizziness (Tullio phenomenon), nystagmus, and neurological disturbance in the absence of hearing loss.

Acoustic resonance Tissues associated with air-filled cavities can be subjected to shear forces when those cavities resonate. Two important factors for resonance are the relationship between the dimensions of the cavity and the wavelength of the sound and the tuning or amplification of the resonance. The latter is described by the Q value, with a high Q indicating greater resonance amplitude. The resonance frequency of beluga and bottlenose whale lungs has been determined to be 30 and 36 Hz, respectively, with relatively low Q values of 2.5 and 3.1, respectively. Thus, at these low frequencies, there is a modest amplification of the resonance magnitude. The resonance characteristics of other air-filled cavities have not been measured but the consensus is that the magnitude of the resonance is not great enough to cause tissue damage.

Rectified diffusion and activation of micro-bubbles Bubble growth either through acoustically driven rectified diffusion or acoustic activation of micro-bubbles could lead to symptoms similar to decompression sickness in human divers. The basic requirement in either model is that the tissues be highly supersaturated. In some deep-diving marine mammals, such as beaked whales, supersaturation has been calculated to exceed 300%. Beaked whales that stranded subsequent to naval sonar activities have shown bubbles in a number of tissues. However, the current understanding of the exposures of the whales that stranded is that the received sound levels were too low to cause bubble growth or activation (demonstrated at 210 dB re 1 μPa *in vitro*). A more likely explanation is that a behavioral response to lower received levels initiated the cascade of events resulting in the stranding. Studies in right whales, a species not involved in naval sonar stranding incidents, have shown that received levels of 133 dB re 1 μPa of an unfamiliar signal will cause abrupt changes in diving behavior.

Blast injury Blast injuries to marine mammals have only rarely been observed. The best-documented case is that of humpback whales that died within 3 days after detonation of 1700–5000-kg Tovex (a trinitrotoluene (TNT) clone) blasts. The mechanical traumas in the whale ears were consistent with classic blast injuries in humans including round window rupture, ossicular chain disruption, bloody effusion in the ear region, and bilateral periotic fractures. These traumas result from eruptive injury during the rarefactive portion of the shock wave when inner ear fluid pressures are much greater than ambient.

Stress Both high-intensity, short-duration stimuli and long-term exposure to much lower levels of noise can result in elevated levels of stress. Some studies conducted in humans have shown that exposure to chronic noise elevates neuroendocrine and cardiovascular indices of stress and results in diminished performance on cognitive tests of reading ability and long-term memory. For most studies in humans, it has been difficult to demonstrate statistically significant effects of chronic noise, although there is a consistent trend toward increased cardiovascular risk if the daytime exposure level exceeds 65 dB(A) (*see* XXX for a discussion of the differences between in-air and underwater decibel reference levels and measurements). Within this context, the 12-dB increase in low-frequency noise due to shipping in the past four decades could be a stress factor in addition to its role in masking communications. Another component of chronic noise, at least in the North Atlantic, is the long-range propagation of the sounds from seismic surveys. Autonomous hydrophones located near the mid-Atlantic ridge frequently, particularly in the summer, recorded sounds of seismic surveys taking place over 3000 km from the recording location. The effects of such long-term increases in anthropogenic sound on the stress response of marine mammals have not been determined.

There have been two studies of the short-term effects of noise on the stress response of marine mammals. One detected no change in behavior or catecholamine levels of captive beluga whales exposed to playbacks of the operating noise from a semisubmersible drilling platform at a source level of 153 dB re 1 µPa at 1 m. In contrast, a beluga whale exposed to high-level (> 100 kPa) impulsive sounds and high-intensity tones had significantly elevated norepinephrine, epinephrine, and dopamine levels. A bottlenose dolphin similarly exposed did not show elevated catecholamines, but did show an increase in aldosterone and a decrease in absolute monocyte levels after exposure to a seismic water gun.

Among the range of possible mechanisms contributing to the stranding of beaked whales exposed to naval sonar is an acute stress phenomenon, that of hemorrhagic diathesis. A precondition for hemorrhagic diathesis is a depletion or lack of clotting factors or platelet dysfunction. Humans with a hereditary deficiency in clotting factors develop subarachnoid and inner ear hemorrhages similar to those seen in the beaked whales. No studies have been conducted on the clotting ability of beaked whale blood, but in the few cetacean species studied to date, all have shown a lack of certain clotting factors. None of the species studied have stranded in association with naval sonar; so if hemorrhagic diathesis is a contributor to beaked whale strandings, the level of stress and the physiological responses to that stress are different in beaked whales.

Magnitude of the Problem

Other than the strandings of primarily beaked whales in association with naval mid-frequency sonar, anthropogenic sound has not been identified as a cause of marine mammal mortality. Even in the case of the beaked whales and naval sonar, the sequence of behavioral and physiological responses leading to stranding and death is unknown. The total number of beaked whales that are known to have died as a result of anthropogenic sound or in association with alleged naval activity is over 3 orders of magnitude fewer than the number of cetaceans killed annually in direct fisheries bycatch.

The preceding paragraph reflects our current understanding but is more likely a reflection of our lack of knowledge than a true assessment of the situation. In order to better gauge the extent of the problem, we need a currency in which to measure the effects of anthropogenic sound other than mortality. As the preceding discussion has shown, anthropogenic sound can affect marine mammals in ways that range from behavioral responses, to inhibited communication and sensing of the environment, to temporary and permanent auditory threshold shifts, to trauma producing morbidity or mortality. The National Research Council (2005) proposed assigning lethality equivalents to sublethal effects of sound. There was recognition that sound which interfered with foraging, displaced animals, masked communication, or caused TTS had the potential to reduce lifetime reproductive fitness and should be given a proportional weighting. For example, if anthropogenic sound interfered with prey detection by passive listening so that it caused a decrease of 0.1% in the lifetime reproductive fitness of each animal affected, then the lethal equivalent would be 0.001, and if 1000 animals were affected, that would be the equivalent of removing one animal from the population.

The zones of influence in **Figure 1** show that the largest number of animals will be affected at the level

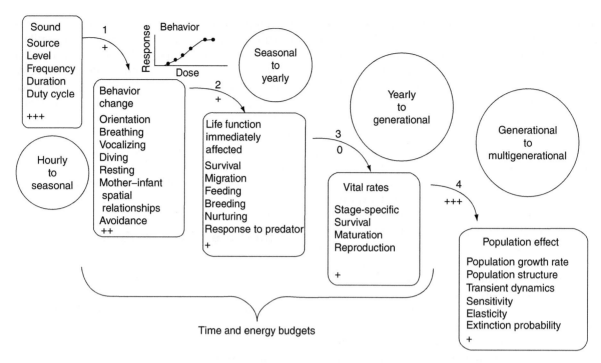

Figure 2   The conceptual 'population consequences of acoustic disturbance' model describes several stages required to relate acoustic disturbance to effects on a marine mammal population. Five groups of variables are of interest, and transfer functions specify the relationships between the variables listed, for example, how sounds of a given frequency affect the vocalization rate of a given species of marine mammal under specified conditions. A typical type of dose–response transfer function is illustrated. Each box lists variables with observable features (sound, behavior change, life function affected, vital rates, and population effect). In most cases, the causal mechanisms of responses are not known. For example, survival is included as one of the life functions that could be affected to account for such situations as the beaked whale strandings in response to naval mid-frequency sonar, in which it is generally agreed that exposure can result in death. The '+' signs at the bottom of the boxes indicate how well the variables can be measured at the present time. The indicators between boxes show how well the 'black box' nature of the transfer functions is understood; these indicators scale from '+ + +' (well known and easily observed) to '0' (unknown). Reproduced from National Research Council (2005) *Marine Mammal Populations and Ocean Noise: Determining When Noise Causes Biologically Significant Effects*. Washington, DC: The National Academies Press, with permission from The National Academies Press.

of a behavioral change. The National Research Council in 2005 presented a model of the population consequences of behavioral change in response to anthropogenic sound (**Figure 2**). This model summarizes what we need to know and indicates how well we know various components leading from behavioral response to population effects. We can measure characteristics of the sound sources well and can do a pretty good job of observing behavioral changes. There is less understanding of how a given sound causes a particular behavioral change under a certain set of conditions. Much of the uncertainty regarding the significance of an observable behavioral response is related to the difficulty in determining how that behavioral change can cause a significant change in a life function. Little is known about the functional response of a behavioral change, and the measurement of the life function altered is difficult for most marine mammal species. Integrating changes in migration, feeding, breeding, etc., over a lifetime in order to determine effects on

vital rates is currently beyond our capabilities. However, when the changes in vital rates are known, the population-level consequences are readily determined. Unfortunately, at least a decade of work will be required before this conceptual model can become a predictive model and can be used to determine the lethal equivalents of sound-induced behavioral changes and subsequent population effects. The uncertainty regarding the magnitude of the problem leaves an extensive gray area between under-regulation resulting in unacceptable harm to marine mammal populations and over-regulation unnecessarily inhibiting essential human activities on and in the ocean.

Further Reading

Cox TM, Ragen TJ, Read AJ, *et al.* (2006) Understanding the impacts of anthropogenic sound on beaked whales. *Journal of Cetacean Research Management* 7: 177–187.

D'Spain GL, D'Amico A, and Fromm DM (2006) Properties of the underwater sound fields during some well documented beaked whale mass stranding events. *Journal of Cetacean Research Management* 7: 223–238.

Fernandez A, Edwards JF, Rodriguez F, *et al.* (2005) 'Gas and fat embolic syndrome' involving a mass stranding of beaked whales (family *Ziphiidae*) exposed to anthropogenic sonar signals. *Veterinary Pathology* 42: 446–457.

Finneran JJ, Schlundt CE, Dear R, Carder DA, and Ridgway SH (2002) Temporary shift in masked hearing thresholds in odontocetes after exposure to single underwater impulses from a seismic watergun. *Journal of the Acoustical Society of America* 111: 2929–2940.

Hildebrand J (2005) Impacts of anthropogenic sound. In: Reynolds JE, III, Perrin WF, Reeves RR, Montgomery S, and Ragen TJ (eds.) *Marine Mammal Research: Conservation beyond Crisis*, pp. 101–123. Baltimore, MD: Johns Hopkins Press.

McDonald MA, Hildebrand JA, and Wiggins SM (2006) Increases in deep ocean ambient noise west of San Nicolas Island, California. *Journal of the Acoustical Society of America* 120: 1–8.

Merrill J (ed.) (2004) *Human-generated ocean sound and the effects on marine life. Marine Technology Society Journal* 37(4).

National Research Council (1994) *Low-Frequency Sound and Marine Mammals: Current Knowledge and Research Needs.* Washington, DC: The National Academies Press.

National Research Council (2000) *Marine Mammals and Low-Frequency Sound.* Washington, DC: The National Academies Press.

National Research Council (2003) *Ocean Noise and Marine Mammals.* Washington, DC: The National Academies Press.

National Research Council (2005) *Marine Mammal Populations and Ocean Noise: Determining When Noise Causes Biologically Significant Effects.* Washington, DC: The National Academies Press.

Potter J and Tyack PL (2003) Special Issue on Marine Mammals and Noise. *IEEE Journal of Oceanic Engineering* 28: 163.

Richardson WJ, Greene CR, Malme CI, and Thompson DH (1995) *Marine Mammals and Noise.* San Diego, CA: Academic Press.

Wartzok D and Ketten DR (1999) Marine mammal sensory systems. In: Reynolds JE, III and Rommel S (eds.) *Biology of Marine Mammals*, pp. 117–175. Washington, DC: Smithsonian Institution Press.

Wartzok D, Popper AN, Gordon J, and Merrill J (2004) Factors affecting the responses of marine mammals to acoustic disturbance. *Marine Technology Society Journal* 37: 6–15.

SEABIRD CONSERVATION

J. Burger, Rutgers University, Piscataway, USA

Introduction

Conservation is the preservation and protection of plants, animals, communities, or ecosystems; for marine birds, this means preserving and protecting them in all of their diverse habitats, at all times of the year. Conservation implies some form of management, even if the management is limited to leaving the system alone, or monitoring it, without intervention or human disturbance. The appropriate degree of management is often controversial. Some argue that we should merely protect seabirds and their nesting habitats from further human influences, leaving them alone to survive or to perish. For many species, however, this solution is not practical because they do not live on remote islands, in inaccessible sites, or places that could be totally ignored by people. For other species, their nesting and foraging habitats have been so invaded by human activities that they must adapt to new, less suitable conditions. For some species, their declines have been so severe that only aggressive intervention will save them. Even species that appear to be unaffected by people have suffered from exotic feral animals and diseases that have come ashore, brought by early seafarers in dugout canoes or later by mariners in larger boats with more places for invading species to hide.

For marine birds there is compelling evidence that the activities of man over centuries have changed their habitats, their nesting biology, and their foraging ecology. Thus, we have a responsibility to conserve the world's marine birds. For conservation to be effective, the breeding biology, natural history, foraging ecology, and interactions with humans must be well understood, and the factors that contribute to their overall reproductive success and survival known.

Marine Bird Biology and Conservation

In this article, seabirds, or marine birds, include both the traditional seabird species (penguins, petrels and shearwaters, albatrosses, tropicbirds, gannets and boobies, frigate-birds, auks, gulls and terns, and pelicans) and closely related species that spend less time at sea (cormorants, skimmers), and also other species that spend a great deal of their life cycle along coasts or at sea, but may nest inland, such as shore birds. Seabirds are distributed worldwide, and nest in a variety of habitats, from remote oceanic islands that are little more than coral atolls, to massive rocky cliffs, saltmarsh islands, sandy beaches or grassy meadows, and even rooftops. While most species of seabirds nest colonially, some breed in loose colonies of scattered nests, and still others nest solitarily. Understanding the nesting pattern and habitat preferences of marine birds is essential to understanding the options for conservation of these birds on their nesting colonies. Without such information, appropriate habitats might not be preserved.

The attention of conservationists is normally directed to protecting seabirds while they are breeding, but marine birds spend most of their lives away from the breeding colonies. Although some terns and gulls first breed at 2 or 3 years of age, other large gulls and other seabirds breed when they are much older. Some albatrosses do not breed until they are 10 years old. Nonbreeders often wander the oceans, bays, and estuaries, and do not return to the nesting colonies until they are ready to breed. They face a wide range of threats during this period, and these often pose more difficult conservation issues because the birds are dispersed, and are not easy to protect. In some cases, such as roseate terns (*Sterna dougallii*), we do not even know where the vast majority of overwintering adults spend their time. Since many species forage over the vast oceans during the nonbreeding season, or before they reach adulthood, conditions at sea are critical for their long-term survival. While landscape ecology has dominated thought for terrestrial systems, few of its tenets have been applied to oceanic ecosystems, or to the conservation needs of marine birds.

A brief description of the factors that affect the success of marine bird populations will be enumerated before discussion of conservation strategies and management options for the protection and preservation of marine bird populations.

Threats to Marine Birds

Marine bird conservation can be thought of as the relationship between hazards or threats, marine bird vulnerabilities, and management. The schematic in **Figure 1** illustrates the major kinds of hazards faced by marine birds, and indeed all birds, and the different kinds of vulnerabilities they face. The outcomes shown in **Figure 1** are the major ones; however, there are many others that contribute to the overall decline in population levels (**Figure 2**). Conservation involves some form of intervention or management for each of these hazards, to preserve and conserve the species.

Marine Bird Vulnerabilities

Factors that affect marine bird vulnerability include the stage in the life cycle, their activity patterns, and their ecosystem (**Figure 3**). Marine birds are differentially vulnerable during different life stages. Many of the life cycle vulnerabilities are reduced by nesting in remote oceanic islands (albatrosses, many petrels, many penguins) or in inaccessible locations, such as cliffs (many alcids, kittiwakes, *Rissa tridactyla*) or tall trees. However, not all marine birds nest in such inaccessible sites, and some sites that were inaccessible for centuries are now inhabited by people and their commensal animals.

For many species, the egg stage is the most vulnerable to predators, since eggs are sufficiently small that a wide range of predators can eat them. Eggs are placed in one location, and are entirely dependent upon parents for protection from inclement weather, accidents, predators, and people. In many cultures, bird eggs, particularly seabird eggs, play a key role, either as a source of protein or as part of cultural traditions. In some cultures, the eggs of particular species are considered aphrodisiacs and are highly prized and sought after.

Egging is still practiced by humans in many places in the world, usually without any legal restrictions. Even where egging is illegal, either the authorities overlook the practice or it is impossible to enforce, or it is sufficiently clandestine to be difficult to apprehend the eggers.

Chicks are nearly as vulnerable as eggs, although many seabirds are semiprecocial at birth and are able to move about somewhat within a few days of hatching. The more precocial, the more likely the chick can move about to hide from predators or people, or seek protection from inclement weather. Nonetheless, chicks are unable to fly, and thus cannot avoid most ground predators and many aerial

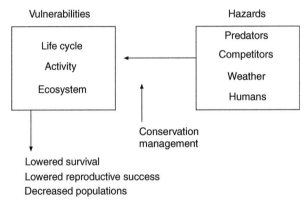

Figure 1 Marine avian conservation is the relationship between the hazards marine birds must face, along with their vulnerabilities.

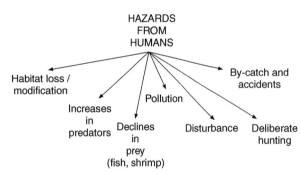

Figure 3 Humans provide a wide range of hazards, including direct and indirect effects.

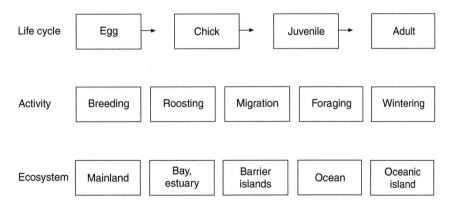

Figure 2 The primary vulnerabilities marine birds face deal with aspects of their life cycles, activity patterns and the ecosystems they inhabit.

predators if they cannot hide sufficiently. The pre-fledging period can last for weeks (small species such as terns) to 6 months for albatrosses.

The relative vulnerability of juveniles and adults is usually the same, at least with respect to body size. Most juveniles are as large as adults and can fly, and are thus able to avoid predators. Juveniles, however, are less experienced with predators and with foraging, and so are less adept at avoiding predators and at foraging efficiently. The relative vulnerability of juveniles and adults depends on their size, habitat, type of predator, and antipredator behavior. For example, species that nest high in trees (Bonaparte's gull, *Larus philadelphia*) or on cliffs (e.g., kittiwakes, some murres, some alcids) are not exposed to ground predators that cannot reach them. Species that nest on islands far removed from the mainland are less vulnerable to ground predators (e.g., rats, foxes), unless these have been introduced or have unintentionally reached the islands. Marine birds that are especially aggressive in their defense of their nests (such as most terns) can sometimes successfully defend their eggs and young from small predators by mobbing or attacks.

Activity patterns also influence their vulnerability. When marine birds are breeding, they are tied to a particular nest site, and must either abandon their eggs or chicks or stay to protect them from inclement weather, predators, or people. At other times of the year, seabirds are not as tied to one location, and can move to avoid these threats. Foraging birds are vulnerable not only to predators and humans, but to accidents from being caught in fishing gill nets, drift nets, or longlines, from which mortality can be massive, especially to petrels and albatrosses.

The choice of habitats or ecosystems also determines their relative vulnerability to different types of hazards. Marine birds nesting on oceanic islands are generally removed from ground predators and most aerial predators but face devastation when such predators reach these islands. Cats and rats have proven to be the most serious threat to seabirds nesting on oceanic and barrier islands. The threat from predators increases the closer nesting islands are to the mainland, a usual source of predators and people. Similarly, the threat from storm and hurricane tides is greater near-shore, particularly for ground-nesting seabirds.

Marine Bird Hazards

The major hazards and challenges to survival of marine birds are from competitors (for mates, food, nesting sites), predators, inclement weather, and humans (**Figure 1**). Of these, humans are the greatest problem for the conservation of marine birds and strongly influence the other three types of hazards. Humans affect marine birds in a wide range of ways, by changing the environment around seabirds (**Figure 4**), ultimately causing population declines. While other hazards, such as competition for food and inclement weather are widespread, marine birds have always faced these challenges.

Habitat loss and modification are the greatest threats to marine birds that nest in coastal regions, and for birds nesting on near-shore islands. Direct loss of habitat is often less severe on remote oceanic islands, although recent losses of habitat on the Galapagos and other islands are causes for concern. Habitat loss can also include a decrease in available foraging habitat, either directly through its loss or through increased activities that decrease prey abundance or their ability to forage within that habitat.

Humans cause a wide range of other problems:

- Introducing predators to remote islands, and increasing the number of predators on islands and coastal habitats. For example, rats and cats have been introduced to many remote islands, either deliberately or accidentally. Further, because of the construction of bridges and the presence of human foods (garbage), foxes, raccoons and other predators have reached many coastal islands.
- Decreasing available prey through overfishing, habitat loss, or pollution. Coastal habitat loss can decrease fish production because of loss of nursery areas, and pollution can further decrease reproduction of prey fish used by seabirds.
- Decreasing reproductive success, causing behavioral deficits, or direct mortality because of pollution. Contaminants, such as lead and mercury, can reduce locomotion, feeding behavior, and parental recognition in young, leading to decreased reproductive success.
- Decreasing survival or causing injuries because birds are inadvertently caught in fishing lines, gillnets, or ropes attached to longlines.
- Decreasing reproductive success or foraging success because of deliberate or accidental disturbance of nesting, foraging, roosting, or migrating marine birds. For some marine birds the presence of recreational fishing boats, personal watercraft, and commercial fishing boats reduces the area in which they can forage.
- Deliberate collection of eggs, and killing of chicks or adults for food, medicine, or other purposes. On many seabird nesting islands in the Caribbean, and elsewhere, the eggs of terns and other species

IMMEDIATE RESPONSES

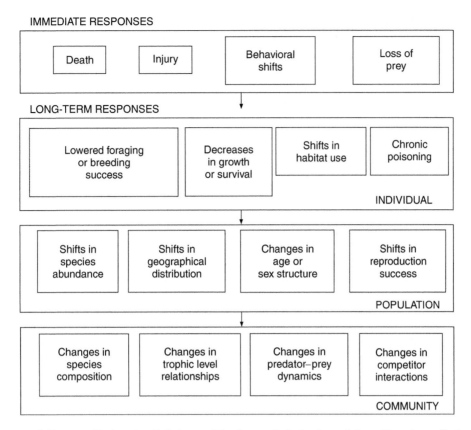

Figure 4 Human activities can affect marine birds in a variety of ways, including immediate and long-term effects.

are still collected for food. Egging of murres and other species is also practiced by some native peoples in the Arctic.

Conservation and Management of Marine Birds

Conservation of marine birds is a global problem, and global solutions are needed. This is particularly true for problems that occur at sea, where the birds roost, migrate, and forage. No one governmental jurisdiction controls the world's population of most species. Education, active protection and management, international treaties and agreements, and international enforcement may be required to solve some of the major threats to marine birds. However, the conservation and management of marine birds also involves intervention in each of the above hazards, and this can often be accomplished locally or regionally with positive results.

Habitat loss can be partly mitigated by providing other suitable habitats or nesting spaces nearby; predators can be eliminated or controlled; fishing can be managed so that stocks are not depleted to a point where there are no longer sufficient resources for marine birds; contamination can be reduced by legal enforcement; human disturbance can be reduced by laws, wardening, and voluntary action; by-catch can be reduced by redesigning fishing gear or changing the spatial or temporal patterns of fishing; and deliberate or illegal hunting can be reduced by education and legal enforcement. Each will be discussed below.

Habitat creation and modification is one of the most useful conservation tools because it can be practiced locally to protect a particular marine bird nesting or foraging area. In many coastal regions, nesting habitat has been created for beach-nesting terns, shore birds, and other species by placing sand on islands that are otherwise unsuitable, extending sandy spits to create suitable habitat, and removing vegetation to keep the appropriate successional stage. In some places grassy meadows are preserved for nesting birds, while in others sand cliffs have been modified, and concrete slabs have been provided for cormorants, gannets, and boobies (albeit to make it easier to collect the guano for fertilizer). Habitat modification can also include creation of nest sites. Burrows have been constructed for petrels; chick shelters have been created for terns; and platforms have been built for cormorants, anhingas, and terns.

The increase in the diversity and number of introduced and exotic predators on oceanic and other islands is a major problem for many marine birds. One of the largest problems marine birds face worldwide is the introduction of cats and rats to remote nesting islands. Since most marine birds on remote islands nest on the ground, their eggs and chicks are vulnerable to cats and rats. Some governments, such as that of New Zealand, have devoted considerable time and resources to removal of these two predators from remote nesting islands, but the effort is enormous.

Many marine birds evolved on remote islands where there were no mammalian predators. These species lack antipredator behaviors that allow them to defend themselves, as seen in albatrosses, which do not leave their nests while rats gnaw them. Most sea birds on remote islands nest on, or under the ground, where they are vulnerable to ground predators, and they do not leave their nests when approached. Rats and cats have proven to be the most significant threat to sea birds worldwide, and their eradication is essential if some marine birds are to survive. New Zealand has invested heavily in eradicating invasive species on some of its offshore islands, allowing sea birds and other endemic species to survive. Cats, however, are extremely difficult to remove, even from small offshore islands, and up to three years were required to remove them completely from some New Zealand islands. Such a program involves a major commitment of time, money, and personnel by local or federal governments.

Simply removing predators, however, does not always result in immediate increases in seabird populations. Sometimes unusual management practices are required, such as use of decoys and playback of vocalizations to attract birds to former colony sites. Steve Kress of National Audubon reestablished Atlantic puffins (*Fratercula arctica*) on nesting colonies in Maine by a long-term program of predator removal, decoys, and the playback of puffin vocalizations.

Increasing observations of chick mortality from starvation or other breeding failures have focused attention on food availability, and the declines in fish stocks. Declines in prey can be caused by sea level changes, water temperature changes, increases in predators and competitors, and other natural factors. However, they can also be caused by overfishing that depletes the breeding stocks, and reducing the production of small fish that serve as prey for seabirds. There are two mechanisms at work: in some cases fishermen take the larger fish, thereby removing the breeding stock, with a resultant decline in small fish for prey. This may have happened in the northern

Atlantic. In other cases, fishermen take small fish, thereby competing directly with the seabirds, as partially happened off the coast of Peru.

Overfishing is a complicated problem that often requires not only local fisheries management but national and international treaties and laws. Even then, fisheries biologists, politicians, importers/exporters, and lawyers see no reasons to maintain the levels of fish stocks necessary to provide for the foraging needs of sea birds. Nonetheless, the involvement of conservationists interested in preserving marine bird populations must extend to fisheries issues, for this is one of the major conservation challenges that seabirds face.

By-catch in gill nets, drift nets, and longlines is also a fisheries problem. With the advent of longlines, millions of seabirds of other species are caught annually in the miles of baited lines behind fishing vessels. The control and reduction in the number of such fishing gear is critical to reducing seabird mortality. Longlines are major problems for seabirds in the oceans of the Southern Hemisphere, although Australia and New Zealand are requiring bird-deterrents on longline boats.

Pollution is another threat to sea birds: Pollutants include heavy metals, organics, pesticides, plastics, and oil, among others. Oil spills have often received the most attention because there are often massive and conspicuous die-offs of sea birds following major oil spills. Usually, however, the carcass counts underestimate the actual mortality because the spills happen at sea or in bad conditions where the carcasses are never found or do not reach land before they are scavenged or decay. Although direct mortality is severe from oil spills, one of the greatest problems following an oil spill is the decline of local breeding populations, as happened following the *Exxon Valdez* in Alaska. Ten years after the spill some seabird species had still not recovered to pre-spill levels. Partially this resulted from a lack of excess reproduction on nearby islands, where predators such as foxes kept reproduction low.

While major oil spills have the potential to cause massive die-offs of birds that are foraging and breeding nearby, or migrating through the area, chronic oil pollution is also a serious threat. Many coastal areas, particularly near major ports, experience chronic oil spillage that accounts for far more oil than the massive oil spills that receive national attention. Chronic pollution can cause more subtle effects such as changes in foraging behavior, deficits in begging, weight loss, and internal lesions.

When there are highly localized population declines as a result of pollution, such as oil spills, or of inclement weather, predators, or other causes, the management options are limited. However, one method to encourage rapid recovery is to manage the breeding colonies outside of the affected area, allowing them to serve as sources for the depleted colonies. In the case of the *Exxon Valdez*, for example, there were numerous active colonies immediately outside of the spill impact zone. However, reproduction on many of these islands was suboptimal owing to the presence of predators (foxes). Fox removal would no doubt increase reproductive success on those islands, providing surplus birds that could colonize the depleted colonies within the spill zone itself.

Management for reductions in marine pollutants, including oil, can be accomplished by education, negotiations with companies, laws and treaties, and sanctions. For example, following the *Exxon Valdez*, the U.S. government passed the Oil Pollution Act that ensured that by 2020 all ships entering U.S. waters would have double hulls and many other safety measures to reduce the possibility of large oil spills.

Another threat to marine birds is through atmospheric deposition of mercury, cadmium, lead, and other contaminants. At present, mercury and other contaminants have been found in the tissues of birds throughout the world, including the Arctic and Antarctic. While atmospheric deposition is greatest in the Northern Hemisphere, contaminants from the north are reaching the Southern Hemisphere. The problem of atmospheric deposition of mercury, and oxides of nitrogen and sulfur, can be managed only by regional, national, and international laws that control emissions from industrial and other sources, although some regional negotiations can be successful.

Marine birds have been very useful as indicators of coastal and marine pollution because they integrate over time and space. While monitoring of sediment and water is costly and time-consuming, monitoring of the tissues of birds (especially feathers) can be used to indicate where there may be a problem. Declines in marine bird populations such as occurred with DDT, were instrumental in regulating contaminants. Marine birds have been especially useful as bioindicators in the Great Lakes for polychlorinated biphenyls (PCBs), on the East Coast of North America and in northern Europe for mercury, and in the Everglades for mercury.

Human disturbance is a major threat to seabirds, both in coastal habitats and on oceanic islands. While the level and kinds of human disturbance to marine birds in coastal regions is much higher than for oceanic islands, the birds that nest on oceanic islands did not evolve with human disturbance and are far less equipped to deal with it. Disturbance to breeding and feeding assemblages can be deliberate or accidental, when people come close without even realizing they are doing so, and fail to notice or be concerned. Sometimes colonial birds mob people who enter their colony, but the people do not see any eggs or chicks (because they are cryptic), and so are unaware they are causing any damage. Chicks and eggs, however, can be exposed to heat or cold stress during these disturbances.

Human disturbance can be managed by education, monitoring (by volunteers, paid wardens, or law enforcement officers), physical barriers (signs, strings, fences, barricades), laws, and treaties. In most cases, however, it is worth meeting with affected parties to figure out how to reduce the disturbance to the birds while still allowing for the human activities. This can be done by limiting access temporally and keeping people away during the breeding season, or by posting the sensitive location but allowing human activities in other regions. Compliance will be far higher if the interested parties are included in the development of the conservation strategy, rather than merely being informed at a later point. Moreover, such people often have creative solutions that are successful.

The deliberate collecting of eggs and marine birds themselves can be managed by education, negotiations, laws and treaties, and enforcement. In places where the collection of eggs or adult birds is needed as a source of protein or for cultural reasons, mutual education by the affected people and managers will be far more successful. In many cases, indigenous peoples have maintained a sustainable harvest of seabird eggs and adults for centuries without ill effect to the seabird populations. However, if the populations of these people increase, the pressure on seabird populations may exceed their reproductive capacity. People normally took only the first eggs, and allowed the birds to re-lay and raise young. Conservation was often accomplished because individuals 'owned' a particular section of the colony, and their 'section' was passed down from generation to generation. There was thus strong incentive to preserve the population and not to overuse the resource, particularly since most seabirds show nest site tenacity and will return to the same place to nest year after year. When 'governments' took over the protection of seabird colonies, no one owned them any longer, and they suffered the fate of many 'commons' resources: they were exploited to the full with devastating results. Whereas subsistence hunting of seabirds and their eggs was successfully managed for centuries, the populations suffered overnight with the advent of government control and the availability of new technologies (snowmobiles, guns). More recently, the use of personal watercraft has increased in some coastal areas, destroying nurseries for fish and shellfish, disturbing foraging activities, and disrupting the nesting activities of terns and other species.

Hunting by nontraditional hunters can also be managed by education, persuasion, laws, and treaties. However, both types of hunting can be managed only when there are sufficient data to provide understanding of the breeding biology, population dynamics, and population levels. Without such information on each species of marine bird, it is impossible to determine the level of hunting that the populations can withstand. Extensive egging and hunting of marine birds by native peoples still occurs in some regions, such as that of the murres in Newfoundland and Greenland.

On a few islands, some seabird populations have both suffered and benefitted at the hands of the military. Some species nested on islands that were used as bombing ranges (Culebra, Puerto Rico) or were cleared for air transport (Midway) or were directly bombed (Midway, during the Second World War). In these cases, conservation could only involve governmental agreements to stop these activities, and of course, in the case of war, it is no doubt out of the hands of conservationists. However, military occupancy may protect colonies by excluding those who would exploit the birds.

Conclusions

Conservation of marine birds is a function of understanding the hazards that a given species or group of species face, understanding the species vulnerabilities and possible outcomes, and devising methods to reduce or eliminate these threats so that the species can flourish. Methods range from preserving habitat and preventing any form of disturbance (including egging and hunting), to more complicated and costly procedures such as wardening, and attracting birds back to former nesting colonies.

The conservation methods that are generally available include education, creation of nesting habitat and nest sites, the elimination of predators, the cessation of overfishing, building of barriers, use of wardens and guards, use of decoys and vocalization, creation of laws and treaties, and the enforcement of these laws. In most cases, the creation of coalitions of people with differing interests in

seabirds, to reach mutually agreeable solutions, will be the most effective and long-lasting. Although ecotourism may pose the threat of increased disturbance or beach development, it can be managed as a source of revenue to sustain conservation efforts.

It is necessary to bear in mind that conservation of seabirds is not merely a matter of protecting and preserving nesting assemblages, but of protecting their migratory and wintering habitat and assuring an adequate food supply. Assuring a sufficient food supply can place marine birds in direct conflict with commercial and recreational fishermen, and with other marine activities, such as transportation of oil and other industrial products, use of personal watercraft and boats, and development of shoreline industries and communities. Conservation of marine birds, like many other conservation problems, is a matter of involving all interested parties in solving a 'commons' issue.

See also

Ecosystem Effects of Fishing. Oil Pollution. Seabird Conservation.

Further Reading

Burger J (2001) *Tourism and ecosystems. Encyclopedia of Global Environmental Change.* Chichester: Wiley.

Burger J and Gochfeld M (1994) Predation and effects of humans on island-nesting seabirds. In: Nettleship DN, Burger J, and Gochfeld M (eds.) *Threats to Seabirds on Islands*, pp. 39–67. Cambridge: International Council for Bird Preservation Technical Publication.

Croxall JP, Evans PGH, and Schreiber RW (1984) *Status and Conservation of the World's Seabirds.* Cambridge: International Council for Bird Preservation Technical Publication No. 2.

Kress S (1982) The return of the Atlantic Puffin to Eastern Egg Rock, Maine. *Living Bird Quarterly* 1: 11–14.

Moors PJ (1985) *Conservation of Island Birds.* Cambridge: International Council for Bird Preservation Technical Publication.

Nettleship DN, Burger J, and Gochfeld M (eds.) (1994) *Threats to Seabirds on Islands.* Cambridge: International Council for Bird Preservation Technical Publication.

Vermeer K, Briggs KT, Morgan KH, and Siegel-Causey D (1993) *The Status, Ecology, and Conservation of Marine Birds of the North Pacific.* Ottawa: Canadian Wildlife Service Special Publication.

SEA TURTLES

F. V. Paladino, Indiana-Purdue University at Fort Wayne, Fort Wayne, IN, USA
S. J. Morreale, Cornell University, Ithaca, NY, USA

Introduction

There are seven living species of sea turtles that include six representatives from the family Cheloniidae and one from the family Dermochelyidae. These, along with two other extinct families, Toxochelyidae and Prostegidae, had all evolved by the early Cretaceous, more than 100 Ma (million years ago). Today the cheloniids are represented by the loggerhead turtle (*Caretta caretta*), the green turtle *Chelonia mydas*, the hawksbill turtle (*Eretmochelys imbricata*), the flatback turtle (*Natator depressus*), and two congeneric turtles, the Kemp's ridley turtle (*Lepidochelys kempii*) and the olive ridley turtle (*Lepidochelys olivacea*). The only remaining member of the dermochelyid family is the largest of all the living turtles, the leatherback turtle (*Dermochelys coriacea*).

Sea turtles evolved from a terrestrial ancestor, and like all reptiles, use lungs to breathe air. Nevertheless, a sea turtle spends virtually its entire life in the water and, not surprisingly, has several extreme adaptations for an aquatic existence. The rear limbs of all sea turtles are relatively short and broadly flattened flippers. In the water, these are used as paddles or rudders to steer the turtle's movements whereas on land they are used to push the turtle forward and to scoop out the cup shaped nesting chamber in the sand (**Figure 1**). The front limbs are highly modified structures that have taken the appearance of a wing. Internally the front limbs have a shortened radius and ulna (forearm bones) and greatly elongated digits that provide the support for the flattened, blade-like wing structure. Functionally the front flippers are nearly identical to bird wings, providing a lift-based propulsion system that is not seen in other turtles.

Other special adaptations help sea turtles live in the marine environment. All marine turtles are well suited for diving, with specialized features in their blood, lungs, and heart that enable them to stay submerged comfortably for periods from 20 min to more than an hour. Since everything they eat and drink comes from the ocean, their kidneys are designed to minimize salt uptake and conserve water and they have highly developed glands in their heads that concentrate and excrete salt in the form of tears.

Despite all their adaptations and highly specialized mechanisms for life in the ocean, sea turtles are inescapably tied to land at some stage of their life. Turtles, like all other reptiles, have amniotic eggs which contain protective membranes that allow for complete embryonic development within the protected environment of the egg. This great advancement in evolution provided many advantages for reptiles over their predecessors: fish and amphibians. However, turtle eggs must develop in a terrestrial environment. Thus, in order to reproduce, female sea turtles need to emerge from the water and come ashore to lay eggs. All seven species lay their eggs on sandy beaches in warmer tropical and subtropical regions of the world.

Eggs are deposited on the beach in nest chambers, which can be as deep as 1 m below the sand surface. The adult female crawls on land, excavates a chamber into which it lays 50–130 eggs, and returns to the water after covering the new nest. Eggs usually take 50–70 days to fully develop and produce hatchlings that clamber to the beach surface at night. The length of the incubation primarily is influenced by the prevailing temperature of the nest during development; warmer temperatures hasten the hatchlings' development rates.

Nest temperature also plays a key role in determining the sex of hatchlings. Sea turtles share with other reptiles a phenomenon known as temperature-dependent sex determination (TSD). In sea turtles, warmer nest temperatures produce females, whereas cooler temperatures generate males. More specifically, it appears that there is a crucial period in the middle trimester of incubation during which temperature acts on the sex-determining mechanism in the developing embryo. Temperatures of $>30°C$ generate female sea turtles, and temperatures cooler than 29°C produce males. There is a pivotal range between these temperatures that can produce either sex. During rainy periods it is very common to have nests that are exclusively male, whereas a sunny dry climate tends to produce many nests of all females. Extended periods of extreme weather can produce an extremely skewed sex ratio for an entire beach over an entire nesting season. It is of much concern that global warming trends could have drastic effects on reptiles for which a balanced sex ratio totally depends on temperature.

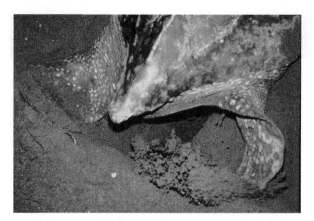

Figure 1 Leatherback digging nest.

During a single nesting season an individual female nests several different times, emerging from the water to lay eggs at roughly two-week intervals. Thus a female's reproductive output can total sometimes more than 1000 eggs by the end of the nesting season. However, once finished, most individuals do not return to nest again for several years. There is much variation in the measured intervals of return, between 1 and 9 years, before subsequent nesting bouts. Alternatively, males may mate at a much more frequent rate. Mating occurs in the water, and has been observed in some areas, usually near the nesting beaches. However, timing and location of mating is not known for many nesting populations, and is virtually unknown for some species.

Sea turtles have evolved a life history strategy that includes long-lived adults, a high reproductive potential, and high mortality as hatchlings or juveniles. As adults they have few predators with the exception of some of the larger sharks and crocodilians. Age to reproductive maturity is estimated at 8–20 years depending on the species and feeding conditions. Their eating habits include herbivores, like the green and black turtle, carnivores like the ridley turtles that eat crustaceans, and even spongivores, like the hawksbill. Almost all sea turtle hatchlings have a pelagic dispersal phase which is poorly documented and understood. Juveniles and adults congregate in feeding areas after the hatchling has attained a certain size. All sea turtles are currently listed in Appendix I of the Convention on International Trade in Endangered Species of Flora or Fauna (CITES Convention) and all species except the flatback (*Natator depressus*) from Australia, are also listed as threatened or endangered by the World Conservation Union (IUCN). Historically populations of sea turtles have declined worldwide over the past 20 years due to loss of nesting beach habitat, harvest of their eggs due to natural predation or human consumption, and the extensive adult mortality brought about by fishing pressure from the expansion of shrimp and drift net or long-line ocean fisheries.

General Sea Turtle Biology

The first fossil turtles appear in the Upper Triassic (210–223 Ma) and the sea turtles of today are descendants of that lineage. The current species of the family Cheloniidae evolved about 2–6 Ma and the sole representative of the Dermochelyidae family is estimated to have evolved about 20 Ma. All sea turtles are characterized by an anapsid skull, in which the region behind the eye socket is completely covered by bone without any openings, and the presence of a bony upper shell (carapace) and a similar bony lower shell (plastron). In all of the cheloniids the carapace and plastron consist of paired bony plates, in contrast to the sole Dermochelyid which has no evident bony plates in the leathery carapace and plastron, but instead interlocking cartilaginous osteoderms.

Like all reptiles, sea turtles have indeterminate growth, i.e., they continue to grow throughout their entire lifetime, which is estimated to be 40–70 years. Growth rates are often rapid in early life stages depending on amount of food available and environmental conditions, but diminish greatly after sexual maturity. As turtles get very old, growth probably becomes negligible. Sea turtles can grow to be quite large; and the leatherback is by far the largest, with most adults ranging between 300 and 500 kg with lengths of over 2 m. The cheloniid turtles are all smaller ranging from the smallest, the Kemp's and olive ridleys, to the flatback, which may weigh as much as 350 kg.

Sea turtles can not retract their necks or limbs into their shell like other turtle species and have evolved highly specialized paddle-like, hind limbs and wing-like front limbs with reduced nails on the forelimbs limited to one or two claw-like growths (**Figure 2**) that are designed to facilitate clasping of the carapace on the female by the male during copulation. Fertilization is internal and sea turtles lay terrestrial nests on sandy beaches in the tropics and subtropics. Adults migrate from feeding areas to nesting beaches at intervals of 1–9 years, distances ranging from hundreds to thousands of nautical kilometers, to lay nests on land. Nest behavior can be solitary or aggregate in phenomena called an 'Arribada' where up to 75 000–350 000 female turtles will emerge to nest on one beach over a period of three to five consecutive evenings (**Figure 3**). Egg clutches of 50–200 eggs are buried in the sand at 20–40 cm below the

Figure 2 Single nail on fore-flipper of a juvenile loggerhead.

Figure 3 Olive ridley arribada at Nancite Costa Rica.

surface. The nests are unattended and hatch 45–70 days later into hatchlings that average 15–30 g in weight. The sex of a hatchling sea turtle is determined by the temperature at which the nest is incubated (temperature dependent sex determination) during the middle trimester of development (critical period). For most sea turtles, temperatures above 29.5 °C result in the development of a female whereas those eggs incubated at temperatures below 29 °C become males. Sea turtles lay cleidoic eggs which are 2–6 cm in diameter and have a typical leathery inorganic shell constructed by the shell gland in the oviduct of the laying females. Nesting is seasonal and controlled by photoperiod cues integrated by the pineal gland and also influenced by the level of nutrition and fat stores available.

Sea turtle physiology is well adapted for deep and shallow diving that can average from 15 min to 2 hours in length. Their tissues contain high levels of respiratory pigments like myoglobin as a reserve oxygen store during the breath hold/dive. Respiratory tidal volumes are quite large (2–6 liters per breath) and allow for rapid washout of carbon dioxide and oxygen uptake during the brief periods spent on the surface breathing (**Figure 4**). Thus in a normal one hour period a sea turtle may spend 50 min under the water and only 10 min at the surface breathing and operating entirely aerobically and accrue no oxygen debt. Cardiovascular adaptations include counter current heat exchangers in the flippers to reduce or enhance heat exchange with the surrounding water. Control of the vascular tree is much higher than the level of arterioles and can permit almost complete restriction of blood flow to all tissues but the heart, brain, central nervous system, and kidney during the deepest dives. They also have evolved temperature and pressure adapted enzymes that will operate well at both the surface and at extreme depths. Leatherbacks are the deepest diving sea turtles; dives of over 750 m in depth have been recorded.

All sea turtles have lacrimal salt glands (**Figure 5**) and reptilian kidneys with short loops of Henle to regulate ion and water levels. Lacrimal salt glands allow sea turtles to drink sea water from birth and maintain water balance despite the high levels of inorganic salts in both their marine diet and the salty ocean water they drink. Diets range from: plant materials like sea grass, *thallasia* and algae for herbivores like the green and black sea turtles; crabs and crustaceans for the ridley turtles; sponges for the loggerhead; and jellyfish and other Cniderians are the sole diet of leatherbacks.

Dermochelyidae (1 genera, 1 species)

The genus *Dermochelys*: *Dermochelys coriacea* (Linne') the leatherback turtle Leatherbacks are the largest (up to 600 kg as adults) and most ancient of the sea turtles diverging from the other turtle families in the Cretaceous. About 20 Ma *Dermochelys* evolved to a body form very similar to that seen today. Their shell consists of cartilaginous osteoderms and they do not have the characteristic laminae and plates found in the plastron and carapace of other sea turtles. The leatherback shell is streamlined with seven cartilaginous narrow ridges on the carapace and five ridges on the plastron that direct water flow in a

Figure 4 Leatherback breathing at surface.

Figure 5   Clear salt gland secretion from base of eye.

Figure 6   Leatherback osteoderm with ridges on carapace.

laminar manner over their entire body (**Figure 6**). The appearance is a distinctive black with white spots, a smooth, scaleless carapace skin and a white smooth-skinned plastron. The head has two saber tooth like projections on the upper beak that overlap the front of the lower beak and serve to pierce the air bladder of floating cniderians that are a large component of their diet when available. Their nesting distribution is worldwide with colonies in Africa, Islands across the Caribbean, Florida, Pacific Mexico, the Pacific and Caribbean coastlines of Central America, South America, Malaysia, New Guinea, and Sri Lanka. Remarkably they are found in subpolar oceans at surface water temperatures of 7–10°C while maintaining a core body temperature above 20°C.

Leatherbacks are pelagic and remain in the open ocean throughout their life feeding on soft-bodied invertebrates such as cniderians, ctenophores, and salps. Average adult females weigh about 300 kg and may take only 8–10 years to attain that size after emerging from their nests at about 24 g. Sex is determined by TSD with a pivotal temperature of 29.5°C. They build a simple cup shaped body pit and lay the largest eggs, about the size of a tennis ball, in clutches of 60–110 yolked eggs. Unlike other sea turtles leatherbacks also deposit 30–60 smaller yolkless eggs that are infertile and only contain albumins. These yolkless eggs tend to be laid late in the nesting process and are primarily on the top of the clutch and may serve as a reservoir for air when hatchlings emerge and congregate prior to the frenzied digging to emerge from their nest chamber.

Unlike other sea turtles leatherbacks crawl on their wrists while on land, rotating both front flippers simultaneously and pushing with both rear flippers in unison. This contrasts with the alternating right to left front and rear flipper crawls of the Cheloniidae. In the ocean their enlarged paddle-like front flippers generate enormous thrust providing excellent propulsion and allowing leatherbacks to migrate an average of 70 km per day when leaving the nesting beaches. Leatherback migrations from Central American beaches are along narrow 'corridors' that are about 100 km wide (**Figure 7**). These oceanic migratory corridors provide important insights into the complex reproductive behavior of these animals. Genetic studies have demonstrated that there is excellent gene flow across all the ocean basins with a strong natal homing and distinct genetic haplotypes in different nesting populations.

Cheloniidae (5 genera, 6 species, 1 race)

The genus *Chelonia*:*Chelonia mydas* (Linne') the green turtle and *Chelonia mydas agassizi* (Bocourt) the black turtle   The genus *Chelonia* contains only one living species the green turtle *Chelonia mydas*

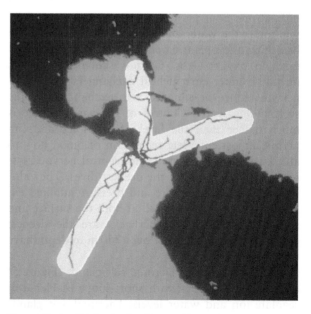

Figure 7   Leatherback migration corridors.

Figure 8 Green turtle.

mydas (Linne') that also has a Pacific–Mexican population that is considered by some researchers as a subspecies or race, the Pacific green turtle or black turtle, *Chelonia mydas agassizi* (Bocourt). The green turtle (**Figure 8**) actually has a brownish colored carapace and scales on the legs, with a yellowish plastron and has one pair of prefrontal head scales with four pairs of lateral laminae. The name 'green turtle' comes from the large greenish fat deposit found under the carapace which is highly desired for the cooking of turtle soup. The black turtle (**Figure 9**) is distinguished by the dark black color of both the carapace and plastron in the adult form. This subspecies or race is found only along the Pacific coastline of Mexico and extends in smaller populations along the Pacific Central American coastline to Panama. *Chelonia* are herbivorous; the green turtle eats marine algae and other marine plants of the genera *Zostera*, *Thallassina*, *Enhaus*, *Posidonia*, and *Halodule* and

Figure 9 Black turtle (with transmitter).

is readily found in the Caribbean in eel-grass (*Zostera*) beds, whereas black turtles rely heavily on red algae and other submerged vegetation. These turtles live in the shallow shoals along the equatorial coastlines as far North as New England in the Atlantic and San Diego in the Pacific and as far south as the Cape of Good Hope in the Atlantic and Chile in the Pacific. The main breeding rookeries include the Coast of Central America, many islands in the West Indies, Ascension Island, Bermuda, the Florida Coast, islands off Sarawak (Malaysia), Vera Cruz coastline in Mexico (center for black turtles), and islands off the coast of Australia. Adult females emerge on sandy beaches at night to construct nests in which they lay 75–200 golf ball-sized leathery eggs. The eggs in these covered nests develop unattended and the young emerge 45–60 days latter. The sex of the hatchlings is determined by the temperature at which the eggs are incubated during the third trimester of development called TSD. Temperatures above 29°C tend to produce all female hatchlings whereas those below 29°C produce males. The hatchlings have a high mortality in the first year and spend a pelagic period before reappearing in juvenile/adult feeding grounds, such as the reefs off Bermuda, the Azores, and Heron Island in the Pacific. As adults they average 150–300 kg with straight line carapace lengths of 65–90 cm. Black turtles tend to be smaller than green turtles. Genetic studies have shown a strong natal homing of the females to the beaches where they were hatched with significant gene flow between rookeries probably due to mating with males from different natal beaches found on common feeding grounds.

The genus *Lepidochelys*: *Lepidochelys kempii* (Garman) the Kemp's ridley, *Lepidochelys olivacea* (Eschsholtz) the olive ridley The Kemp's or Atlantic Ridley and the olive or Pacific Ridley turtles are the two distinct species of this genus. The Kemp's ridley (**Figure 10**) is the rarest of the sea turtles whereas the olive ridley is the most abundant. Anatomically ridleys are the smallest sea turtles. Adult Kemp's usually have five pairs of grey-colored coastal scutes and olive ridleys have 5–9 olive-colored pairs. The adults have a straight line carapace length of 55–70 cm and weigh on average 100–200 kg. Both species have an interesting reproductive behavior in that they have communal nesting called 'arribadas' (Spanish for arrival). There are a number of arribada beaches where females come out by the thousands on sandy beaches 2–6 km long on three or four successive evenings once a month during the nesting seasons.

Figure 10 Juvenile Kemp's ridley.

A percentage of the total population are solitary nesters that emerge on other nearby beaches or on arribada beaches on nights other than the arribadas. These solitary nesters have a very high hatching success of their individual nests (about 80%). It is unknown what proportion of the total population of females nest in this solitary manner and what role they play in the contribution of new recruits into the reproductive adult numbers. Arribadas in Gahirmatha, India have been described with 100 000 turtles emerging in one night to nest communally only inches apart. Many nests are dug up by successive nesting females in the same or subsequent nights of the arribada. The hatching success of these arribada nests is about 4–8% which is the lowest among all sea turtles. Kemp's ridley turtles have arribadas on only one nesting beach in the world, Rancho Nuevo (Caribbean), Mexico. Historically Kemp's arribadas were estimated at 20 000–40 000 female turtles per night in the 1950s, but recent arribadas average only 300–400 individuals per night. This dramatic decline has been attributed to the numbers of adults and juveniles killed in the shrimping nets of the fisheries in the Gulf of Mexico and the American Atlantic coastline. These genera feed primarily on crustaceans, crabs and shrimp, and as a result have been in direct competition with these well-developed fisheries for many years and have suffered dire consequences. Olive ridleys have arribada beaches in Pacific Mexico, Nicaragua, Costa Rica, and along the Bay of Bengal in India. Very little is known about the ocean life of these sea turtle species and it is believed that after hatching they also spend 1–4 years in a pelagic phase associated with floating *Sargassum* and then reappear in the coastal estuaries and neretic zones worldwide at a straight carapace length of about 20–30 cm. Although Kemp's ridley turtles feed primarily on crabs in inshore areas, olive ridley turtles have a more varied diet that includes salps (*Mettcalfina*), jellyfish, fish, benthic invertebrates, mollusks, crabs, shrimp and bryozoans. This more varied diet may account for the different deep ocean and nearshore habitats in which the olive ridleys are found.

The genus *Eretmochelys*: *Eretmochelys imbricata* (Linne'), the hawksbill This species is the most sought after sea turtle for the beauty of the carapace. Historically eyeglass frames and hair combs were made from the carapace of hawksbills. The head is distinguished by a narrow, elongated snout-like mouth and jaw that resembles the beak of a raptor (**Figure 11**). They have four pairs of thick laminae and 11 peripheral bones in the carapace. They tend to be more solitary in their nesting behavior than the other sea turtles but this may be due to their reduced populations. Other than the kemp's ridley turtles they tend to be the smallest turtles averaging 75–150 kg as adults. They are common residents of coral reefs worldwide and appear as juveniles at about 20–25 cm straight length carapace after a pelagic developmental phase in the floating *Sargassum*. What is impressive about the adults is that their diet is 90% sponges. This is one of the few spongivorous animals and electron micrographs of their gastrointestinal tract has shown the microvilli with millions of silica spicules imbedded in the tissue. Tunicates, sea anemone, bryozoans, coelenterates, mollusks, and marine plants have also been found to be important components of the hawksbill diet.

Hawksbills have the largest clutch size of any of the chelonids averaging 130 eggs per nest. They also lay the smallest eggs with a mean diameter of 37.8 mm and mean average weight of 26.6 g. This results in the smallest hatchling of all sea turtles with an average weight of 14.8 g. Other than possibly the

Figure 11 Raptor-like beak of hawksbill.

olive ridley turtle they have the lowest clutch frequency of only 2.74 clutches per nesting season. Hawksbills also tend to be the quickest to construct their nest when out on a nesting beach; they spend only about 45 min to complete the process whereas other turtles take between one to two hours.

Habitats include shallow costal waters and they are readily found on muddy bottoms or on coral reefs. The genetic structure of Atlantic and Pacific populations indicate that hawksbills like other sea turtles have strong natal homing and form distinct nesting populations. They also have TSD and hatchlings also have a pelagic phase before appearing in coastal waters and feeding grounds at about the size of a dinner plate.

Cuba has requested that these turtles be upgraded to CITES Appendix II status which would allow farming of 'local resident populations.' Genetic studies, however, have conclusively demonstrated that these Cuban populations are not local and include turtles from other regions of the Caribbean. These kinds of controversies will increase as human pressures for turtle products, competition for the same food resources such as shrimp, incidental capture due to longline fishing practices, and beach alteration and use by humans increases.

The genus *Caretta*: *Caretta caretta* (Linne'): the loggerhead turtle The loggerhead, like olive ridley turtles, are distinguished by two pairs of prefrontal scales, three enlarged poreless inframarginal laminae, and more than four pairs of lateral laminae. They have a beak like snout that is very broad and not narrow like the hawksbill and they have the largest and broadest head and jaw of the chelonids (**Figure 12**). They nest throughout the Atlantic, Caribbean, Central America, South America, Mediterranean, West Africa, South Africa, through the Indian Ocean, Australia, Eastern and Western Pacific Coastlines.

Loggerheads are primarily carnivorous but have a varied diet that includes crabs (including horseshoe crabs), mollusks, tube worms, sea pens, fish, vegetation, sea pansies, whip corals, sea anemonies, and barnacles and shrimp. Their habitats appear to be quite diverse and they will shift between deeper continental shelf areas and up into shallow river estuaries and lagoons. This sea turtle is the only turtle that has large resident and nesting populations across the North American coastline from Virginia to Florida and the nesting rookeries in Georgia and Florida are currently doing well despite severe human pressures.

Loggerheads that nest on islands near Japan have been shown to develop and grow to adult size along

Figure 12 Broad head of juvenile loggerhead.

the Mexican coastline and then migrate 10 000 km across the Pacific to nest. Genetic studies have confirmed that the Baja Mexico haplotypes are the same as the female adults nesting on these Japanese islands confirming strong natal homing in this genus. Loggerheads make a simple nest at night and lay about 100–120 golf ball-sized eggs. The hatchlings spend a pelagic phase and reappear at 25–30 cm straight carapace length in coastal bays, estuaries and lagoons as well as along the continental shelf and open oceans (**Figure 13**). The orientation and homing of loggerhead hatchlings has been extensively studied and it has been demonstrated that these turtles can detect the direction and intensity of magnetic fields and magnetic inclination angles. This ability together with the use of olfactory cues and chemical imprinting on olfactory cues from natal beaches may account for the natal homing ability of all sea turtles.

The genus *Natator*: *Natator depressus* the flatback This genus has a very limited distribution and is only found in the waters off Australia yet it appears that populations are not endangered at this time. Their nesting beaches are primarily in northern and south-central Queensland. Most of the nesting beaches are quite remote which has protected this species from severe impact by

Figure 13 Juvenile loggerhead turtle.

humans. Apart from the Kemp's ridley this is the only sea turtle to nest in significant numbers during the daytime. It is thought that nighttime nesting has evolved as a behavior to reduce predation and detection but there may also be thermoregulatory considerations. During the daytime in the tropics and subtropics both radiant heat loads from the sun and thermal heat loads from the hot sands may significantly heat the turtles past their critical thermal tolerance. In fact, a number of daytime-nesting flatbacks may die due to overheating if the females do not time their emergence and return to the water to coincide closely with the high tides. If individuals are stranded on the nesting beach when the tides are low and the females must traverse long expanses of beach to emerge or return to the sea during hot and sunny daylight hours, there is a higher potential to overheat and die. On the other hand green turtles in the French Frigate Shoals area of the Pacific are known to emerge on beaches or remain exposed during daylight in shallow tidal lagoons during nonnesting periods and appear to heat up to either aid in digestion or destroy ectoparasites.

Flatback turtles are characterized by a compressed appearance and profile of the carapace with fairly thin and oily scutes. Flatbacks are also distinguished by four pairs of laminae on the carapace and the rim of the shell tends to coil upwards toward the rear. Flatbacks have a head that is very similar to the Kemp's ridley with the exception of a pair of pre-ocular scales between the maxilla and prefrontal scales on the head. These turtles tend to be the largest chelonid, weighing up to 400 kg as adults, and lay the second largest egg with diameters of about 51.5 mm weighing about 51.5 g, smaller only than the leatherback which has a mean egg diameter of 53.4 mm and mean egg weights of 75.9 g. Flatbacks have the smallest clutch size of the chelonids, laying a mean of 53 eggs per clutch, and will lay about three clutches per nesting season. There is only one readily accessible nesting beach for this species at Mon Repos in Queensland, Australia. Other isolated rookeries like Crab Island are found along the Gulf of Carpentaria and Great Barrier Reef.

It is believed that flatbacks may be the only sea turtle that does not have an extended pelagic period in the open ocean. Hatchlings and juveniles spend the early posthatchling stage in shallow, protected coastal waters on the north-eastern Australian continental shelf and Gulf of Carpentaria. Their juvenile and adult diet is poorly known but appears to include snails, soft corals, mollusks, bryozoans, and sea pens.

See also

International Organizations.

Further Reading

Bustard R (1973) *Sea Turtles: Natural History and Conservation.* New York: Taplinger Publishing.

Carr A (1986) *The Sea Turtle: So Excellent a Fishe.* Austin: University of Texas Press.

Carr A (1991) *Handbook of Turtles.* Ithaca, NY: Comstock Publishing Associates of Cornell University Press.

Gibbons JW (1987) Why do turtles live so long? *BioScience* 37(4): 262–269.

Lutz PL and Musick J (eds.) (1997) *The Biology of Sea Turtles.* Boca Raton: CRC Press.

Rieppel O (2000) Turtle origins. *Science* 283: 945–946.

MARINE AQUACULTURE

MARICULTURE OVERVIEW

M. Phillips, Network of Aquaculture Centres in Asia-Pacific (NACA), Bangkok, Thailand

Introduction: Mariculture – A Growing Ocean Industry of Global Importance

Global production from aquaculture has grown substantially in recent years, contributing in evermore significant quantities to the world's supply of fish for human consumption. According to FAO statistics, in 2004, aquaculture production from mariculture was 30.2 million tonnes, representing 50.9% of the global total of farmed aquatic products. Freshwater aquaculture contributed 25.8 million tonnes, or 43.4%. The remaining 3.4 million tonnes, or 5.7%, came from production in brackish environments (**Figure 1**). Mollusks and aquatic plants (seaweeds), on the other hand, almost evenly make up most of mariculture at 42.9% and 45.9%, respectively. These statistics, while accurately reflecting overall trends, should be viewed with some caution as the definition of mariculture is not adopted consistently across the world. For example, when reporting to FAO, it is known that some countries report penaeid shrimp in brackish water, and some in mariculture categories. In this article, we focus on the culture of aquatic animals in the marine environment.

Nevertheless, the amount of aquatic products from farming of marine animals and plants is substantial, and expected to continue to grow. The overall production of mariculture is dominated by Asia, with seven of the top 10 producing countries within Asia. Other regions of Latin America and Europe however also produce significant and growing quantities of farmed marine product. The largest producer of mariculture products by far is China, with nearly 22 million tonnes of farmed marine species. A breakdown of production among the top 15 countries, and major commodities produced, is given in **Table 1**.

Commodity and System Descriptions

A wide array of species and farming systems are used around the world for farming aquatic animals and plants in the marine environment. The range of marine organisms produced through mariculture currently include seaweeds, mollusks, crustaceans, marine fish, and a wide range of other minor species such as sea cucumbers and sea horses. Various containment or holding facilities are common to marine ecosystems, including sea-based pens, cages, stakes, vertical or horizontal lines, afloat or bottom set, and racks, as well as the seabed for the direct broadcast of clams, cockles, and similar species.

Mollusks

Many species of bivalve and gastropod mollusks are farmed around the world. Bivalve mollusks are the major component of mariculture production, with most, such as the commonly farmed mussel, being high-volume low-value commodities. At the other end of the value spectrum, there is substantial production of pearls from farming, an extremely low-volume but high-value product.

Despite the fact that hatchery production technologies have been developed for many bivalves and gastropods, much bivalve culture still relies on collection of seedstock from the wild. Artificial settlement substrates, such as bamboo poles, wooden stakes, coconut husks, or lengths of frayed rope, are used to collect young bivalves, or spat, at settlements. The spat are then transferred to other grow-out substrates ('relayed'), or cultured on the settlement substrate.

Some high-value species (such as the abalone) are farmed in land-based tanks and raceways, but most mollusk farming takes place in the sea, where three major systems are commonly used:

- Within-particulate substrates – this system is used to culture substrate-inhabiting cockles, clams, and other species. Mesh covers or fences may be used to exclude predators.

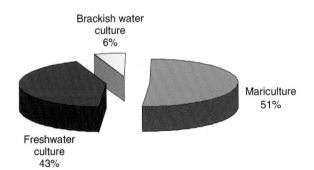

Figure 1 Aquaculture production by environment in 2004. From FAO statistics for 2004.

Brackish water culture 6%

Mariculture 51%

Freshwater culture 43%

Table 1   Top 15 mariculture producers

Country	Production (tonnes)	Major species/commodities farmed in mariculture
China	21 980 595	Seaweeds (kelp, wakame), mollusks (oysters, mussels, scallops, cockles, etc.) dominate, but very high diversity of species cultured, and larger volumes of marine fish and crustaceans
Philippines	1 273 598	Seaweeds (*Kappaphycus*), with small quantities of fish (milkfish and groupers) and mollusks
Japan	1 214 958	Seaweeds, mollusks, marine fish
Korea, Republic of	927 557	Seaweeds, mollusks, marine fish
Chile	688 798	Atlantic salmon, other salmonid species, smaller quantities of mollusks, and seaweed
Norway	637 993	Atlantic salmon, other salmonid species
Korea, Dem. People's Rep.	504 295	Seaweeds, mollusks
Indonesia	420 919	Seaweeds, smaller quantities of marine fish, and pearls
Thailand	400 400	Mollusks (cockle, oyster, green mussel), small quantities of marine fish (groupers)
Spain	332 062	Blue mussel dominates, but high diversity of fish and mollusks
United States of America	242 937	Mollusks (oysters) with small quantities of other mollusks and fish (salmon)
France	198 625	Mollusks (mussel and oyster) with small quantities of other mollusks and fish (salmon)
United Kingdom	192 819	Atlantic salmon and mussels
Vietnam	185 235	Seaweeds (*Gracilaria*) and mollusks
Canada	134 699	Atlantic salmon and other salmonid species, and mollusks
Total (all countries)	30 219 472	

From FAO statistics for 2004.

- On or just above the bottom – this culture system is commonly used for culture of bivalves that tolerate intertidal exposure, such as oysters and mussels. Rows of wooden or bamboo stakes are arranged horizontally or vertically. Bivalves may also be cultured on racks above the bottom in mesh boxes, mesh baskets, trays, and horizontal wooden and asbestos-cement battens.
- Surface or suspended culture – bivalves are often cultured on ropes or in containers, suspended from floating rafts or buoyant long-lines.

Management of the mollusk cultures involves thinning the bivalves where culture density is too high to support optimal growth and development, checking for and controlling predators, and controlling biofouling. Mollusk production can be very high, reaching 1800 tonnes per hectare annually. With a cooked meat yield of around 20%, this is equivalent to 360 tonnes of cooked meat per hectare per year, an enormous yield from a limited water area. Farmed bivalves are commonly sold as whole fresh product, although some product is simply processed, for example, shucked and sold as fresh or frozen meat. There has been some development of longer-life products, including canned and pickled mussels.

Because of their filter-feeding nature, and the environments in which they are grown, edible bivalves are subject to a range of human health concerns, including accumulation of heavy metals, retention of human-health bacterial and viral pathogens,

and accumulation of toxins responsible for a range of shellfish poisoning syndromes. One option to improve the product quality of bivalves is depuration, which is commonly practiced with temperate mussels, but less so in tropical areas.

Seaweeds

Aquatic plants are a major production component of mariculture, particularly in the Asia-Pacific region. About 13.6 million tonnes of aquatic plants were produced in 2004. China is the largest producer, producing just less than 10 million tonnes. The dominant cultured species is Japanese kelp *Laminaria japonica*. There are around 200 species of seaweed used worldwide, or which about 10 species are intensively cultivated – including the brown algae *L. japonica* and *Undaria pinnatifida*; the red algae *Porphyra*, *Eucheuma*, *Kappaphycus*, and *Gracilaria*; and the green algae *Monostrema* and *Enteromorpha*.

Seaweeds are grown for a variety of uses, including direct consumption, either as food or for medicinal purposes, extraction of the commercially valuable polysaccharides alginate and carrageenan, use as fertilizers, and feed for other aquaculture commodities, such as abalone and sea urchins.

Because cultured seaweeds reproduce vegetatively, seedstock is obtained from cuttings. Grow-out is undertaken using natural substrates, such as long-lines, rafts, nets, ponds, or tanks.

Production technology for seaweeds is inexpensive and requires only simple equipment. For this reason, seaweed culture is often undertaken in relatively undeveloped areas where infrastructure may limit the development of other aquaculture commodities, for example, in the Pacific Island atolls.

Seaweeds can be grown using simple techniques, but are also subject to a range of physiological and pathological problems, such as 'green rot' and 'white rot' caused by environmental conditions, 'ice-ice' disease, and epiphyte growth. In addition, cultured seaweeds are often consumed by herbivores, particularly sea urchins and herbivorous fish species, such as rabbitfish.

Selective breeding for specific traits has been undertaken in China to improve productivity, increase iodine content, and increase thermal tolerance to better meet market demands. More recently, modern genetic manipulation techniques are being used to improve temperature tolerance, increase agar or carrageenan content, and increase growth rates. Improved growth and environmental tolerance of cultured strains is generally regarded as a priority for improving production and value of cultured seaweeds in the future.

Seaweed aquaculture is well suited for small-scale village operations. Seaweed fisheries are traditionally the domain of women in many Pacific island countries, so it is a natural progression for women to be involved in seaweed farming. In the Philippines and Indonesia, seaweed provides much-needed employment and income for many thousands of farmers in remote coastal areas.

Marine Finfish

Marine finfish aquaculture is well established globally, and is growing rapidly. A wide range of species is cultivated, and the diversity of culture is also steadily increasing.

In the Americas and northern Europe, the main species is the Atlantic salmon (*Salmo salar*), with smaller quantities of salmonids and species. Chile, in particular, has seen the most explosive growth of salmon farming in recent years, and is poised to become the number-one producer of Atlantic salmon. In the Mediterranean, a range of warmer water species are cultured, such as seabass and seabream.

Asia is again the major producer of farmed marine fish. The Japanese amberjack *Seriola quinqueradiata* is at the top of the production tables, with around 160 000 tonnes produced in 2004, but the region is characterized by the extreme diversity of species farmed, in line with the diverse fish-eating habits of the people living in the region. Seabreams are also common, with barramundi or Asian seabass (*Lates calcarifer*) cultured in both brackish water and mariculture environments. Grouper culture is expanding rapidly in Asia, driven by high prices in the live-fish markets of Hong Kong SAR and China, and the decreasing availability of wild-caught product due to overfishing. Southern bluefin tuna (*Thunnus mccoyii*) is cultured in Australia using wild-caught juveniles. Although production of this species is relatively small (3500–4000 tonnes per annum in 2001–03), it brings very high prices in the Japanese market and thus supports a highly lucrative local industry in South Australia. The 2003 production of 3500 tonnes was valued at US$65 million.

Hatchery technologies are well developed for most temperate species (such as salmon and seabream) but less well developed for tropical species such as groupers where the industry is still reliant on collection of wild fingerlings, a concern for future sustainability of the sector.

The bulk of marine fish are presently farmed in net cages located in coastal waters. Most cultured species are carnivores, leading to environmental concerns over the source of feed for marine fish farms, with most still heavily reliant on wild-caught so-called 'trash' fish. Excessive stocking of cages in coastal waters also leads to concern over water and sediment pollution, as well as impacts from escapes and disease transfer on wild fish populations.

Crustaceans

Although there is substantial production of marine shrimps globally, this production is undertaken in coastal brackish water ponds and thus does not meet the definition of mariculture. There has been some experimental culture of shrimp in cages in the Pacific, but this has not yet been commercially implemented. Tropical spiny rock lobsters and particularly the ornate lobster *Panulirus ornatus* are cultured in Southeast Asia, with the bulk of production in Vietnam and the Philippines. Lobster aquaculture in Vietnam produces about 1500 tonnes valued at around US$40 million per annum. Tropical spiny rock lobsters are cultured in cages and fed exclusively on fresh fish and shellfish.

In the medium to long term, it is necessary to develop hatchery production technology for seedstock for tropical spiny rock lobsters. There is currently considerable research effort on developing larval-rearing technologies for tropical spiny rock lobsters in Southeast Asia and in Australia. As in the case of tropical marine fish farming, there is also a need to develop less-wasteful and less-polluting diets to replace the use of wild-caught fish and shellfish as diets.

Other Miscellaneous Invertebrates

There are a range of other invertebrates being farmed in the sea, such as sea cucumbers, sponges, corals, sea horses, and others. Farming of some species has been ongoing for some time, such as the well-developed sea cucumber farming in northern China, but others are more recent innovations or still at the research stage. Sponge farming, for example, is generating considerable interest in the research community, but commercial production of farmed sponges is low, mainly in the Pacific islands. This farming is similar to seaweed culture as sponges can be propagated vegetatively, with little infrastructure necessary to establish farms. The harvested product, bath sponges, can be dried and stored and, like seaweed culture, may be ideal for remote communities, such as those found among the Pacific islands.

Environmental Challenges

Environmental Impacts

Mariculture is an important economic activity in many coastal areas but is facing a number of environmental challenges because of the various environmental 'goods' and 'services' required for its development. The many interactions between mariculture and the environment include impacts of: (1) the environment on mariculture; (2) mariculture on the environment; and (3) mariculture on mariculture.

The environment impacts on mariculture through its effects on water, land, and other resources necessary for successful mariculture. These impacts may be negative or positive, for example, water pollution may provide nutrients which are beneficial to mariculture production in some extensive culture systems, but, on the other hand, toxic pollutants and pathogens can be extremely damaging. An example is the farming of oysters and other filter-feeding mollusks which generally grow faster in areas where nutrient levels are elevated by discharge of wastewater from nearby centers of human population. However, excessive levels of human and industrial waste cause serious problems for mollusk culture, such as contamination with pathogens and toxins from dinoflagellates. Aquaculture is highly sensitive to adverse environmental changes (e.g., water quality and seed quality) and it is therefore in the long-term interests of mariculture farmers and governments to work toward protection and enhancement of environmental quality. The effects of global climate change, although poorly understood in the fishery sector, are likely to have further significant influences on future mariculture development.

The impacts of mariculture on the environment include the positive and negative effects farming operations may have on water, land, and other resources required by other aquaculturists or other user groups. Impacts may include loss or degradation of natural habitats, water quality, and sediment changes; overharvesting of wild seed; and introduction of disease and exotic species and competition with other sectors for resources. In increasingly crowded coastal areas, mariculture is running into more conflicts with tourism, navigation, and other coastal developments.

Mariculture can have significant positive environmental impacts. The nutrient-absorbing properties of seaweeds and mollusks can help improve coastal water quality. There are also environmental benefits from restocking of overfished populations or degraded habitats, such as coral reefs. For example, farming of high-value coral reef species is being seen as one means of reducing threats associated with overexploitation of threatened coral reef fishes traditionally collected for food and the ornamental trade.

Finally, mariculture development may also have an impact on itself. The rapid expansion in some areas with limited resources (e.g., water and seed) has led to overexploitation of these resources beyond the capacity of the environment to sustain growth, followed by an eventual collapse. In mariculture systems, such problems have been particularly acute in intensive cage culture, where self-pollution has led to disease and water-quality problems which have undermined the sustainability of farming, from economic and environmental viewpoints. Such problems emphasize the importance of environmental sustainability in mariculture management, and the need to minimize overharvesting of resources and hold discharge rates within the assimilative capacity of the surrounding environment.

The nature and the scale of the environmental interactions of mariculture, and people's perception of their significance, are also influenced by a complex interaction of different factors, such as follows:

- The technology, farming and management systems, and the capacity of farmers to manage technology. Most mariculture technology, particularly in extensive and semi-intensive farming systems, such as mollusk and seaweed farming, and well-managed intensive systems, is environmentally neutral or low in impact compared to other food production sectors.

- The environment where mariculture farms are located (i.e., climatic, water, sediment, and biological features), the suitability of the environment for the cultured animals and the environmental conditions under which animals and plants are cultured.
- The financial and economic feasibility and investment, such as the amount invested in proper farm infrastructure, short- versus long-term economic viability of farming operations, and investment and market incentives or disincentives, and the marketability of products.
- The sociocultural aspects, such as the intensity of resource use, population pressures, social and cultural values, and aptitudes in relation to aquaculture. Social conflicts and increasing consumer perceptions all play an important role.
- The institutional and political environment, such as government policy and the legal framework, political interventions, plus the scale and quality of technical extension support and other institutional and noninstitutional factors.

These many interacting factors make both understanding environmental interactions and their management (as in most sectors – not just mariculture) both complex and challenging.

Environmental Management of Mariculture

The sustainable development of mariculture requires adoption of management strategies which enhance positive impacts (social, economic, and environmental impacts) and mitigate against environmental impacts associated with farm siting and operation. Such management requires consideration of: (1) the farming activity, for example, in terms of the location, design, farming system, investment, and operational management; (2) the 'integration' of mariculture into the surrounding coastal environment; and (3) supporting policies and legislation that are favorable toward sustainable development.

Technology and Farming Systems Management

The following factors are of crucial importance in environmental management at the farm level:

- *Farm siting.* The sites selected for aquaculture and the habitat at the farm location play one of the most important roles in the environmental and social interactions of aquaculture. Farm siting is also crucial to the sustainability of an investment; incorrect siting (e.g., cages located in areas with unsuitable water quality) often lead to increased

investment costs associated with operation and amelioration of environmental problems. Farms are better sited away from sensitive habitats (e.g., coral reefs) and in areas with sufficient water exchange to maintain environmental conditions. Problems of overstocking of mollusk culture beds are recognized in the Republic of Korea, for example, where regulations have been developed to restrict the areas covered by mollusk culture. For marine cage culture, one particularly interesting aspect of siting is the use of offshore cages, and new technologies developed in European countries are now attracting increasing interest in Asia.

- *Farm construction and design features.* Farm construction and system design has a significant influence on the impact of mariculture operations on the environment. Suitable design and construction techniques should be used when establishing new farms, and as far as possible seek to cause minimum disturbance to the surrounding ecosystems. The design and operation of aquaculture farms should also seek to make efficient use of natural resources used, such as energy and fuel. This approach is not just environmentally sound, but also economic because of increasing energy costs.

- *Water and sediment management.* Development of aquaculture should minimize impacts on water resources, avoiding impacts on water quality caused by discharge of farm nutrients and organic material. For sea-based aquaculture, where waste materials are discharged directly into the surrounding environment, careful control of feed levels and feed quality is the main method of reducing waste discharge, along with good farm siting. In temperate aquaculture, recent research has been responsible for a range of technological and management innovations – low-pollution feeds and novel self-feeding systems, lower stocking densities, vaccines, waste-treatment facilities – that have helped reduce environmental impacts. Complex models have also been developed to predict environmental impacts, and keep stocking levels within the assimilative capacity of the surrounding marine environment. In mariculture, there are also examples of integrated, polyculture, and alternate cropping farming systems that help to reduce impacts. For example in China and Korea, polyculture on sea-based mollusk and seaweed farms is practiced and for more intensive aquaculture operations, effluent rich in nutrients and microorganisms, is potentially suitable for culturing fin fish, mollusks, and seaweed.

- *Suitable species and seed.* A supply of healthy and quality fish, crustacean, and mollusk seed is essential for the development of mariculture. Emphasis should be given to healthy and quality hatchery-reared stock, rather than collection from the wild. Imports of alien species require import risk assessment and management, to reduce risks to local aquaculture industries and native biodiversity.

- *Feeds and feed management.* Access to feeds, and efficient use of feeds is of critical importance for a cost-effective and environmentally sound mariculture industry. This is due to many factors, including the fact that feeds account for 50% or more of intensive farming costs. Waste and uneaten feed can also lead to undesirable water pollution. Increasing concern is also being expressed about the use of marine resources (fish meal as ingredients) for aquaculture feeds. One of the biggest constraints to farming of carnivorous marine fish such as groupers is feed. The development of sustainable supplies of feed needs serious consideration for future development of mariculture at a global level.

- *Aquatic animal health management.* Aquatic animal and plant diseases are a major cause of unsustainability, particularly in more intensive forms of mariculture. Health management practices are necessary to reduce disease risks, to control the entry of pathogens to farming systems, maintain healthy conditions for cultured animals and plants, and avoid use of harmful disease control chemicals.

- *Food safety.* Improving the quality and safety of aquaculture products and reducing risks to ecosystems and human health from chemical use and microbiological contamination is essential for modern aquaculture development, and marketing of products on domestic and international markets. Normally, seafood is considered healthy food but there are some risks associated with production and processing that should be minimized. The two food-safety issues, that can also be considered environmental issues, are chemical and biological. The chemical risk is associated with chemicals applied in aquaculture production and the biological is associated with bacteria or parasites that can be transferred to humans from the seafood products. Increasing calls for total traceability of food products are also affecting the food production industry such that consumers can be assured that the product has been produced without addition of undesirable or harmful chemicals or additives, and that the environments and ecosystems affected by the production facilities have not been compromised in any way.

- *Economic and social/community aspects.* The employment generated by mariculture can be highly significant, and globally aquaculture has become an important employer in remote and poor coastal communities. Poorly planned mariculture can also lead to social conflicts, and the future development and operation of mariculture farms must also be done in a socially responsible manner, as far as possible without compromising local traditions and communities, and ideally contributing positively. The special traditions of many coastal people and their relation with the sea in many places deserve particularly careful attention in planning and implementation of mariculture.

Planning, Policy, and Legal Aspects

Integrated Coastal Area Management

Effective planning processes are essential for sustainable development of mariculture in coastal areas. Integrated coastal area management (ICAM) is a concept that is being given increasing attention as a result of pressures on common resources in coastal areas arising from increasing populations combined with urbanization, pollution, tourism, and other changes. The integration of mariculture into the coastal area has been the subject of considerable recent interest, although practical experience in implementation is still limited in large measure because of the absence of adequate policies and legislation and institutional problems, such as the lack of unitary authorities with sufficiently broad powers and responsibilities.

Zoning of aquaculture areas within the coastal area is showing some success. In China, Korea, Japan, Hong Kong, and Singapore, there are now well-developed zoning regulations for water-based coastal aquaculture operations (marine cages, mollusks, and seaweeds). For example, Hong Kong has 26 designated 'marine fish culture zones' within which all marine fish-culture activities are carried out. In the State of Hawaii, 'best areas' for aquaculture have been identified, and in Europe zoning laws are being strictly applied to many coastal areas where aquaculture is being developed. Such an approach allows for mariculture to be developed in designated areas, reducing risks of conflicts with other coastal zone users and uses.

Policy and Legal Issues

While much can be done at farm levels and by integrated coastal management, government involvement through appropriate policy and legal instruments is important in any strategy for mariculture sustainability. Some of the important issues include legislation, economic incentives/disincentives, private sector/community participation in policy formulation, planning processes, research and knowledge transfer, balance between food and export earnings, and others.

While policy development and most matters of mariculture practice have been regarded as purely national concerns, they are coming to acquire an increasingly international significance. The implication of this is that, while previously states would look merely to national priorities in setting mariculture policy, particularly legislation/standards, for the future it will be necessary for such activities to take account of international requirements, including various bilateral and multilateral trade policies. International standards of public health for aquaculture products and the harmonization of trade controls are examples of this trend.

Government regulations are an important management component in maintaining environmental quality, reducing negative environmental impacts, and allocating natural resources between competing users and integration of aquaculture into coastal area management. Mariculture is a relative newcomer among many traditional uses of natural resources and has commonly been conducted within an amalgam of fisheries, water resources, and agricultural and industrial regulations. It is becoming increasingly clear that specific regulations governing aquaculture are necessary, not least to protect aquaculture development itself. Key issues to be considered in mariculture legislation are farm siting, use of water area and bottom in coastal and offshore waters; waste discharge, protection of wild species, introduction of exotic or nonindigenous species, aquatic animal health; and use of drugs and chemicals.

Environmental impact assessment (EIA) can also be an important legal tool which is being more widely applied to mariculture. The timely application of EIA (covering social, economic, and ecological issues) to larger-scale coastal mariculture projects can be one way to properly identify environmental problems at an early phase of projects, thus enabling proper environmental management measures to be incorporated in project design and management. Such measures will ultimately make the project more sustainable. A major difficulty with EIAs is that they are difficult (and generally impractical) to apply to smaller-scale mariculture developments, common throughout many parts of Asia, and do not easily take account of the potential cumulative effects of many small-scale farms. Strategic environmental assessment (SEA) can provide a broader means of assessing impacts.

Conclusion

Mariculture is and will increasingly become an important producer of aquatic food in coastal areas, as well as a source of employment and income for many coastal communities. Well-planned and -managed mariculture can also contribute positively to coastal environmental integrity. However, mariculture's future development will occur, in many areas, with increasing pressure on coastal resources caused by rising populations, and increasing competition for resources. Thus, considerable attention will be necessary to improve the environmental management of aquaculture through environmentally sound technology and better management, supported by effective policy and planning strategies and legislation.

See also

Mariculture, Economic and Social Impacts.

Further Reading

Clay J (2004) *World Aquaculture and the Environment. A Commodity by Commodity Guide to Impacts and Practices*. Washington, DC: Island Press.

FAO/NACA/UNEP/WB/WWF (2006) *International Principles for Responsible Shrimp Farming*, 20pp. Bangkok, Thailand: Network of Aquaculture Centres in Asia-Pacific (NACA). http://www.enaca.org/uploads/international-shrimp-principles-06.pdf (accessed Apr. 2008).

Hansen PK, Ervik A, Schaanning M, et al. (2001) Regulating the local environmental impact of intensive, marine fish farming-II. The monitoring programme of the MOM system (Modelling-Ongrowing fish farms-Monitoring). *Aquaculture* 194: 75–92.

Hites RA, Foran JA, Carpenter DO, Hamilton MC, Knuth BA, and Schwager SJ (2004) Global assessment of organic contaminants in farmed salmon. *Science* 303: 226–229.

Joint FAO/NACA/WHO Study Group (1999) Food safety issues associated with products from aquaculture. *WHO Technical Report Series 883*. http://www.who.int/foodsafety/publications/fs_management/en/aquaculture.pdf (accessed Apr. 2008).

Karakassis I, Pitta P, and Krom MD (2005) Contribution of fish farming to the nutrient loading of the Mediterranean. *Scientia Marina* 69: 313–321.

NACA/FAO (2001) Aquaculture in the third millennium. In: Subasinghe RP, Bueno PB, Phillips MJ, Hough C, McGladdery SE, and Arthur JR (eds.) *Technical*

Proceedings of the Conference on Aquaculture in the Third Millennium. Bangkok, Thailand, 20–25 February 2000, 471pp. Bangkok, NACA and Rome: FAO.

Naylor R, Hindar K, Flaming IA, *et al.* (2005) Fugitive salmon: Assessing the risks of escaped fish from net-pen aquaculture. *BioScience* 55: 427–473.

Neori A, Chopin T, Troell M, *et al.* (2004) Integrated aquaculture: Rationale, evolution and state of the art emphasizing sea-weed biofiltration in modern mariculture. *Aquaculture* 231: 361–391.

Network of Aquaculture Centres in Asia-Pacific (2006) Regional review on aquaculture development. 3. Asia and the Pacific – 2005. *FAO Fisheries Circular No. 1017/3,* 97pp. Rome: FAO.

Phillips MJ (1998) Tropical mariculture and coastal environmental integrity. In: De Silva S (ed.) *Tropical Mariculture,* pp. 17–69. London: Academic Press.

Pillay TVR (1992) *Aquaculture and the Environment,* 158pp. London: Blackwell.

Secretariat of the Convention on Biological Diversity (2004) Solutions for sustainable mariculture – avoiding the adverse effects of mariculture on biological diversity, *CBD Technical Series No. 12.* http://www.biodiv.org/doc/publications/cbd-ts-12.pdf (accessed Apr. 2008).

Tacon AJC, Hasan MR, and Subasinghe RP (2006) Use of fishery resources as feed inputs for aquaculture development: Trends and policy implications. *FAO Fisheries Circular No. 1018.* Rome: FAO.

World Bank (2006) *Aquaculture: Changing the Face of the Waters. Meeting the Promise and Challenge of Sustainable Aquaculture.* Report no. 36622. Agriculture and Rural Development Department, the World Bank. http://siteresources.worldbank.org/INTARD/Resources/Aquaculture_ESW_vGDP.pdf (accessed Apr. 2008).

Relevant Websites

http://www.pbs.org
– Farming the Seas, Marine Fish and Aquaculture Series, PBS.

http://www.fao.org/fi
– Food and Agriculture Organisation of the United Nations.

http://www.cbd.int
Jakarta Mandate, Marine and Coastal Biodiversity: Mariculture, Convention on Biological Diversity.

http://www.enaca.org
– Network of Aquaculture Centres in Asia-Pacific.

http://www.seaplant.net
– The Southeast Asia Seaplant Network.

http://www.oceansatlas.org
– UN Atlas of the Oceans.

MARICULTURE DISEASES AND HEALTH

A. E. Ellis, Marine Laboratory, Aberdeen, Scotland,
UK

Introduction

As with all forms of intensive culture where a single
species is reared at high population densities, infectious disease agents are able to transmit easily
between host individuals and large economic losses
can result from disease outbreaks. Husbandry
methods are designed to minimize these losses by
employing a variety of strategies, but central to all of
these is providing the cultured animal with an optimal environment that does not jeopardize the animal's health and well-being. All animals have innate
and acquired defenses against infectious agents and
when environmental conditions are good for the
host, these defense mechanisms will provide protection against most infections. However, animals
under stress have less energy available to combat
infections and are therefore more prone to disease.
Although some facilities on a farm may be able to
exclude the entry of pathogens, for example hatcheries with disinfected water supplies, it is impossible
to exclude pathogens in an open marine situation.
Under these conditions, stress management is para-
mount in maintaining the health of cultured animals.
Even then, because of the close proximity of individuals in a farm, if certain pathogens do gain entry
they are able to spread and multiply extremely rapidly and such massive infectious burdens can overcome the defenses of even healthy animals. In such
cases some form of treatment, or even better,
prophylaxis, is required to prevent crippling losses.
This article describes some of the management
strategies available to fish and shellfish farmers in
avoiding or reducing the losses from infectious diseases and some of the prophylactic measures and
treatments. The most important diseases encountered in mariculture are summarized in **Table 1.**

Health Management

Facility Design

Farms and husbandry practices can be designed in
such a way as to avoid the introduction of pathogens
and to restrict their spread within a farm in a variety
of ways.

Isolate the hatchery Infectious agents can be
excluded from hatcheries by disinfecting the
incoming water using filters, ultraviolet lamps or
ozone treatments. It is also important not to
introduce infections from other parts of the farm
that may be contaminated. The hatchery then
should stand apart and strict hygiene standards
applied to equipment and personnel entering the
hatchery. Some diseases cause major mortalities in
young fry while older fish are more resistant. For
example, infectious pancreatic necrosis virus
(IPNV) causes mass mortality in halibut fry, but
juveniles are much more resistant. It is vitally
important therefore, to exclude the entry of IPNV
into the halibut hatchery and as this virus has a
widespread distribution in the marine environment,
disinfection of the water supply may be necessary.

Hygiene practice Limiting the spread of disease
agents on a farm include having hand nets for each
tank and disinfecting the net after each use,
disinfectant foot-baths at the farm entrance and
between buildings, and restricted movement of staff,
their protective clothing and equipment. Prompt
removal of dead and moribund stock is essential as
large numbers of pathogens are shed into the water
from such animals. In small tanks this can easily be
done using a hand net. In large sea cages, lifting the
net can be stressful to fish and divers are expensive.
Special equipment such as air-lift pumps, or specially
designed cages with socks fitted in the bottom in
which dead fish collect and which can be hoisted out
on a pulley are more practical. Proper disposal of
dead animals is essential. Methods such as
incineration, rendering, ensiling and, on a small scale,
in lime pits are recommended.

Husbandry and Minimizing Stress

Animals under stress are more prone to infectious
diseases. However, it is not possible to eliminate all
the procedures that are known to induce stress in
aquaculture animals, as many are integral parts of
aquaculture, e.g., netting, grading, and transport.
Nevertheless, it is possible for farming practices to
minimize the effects of these stressors and others,
e.g., overcrowding and poor water quality, can be
avoided by farmers adhering to the recommended

Table 1 Principal diseases of fish and shellfish in mariculture

Disease agent	Host	Prevention/treatment
Shellfish pathogens		
Protozoa		
Bonamia ostreae	European flat oyster	Exclusion
Marteilia refringens	Oyster, mussel	Exclusion
Bacteria:		
Aerococcus viridans (Gaffkaemia).	Lobsters	Improve husbandry
Fin-fish pathogens		
Viruses		
Infectious pancreatic necrosis	Salmon, turbot, halibut, cod, sea bass, yellowtail	Sanitary precautions, vaccinate
Infectious salmon anemia	Atlantic salmon	Sanitary precautions, eradicate
Pancreas disease	Atlantic salmon	Avoid stressors, vaccinate
Viral hemorrhagic septicemia	Turbot	Sanitary precautions
Viral nervous necrosis	Stripped jack, Japanese flounder, barramundi, sea bass, sea bream, turbot, halibut	Sanitary precautions
Bacteria		
Vibriosis (*Vibrio anguillarum*)	All marine species	Vaccinate, antibiotics
Cold water vibriosis (*Vibrio salmonicida*)	Salmon, trout, cod	Vaccinate, antibiotics
Winter ulcers (*Vibrio viscosus*)	Salmon	Avoid stress, vaccinate
Typical furunculosis (*Aeromonas salmonicida*)	Salmon	Vaccinate, antibiotics
Atypical furunculosis (*Aeromonas salmonicida achromogenes*)	Turbot, halibut, flounder, salmon	Vaccinate, antibiotics, avoid stress
Flexibacter maritimus	Sole, flounder, turbot, salmon, sea bass	Avoid stress, antibiotics
Enteric redmouth (*Yersinia ruckeri*)	Atlantic salmon	Vaccinate
Bacterial kidney disease (*Renibacterium salmoninarum*)	Pacific salmon	Exclude
Mycobacteriosis (*Mycobacterium marinum*)	Salmonids, sea bass, sea bream, stripped bass, cod, red drum, tilapia	Sanitary
Pseudotuberculosis (*Photobacterium damselae piscicida*)	Yellowtail, sea bass, sea bream	Vaccinate, sanitary
Piscirickettsiosis (*Piscirickettsia salmonis*)	Salmonids	Sanitary
Parasites		
Protozoan: many species; *Amyloodinium, Cryptobia, Ichthyobodo, Cryptocaryon, Tricodina*	Most species	Avoid stress, sanitary, chemical baths, e.g., formalin
Paramebic gill disease	Salmonids	Low salinity bath, H_2O_2
Crustacea: Sea lice	Salmonids, sea bass, sea bream	In-feed insecticides, H_2O_2, cleaner fish

limits for stocking densities, water flow rates and feeding regimes. In cases where stressors are unavoidable, farmers can adopt certain strategies to minimize the stress.

Withdrawal of food prior to handling Following feeding the oxygen requirement of fish is increased. Withdrawal of food two or three days prior to handling the fish will therefore minimize respiratory stress. It also avoids fouling of the water during transportation with fecal material and regurgitated food.

Use of anesthesia Although anesthetics can disturb the physiology of fish, light anesthesia can have a calming effect on fish during handling and transport and so reduce the stress resulting from these procedures.

Avoidance of stressors at high temperatures High temperatures increase the oxygen demand of animals and stress-induced mortality can result from respiratory failure at high temperatures. It is therefore safer to carry out netting, grading and transport at low water temperatures.

Avoidance of multiple stressors and permitting recovery The effects of multiple stressors can be additive or even synergistic so, for instance, sudden changes of temperature should be avoided during or after transport. Where possible, recovery from stress should be facilitated. Generally, the duration of the recovery period is proportional to the duration of the stressor. Thus, reducing the time of netting, grading or transport will result in recovery in a shorter time. The duration required for recovery to occur may be from a few days to two weeks.

Selective breeding In salmonids, it is now established that the magnitude of the stress response is a heritable characteristic and programs now exist for selecting broodstock which have a low stress response to handling stressors. This accelerates the process of domestication to produce stocks, which are more tolerant of aquaculture procedures with resultant benefits in increased health, survival and productivity. Such breeding programs have been conducted in Norway for some years and have achieved improvements in resistance to furunculosis and infectious salmon anemia virus (ISAV) in Atlantic salmon.

Management of the Pathogen

Breaking the pathogen's life cycle If a disease is introduced on to a farm, it is important to restrict the horizontal spread especially to different year classes of fish/shellfish. Hence, before a new year class of animals is introduced to a part of the farm, all tanks, equipment etc. should be thoroughly cleaned and disinfected. In sea-cage sites it is a useful technique to physically separate year classes to break the infection cycle.

Many pathogens do not survive for long periods of time away from their host and allowing a site to be fallow for a period of time may eliminate or drastically reduce the pathogen load. This practice has been very effective in controlling losses from furunculosis in marine salmon farms and also significantly reduces the salmon lice populations particularly in the first year after fallowing.

Eliminate vertical transmission Several pathogens may persist as an asymptomatic carrier state and be present in the gonadal fluids of infected broodstock and can infect the next generation of fry. In salmon farming IPNV may persist in or on the ova and disinfection of the eggs with iodine-based disinfectants (Iodophores) is recommended immediately after fertilization.

The testing of gonadal fluids for the presence of IPNV can also be carried out and batches of eggs from infected parents destroyed. IPNV has been associated with mass mortalities in salmon and halibut fry and has been isolated from a wide range of marine fish and shellfish, including sea bass, turbot, striped bass, cod, and yellowtail.

Avoid infected stock Many countries employ regulations to prevent the movement of eggs or fish that are infected with certain 'notifiable' diseases from an infected to a noninfected site. These policies are designed to limit the spread of the disease but require specialized sampling procedures and laboratory facilities to perform the diagnostic techniques. By testing the stock frequently they can be certified to be free from these diseases. Such certification is required for international trade in live fish and eggs but within a country it is widely practiced voluntarily because stock certified to be 'disease-free' command premium prices.

Eradication of infected stock Commercially this is a drastic step to take especially when state compensation is not usually available even when state regulations might require eradication of stock. This policy is usually only practiced rarely and when potential calamitous circumstances may result, for instance, the introduction of an important exotic pathogen into an area previously free of that disease. This has occurred in Scotland where the European Commission directives have required compulsory slaughter of turbot infected with viral hemorrhagic septicemia virus (VHSV) and Atlantic salmon infected with ISAV.

Treatments

Viral Diseases

There are no treatments available for viral diseases in aquaculture. These diseases must be controlled by husbandry and management strategies as described above, or by vaccination (see below).

Bacterial Diseases

Antibiotics can be used to treat many bacterial infections in aquaculture. They are usually mixed into the feed. Before the advent of vaccines against many bacterial diseases of fin fish, antibiotic treatments were commonly used. However, after a few years, the bacterial pathogens developed resistance to the antibiotics. Furthermore, there was a growing concern that the large amounts of antibiotics being used in aquaculture would have damaging effects on the environment and that antibiotic residues in fish flesh may have dangerous consequences for consumers by promoting the development of antibiotic resistant strains of human bacterial pathogens. These concerns have led to many restrictions on the use of antibiotics in aquaculture, especially in defining long withdrawal periods to ensure that carcasses for consumption are free of residues. These regulations have made the use of antibiotic treatments impractical for fish that are soon to be harvested for consumption but their use in the hatchery is still an important method of controlling losses from bacterial pathogens.

For most bacterial diseases of fish, vaccines have become the most important means of control (see below) and this has led to drastic reductions in the use of antibiotics in mariculture.

Parasite Diseases

Sea lice   The most economically important parasitic disease in mariculture of fin fish is caused by sea lice infestation of salmon. These crustacean parasites normally infest wild fish and when they enter the salmon cages they rapidly multiply. The lice larvae and adults feed on the mucus and skin of the salmon and heavy infestations result in large haemorrhagic ulcers especially on the head and around the dorsal fin. These compromise the fish's osmoregulation and allow opportunistic bacterial pathogens to enter the tissues. Without treatment the fish will die.

A range of treatments are available and recently very effective and environmental friendly in-feed treatments such as 'Slice' and 'Calicide' have replaced the highly toxic organophosphate bath treatments. Hydrogen peroxide is also used. As a biological control method, cleaner fish are used but they are not a complete method of control.

Vaccination

Use of Vaccines

Vertebrates can be distinguished from invertebrates in their ability to respond immunologically in a specific manner to a pathogen or vaccine. Invertebrates, such as shellfish, only possess nonspecific defense mechanisms. In the strict definition, vaccines are used only as a prophylactic measure in vertebrates because a particular vaccine against a particular disease induces protection that is specific for that particular disease and does not protect against other diseases. Vaccines induce long-term protection and in aquaculture a single administration is usually sufficient to induce protection until the fish are harvested. However, vaccines also have nonspecific immunostimulatory properties that can also activate many nonspecific defense mechanisms. These can increase disease resistance levels but only for a short period of time. Thus, in their capacity to induce such responses they are also used in shellfish culture, especially of shrimps.

Current Status of Vaccination

In Atlantic salmon mariculture, vaccination has been very successful in controlling many bacterial diseases and has almost replaced the need for antibiotic treatments. In recent years vaccination of sea bass and sea bream has become common practice. Most of the commercial vaccines are against bacterial diseases because these are relatively cheap to produce. Obviously the cost per dose of vaccine for use in aquaculture must be very low and it is inexpensive to culture most bacteria in large fermenters and to inactivate the bacteria and their toxins chemically (usually with formalin). It is much more expensive to culture viruses in tissue culture and this has been a major obstacle in commercializing vaccines against virus diseases of fish. However, modern molecular biology techniques have made it possible to transfer viral genes to bacteria and yeasts, which are inexpensive to culture and produce large amounts of viral vaccine cheaply. A number of vaccines against viral diseases of Atlantic salmon are now becoming available. Currently available commercial vaccines for use in mariculture are summarized in **Table 2**.

Methods of Vaccination

There are two methods of administering vaccines to fish: immersion in a dilute suspension of the vaccine or injection into the body cavity. For practical

Table 2   Vaccines available for fish species in mariculture

Vaccines against	Maricultured species
Bacterial diseases	
Vibriosis	Salmon, sea bass, sea bream, turbot, halibut
Winter ulcers	Salmon
Coldwater Vibriosis	Salmon
Enteric redmouth	Salmon
Furunculosis	Salmon
Pseudotuberculosis	Sea bass
Viral diseases	
Infectious pancreatic necrosis (IPNV)	Salmon
Pancreas disease	Salmon
Infectious salmon anemia (ISAV)	Salmon

reasons the latter method requires the fish to be over about 15 g in weight.

Immersion vaccination is effective for some, but not all vaccines. The vaccine against the bacterial disease vibriosis is effective when administered by immersion. It is used widely in salmon and sea bass farming and probably could be administered by this route to most marine fish species. The vaccine against pseudotuberculosis can also be administered by immersion to sea bass. With the exception of the vaccine against enteric redmouth, which is delivered by immersion to fish in freshwater hatcheries, all the other vaccines must be delivered by injection in order to achieve effective protection.

Injection vaccination induces long-term protection and the cost per dose is very small. However, it is obviously very labor intensive. Atlantic salmon are usually vaccinated several months before transfer to sea water so that the protective immunity has time to develop before the stress of transportation to sea and exposure to the pathogens encountered in the marine environment.

Conclusions

It is axiomatic that intensive farming of animals goes hand in hand with culture of their pathogens. The mariculture of fish and shellfish has had severe problems from time to time as a consequence of infectious diseases. During the 1970s, *Bonamia* and *Marteilia* virtually eliminated the culture of the European flat oyster in France and growers turned to production of the more resistant Pacific oyster. In Atlantic salmon farming, Norway was initially plagued with vibriosis diseases and Scotland suffered badly from furunculosis in the late 1980s. These bacterial diseases have been very successfully brought under control by vaccines. However, there are still many diseases for which vaccines are not available and the susceptibility of Pacific salmon to bacterial kidney disease has markedly restricted the development of the culture of these fish species on the Pacific coast of North America. As new industries grow, new diseases come to the foreground, for instance piscirikettsia in Chilean salmon culture, paramoebic gill disease in Tasmanian salmon culture, pseudotuberculosis in Mediterranean sea bass and

sea bream and Japanese yellowtail culture. Old diseases find new hosts, for example IPNV long known to affect salmon hatcheries, has in recent years caused high mortality in salmon postmolts and has devastated several halibut hatcheries.

To combat these diseases and to ensure the sustainability of aquaculture great attention must be paid to sanitation and good husbandry (including nutrition). In some cases these are insufficient in themselves and the presence of certain enzootic diseases, or following their introduction, have made it impossible for certain species to be cultured, for example, the European flat oyster in France. The treatment of disease by chemotherapy, which was performed widely in the 1970s and 1980s, resulted in the induction of antibiotic-resistant strains of bacteria and chemoresistant lice. Furthermore, the growing concern for the environment and the consumer about the increasing usage of chemicals and antibiotics in aquaculture, led to increasing control and restrictions on their usage. This stimulated much research in the 1980s and 1990s into development of more environmentally and consumer friendly methods of control such as vaccines and immunostimulants. These have achieved remarkable success and the pace of current research in this area using biotechnology to produce vaccines more cheaply, suggests that this approach will allow continued growth and sustainability of fin-fish mariculture into the future.

See also

Mariculture Overview.

Further Reading

Bruno DW and Poppe TT (1996) *A Colour Atlas of Salmonid Diseases*. London: Academic Press.

Ellis AE (ed.) (1988) *Fish Vaccination*. London: Academic Press.

Lightner DV (1996) *A Handbook of Pathology and Diagnostic Procedures for Diseases of Cultured Penaeid Shrimps*. The World Acquaculture Society.

Roberts RJ (ed.) (2000) *Fish Pathology*, 3rd edn. London: W.B. Saunders.

SALMONID FARMING

L. M. Laird[†], Aberdeen University, Aberdeen, UK

Introduction

All salmonids spawn in fresh water. Some of them complete their lives in streams, rivers, or lakes but the majority of species are anadromous, migrating to sea as juveniles and returning to spawn as large adults after one or more years feeding. The farmed process follows the life cycle of the wild fish; juveniles are produced in freshwater hatcheries and smolt units and transferred to sea for ongrowing in floating sea cages. An alternative form of salmonid mariculture, ocean ranching, takes advantage of their accuracy of homing. Juveniles are released into rivers or estuaries, complete their growth in sea water and return to the release point where they are harvested.

The salmonids cultured in seawater cages belong to the genera *Salmo*, *Oncorhynchus*, and *Salvelinus*. The last of these, the charrs are currently farmed on a very small scale in Scandinavia; this article concentrates on the former two genera. The Atlantic salmon, *Salmo salar* is the subject of almost all production of fish of the genus *Salmo* (1997 worldwide production 640 000 tonnes) although a small but increasing quantity of sea trout (*Salmo trutta*) is produced (1997 production 7000 tonnes). Three species of *Oncorhynchus*, the Pacific salmon are farmed in significant quantities in cages, the chinook salmon (also known as the king, spring or quinnat salmon), *O. tshawytscha* (1997, 10 000 tonnes), the coho (silver) salmon, *O. kisutch* (1997, 90 000 tonnes) and the rainbow trout, *O. mykiss*. The rainbow trout (steelhead) was formerly given the scientific name *Salmo gairdneri* but following studies on its genetics and native distribution was reclassified as a Pacific salmon species. Much of the world rainbow trout production (1997, 430 000 tonnes) takes place entirely in fresh water although in some countries such as Chile part-grown fish are transferred to sea water in the same way as the salmon species.

Here, the history of salmonid culture leading to the commercial mariculture operations of today is reviewed. This is followed by an overview of the requirements for successful operation of marine salmon farms, constraints limiting developments and prospects for the future.

[†] Deceased

History

Salmonids were first spawned under captive conditions as long ago as the fourteenth century when Dom Pinchon, a French monk from the Abbey of Reome stripped ova from females, fertilized them with milt from males, and placed the fertilized eggs in wooden boxes buried in gravel in a stream. At that time, all other forms of fish culture were based on the fattening of juveniles captured from the wild. However, the large (4–7 mm diameter) salmonid eggs were much easier to handle than the tiny, fragile eggs of most freshwater or marine fish. By the nineteenth century the captive breeding of salmonids was well established; the main aim was to provide fish to enhance river stocks or to transport around the world to provide sport in countries where there were no native salmonids. In this way, brown trout populations have become established in every continent except Antarctica, sustaining game fishing in places such as New Zealand and Patagonia.

A logical development of the production of eggs and juveniles for release was to retain the young fish in captivity, growing them until a suitable size for harvest. The large eggs hatch to produce large juveniles that readily accept appropriate food offered by the farmer. In the early days, the fish were first fed on finely chopped liver and progressed to a diet based on marine fish waste. The freshwater rainbow trout farming industry flourished in Denmark at the start of the twentieth century only to be curtailed by the onset of World War I when German markets disappeared. The success of the Danish trout industry encouraged a similar venture in Norway. However, when winter temperatures in fresh water proved too low, fish were transferred to pens in coastal sea water. Although these pens broke up in bad weather, the practice of seawater salmonid culture had been successfully demonstrated. The next major steps in salmonid mariculture came in the 1950s and 1960s when the Norwegians developed the commercial rearing of rainbow trout and then Atlantic salmon in seawater enclosures and cages. Together with the development of dry, manufactured fishmeal-based diets this led to the industry in its present form.

Salmonid Culture Worldwide

Fish reared in seawater pens are subject to natural conditions of water quality and temperature. Optimum water temperatures for growth of most

Table 1 Production of four species of salmonids reared in seawater cages

	1988 production (t)	1997 production (t)
Atlantic salmon	112 377	638 951
Rainbow trout[a]	248 010	428 963
Coho salmon	25 780	88 431
Chinook salmon	4 698	9 774

[a]Includes freshwater production.

salmonid species are in the range 8–16°C. Such temperatures, together with unpolluted waters are found not only around North Atlantic and North Pacific coasts within their native range but also in the southern hemisphere along the coastlines of Chile, Tasmania, and New Zealand. Salmon and trout are thus farmed in seawater cages where conditions are suitable both within and outwith their native ranges.

Seawater cages are used for almost the entire sea water production of salmonids. A very small number of farms rear fish in large shore-based silo-type structures into which sea water is pumped. Such structures have the advantage of better protection against storms and predators and the possibility of control of environmental conditions and parasites such as sea lice. However, the high costs of pumping outweigh these advantages and such systems are generally now used only for broodfish that are high in value and benefit from controlled conditions.

The production figures for the four species of salmonid reared in seawater cages are shown in **Table 1**

Norway, the pioneering country of seawater salmonid mariculture, remains the biggest producer of Atlantic salmon. The output figures for 1997 show the major producing countries to be Norway (331 367 t), Scotland, UK (99 422 t), Chile (96 675 t), Canada (51 103 t), USA (18 005 t). Almost the entire farmed production of chinook salmon comes from Canada and New Zealand with Chile producing over 70 000 tonnes of coho salmon (1997 figures). Most of the Atlantic salmon produced in Europe is sold domestically or exported to other European countries such as France and Spain. Production in Chile is exported to North America and to Japan.

Seawater Salmonid Rearing

Smolts

Anadromous salmonids undergo physiological, anatomical and behavioral changes that preadapt them for the transition from fresh water to sea water. At this stage one of the most visible changes in the young fish is a change in appearance from mottled brownish to silver and herring-like. The culture of farmed salmonids in sea water was made possible by the availability of healthy smolts, produced as a result of the technological progress in freshwater units. Hatcheries and tank farms were originally operated to produce juveniles for release into the wild for enhancement of wild stocks, often where there had been losses of spawning grounds or blockage of migration routes by the construction of dams and reservoirs. One of the most significant aspects of the development of freshwater salmon rearing was the progress in the understanding of dietary requirements and the production of manufactured pelleted feed. The replacement of a diet based on wet trash fish with a dry diet also benefitted the freshwater rainbow trout farming industry, by improving growth and survival and reducing disease and the pollution of the watercourses receiving the outflow water from earth ponds.

It was found possible to transfer smolts directly to cages moored in full strength sea water. If the smolts are healthy and the transfer stress-free, survival after transfer is high and feeding begins within 1–3 days (Atlantic salmon).

The smolting process in salmonids is controlled by day length; natural seasonal changes regulate the physiological processes, resulting in the completion of smolting and seaward migration of wild fish in spring. For the first two decades of seawater Atlantic salmon farming, producers were constrained by the annual seasonal availability of smolts. These fish are referred to as 'S1s', being approximately one year post-hatching. This had consequences for the use of equipment (nonoptimal use of cages) and for the timing of harvest. Most salmon reached their optimum harvest size or began to show signs of sexual maturation at the same time of year; thus large quantities of fish arrived on the market together for biological rather than economic reasons. These fish competed with wild salmonids and missed optimum market periods such as Christmas and Easter.

Research on conditions controlling the smolting process enabled smolt producers to alter the timing of smolting by manipulating photoperiod. Compressing the natural year by shortening day length and giving the parr an early 'winter' results in S1/2s or S3/4s, smolting as early as six months after hatch. Similarly, by delaying winter, smolting can be postponed. Thus it is now possible to have Atlantic salmon smolts ready for transfer to sea water throughout the year. Although this benefits marketing it makes site fallowing (see below) more difficult than when smolt input is annual.

The choice of smolts for seawater rearing is becoming increasingly important with the establishment of controlled breeding programs. Few species of fish can be said to be truly domesticated. The only examples approaching domestication are carp species and, to a lesser degree, rainbow trout. Other salmon species have been captive bred for no more (and usually far less than) ten generations; the time between successive generations of Atlantic salmon is usually a minimum of three years which prevents rapid progress in selection for preferred characters although this is countered by the fact that many thousand eggs are produced by each female. Trials carried out mainly in Norway have demonstrated that several commercially important traits can be improved by selective breeding. These include growth rate, age at sexual maturity, food conversion efficiency, fecundity, egg size, disease resistance and survival, adaptation to conditions in captivity and harvest quality, including texture, fat, and color. All of these factors can also be strongly influenced by environmental factors and husbandry.

Sexual maturation before salmonids have reached the desired size for harvest has been a problem for salmonid farmers. Pacific salmon species (except rainbow trout) die after spawning; Atlantic salmon and rainbow trout show increased susceptibility to disease, reduced growth rate, deterioration in flesh quality and changes in appearance including coloration. Male salmonids generally mature at a smaller size and younger age than the females. One solution to this problem, routinely used in rainbow trout culture, is to rear all-female stocks, produced as a result of treating eggs and fry of potential broodstock with methyl testosterone to give functional males which are in fact genetically female. When crossed with normal females, all-female offspring are produced as the Y, male, sex chromosome has been eliminated. Sexual maturation can be eliminated totally by subjecting all-female eggs to pressure or heat shock to produce triploid fish. This is common practice for rainbow trout but used little for other salmonids, partly because improvements in stock selection and husbandry are overcoming the problem but also because of adverse press comment on supposedly genetically modified fish. This same reaction has limited the commercial exploitation of fast-growing genetically modified salmon, produced by the incorporation into eggs of a gene from ocean pout.

Site Selection

The criteria for the ideal site for salmonid cage mariculture have changed with the development of stronger cage systems, use of automatic feeders with a few days storage capacity and generally bigger and stronger boats, cranes and other equipment on the farm. The small, wooden-framed cages (typically 6 m × 6 m frame, 300 m^3 capacity) with polystyrene flotation required sheltered sites with protection from wind and waves greater than 1–2 m high. Recommended water depth was around three times the depth of the cage to ensure dispersal of wastes. This led to the siting of cages in inshore sites such as inner sea lochs and fiords. These sheltered sites had several disadvantages, notably variable water quality caused by runoff of fresh water, silt, and wastes from the land and susceptibility to the accumulation of feces and waste feed on the seabed because of poor water exchange. In addition, cage groups were often sited near public roads in places valued for their scenic beauty, attracting adverse public reaction to salmon farming.

Cages in use today are far larger (several thousand m^3 volume) and stronger. Frames are made from either galvanized steel or flexible plastic or rubber and can be designed to withstand waves of 5 m or more. Flotation collars are stronger and mooring systems designed to match cages to sites. Such sites are likely to provide more constant water quality than inshore sites; an ideal salmonid rearing site has temperatures of 6–16°C and salinities of 32–35‰ (parts per thousand). Rearing is thus moving into deeper water away from sheltered lochs and bays. However, there are still advantages to proximity to the coast; these include ease of access from shore bases, proximity to staff accommodation, reduction in costs of transport of feed and stock and ability to keep sites under regular surveillance. Other factors to be taken into account in siting cage groups are the avoidance of navigation routes and the presence of other fish or shellfish farms. Maintaining a minimum separation distance from other fish farms is preferred to minimize the risk of disease transfer; if this is not possible, farms should enter into agreements to manage stock in the same way to reduce risk. Models have been developed in Norway and Scotland to determine the carrying capacity of cage farm sites.

Current speed is an important factor in site selection. Water exchange through the cage net ensures the supply of oxygen to the stock and removal of dissolved wastes such as ammonia as well as feces and waste feed. Salmon have been shown to grow and feed most efficiently in currents with speeds equivalent to 1–2 body lengths per second. In an ideal site this current regime should be maintained for as much of the tidal cycle as possible. At faster current speeds the salmon will use more energy in swimming and cage nets will tend to twist, sometimes forming pockets and trapping fish, causing scale removal.

Some ideal sites may be situated near offshore islands; access from the mainland may require crossing open water with strong tides and currents making access difficult on stormy days. However, modern workboats and feeding barges with the capacity to store several days supply of feed make the operation of such sites possible.

The presence of predators in the vicinity is often taken as a criterion for site selection. Unprotected salmon cage farms are likely to be subject to predation from seals or, in Chile, sea lions, and birds, such as herons and cormorants. Such predators not only remove fish but also damage others, tear holes in nets leading to escapes and stress stock making it more susceptible to disease. Protection systems to guard against predators include large mesh nets surrounding cages or cage groups, overhead nets and acoustic scaring devices. When used correctly these can all be effective in preventing attacks. Attacks from predators are frequently reported to involve nonlocal animals, attracted to a food source. Because of this and the possibility of excluding and deterring predators, it seems that proximity to colonies is not necessarily one of the most important factors in determining site selection.

A further factor, which must be taken into account in the siting of cage salmonid farms, is the occurrence of phytoplankton blooms. Phytoplankton can enter surface-moored cages and can physically damage gills, cause oxygen depletion or produce lethal toxins that kill fish. Historic records may indicate prevalence of such blooms and therefore sites to be avoided although some cages are now designed to be lowered beneath the surface and operated as semi-submersibles, keeping the fish below the level of the bloom until it passes.

Farm Operation

Operation of the marine salmon farm begins with transfer of stock from freshwater farms. Where possible, transfer in disinfected bins suspended under helicopters is the method of choice as it is quick and relatively stress-free. For longer journeys, tanks on lorries or wellboats are used. The latter require particular vigilance as they may visit more than one farm and have the potential to transfer disease. Conditions in tanks and wellboats should be closely monitored to ensure that the supply of oxygen is adequate (minimum 6 mg l^{-1}).

The numbers of smolts stocked into each cage is a matter for the farmer; some will introduce a relatively small number, allowing for growth to achieve a final stocking density of 10–15 kg^{-3} whereas others stock a greater number and split populations between cages during growth. This latter method makes better use of cage space but increases handling and therefore stress. Differential growth may make grading into two or three size groups necessary.

Stocking density is the subject of debate. It is essential that oxygen concentrations are maintained and that all fish have access to feed when it is being distributed. Fish may not distribute themselves evenly within the water column; because of crowding together the effective stocking density may therefore be a great deal higher than the theoretical one.

As with all farmed animals the importance of vigilance of behavior and health and the maintenance of accurate, useful records cannot be overemphasized. When most salmon farms were small, producing one or two hundred tonnes of salmon a year rather than thousands, hand feeding was normal; observation of stock during feeding provided a good indication of health. Today, fish are often fed automatically using blowers attached to feed storage systems. The best of these systems incorporate detectors to monitor consumption of feed and underwater cameras to observe the stock.

All of the nutrients ingested by cage-reared salmonids are supplied in the feed distributed. Typically, manufactured diets for salmonids will contain 40% protein (mainly obtained from fishmeal) and up to 30% oil, providing the source of energy, sparing protein for growth. Although very poorly digested by salmonids, carbohydrate is necessary to bind other components of the diet. Vitamins and minerals are also added, as are carotenoid pigments such as astaxanthin, necessary to produce the characteristic pink coloration of the flesh of anadromous salmonids. The feed used on marine salmon farms is nowadays almost exclusively a pelleted or extruded fishmeal-based diet manufactured by specialist companies. Feed costs make up the biggest component of farm operating costs, sometimes reaching 50%. It is therefore important to make optimum use of this valuable input by minimizing wastes. This is accomplished by ensuring that feed is delivered to the farm in good condition and handled with care to prevent dust formation, increasing the size of pellets as the fish grow and distributing feed to satisfy the appetites of the fish. Improvements in feed manufacture and in feeding practices have reduced feed conversion efficiency (feed input : increase in weight of fish) from 2 : 1 to close to 1 : 1. Such figures may seem improbable but it must be remembered that they represent the conversion of a nearly dry feed to wet fish flesh and other tissues.

The importance of maintaining a flow of water through the net mesh of the cages has been emphasized. Mesh size is generally selected to be the maximum capable of retaining all fish and preventing escapes. Any structure immersed in the upper few meters of coastal or marine waters will quickly be subjected to colonization by fouling organisms including bacteria, seaweeds, mollusks and sea squirts. Left unchecked, such fouling occludes the mesh, reducing water exchange and may place a burden on the cage reducing its resistance to storm damage. One of the most effective methods of preventing fouling of nets and moorings is to treat them with antifouling paints and chemicals prior to installation. However, one particularly effective treatment used in the early 1980s, tributyl tin, has been shown to have harmful effects on marine invertebrates and to accumulate in the flesh of the farmed fish; its use in aquaculture is now banned. Other antifoulants are copper or oil based; alternative, preferred methods of removing fouling organisms include lifting up sections of netting to dry in air on a regular basis or washing with high pressure hoses or suction devices to remove light fouling.

The aim of the salmonid farmer is to produce maximum output of salable product for minimum financial input. To do this, fish must grow efficiently and a high survival rate from smolt input to harvest must be achieved. Minimizing stress to the fish by reducing handling, maintaining stable environmental conditions and optimizing feeding practices will reduce mortalities. Causes of mortality in salmonid and other farms are reviewed elsewhere (see Mariculture Diseases and Health). It is vital to keep accurate records of mortalities; any increase may indicate the onset of an outbreak of disease. It is also important that dead fish are removed; collection devices installed in the base of cages are often used to facilitate this. Treatment of diseases or parasitic infestations such as sea lice (*Lepeophtheirus salmonis, Caligus elongatus*) is difficult in fish reared in sea cages because of their large volumes and the high numbers of fish involved. Some treatments for sea lice involve reducing the cage volume and surrounding with a tarpaulin so that the fish can be bathed in chemical. After the specified time the tarpaulin is removed and the chemical disperses into the water surrounding the cage. Newer treatments incorporate the chemicals in feed and are therefore simpler to apply. In the future, vaccines are increasingly likely to replace chemicals.

The health of cage-reared salmonids can be maintained by a site management system incorporating a period of fallowing when groups of cages are left empty for a period of at least three months and preferably longer. This breaks the life cycle of parasites such as sea lice and allows the seabed to recover from the nutrient load falling from the cages. Ideally a farmer will have access to at least three sites; at any given time one will be empty and the other two will contain different year classes, separated to prevent cross-infection.

Harvesting

Most of the farmed salmonids reared in sea water reach the preferred harvest size (3–5 kg) 10 months or more after transfer to sea water. Poor harvesting and handling methods can have a devastating effect on flesh quality, causing gaping in muscle blocks and blood spotting. After a period of starvation to ensure that guts are emptied of feed residues the fish are generally killed by one of two methods. One of these involves immersion in a tank of sea water saturated with carbon dioxide, the other an accurate sharp blow to the cranium. Both methods are followed by excision of the gill arches; the loss of blood is thought to improve flesh quality. It is important that water contaminated with blood is treated to kill any pathogens which might infect live fish.

Ocean Ranching

The anadromous behavior of salmonids and their ability to home to the point of release has been exploited in ocean ranching programs which have been operated successfully with Pacific salmon. Some of these programs are aimed at enhancing wild stocks and others are operated commercially. The low cost of rearing Pacific salmon juveniles, which are released into estuaries within weeks of hatching, makes possible the release of large numbers. In Japan over two billion juveniles are released annually; overall return rates have increased to 2%, 90% of which are chum (*Oncorhynchus keta*) and 8% pink (*Oncorhynchus gorbuscha*) salmon. The success of the operation depends on cooperation between those operating and financing the hatcheries and those harvesting the adult fish. The relatively high cost of producing Atlantic salmon smolts and the lack of control over harvest has restricted ranching operations.

See also

Mariculture Diseases and Health. Ocean Ranching.

Further Reading

Anon (ed.) (1999) *Aquaculture Production Statistics 1988–1997.* Rome: Food and Agriculture Organization.

Black KD and Pickering AD (eds.) (1998) *Biology of Farmed Fish*. Sheffield Academic Press.

Heen K, Monahan RL, and Utter F (eds.) (1993) *Salmon Aquaculture*. Oxford: Fishing News Books.

Pennell W and Barton BA (eds.) (1996) *Principles of Salmonid Culture*. Amsterdam: Elsevier.

Stead S and Laird LM (In press) *Handbook of Salmon Farming Praxis*. Chichester: Springer-Praxis.

Willoughby S (1999) *Manual of Salmonid Farming*. Oxford: Blackwell Science.

OYSTERS – SHELLFISH FARMING

I. Laing, Centre for Environment Fisheries and Aquaculture Science, Weymouth, UK

As I ate the oysters with their strong taste of the sea and their faint metallic taste that the cold white wine washed away, leaving only the sea taste and the succulent texture, and as I drank their cold liquid from each shell and washed it down with the crisp taste of the wine, I lost the empty feeling and began to be happy and to make plans. (Ernest Hemingway in *A Moveable Feast*).

A Long History

Oysters have been prized as a food for millennia. Carbon dating of shell deposits in middens in Australia show that the aborigines took the Sydney rock oyster for consumption in around 6000 BC. Oyster farming, nowadays carried out all over the world, has a very long history, although there are various thoughts as to when it first started. It is generally believed that the ancient Romans and the Greeks were the first to cultivate oysters, but some maintain that artificial oyster beds existed in China long before this.

It is said that the ancient Chinese raised oysters in specially constructed ponds. The Romans harvested immature oysters and transferred them to an environment more favorable to their growth. Greek fishermen would toss broken pottery dishes onto natural oyster beds to encourage the spat to settle.

All of these different cultivation methods are still practiced in a similar form today.

A Healthy Food

Oysters (**Figure 1**) have been an important food source since Neolithic times and are one of the most nutritionally well balanced of foods. They are ideal for inclusion in low-cholesterol diets. They are high in omega-3 fatty acids and are an excellent source of vitamins. Four or five medium-size oysters supply the recommended daily allowance of a whole range of minerals.

The Oyster

Oysters are one among a number of bivalve mollusks that are cultivated. Others include clams, cockles, mussels, and scallops.

Current Status

Oysters form the second most important group, after cyprinid fishes, in world aquaculture production.

Figure 1 A dish of Pacific oysters.

In 2004, 4.6 million tonnes were produced (Food and Agriculture Organization (FAO) data). Asia and the Pacific region produce 93.4% of this total production.

The FAO of the United Nations lists 15 categories of cultivated oyster species. The Pacific oyster (*Crassostrea gigas*) is by far the most important. Annual production of this species now exceeds 4.4 million metric tonnes globally, worth US$2.7 billion, and accounts for 96% of the total oyster production (see **Table 1**). China is the major producer.

The yield from wild oyster fisheries is small compared with that from farming and has declined steadily in recent years. It has fallen from over 300 000 tonnes in 1980 to 152 000 tonnes in 2004. In contrast to this, Pacific oyster production from aquaculture has increased by 51% in the last 10 years (see **Figure 2**).

General Biology

As might be expected, given a long history of cultivation, there is a considerable amount known about the biology of oysters.

The shell of bivalves is in two halves, or valves. Two muscles, called adductors, run between the inner surfaces of the two valves and can contract rapidly to close the shell tightly. When exposed to the air, during the tidal cycle, oysters close tightly to prevent desiccation of the internal tissues. They can respire anaerobically (i.e., without oxygen) when out of water but have to expel toxic metabolites when reimmersed as the tide comes in. They are known to be able to survive for long periods out of water at low temperatures such as those used for storage after collection.

Within the shell is a fleshy layer of tissue called the mantle; there is a cavity (the mantle cavity) between the mantle and the body wall proper. The mantle secretes the layers of the shell, including the inner nacreous, or pearly, layer. Oysters respire by using both gills and mantle. The gills, suspended within the mantle cavity, are large and function in food gathering (filter feeding) as well as in respiration. As water passes over the gills, organic particulate material, especially phytoplankton, is strained out and is carried to the mouth.

Table 1 The six most important oyster species produced worldwide

Oyster species	Number of producing countries	Major producing countries (% of total)	Production (metric tonnes)
Pacific oyster	30	China (85), Korea (5), Japan (5), France (2.5)	4 429 337
American oyster	4	USA (95), Canada (4.5)	110 770
Slipper oyster	1	Philippines (100)	15 915
Sydney oyster	1	Australia (100)	5600
European oyster	17	Spain (50), France (29), Ireland (8)	5071
Mangrove oyster	3	Cuba (99)	1184

Tonnages are 2004 figures (FAO data). All are cupped oyster species, apart from the European oyster, which is a flat oyster species.

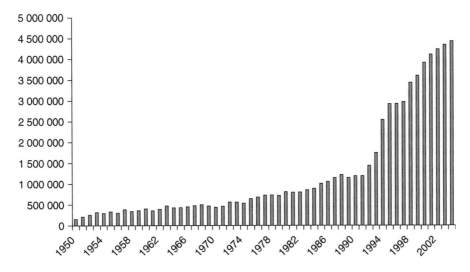

Figure 2 The increase in world production, in metric tonnes, of the Pacific oysters (1950–2004).

Oysters have separate sexes, but they may change sex one or more times during their life span, being true hermaphrodites. In most species, the eggs and sperm are shed directly into the water where fertilization occurs. Larvae are thus formed and these swim and drift in the water, feeding on natural phytoplankton. After 2–3 weeks, depending on local environmental conditions, the larvae are mature and they develop a foot. At this stage they sink to the seabed and explore the sediment surface until they find a suitable surface on which to settle and attach permanently, by cementation. Next, they go through a series of morphological and physiological changes, a process known as metamorphosis, to become immature adults. These are called juveniles, spat, or seed.

Cultivated species of flat oysters brood the young larvae within the mantle cavity, releasing them when they are almost ready to settle. Fecundity is usually related to age, with older and larger females producing many more larvae.

Growth of juveniles is usually quite rapid initially, before slowing down in later years. The length of time that oysters take to reach a marketable size varies considerably, depending on local environmental conditions, particularly temperature and food availability. Pacific oysters may reach a market size of 70–100 g live weight (shell-on) in 18–30 months.

Methods of Cultivation – Seed Supply

Oyster farming is dependent on a regular supply of small juvenile animals for growing on to market size.

These can be obtained primarily in one of two ways, either from naturally occurring larvae in the plankton or artificially, in hatcheries.

Wild Larvae Collection

Most oyster farmers obtain their seed by collecting wild set larvae. Special collection materials, generically known as 'cultch', are placed out when large numbers of larvae appear in the plankton. Monitoring of larval activity is helpful to determine where and when to put out the cultch. It can be difficult to discriminate between different types of bivalve larvae in the plankton to ensure that it is the required species that is present but recently, modern highly sensitive molecular methods have been developed for this.

Various materials can be used for collection. Spat collected in this way are often then thinned prior to growing on (**Figure 3**). In China, coir rope is widely used, as well as straw rope, flax rope, and ropes woven by thin bamboo strips. Shells, broken tiles, bamboo, hardwood sticks, plastics, and even old tires can also be used. Coatings are sometimes applied.

Hatcheries

Restricted by natural conditions, the amount of wild spat collected may vary from year to year. Also, where nonnative oyster species are cultivated, there may be no wild larvae available. In these circumstances, hatchery cultivation is necessary. This also allows for genetic manipulation of stocks, to rear and maintain lines specifically adapted for certain

Figure 3 A demonstration of the range of oyster spat collectors used in France.

traits. Important traits for genetic improvement include growth rate, environmental tolerances, disease resistance, and shell shape.

Methods of cryopreservation of larvae are being developed and these will contribute to maintaining genetic lines. Furthermore, triploid oysters, with an extra set of chromosomes, can only be produced in hatcheries. These oysters often have the advantage of better growth and condition, and therefore marketability, during the summer months, as gonad development is inhibited.

Techniques for hatchery rearing were first developed in the 1950s and today follow well-established procedures.

Adult broodstock oysters are obtained from the wild or from held stocks. Depending on the time of year, these may need to be bought into fertile condition by providing a combination of elevated temperature and food (cultivated phytoplankton diets) over several weeks (see **Figure 4**). Selection of the appropriate broodstock conditioning diet is very important. An advantage of hatchery production is that it allows for early season production in colder climates and this ensures that seeds have a maximum growing period prior to their first overwintering.

For cultivated oyster species, mature gametes are usually physically removed from the gonads. This involves sacrificing some ripe adults. Either the gonad can be cut repeatedly with a scalpel and the gametes washed out with filtered seawater into a part-filled container, or a clean Pasteur pipette can be inserted into the gonad and the gametes removed by exerting gentle suction.

Broodstock can also be induced to spawn. Various stimuli can be applied. The most successful methods are those that are natural and minimize stress. These include temporary exposure to air and thermal cycling (alternative elevated and lowered water temperatures). Serotonin and other chemical triggers can also be used to initiate spawning but these methods are not generally recommended as eggs liberated using such methods are often less viable.

Flat oysters, of the genera *Ostrea* and *Tiostrea*, do not need to be stimulated to spawn. They will spawn of their own accord during the conditioning process as they brood larvae within their mantle cavities for varying periods of time depending on species and temperature.

The fertilized eggs are then allowed to develop to the fully shelled D-larva veliger stage, so called because of the characteristic 'D' shape of the shell valves (**Figure 5**).

These larvae are then maintained in bins with gentle aeration. Static water is generally used and this is exchanged daily or once every 2 days. Through-flow systems, with meshes to prevent loss of larvae, are also employed. Cultured microalgae are added into the tanks several times per day, at an appropriate

Figure 4 The SeaCAPS continuous culture system for algae. An essential element for a successful oyster hatchery is the means to cultivate large quantities of marine micro algae (phytoplankton) food species.

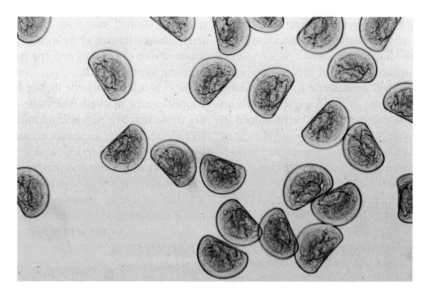

Figure 5 Pacific oyster D larvae. At this stage shell length is about 70 microns.

daily ration according to the number and size of the larvae.

The larvae eventually become competent to settle and for this, surfaces must be provided. The area of settlement surface is important. Types of materials in common usage to provide large surface areas for settlement include sheets of slightly roughened polyvinyl chloride (PVC), layers of shell chips and particles prepared by grinding aged, clean oyster shell spread over the base of settlement trays or tanks, bundles, bags, or strings of aged clean oyster shells dispersed throughout the water column, usually in settlement tanks or various plastic or ceramic materials coated with cement (lime/mortar mix).

For some of these methods, the oysters are subsequently removed from the settlement surface to produce 'cultchless' spat. These can then be grown as separate individuals through to marketable size and sold as whole or half shell, usually live.

Provision of competence to settle Pacific oyster larvae for remote setting at oyster farms is common practice on the Pacific coast of North America. Hatcheries provide the mature larvae and the farmers themselves set them and grow the spat for seeding oyster beds or in suspended culture.

In other parts of the world, hatcheries set the larvae, as described above, and grow the spat to a size that growers are able to handle and grow.

Oyster juveniles from hatchery-reared larvae perform well in standard pumped upwelling systems and survival is usually good, although some early losses may occur immediately following metamorphosis. Diet, ration, stocking density, and water flow rate are all important in these systems (**Figure 6**). They are only suitable for initial rearing of small seed. As the spat grow, food is increasingly likely to become limiting in these systems and they must be transferred to the sea for on-growing.

The size at which spat is supplied is largely dictated by the requirements and maturity of the grow-out industry. Seed native oysters are made available from commercial hatcheries at a range of sizes up to 25–30 mm. The larger the seed, the more expensive they are but this is offset by the higher survival rate of larger seed. Larger seed should also be more tolerant to handling.

Ponds

Pond culture offers a third method to provide seed. This was the method that was originally developed in Europe in early attempts to stimulate production following the decline of native oyster stocks in the late nineteenth century. Ponds of 1–10 ha in area and 1–3 m deep were built near to high water spring tides, filled with seawater, and then isolated for the period of time during which the oysters are breeding naturally, usually May to July. Collectors put into the ponds encourage and collect the settlement of juvenile oysters. There is an inherent limited amount of control over the process and success is very variable. In France, where spat collectors were also deployed in the natural environment, it was relatively successful and became an established method for a time. In the UK, spat production from ponds built in the early 1900s was insufficiently regular to provide a reliable supply of seed to the industry and the method was largely abandoned in the middle of the twentieth century in favor of the more controlled conditions available in hatcheries.

(a)

(b)

Figure 6 Indoor (a) and outdoor (b) oyster nursery system, in which seawater and algae food are pumped up through cylinders fitted with mesh bases and containing the seed.

Methods of Cultivation – On-Growing

Various methods are available for on-growing oyster seed once this has been obtained, from either wild set larvae, hatcheries, or ponds.

Oysters smaller than 10 g need to be held in trays or bags attached to a superstructure on the foreshore until they are large enough to be put directly on to the substrate and be safe from predators, strong tidal and wave action, or siltation (**Figure 7**).

Tray cultivation of oysters can be successful in water of minimal flow, where water exchange is driven only by the rise and fall of the tide and gentle wave action. In such circumstances it may even be possible in open baskets.

Figure 7 The traditional bag and trestle method for on-growing Pacific oysters.

Figure 8 Open baskets for on-growing, as developed in Tasmania for Pacific oysters.

In more exposed sites, systems developed in Australia and employing plastic mesh baskets attached to or suspended from wires (see **Figure 8**) are becoming increasingly popular. It is claimed that an oyster with a better shape, free from worm (*Polydora*) infestation, will result from using these systems. *Polydora* can cause unsightly brown blemishes on the inner surface of the shell and decrease marketability of the stock. Less labor is required with these systems but they are more expensive to purchase initially.

In all of the above systems, the mesh size can be increased as the oysters grow, to improve the flow of water and food through the animals.

Holding seed oysters in trays or attached to the cultch material onto which they were settled, and suspended from rafts, pontoons, or long lines is an alternative method in locations where current speed will allow. The oysters should grow more quickly because they are permanently submerged but the shell may be thinner and therefore more susceptible to damage. Early stages of predator species such as crabs and starfish can settle inside containers, where they can cause significant damage unless containers are opened and checked on a regular basis. In the longer term the oysters will perform better on the seabed, although they can be reared to market size in the containers.

Protective fences can be put up around ground plots to give some degree of protection to smaller oysters from shore crabs. Potting crabs in the area of the lays is another method of control. The steps in the oyster cultivation process are shown as a diagram in **Figure 9**.

The correct stocking density is important so as not to exceed the carrying capacity of a body of water. When carrying capacity is exceeded, the algal population in the water is insufficient and growth declines. Mortalities also sometimes occur.

Yields in extensively cultivated areas are usually about 25 tonnes per hectare per year but this can increase to 70 tonnes where individual plots are well separated.

In France, premium quality oysters are sometimes finished by holding them in fattening ponds known as claires (see **Figure 10**), in which a certain type of algae is encouraged to bloom, giving a distinctive green color to the flesh.

In the UK, part-grown native flat oysters from a wild fishery are relayed in spring into areas in which conditions are favorable to give an increase in meat yield over a period of a few months, prior to marketing in the winter.

There are proposals to establish standards for organic certification of mollusk cultivation but many oyster growers consider that the process is intrinsically organic.

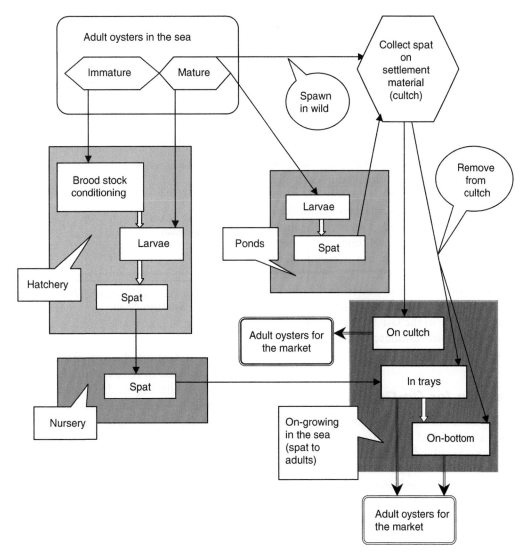

Figure 9   The steps and processes of oyster farming.

Figure 10 Oyster fattening ponds, or Claires, in France.

Pearl Oysters

There is an important component of the oyster farming industry, located mainly in Japan, Australia, and the South Sea Islands, devoted to the production of pearls. The process involves inserting into the oyster a nucleus, made with shell taken from a North American mussel, and a tiny piece of mantle cut from another oyster. The oysters are cultivated in carefully tended suspended systems while pearls develop around the nucleus. Once the pearls have been taken out of the oysters, they are seeded anew. A healthy oyster can be reseeded as many as 4 times with a new nucleus. As the oyster grows, it can accommodate progressively bigger pearl nuclei. Therefore, the biggest pearls are most likely to come from the oldest oysters. There has been a significant increase in demand for pearl jewelry in the last 10 years. The US is the biggest market, with estimated sales of US$1.5 billion of pearl jewelry per year (**Table 2**).

Site Selection for On-Growing

A range of physical, biological, and chemical factors will influence survival and growth of oysters. Many of these factors are subject to seasonal and annual variation and prospective oyster farmers are usually advised to monitor the conditions and to see how well oysters grow and survive at their chosen site for at least a year before any commercial culture begins.

Seawater temperature has a major effect on seasonal growth and may be largely responsible for any differences in growth between sites. Changes in

Table 2 Estimated values of pearl production (2004)

	Market share in value (%)	Production value (2004)
White South Sea cultured pearls (Australia, Indonesia, the Philippines, Myanmar)	35	US$220 million
Freshwater cultured pearls (China)	24	US$150 million
Akoya cultured pearls (Japan, China)	22	US$135 million
Tahitian cultured pearls (French Polynesia)	19	US$120 million
Total		US$625 million

Source: Pearl World, The International Pearling Journal (January/February/March 2006).

salinity do not affect the growth of bivalves by as much as variation in temperature. However, most bivalves will usually only feed at higher salinities. Pacific oysters prefer salinity levels nearer to 25 psu, conditions typical of many estuaries and inshore waters.

Growth rates are strongly influenced by the length of time during which the animals are covered by the tide. Growth of oysters in trays stops when they are exposed to air for more than about 35% of the time. However, this fact can be used to advantage by the cultivator who may wish, for commercial reasons, to slow down or temporarily stop the growth of the stock. This can be achieved by moving the stock higher up the beach. It is a practice that is routinely adopted in Korea and Japan for 'hardening off' wild-caught spat prior to sale.

Other considerations are related to access and harvesting. It is important to consider the type of equipment likely to be used for planting, maintenance, and harvesting, particularly at intertidal sites. Some beaches will support wheeled or tracked vehicles, while others are too soft and will require the use of a boat to transport equipment.

Risks

Oyster farming can be at risk from pollutants, predators, competitors, diseases, fouling organisms, and toxic algae species.

Pollutants

Some pollutants can be harmful to oysters at very low concentrations. During the 1980s, it was found that tributyl tin (TBT), a component of marine antifouling paints, was highly toxic to bivalve mollusks at extremely low concentrations in the seawater. Pacific oysters cultivated in areas in which large numbers of small vessels were moored showed stunted growth and thickening of the shell, and natural populations of flat oysters failed to breed (see **Figure 11**). The use of this compound on small vessels was widely banned in the late 1980s, and since then the oyster industry has recovered. The International Maritime Organization has since announced a ban for larger vessels as well. When marketed for consumption, oysters must meet a number of 'end product' standards. These include a requirement that the shellfish should not contain toxic or objectionable compounds such as trace metals, organochlorine compounds, hydrocarbons, and polycyclic aromatic hydrocarbons (PAHs) in such quantities that the calculated dietary intake exceeds the permissible daily amount.

Predators

In some areas oyster drills or tingles, which are marine snails that eat bivalves by rasping a hole through the shell to gain access to the flesh, are a major problem.

Competitors

Competitors include organisms such as slipper limpets (*Crepidula fornicata*), which compete for food and space. In silt laden waters, they also produce a muddy substrate, which is unsuitable for cultivation. They can be a significant problem. An example is around the coast of Brittany, where slipper limpets have proliferated in the native flat oyster areas during the last 30 years. In order to try to control this problem, approximately 40 000–50 000 tonnes of slipper limpets are harvested per year. These are taken to a factory where they are converted to a calcareous fertilizer.

Diseases

The World Organisation for Animal Health (OIE) lists seven notifiable diseases of mollusks and six of

Figure 11 Compared with a normal shell (upper shell section), TBT from marine antifouling paints causes considerable thickening of Pacific oyster shells (lower section). These compounds are now banned.

these can infect at least one of the cultivated oyster species. Some of these diseases, often spread through national and international trade, have had a devastating effect on oyster stocks worldwide. For example, Bonamia, a disease caused by the protist parasitic organism *Bonamia ostreae*, has had a significant negative impact on *Ostrea edulis* production throughout its distribution range in Europe. Mortality rates in excess of 80% have been noted. The effect this can have on yields can be seen in the drastic (93%) drop in recorded production in France, from 20 000 tonnes per year in the early 1970s to 1400 tonnes in 1982. A major factor in the introduction and success of the Pacific oyster as a cultivated species has been that it is resistant to major diseases, although it is not completely immune to all problems.

There are reports of summer mortality episodes, especially in Europe and the USA, and infections by a herpes-like virus have been implicated in some of these. *Vibrio* bacteria are also associated with larval mortality in hatcheries.

The Fisheries and Oceans Canada website lists 53 diseases and pathogens of oysters. Juvenile oyster disease is a significant problem in cultivated *Crassostrea virginica* in the Northeastern United States. It is thought to be caused by a bacterium (*Roseovarius crassostreae*).

Fouling

Typical fouling organisms include various seaweeds, sea squirts, tubeworms, and barnacles. The type and degree of fouling varies with locality. The main effect is to reduce the flow of water and therefore the supply of food to bivalves cultivated in trays or on ropes and to increase the weight and drag on floating installations. Fouling organisms grow in response to the same environmental factors as are desirable for good growth and survival of the cultivated stock, so this is a problem that must be controlled rather than avoided.

Toxic Algae

Mortalities of some marine invertebrates, including bivalves, have been associated with blooms of some alga species, including *Gyrodinium aureolum* and *Karenia mikimoto*. These so-called red tides cause seawater discoloration and mortalities of marine organisms. They have no impact on human health.

Stock Enhancement

If we accept the FAO definition of aquaculture as "The farming of aquatic organisms in inland and coastal areas, involving intervention in the rearing process to enhance production and the individual or corporate ownership of the stock being cultivated," then in many cases stock enhancement can be included as a type of oyster farming.

Natural beds can be managed to encourage the settlement of juvenile oysters and sustain a fishery. Beds can be raked and tilled on a regular basis to remove silt and ensure that suitable substrates are available for the attachment of the juvenile stages. Adding settlement material (cultch) is also beneficial.

In some areas of the world, there has been a dramatic reduction in stocks of the native oyster species. This is attributed mainly to overexploitation, although disease is also implicated. Native oyster beds form a biotope, with many associated epifaunal and infaunal species. Loss of this habitat has resulted in a major decline in species richness in the coastal environment.

Considerable effort has been put into restoring these beds, using techniques that might fall under the definition of oyster farming. A good example is the Chesapeake Bay Program for the native American oyster *Crassostrea virginica*. Overfishing followed by the introduction of two protozoan diseases, believed to be inadvertently introduced to the Chesapeake through the importation of a nonnative oyster, *Crassostrea gigas*, in the 1930s, has combined to reduce oyster populations throughout Chesapeake Bay to about 1% of historical levels. The Chesapeake Bay Program is committed to the restoration and creation of aquatic reefs. A key component of the strategy to restore oysters is to designate sanctuaries, that is, areas where shellfish cannot be harvested. It is often necessary within a sanctuary to rehabilitate the bottom to make it a suitable oyster habitat. Within sanctuaries aquatic reefs are created primarily with oyster shell, the preferred substrate of spat. Alternative materials including concrete and porcelain are also used. These permanent sanctuaries will allow for the development and protection of large oysters and therefore more fecund and potentially disease-resistant oysters. Furthermore, attempts have been made at seeding with disease-free hatchery oysters.

Nonnative Species

The International Council for the Exploration of the Sea (ICES) Code of Practice sets forth recommended procedures and practices to diminish the risks of detrimental effects from the intentional introduction and transfer of marine organisms. The Pacific oyster, originally from Asia, has been introduced around the world and has become invasive in parts of Australia and New Zealand, displacing the native rock oyster in some areas. It is also increasingly becoming a problem in northern European coastal regions. In parts of France naturally recruited stock is competing with cultivated stock and has to be controlled.

The possibility of farming the nonnative Suminoe oyster (*Crassostrea ariakensis*) to restore oyster stocks in Chesapeake Bay is being examined and there are differing opinions about the environmental risks involved, with concerns of potential harmful effects on the local ecology.

It should also be noted that aquaculture has been responsible for the introduction of a whole range of passenger species, including pest species such as the slipper limpet and oyster drills, throughout the world.

Food Safety

Bivalve mollusks filter phytoplankton from the seawater during feeding; they also take in other small particles, such as organic detritus, bacteria, and viruses. Some of these bacteria and viruses, especially those originating from sewage outfalls, can cause serious illnesses in human consumers if they remain in the bivalve when it is eaten. The stock must be purified of any fecal bacterial content in cleansing (depuration) tanks (**Figure 12**) before sale for consumption. There are regulations governing this, which are based on the level of contamination of the mollusks.

Viruses are not all removed by normal depuration processes and they can cause illness if the bivalves are eaten raw or only lightly cooked, as oysters often are. These viruses can only be detected by using sophisticated equipment and techniques, although research is being carried out to develop simpler methods.

Finally, certain types of naturally occurring algae produce toxins, which can accumulate in the flesh of oysters. People eating shellfish containing these toxins can become ill and in exceptional cases death can result. Cooking does not denature the toxins responsible nor does cleansing the shellfish in depuration tanks eliminate them. The risks to consumers are usually minimized by a requirement for samples to be tested regularly. If the amount of toxin exceeds a certain threshold, the marketing of shellfish for consumption is prohibited until the amount falls to a safe level.

Further Reading

Andersen RA (ed.) (2005) *Algal Culturing Techniques.* New York: Academic Press.

Bueno P, Lovatelli A, and Shetty HPC (1991) Pearl oyster farming and pearl culture, Regional Seafarming Development and Demonstration Project. *FAO Project Report No. 8.* Rome: FAO.

Dore I (1991) *Shellfish – A Guide to Oysters, Mussels, Scallops, Clams and Similar Products for the Commercial User.* London: Springer.

Gosling E (2003) *Bivalve Mollusks; Biology, Ecology and Culture.* London: Blackwell.

Helm MM, Bourne N, and Lovatelli A (2004) Hatchery culture of bivalves; a practical manual. *FAO Fisheries Technical Paper 471.* Rome: FAO.

Huguenin JE and Colt J (eds.) (2002) *Design and Operating Guide for Aquaculture Seawater Systems,* 2nd edn. Amsterdam: Elsevier.

Lavens P and Sorgeloos P (1996) Manual on the production and use of live food for aquaculture. *FAO Fisheries Technical Paper 361.* Rome: FAO.

Mann R (1979) Exotic species in mariculture. Proceedings of Symposium on Exotic Species in Mariculture: Case Histories of the Japanese Oyster, *Crassostrea gigas* (Thunberg), with implications for other fisheries. Woods Hole Oceanographic Institution, Cambridge Press.

Matthiessen G (2001) *Fishing News Books Series: Oyster Culture.* New York: Blackwell.

Spencer BE (2002) *Molluscan Shellfish Farming.* New York: Blackwell.

Figure 12 A stacked depuration system, suitable for cleansing oysters of microbiological contaminants. The oysters are held in trays in a cascade of UV-sterlized water.

Relevant Websites

http://www.crlcefas.org
– Cefas is the EU community reference laboratory for bacteriological and viral contamination of oysters.

http://www.oie.int

  – Definitive information on oyster diseases can be found on the website of The World Organisation for Animal Health (OIE).

http://www.pac.dfo-mpo.gc.ca

  – Further information on diseases can be found on the Fisheries and Oceans Canada website.

http://www.cefas.co.uk

  – The online magazine 'Shellfish News', although produced primarily for the UK industry, has articles of general interest on oyster farming. It can be found on the News/Newsletters area of the Cefas web site. A booklet on site selection for bivalve cultivation is also available at this site.

http://www.vims.edu

  – The Virginia Institute of Marine Science has information on oyster genetics and breeding programs and the proposals to introduce the non-native *Crassostrea ariakensis* to Chesapeake Bay.

http://www.was.org

  – The World Aquaculture Society website has a comprehensive set of links to other relevant web sites, including many national shellfish associations.

http://www.fao.org

  – There is a great deal of information on oyster farming throughout the world on the website of The Food and Agriculture Organization of the United Nations (FAO). This website has statistics on oyster production, datasheets for various cultivated species, including Pacific oysters, and online manuals on the production and use of live food for aquaculture (*FAO Fisheries Technical Paper 361*), Hatchery culture of bivalves (*FAO Fisheries Technical Paper. No. 471*) and Pearl oyster farming and pearl culture (*FAO Project Report No. 8*).

MARICULTURE OF AQUARIUM FISHES

N. Forteath, Inspection Head Wharf, Tasmania,
Australia

Introduction

Marine fishes and invertebrates have been kept in aquaria for decades. However, attempts to maintain marine species in a captive environment have been dependent on trial and error for the most part but it has been mainly through the attention and care of aquarists that our knowledge about many marine species has been obtained. This is particularly true of the charismatic syngnathids, which includes at least 40 species of sea horses.

During the past 30 years, technological advances in corrosion resistant materials together with advances in aquaculture systems have brought about a rapid increase in demand for large public marine aquarium displays, oceanariums and hobby aquaria suitable for colorful and exotic ornamental species. These developments have led to the establishment of important export industries for live fishes, invertebrates and so-called 'living rocks'. Attempts to reduce dependence on wild harvesting through the development of marine fish and invertebrate hatcheries met with limited success.

The availability of equipment, which greatly assists in meeting the water quality requirements for popular marine organisms, has turned the attention of aquarists towards maintaining increasingly complex living marine ecosystems and more exotic species. A fundamental requirement for the success of such endeavors is the need to understand species biology and interspecific relationships within the tank community. Modern marine aquarists must draw increasingly on scientific knowledge and this is illustrated below with reference to sea horse and coral reef aquaria, respectively.

History

The first scientific and public aquarium was built in the London Zoological Gardens in 1853. This facility was closed within a few years and another attempt was not undertaken until 1924. By the 1930s several public aquaria were built in other European capitals but by the end of World War II only that of Berlin remained. During the 1970s the scene was set for a new generation of public aquaria, several specializing in marine displays, and others becoming more popularly known as oceanariums due to the presence of marine mammals, displays of large marine fish and interactive educational activities.

Themes have added to the public interest. For example, Monterey Bay Aquarium exhibits a spectacular kelp forest, a theme repeated by several world class aquaria and Osaka Aquarium sets out to recreate the diverse environments found around the Pacific Ocean. Some of these aquaria have found that exhibits of species native to their location alone are not successful in attracting visitor numbers. The New Jersey Aquarium, for example, has been forced to build new facilities and tanks housing over 1000 brightly colored marine tropical fish with other ventures having to rely on the lure of sharks and touch pools.

The history of public aquaria has evolved from stand alone tank exhibits to massive 2–3 million liter tanks through which pass viewing tunnels. Once, visitors were content to be mere observers of the fishes and invertebrates but by the end of the millennium the emphasis changed to ensuring the public became actual participants in the aquarium experience. The modern day visitors seek as near an interactive experience as possible and hope to be transformed into the marine environment and witness for themselves the marine underwater world.

The concept of modern marine aquarium-keeping in the home has its origins in the United Kingdom and Germany. The United States is now the world's most developed market in terms of households maintaining aquaria, especially those holding exotic marine species: there are about 2.5 million marine hobby aquariums in the USA. In Holland and Germany, the emphasis has been on reef culture, a hobby which is becoming more widespread. The manufacture of products designed specifically for the ornamental fish trade first began in 1954 and scientific and technical advances during the 1960s brought aquarium keeping a very long way from the goldfish and goldfish bowl. The development of suitable materials for marine aquaria has been hampered by the corrosive nature and the toxicity of materials when immersed in sea water. One of the first authoritative texts on materials and methods for marine aquaria was written by Spotte in 1970, followed by Hawkins in 1981.

The volume of marine ornamental fish involved in the international trade is difficult to calculate accurately since records are poor. Current estimates are between 100 and 200 tonnes per annum which probably corresponds to more than 20 million individual specimens. The trade is highly dependent on harvesting from the wild. In Sri Lanka, Indonesia, the Philippines, Fiji and Cook Islands, the export of tropical reef species is now one of the most important export industries employing significant numbers of village people.

Members of the family Pomacentridae, in particular clown fishes, *Amphiprion* spp., and blue–green chromis, *Chromis viridis*, are central to the industry and cleaner wrasse, *Labroides dimidiatus*, flame angels, *Centropyge loriculus*, red hawks, *Neocirrhites armatus*, tangs, *Acanthus* spp., and seahorses, *Hippocampus* spp. are important. The fire shrimp, *Lysmata debelius*, is a major species with respect to invertebrates.

Historically, culture of marine ornamentals has not been in competition with the wild harvest industry. However, recent advances in aquaculture technology will undoubtedly enable more marine ornamentals to be farmed. One European hatchery already has an annual production of *Amphiprion* spp. which is 15 times that exported from Sri Lanka, whereas in Australia seahorse farming is gaining momentum. In both Europe and USA, commercial production of fire shrimp, *L. debelius*, is being attempted. To date the greatest impediment to culture lies in the fact that many popular marine ornamentals produce tiny, free-floating eggs and the newly hatched larvae either do not accept traditional prey used for rearing such as rotifers and brine shrimp, or the prey has proved nutritionally inadequate.

During the 1980s, the Dutch and German aquarists pioneered the development of miniature reef aquaria. Their success among other things, depended on efficient means of purifying the water. The Dutch developed a 'wet and dry' or trickle filter. These filters acted as both mechanical and biological systems. More compact and efficient trickle filter systems have been developed with the advent of Dupla bioballs

during the 1990s. These aquaria are filled with so-called 'living rock' which is initially removed from coral reefs.

Coral culture *per se* has recently been established in the Philippines, Solomon Islands, Palau, Guam, and the United States and is aimed at reducing the need to remove living coral from natural reefs.

The upsurge in popularity in marine ornamentals over the past 30 years for both public and hobby aquaria has raised serious conservation concerns. There are calls for sustainable management of coral reefs worldwide and even bans on harvesting of all organisms including 'living rock'.

The sea horses in particular have received attention from aquarists and conservationists. Sea horses are one of the most popular of all marine species and are probably responsible for converting more aquarists to marine aquarium-keeping than any other fish. Unlike most other marine ornamentals sea horses have been bred in captivity for many years.

In 1996, the international conservation group TRAFFIC (the monitoring arm of the World Wide Trust for Nature) claimed sea horses were under threat from overfishing for use in traditional medicines, aquaria, and as curios. According to TRAFFIC at least 22 countries export sea horses, the largest known exporters being India, the Philippines, Thailand and Vietnam. Importers for aquaria include Australia, Canada, Germany, Japan, The Netherlands, United Kingdom, and the United States. The species commonly in demand are shown in **Table 1**.

The greatest problem for both sides in the debate over the trade in wild-caught marine ornamental fishes in general is a historical lack of scientific data on the biology of these animals both in their natural and captive environments. It is true to say that most of the information about marine ornamentals is derived from intelligence gathering by aquarists and scientific rigor has been applied in the case of only a few species. One such species is the pot-bellied sea horse, *Hippocampus abdominalis*, which has been studied both in the wild and captive environment for several years. Data on this species serve as useful

Table 1 Sea horse species commonly kept in aquaria

Aquarium common name	Species name	Geographic reference	Size (cm)	Color
Dwarf sea horse	*Hippocampus zosterae*	Florida coast	5	Green, gold, black, white
Northern giant	*H. erectus*	New Jersey coast	20	Mottled, yellow, red, black, white
Spotted sea horse	*H. reidi*	Florida coast	18	Mottled, white, black
Short-snouted	*H. hippocampus*	Mediterranean	15	Red/black
Mediterranean	*H. ramulosus*	Mediterranean	15	Yellow/green
Golden or Hawaiian	*H. kuda*	Western Pacific	30	Golden
Pot-bellied sea horse	*H. abdominalis*	Southern Australian coast	30	Mottled, gold, white, black

comparative tools for knowledge about other ornamental fishes.

The Pot-bellied Sea Horse and Other Aquarium Species

Species Suitability for Aquaria

Table 2 sets out major parameters affecting life support in marine aquaria. Factors about a species that it would be advantageous to know prior to selection for the aquarium are:

- water temperature range,
- water quality requirements,
- behavior and habitat (territorial, aggressive, cannibalistic, pelagic, benthic),
- diet,
- breeding biology,
- size and age,
- ability to withstand stress,
- health (resistance and susceptibility).

Water Temperature Range

The pot-bellied seahorse *H. abdominalis* has a broad distribution along the coastal shores of Australia, being found from Fremantle in Western Australia eastwards as far as Central New South Wales, all around Tasmania and also much of New Zealand. Within its range water temperatures may reach 28°C for several months and fall to 9°C for a few weeks. Acclimation trials in the laboratory and in home aquaria have shown that this species will live at water temperatures as high as 30°C and as low as 8°C. Unlike many marine ornamental fishes, *H. abdominalis* is eurythermal.

Many marine aquaria are kept between 24 and 26°C which is considered satisfactory for a number of marine ornamentals, however some species require higher temperatures, for example some butterfly fishes (Chaetodontidae) survive best at 29°C. Many tropical coral reef species are stenothermous and are difficult to maintain in temperate climates without accurate thermal control. The more eurythermous a species, the easier it will be to acclimate to a range of water temperature fluctuations.

Water Quality Requirements

Salinity The salinity of sea water may alter due to freshwater run-off or evaporation. Some marine species are less tolerant than others to salinity changes, and it is important to determine whether or not a species will survive even relatively minor changes. *H. abdominalis* is euryhaline, growing and

Table 2 Parameters affecting life support in marine aquaria

Parameter	Factor
Physical	
Temperature	
Salinity	
Particular matter	Composition
	Size
	Concentration
Light	Artificial/natural
	Photoperiod
	Spectrum
	Intensity
Water motion	Surge
	Laminar
	Turbulence
Chemical	
pH and alkalinity	
Gases	Total gas pressure
	Dissolved oxygen
	Un-ionized ammonia
	Hydrogen sulfide
	Carbon dioxide
Nutrients	Nitrogen compounds
	Phosphorus compounds
	Trace metals
Organic compounds	Biodegradable
	Nonbiodegradable
Toxic compounds	Heavy metals
	Biocides
Biological	
Bacteria	
Virus	
Fungi	
Others	

breeding at salinities between 15–37 parts per thousand (‰). It is a coastal dwelling species being recorded at depths of 1–15 m and may be present in a range of habitats from estuaries, open rocky substrates and artificial harbors. Often, pot-bellied sea horses can be found attached to nets and cages used to farm other fish species. The somewhat euryecious behavior of this sea horse has probably resulted in its broad tolerance of salinities.

Many marine aquaria depend on artificial sea water which is purchased as a salt mixture and added to dechlorinated fresh water. Natural sea water is a complex chemical mixture of salts and trace elements. It has been shown that some artificial seawater mixtures are unsuitable for marine plants and even particular life stages of some fishes. Particular attention must be paid to trace elements in coral reef aquaria when using either natural or artificial salt water. Coral reef aquaria are also more sensitive to salinity changes than general fish aquaria. **Table 3** gives a useful saltwater recipe.

Table 3 The Wiedermann–Kramer saltwater formula

In 100 liters of distilled water:	
Sodium chloride (NaCl)	2765 g
Magnesium sulfate crystals (MgSO$_4 \cdot$7H$_2$O)	706 g
Magnesium chloride crystals (MgCl$_2 \cdot$6H$_2$O)	558 g
Calcium chloride crystals[a] (Cacl$_2 \cdot$6H$_2$O)	154 g
Potassium chloride[a] (KCl)	69.7 g
Sodium bicarbonate (NaHCO$_3$)	25 g
Sodium bromide (NaBr)	10 g
Sodium bicarbonate (NaCO$_3$)	3.5 g
Boric acid (H$_3$BO$_3$)	2.6 g
Strontium chloride (SrCl$_2$)	1.5 g
Potassium iodate (KIO$_3$)	0.01 g

[a]The potassium chloride should be dissolved separately with some of the 100 liters of distilled water as should the calcium chloride. Add these after the other substances have been dissolved.

Other Water Quality Guidelines

Various tables have been provided setting out water quality guidelines for mariculture. **Table 4** is useful for well-stocked marine fish and coral reef aquaria but is possibly too rigid for lightly stocked fish tanks.

The toxicity of several parameters given in **Table 3** may be reduced by ensuring the water is always close to saturation with respect to dissolved oxygen concentration (DOC). Studies on *H. abdominalis* have indicated that this species becomes stressed when DOC falls below 85% saturation. Furthermore, the species is tolerant of much higher un-ionized ammonia (NH$_3$-N) levels when the water is close to DOC saturation. A combination of low dissolved oxygen (<80%) and NH$_3$N greater than 0.02 mg l^{-1} may result in high mortalities.

Ammonia is the major end product of protein metabolism in sea horses and most aquatic animals. It is toxic in the un-ionized form (NH$_3$). Ammonia concentration expressed as the NH$_3$ compound is converted into a nitrogen basis by multiplying by 0.822. The concentration of un-ionized ammonia depends on total ammonia, pH, temperature, and salinity. The concentration of un-ionized ammonia is equal to:

$$\text{Un-ionized ammonia}\left(\text{mgl}^{-1} \text{ as } \text{NH}_3\text{-N}\right) = (a)(\text{TAN})$$

where a = mole fraction of un-ionized ammonia and TAN = total ammonia nitrogen (mg l^{-1} as N).

Table 5 gives the mole fraction for given temperatures and pH in sea water.

The concentration of un-ionized ammonia is about 40% less in sea water than fresh water, but its toxicity is increased by the generally higher pH in the former. **Figure 1(A)** shows the operation of a simple

Table 4 Water quality levels for the aquarium

Parameter	Level
Dissolved oxygen	90–100% saturation (>6 mg l^{-1})
Ammonia	<0.02 mg l^{-1} NH$_3$-N
Nitrite	<0.1 mg l^{-1} NO$_2$-N
Hydrogen sulfide	<0.001 mg l^{-1} as H$_2$S
Chlorine residual	<0.001 mg l^{-1}
pH	7.8–8.2
Copper	<0.003 mg l^{-1}
Zinc	<0.0025 mg l^{-1}

Table 5 Mole fraction of un-ionized ammonia in sea water[a]

Temp (°C)	pH				
	7.8	7.9	8.0	8.1	8.2
20	0.0136	0.0171	0.0215	0.0269	0.0336
25	0.0195	0.0244	0.0305	0.0381	0.0475
30	0.0274	0.0343	0.0428	0.0532	0.0661

[a]Modified from Huguenin and Colt (1989).

subgravel filter which removes ammonia and nitrite by nitrification and **Figure 1(B)** shows the configuration of a power filter using both nitrification and absorption to remove ammonia. Both methods have been used to maintain water quality in marine ornamental aquaria.

Unfortunately, information on toxicity levels of ammonia for marine ornamentals is poorly documented but **Table 4** is probably a useful guide given the data on cultured species.

The pot-bellied sea horse is intolerant to even low levels of hydrogen sulfide (H$_2$S) (**Table 4**). This gas is difficult to measure at low levels thus care is required to avoid anoxic areas in aquaria. H$_2$S is almost certainly toxic to other marine ornamentals also, particularly reef dwellers.

Chlorine, copper and zinc have all proved toxic to *H. abdominalis* at levels exceeding those shown in **Table 4**. Chorine and copper are often used in aquaria: the former to sterilize equipment and the latter in treatment of various diseases. Furthermore, chlorine may be present in tap water when mixed with artificial seawater mixtures. Great care is required in the use of these chemicals. Available aquarium test kits are seldom sensitive enough to detect chronic chlorine concentrations. Often 1–5 mg l^{-1} sodium thiosulfate or sodium sulfite are used to remove chlorine but for some marine species these too may prove toxic. Bioassays for chlorine toxicity using marine ornamentals have not been carried out.

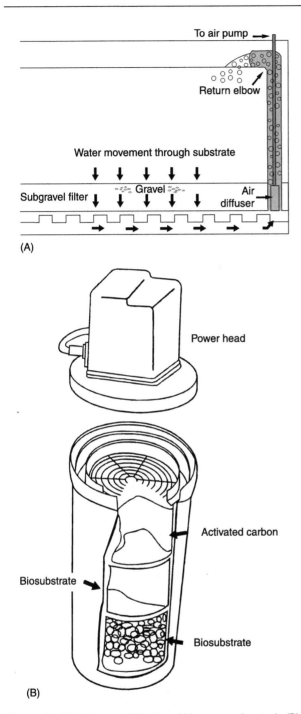

Figure 1 (A) Undergravel filtration within an aquarium tank. (B) Canister filter with power head and filter media chambers.

Copper toxicity can be significantly reduced with the addition of 1–10 mg l^{-1} of EDTA. EDTA is also a good chelating agent for zinc.

Behavior and Habitat

It is widely believed that sea horses spend their time anchored by their prehensile tails to suitable objects. This is not necessarily true. *H. abdominalis* is an active species feeding both in the water column and over the substratum. However, at night the fish 'roosts' often in association with other specimens. Furthermore, this species is remarkably gregarious and stocking levels as high as 10 fish, 6 cm in length per liter, have been regularly maintained in hatchery trials.

Although sea horses may tolerate the presence of several of their own species, their slow feeding behavior puts them at a competitive disadvantage in a mixed species aquarium, where faster feeders will ingest the sea horses' food.

Predation is a serious problem in marine aquaria. Sea horses are known to be prey for other fishes both in the wild and in aquaria. Members of the antennariids, particularly the sargassum fish, *Histrio histrio*, are known to feed on sea horses as are groupers (Serranidae) and trigger Fishes (Balistinae), flatheads (Platycephalidae) and cod (Moridae).

Territorial species are common among the coral-dwelling fishes and such behavior makes them difficult to keep in mixed-species aquaria. The blue damsel, *Pomacentrus coelestris*, shoals in its natural habitat but becomes pugnacious in the confines of the aquarium.

Several marine ornamentals seek protection or are cryptic. The majority of clown or anemone fish (*Amphiprion*) retreat into sea anemones if threatened, in particular, the anemones *Stoichaetis* spp., *Radianthus* spp., and *Tealia* spp., whereas some wrasse dive beneath sand when frightened. Cryptic coloration is seen in the sea horses and color changes have often been reported.

The cleaner wrasse, *Labroides dimidiatus*, lives in shoals over reefs but is successfully maintained singly in the aquarium, where even the most aggressive fish species welcome its attention. Other wrasses mimic the coloration and shape of *L. dimidiatus* simply to lure potential prey towards them.

Behavior and habitat of many marine ornamentals has been gleaned from observation only, but failure to understand these factors make fish-keeping difficult.

Diet

The dietary requirements are poorly known for marine ornamentals and many artificial feeds may do little more than prevent starvation without live or frozen feed supplements. Furthermore, a given species may require different foods at various life stages. Several stenophagic species are known in the aquarium mainly consisting of algal and live coral feeders, for example the melon butterfish *Chaetodon trifasciatus*. The diet of others may not even be known in spite of such fish being sold for the aquarium. The regal angelfish, *Pygoplites diacanthus*, seldom lives for long in the aquarium and dies from starvation.

Sea horses are easy to feed but require either live or frozen crustacea or small fish. The pot-bellied seahorse has been reared through all growth stages using diets of enriched brine shrimp, *Artemia salina*, live or frozen amphipods, small krill species, and fish fry. The ready acceptance and good growth rates recorded in hatchery-produced pot-bellied sea horses using 48-h-old enriched brine shrimp have resulted in significant numbers of sea horses being raised in at least one commercial farm. Apart from hatcheries for clown fishes and pot-bellied sea horses, the intensive culture of marine ornamentals has proved difficult due to a lack of suitable prey species for fry.

Breeding Behavior

Breeding behavior can be induced in several ornamental fishes with appropriate stimuli and environments. The easier species are sequential hermaphrodites such as serranids. The pomacentrids of the genus *Amphiprion* are a further good example. However, sea horses have been extensively studied.

The sea horses are unique in that the male receives, fertilizes and broods the eggs in an abdominal pouch following a ritualized dance. Much has been made of monogamy but as further studies are undertaken scientific support for such breeding behavior is being questioned. *H. abdominalis* is polygamous both in the wild and captivity.

The pot-bellied sea horse in captivity, at least, shows breeding behavior as early as four months of age and males may give birth to a single offspring; by one year of age males may give birth to as many as 80 fry and at two years of age 500 fry. Precocity in other marine ornamentals is not recorded but may exist.

Size and Age

The size and age of aquarium fish have been seldom studied scientifically but has been observed. Groupers (Serranidae) and triggerfishes (Balistinae) quickly outgrow aquaria, and some angelfishes and emperors may show dramatic color changes with age, becoming more or less pleasing to aquarists.

Longevity likewise is unknown for most marine ornamentals but the pygmy sea horse *H. zosterae* lives for no more than two years, whereas *H. abdominalis* may live for up to nine years.

Stress

Stress probably plays a pivotal role in the health of marine ornamentals but scientific studies have not been undertaken. The aquarist would do well to remember that stress suppresses aspects of the immune response of fishes and that studies on cultured species demonstrate that capture, water changes, crowding, transport, temperature changes, and poor water quality induce stress responses. Furthermore, stress can be cumulative and some species may be more responsive than others. Farmed species tend to show a higher stress threshold than wild ones. The potential advantage of purchasing hatchery-reared ornamentals (if available) are obvious, since survival in farmed stock should be greater than in wild fish held in aquaria.

Health Management

Good health management results from an understanding of the biological needs of a species. Treatment with chemicals is a short-term remedy only and the use of antibiotics may exacerbate problems through bacterial resistance.

A considerable number of pathogens have been recorded in marine aquarium fishes and include viruses, bacteria, protozoa, and metazoa. Most diseases have been shown not to be peculiar to a given species but epizootic. For example, several of the disease organisms recorded in sea horses, in particular *Vibrio* spp., protozoa and microsporidea, are known to infect other fish species also.

In coral reef aquaria, nonpathogenic diseases due to poor water quality may be common and in-depth knowledge pertaining to the husbandry of such systems is essential for their well-being.

Coral Reef Aquaria

The Challenge

The coral reef is one of the best-adapted ecosystems to be found in the world. Such reefs are biologically derived and the organisms which contribute substantially to their construction are hermatypic corals although ahermatypic species are present. Coral reefs support communities with a species diversity that far exceeds those of neighboring habitats and the symbiotic relationship between zooxanthellae and the scleractinian corals are central to the reef's well-being. Zooxanthellae are also present in many octocorallians, zoanthids, sea anemones, hydrozoans and even giant clams. As zooxanthellae require light for photosynthesis, the reef is dependent on clear water.

Coral reefs are further restricted by their requirement for warm water at 20–28 °C, and the great diversity of life demands a plentiful supply of oxygen. The challenge for the aquarist lies in the need to match the physical parameters of the water in the aquarium as closely as possible with sea water of the reef itself.

Physical Considerations

Temperature The recommended water temperature for coral reef aquaria is a stable 24°C. The greater the temperature fluctuations the less the diversity of life the aquarium will support. At temperatures less than 18°C the reef will die and above 30°C increasing mortalities among zooxanthellae will occur leading to the death of hermatypic corals.

Light Water bathing coral reefs has a blue appearance which has been called the color of ocean deserts. Most of the primary production is the result of photosynthesis by benthic autotrophs (zooxanthellae) rather than drifting plankton. Photosynthetic pigments of zooxanthellae absorb maximally within the light wavelength bands that penetrate furthest into sea (400–750 μm) and therefore clear oceanic water is essential.

In the aquarium, both fluorescent and metal halides are available which will supply light at the correct wavelength. Actimic-03 fluorescents combined with white fluorescents are suitable. Typically three tubes, two actinic and one white, will be needed for a 200 liter tank. Metal halide lamps cannot be placed close to the tank because of heat and such lamps produce UV light which will destroy some organisms. A glass sheet placed over the aquarium will prevent UV penetration. One 175 W lamp is recommended per 60 cm of aquarium length.

Water movement Coral reefs are subjected to various types of water movements, namely surges, laminar currents, and turbulence. These water motions play an essential role by bringing oxygen to the corals and plants, removing detritus and, in the case of ahermatypic corals, transporting their food.

In the aquarium these necessary water movements must be present. Power head filters are available which produce acceptable surges and currents.

Biological Considerations

Nutrients Coral reefs are limited energy and nutrient traps: rather than being lost to deep water sediments, some organic compounds and nutrients are retained and recycled. However, water movements rid the reef of dangerous excess nutrients which might promote major macrophyte growth.

The coral reef aquarium soon becomes a nutritional soup if both organic compounds and nutrients are not recycled or removed. Although skilled and knowledgeable aquarists are able to use the

biological components of the reef itself to produce an autotrophic system, most employ protein skimmers, mechanical filters and biological filters to prevent poisoning of the system. **Figure 2** represents nutrient cycling over a coral reef and **Figure 1** shows a suitable filter for reef aquaria. Removal of nitrate can be achieved through the use of specialized filters which grow denitrifying bacteria.

Living rock Living rock is dead compacted coral which has been colonized by various invertebrates. In addition, there will be algae and bacteria. Different sources of living rock will provide different populations of organisms. Over time the organisms which survived the transfer from reef to aquarium become established

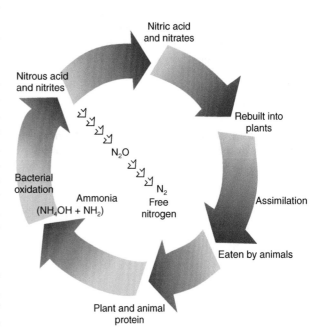

Figure 2 Nutrient cycling over a coral reef.

Table 6 Some additional faunal components for coral reef aquaria

Niche	Common name	Scientific name
Detritus feeder	Anemone shrimp	*Periclimenea brevicarpalis*
	Shrimp	*P. pedersoni*
	Fiddler crab	*Uca* spp.
Algal feeder	Tiger cowrie	*Cypraea tigris*
	Money cowrie	*C. moneta*
	Starfish	*Patiria* spp.
Plankton feeders	Mandarin fish	*Synchiropus splendidus*
	Psychedelic fish	*S. picturatus*
	Midas blenny	*Escenius midas*

and the aquarium is ready for corals. By the time the corals are placed in the aquarium all filter systems must be operating efficiently. Ahermatypic corals should be introduced first and placed in the darker regions of the tank since they feed on plankton. Once hermatypic corals are introduced appropriate blue light for at least 12 hours each day must be available, and strict water quality maintained (**Table 4**). Nitrate must not rise above $15 \, mg \, l^{-1}$ and pH fall below 8.

Additional faunal components The living rock will introduce various invertebrates. Additional species must be selected carefully and on a scientific rather than an esthetic basis. The introduction of detritus and algal feeders will probably be essential and coral eaters must be avoided. Fish species require high protein diets which will necessitate further reductions of ammonia, nitrite, and nitrate from the aquarium. The species given in **Table 6** might be considered but there are many others. Selection will depend on the inhabitants.

See also

Corals and Human Disturbance. Mariculture Diseases and Health. Mariculture Overview.

Further Reading

Barnes RSK and Mann KH (1995) *Fundamentals of Aquatic Ecology.* Oxford: Blackwell Scientific.

Emmens CW (1995) *Marine Aquaria and Miniature Reefs.* Neptune City: T.F.H. Publications.

Hawkins AD (1981) *Aquarium Systems.* New York: Academic Press.

Huguenin JE and Colt J (1989) *Design and Operating Guide for Aquaculture Seawater Systems.* Amsterdam: Elsevier.

Lawson TB (1995) *Fundamentals of Aquacultural Engineering.* New York: Chapman & Hall.

Spottle S (1979) *Seawater Aquariums – the Captive Environment.* New York: John Wiley.

Timmons MB and Losordo TM (1994) *Aquaculture Water Reuse Systems. Engineering Design and Management.* Amsterdam: Elsevier.

Untergasser D (1989) *Handbook of Fish Diseases.* Neptune City: T.F.H. Publications.

MARICULTURE OF MEDITERRANEAN SPECIES

G. Barnabé, Université de Montpellier II, France
F. Doumenge, Musée Océanographique de Monaco, Monaco

Basic Requirements

Obtaining Stock for Ongrowing

The starting point of any farming operation is the acquisition of stock for rearing; these may be spat for mollusks or alevins, fry or juveniles for fish.

From the wild For the Mediterranean mussel (*Mytilus galloprovincialis*), spat is always collected from the wild, from rocky shores or shallow harbors where they are abundant. Conditions are less favorable for oyster culture; the native (flat) oyster (*Ostrea edulis*) is captured only in the Adriatic and the other remnants of natural stocks are unable to support intensive culture. Spat from Japanese (cupped) oysters (*Crassostrea gigas*) has to be imported from the Atlantic coast. Clam culture utilizes both spat from Mediterranean species (*Tapes decussatus*) and a species originating in Japan (*Tapes philippinarum*) which has spread very rapidly, especially in the Adriatic.

Juvenile marine fish such as eels (*Anguilla anguilla*), mullets (*Mugil* sp.), gilthead sea bream (*Sparus aurata*) and sea bass (*Dicentrarchus labrax*) are traditionally captured in spring in the mouths of rivers or in traps or other places in protected lagoons in Italy; these form the basis of the valliculture of the northern Adriatic, a type of extensive fish culture. In practice, elvers and yellow eels are supplied mainly from fisheries in other Mediterranean lagoons, especially in France.

Bluefin tuna (*Thunnus thynnus*) for ongrowing are taken from the spring and summer fishery for juveniles weighing a few tens of kilograms by Spanish seine netters in the western Mediterranean and Croatians in the Adriatic.

Controlled reproduction in hatcheries The transition to intensive fish culture has been made possible by the control of reproduction in sea bass and sea bream (*see* Mariculture Overview). In 1999 around 100 hatcheries produced 209 million sea bass fry and 242 million gilthead sea bream fry; this represents respectively a doubling and a 50% increase on production in 1997. Hatcheries are very different from the structures used for ongrowing; France exports tens of millions of juveniles to all the Mediterranean countries.

Access to Technology

As an activity, aquaculture is becoming increasingly complex with regard to the technology associated with rearing as well as the interactions with the physical and socio-economic environment. It requires more specialized training than can be obtained in the traditional workplace. Management of a hatchery, monitoring of water quality, and genetic research are examples of the new requirements for training.

The transition to cage culture took place using cages designed and manufactured for salmonid culture. Manufacturing feed granules has developed only recently in countries such as Greece.

Feeding

Mollusk culture exploits the natural production of plankton and thus follows natural changes in productivity.

The extensive valliculture systems used in northern Italy in managed protected lagoons are based on natural production improved through control of the water (fish trapped in the channels communicating with the sea, overwintering in deep areas) and input of juveniles. Production remains low ($20\,\mathrm{kg\,ha^{-1}\,y^{-1}}$).

Intensive fish culture in cages uses pelleted protein-rich diets where fishmeal comprises 60% of the dry weight. However, bluefin tuna only consume whole fish or fresh or frozen cephalopods during ongrowing. Their conversion rate is exceptionally poor, of the order of 15–20% depending on water temperature.

Transport

The aquaculture cycle requires dependable and rapid transport to take juveniles to ongrowing facilities and, especially, to move the final production output which is generally valuable and perishable to the consumer market, and must remain chilled at all times.

For isolated sites, particularly small islands, there is always the problem that the transport of feed and various items of equipment is expensive.

Manpower Requirements

Hatcheries require a specialized and motivated workforce, as the production cycle for fry must not be interrupted. Sea sites are becoming ever more highly mechanized, even in countries such as Turkey where labor is cheap, because the volumes to be handled are increasing continuously.

Capital

Mediterranean aquaculture requires heavy investment both for setting up and for operating. It is dependent on international capital. Scandinavian, British, and Japanese interests play a considerable part, alongside local entrepreneurs.

Distinctive Aspects

Favorable

Sheltered waters   There are numerous large areas of sheltered water around the Mediterranean, the shoreline having been submerged and straightened out by the rapid rise in sea level following the last glaciation, 12 000 years ago. Nowadays, many bays provide big expanses of water which are both deep and sheltered (Carthagena, Toulon, La Spezia, Gaete, Naples, Trieste, Thessaloniki, etc.). These waters are productive, but are threatened by pollution.

Channels between the islands of the Dalmatian archipelago or in the Aegean Sea or the large gulfs surrounding the Hellenic peninsular (Patras, Corinth, Thessaloniki, Argolis, Evvoia, etc.), and to the north of Entail are perfectly suited to the types of aquaculture systems that benefit from a rapid turnover of water.

Above all, the input of sediment from deltas or the effects of littoral erosion have led to the build-up of sand bars forming lido-type complexes of impounded lagoons where the waters experience strong variations in salinity and high productivity because of the movements through gaps in the barriers.

High demand from local markets   The Latin countries of the north-west Mediterranean and Adriatic have a culinary tradition based on the consumption of large quantities of seafood. Markets in Spain, Mediterranean France, and Italy pay high prices for the fresh products of mollusk culture and marine fish farming.

There is also the strong seasonal demand from more than 100 million tourists who travel around all the coasts and particularly the islands of the Mediterranean, except where there are political problems.

Unfavorable

Eutrophication, toxic blooms, bacterial pollution   As elsewhere in the world, Mediterranean aquaculture cannot escape problems of disease. Nodovirus has hit fish in cages and two protozoans have decimated cultured stocks of the native oyster, *Ostrea edulis*.

Harmful blooms frequently affect lagoons where shellfish are cultured and the presence of serious pathogens (cholera, salmonella) has prohibited the use of certain zones for production.

Extreme temperatures   In sheltered lagoons or enclosed waters (e.g., the Bougara Sea, Tunisia), temperatures may exceed 30°C in summer, causing harmful blooms which are prejudicial to the success of aquaculture. In the north-west Mediterranean and the northern Adriatic, the water temperature in the lagoons may drop to below + 5°C and ice may develop on the surface of the lagoons. In general, thermal conditions favor the eastern Mediterranean.

Competition for space (urbanization, tourism)   Littoral space is under pressure from several users; tourism, navigation, and especially the extension of industrial and urban developments. Land bases are essential for mariculture. Lack of sites is further aggravated by complex regulations which are applied rigorously. Finally, the tendency to designate large areas of wetland and similar areas of sea for nature conservation removes significant areas from the expansion of aquaculture.

Dependence on fisheries   When juvenile mollusks or fish for mariculture rearing are taken from the natural environment they are supplied by fisheries. This dependence is a major risk to the effective operation of the production process. In the absence of industrial pelagic fisheries there is no fishmeal production in the Mediterranean. In order to satisfy the demand linked to the expansion of fish culture, almost all components of the feed must be imported. This places production further under external control. Substitution of plant for animal protein in the diet is becoming necessary.

Cultural limits of the markets   While the culture of Roman Catholic Christianity provides for a boosted local market, the Orthodox Christian culture is far less demanding in terms of seafood products. The Muslim countries of the southern coast and the Levantine Basin do not have the monetary resources nor, significantly, a dietary tradition adapted to the products of marine aquaculture. In contrast, the Mediterranean bluefin tuna achieves high prices on the Japanese sashimi market.

Production Systems

Mollusk Culture

The oyster reared everywhere (although sometimes in small quantities) is the cupped or Japanese oyster (*Crassostrea gigas*) introduced to Europe from Japan in the 1960s to replace the Portuguese oyster (*Crassostrea angulata*), stocks of which had been drastically reduced by disease. One single major center, the Etang de Thau, produces around 12 000 tonnes each year using spat taken from the Atlantic coast.

The native mussel (*Mytilus galloprovincialis*) is reared in small quantities around the Spanish coast (Ebro Delta, Mar Menor). In contrast, in Italy annual production is between 100 000 and 130 000 tonnes; the main sites are the Gulf of Taranto and the northern and mid-Adriatic, as well as several bays in the Tyrrhenian Sea (La Spezia, Gaete, and Naples). Along the French coast, only the Etang de Thau has a production of a few thousand tonnes. In Greece, the Gulf of Thessaloniki has a production of the same order.

For around 20 years, Japanese-type long lines have been installed in the open sea off Languedoc Rousillon in waters between 15 and 35 m deep. These structures resist the forces of the sea and production can reach tens of thousands of tonnes of mussels, but is strongly limited by storms and predation by sea bream.

Oysters and mussels are always cultured in suspension on ropes either in lagoons or in the sea. The spat of oyster is captured on the Atlantic coast on various substrates and is often attached with the help of quick-setting cement on the rearing bars, as this technique yields the best results. Other farmers leave the spat to develop on the mollusc shells where they have attached; the shells are placed in nets, spaced regularly, on the rearing ropes.

Mussel spat, collected from the wild, is placed in a tubular net, which is then attached to the rearing ropes. Mussels attach themselves to the artificial substrate with the aid of their byssus. It is therefore necessary to detach and clean the mussels once or twice during their growth.

Clams are especially abundant around the Italian Adriatic coast. The Japanese species, introduced as hatchery-reared spat at the beginning of the 1980s to seed protected areas, has spread very rapidly and invaded neighboring sectors. The density of these mollusks can reach 4000 individuals per square meter and production is around 40 000 tonnes per annum. A large part of the production is exported to Spain.

Culture of Sea Bass and Gilthead Sea Bream

Control and management of the spawning of sea bass and sea bream has made it possible to respond to the demand for high quality aquatic products that cannot be supplied on a regular basis from fisheries based on wild stocks. This production is based on hatcheries (*see* Mariculture Overview). Broodstock can be held in cages or in earth ponds. In such ponds, control of the photoperiod and temperature allows fertilized eggs to be obtained through almost the whole year. Larvae are reared in the same way as those of other marine fish (*see* Marine Fish Larvae) although sea bass can be fed on *Artemia nauplii* from first feeding; this does away with the burden of rearing rotifers in these hatcheries. The trend is towards enlargement to bring about economies of scale; some hatcheries produce 20 million juveniles each year; those producing fewer than 1–2 million juveniles per year are unlikely to be profitable.

Water is generally recycled within the hatchery to save energy and to maintain the stability of physicochemical characteristics. The quality of the water available in the natural environment determines the suitability of a site for a hatchery.

Almost all ongrowing is carried out in sea cages. All Mediterranean countries have contributed to the development and there is a tendency to move further and further east. Cages have been installed in sheltered coastal waters, but the scarcity of such sites and progress in cage technology are encouraging the development of exposed sites in the open sea. Techniques used in this type of farming resemble those used in the cage farming of salmonids. (*see* Salmonid Farming) and identical cages are used. These have a diameter of up to 20 m and are up to 10 m deep. Dry granular feed is used and the conversion rate is continuously improving (between 1.3 and 2).

Growth rate is determined by water temperature; in Greece a sea bass reaches a weight of 300 g in 11 months, but would take twice as long to reach the same weight on the French coast. This demonstrates the advantage of the eastern Mediterranean.

Other species, for which larval rearing is more difficult, are reared on a small scale using different techniques. These include the dentex (*Dentex dentex*), common (Couch's) sea bream (*Pagrus pagrus*) produced in mesocosms in Greece, and the greater amberjack (*Seriola dumerilii*), ongrown in Spain. Diversification of species offers the potential for increasing markets; this is becoming true for the meagre (*Argyrosomos regius*) and the red drum (*Scianops ocellatus*), and many other trials are taking place.

Ongrowing Bluefin Tuna

Japanese attempts at rearing larval bluefin tuna have not yet produced economically viable results. Mediterranean aquaculture depends on juveniles or sub-adults captured in the spring fishery, which are transported in nets supported by rafts to large cages where they will be kept for a few months. The towage, which must be at speeds of between 1 and 2 knots, may take several weeks.

The fish are harvested at the end of autumn and beginning of winter, before periods of low temperature and bad weather. In general, the tuna double their weight after 6–7 months of ongrowing and their flesh acquires the color and quality which puts them in demand for the Japanese sashimi market.

Regionalization

During the last 20 years, regional specialization has developed progressively, based on a balance of favorable and unfavorable factors for each of the types of mariculture.

The development of Mediterranean aquaculture production has thus been subject to an evolutionary process which has taken account not only of the major factors previously described but also conditions peculiar to each nation. This has produced contrasting situations, as decribed below.

France and Spain – Relatively Low Level of Aquaculture Development

This is due to a coincidence of relatively high levels of aquaculture development on the Atlantic coasts of both countries (mussels and fish in Galicia, mussels and oysters in the Bay of Biscay and the English Channel) and large quantities of imports supplementing regional production. Aquaculture developments remain small in size and are very spread out. One exception, demonstrating the possibilities that exist, is the success of the ongrowing of bluefin tuna in Spain in the Bay of Carthagena, from which 6000–7000 tonnes were sold in 1999–2000, for a revenue of over US $100 000 000.

The Size of the Italian Market

In the year 2000 only one third of the market was supplied by national production, in spite of rapid increases; production of sea bass (8800 tonnes) and gilthead sea bream (6200 tonnes) have both doubled from 1997. Italy absorbs almost all of the Maltese production (600 tonnes of sea bass and 1600 tonnes of sea bream in 2000); most of this comes from stock from eastern Mediterranean hatcheries.

The 40 000 tonnes of clams produced in the Adriatic saturate the market and the excess is exported to Spain. In spite of health problems, Italian mussel culture supports an annual market of 100 000–130 000 tonnes. In addition, the upper Adriatic has preserved the tradition of exploiting the valli and the lagoons.

In 2000 the Italian aquaculture market (by value) was supplied 40% by sea bass and sea bream (200 million lire), 33% by clams (165 million lire), and 27% by mussels (135 million lire).

The Dalmatian archipelago belonging to Croatia retains a sector of high quality traditional oyster culture. However, new possibilities have opened up with the transfer of technology and finance from Croatian emigrants to south Australia who have developed the ongrowing of bluefin tuna in the Ile de Kali, using the techniques they practiced in Port Lincoln. However, biological and logistic constraints are currently holding back development. Production from the Croatian businesses in 1999 and 2000 was limited to 1000 tonnes of small tuna (20–25 kg each); these receive relatively poor prices in Japan.

The Pioneering Front for Culture of Sea Bass and Sea Bream in the Eastern Mediterranean

This has reached Greece where production has gone from 1600 tonnes in 1990 to 6000 tonnes in 1992, 18 000 tonnes in 1996, 36 000 tonnes in 1998 and 56 000 tonnes in 2000. The movement then reached neighbouring Turkey, going from 1200 tonnes in 1992 to 3500 tonnes in 1994, reaching 12 500 tonnes in 1998 and 14 000 tonnes in 2000. Cyprus has also recently developed production which reached 1500 tonnes in 2000.

As part of this conquest of space, the first Greek seafish farms were installed between 1985 and 1995 in the west, in the Ionian islands and in Arcadia (center of the Peloponese). Then, via the Gulf of Corinth, they were joined from 1990 to 1995 by an active center in Argolis (east of the Peloponese). In addition, suitable sites for cages were found in the bays behind the barrier of the Isle of Evvoia. Finally, since 1990, developments have progressed to the Aegean around the archipelagos fringing the Peloponese. Between 1995 and 2000 Greek sites appear to have become saturated and mariculture has moved to the Turkish coast and the Anatolian bays of the Aegean coast.

Perspectives and Problems

Lack of Sites

The area dedicated to intensive cage mariculture remains modest: the whole of the French marine fish culture, around 6000 tonnes annual production, occupies no more than 10 ha of sea and 5 ha of land.

Pollution Ascribed to Aquaculture

It is politically correct to speak of the pollution derived from intensive cage rearing. When this alleged pollution (from fish excreting mainly nitrogen and phosphorus) is ejected into oligotrophic open waters such as the Mediterranean it can be seen as a benefit rather than a nuisance. The FAO, elsewhere, has suggested that significant increases in fish catches occur in areas where human-derived wastes have increased in the Mediterranean.

Health Limits and Shellfish

Production of mussels fluctuates widely: pollution of coastal and lagoon waters (toxic plankton and bacterial pollution) regularly prevents their sale. Regular consumption of mussels could be dangerous as the species concentrates okadaic acid (a strong carcinogen) produced by toxic phytoplankton.

Marketing Problems

Overproduction of sea bass and gilthead sea bream has led to a periodic collapse in prices. This phenomenon appears to be due to lack of planning, organization, and commercial astuteness by the producers, as well as competition between countries operating within different economic frameworks (cost of manpower). The salmon market is characterized by identical examples. For bluefin tuna, dependence on a single, distant market (the sashimi market in Japan) makes ongrowing a risky activity but very profitable in economic terms.

Conclusions

Traditional mollusk culture maintains its position but is encountering problems of limited availability of water. Transfer out to the open sea which is less polluted has been piloted in Languedoc (France) for two decades, but has demonstrated neither the suitability of the techniques nor their profitability, and production has stagnated.

The explosive growth of the production of marine fish in cages can be said to demonstrate the true revolution in Mediterranean mariculture. This is based entirely on species with a high commercial value. This type of rearing has expanded eastwards from the European Mediterranean countries but has not yet reached the southern shore.

Markets, particularly the huge European market of 360 million inhabitants, are not yet saturated. Diversification of the species produced may open up new markets. The expansion of cage-based mariculture has not yet finished, while progress in technology is unpredictable.

The major missing element in Mediterranean aquaculture is the rearing of penaeids, in spite of several sporadic but insignificant attempts at production in Southern Italy and Morocco.

See also

Mariculture Overview. Salmonid Farming.

Further Reading

Barnabe G (1974) Mass rearing of the bass *Dicentrarchus labrax* L. In: Blaxter JHS (ed.) *The Early Life History of Fish*, pp. 749–753. Berlin: Springer-Verlag.

Barnabe G (1976) Ponte induite ET élevage des larves du Loup *Dicentrarchus labrax* (L.) et de la Dorade *Sparus aurata* (L.). *Stud. Rev. C.G.P.M. (FAO)* 55: 63–116.

Barnabe G (1990) Open sea aquaculture in the Mediterranean. In: Barnabe G (ed.) *Aquaculture*, pp. 429–442. New York: Ellis Horwood.

Barnabe G (1990) Rearing bass and gilthead bream. In: Barnabe G (ed.) *Aquaculture*, pp. 647–683. New York: Ellis Horwood.

Barnabe G and Barnabe-Quet R (1985) Avancement et amélioration de la ponte induite chez le Loup Dicentrarchus labrax (L) à l'aide d'un analogue de LHRH injecté. *Aquaculture* 49: 125–132.

Doumenge F (1991) Meditérranée. In: *Encyclopaedia Universalis* pp. 871–873.

Doumenge F (1999) L'aquaculture des thons rouges et son développement économique. *Biol Mar Medi* 6: 107–148.

Ferlin P and Lacroix D (2000) Current state and future development of aquaculture in the Mediterranean region. *World Aquaculture* 31: 20–23.

Heral M (1990) Traditional oyster culture in France. In: Barnabé G (ed.) *Aquaculture*, pp. 342–387. New York: Ellis Horwood.

SEAWEEDS AND THEIR MARICULTURE

T. Chopin and M. Sawhney, University of New Brunswick, Saint John, NB, Canada

Introduction: What are Seaweeds and their Significance in Coastal Systems

Before explaining how they are cultivated, it is essential to try to define this group of organisms commonly referred to as 'seaweeds'. Unfortunately, it is impossible to give a short definition because this heterogeneous group is only a fraction of an even less natural assemblage, the 'algae'. In fact, algae are not a closely related phylogenetic group but a diverse group of photosynthetic organisms (with a few exceptions) that is difficult to define, by either a lay person or a professional botanist, because they share only a few characteristics: their photosynthetic system is based on chlorophyll *a*, they do not form embryos, they do not have roots, stems, leaves, nor vascular tissues, and their reproductive structures consist of cells that are all potentially fertile and lack sterile cells covering or protecting them. Throughout history, algae have been lumped together in an unnatural group, which now, especially with the progress in molecular techniques, is emerging as having no real cohesion with representatives in four of the five kingdoms of organisms. During their evolution, algae have become a very diverse group of photosynthetic organisms, whose varied origins are reflected in the profound diversity of their size, cellular structure, levels of organization and morphology, type of life history, pigments for photosynthesis, reserve and structural polysaccharides, ecology, and habitats they colonize. Blue-green algae (also known as Cyanobacteria) are prokaryotes closely related to bacteria, and are also considered to be the ancestors of the chloroplasts of some eukaryotic algae and plants (endosymbiotic theory of evolution). The heterokont algae are related to oomycete fungi. At the other end of the spectrum (one cannot presently refer to a typical family tree), green algae (Chlorophyta) are closely related to vascular plants (Tracheophyta). Needless to say, the systematics of algae is still the source of constant changes and controversies, especially recently with new information provided by molecular techniques. Moreover, the fact that the roughly 36 000 known species of algae represent only about 17% of the existing algal species is a measure of our still rudimentary knowledge of this group of organisms despite their key role on this planet: indeed, approximately 50% of the global photosynthesis is algal derived. Thus, every second molecule of oxygen we inhale was produced by an alga, and every second molecule of carbon dioxide we exhale will be re-used by an alga.

Despite this fundamental role played by algae, these organisms are routinely paid less attention than the other inhabitants of the oceans. There are, however, multiple reasons why algae should be fully considered in the understanding of oceanic ecosystems: (1) The fossil record, while limited except in a few phyla with calcified or silicified cell walls, indicates that the most ancient organisms containing chlorophyll *a* were probably blue-green algae 3.5 billion years ago, followed later (900 Ma) by several groups of eukaryotic algae, and hence the primacy of algae in the former plant kingdom. (2) The organization of algae is relatively simple, thus helping to understand the more complex groups of plants. (3) The incredible diversity of types of sexual reproduction, life histories, and photosynthetic pigment apparatuses developed by algae, which seem to have experimented with 'everything' during their evolution. (4) The ever-increasing use of algae as 'systems' or 'models' in biological or biotechnological research. (5) The unique position occupied by algae among the primary producers, as they are an important link in the food web and are essential to the economy of marine and freshwater environments as food organisms. (6) The driving role of algae in the Earth's planetary system, as they initiated an irreversible global change leading to the current oxygen-rich atmosphere; by transfer of atmospheric carbon dioxide into organic biomass and sedimentary deposits, algae contribute to slowing down the accumulation of greenhouse gases leading to global warming; through their role in the production of atmospheric dimethyl sulfide (DMS), algae are believed to be connected with acidic precipitation and cloud formation which leads to global cooling; and their production of halocarbons could be related to global ozone depletion. (7) The incidence of algal blooms, some of which are toxic, seems to be on the increase in both freshwater and marine habitats. (8) The ever-increasing use of algae in pollution control, waste treatment, and biomitigation by developing balanced management practices such as integrated multi-trophic aquaculture (IMTA).

This chapter restricts itself to seaweeds (approximately 10 500 species), which can be defined as marine benthic macroscopic algae. Most of them are members of the phyla Chlorophyta (the green seaweeds of the class Ulvophyceae (893 species)), Ochrophyta (the brown seaweeds of the class Phaeophyceae (1749 species)), and Rhodophyta (the red seaweeds of the classes Bangiophyceae (129 species) and Florideophyceae (5732 species)). To a lot of people, seaweeds are rather unpleasant organisms: these plants are very slimy and slippery and can make swimming or walking along the shore an unpleasant experience to remember! To put it humorously, seaweeds do not have the popular appeal of 'emotional species': only a few have common names, they do not produce flowers, they do not sing like birds, and they are not as cute as furry mammals! One of the key reasons for regularly ignoring seaweeds, even in coastal projects is, in fact, this very problem of identification, as very few people, even among botanists, can identify them correctly. Reasons for this include: a very high morphological plasticity; taxonomic criteria that are not always observable with the naked eye but instead are based on reproductive structures, cross sections, and, increasingly, ultrastructural and molecular arguments; an existing classification of seaweeds that is in a permanent state of revisions; and algal communities with very large numbers of species from different algal taxa that are not always well defined. The production of benthic seaweeds has probably been underestimated, since it may approach 10% of that of all the plankton while only occupying 0.1% of the area used by plankton; this area is, however, crucial, as it is the coastal zone.

The academic, biological, environmental, and economic significance of seaweeds is not always widely appreciated. The following series of arguments emphasizes the importance of seaweeds, and why they should be an unavoidable component of any study that wants to understand coastal biodiversity and processes: (1) Current investigations about the origin of the eukaryotic cell must include features of present-day algae/seaweeds to understand the diversity and the phylogeny of the plant world, and even the animal world. (2) Seaweeds are important primary producers of oxygen and organic matter in coastal environments through their photosynthetic activities. (3) Seaweeds dominate the rocky intertidal zone in most oceans; in temperate and polar regions they also dominate rocky surfaces in the shallow subtidal zone; the deepest seaweeds can be found to depths of 250 m or more, particularly in clear waters. (4) Seaweeds are food for herbivores and, indirectly, carnivores, and hence are part of the foundation of the food web. (5) Seaweeds participate naturally in nutrient recycling and waste treatment (these properties are also used 'artificially' by humans, for example, in IMTA systems). (6) Seaweeds react to changes in water quality and can therefore be used as biomonitors of eutrophication. Seaweeds do not react as rapidly to environmental changes as phytoplankton but can be good indicators over a longer time span (days vs. weeks/months/years) because of the perennial and benthic nature of a lot of them. If seaweeds are 'finally' attracting some media coverage, it is, unfortunately, because of the increasing report of outbreaks of 'green tides' (as well as 'brown and red tides') and fouling species, which are considered a nuisance by tourists and responsible for financial losses by resort operators. (7) Seaweeds can be excellent indicators of natural and/or artificial changes in biodiversity (both in terms of abundance and composition) due to changes in abiotic, biotic, and anthropogenic factors, and hence are excellent monitors of environmental changes. (8) Around 500 species of marine algae (mostly seaweeds) have been used for centuries for human food and medicinal purposes, directly (mostly in Asia) or indirectly, mainly by the phycocolloid industry (agars, carrageenans, and alginates). Seaweeds are the basis of a multibillion-dollar enterprise that is very diversified, including food, brewing, textile, pharmaceutical, nutraceutical, cosmetic, botanical, agrichemical, animal feed, bioactive and antiviral compounds, and biotechnological sectors. Nevertheless, this industry is not very well known to Western consumers, despite the fact that they use seaweed products almost daily (from their orange juice in the morning to their toothpaste in the evening!). This is due partly to the complexity at the biological and chemical level of the raw material, the technical level of the extraction processes, and the commercial level with markets and distribution systems that are difficult to understand and penetrate. (9) The vast majority of seaweed species has yet to be screened for various applications, and their extensive diversity ensures that many new algal products and processes, beneficial to mankind, will be discovered.

Seaweed Mariculture

Seaweed mariculture is believed to have started during the Tokugawa (or Edo) Era (AD 1600–1868) in Japan. Mariculture of any species develops when society's demands exceed what natural resources can supply. As demand increases, natural populations frequently become overexploited and the need for the cultivation of the appropriate species emerges. At present, 92% of the world seaweed supply comes from cultivated species. Depending on the selected species, their biology, life history, level of tissue specialization, and the socioeconomic situation of the region where it is developed, cultivation technology (**Figures 1–6**) can be low-tech (and still extremely successful with highly efficient and simple culture techniques, coupled with intensive labour at low costs) or can become highly advanced and mechanized, requiring on-land cultivation systems for seeding some phases of the life history before growth-out at open-sea aquaculture sites. Cultivation and seed-stock improvement techniques have been refined over centuries, mostly in Asia, and can now be highly sophisticated. High-tech on-land cultivation systems (**Figure 7**) have been developed in a few rare cases, mostly in the Western World; commercial viability has only been reached when high value-added products have been obtained, their markets secured (not necessarily in response to a local demand, but often for export to Asia), and labor costs reduced to balance the significant technological investments and operational costs.

Because the mariculture of aquatic plants (11.3 million tonnes of seaweeds and 2.6 million tonnes of unspecified 'aquatic plants' reported by the Food and Agriculture Organization of the United Nations) has developed essentially in Asia, it remains mostly unknown in the Western World, and is often neglected or ignored in world statistics ... a situation we can only explain as being due to a deeply rooted zoological bias in marine academics, resource managers, bureaucrats, and policy advisors! However, the seaweed aquaculture sector represents 45.9% of the biomass and 24.2% of the value of the world mariculture production, estimated in 2004 at 30.2 million tonnes, and worth US$28.1 billion (**Table 1**). Mollusk aquaculture comes second at 43.0%, and the finfish aquaculture, the subject of many debates, actually only represents 8.9% of the world mariculture production.

The seaweed aquaculture production, which almost doubled between 1996 and 2004, is estimated at 11.3 million tonnes, with 99.7% of the biomass being

Figure 1 Long-line aquaculture of the brown seaweed, *Laminaria japonica* (kombu), in China. Reproduced by permission of Max Troell.

Figure 2   Long-line aquaculture of the brown seaweed, *Undaria pinnatifida* (wakame), in Japan. Photo by Thierry Chopin.

Figure 3   Net aquaculture of the red seaweed, *Porphyra yezoensis* (nori), in Japan. Photo by Thierry Chopin.

cultivated in Asia (**Table 2**). Brown seaweeds represent 63.8% of the production, while red seaweeds represent 36.0%, and the green seaweeds 0.2%. The seaweed aquaculture production is valued at US$5.7 billion (again with 99.7% of the value being provided by Asian countries; **Table 3**). Brown seaweeds dominate with 66.8% of the value, while red seaweeds contribute 33.0%, and the green seaweeds 0.2%. Approximately 220 species of seaweeds are cultivated worldwide; however, six genera (*Laminaria* (kombu; 40.1%), *Undaria* (wakame; 22.3%), *Porphyra* (nori; 12.4%), *Eucheuma/Kappaphycus* (11.6%), and

Gracilaria (8.4%)) provide 94.8% of the seaweed aquaculture production (**Table 4**), and four genera (*Laminaria* (47.9%), *Porphyra* (23.3%), *Undaria* (17.7%), and *Gracilaria* (6.7%)) provide 95.6% of its value (**Table 5**). Published world statistics, which regularly mention 'data exclude aquatic plants' in their tables, indicate that in 2004 the top ten individual species produced by the global aquaculture (50.9% mariculture, 43.4% freshwater aquaculture, and 5.7% brackishwater aquaculture) were Pacific cupped oyster (*Crassostrea gigas* – 4.4 million tonnes), followed by three species of carp – the silver

Figure 4 Off bottom-line aquaculture of the red seaweed, *Eucheuma denticulatum*, in Zanzibar. Photo by Thierry Chopin.

Figure 5 Bottom-stocking aquaculture of the red alga, *Gracilaria chilensis*, in Chile. Photo by Thierry Chopin.

carp (*Hypophthalmichthys molitrix* – 4.0 million tonnes), the grass carp (*Ctenopharyngodon idellus* – 3.9 million tonnes), and the common carp (*Cyprinus carpio* – 3.4 million tonnes). However, in fact, the kelp, *Laminaria japonica*, was the first top species, with a production of 4.5 million tonnes.

Surprisingly, the best-known component of the seaweed-derived industry is that of the phyco-colloids, the gelling, thickening, emulsifying, binding, stabilizing, clarifying, and protecting agents known as carrageenans, alginates, and agars. However, this component represents only a minor volume (1.26 million tonnes or 11.2%) and value (US$650 million or 10.8%) of the entire seaweed-derived

industry (**Table 6**). The use of seaweeds as sea vegetable for direct human consumption is much more significant in tonnage (8.59 million tonnes or 76.1%) and value (US$5.29 billion or 88.3%). Three genera – *Laminaria* (or kombu), *Porphyra* (or nori), and *Undaria* (or wakame) – dominate the edible seaweed market. The phycosupplement industry is an emerging component. Most of the tonnage is used for the manufacturing of soil additives; however, the agrichemical and animal feed markets are comparatively much more lucrative if one considers the much smaller volume of seaweeds they require. The use of seaweeds in the development of pharmaceuticals and nutraceuticals, and as a source of pigments and

Figure 6 Net aquaculture of the green alga, *Monostroma nitidum*, in Japan. Photo by Thierry Chopin.

Figure 7 Land-based tank aquaculture of the red alga, *Chondrus crispus* (Irish moss), in Canada for high value-added sea-vegetable (Hana-nori) production. Photo by Acadian Seaplants Limited.

bioactive compounds is in full expansion. Presently, that component is difficult to evaluate accurately; the use of 3000 tonnes of raw material to obtain 600 tonnes of products valued at US$3 million could be an underestimation.

The Role of Seaweeds in the Biomitigation of Other Types of Aquaculture

One may be inclined to think that, on the world scale, the two types of aquaculture, fed and extractive,

environmentally balance each other out, as 45.9% of the mariculture production is provided by aquatic plants, 43.0% by mollusks, 8.9% by finfish, 1.8% by crustaceans, and 0.4% by other aquatic animals. However, because of predominantly monoculture practices, economics, and social habits, these different types of aquaculture production are often geographically separate, and, consequently, rarely balance each other out environmentally, on either the local or regional scale (**Figure 8**). For example, salmon aquaculture in Canada represents 68.2% of the tonnage of the aquaculture industry and 87.2% of its farmgate value. In Norway, Scotland, and Chile,

Table 1 World mariculture production and value from 1996 to 2004 according to the main groups of cultivated organisms

	Production (%)		% of value in	
	1996	2000	2004	2004
Mollusks	48	46.2	43.0	34.0
Aquatic plants	44	44.0	45.9	24.2
Finfish	7	8.7	8.9	34.0
Crustaceans	1	1.0	1.8	6.8
Other aquatic animals		0.1	0.4	1.0

Source: FAO (1998) *The State of World Fisheries and Aquaculture 1998*. Italy: Food and Agriculture Organization of United Nations. http://www.fao.org/docrep/w9900e/w9900e00.htm; FAO (2002) *The State of World Fisheries and Aquaculture 2002*. Italy: Food and Agriculture Organization of the United Nations. http://www.fao.org/docrep/005/y7300e/y7300e00.htm; and FAO (2006) *The State of World Fisheries and Aquaculture 2006*. Italy: Food and Agriculture Organization of the United Nations. http://www.fao.org/docrep/009/a0699e/A0699E00.htm (accessed Mar. 2008).

Table 2 Seaweed aquaculture production (tonnage) from 1996 to 2004 according to the main groups of seaweeds (the brown, red, and green seaweeds) and contribution from Asian countries

	Production (tonnes)		
	1996	2000	2004
Brown seaweeds			
World	4 909 269	4 906 280	7 194 316
Asia (%)	4 908 805	4 903 252	7 194 075
	(99.9)	(99.9)	(99.9)
Red seaweeds			
World	1 801 494	1 980 747	4 067 028
Asia (%)	1 678 485	1 924 258	4 035 783
	(93.2)	(97.2)	(99.2)
Green seaweeds			
World	13 418	33 584	19 046
Asia (%)	13 418	33 584	19 046
	(100)	(100)	(100)
Total			
World	6 724 181	6 920 611	11 280 390
Asia (%)	6 600 708	6 861 094	11 248 904
	(98.2)	(99.1)	(99.7)

Source: FAO (1998) *The State of World Fisheries and Aquaculture 1998*. Italy: Food and Agriculture Organization of United Nations. http://www.fao.org/docrep/w9900e/w9900e00.htm; FAO (2002) *The State of World Fisheries and Aquaculture 2002*. Italy: Food and Agriculture Organization of the United Nations. http://www.fao.org/docrep/005/y7300e/y7300e00.htm; and FAO (2006) *The State of World Fisheries and Aquaculture 2006*. Italy: Food and Agriculture Organization of the United Nations. http://www.fao.org/docrep/009/a0699e/A0699E00.htm (accessed Mar. 2008).

salmon aquaculture represents 88.8%, 93.3%, and 81.9% of the tonnage of the aquaculture industry, and 87.3%, 90.9%, and 95.5% of its farmgate value, respectively. Conversely, while Spain (Galicia) produces only 8% of salmon in tonnage (16% in farmgate

Table 3 Seaweed aquaculture production (value) from 1996 to 2004 according to the main groups of seaweeds (the brown, red, and green seaweeds) and contribution from Asian countries

	Value (×US$1000)		
	1996	2000	2004
Brown seaweeds			
World	3 073 255	2 971 990	3 831 445
Asia (%)	3 072 227	2 965 372	3 831 170
	(99.9)	(99.8)	(99.9)
Red seaweeds			
World	1 420 941	1 303 751	1 891 420
Asia (%)	1 367 625	1 275 090	1 875 759
	(96.3)	(97.8)	(99.2)
Green seaweeds			
World	7263	5216	12 751
Asia (%)	7263	5216	12 751
	(100)	(100)	(100)
Total			
World	4 501 459	4 280 957	5 735 615
Asia (%)	4 447 115	4 245 678	5 719 680
	(98.8)	(99.2)	(99.7)

Source: FAO (1998) *The State of World Fisheries and Aquaculture 1998*. Italy: Food and Agriculture Organization of United Nations. http://www.fao.org/docrep/w9900e/w9900e00.htm; FAO (2002) *The State of World Fisheries and Aquaculture 2002*. Italy: Food and Agriculture Organization of the United Nations. http://www.fao.org/docrep/005/y7300e/y7300e00.htm; and FAO (2006) *The State of World Fisheries and Aquaculture 2006*. Italy: Food and Agriculture Organization of the United Nations. http://www.fao.org/docrep/009/a0699e/A0699E00.htm (accessed Mar. 2008).

value), it produces 81% of its tonnage in mussels (28% in farmgate value). Why should one think that the common old saying "Do not put all your eggs in one basket", which applies to agriculture and many other businesses, would not also apply to aquaculture? Having too much production in a single species leaves a business vulnerable to issues of sustainability because of low prices due to oversupply, and the possibility of catastrophic destruction of one's only crop (diseases, damaging weather conditions). Consequently, diversification of the aquaculture industry is imperative to reducing the economic risk and maintaining its sustainability and competitiveness.

Phycomitigation (the treatment of wastes by seaweeds), through the development of IMTA systems, has existed for centuries, especially in Asian countries, through trial and error and experimentation. Other terms have been used to describe similar systems (integrated agriculture-aquaculture systems (IAAS), integrated peri-urban aquaculture systems (IPUAS), integrated fisheries-aquaculture systems (IFAS), fractionated aquaculture, aquaponics); they can, however, be considered to be variations on the IMTA concept. 'Multi-trophic' refers to the incorporation of species from different trophic or nutritional levels in

Table 4 Production, from 1996 to 2004, of the eight genera of seaweeds that provide 96.1% of the biomass for the seaweed aquaculture in 2004

Genus	Production (tonnes)			% of production in 2004
	1996	2000	2004	
Laminaria (kombu)[a]	4 451 570	4 580 056	4 519 701	40.1
Undaria (wakame)[a]	434 235	311 125	2 519 905	22.3
Porphyra (nori)[b]	856 588	1 010 778	1 397 660	12.4
Kappaphycus/ Eucheuma[b]	665 485	698 706	1 309 344	11.6
Gracilaria[b]	130 413	65 024	948 292	8.4
Sargassum[a]	0	0	131 680	1.2
Monostroma[c]	8277	5288	11 514	0.1

[a]Brown seaweeds.
[b]Red seaweeds.
[c]Green seaweeds.
Source: FAO (1998) *The State of World Fisheries and Aquaculture 1998*. Italy: Food and Agriculture Organization of United Nations. http://www.fao.org/docrep/w9900e/w9900e00.htm; FAO (2002) *The State of World Fisheries and Aquaculture 2002*. Italy: Food and Agriculture Organization of the United Nations. http://www.fao.org/docrep/005/y7300e/y7300e00.htm; and FAO (2006) *The State of World Fisheries and Aquaculture 2006*. Italy: Food and Agriculture Organization of the United Nations. http://www.fao.org/docrep/009/a0699e/A0699E00.htm (accessed Mar. 2008).

Table 5 Value, from 1996 to 2004, of the eight genera of seaweeds that provide 99.0% of the value of the seaweed aquaculture in 2004

Genus	Value ($\times$ US$1000)			% of value in 2004
	1996	2000	2004	
Laminaria (kombu)[a]	2 875 497	2 811 440	2 749 837	47.9
Porphyra (nori)[b]	1 276 823	1 183 148	1 338 995	23.3
Undaria (wakame)[a]	178 290	148 860	1 015 040	17.7
Gracilaria[b]	60 983	45 801	385 794	6.7
Kappaphycus/ Eucheuma[b]	67 883	51 725	133 324	2.3
Sargassum[a]	0	0	52 672	0.9
Monostroma[c]	6622	1849	9937	0.2

[a]Brown seaweeds.
[b]Red seaweeds.
[c]Green seaweeds.
Source: FAO (1998) *The State of World Fisheries and Aquaculture 1998*. Italy: Food and Agriculture Organization of United Nations. http://www.fao.org/docrep/w9900e/w9900e00.htm; FAO (2002) *The State of World Fisheries and Aquaculture 2002*. Italy: Food and Agriculture Organization of the United Nations. http://www.fao.org/docrep/005/y7300e/y7300e00.htm; and FAO (2006) *The State of World Fisheries and Aquaculture 2006*. Italy: Food and Agriculture Organization of the United Nations. http://www.fao.org/docrep/009/a0699e/A0699E00.htm (accessed Mar. 2008).

the same system. This is one potential distinction from the age-old practice of aquatic polyculture, which could simply be the co-culture of different fish species from the same trophic level. In this case, these organisms may all share the same biological and chemical processes, with few synergistic benefits, which could potentially lead to significant shifts in the ecosystem. Some traditional polyculture systems may, in fact, incorporate a greater diversity of species, occupying several niches, as extensive cultures (low intensity, low management) within the same pond. The 'integrated' in IMTA refers to the more intensive cultivation of the different species in proximity to each other (but not necessarily right at the same location), connected by nutrient and energy transfer through water.

The IMTA concept is very flexible. IMTA can be land-based or open-water systems, marine or freshwater systems, and may comprise several species combinations. Ideally, the biological and chemical processes in an IMTA system should balance. This is achieved through the appropriate selection and proportioning of different species providing different ecosystem functions. The co-cultured species should be more than just biofilters; they should also be organisms which, while converting solid and soluble nutrients from the fed organisms and their feed into

biomass, become harvestable crops of commercial value and acceptable to consumers. A working IMTA system should result in greater production for the overall system, based on mutual benefits to the co-cultured species and improved ecosystem health, even if the individual production of some of the species is lower compared to what could be reached in monoculture practices over a short-term period.

While IMTA likely occurs due to traditional or incidental, adjacent culture of dissimilar species in some coastal areas (mostly in Asia), deliberately designed IMTA sites are, at present, less common. There has been a renewed interest in IMTA in Western countries over the last 30 years, based on the age-old, common sense, recycling and farming practice in which the by-products from one species become inputs for another: fed aquaculture (fish or shrimp) is combined with inorganic extractive (seaweed) and organic extractive (shellfish) aquaculture to create balanced systems for environmental sustainability (biomitigation), economic stability (product diversification and risk reduction), and social acceptability (better management practices). They are presently simplified systems, like fish/seaweed/shellfish. Efforts to develop such IMTA systems are currently taking place in Canada, Chile, China, Israel, South Africa, the USA, and several European countries. In the future, more advanced

Table 6 Biomass, products, and value of the main components of the world's seaweed-derived industry in 2006

Industry component	Raw material (wet tonnes)	Product (tonnes)	Value (US$)
Sea vegetables	8.59 million	1.42 million	5.29 billion
Kombu (*Laminaria*)[a]	4.52 million	1.08 million	2.75 billion
Nori (*Porphyra*)[b]	1.40 million	141 556	1.34 billion
Wakame (*Undaria*)[a]	2.52 million	166 320	1.02 billion
Phycocolloids	1.26 million	70 630	650 million
Carrageenans[b]	528 000	33 000	300 million
Alginates[a]	600 000	30 000	213 million
Agars[b]	127 167	7630	137 million
Phycosupplements	1.22 million	242 600	53 million
Soil additives	1.10 million	220 000	30 million
Agrichemicals (fertilizers, biostimulants)	20 000	2000	10 million
Animal feeds (supplements, ingredients)	100 000	20 000	10 million
Pharmaceuticals nutraceuticals, botanicals, cosmeceuticals, pigments, bioactive compounds, antiviral agents, brewing, etc.	3000	600	3 million

[a]Brown seaweeds.
[b]Red seaweeds.
Source: McHugh (2003); Chopin and Bastarache (2004); and FAO (2004) *The State of World Fisheries and Aquaculture 2004*. Italy: Food and Agriculture Organization of the United Nations. http://www.fao.org/docrep/007/y5600e/y5600e00.htm (accessed Mar. 2008); FAO (2006) *The State of World Fisheries and Aquaculture 2006*. Italy: Food and Agriculture Organization of the United Nations. http://www.fao.org/docrep/009/a0699e/A0699E00.htm (accessed Mar. 2008).

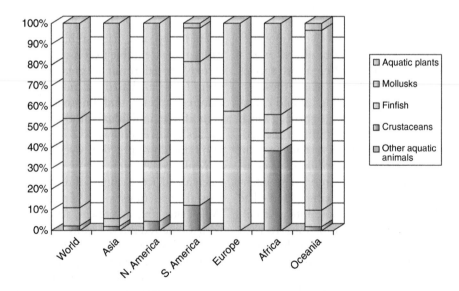

Figure 8 Distribution (%) of mariculture production among the main farmed groups of organisms, worldwide and on the different continents. The world distribution is governed by the distribution in Asia; however, large imbalances between fed and extractive aquacultures exist on the other continents. Source: FAO (2006) *FAO Fisheries Technical Paper 500: State of World Aquaculture 2006*. Rome: Food and Agriculture Organization of the United Nations. http://www.fao.org/docrep/009/a0874e/a0874e00.htm (accessed Mar. 2008).

systems with several other components for different functions, or similar functions but different size brackets of particles, will have to be designed.

Recently, there has been a significant opportunity for repositioning the value and roles seaweeds have in coastal ecosystems through the development of IMTA systems. Scientists working on seaweeds, and industrial companies producing and processing them, have an important role to play in educating the animal-dominated aquaculture world, especially in the Western World, about how to understand and take advantage of the benefits of such extractive organisms, which will help bring a balanced ecosystem approach to aquaculture development.

It is difficult to place a value on the phycomitigation industry, inasmuch as no country has yet implemented guidelines and regulations regarding nutrient discharge into coastal waters. Because the 'user pays' concept is expected to gain momentum as a tool in integrated coastal management, one should soon be able to put a value on the phycomitigation services of IMTA systems for improving water quality and coastal health. Moreover, the conversion of fed aquaculture by-products into the production of salable biomass and biochemicals used in the sea-vegetable, phycocolloid, and phycosupplement sectors should increase the revenues generated by the phycomitigation component.

See also

Mariculture Overview.

Further Reading

Barrington K, Chopin T, and Robinson S (in press) Integrated multi-trophic aquaculture in marine temperate waters. FAO case study review on integrated multi-trophic aquaculture. Rome: Food and Agriculture Organization of the United Nations.

Chopin T, Buschmann AH, Halling C, et al. (2001) Integrating seaweeds into marine aquaculture systems: A key towards sustainability. *Journal of Phycology* 37: 975–986.

Chopin T and Robinson SMC (2004) Proceedings of the integrated multi-trophic aquaculture workshop held in Saint John, NB, 25–26 March 2004. *Bulletin of the Aquaculture Association of Canada* 104(3): 1–84.

Chopin T, Robinson SMC, Troell M, et al. (in press) Multi-trophic integration for sustainable marine aquaculture. In: Jorgensen SE (ed.) *Encyclopedia of Ecology.* Oxford, UK: Elsevier.

Costa-Pierce BA (2002) *Ecological Aquaculture: The Evolution of the Blue Revolution*, 382pp. Oxford, UK: Blackwell Science.

Critchley AT, Ohno M, and Largo DB (2006) *World Seaweed Resources An Authoritative Reference System.* Amsterdam: ETI BioInformatics Publishers (DVD ROM).

FAO (1998) *The State of World Fisheries and Aquaculture 1998.* Italy: Food and Agriculture Organization of the United Nations. http://www.fao.org/docrep/w9900e/w9900e00.htm (accessed Jul. 2008).

FAO (2002) *The State of World Fisheries and Aquaculture 2002.* Italy: Food and Agriculture Organization of the United Nations. http://www.fao.org/docrep/005/y7300e/y7300e00.htm (accessed Jul. 2008)

FAO (2004) *The State of World Fisheries and Aquaculture 2004.* Italy: Food and Agriculture Organization of the United Nations. http://www.fao.org/docrep/007/y5600e/y5600e00.htm (accessed Mar. 2008).

FAO (2006) *FAO Fisheries Technical Paper 500: State of World Aquaculture 2006.* Rome: Food and Agriculture Organization of the United Nations. http://www.fao.org/docrep/009/a0874e/a0874e00.htm (accessed Mar. 2008).

FAO (2006) *The State of World Fisheries and Aquaculture 2006.* Italy: Food and Agriculture Organization of the United Nations. http://www.fao.org/docrep/009/a0699e/A0699E00.htm (accessed Mar. 2008).

Graham LE and Wilcox LW (2000) *Algae*, 699pp. Upper Saddle River, NJ: Prentice-Hall.

Lobban CS and Harrison PJ (1994) *Seaweed Ecology and Physiology*, 366pp. Cambridge, UK: Cambridge University Press.

McHugh DJ (2001) *Food and Agriculture Organization Fisheries Circular No. 968 FIIU/C968 (En): Prospects for Seaweed Production in Developing Countries.* Rome: FAO. http://www.fao.org/DOCREP/004/Y3550E/Y3550E00.HTM (accessed Mar. 2008).

McVey JP, Stickney RR, Yarish C, and Chopin T (2002) Aquatic polyculture and balanced ecosystem management: New paradigms for seafood production. In: Stickney RR and McVey JP (eds.) *Responsible Marine Aquaculture*, pp. 91–104. Oxon, UK: CABI Publishing.

Neori A, Chopin T, Troell M, et al. (2004) Integrated aquaculture: Rationale, evolution and state of the art emphasizing seaweed biofiltration in modern mariculture. *Aquaculture* 231: 361–391.

Neori A, Troell M, Chopin T, Yarish C, Critchley A, and Buschmann AH (2007) The need for a balanced ecosystem approach to blue revolution aquaculture. *Environment* 49(3): 36–43.

Ridler N, Barrington K, Robinson B, et al. (2007) Integrated multi-trophic aquaculture. Canadian project combines salmon, mussels, kelps. *Global Aquaculture Advocate* 10(2): 52–55.

Ryther JH, Goldman CE, Gifford JE, et al. (1975) Physical models of integrated waste recycling-marine polyculture systems. *Aquaculture* 5: 163–177.

Troell M, Halling C, Neori A, et al. (2003) Integrated mariculture: Asking the right questions. *Aquaculture* 226: 69–90.

Troell M, Rönnbäck P, Kautsky N, Halling C, and Buschmann A (1999) Ecological engineering in aquaculture: Use of seaweeds for removing nutrients from intensive mariculture. *Journal of Applied Phycology* 11: 89–97.

Relevant Website

http://en.wikipedia.org
– Integrated Multi-Trophic Aquaculture, Wikipedia.

OCEAN RANCHING

A. G. V. Salvanes, University of Bergen, Bergen, Norway

Introduction

Ocean ranching is most often referred to as stock enhancement. It involves mass releases of juveniles which feed and grow on natural prey in the marine environment and which subsequently become recaptured and add biomass to the fishery. Releases of captive-bred individuals are common actions when critically low levels of fish species or populations occur either due to abrupt habitat changes, overfishing or recruitment failure from other causes. Captive-bred individuals are also introduced inside or outside their natural geographic range of the species to build up new fishing stocks.

At present 27 countries (excluding Japan) have been involved with ranching of over 65 marine or brackish-water species. Japan leads the world with approximately 80 species being ranched or researched for eventual stocking. This includes 20 shared with other nations and 60 additional species. **Table 1** shows an overview of the most important species worldwide. Many marine ranching projects are in the experimental or pilot stage. Around 60% of the release programs are experimental or pilot, 25% are strictly commercial, and 12% have commercial and recreational purposes. Only a few are dedicated solely to sport fish enhancement.

The success of ocean ranching relies on a knowledge of basic biology of the species that are captive bred, but also on how environmental factors and wild conspecifics and other species interact with the released. This article provides general information on life histories of the three major groups of animals that are being stocked, salmon, marine fish and invertebrates, and an overview of the status and success of ocean ranching programs. It also devotes a short section to the history of ocean ranching and a larger section on how success of stock enhancement is measured.

History

Ocean ranching has a long history going back to 1860–1880 that commenced with the anadromous salmonids in the Pacific. In order to restore populations that had been reduced or eliminated due to factors such as hydroelectric development or pollution, large enhancement programs were initiated on various Pacific salmon species mainly within the USA, Canada, USSR and Japan. In addition, transplantation of both Atlantic and Pacific salmonids to other parts of the world (e.g. Australia, New Zealand and Tasmania) with no native salmon populations was attempted.

Around 1900, ocean ranching was extended to coastal populations of marine fish. Because of large fluctuations in the landings of these, release programs of yolk-sac larvae of cod, haddock, pollack, place, and flounder were initiated in the USA, Great Britain, and Norway. It was intended that such releases should stabilize the recruitment to the populations and, thus, stabilize the catches in the coastal fisheries. There was, however, a scientific controversy of whether releases of yolk-sac larvae could have positive effects on the recruitment to these populations. In the USA the releases ceased by World War II without evaluation. In Norway evaluation was conducted in 1970 when it was shown to be impossible to separate the effect of releases of yolk-sac larvae from natural fluctuations in cod recruitment, a conclusion that led to termination of the program. Recent field estimates of the mortality of early life stages of cod suggest that only a handful of the 33 million larvae, an amount normally released, survive the three first months.

Larvae and juveniles of European lobster (*Homarus gammarus*) have been cultured and released along the coast of Norway for over 100 years. In 1889 newly hatched lobster eggs and newly settled juveniles were released in Southern Norway on an island which has its own continental shelf separated from the mainland. Because the larvae were too small to be tagged, no recapture could be registered. The first attempts at scallop enhancement occurred in 1900, but the activity ceased after a short time. From 1970, stock enhancement started on scallops on a commercial scale in Europe and Japan and from 1980 on giant clams in Indo-Pacific countries. Other invertebrate enhancement programs started on a commercial basis in 1963 when the government of Japan instituted such actions as a national policy to augment both marine fish populations and commercially interesting invertebrate populations (e.g. abalone, clams, sea urchins, shrimps and prawns).

Table 1 Main species which are captive bred and released in ocean ranching programs

English name	Scientific name	Country
Salmonids		
Atlantic salmon	*Salmo salar*	Norway, Iceland, UK
Pink salmon	*Oncorhynchus gorbuscha*	USA, Canada, Japan
Chinook	*O. tschawytcha*	USA, Canada
Chum salmon	*O. keta*	USA, Canada, Japan
Coho salmon	*O. kisutch*	USA, Canada, Japan
Sockeye salmon	*O. nerka*	USA, Canada, Japan
Masu salmon	*O. masou*	Japan
Marine fish		
Pacific herring	*Clupea pallasi*	Japan
Black sea bream	*Acanthopagrus schlegeli*	Japan
Red sea bream	*Pagrus major*	Japan
Sandfish	*Arctoscopus japonicus*	Japan
Jacopever	*Sebastes schlegeli*	Japan
Japanese flounder	*Paralichthys olivaceus*	Japan
Mud dab	*Limanda yokohamae*	Japan
Ocellate puffer	*Takifugu rubripes*	Japan
Striped jack	*Pseudocaranx dentex*	Japan
Yellow tail	*Seriola quinqueradiata*	Japan
Sea bass	*Lateolabrax japonicus*	Japan
Red drum	*Sciaenops ocellatus*	USA
Spotted sea trout	*Cynoscion nebulosus*	USA
Striped bass	*Morone saxatilis*	USA
Mullet	*Mugil cephalus*	Hawaii
Threadfin	*Polydactylus sexfilis*	Hawaii
Turbot	*Scophthalmus maximus*	Denmark, Spain
White sea bass	*Atractoscion nobilis*	USA
Whitefish	*Coregonus lavaretus*	Baltic
Cod	*Gadus morhua* L.	Norway, Sweden, Denmark, Faeroe Islands, USA
Invertebrates		
Kuruma shrimps	*Penaeus japonicus*	Japan
Chinese shrimps	*Penaeus chinensis*	Japan
Speckled shrimp	*Metapenaeus monoceros*	Japan
Mangrove crab	*Scylla serrata*	Japan
Swimming crab	*Portunus trituberculatus*	Japan
Blue crab	*Portunus pelagricus*	Japan
Japanese abalone	*Sulculus diversicolor*	Japan
Disk abalone	*Haliotis discus*	Japan
Yezo abalone	*Haliotis discus hannai*	Japan
Giant abalone	*Haliotis gigantea*	Japan
Spiny top shell	*Batillus cornutus*	Japan
Ark shell	*Scapharca broughtonii*	Japan
Scallop	*Patinopecten yessoensis*	Japan
Scallop	*Pecten maximus*	France, Ireland, Norway, UK
Hard clam	*Meretrix lusoria*	Japan
Hard clam	*Meretrix lamarckii*	Japan
Giant clam	*Tridacna maxima*	Indo-Pacific
Giant clam	*Tridacna derasa*	Indo-Pacific
Surf clam	*Spisula sachalinensis*	Japan
European lobster	*Homarus gammarus*	Norway, France, Ireland
Sea urchin	*Tripneustes gratilla*	Japan
Red sea urchin	*Pseudocentrotus depressus*	Japan
Sea urchin	*Strongylocentrotus intermedius*	Japan
Sea urchin	*Strongylocentrotus nudus*	Japan
Sea cucumber	*Stichopus japonicus*	Japan

Salmonids

Ocean ranching programs of salmon have the longest history and have been most comprehensive. Large programs occur in Japan, along the west coast of North America, and also in the Northern Atlantic. Seven species of salmon, which are all anadromous, have been used for releases of captive-bred individuals: Atlantic salmon (*Salmo salar*) which inhabit the northern Atlantic area and the northern Pacific species: pink (*Oncorhynchus gorbuscha*); Chinook (*O. tschawytcha*); chum (*O. keta*); coho (*O. kisutch*); sockeye (*O. nerka*); and the masu salmon (*O. masou*). Atlantic salmon juveniles remain in fresh water for 1–5 years before they smolt and migrate to the marine environment, whereas the pink and chum leave the rivers soon after they have resorbed the yolk-sac.

The ocean ranching program for chum salmon (*Oncorhynchus keta*) in Japan is typical of the way salmon programs are conducted and is the only program that is considered as economically successful to the operators. Chum fry are easy to produce, their survival is good and the program produces 40–50 million fish annually for the fishermen. The number of chum salmon released during the 15-year period 1980–1995 was stable and was about 2 billion per year, whereas the catches tripled during the same period. More than 90% of the chum salmon catches originated from released juveniles, suggesting an increase in the recapture rate from 1% to 3%. The success of ocean ranching of salmon in Japan was due to favorable environmental conditions over the Northern Pacific, but also due to improvements in marine ranching techniques such as egg collection, artificial diets and timing of release, improvements that also were supported by dietary, physiological and behavioral research.

Salmonids show a wide range of lifestyles and a high phenotypic plasticity. The degree of variation depends on the species. The pink salmon has few life history variants and spawns generally in the gravel of estuaries or the lower parts of North Pacific rivers and then dies. After emergence from the gravel the fry leave immediately for the ocean where they grow to maturity, then return to spawn in their natal gravel at 2 years old. The most complex lifestyles are those of steelhead trout (*Oncorhynchus mykiss*) ranging from anadromous to landlocked. The species spawn well upriver in the gravel of a stream. On emergence from the gravel, their fry may spend 1–6 years in the river before emigrating to the sea, or they may not emigrate at all, but mature in the juvenile freshwater habitat and reproduce there (in this case they are referred to as rainbow trout). The steelhead is generally iteroparous. The individuals that emigrate to the sea spend 1–6 years there feeding on prey organisms that are not harvested such as krill, copepods, amphipods, mesopelagic fish, but also on commercial species such as capelin, herring and squid before maturing and returning to breed. After spawning they may return to the sea again, mature, and return the next or later years. Through a combination of all these different developmental possibilities, a single cohort of steelhead may give rise to over 30 different life history types.

The length of the spawning period in anadromous salmonids varies and may exceed 3 months in wild populations. The individuals that return early in the spawning season generally produce offspring that hatch early, and grow faster than the fry of those that arrive later. The timing of the returns reflects individual differences in physiological and behavioral strategies for energy storage and usage, and differences in genetic control between individuals with different seasonal runs. The variation in seasonal runs in a population will have implications for survival probabilities of new recruits in fluctuating natural environments and thus for the stability of population sizes over generations. Populations that have long spawning periods exhibit a high degree of phenotypic variation whereas those that have a short spawning period have lower. The more variation there is, the more flexible will the population be in its adaptations to environmental fluctuations. The smaller the variation, the higher the possibility will be for a recruitment failure and for extinction if individuals are semelparous.

When entering the marine environment wild salmonids migrate to highly productive high sea areas where they grow fast. At the onset of sexual maturation most individuals return to the stream where they were spawned after migrating for several years and thousands of kilometers. Exceptions are jacks (males maturing at age 1 year) that remain at sea for a few months; some species remain in coastal waters rather than undergoing long oceanic migrations. Atlantic salmon can return after one sea winter (grilse), two sea-winters and to a lesser extent after three sea-winters. It has been shown that salmon have the ability to recognize distinctive odors of their home stream and that they become imprinted to this odor as they transform into smolts, just prior to their outward migration. Furthermore, there is evidence that salmon may respond to chemicals (pheromones) given off by conspecifics, such as juveniles inhabiting a stream, and that they are able to discriminate between water that contains fish from their own populations and those from other populations. These abilities allow a great deal of precision in homing. Most commercial harvesting occurs on the coast and recreational and native fishery in the rivers and streams.

Under the salmon enhancement programs it is assumed that the captive-bred released individuals have the same biological and genetic characteristics as wild individuals and that they undertake the same migration routes as their wild conspecifics, and return to their native rivers to spawn. However, it has not been possible to study whether released salmon migrate to the same high seas areas as wild individuals because the main fishery is on the coast and not offshore. The return rates of tagged wild and captive-bred individuals to the various rivers and streams have been compared and show that the homing instinct is not always 100% in any of the groups; the straying rate is three times as high for captive-bred fish with up to 13% recaptured in a different location than where the individuals were released. This suggests different homing instinct abilities in captive-bred and wild salmonids.

The nature of genetic variation within species that are part of ocean ranching programs was not appreciated when they started. The goal of ocean ranching was to restore wild stocks to their former abundance and so to restore the fishery to its former levels. No attention was paid to the number of spawners that were required in order to obtain sufficient genetic variation among captive-bred fish. Often the effective number of spawners was low, resulting in reduced genetic diversity of the population. Moreover, individuals from the broodstock were often taken from the first adults to return, and fry from early-spawning adults were hatched in captivity and released on the spawning grounds together with offspring from fish that had spawned in the wild. Such actions may have had a negative influence on the genetic diversity of the mixed population of wild and captive-bred fish because the fry that hatched early, grow faster, and can displace fry of later-spawning fish. For example, in the USA a decrease in the spawning season from 13 to 3 weeks over just 13 years has been demonstrated for coho salmon. It has also been reported that offspring of captive-bred steelhead that spawned naturally had much lower survival than those from wild fish. This could be a contributing factor to the continuing decline in enhanced salmon stocks despite large-scale releases.

Marine Fish

Ocean ranching is practiced on many marine fish in several countries; the Japanese programs are the largest. Captive-bred individuals of eleven marine fish species are released on a commercial basis in Japan (**Table 1**) with a total yearly release of 68 million fry or juveniles. Red sea bream (*Pagrus major*) and Japanese flounder (*Paralichthys olivaceus*) are released in highest quantities (22 million of each per year).

Juvenile Japanese flounder are 4–12 cm long when released. Those that are larger than 9 cm at release survive to commercial size. Captive-bred individuals show a high degree of partial or complete albinism

on the ocular side and characteristic melanin deposits on the other side. These juveniles suffer from high mortality immediately after release (**Figure 1**). The proportion of the catches that have a captive origin is used to quantify impact on catches. Between 10 and 40% of the catches consist of released fish; but this program cannot be considered as a success. Despite releases having increased from 5 million to 22 million juveniles from 1985 to 1995, the total catch of flounder is decreasing (**Figure 2**).

There is also a size-dependent survival of red sea bream; up to 50% are reported in catches if fish were 4 cm at release. Enhancement of red sea bream has been conducted since 1974 and mostly 4 cm-long individuals are now released. However, the population continues to decline despite the releases (**Figure 3**).

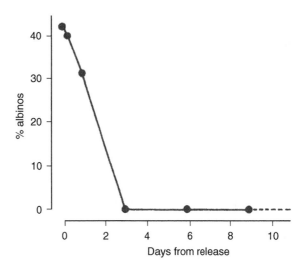

Figure 1 Change in the proportion of albino Japanese flounder juveniles in the catch for successive days after release. (After Blaxter, 2000.)

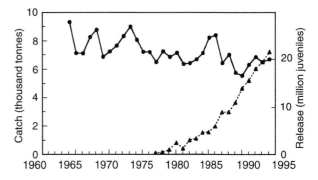

Figure 2 The catch (●) and number of released juveniles (▲) of flounder in Japan. (After Masuda and Tsukamoto, 1998; in Coleman *et al.*, 1998.)

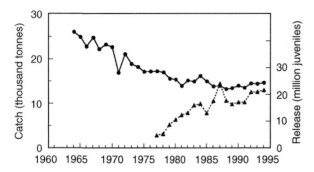

Figure 3 The catch (●) and number of released juveniles (▲) of red sea bream in Japan. (After Masuda and Tsukamoto, 1998; in Coleman *et al.*, 1998.)

Ten other marine fish species (**Table 1**) are used for ocean ranching, but most can be considered to be at an experimental stage. Only the red drum (*Sciaenops ocellatus*) and spotted sea trout (*Cynoscion nebulosus*), which are being released in Texas lagoons of the Gulf of Mexico, have reached a commercial scale. Since the early 1980s there have been yearly releases of 16–30 millions of red drum fingerlings and 5 million sea trout. The adult populations of red drum live offshore and the eggs and larvae are swept into the inshore areas and lagoon where they grow to juveniles for up to six years, and where they support a recreational, and sometimes a commercial fishery. The proportion of released individuals in the catches has been as high as 20%.

The First Scientific Evaluation of Stock Enhancement; The Norwegian Cod Study

The most comprehensive scientific study on marine fish stock enhancement has been conducted on Atlantic cod. Average age at maturity in coastal cod is 3 years. Spawning occurs at grounds located at *c.* 50 m depth and the spawning period is February–April. In western Norway juveniles settle in the shallow nearshore areas during summer and early fall and inhabit mainly 0–20 m depth. There is a commercial and recreational fishery for cod. Cod enter the fishery as by-catch when 6 months old, but most are harvested when two years or older. Coastal cod remain localized in the fiords at least until maturation. Tagging experiments have shown that very few individuals migrate to other areas. Because of this resident life history, cod need to feed on local prey or on prey that are advected into the fiords. This is very different from salmonids that move to high sea areas and bring biomass 'home' to the fishery.

Despite the failure of releases of yolk-sac larvae, new attempts with ocean ranching of cod commenced in Norway in 1983 first through large-scale

experiments, but later as small-scale experiments also in Sweden, Denmark, the Faroe Islands and USA. The new approach was to release 6–9-month-old juveniles, 10–20 cm long. This was decided because a positive correlation had been documented between abundance at 6–9 months old and subsequent recruitment to the fishable population, and because large-scale production techniques for juvenile cod had been developed by the Institute of Marine Research in Norway.

The very extensive work on cod by the Norwegians is thought to represent the most thorough investigation of the possibilities and limitations of stock enhancement. A large governmental interdisciplinary research program was initiated in 1983 with experiments being carried out in a number of fiords along the coast. The aim was to develop full-scale production of juveniles, to develop techniques for mass marking and to design and carry out large-scale release experiments on the Norwegian coast. These were to be in association with wide-ranging field studies which were intended to evaluate the potential and limitations of sea ranching of cod from an ecological perspective, before ocean ranching was initiated on a commercial scale. Masfjorden was selected for the study of the whole ecosystem and its dynamic fluctuation in carrying capacity. Research here involved field studies before and after large-scale releases. In addition dynamic ecosystem simulation models were developed. The models integrated all relevant field data and were used to study the effect on fish production and carrying capacity of environmental and biological factors. It was shown that although juvenile wild cod populations were augmented after release of captive-bred cod, there was no sign of stock enhancement of 2-year-old cod when these should have recruited to the fishery. A recent time-series analysis of the survival of captive-bred cod from this fiord shows a high mortality rate in the spring at one year old. It is uncertain what causes this. One possibility is a higher mortality risk in captive-bred than wild cod in spring when there is a massive immigration of spurdog and also higher abundance of other predators. The modeling predicted that density-dependent growth, predation, and cannibalism restricted cod productivity, and that the most important environmental factor was non-local wind-driven fiord-coast advection of organisms at lower trophic levels. This limited the carrying capacity for fish at higher trophic levels in the fiord. The zooplankton *Calanus finmarchicus*, which in spring and summer occur in high abundance in the coastal waters of Norway, were exchanged between coast and fiord via advection of intermediate watermasses. These are the main prey of gobies, and gobies are the main prey of juvenile cod. If strong southerly winds occur frequently, they transport planktonic organisms into the fiord, whereas if strong northerly winds occur most frequently, they may reduce the abundance of planktonic organisms within the fiord and thus limit the carrying capacity for fish at higher trophic levels. This means that the carrying capacity of resident fish in local habitats is highly dependent on environmental processes that occur on a larger scale. The modeling also predicted that the distance from the coast affected the carrying capacity and individual growth of fish in the fiords with lower production and growth within the inner fiords than on the coast (**Figure 4**) because the advection rate and zooplankton density become damped with increasing distance from the coast (**Figure 5**). This was confirmed by empirical estimates of growth curves for cod (**Figure 6**).

As a consequence of the negative conclusion of the Norwegian experimental ocean ranching on cod, the releases were terminated and it was decided not to scale up to commercial level. The research has, however, not been wasted. New insights into fiord

ecology and particularly the influence of environmental factors on ecosystem dynamics have been achieved.

Marine Invertebrates

Enhancement programs on invertebrates are most comprehensive in Japan, including shrimps, crabs, abalone, clams, scallops, sea urchins and sea cucumber (**Table 1**). The present program started on a commercial basis around 1970. Annual releases of seedlings are 3.47×10^9, mostly of scallop (*Patinopecten yessoensis*) (3×10^9).

Scallop seabed culture of *Pecten maximus* started in Europe in France in 1980 and extended to Ireland, Scotland, and Norway. Each country produces 25 million juveniles per year for release. Juveniles are

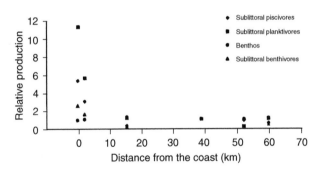

Figure 4 Simulated yearly production for five west Norwegian fiords located at different distances from the outer coast. Sublittoral planktivores refers to gobies, sublittoral piscivores refers to cod and other fish at the same trophic level. (After Salvanes *et al.*, 1995.)

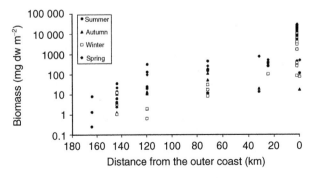

Figure 5 The density of *Calanus finmarchicus* as a function of the distance from the coast by season. Note the logarithmic scale on the biomass axis. (After Salvanes *et al.*, 1995.)

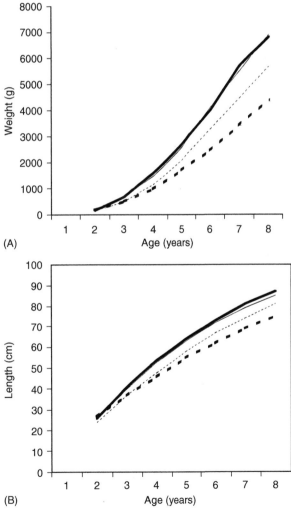

Figure 6 Growth curves ((A) weight; (B) length) for released and wild cod. ——, wild cod, outer coast; - - -, wild cod, inner fiord: ——, released cod, outer coast; - - -, released cod, inner fiord. (Modified from Svsånd *et al.*, 2000.)

first kept in intermediate culture for 1–2 years before release. In France 3 cm juveniles are released on seabeds, whereas in Norway they are 5–6 cm before being released. Ocean ranching with scallops in Japan is considered successful because the landings have increased from 5000 tonnes around 1970 to 200 000 tonnes in the mid 1990s (**Figure 7**). Other invertebrate populations in Japan are, however, still declining despite the high numbers of released seedlings. For example, the populations of abalone have more than halved over the last 25 years (**Figure 8**).

The benefits of releasing invertebrates is that they are either stationary or do not move far from the release point and that they feed lower in the food web than the fish and therefore utilize a larger proportion of the primary production. For example bivalves benefit from transferring locally produced algae to highly valuable biomass by filter feeding; others such as prawns, shrimps and crabs feed on locally produced or advected secondary producers. However, the drawback is that often juveniles that are being released are small and grow much more slowly than fish of the same age and suffer from high predation mortality. Moreover, some species do not move very much, and if juveniles are released in high densities, this could result in unwanted density-dependent mortality (predation, food competition). In addition, the production costs of viable seedlings are often high.

One example of a slow grower is the European lobster. In the 1970s, Norway succeeded in producing one-year-old juveniles for release, and soon thereafter a commercial lobster hatchery with a production capacity of 120 000 juveniles per year was built. These have been released along the Norwegian coast in attempts to rebuild depleted populations. A large-scale experiment has been conducted at Kvitsøy, a small island with its own continental shelf, located on the south-western coast of Norway. It took 5–8 years after the first large releases in 1985 and 1986 before the lobster recruited to the regulated commercial fishery. Although captive-bred individuals were not tagged, it was possible to distinguish them from the wild because most had developed two scissors claws instead of the normal one scissors and one crusher claw. From 1990 to 1994 the released juveniles were microtagged. In 1997 43% of the landings were of released lobster. The landings of lobster increased from 1995 to 1997, but catches of the wild stock showed a weak decrease. It is not known whether released lobsters replace wild stock.

Measuring the Success of Ocean Ranching

All enhancement programs rely on the assumptions that human operations have reduced populations to sizes below the carrying capacities, or that there are possibilities for increased production of the target species, and that releases of captive-bred juveniles will increase the number of adults and thus subsequent harvests. However, few attempts have been made to test these assumptions scientifically due to their complexity, and many questions are therefore still unanswered.

Ocean ranching programs involve enormous investments, but the pay-off has been difficult to evaluate and therefore generally ignored. The evaluation should be done in two steps; first biologically and then economically. The measure of the success of an enhancement program depends on the objectives: the fishermen consider it successful if they catch more fish; biologists at the hatcheries may consider the production and release of viable juveniles as a success; a conservation biologist will be happy if previously nearly extinct populations increase again; an ecologist will also demand that

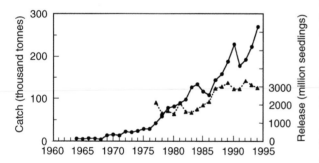

Figure 7 The catch (●) and the number of released seedling (▲) of scallop in Japan. (After Masuda and Tsukamoto, 1998; in Coleman *et al.*, 1998.)

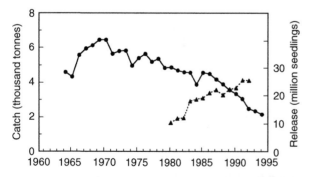

Figure 8 The catch (●) and the number of released seedlings (▲) of abalone in Japan. (After Masuda and Tsukamoto, 1998; in Coleman *et al.*, 1998.)

increased biological production on the target species does not have negative influence on wild conspecifics or on other parts of the ecosystem; and an economist will only consider it a success if there is a positive net pay-off of investments.

The first stage of an evaluation is to determine if there is a net biological benefit by: (1) estimating survival of released fish; (2) verifying survival over many generations; (3) measuring eventual negative effects on wild fish or on the ecosystem as a whole by covering questions such as: does individual variation in growth and abundance change? The final stage would then be to evaluate eventual economic benefit. The programs that occur today usually have immediacy and an applied aspect, and investments into experimental ecology to test major underlying assumptions for increased biological production have therefore been ignored. From a biological point of view this is very shortsighted as the absence of proper evaluation means that ocean ranching programs are conducted in an *ad hoc* way and this will result in a high chance of investment for nothing.

Fish populations have fluctuated in the periods before fishing technology allowed high fishing pressure. It has been reported that some of the marked shifts in sockeye salmon abundance over the last 300 years were associated with climatic changes long before humans were able to overfish. The way climate affects cycles in population sizes is, however, difficult to separate from human made effects. If fish are released because a population declines, and if such releases are conducted during a time period when there are unfavorable climatic conditions, the negative climatic factors may counteract positive enhancement effects. For example, the simulation modeling of the Masfjorden ecosystem showed that nonlocal wind-driven coast–fiord advection of organisms at lower trophic levels had a large impact on the carrying capacity of the cod populations. This means that carrying capacities for fish in coastal ecosystems fluctuate in an unpredictable manner. Simulations also suggest that with imperfect knowledge of the carrying capacity for juvenile cod – which represents more or less the current situation – it is impossible to conduct 'perfect releases' that match the carrying capacity. This means that there may be no payoff from releasing cod to increase a stock.

Other major questions concern the ecology of wild stocks, and also whether the captive-bred individual's genetic, physiological, morphological and behavioral traits are similar to those of wild conspecifics. If not, are these factors deviating so much that released individuals have poorer or better survival than wild conspecifics? Hence, will enhancement effects disappear soon after release or will released animals just

replace wild and therefore not enhance total population sizes (released and wild) and harvests? For example in salmon groups size hierarchies among juveniles soon develop in their nursery streams; this also happens among adults in the spawning habitats. Large juveniles tend to monopolize the best feeding spots, and on the spawning beds the large adults get access to the best gravel beds and large males have higher mating success with females. Mature escaped farmed salmon enter Norwegian rivers. Here they often replace wild fish on the spawning grounds. This can be a long-term problem due both to the way the farmed salmon has been domesticated and selectively bred for rapid growth and late maturation, and to a life history that would not necessarily have any benefit under all environmental situations if they escape to the wild. Moreover, offspring of captive-bred steelhead that spawned naturally had much lower survival than those from the wild. It is possibile that instead of enhancing a population, large-scale releases of captive-bred individuals can first replace wild conspecifics, and thereafter become extinct over a few generations in extreme environmental conditions to which the released fish are not adapted. Hence, differences in the life history of captive-bred and wild fish can be a contributing factor to the continuing decline in enhanced salmon stocks despite large-scale releases. If enhancement programs are initiated without taking into consideration the genetic diversity the wild populations have evolved through generations and that is required for survival under extreme environmental fluctuations, the program may fail.

Another documented difference between captive-bred and wild fish is the capability of the latter to show flexible behavior under changing environmental conditions. Captive-bred cod and salmonids adapt more slowly than wild fish to novel food or to food encountered in a different way than previously. They also tend to take more risks than do wild fish. When captive-bred cod were released for sea ranching in Masfjorden off western Norway, it was evident from field samples that these fish had a different diet and higher mortality rate than wild cod at least for the first three days after release. Moreover, these captive-bred cod were released into the natural habitat in numbers that greatly exceeded that of wild cod of the same cohort. Despite this, released cod abundance declined sharply during the spring when the fish were one-year-old, six months after release. At release these fish were naive to the heterogeneous marine environment, to predators, and also to encounters with natural prey. It is possible that feeding behavior, and perhaps also other behavior, of captive-bred animals differed from that of wild cod over a much longer time than the first three days after release and that this could have made them more vulnerable toward piscivorous predators. Released salmonids also do less well than wild. Although they were able to adapt to novel and/or live prey by learning for some days or weeks before they were released, they took less prey, grew more slowly, had a higher mortality rate than wild, had a narrower range of dietary items and frequently lagged behind wild fish in switching to new prey items as these became abundant.

In the literature there is hardly any ocean ranching program that can be considered successful on both ecological and economic grounds. Only the chum salmon ranching program in Japan is considered economically successful, at least for the operators, because the cost of fry production is low and the fishery catches more fish (**Figure 9A**). There are, however, two possible drawbacks even for this program: the average size of individual chum has been decreasing (**Figure 9B**), and the change is correlated with the Japanese chum releases, and with the abundance of chum in the North Pacific. The

economic value of individual fish has therefore decreased. Moreover, total Pacific chum production has only increased by 50% since 1970, and the North American production (largely natural) increased by about the same amount. It might thus be that the apparent increase in Japanese chum production could be due to favorable climatic conditions in combination with replacement of wild fish. It is, however, very difficult to separate enhancement and environmental effects. Nevertheless, cyclical changes in Pacific salmon populations have occurred earlier and long before humans were able to overfish or conduct captive breeding for release. This means that environmental factors would mask any cause of change in populations that are seen in populations chosen for ocean ranching. Another study, on chinook salmon of the Columbian river basin showed that the population continues to decline towards extinction despite the release of captive-bred individuals. It was concluded through simulation studies that the only way to reverse the decline would be to increase the survival of first-year fish, and the only way of doing this might be to remove the dams that had been constructed for hydropower production.

The prospects of marine ranching should be critically evaluated biologically and economically before commercial large-scale programs are initiated on a target species. The optimal species for stock enhancement should be easy and cheap to rear to release size, it should grow fast, have a low mortality rate and it should feed low in the food chain. It should also have a behavior that does not deviate from wild conspecifics or lead to negative effects on other species. However, the case studies reported here illustrate clear difficulties in increasing population sizes by releasing captive-bred individuals and show that hardly any commercial enhancement program can be regarded as clearly successful. Model simulations suggest, however, that stock-enhancement may be possible if releases can be made that match closely the current ecological and environmental conditions. However, this requires improvements of assessment methods of these factors beyond present knowledge. Marine systems tend to have strong nonlinear dynamics, and unless one is able to predict these dynamics over a relevant time horizon, release efforts are not likely to increase the abundance of the target population.

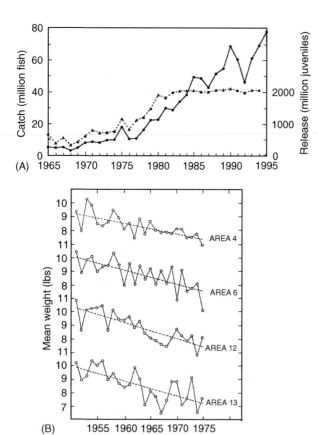

Figure 9 (A) The catch (●) and the number released of juveniles (▲) of chum salmon in Japan. (B) Changes in the mean weights of coho salmon harvested from the Pacific ocean between 1950 and 1981. (After Masuda and Tsukamoto, 1998; in Coleman *et al.*, 1998.)

See also

Mariculture, Economic and Social Impacts. Salmonid Farming.

Further Reading

Asplin L, Salvanes AGV, and Kristoffersen JB (1999) Non-local wind-driven fjord–coast advection and its potential effect on pelagic organisms and fish recruitment. *Fisheries Oceanography* 8: 255–263.

Blaxter JHS (2000) The enhancement of marine fish stocks. *Advances in Marine Biology* 38: 2–54.

Coleman F, Travis J and Thistle AB (1998) *Marine Stock Enhancement: A New Perspective.* Bulletin of Marine Science 62.

Danielssen DS, Howell BR, and Moksness E (1994) An International Symposium on Sea Ranching of Cod and other Marine Fish Species. *Aquaculture and Fisheries Management* 25 (Suppl. 1).

Finney B, Gregory-Eaves I, Sweetman J, Douglas MSV, and Smol JP (2000) Impacts of climatic change and fishing on Pacific salmon over the past 300 years. *Science* 290: 795–799.

Giske J and Salvanes AGV (1999) A model for enhancement potentials in open ecosystems. In: Howell BR, Moksness E, and Svåsand T (eds.) *Stock Enhancement and Sea Ranching.* Blackwell: Fishing, News Books.

Howell BR, Moksness E, and Svåsand T (1999) *Stock Enhancement and Sea Ranching.*

Kareiva P, Marvier M, and McClure M (2000) Recovery and management options for spring/summer chinnook salmon in the Columbia River basin. *Science* 290: 977–979.

Mills D (1989) *Ecology and Management of Atlantic Salmon.* London: Chapman Hall.

Ricker WE (1981) Changes in the average size and average age of Pacific salmon. *Canadian Journal of Fisheries and Aquatic Science* 38: 1636–1656.

Salvanes AGV, Aksnes DL, Foss JH, and Giske J (1995) Simulated carrying capacities of fish in Norwegian fjords. *Fisheries Oceanography* 4(1): 17–32.

Salvanes AGV, Aksnes DL, and Giske J (1992) Ecosystem model for evaluating potential cod production in a west Norwegian fjord. *Marine Ecology Progress Series* 90: 9–22.

Salvanes AGV and Baliño B (1998) Production and fitness in a fjord cod population: an ecological and evolutionary approach. *Fisheries Research* 37: 143–161.

Shelbourne JE (1964) The artificial propagation of marine fish. *Advances in Marine Biology* 2: 1–76.

Svåsand T, Kristiansen TS, Pedersen T, *et al.* (2000) The biological and economical basis of cod stock enhancement. *Fish and Fisheries* 1: 173–205.

Thorpe JE (1980) *Salmon Ranching.* London: Academic Press.

MARICULTURE, ECONOMIC AND SOCIAL IMPACTS

C. R. Engle, University of Arkansas at Pine Bluff, Pine Bluff, AR, USA

Introduction

Mariculture is a broad term that encompasses the cultivation of a wide variety of species of aquatic organisms, including both plants and animals. These different products are produced across the world with a wide array of technologies. Each technology in each situation will entail various price and cost structures within different social contexts. Thus, each aquaculture enterprise will have distinct types and levels of economic and social impacts. Moreover, there are as many management philosophies, strategies, and business plans as there are aquaculture entrepreneurs. Each of these will result in different economic and social impacts. As an example, consider two shrimp farms located in a developing country that utilize the same production technology. One farm hires and trains local people as both workers and managers, invests in local schools and health centers, while paying local and national taxes. This farm will likely have a large positive social and economic impact. On the other hand, another farm that imports managers, pays the lowest wages possible, displaces local families through land acquisitions, pays few taxes, and exports earnings to developed countries may have negative social and perhaps even economic impacts.

Economic and social structure, interactions, and impacts are complex and interconnected, even in rural areas with seemingly simple economies. This article discusses a variety of types of impacts that can occur from mariculture enterprises.

Economic Impacts

Economic impacts begin with the direct effects from the sale of product produced by the mariculture operation. However, the impacts extend well beyond the effect of sales revenue to the farm. As the direct output of marine fish, shellfish, and seaweed production increases, the demand for supply inputs such as feed, fingerlings, equipment, repairs, transportation services, and processing services also increases.

These activities represent indirect effects. Subsequent increased household spending will follow. As the industry grows, employment in all segments of the industry also grows and these new jobs create more income that generates additional economic activity. Thus, growth of the mariculture industry results in greater spending that multiplies throughout the economy.

Mariculture is an important economic activity in many parts of the world. **Table 1** lists the top 15 mariculture-producing countries in the world, both in terms of the volume of metric tons produced and the value in 2005. Its economic importance can be measured in total sales volume, total employment, or total export volume for large aquaculture industry sectors. Macroeconomic effects include growth that promotes trade and domestic resource utilization. Fish production in the Philippines, for example, accounted for 3.9% of gross domestic product (GDP) in 2001. In India, it contributed 1.4% of national GDP and 5.4% of agricultural GDP.

On the microlevel, incomes and livelihoods of the poor are enhanced through mariculture production. Small-scale and subsistence mariculture provides high-quality protein for household consumption,

Table 1 Top 15 mariculture-producing countries, with volumes (metric tons) produced in 2005 and value (in US)

Country	Volume of production (metric ton)	Value of production ($1000 US)
China	22 677 724	14 981 801
The Philippines	1 419 727	165 335
Japan	1 211 959	3 848 906
Korea, Rep.	1 042 142	1 317 250
Indonesia	950 819	338 093
Chile	889 867	3 069 169
Korea, Dem.	703 292	283 362
Norway	504 295	2 072 562
Thailand	656 636	58 587
France	347 750	571 543
Canada	216 103	503 974
Spain	136 724	262 394
United Kingdom	190 426	584 152
United States	161 339	235 912
Vietnam	134 937	158 800
Others	173 800	3 268 283
Total	31 417 540	31 720 123

Source: FishStat Plus (http://www.fao.org).

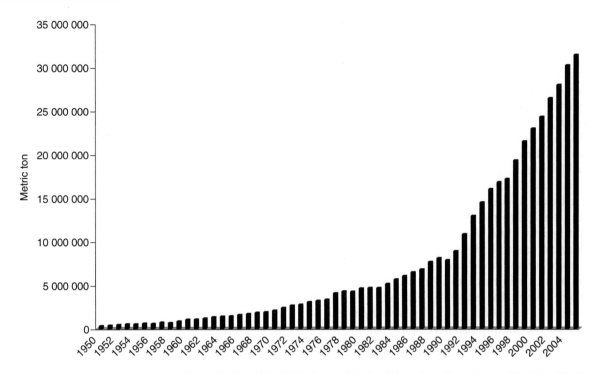

Figure 1   Growth of mariculture production worldwide, 1950–2005. Source: Food and Agriculture Organization of the United Nations.

generates supplemental income from sales to local markets, and can serve as a savings account to meet needs for cash during difficult financial times.

Income from Sales Revenue

According to the Food and Agriculture Organization of the United Nations, mariculture production grew from just over 300 000 metric ton in 1950 to 31.4 million metric ton valued at $31.7 billion in 2005 (**Figure 1**). **Figure 1** also shows that the total volume of mariculture production has tripled over the past decade alone. Sales revenue received by operators of mariculture businesses has increased rapidly over this same time period. The top mariculture product category in 2005, based on quantity produced, was Japanese kelp, followed in descending order of quantities produced by oysters, Japanese carpetshell, wakame, Yesso scallop, and salmon (**Table 2**). In terms of value, salmon was the top mariculture product produced in 2005, followed by oysters, kelp, and Japanese carpetshell. The top five marine finfish raised in 2005 (in terms of value) were Atlantic salmon, Japanese amberjack, rainbow trout, seabreams, and halibut.

The increase in production and sales of mariculture products has resulted not only from the expansion of existing products and technologies (i.e., salmon, Japanese amberjack) but also from the rapid development of technologies to culture new species

Table 2   Top 15 mariculture species cultured, volume (metric tons), and value ($1000 US), 2005

Species	Volume (metric ton)	Value ($1000 US)
Atlantic salmon	1 216 791	4 659 841
Blood cockle	436 924	436 772
Blue mussel	391 210	385 131
Constricted tagelus	713 846	589 836
Green mussel	280 267	26 679
Japanese carpetshell	2 880 687	2 358 586
Japanese kelp	4 911 256	2 941 148
Laver (nori)	1 387 990	1 419 130
Pacific cupped oyster	4 496 196	3 003 831
Rainbow trout	183 575	762 707
Red seaweeds	866 383	121 294
Sea snails	238 331	131 188
Wakame	2 739 753	1 101 507
Warty gracilaria	985 667	394 257
Yesso scallop	1 239 811	1 677 870
Others	8 448 853	11 710 346
Total	31 417 540	31 720 123

Source: FishStat Plus (http://www.fao.org).

(i.e., cobia, grouper, and tuna). New mariculture farms have created new markets and opportunities for local populations, such as the development of backyard hatcheries in countries such as Indonesia. In Bali, for example, the development of small-scale hatcheries has resulted in substantial increases in income as compared to the more traditional coconut crops.

The immediate benefit to a mariculture entrepreneur is the cash income received from sale of the aquatic products produced. This income then is spent to pay for cash expenditures related to feed, fingerlings, repairs to facilities and equipment, fuel, labor, and other operating costs. Payments on loans will be made to financial lenders involved in providing capital for the facilities, equipment, and operation of the business. Profit remaining after expenses can be spent by the owner on other goods and services, invested back into the business, or invested for the future benefit of the owner.

Mariculture of certain species has begun to exceed the value of the same species in capture fisheries. Salmon is a prime example, in which the value of farmed salmon has exceeded that of wild-caught salmon for a number of years. A similar trend can be observed for the rapidly developing capture-based mariculture technologies. In the Murcia region of Spain, for example, the economic value of tuna farming now represents 8 times the value of regional fisheries.

Employment

Mariculture businesses generate employment for the owner of the business, for those who serve as managers and foremen, and those who constitute the principal workforce for the business. Studies in China, Vietnam, Philippines, Indonesia, and Thailand suggest that shrimp farming uses more labor per hectare than does rice farming.

Employment is generated throughout the market supply chain. Development of mariculture businesses increases demand for fry and fingerlings, feeds, construction, equipment used on the farm (trucks, tractors, aerators, nets, boats, etc.), and repairs to facilities and equipment. In Bangladesh, it is estimated that about 300 000 people derive a significant part of their annual income from shrimp seed collection.

Mariculture businesses also create jobs for fish collectors, brokers, and vendors. These intermediaries provide important marketing functions that are difficult for smaller-scale producers to handle. With declines in catch of some previously important fish products, fish vendors need new fish products to sell. Increasing volumes from mariculture can keep these marketers in business.

The employment stimulated by development of mariculture businesses is especially significant in economically depressed areas with high rates of unemployment and underemployment. Many mariculture activities develop in rural coastal communities where jobs are limited and mariculture can constitute a critical source of employment. Many jobs in fishing communities have been reduced as fishing opportunities have become more scarce. Given its relationship to the fishing industry and some degree of similarity in the skills required, mariculture businesses are often a welcome alternative for the existing skilled workforce.

Capture-based mariculture operations in particular provide job opportunities that require skill sets that are easily met by those who have worked in the fishing industry. The rapid development of capture-based tuna farms in the Mediterranean Sea provides a good example. In Spain, fishermen have become active partners in tuna farms. As a result, the number of specialized bluefin tuna boats has increased to supply the capture-based tuna farms in the area. These boats capture the small pelagic fish that are used as feed. In Croatia, trawlers transport live fish or feed to tuna farms off shore. This has generated important sources of employment in heavily depopulated Croatian islands. Tuna farming is labor-intensive, but offers opportunities for younger workers to develop new skills. Many of the employees on the tuna farms are young, 25–35 years old, who work as divers. The divers are used to crowd the tuna for hand-harvesting without stressing the fish, inspect for mortalities, and check the integrity of moorings. Working conditions on tuna farms are preferable to those on tuna boats because the hours and salaries are more regular. Workers can spend weekends on shore as compared to spending long periods of time at sea on tuna boats. The improved working conditions have improved social stability.

Formal studies of economic impacts have measured the effects on employment for fish farm owners and from the secondary businesses such as supply companies, feed mills, processing plants, transportation services, etc. In the Philippines, fish production accounted for 4.4% of total national employment in 2001. In the United States, catfish production accounted for nearly half of all employment in one particular county.

Tax Revenue

Tax policies vary considerably from country to country. Nevertheless, there is some form of tax structure in most countries that is used to finance local, regional, and national governments. Tax structures typically include some form of business tax in addition to combinations of property, income, value-added, or sales taxes. Tax revenue is the major source of revenue used for investments in roads and bridges, schools, health, and public sanitation facilities.

Mariculture businesses contribute to the tax base in that particular region or market. Moreover, the wages paid to managers and workers in the business are subject to income, sales, and property taxes. As the business grows, it pays more taxes. Increasing payrolls from increased employment by the company results in greater tax revenue from increased incomes, increased spending (sales taxes), and increased property taxes paid (as people purchase larger homes and/or more land).

Export Revenue and Foreign Exchange

The largest volume of mariculture production is in developing countries whereas the largest markets for mariculture products are in the more developed nations of the world. Thus, much of the flow of international trade in mariculture products is from the developing to the developed world. This trade results in export revenue for the companies involved. Exporters in the live grouper trade in Asia have been shown to earn returns on total costs as high as 94%.

Export volumes and revenue further generate foreign exchange for the exporting country. This is particularly important for countries that are dependent on other countries for particular goods and services. In order to import products not produced domestically, the country needs sufficient foreign exchange. Moreover, if the country exports products to countries whose currency is strong (i.e., has a high value relative to other forms of currency), the country can then import a relatively greater amount of product from countries whose currency does not have as high a value. Many low-income food-deficient countries use fish exports to pay for imports of low-value food commodities. For example, in 2000, Indonesia, China, the Philippines, India, and Bangladesh paid off 63%, 62%, 22%, 54%, and 32%, respectively, of their food import bills by exporting fish and fish products.

Economic Growth

Economic growth is generated from increased savings, investment, and money supply. The process starts with profitable businesses. Profits can either be saved by the owner or invested back into the business. Savings deposited in a bank or invested in stocks or bonds increase the amount of capital available to be loaned out to other potential investors. With increasing amounts of capital available, the cost of capital (the interest rate) is reduced, making it easier for both individuals and businesses to borrow capital. As a result, spending on housing, vehicles, property, and new and expanded businesses increases. This in turn increases demand for housing, vehicles, and other goods which increases demand for employment. As the economy grows (as measured by the GDP), the standard of living rises as income levels rise and citizens can afford to purchase greater varieties of goods and services. The availability of goods and services also increases with economic growth. Economic growth increases demand which also tends to increase price levels. As prices increase, businesses become more profitable and wages rise.

Economic growth is necessary for standards of living to increase. However, continuous economic growth means that prices also rise continuously. Continuous increases in prices constitute inflation that, if excessive, can create economic difficulties. It is beyond the scope of this article to discuss the mechanisms used by different countries to manage the economy on a national scale and how to manage both economic growth and inflation.

Economic Development

Economic development occurs as economies diversify, provide a greater variety of goods and services, and as the purchasing power of consumers increases. This combination results in a higher standard of living for more people in the country. Economic development results in higher levels of education, greater employment opportunities, and higher income levels. Communities are strengthened with economic development because increasing numbers of jobs result in higher income levels. Higher standards of living provide greater incentives for young people to stay in the area rather than outmigrate in search of better employment and income opportunities.

Studies of the economic impacts of aquaculture have highlighted the importance of backwardly linked businesses. These are the businesses that produce fingerlings, feed, and sell other supplies to the aquaculture farms. Studies of the linkages in the US catfish industry show that economic output of and the value added by the secondary businesses are greater than those of the farms themselves.

Economic Multiplier Effects

Each time a dollar exchanges hands in an economy, its impact 'multiplies'. This occurs because that same dollar can then be used to purchase some other item. Each purchase represents demand for the product being purchased. Each time that same dollar is used to purchase another good or service, it generates additional demand in the economy. Those businesses that have higher expenditures, particularly those with high initial capital expenditures, tend to have higher multiplier effects in the economy.

Aquaculture businesses tend to have high capital expenditures and, thus, tend to have relatively higher multipliers than other types of businesses, particularly other types of agricultural businesses. These high capital expenditures come from the expense of building ponds, constructing net pens or submerged cages, and the equipment costs associated with aerators, boats, trucks, tractors, and other types of equipment. As an example of how economic multipliers occur, consider a new shrimp farm enterprise. Construction of the ponds will require either purchasing bulldozers and bulldozer operators or contracting with a company that has the equipment and expertise to build ponds. The shrimp farm owner pays the pond construction company for the ponds that were built. The dollars paid to the construction company are used to pay equipment operators, fuel companies, for water supply and control pipes and valves, and repairs to equipment. The pond construction workers use the dollars received as wages to purchase more food, clothing for the family, school costs, healthcare costs, and other items. The dollars spent for food in the local market become income for the vendors who pay the farmers who raised the food. Those farmers can now pay for their seed and fertilizer, buy more clothing, and pay school and healthcare expenses, etc. In this expenditure chain, the dollar 'turned over' five times, creating additional economic demand each time.

Several formal studies of economic impacts from aquaculture have been conducted. These studies have shown large amounts of total economic output, employment, and value-added effects from aquaculture businesses. On the local area, aquaculture can generate the majority of economic output in a given district. Multipliers as high as 6.1 have been calculated for aquaculture activities on the local level.

Potential Negative Economic Impacts

Mariculture has been criticized by some for what economists call 'externalities'. Externalities refer to an effect on an individual or community other than the individual or community that created the problem. Pollution is often referred to as an externality. For example, if effluents discharged from a mariculture business create water-quality problems for another farm or community downstream, that farm or community will have to spend more to clean up that pollution. Yet, those costs are 'external' to the business that created the problem.

To the extent that a mariculture business creates an externality that increases costs for another business or community, there will be negative economic impacts. The primary potential sources of negative economic impacts from mariculture include: (1) discharge of effluents that may contain problematic levels of nutrients, antibiotics, or nonnative species; (2) spread of diseases; (3) use of high levels of fish meal in the diets fed to the species cultured; (4) clearing mangroves from coastal areas; and (5) consequences of predator control measures. The history of the development of mariculture includes cases in which unregulated discharge of untreated effluents from shrimp farms resulted in poor water-quality conditions in bays and estuaries. Since this same water also served as influent for other shrimp farms, its poorer quality stressed shrimp and facilitated spread of viral diseases. Economic losses resulted.

Concerns have been noted over the increasing demand for fish meal as mariculture of carnivorous species grows. The fear is that this increasing use will result in a decline in the stocks of the marine species that serve as forage for wild stocks and that are used in the production of fish meal. If this were to occur, economic losses could occur from declines in capture fisheries.

Loss of mangrove forests in coastal areas results in loss of important nursery grounds for a number of species and less protection during storms. The loss of mangrove areas is due to many uses, including for construction, firewood, for salt production, and others. If the construction of ponds for mariculture reduces mangrove areas, additional negative impacts will occur.

Control of predators on net pens and other types of mariculture operations often involves lethal methods. The use of lethal methods raises concerns related to biodiversity and viability of wild populations. While there is no direct link to other economic activities in many of these cases, the loss of ecosystem services from reductions in biodiversity may at some point result in other economic problems.

Social Impacts

Food Security

Aquaculture is an important food security strategy for many individuals and countries. Domestic production of food provides a buffer against interruptions in the food supply that can result from international disputes, trade embargoes, war, transportation accidents, or natural disasters. There are many examples of fish ponds being used to feed households during times of strife. Examples include fish pond owners in Vietnam during the war with the United States and during the Contra War in Nicaragua, among others.

Aquaculture has been shown to function as a food reserve for many subsistence farming households. Many food crops are seasonal in nature while fish in ponds or cages can be available year-round and used particularly during periods of prolonged drought or in between different crops. Pond levees also can provide additional land area to raise vegetable crops for household consumption and sale.

Nutritional Benefits

Fish are widely recognized to be a high-quality source of animal protein. Fish have long been viewed as 'poor people's food'. Subsistence farmers who raise fish, shellfish, or seaweed have a ready source of nutritious food for home consumption throughout the year. Particularly in Asia, fish play a vital role in supplying inexpensive animal protein to poorer households. This is important because the proportion of the food budget allocated to fish is higher among low-income groups. Moreover, it has been shown that rural people consume more fish than do urban dwellers and that fish producers consume more fish than do nonproducers. Those farmers who also supply the local market with aquatic products provide a supply of high-quality protein to other households in the area. Increasing supplies of farmed products result in lower prices in seafood markets. Tuna prices in Japan, for example, have been falling as a result of the increased farmed supply. This has resulted in making tuna and its nutritional benefits more readily available to middle-income buyers in Japan.

Health Benefits

In addition to the protein content, a number of aquatic products provide additional health benefits. Farmed fish such as salmon have high levels of omega-3 fatty acids that reduce the risk of heart disease. Products such as seaweeds are rich in vitamins. Mariculture can enable the poorest of the poor and even the landless to benefit from public resources. For example, cage culture of fish in public waters, culture of mollusks and seaweeds along the coast, and culture-based fisheries in public water bodies provide a source of healthy food for subsistence families.

Poverty Alleviation

As aquaculture has grown, its development has frequently occurred in economically depressed areas.

Poverty alleviation is accompanied by a number of positive social impacts. These include improved access to food (that results in higher nutritional and health levels), improved access to education (due to higher income levels and ability to pay for fees and supplies), and improved employment opportunities. In Sumatra, Indonesia, profits from grouper farming enable members of Moslem communities to make pilgrimages to Mecca, enriching both the individuals and the community. In Vietnam, income from grouper hatcheries contributes 10–50% of the annual income of fishermen.

Enhancement of Capture Fisheries

Marine aquaculture can improve the condition of fisheries through supplemental stockings from aquaculture. This will help to meet the global demand for fish products. Increased mariculture supply relieves pressure on traditional protein sources for species such as live cod and haddock. Given that fish supplies from capture fisheries are believed to have reached or be close to maximum sustainable yield, mariculture can reduce the expected shortage of fish.

Potential Negative Social Impacts

Mariculture has been criticized for creating negative social impacts. These criticisms have tended to focus on displacement of people if large businesses buy land and move people off that land. Economic development is accompanied by social change and individuals and households frequently are affected in various ways by construction of new infrastructure and production and processing facilities. In developed countries, federal laws typically provide for some degree of compensation for land taken through eminent domain for construction of highways or power lines. However, developing countries rarely have similar mechanisms for compensating those who lose land. In situations in which the resources involved are in the public domain (such as coastal areas), granting a concession for a mariculture operation along the coast may result in reduced access for poor and landless individuals to collect shellfish and other foods for their households.

Resource ownership or use rights in coastal zones frequently are ambiguous. Many of these areas traditionally have been common access areas, but are now under pressure for development for mariculture activities. Displaced and poor migrant people are frequently marginalized in coastal lands and often depend to some degree on common resources. However, the extent of conflict over the use of resources is variable. Surveys in Asia reported less than 10% of farms experiencing conflicts in most countries, but higher incidents of conflict were reported in India (29%) and China (94%).

In the Mediterranean, capture-based mariculture has resulted in conflicts between local tuna fishermen who fish with long lines, and cage towing operations. Other conflicts have occurred in Croatia due to the smell and pollution during the summer from bluefin tuna farms. Uncollected fat skims on the sea surface that results from feeding oily trash fish spread outside the licensed zones onto beaches frequented by tourists. On the positive side, tourism in Spain was enhanced by offering guided tours to offshore cages.

Additional negative impacts could potentially include: (1) marine mammal entanglements; (2) threaten genetic makeup of wild fish stocks; (3) privatize what some think of as free open-access resource; (4) esthetically undesirable; and (5) negative impacts on commercial fishermen. For example, excessive collection of postlarvae and egg-laden female shrimp can result in loss of income for fishermen and reduction of natural shrimp and fish stocks.

Conclusions

The economic and social impacts of mariculture are variable throughout the world and are based on the range of technologies, cultures, habitats, land types, and social, economic, and political differences. Positive economic impacts include increased revenue, employment, tax revenue, foreign exchange, economic growth and development, and multiplicative effects through the economy. Negative economic effects could occur through excessive discharge of effluents, spread of disease, and through excessive use of fish meal. Mariculture enhances food security of households and countries and provides nutritional and health benefits while alleviating poverty. However, if earnings are exported and access to public domain resources restricted to poorer classes, negative social impacts can occur.

See also

Dynamics of Exploited Marine Fish Populations. Global Marine Pollution. Mariculture Diseases and Health. Mariculture of Aquarium Fishes. Mariculture of Mediterranean Species. Mariculture Overview. Marine Chemical and Medicine Resources. Marine Fishery Resources, Global State of. Marine Mammals, History of Exploitation. Marine Protected Areas.

Further Reading

Aguero M and Gonzalez E (1997) Aquaculture economics in Latin America and the Caribbean: A regional assessment. In: Charles AT, Agbayani RF, Agbayani EC, et al. (eds.) Aquaculture Economics in Developing Countries: Regional Assessments and an Annotated Bibliography. FAO Fisheries Circular No. 932. Rome: FAO.

Dey MM and Ahmed M (2005) Special Issue: Sustainable Aquaculture Development in Asia. Aquaculture Economics and Management 9(1/2): 286pp.

Edwards P (2000) Aquaculture, poverty impacts and livelihoods. Natural Resource Perspectives 56 (Jun. 2000).

Engle CR and Kaliba AR (2004) Special Issue: The Economic Impacts of Aquaculture in the United States. Journal of Applied Aquaculture 15(1/2): 172pp.

Ottolenghi F, Silvestri C, Giordano P, Lovatelli A, and New MB (2004) Capture-Based Aquaculture. The Fattening of Eels, Groupers, Tunas and Yellowtails, 308pp. Rome: FAO.

Tacon GJ (2001) Increasing the contribution of aquaculture for food security and poverty alleviation. In: Subasinghe RP, Bueno P, Phillips MJ, Hough C, McGladdery SE, and Arthur JR (eds.) Aquaculture in the Third Millennium. Technical Proceedings of the Conference on Aquaculture in the Third Millennium, pp. 63–72, Bangkok, Thailand, 20–25 February 2000. Bangkok and Rome: NACA and FAO.

Relevant Websites

http://www.fao.org
  – Food and Agriculture Organization of the United Nations.
http://www.worldfishcenter.org
  – WorldFish Center.

EXPLOITATION OF NATURAL AND CULTURAL RESOURCES

SHIPPING AND PORTS

H. L. Kite-Powell, Woods Hole Oceanographic
Institution, Woods Hole, MA, USA

Introduction

Ships and ports have been an important medium for trade and commerce for thousands of years. Today's maritime shipping industry carries 90% of the world's 5.1 billion tons of international trade. Total seaborne cargo movements exceeded 21 trillion ton-miles in 1998. Shipping is an efficient means of transport, and becoming more so. World freight payments as a fraction of total import value were 5.24% in 1997, down from 6.64% in 1980. Modern container ships carry a 40-foot container for <10 cents per mile – a fraction of the cost of surface transport.

Segments of the shipping industry exhibit varying degrees of technological sophistication and economic efficiency. Bulk shipping technology has changed little in recent decades, and the bulk industry is economically efficient and competitive. Container shipping has advanced technologically, but economic stability remains elusive. The shipping industry has a history of national protectionist measures and is governed by a patchwork of national and international regulations to ensure safety and guard against environmental damage.

Units

Two standard measures of ship size are deadweight tonnage (dwt) and gross tonnage (gt). Deadweight describes the vessel's cargo capacity; gross tonnage describes the vessel's enclosed volume, where 100 cubic feet equal one gross ton. In container shipping, the standard unit of capacity is the twenty-foot equivalent unit (TEU), which is the cargo volume available in a container 20 feet long, 8 feet wide, and 8 feet (or more) high. These containers are capable of carrying 18 metric tons but more typically are filled to 13 tons or less, depending on the density of the cargo.

World Seaborne Trade

Shipping distinguishes between two main types of cargos: bulk (usually shiploads of a single commodity) and general cargos (everything else). Major dry bulk trades include iron ore, coal, grain, bauxite, sand and gravel, and scrap metal. Liquid bulk or tanker cargos include crude oil and petroleum products, chemicals, liquefied natural gas (LNG), and vegetable oil. General cargo may be containerized, break-bulk (non-container packaging), or 'neo-bulk' (automobiles, paper, lumber, etc.). **Table 1** shows that global cargo weight has grown by a factor of 10 during the second half of the twentieth century, and by about 4% per year since 1986. Tanker cargos (mostly oil and oil products) make up about 40% of all cargo movements by weight.

Both the dry bulk and tanker trades have grown at an average rate of about 2% per year since 1980. Current bulk trades total some 9 trillion ton-miles for dry bulk and 10 trillion ton-miles for tanker cargos.

Before containerization, general cargo was transported on pallets or in cartons, crates, or sacks (in general cargo or break-bulk vessels). The modern container was first used in US domestic trades in the 1950s. By the 1970s, intermodal services were developed, integrating maritime and land-based modes (trains and trucks) of moving containers. Under the 'mini land-bridge' concept, for example, containers bound from Asia to a US east coast port might travel across the Pacific by ship and then across the USA by train. In the 1980s, intermodal services were streamlined further with the introduction of double-stack trains, container transportation became more

Table 1 World seaborne dry cargo and tanker trade volume (million tons) 1950–1998

Year	Dry cargo	Tanker cargo	Total	% change
1950	330	130	460	—
1960	540	744	1284	—
1970	1165	1440	2605	—
1986	2122	1263	3385	3
1987	2178	1283	3461	2
1988	2308	1367	3675	6
1989	2400	1460	3860	5
1990	2451	1526	3977	3
1991	2537	1573	4110	3
1992	2573	1648	4221	3
1993	2625	1714	4339	3
1994	2735	1771	4506	4
1995	2891	1796	4687	4
1996	2989	1870	4859	4
1997	3163	1944	5107	5
1998	3125	1945	5070	− 1

Table 2   Cargo volume on major shipping routes (million tons) 1998 (for containers: millions of TEU, 1999)

Cargo type	Route	Cargo volume	% of total for cargo
Petroleum	Middle East–Asia	525.0	26
	Intra-Americas	326.5	16
	Middle East–Europe	229.5	12
	Intra-Europe	142.7	7
	Africa–Europe	141.4	7
	Middle East–Americas	137.7	7
	Africa–Americas	106.1	5
Dry bulk: iron ore	Australia–Asia	116.8	28
	Americas–Europe	87.1	21
	Americas–Asia	63.3	15
Dry bulk: coal	Australia–Asia	118.4	25
	Americas–Europe	61.3	13
	S. Africa–Europe	37.8	8
	Americas–Asia	35.2	7
	Intra-Asia	29.3	6
	Australia–Europe	27.3	6
Dry bulk: grain	Americas–Asia	60.2	31
	Americas–Europe	20.6	10
	Intra-Americas	33.1	17
	Americas–Africa	15.2	8
	Australia–Asia	15.1	8
Container	Trans-Pacific (Asia/N. America)	7.3	14
	Asia/Europe	4.5	8
	Trans-Atlantic (N. America/Europe)	4.0	8

standardized, and shippers began to treat container shipping services more like a commodity. Today, >70% of general cargo moves in containers. Some 58.3 million TEU of export cargo moved worldwide in 1999; including empty and domestic movement, the total was over 170 million TEU. The current total container trade volume is some 300 billion TEU-miles. Container traffic has grown at about 7% per year since 1980.

Table 2 lists the major shipping trade routes for the most significant cargo types, and again illustrates the dominance of petroleum in total global cargo volume. The most important petroleum trade routes are from the Middle East to Asia and to Europe, and within the Americas. Iron ore and coal trades are dominated by Australian exports to Asia, followed by exports from the Americas to Europe and Asia. The grain trades are dominated by American exports to Asia and Europe, along with intra-American routes. Container traffic is more fragmented; some of the largest routes connect Asia with North America to the east and with Europe to the west, and Europe and North America across the Atlantic.

The World Fleet

The world's seaborne trade is carried by an international fleet of about 25 000 ocean-going commercial cargo ships of more than 1000 gt displacement. As trade volumes grew, economies of scale (lower cost per ton-mile) led to the development of larger ships: ultra-large crude oil carriers (ULCCs) of >550 000 dwt were launched in the 1970s. Structural considerations, and constraints on draft (notably in US ports) and beam (notably in the Panama Canal) have curbed the move toward larger bulk vessels. Container ships are still growing in size. The first true container ships carried around 200 TEU; in the late 1980s, they reached 4000 TEU; today they are close to 8000 TEU. The world container fleet capacity was 4.1 million TEU in 1999.

Table 3 shows the development of the world fleet in dwt terms since 1976. Total fleet capacity has grown by an average of 1.5% per year, although it contracted during the 1980s. The significant growth in the 'other vessel' category is due largely to container vessels; container vessel dwt increased faster than any other category (nearly 9% annually) from 1985 to 1999. Specialized tankers and auto carriers have also increased rapidly, by around 5% annually. Combos (designed for a mix of bulk and general cargo) and general cargo ships have decreased by >5% per year.

Four standard vessel classes dominate both the dry bulk and tanker fleets, as shown in **Table 4**. Panamax bulkers and Suezmax tankers are so named because they fall just within the maximum dimensions for the

Table 3 World fleet development (million dwt) 1976–1998

Year	Oil tankers	Dry bulk carriers	Other vessels	Total world fleet	% change
1976	320.0	158.1	130.3	608.4	—
1977	335.3	174.4	139.1	648.8	7
1978	339.1	184.5	146.8	670.4	3
1979	338.3	188.5	154.7	681.5	2
1980	339.8	191.0	160.1	690.9	1
1981	335.5	199.5	162.2	697.2	1
1982	325.2	211.2	165.6	702.0	1
1983	306.1	220.6	167.8	694.5	−1
1984	286.8	228.4	168.1	683.3	−2
1985	268.4	237.3	168.0	673.7	−1
1986	247.5	235.2	164.9	647.6	−4
1987	245.5	231.8	163.5	640.8	−1
1988	245.0	230.1	162.0	637.1	−1
1989	248.4	231.4	166.9	646.7	2
1990	257.4	238.9	170.5	666.8	3
1991	264.2	244.0	176.1	684.3	3
1992	270.6	245.7	178.3	694.6	2
1993	270.2	251.3	194.4	715.9	3
1994	269.4	254.3	220.3	744.0	4
1995	265.8	259.8	241.5	767.1	3
1996	271.2	269.6	252.2	793.0	4
1997	272.5	277.8	263.0	813.3	3
1998	279.7	271.6	274.0	825.3	2

Table 4 Major vessel categories in the world's ocean-going cargo ship fleet, 2000

Cargo type	Category	Typical size	Number in world fleet
Dry bulk	Handysize	27 000 dwt	2700
	Handymax	43 000 dwt	1000
	Panamax	69 000 dwt	950
	Capesize	150 000 dwt	550
Tanker	Product tanker	45 000 dwt	1400
	Aframax	90 000 dwt	700
	Suezmax	140 000 dwt	300
	VLCC	280 000 dwt	450
General cargo	Container	>2000 TEU	800
	Container feeder	<2000 TEU	1800
	Ro/Ro	<2000 TEU	900
	Semi-container	<1000 TEU	2800

VLCC, very large crude carrier; RoRo, roll-on/roll-off.

Panama and Suez Canals, respectively. Capesize bulkers and VLCCs (very large crude carriers) are constrained in size mostly by the limitations of port facilities (draft restrictions).

In addition to self-propelled vessels, ocean-going barges pulled by tugs carry both bulk and container cargos, primarily on coastal routes.

Dry Bulk Carriers

The dry bulk fleet is characterized by fragmented ownership; few operators own more than 100 vessels. Most owners charter their vessels to shippers through a network of brokers, under either long-term or single-voyage ('spot') charter arrangements (see **Table 5**). The top 20 dry bulk charterers account for about 30% of the market. The dry bulk charter market is, in general, highly competitive.

Capesize bulkers are used primarily in the Australian and South American bulk export trades. Grain shipments from the USA to Asia move mainly on Panamax vessels. Some bulk vessels, such as oil/bulk/ore (OBO) carriers, are designed for a variety of cargos. Others are 'neo-bulk' carriers, specifically

Table 5 Average daily time charter rates, 1980–2000

Cargo type	Vessel type	$/day
Dry bulk	Handysize	6000
	Handymax	8000
	Panamax	9500
	Capesize	14 000
Tanker	Product tanker	12 000
	Aframax	13 000
	Suezmax	16 500
	VLCC	22 000
Container	400 TEU geared	5000
	1000 TEU geared	9000
	1500 TEU geared	13 500
	2000 TEU gearless	18 000

VLCC, very large crude carrier.

designed to carry goods such as automobiles, lumber, or paper.

Tankers

Ownership of the tanker fleet is also fragmented; the average tanker owner controls fewer than three vessels; 70% of the fleet is owned by independent owners and 25% by oil companies. Like dry bulk ships, tankers are chartered to shippers on longer time charters or spot voyage charters (see **Table 5**). The top five liquid bulk charterers account for about 25% of the market. Like the dry bulk market, the tanker charter market is highly competitive.

VLCCs are used primarily for Middle East exports to Asia and the USA. In addition to liquid tankers, a fleet of gas carriers (world capacity: 17 million gt) moves liquefied natural gas (LNG) and petroleum gas (LPG).

General Cargo/Container Ships

Ownership is more concentrated in the container or 'liner' fleet. ('Liner' carriers operate regularly scheduled service between specific sets of ports.) Of some 600 liner carriers, the top 20 account for 55% of world TEU capacity. The industry is expected to consolidate further. Despite this concentration, liner shipping is competitive; barriers to entry are low, and niche players serving specialized cargos or routes have been among the most profitable. Outside ownership of container vessels and chartering to liner companies increased during the 1990s.

Large ships (>4000 TEU) are used on long ocean transits between major ports, while smaller 'feeder' vessels typically carry coastwise cargo or serve niche markets between particular port pairs. In addition to container vessels, the general cargo trades are served by roll-on/roll-off (RoRo) ships (world fleet capacity: 23 million gt) that carry cargo in wheeled vehicles, by refrigerated ships for perishable goods, and by a diminishing number of multi-purpose general cargo ships.

Passenger Vessels

The world's fleet of ferries carries passengers and automobiles on routes ranging from urban harbor crossings to hundreds of miles of open ocean. Today's ferry fleet is increasingly composed of so-called fast ferries, >1200 of which were active in 1997. Many of these are aluminum-hull catamarans, the largest capable of carrying 900 passengers and 240 vehicles at speeds between 40 and 50 knots. Fast ferries provide an inexpensive alternative to land and air travel along coastal routes in parts of Asia, Europe, and South America.

The cruise ship industry was born in the 1960s, when jet aircraft replaced ships as the primary means of crossing oceans, and underutilized passenger liners were converted for recreational travel. The most popular early cruises ran from North America to the Caribbean. Today, the Caribbean remains the main cruise destination, with more than half of all passengers; others are Alaska and the Mediterranean. The US cruise market grew from 500 000 passengers per year in 1970 to 1.4 million in 1980 and 3.5 million in 1990. Today, more than 25 cruise lines serve the North American cruise market. The three largest lines control over 50% of the market, and further consolidation is expected; 149 ships carried 6 million passengers and generated $18 billion in turnover in 1999. The global market saw 9 million passengers. Cruise vessels continue to grow larger due to scale economies; the largest ships today carry more than 2000 passengers. Most cruise lines are European-owned and fly either European or open registry flags (see below).

Financial Performance

Shipping markets display cyclical behavior. Charter rates (see **Table 5**) are set by the market mechanism of supply and demand, and are most closely correlated with fleet utilization (rates rise sharply when utilization exceeds 95%). Shipping supply (capacity) is driven by ordering and scrapping decisions, and sometimes by regulatory events such as the double hull requirement for tankers (see below). Demand (trade volume) is determined by national and global business cycles, economic development, international trade policies, and trade disruptions such as wars/conflicts, canal closures, or oil price shocks. New vessel deliveries typically lag orders by 1–2 years.

Excessive ordering of new vessels in times of high freight rates eventually leads to overcapacity and a downturn in rates; rates remain low until fleet contraction or trade growth improves utilization, and the cycle repeats.

Average annual returns on shipping investments ranged from 8% (dry bulk) to 11% (container feeder ships) and 13% (tankers) during 1980–2000, with considerable volatility. Second-hand ship asset prices fluctuate with charter rates, and many ship owners try to improve their returns through buy-low and sell-high strategies.

Law and Regulation

Legal Regimes

International Law of the Sea   The United Nations Convention on the Law of the Sea (UNCLOS) sets out an international legal framework governing the oceans, including shipping. UNCLOS codifies the rules underlying the nationality (registry) of ships, the right of innocent passage for merchant vessels through other nations' territorial waters, etc. (see the entry in this volume on the Law of the Sea for details).

International Maritime Organization   The International Maritime Organization (IMO) was established in 1948 and became active in 1959. Its 158 present member states have adopted some 40 IMO conventions and protocols governing international shipping. Major topics include maritime safety, marine pollution, and liability and compensation for third-party claims. Enforcement of these conventions is the responsibility of member governments and, in particular, of port states. The principle of port state control allows national authorities to inspect foreign ships for compliance and, if necessary, detain them until violations are addressed.

In 1960, IMO adopted a new version of the International Convention for the Safety of Life at Sea (SOLAS), the most important of all treaties dealing with maritime safety. IMO next addressed such matters as the facilitation of international maritime traffic, load lines, and the carriage of dangerous goods. It then turned to the prevention and mitigation of maritime accidents, and the reduction of environmental effects from cargo tank washing and the disposal of engine-room waste.

The most important of all these measures was the Marine Pollution (MARPOL) treaty, adopted in two stages in 1973 and 1978. It covers accidental and operational oil pollution as well as pollution by chemicals, goods in packaged form, sewage, and garbage. In the 1990s, IMO adopted a requirement for all new tankers and existing tankers over 25 years of age to be fitted with double hulls or a design that provides equivalent cargo protection in the event of a collision or grounding.

IMO has also dealt with liability and compensation for pollution damage. Two treaties adopted in 1969 and 1971 established a system to provide compensation to those who suffer financially as a result of pollution.

IMO introduced major improvements to the maritime distress communications system. A global search and rescue system using satellite communications has been in place since the 1970s. In 1992, the Global Maritime Distress and Safety System (GMDSS) became operative. Under GMDSS, distress messages are transmitted automatically in the event of an accident, without intervention by the crew.

An International Convention on Standards of Training, Certification and Watchkeeping (STCW), adopted in 1978 and amended in 1995, requires each participating nation to develop training and certification guidelines for mariners on vessels sailing under its flag.

National Control and Admiralty Law   The body of private law governing navigation and shipping in each country is known as admiralty or maritime law. Under admiralty, a ship's flag (or registry) determines the source of law. For example, a ship flying the American flag in European waters is subject to American admiralty law. This also applies to criminal law governing the ship's crew.

By offering advantageous tax regimes and relatively lax vessel ownership, inspection, and crewing requirements, so-called 'flags of convenience' or 'open registries' have attracted about half of the world's tonnage (see **Table 6**). The open registries include Antigua and Barbuda, Bahamas, Bermuda, Cayman Islands, Cyprus, Gibraltar, Honduras, Lebanon, Liberia, Malta, Mauritius, Oman, Panama, Saint Vincent, and Vanuatu. Open registries are the flags of choice for low-cost vessel operation, but in some instances they have the disadvantage of poor reputation for safety.

Protectionism/Subsidies

Most maritime nations have long pursued policies that protect their domestic flag fleet from foreign competition. The objective of this protectionism is usually to maintain a domestic flag fleet and cadre of seamen for national security, employment, and increased trade.

Table 6 Fleets of principal registries (in 1000s of gt) 1998 (includes ships of 100 gt)

Country of registry	Tankers	Dry bulk	General cargo	Other	Total	% of world
Panama	22 680	40 319	26 616	8608	98 223	18.5
Liberia	26 361	16 739	8803	8590	60 493	11.4
Bahamas	11 982	4990	7143	3601	27 716	5.2
Greece	12 587	8771	1943	1924	25 225	4.7
Malta	9848	8616	4658	952	24 074	4.5
Cyprus	3848	11 090	6981	1383	23 302	4.4
Norway	8994	4041	4153	5949	23 137	4.4
Singapore	8781	4585	5596	1408	20 370	3.8
Japan	5434	3869	3111	5366	17 780	3.3
China	2029	6833	6199	1443	16 504	3.1
USA	3436	1268	3861	3286	11 851	2.2
Russia	1608	1031	3665	4786	11 090	2.1
Others	33 448	46 414	57 350	34 917	172 129	32.4
World total	151 036	158 566	140 079	82 213	531 894	100.0

Laws that reserve domestic waterborne cargo (cabotage) for domestic flag ships are common in many countries, including most developed economies (although they are now being phased out within the European Union). The USA has a particularly restrictive cabotage system. Cargo and passengers moving by water between points within the USA, as well as certain US government cargos, must be carried in US-built, US-flag vessels owned by US citizens. Also, US-flag ships must be crewed by US citizens.

Vessels operating under restrictive crewing and safety standards may not be competitive in the international trades with open registry vessels due to high operating costs. In some cases, nations have provided operating cost subsidies to such vessels. The US operating subsidy program was phased out in the 1990s.

Additional subsidies have been available to the shipping industry through government support of shipbuilding (see below).

Liner Conferences

The general cargo trades have traditionally been served by liner companies that operate vessels between fixed sets of ports on a regular, published schedule. To regulate competition among liner companies and ensure fair and consistent service to shippers, a system of 'liner conferences' has regulated liner services and allowed liner companies, within limits, to coordinate their services.

Traditionally, so-called 'open' conferences have been open to all liner operators, published a single rate for port-to-port carriage of a specific commodity, and allowed operators within the conference to compete for cargo on service. The US foreign liner trade has operated largely under open conference rules. By contrast, many other liner trades have been governed by closed conferences, participation in which is restricted by governments. In 1974, the United Nations Conference on Trade and Development (UNCTAD) adopted its Code of Conduct for Liner Conferences, which went into effect in 1983. Under this code, up to 40% of a nation's trade can be reserved for its domestic fleet.

In the 1970s, price competition began within conferences as carriers quoted lower rates for intermodal mini land-bridge routes than the common published port-to-port tariffs. By the 1980s, intermodal rates were incorporated fully in the liner conference scheme. The 1990s saw growing deregulation of liner trades (for example, the US 1998 Ocean Shipping Reform Act), which allowed carriers to negotiate confidential rates separately with shippers. The effect has been, in part, increased competition and a drive for consolidation among liner companies that is expected to continue in the future.

Environmental Issues

With the increased carriage of large volumes of hazardous cargos (crude oil and petroleum products, other chemicals) by ships in the course of the twentieth century, and especially in the wake of several tanker accidents resulting in large oil spills, attention has focused increasingly on the environmental effects of shipping. The response so far has concentrated on the operational and accidental discharge of oil into the sea. Most notably, national and international regulations (the Oil Pollution Act (OPA) of 1990 in the US; IMO MARPOL 1992 Amendments, Reg. 13F and G) are now forcing conversion of the world's tanker fleet to double hulls. Earlier, MARPOL 1978 required segregated ballast tanks on tankers to avoid the discharge of oil residue following the use of cargo tanks to hold ballast water.

Recently, ballast and bilge water has also been identified as a medium for the transport of non-indigenous species to new host countries. A prominent example is the zebra mussel, which was brought to the USA from Europe in this way.

Liability and Insurance

In addition to design standards such as double hulls, national and international policies have addressed the compensation of victims of pollution from shipping accidents through rules governing liability and insurance. The liability of ship owners and operators for third-party claims arising from shipping accidents historically has been limited as a matter of policy to encourage shipping and trade. With the increased risk associated with large tankers, the international limits were raised gradually in the course of the twentieth century. Today, tanker operators in US waters face effectively unlimited liability under OPA 90 and a range of state laws.

Most liability insurance for the commercial shipping fleet is provided through a number of mutual self-insurance schemes known as P&I (protection and indemnity) clubs. The International Group of P&I Clubs includes 19 of the largest clubs and collectively provides insurance to 90% of the world's ocean-going fleet. Large tankers operating in US waters today routinely carry liability insurance coverage upward of $2 billion.

Shipbuilding

Major technical developments in twentieth century shipbuilding began with the introduction of welding in the mid-1930s, and with it, prefabrication of steel sections. Prefabrication of larger sections, improved welding techniques, and improved logistics were introduced in the 1950s. Mechanized steel prefabrication and numerically controlled machines began to appear in the 1970s. From 1900 to 2000, average labor hours per ton of steel decreased from 400–500 to fewer than 100; and assembly time in dock/berth decreased from 3–4 years to less than 6 months.

South Korea and Japan are the most important commercial shipbuilding nations today. In the late 1990s, each had about 30% of world gross tonnage on order; China, Germany, Italy, and Poland each accounted for between about 4 and 6%; and Taiwan and Romania around 2%. The US share of world shipbuilding output fell from 9.5% in 1979 to near zero in 1989.

Shipbuilding is extensively subsidized in many countries. Objectives of subsidies include national security goals (maintenance of a shipbuilding base), employment, and industrial policy goals (shipyards are a major user of steel industry output). Direct subsidies are estimated globally around $5–10 billion per year. Indirect subsidies include government loan guarantees and domestic-build requirements (for example, in the US cabotage trade).

An agreement on shipbuilding support was developed under the auspices of the Organization for Economic Cooperation and Development (OECD) and signed in December 1994 by Japan, South Korea, the USA, Norway, and the members of the European Union. The agreement would limit loan guarantees to 80% of vessel cost and 12 years, ban most direct subsidy practices, and limit government R&D support for shipyards. Its entry into force, planned for 1996, has been delayed by the United States' failure to ratify the agreement.

Ports

Commercial ships call on thousands of ports around the world, but global cargo movement is heavily concentrated in fewer than 100 major bulk cargo ports and about 24 major container ports. The top 20 general cargo ports handled 51% of all TEU movements in 1998 (see **Table 7**).

Most current port development activity is in container rather than bulk terminals. Port throughput efficiency varies greatly. In 1997, some Asian container ports handled 8800 TEU per acre per year, while European ports averaged 3000 and US ports only 2100 TEU per acre per year. The differences are due largely to the more effective use of automation and lesser influence of organized labor in Asian ports. Because container traffic is growing rapidly and container ships continue to increase in size, most port investment today is focused on the improvement and new development of container terminals.

In many countries, the public sector plays a more significant role in port planning and development than it does in shipping. Most general cargo or 'commercial' ports are operated as publicly controlled or semi-public entities, which may lease space for terminal facilities to private terminal operators. Bulk cargo or 'industrial' ports for the export or import/processing of raw materials traditionally have been built by and for a specific industry, often with extensive public assistance. The degree of national coordination of port policy and planning varies considerably. The USA has one of the greatest commercial port densities in the world, particularly along its east coast. US commercial ports generally compete among each other as semi-private entities run in the interest of local economic development objectives. The US federal government's primary role in

Table 7 Top 20 container ports by volume, 1998

Port	Country	Throughput (1000s TEU)
Singapore	Singapore	15 100
Hong Kong	China	14 582
Kaohsiung	Taiwan	6271
Rotterdam	Netherlands	6011
Busan	South Korea	5946
Long Beach	USA	4098
Hamburg	Germany	3547
Los Angeles	USA	3378
Antwerp	Belgium	3266
Shanghai	China	3066
Dubai	UAE	2804
Manila	Philippines	2690
Felixstowe	UK	2524
New York/New Jersey	USA	2500
Tokyo	Japan	2169
Tanjung Priok	Indonesia	2131
Gioia Tauro	Italy	2126
Yokohama	Japan	2091
San Juan	Puerto Rico	2071
Kobe	Japan	1901

port development is in the improvement and maintenance of navigation channels, since most US ports (apart from some on the west coast) are shallow and subject to extensive siltation.

Future Developments

A gradual shift of the 'centroid' of Asian manufacturing centers westward from Japan and Korea may in the future shift more Asia–America container cargo flows from trans-Pacific to Suez/trans-Atlantic routes. Overall container cargo flows are expected to increase by 4–7% per year. Bulk cargo volumes are expected to increase as well, but at a slower pace.

Container ships will continue to increase in size, reaching perhaps 12 000 TEU capacity in the course of the next decade. Smaller, faster cargo ships may be introduced to compete for high-value cargo on certain routes. One concept calls for a 1400 TEU vessel capable of 36–40 knots (twice the speed of a conventional container ship), making scheduled Atlantic crossings in 4 days.

Port developments will center on improved container terminals to handle larger ships more efficiently, including berths that allow working the ship from both sides, and the further automation and streamlining of moving containers to/from ship and rail/truck terminus.

Conclusions

Today's maritime shipping industry is an essential transportation medium in a world where prosperity is often tied to international trade. Ships and ports handle 90% of the world's cargo and provide a highly efficient and flexible means of transport for a variety of goods. Despite a history of protectionist regulation, the industry as a whole is reasonably efficient and becoming more so. The safety of vessels and their crews, and protection of the marine environment from the results of maritime accidents, continue to receive increasing international attention. Although often conservative and slow to adopt new technologies, the shipping industry is poised to adapt and grow to support the world's transportation needs in the twenty-first century.

See also

International Organizations. Law of the Sea. Marine Policy Overview.

Further Reading

Containerization Yearbook 2000. London: National Magazine Co.

Shipping Statistics Yearbook. Bremen: Institute of Shipping Economics and Logistics.

Lloyd's Shipping Economist. London: Lloyd's of London Press.

United Nations Conference on Trade and Development (UNCTAD) (Annual) *Review of Maritime Transport.* New York: United Nations.

RIGS AND OFFSHORE STRUCTURES

C. A. Wilson III, Department of Oceanography and Coastal Sciences, and Coastal Fisheries Institute, CCEER Louisiana State University, Baton Rouge, LA, USA
J. W. Heath, Coastal Fisheries Institute, CCEER Louisiana State University, Baton Rouge, LA, USA

Introduction

The rapid advance in offshore engineering and physical oceanography has provided mankind with the ability to emplace and maintain structures that support what are effectively small cities in some of the ocean's harshest environments. The majority of these offshore structures are associated with the development of hydrocarbon extraction off the world's continents. Offshore structures have been focal points for controversies concerning pollution and hazard to navigation, yet serve as a basis for many coastal economies. Oil spills, drilling muds, and produced water have been continuous public and regulatory issues. Ironically, oil and gas platforms have also been the target of the locally popular 'rigs to reefs' program since the early 1980s. State and federal governments of the United States have encouraged states and private sector organizations to maintain these structures as artificial reefs as it is perceived that they improve fishing on a local scale.

In addition to their ecological value as 'artificial reefs' these platforms are now recognized as excellent facilities for research as we explore our Earth's final frontier.

History

Since the mid 1950s approximately 7000 structures associated with oil and gas development have been emplaced on the continental shelf and slope, and are now approaching the ocean basins (**Figure 1**). The majority of these structures (>65%) are along United States Gulf Coast, and the remainder are concentrated in the North Sea, Middle East, Africa, Australia, Asia, and South America.

The oldest, most well developed, and largest concentration of structures is in the US Gulf of Mexico (GOM). There are currently about 4000 oil and gas platforms off the coasts of Louisiana and Texas, and industry is now challenging depths in excess of 900 m and over 220 km from shore. The GOM now hosts over 99% of structures associated with US offshore petroleum production. The earliest of these structures were installed in the late 1920s, and were wooden structures erected with steam power and human muscle in the shallow bays and near-shore coastal waters. They were later followed by single piled caissons and multiple leg concrete structures. As technology improved, large steel towers supported by multiple, pipe-like legs approaching 200 cm in diameter were used on the continental shelf by the 1950s. Today, engineers use wire-supported towers

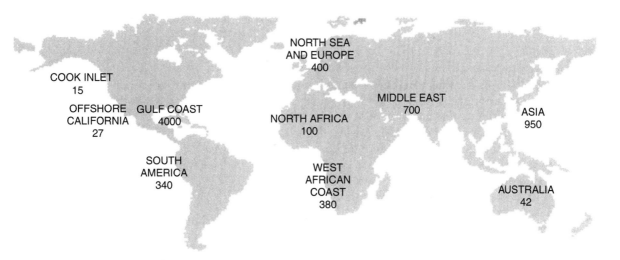

Figure 1 Distribution of the world's offshore platforms in the year 2000. (Information provided by William Griffin.)

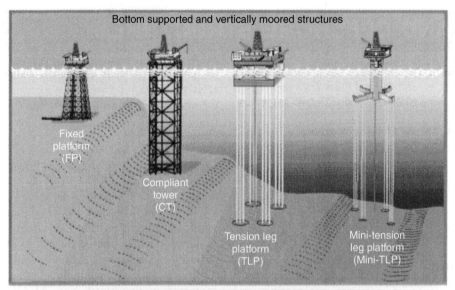

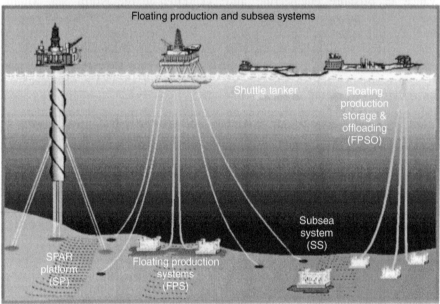

Figure 2 Various shapes and sizes of common offshore platforms (modified from the http://www.gomr.mnms.gov/homepg/offshore/deepwater/options.html/deepwater info).

and floating platforms in water depths not suitable for standard steel frame structures.

Oil and gas structures come in a variety of shapes and sizes (**Figure 2**) that reflect engineering demands depending on water depth, range of sea state conditions, advancements in engineering design, metallurgy, and ocean physics (e.g. currents). Several of the larger platforms have set engineering records for their time, and current structures reach 3.5 km tall. The largest of these structures are found in the North Sea, where ocean conditions are amongst the harshest in the world – winter storm waves can exceed 45 m in height and last for many days. These types of environments demand tremendous feats in engineering. For example, the North Sea 'Troll' weighs over 1 million tons, is over 450 m tall, and is expected to have 70 year lifetimes to exhaust the mineral resources beneath its legs. North Sea structures dwarf those found in the rest of the world.

Environmental Issues

Public perception of oil and gas development has been greatly influenced by oil spills, visual pollution, drilling discharges, and unrelated but dramatic oil tanker accidents. The greatest wealth of information concerning oil and gas environmental issues comes from regulations enforced by the Environmental Protection Agency of the United States and regulations of the UK government for operations in the North Sea. Otherwise, international treaties guide the environmental conscience of industry such as the International Maritime Organization and the Law of the Treaty of the Sea. During the early exploratory years of offshore drilling there were several recorded instances of oil spills and the occasional 'blow out'. However, over the past 25 years strict regulations, put in place by the developed countries (e.g. the US Minerals Management Service and the Environmental Protection Agency), have provided safeguards to prevent such ecological disasters.

Environmental concern of the 1980s focused on the fate of discharged fluids and drilling mud. Several books have been written about produced waters; these are the hydrocarbon-permeated waters pumped to the surface with oil and gas. Recently US regulations were enacted that require operators of platforms located in 60 m of water or less to re-inject produced waters back into the wells. Farther offshore these fluids are discharged in association with large compressor-driven bubblers that mix produced water with the surrounding water. The fate of the re-injected produced waters is unknown. Drilling muds have also caused a major amount of concern due to the heavy metals present in them. A number of studies have been funded by the US Mineral Management Service to investigate the impact of drilling muds. The general conclusions of these studies were that environmental impacts of drilling muds are confined to the area immediately around the drilling site.

The oil and gas industry in the Gulf of Mexico has enjoyed a much different public relationship with the community than off California and elsewhere. Since the first platform was placed off California there has been strong public objection concerning visual pollution and the potential for chemical pollution. However, the oil and gas industries of Louisiana and Texas have notably had a dramatic impact on coastal economies, and enjoyed a very positive relationship with the recreational and commercial fisherman. These user groups have come to depend upon the industry for infrastructure support as well as fishing opportunity. The oil and gas industry has been friendly toward helping fishermen in distress, providing weather updates, and serving as a visual reference point for navigation prior to the advent of Loran or GPS technology. Most noteworthy is the common knowledge that fishing around platforms is exceptional and that the Gulf Coast oil and gas

industry does not object to fishermen mooring to a structure.

The appearance of oil and gas platforms has inadvertently impacted on historic or developing fishing fleets along the world's coastlines. High coastal primary and secondary production serves as the basis for production of commercial and recreational fisheries. Coincident with the large oil and gas fields, the crustacean fishery in the Gulf of Mexico is amongst the world's largest. Likewise, the North Sea has historically been one of the richest trawling fisheries in the world. Offshore structures interfere with net fisheries through their physical presence and the materials that were discarded from the oil and gas platforms during the earlier years of development. These issues have plagued Gulf of Mexico shrimpers for years. In the United States there are federal and state regulations and efforts by major oil and gas companies to totally eliminate all debris associated with oil- and gas-related activities. However, fishermen are still catching artifacts which are damaging to their gear, and oil and gas companies have gone to great lengths to provide funding mechanisms to replace the various gear that are damaged. Nevertheless, industries still coexist in most areas.

'Rigs to Reefs'

It is the observation of many federal and state fishery managers throughout the world that fisheries are in a serious state of decline. The last assessment by the FAO indicated that well over 60% of the world's fisheries were in a state of overfishing or overfished and in decline and that the world's fishery harvest is at maximum. As mentioned above many of these fisheries are located in areas of oil and gas development. It has been suggested by some that the presence of oil and gas platforms actually provides a 'safe haven' for many of these fish species that prevents them from being captured. On the contrary, scientists and managers have also proposed that platforms in fact make fish easier to catch for the hook and line fisherman.

Since the first platforms were emplaced in the Gulf of Mexico, off California, and other parts of the world where hook and line fishing activities occur, fishers have realized that these tall steel structures serve as fish havens or 'artificial reefs'. Scientists have documented that catch rates around platforms in the Gulf of Mexico (number of fish caught per hour per person by hook and line) are amongst the highest in the world. Popular recreational and commercial species such as red snapper, gray snapper, amberjack, mackerel, tuna, marlin, sheepshead, and triggerfish are abundant around the platforms on the continental shelf of the Gulf of Mexico. Charter boat fisheries and commercial fisheries now depend upon many of these structures for income. Pursuant to the public's perception of the fishery value of oil and gas platforms, several states including Louisiana, Texas, Mississippi, and Florida have established artificial reef programs that encourage the use of oil and gas platforms as artificial reefs. In most cases these are extremely positive and cooperative efforts between state regulators and industry to establish specific areas for the long-term maintenance of fishing opportunity.

However, these programs have their opponents. Scientists have yet to establish whether artificial reefs in general, and specifically oil and gas platforms, concentrate fish and make them easier to catch or if artificial reefs increase productivity and benefit some life stages of a fish species. Most scientists have come to an agreement that artificial reefs and specifically platforms increase productivity for some fish species and concentrate others. However, the extent and value of that habitat has yet to be determined. A major concern for species such as red snapper is that the benefit of increased production and survival may be outweighed by increased vulnerability to human capture. Due to the hook and line vulnerability of some fish species, such as red snapper, regulators continue to pursue the use of obsolete platforms as artificial reefs, 'rigs to reefs', with caution.

The Minerals Management Service of the United States has funded a number of ecological investigations off California and in the Gulf of Mexico to explore the ecology and habitat value of platforms. Scientists at Louisiana State University have established that an average of 25 000 fish 12 cm in length inhabit the area immediately around and under oil and gas platforms; such density is 50–1000 times greater than adjacent soft bottom sediment. The fish community associated with offshore platforms includes a number of important recreational and commercial species. Research in the Gulf of Mexico indicates that species composition for near-shore platforms (typically within 50 km offshore) is dominated mostly by estuarine species, while platforms farther offshore contain more pelagic and tropical species. However, this is not the case in the North Sea, where species composition is dominated predominantly by pelagic species. Divers in the GOM have also reported large schools of larval and juvenile species of fish that obviously depend upon the hard steel substrate for protection, orientation, and perhaps food. Scientists in the North Sea report fish densities around large steel towers to be two to three times higher than adjacent water bottoms. Hence, there is little debate regarding the higher numbers of fish living around offshore structures than adjacent water bottoms.

The Rigs to Reef issue is fairly unique to the United States, although there has been limited interest in other parts of the world such as Australia, Japan, and in the Caribbean. In reality the utility of oil and gas platforms as artificial reefs will be limited to the most developed countries where recreational and commercial fishing command attention.

Removal

Federal law in the United States, law in the UK, and EC, as well as international treaty through the International Maritime Organization, requires that once an oil and gas company has completed extraction of oil and gas hydrocarbon reserves the platform must be removed and the bottom returned to its original state. This law was developed through international agreements to protect navigation and military interests, and to preserve ocean ecology.

During the early years of oil and gas development, when the industry was operating in shallow water, technology existed for the removal of these massive structures. Since then the semi-permanent placement of offshore structures has challenged engineering and led to several new disciplines, the most formidable of which is decommissioning. Currently the technology does not exist to remove deep-water structures like the Troll, Mars, and others (**Table 1**). The majority of the structures in shallower waters will be subject to removal over the next 50 years.

The oldest platforms, particularly those in the Gulf of Mexico and off California, are 20–30 years old and are nearing the depletion of their hydrocarbon reserves. Several international organizations continue to debate removal technology, and to explore ocean dumping as an alternative. Most of the environmental groups throughout the world strongly oppose this activity, as the concept of ocean dumping is not palatable to most people. Shell's North Sea Brent Spar incident in 1995 illustrates the public's sensitivity to ocean dumping. Greenpeace blocked the towing of the Brent Spar out to sea, and gas stations were burned in Germany in association with this highly publicized event. The British Government conceded to public pressure and reversed its decision to allow ocean dumping, and the fate of the Brent Spar is yet undecided. It is however a reality that the oil and gas industry will be faced with some serious engineering questions as these deep-water platforms are retired.

Table 1 Company, structure name, water depth, and date of installation of deep-water structures in the Gulf of Mexico, USA[a]

Operator	Structure name	Water depth (m)	Date installed
Shell Offshore Inc.	COGNAC	312	1 Jan. 1978
Shell Offshore Inc.	BULLWINKLE	412	1 Jan. 1988
Conoco Inc.	JOLLIET	536	1 Jan. 1989
BP Exploration & Oil Inc.	A	335	1 Jan. 1991
Exxon Mobil Corporation	A	423	3 Oct. 1991
Shell Deepwater Production Inc.	A	872	5 Feb. 1994
BP Exploration & Oil Inc.	A	393	19 Aug. 1994
EEX Corporation	COOPER	639	7 Jul. 1995
Shell Deepwater Production Inc.	MARS	894	18 Jul. 1996
Kerr-McGee Oil & Gas Corporation	NEPTUNE	588	19 Nov. 1996
Shell Deepwater Production Inc.	RAM-POWELL	980	21 May 1997
Amerada Hess Corporation	BALDPATE	502	31 May 1998
Chevron USA Inc.	GENESIS	789	21 July 1998
British-Borneo Exploration, Inc.	MORPETH EAST	518	10 Aug. 1998
British-Borneo Exploration, Inc.	ALLEGHENY SEA	1004	19 Aug. 1999
Elf Exploration, Inc.	VIRGO	344	17 Sep. 1999

[a] From http://www.gomr.mms.gov/homepg/offshore/deepwater/options.html.

The accepted removal technology along the continental shelf involves the use of explosives where the platform legs are severed 5 m below the mud line. Not only are explosives extremely damaging to the marine environment and the associated explosion kills most of the 10 000–25 000 fish that live around a platform, but also the activity involves risk of human life as divers are frequently lowered down inside the legs to place the explosives. More recently mechanical cutting and cryogenics have been used. Cryogenics is still improving and mechanical cutting is practical only to depths of about 60 m. There are many challenges to removing a large steel platform that is some 3.5 km long. The engineers of deep-water platforms of this size need to provide designs or methods for removal when the structure depletes the petroleum stores.

Research

Not only do the oil and gas platforms serve as a support structure for operations involved in the extraction and transmission in oil and gas; these platforms also serve as excellent research stations. The ocean is a final frontier for exploration, as we know more about the surface of the moon than we do about our continental slopes and ocean basins. To aid in the quest for knowledge, many major and minor oil and gas companies have allowed scientists on their platforms for the purpose of research over the past 20 years.

The US Minerals Management Service has funded a number of ecological investigations to explore the physical, chemical, biological, and meteorological events that occur around oil and gas platforms. These permanent steel structures provide us with the opportunity to collect long-term databases on changes in weather pattern and shallow and deep water currents, to track bird migration, and to monitor the aquatic biological community in many parts of the world. This information will be invaluable in predicting the path and strength of hurricanes, to follow changes in sea life due to global warming, and improve general knowledge of the world's oceans and resolve the impacts of offshore structures and the marine environment. This research activity is carried out at a fraction of the cost that it would be if larger vessels were used to collect this data, and under weather conditions that prevent research vessel operations. Oil and gas companies are often cooperative in providing logistical support (lodging and meals) and transportation for scientists who visit these platforms for such investigations. Obviously information such as hurricane path and current patterns are of extreme interest to the platform engineers as well as to coastal resource managers in state and federal government. The advent of microwave, satellite, and cellular technology allow for the transmission of real-time data collected at these platforms and much of this information is being made online to students around the world, in addition to the scientists that use the data.

Summary

Offshore structures are concentrated in areas of oil and gas development off most coastal countries of the world. Platforms will be a common sight along the world's coastlines for as long as hydrocarbon resources are present. Although the industry provides the energy to fuel the world, the presence of offshore structures has both positive and negative impacts concerning pollution, navigation, and fisheries production. Current development and research is being implemented for the best use of offshore structures concerning these issues. Offshore structures are the result of remarkable feats of engineering, and engineers are continually challenged in the design and in the removal guidelines within the constraints of current technology.

See also

Oil Pollution.

Further Reading

American Petroleum Institute (1995) *Impact on U.S. Companies of a Worldwide Ban on Disposal of Offshore Oil and Gas Platforms at Sea.* IFC Resources Inc.

Boesch DF and Rabalais NW (1987) *Long-term Environmental Effects of Offshore Petroleum Development.* New York: Elsevier Applied Science.

Love MS, Nishimoto M, Schroeder D, and Caselle J (1999) *The Ecological Role of Natural Reefs and Oil and Gas Production Platforms on Rocky Reef Fishes in Southern California: Final Interim Report.* US Department of the Interior, US Geological Survey, Biological Resources, Division, USGS/BRD/CR-1999-007.

Pulsipher AG (ed.) (1996) *Proceedings: An International Workshop on Offshore Lease Abandonment and Platform Disposal: Technology, Regulation and Environmental Effects.* MMS contract 14-35-0001-30794. Baton Rouge, LA: Center for Energy Studies, Louisiana State University.

Reggio VC Jr (1987) *Rigs to Reefs: The Use of Obsolete Petroleum Structures as Artificial Reefs.* OCS Report/MMS 87-0015. New Orleans, LA: US Department of the Interior. Minerals Management Service. Gulf of Mexico OCS Region.

Reggio VC Jr (1989) *Petroleum Structures as Artificial Reefs: A Compendium.* Fourth International Conference on Artificial Habitats for Fisheries, Rigs-to-Reefs Special Session, Miami, FL, November 4, 1987. New Orleans, LA: US Department of the Interior, Minerals Management Service, Gulf of Mexico OCS Regional Office.

Wilson CA and Stanley DR (1990) A fishery-dependent based study of fish species composition and associated catch rates around oil and gas structures off Louisiana. *Fishery Bulletin US* 88: 719–730.

Wilson CA, Stickle VR and Pope DL (1987) *Louisiana Artificial Reef Plan.* Louisiana Department of Wildlife and Fisheries Technical Bulletin No. 41. Baton Rouge, LA: Louisiana Sea Grant College Program.

TIDAL ENERGY

A. M. Gorlov, Northeastern University, Boston, Massachusetts, USA

Introduction

Gravitational forces between the moon, the sun and the earth cause the rhythmic rising and lowering of ocean waters around the world that results in Tide Waves. The moon exerts more than twice as great a force on the tides as the sun due to its much closer position to the earth. As a result, the tide closely follows the moon during its rotation around the earth, creating diurnal tide and ebb cycles at any particular ocean surface. The amplitude or height of the tide wave is very small in the open ocean where it measures several centimeters in the center of the wave distributed over hundreds of kilometers. However, the tide can increase dramatically when it reaches continental shelves, bringing huge masses of water into narrow bays and river estuaries along a coastline. For instance, the tides in the Bay of Fundy in Canada are the greatest in the world, with amplitude between 16 and 17 meters near shore. High tides close to these figures can be observed at many other sites worldwide, such as the Bristol Channel in England, the Kimberly coast of Australia, and the Okhotsk Sea of Russia. **Table 1** contains ranges of amplitude for some locations with large tides.

On most coasts tidal fluctuation consists of two floods and two ebbs, with a semidiurnal period of about 12 hours and 25 minutes. However, there are some coasts where tides are twice as long (diurnal tides) or are mixed, with a diurnal inequality, but are still diurnal or semidiurnal in period. The magnitude of tides changes during each lunar month. The highest tides, called spring tides, occur when the moon, earth and sun are positioned close to a straight line (moon syzygy). The lowest tides, called neap tides, occur when the earth, moon and sun are at right angles to each other (moon quadrature). Isaac Newton formulated the phenomenon first as follows: 'The ocean must flow twice and ebb twice, each day, and the highest water occurs at the third hour after the approach of the luminaries to the meridian of the place'. The first tide tables with accurate prediction of tidal amplitudes were published by the British Admiralty in 1833. However, information about tide fluctuations was available long before that time from a fourteenth century British atlas, for example.

Rising and receding tides along a shoreline area can be explained in the following way. A low height tide wave of hundreds of kilometers in diameter runs on the ocean surface under the moon, following its rotation around the earth, until the wave hits a continental shore. The water mass moved by the moon's gravitational pull fills narrow bays and river estuaries where it has no way to escape and spread over the ocean. This leads to interference of waves and accumulation of water inside these bays and estuaries, resulting in dramatic rises of the water level (tide cycle). The tide starts receding as the moon continues its travel further over the land, away from the ocean, reducing its gravitational influence on the ocean waters (ebb cycle).

The above explanation is rather schematic since only the moon's gravitation has been taken into account as the major factor influencing tide fluctuations. Other factors, which affect the tide range are the sun's pull, the centrifugal force resulting from the earth's rotation and, in some cases, local resonance of the gulfs, bays or estuaries.

Energy of Tides

The energy of the tide wave contains two components, namely, potential and kinetic. The potential energy is the work done in lifting the mass of water above the ocean surface. This energy can be calculated as:

$$E = \mathbf{g}\rho A \int z \, \mathrm{d}z = 0.5\mathbf{g}\rho A b^2,$$

where E is the energy, $\mathbf{g}$ is acceleration of gravity, ρ is the seawater density, which equals its mass per unit

Table 1 Highest tides (tide ranges) of the global ocean

Country	Site	Tide range (m)
Canada	Bay of Fundy	16.2
England	Severn Estuary	14.5
France	Port of Ganville	14.7
France	La Rance	13.5
Argentina	Puerto Rio Gallegos	13.3
Russia	Bay of Mezen (White Sea)	10.0
Russia	Penzhinskaya Guba (Sea of Okhotsk)	13.4

volume, A is the sea area under consideration, z is a vertical coordinate of the ocean surface and h is the tide amplitude. Taking an average $(\mathbf{g}\rho) = 10.15\,\text{kN}\,\text{m}^{-3}$ for seawater, one can obtain for a tide cycle per square meter of ocean surface:

$$E = 1.4h^2, \text{watt-hour}$$

or

$$E = 5.04h^2, \text{kilojoule}$$

The kinetic energy T of the water mass m is its capacity to do work by virtue of its velocity V. It is defined by $T = 0.5\,m\,V^2$. The total tide energy equals the sum of its potential and kinetic energy components.

Knowledge of the potential energy of the tide is important for designing conventional tidal power plants using water dams for creating artificial upstream water heads. Such power plants exploit the potential energy of vertical rise and fall of the water. In contrast, the kinetic energy of the tide has to be known in order to design floating or other types of tidal power plants which harness energy from tidal currents or horizontal water flows induced by tides. They do not involve installation of water dams.

Extracting Tidal Energy: Traditional Approach

People used the phenomenon of tides and tidal currents long before the Christian era. The earliest navigators, for example, needed to know periodical tide fluctuations as well as where and when they could use or would be confronted with a strong tidal current. There are remnants of small tidal hydro-mechanical installations built in the Middle Ages around the world for water pumping, watermills and other applications. Some of these devices were exploited until recent times. For example, large tidal waterwheels were used for pumping sewage in Hamburg, Germany up to the nineteenth century. The city of London used huge tidal wheels, installed under London Bridge in 1580, for 250 years to supply fresh water to the city. However, the serious study and design of industrial-size tidal power plants for exploiting tidal energy only began in the twentieth century with the rapid growth of the electric industry. Electrification of all aspects of modern civilization has led to the development of various converters for transferring natural potential energy sources into electric power. Along with fossil fuel power systems and nuclear reactors, which create huge new environmental pollution problems, clean renewable energy sources have attracted scientists

and engineers to exploit these resources for the production of electric power. Tidal energy, in particular, is one of the best available renewable energy sources. In contrast to other clean sources, such as wind, solar, geothermal etc., tidal energy can be predicted for centuries ahead from the point of view of time and magnitude. However, this energy source, like wind and solar energy is distributed over large areas, which presents a difficult problem for collecting it. Besides that, complex conventional tidal power installations, which include massive dams in the open ocean, can hardly compete economically with fossil fuel (thermal) power plants, which use cheap oil or coal, presently available in abundance. These thermal power plants are currently the principal component of world electric energy production. Nevertheless, the reserves of oil and coal are limited and rapidly dwindling. Besides, oil and coal cause enormous atmospheric pollution both from emission of green house gases and from their impurities such as sulfur in the fuel. Nuclear power plants produce accumulating nuclear wastes that degrade very slowly, creating hazardous problems for future generations. Tidal energy is clean and not depleting. These features make it an important energy source for global power production in the near future. To achieve this goal, the tidal energy industry has to develop a new generation of efficient, low cost and environmentally friendly apparatus for power extraction from free or ultra-low head water flow.

Four large-scale tidal power plants currently exist. All of them were constructed after World War II. They are the La Rance Plant (France, 1967), the Kislaya Guba Plant (Russia, 1968), the Annapolis Plant (Canada, 1984), and the Jiangxia Plant (China, 1985). The main characteristics of these tidal power plants are given in **Table 2**. The La Rance plant is shown in **Figure 1**.

All existing tidal power plants use the same design that is accepted for construction of conventional river hydropower stations. The three principal structural and mechanical elements of this design are: a water dam across the flow, which creates an artificial water basin and builds up a water head for operation of hydraulic turbines; a number of turbines coupled with electric generators installed at the lowest point of the dam; and hydraulic gates in the dam to control the water flow in and out of the water basin behind the dam. Sluice locks are also used for navigation when necessary. The turbines convert the potential energy of the water mass accumulated on either side of the dam into electric energy during the tide. The tidal power plant can be designed for operation either by double or single action. Double action means that the turbines work in both water

Table 2 Extant large tidal power plants

Country	Site	Installed power (MW)	Basin area (km²)	Mean tide (m)
France	La Rance	240	22	8.55
Russia	Kislaya Guba	0.4	1.1	2.3
Canada	Annapolis	18	15	6.4
China	Jiangxia	3.9	1.4	5.08

flows, i.e. during the tide when the water flows through the turbines, filling the basin, and then, during the ebb, when the water flows back into the ocean draining the basin. In single-action systems, the turbines work only during the ebb cycle. In this case, the water gates are kept open during the tide, allowing the water to fill the basin. Then the gates close, developing the water head, and turbines start operating in the water flow from the basin back into the ocean during the ebb.

Advantages of the double-action method are that it closely models the natural phenomenon of the tide, has least effect on the environment and, in some cases, has higher power efficiency. However, this method requires more complicated and expensive reversible turbines and electrical equipment. The

Figure 1 Aerial view of the La Rance Tidal Power Plant (Source: Electricitéde France).

single action method is simpler, and requires less expensive turbines. The negative aspects of the single action method are its greater potential for harm to the environment by developing a higher water head and causing accumulation of sediments in the basin. Nevertheless, both methods have been used in practice. For example, the La Rance and the Kislaya Guba tidal power plants operate under the double-action scheme, whereas the Annapolis plant uses a single-action method.

One of the principal parameters of a conventional hydropower plant is its power output P (energy per unit time) as a function of the water flow rate Q (volume per time) through the turbines and the water head h (difference between upstream and downstream water levels). Instantaneous power P can be defined by the expression: $P = 9.81\,Qh$, kW, where Q is in $m^3 s^{-1}$, h is in meters and 9.81 is the product (ρg) for fresh water, which has mass density $\rho = 1000$ kg m^{-3} and $g = 9.81$ m s^{-2}. The (ρg) component has to be corrected for applications in salt water due to its different density (see above).

The average annual power production of a conventional tidal power plant with dams can be calculated by taking into account some other geophysical and hydraulic factors, such as the effective basin area, tidal fluctuations, etc. **Tables 2** and **3** contain some characteristics of existing tidal power plants as well as prospects for further development of traditional power systems in various countries using dams and artificial water basins described above.

Extracting Tidal Energy: Non-traditional Approach

As mentioned earlier, all existing tidal power plants have been built using the conventional design developed for river power stations with water dams as

their principal component. This traditional river scheme has a poor ecological reputation because the dams block fish migration, destroying their population, and damage the environment by flooding and swamping adjacent lands. Flooding is not an issue for tidal power stations because the water level in the basin cannot be higher than the natural tide. However, blocking migration of fish and other ocean inhabitants by dams may represent a serious environmental problem. In addition, even the highest average global tides, such as in the Bay of Fundy, are small compared with the water heads used in conventional river power plants where they are measured in tens or even hundreds of meters. The relatively low water head in tidal power plants creates a difficult technical problem for designers. The fact is that the very efficient, mostly propeller-type hydraulic turbines developed for high river dams are inefficient, complicated and very expensive for low-head tidal power application.

These environmental and economic factors have forced scientists and engineers to look for a new approach to exploitation of tidal energy that does not require massive ocean dams and the creation of high water heads. The key component of such an approach is using new unconventional turbines, which can efficiently extract the kinetic energy from a free unconstrained tidal current without any dams. One such turbine, the Helical Turbine, is shown in **Figure 2**. This cross-flow turbine was developed in 1994. The turbine consists of one or more long helical blades that run along a cylindrical surface like a screwthread, having a so-called airfoil or 'airplane wing' profile. The blades provide a reaction thrust that can rotate the turbine faster than the water flow itself. The turbine shaft (axis of rotation) must be perpendicular to the water current, and the turbine can be positioned either horizontally or vertically. Due to its axial symmetry, the turbine always

Table 3 Some potential sites for tidal power installations (traditional approach)

Country	Site	Potential power (MW)	Basin area (km²)	Mean tide (m)
USA	Passamaquoddy	400	300	5.5
USA	Cook Inlet	Up to 18 000	3100	4.35
Russia	Mezen	15 000	2640	5.66
Russia	Tugur	6790	1080	5.38
UK	Severn	6000	490	8.3
UK	Mersey	700	60	8.4
Argentina	San Jose	7000	780	6.0
Korea	Carolim Bay	480	90	4.7
Australia	Secure	570	130	8.4
Australia	Walcott	1750	260	8.4

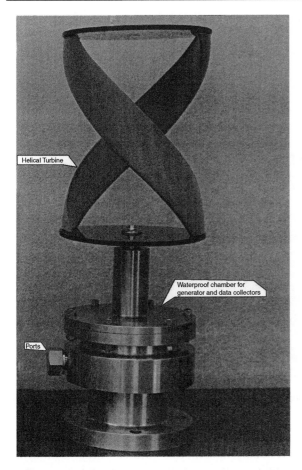

Figure 2 Double-helix turbine with electric generator for underwater installation.

develops unidirectional rotation, even in reversible tidal currents. This is a very important advantage, which simplifies design and allows exploitation of the double-action tidal power plants. A pictorial view of a floating tidal power plant with a number of vertically aligned triple-helix turbines is shown in **Figure 3**. This project has been proposed for the Uldolmok Strait in Korea, where a very strong reversible tidal current with flows up to 12 knots (about 6 m s^{-1}) changes direction four times a day.

The following expression can be used for calculating the combined turbine power of a floating tidal plant (power extracted by all turbines from a free, unconstrained tidal current): $P_t = 0.5\eta\rho AV^3$, where P_t is the turbine power in kilowatts, η is the turbine efficiency ($\eta = 0.35$ in most tests of the triple-helix turbine in free flow), ρ is the mass water density, A is the total effective frontal area of the turbines in m^2 (cross-section of the flow where the turbines are installed) and V is the tidal current velocity in m s^{-1}. Note, that the power of a free water current through a cross-flow area A is $P_w = 0.5\rho AV^3$. The turbine efficiency η, also called power coefficient, is the ratio of the turbine power output P_t to the power of either

the water head for traditional design or unconstrained water current P_w, i.e. $\eta = P_t/P_w$.

The maximum power of the Uldolmok tidal project shown in **Figure 3** is about 90 MW calculated using the above approach for $V = 12$ knots, $A = 2100$ m^2 and $\eta = 0.35$.

Along with the floating power farm projects with helical turbines described, there are proposals to use large-diameter propellers installed on the ocean floor to harness kinetic energy of tides as well as other ocean currents. These propellers are, in general, similar to the well known turbines used for wind farms.

Utilizing Electric Energy from Tidal Power Plants

A serious issue that must be addressed is how and where to use the electric power generated by extracting energy from the tides. Tides are cyclical by their nature, and the corresponding power output of a tidal power plant does not always coincide with the peak of human activity. In countries with a well-developed power industry, tidal power plants can be a part of the general power distribution system. However, power from a tidal plant would then have to be transmitted a long distance because locations of high tides are usually far away from industrial and urban centers. An attractive future option is to utilize the tidal power *in situ* for year-round production of hydrogen fuel by electrolysis of the water. The hydrogen, liquefied or stored by another method, can be transported anywhere to be used either as a fuel instead of oil or gasoline or in various fuel cell energy systems. Fuel cells convert hydrogen energy directly into electricity without combustion or moving parts, which is then used, for instance, in electric cars. Many scientists and engineers consider such a development as a future new industrial revolution. However, in order to realize this idea worldwide, clean hydrogen fuel would need to be also available everywhere. At present most hydrogen is produced from natural gases and fossil fuels, which emit greenhouse gases into the atmosphere and harm the global ecosystem. From this point of view, production of hydrogen by water electrolysis using tidal energy is one of the best ways to develop clean hydrogen fuel by a clean method. Thus, tidal energy can be used in the future to help develop a new era of clean industries, for example, to clean up the automotive industry, as well as other energy-consuming areas of human activity.

Conclusion

Tides play a very important role in the formation of global climate as well as the ecosystems for ocean

Electric generators sit above the water

Figure 3 Artist rendition of the floating tidal power plant with vertical triple-helix turbines for Uldolmok Strait (Korean Peninsula).

habitants. At the same time, tides are a substantial potential source of clean renewable energy for future human generations. Depleting oil reserves, the emission of greenhouse gases by burning coal, oil and other fossil fuels, as well as the accumulation of nuclear waste from nuclear reactors will inevitably force people to replace most of our traditional energy sources with renewable energy in the future. Tidal energy is one of the best candidates for this approaching revolution. Development of new, efficient, low-cost and environmentally friendly hydraulic energy converters suited to free-flow waters, such as triple-helix turbines, can make tidal energy available worldwide. This type of machine, moreover, can be used not only for multi-megawatt tidalpower farms but also for mini-power stations with turbines generating a few kilowatts. Such power stations can provide clean energy to small communities or even individual households located near continental shorelines, straits or on remote islands with strong tidal currents.

Further Reading

Bernshtein LB (ed.) (1996) *Tidal Power Plants*. Seoul: Korea Ocean Research and Development Institute (KORDI).

Gorlov AM (1998) Turbines with a twist. In: Kitzinger U and Frankel EG (eds.) *Macro-Engineering and the Earth: World Projects for the Year 2000 and Beyond*, pp. 1–36. Chichester: Horwood Publishing.

Charlier RH (1982) *Tidal Energy*. New York: Van Nostrand Reinhold.

OFFSHORE SAND AND GRAVEL MINING

E. Garel, CIACOMAR, Algarve University,
Faro, Portugal
W. Bonne, Federal Public Service Health, Food Chain
Safety and Environment, Brussels, Belgium
M. B. Collins, National Oceanography Centre,
Southampton, UK

Introduction

Offshore sand and gravel extraction involves the abstraction of sediments from a bed which is always covered with seawater. This activity started in the early twentieth century (in the mid-1920s, in the UK), but did not reach a significant scale until the 1960s and 1970s, when markets for marine sand and gravel expanded and dredging technology improved (**Figure 1**). In the mining industry, the term 'aggregates' describes a variety of diverse particulate materials, which are provided in bulk, that is, sand and gravel, together with crushed rocks. The distinction between sand and gravel varies, on the basis of the classification adopted. Hence, the nomenclature may change between the end-users or countries (or even regions within a particular country). For example, in the UK industry, 'sand' refers to (noncohesive) minerals with a mean grain size lying between 0.63 and 5 mm, and 'gravel' is reserved for coarser material; in Belgium, the industry considers as gravel the sediment with a mean grain size larger than 4 mm. For comparison, within the scientific literature, the limit between sand and gravel is generally at 2 mm.

Resource Origin

Marine aggregate deposits from the continental shelf can be either relict or modern. Relict deposits have been formed under hydrodynamic and sedimentological regimes that no longer exist. They consist of river or coastal bank deposits, formed during the lower seawater stands, induced by the glacial periods affecting the Earth over the past 2 My. At that time, the rivers extended widely across the continental

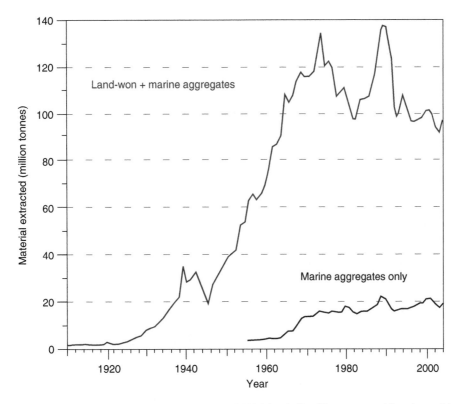

Figure 1 Production of aggregates in the UK between 1900 and 2004 (excluding fill contract and beach nourishment, for marine aggregates). Note the rapid growth of extraction during the late 1960s and the early 1970s, due to the boom in construction in southern England. Sources: Crown-Estate, British Geological Survey.

shelf, transporting and depositing large amounts of sand and gravel in their valleys and coastal waters. This material has been submerged and possibly reworked, especially the mobile sandy deposits, during subsequent warmer interglacial periods, due to the retreating and melting of the ice. Concentration of deposits has resulted from the numerous repetitions of this warm–cold cycle. These sedimentary bodies have remained undisturbed since their formation and are immobile within the context of prevailing hydrodynamic conditions. Relict sand and gravel deposits include mostly thin sheets, banks, drowned beaches and spits, and infilled channel systems, such as river courses and terraces. In contrast, modern deposits have been formed and are controlled by the present prevailing hydrodynamic and sedimentological regime. Therefore, they are related to present-day coastal sediment budgets and sediment dynamics on the seabed. These deposits include, exclusively, sandbanks and sand bedform fields.

The composition, roughness, and grain size of marine aggregate particles vary considerably, depending upon their mode of formation, depositional processes, and history of reworking. Although dependent upon the intensity of post-erosional transport, the grains consist usually of fragments derived from rocks of the surrounding emerged regions. Due to their glacial origin, relict deposits have a composition which is similar to those quarried on land (which are their upstream equivalent), having a low content of shell fragments. In comparison, modern deposits may include high concentration of shelly sediments, which affect the commercial quality of the deposit. In addition, marine aggregates tend to have an higher chloride content. As such, some objections have been raised on their use, in relation to potential alkali–silica reactions, which could affect concrete, or chloride attack on steel reinforcement. However, experience and some experiments have shown that marine aggregates are as suitable as on-land quarried aggregates, for construction purposes.

Production and Usage

Marine aggregates are used mostly in the construction industry, but also for beach replenishment and shore protection, for land reclamation, and in other fill-related uses (such as drainage and capping material). The quality required is governed generally by the usage of the product. High-quality marine aggregates (i.e., well sorted and free from impurities such as clay) are used for mixing with a cementing agent, to produce hard building material (e.g., concrete, mortar, and plaster), or are coated with bitumen for road surfacing. For other usages, such as base material under foundations, roads, and railways, or as the construction of embankments, lower-grade materials can be used. The quality of the material used for beach nourishment may also show strong variability, with location and local/regional policies. Nonetheless, not only should the grain size of the nourished material be similar to (or greater than) that eroded, but it should also not contain fresh biogenic material or contaminants.

Production and usage of marine aggregates vary considerably between countries, with the largest global producers being Japan, the Netherlands, Hong Kong, Korea, and the UK (**Figure 2**). Clusters of extraction areas lie in the vicinity of these countries (e.g., in the North Sea, the Channel, and the Baltic Sea), although aggregate extraction takes place

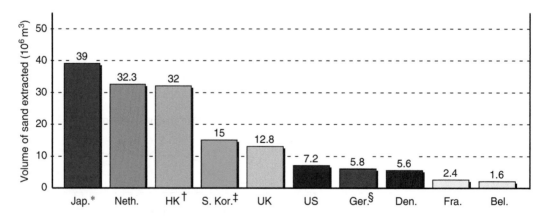

Figure 2 Overview of sand extracted from coastal waters around the world (2002). Annual volumes in million cubic meters per year, for different countries (Japan, the Netherlands, Hong Kong, Korea, United Kingdom, United States, Germany, Denmark, France, Belgium). Key: * data from 1998; † Hong Kong, averaged over 1990–98; ‡ averaged over 1993–95; and § from 2000. Source: Roos PC (2004) *Seabed Pattern Dynamics and Offshore Sand Extraction*, 166pp. PhD Thesis, University of Twente, The Netherlands (ISBN 90-365-2067-3).

in shallow continental seas all over the world. The largest producers present several – if not all – of the following characteristics: limited, or depleted, on-land aggregate resources, increasing environmental pressure on quarries, and large consumption of material. Besides, the yearly national productions may rise drastically as a result of major public works involving, temporarily, a huge volume of extracted material (e.g., Belgium, in 1997, for the construction of pipelines; and the Netherlands, in 2001, for major land reclamation schemes). Countries producing large quantities of marine aggregates extract different types of material, and for distinct purposes (e.g., concrete, filling, or beach recharge). In addition, they may import and export material to and from other countries. For example, in 1998, almost 90% of the marine aggregates production in the UK was for concrete (gravel and coarse sand), but a third of this (c. 7 Mt) was exported to other northwestern European countries. In Spain, by contrast, the dredging of marine aggregates is authorized only for beach replenishment.

Prospecting

The identification of marine aggregate resources is based upon both desk studies and offshore prospecting surveys. The purpose of the desk study is to locate suitable deposits, based upon a scientific literature review and other considerations. Prospecting surveys involve geophysical data acquisition and sediment sampling. In addition to an accurate vessel positioning system (e.g., GPS), the geophysical instrumentation includes usually: an echo sounder, to investigate the seafloor bathymetry; a high-resolution seismic system (e.g., Boomer, Pinger, and Sparker), to produce reflection profiles of the first (c. 10) few meters of the seabed layers; and, a side scan sonar, to characterize the seabed hardness and roughness (**Figure 3**). The collected data set provides information about the thickness, horizontal distribution, and surficial character of the deposit. Vertical and horizontal ground-truth information is provided by core and grab samples, respectively. In addition, grab sampling is useful to examine the surficial sediment grain-size distribution (and also to identify the benthic biological communities present within the area). The selection of the sampling equipment depends upon the difficulty of penetration into the various sediment layers. Large hydraulic grabs can be used to obtain surficial sediment samples, while vibrocorers (up to 10-m long) are capable of coring and recovering coarse-grained material. The level of accuracy of deposit recognition, dependent upon the resolution of both the instruments and the sampling spatial coverage, is governed by the cost of the surveys.

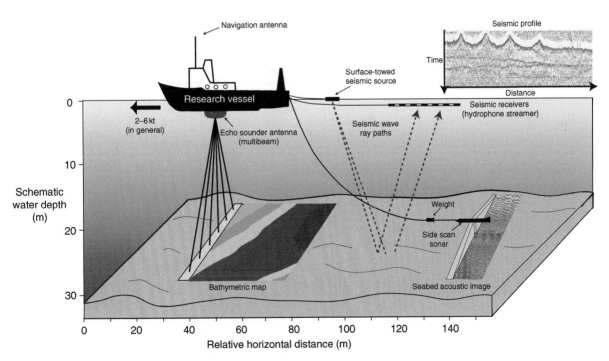

Figure 3 Schematic representation of a survey vessel undertaking geophysical prospecting operations: (multibeam) echo sounder mapping; side scan sonar imaging; and seismic reflection survey.

Extraction Process

Offshore sand and gravel extraction is performed by anchor (or static) suction dredging, and trailer suction dredging. Both techniques utilize powerful centrifugal pumps to draw up the seabed material into the hopper dredger, through pipes of up to 1 m diameter. The dredged material displaces seawater within the hold of the ship, loaded previously as ballast. Anchor suction dredging involves a vessel anchoring or remaining stationary over a deposit, together with forward suction of the bed material through the pipe. As such, the technique is effective for small spatially restricted or, locally, thick deposits. This abstraction procedure creates crater-like pits, or a series of pits, up to 25 m deep and 200 m in diameter, with a slope of about 5° (**Figure 4**). In contrast, trailer suction dredging requires the dredger to drag the lower end of its rear-facing pipe(s) slowly along the seabed, while the ship is underway. This technique permits relatively large areas with thin and more evenly distributed deposits to be worked. It is used also for thicker deposits, such as sandbanks, to limit the harmful environmental impacts to the superficial layers. At the bottom, the head of the pipe creates linear furrows, of 1–3 m in width and up to 50 cm in depth (**Figure 5**). However, repeated trailer suction hopper dredging over an area can lead to large dredged depressions (**Figure 6**).

The total dredged cargo contains generally a mix of different grain sizes. In some cases, the dredgers

(a)

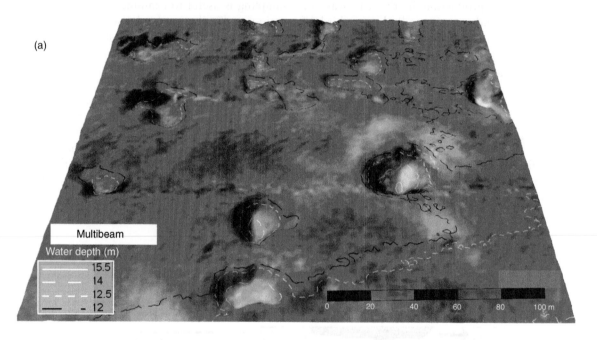

Multibeam

Water depth (m)

15.5
14
12.5
12

0   20   40   60   80   100 m

(b)

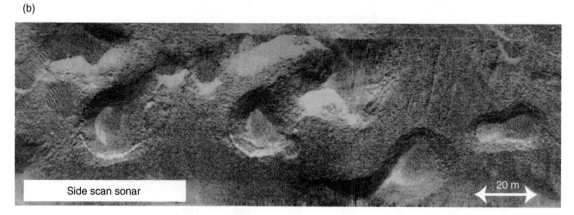

Side scan sonar

20 m

Figure 4 Examples of gravel pits created by anchor hopper dredging (Tromper Wiek Area, Baltic Sea): (a) bathymetry draped with backscatter (illuminated from the west), obtained from a multibeam system; (b) side scan sonar. Source: Manso F, Diesing M, Schwarzer K, and Garel E (2004) Monitoring of dredged areas by combination of multibeam echosounder and sidescan sonar data. *Conference Littoral 2004*, Aberdeen, UK, 20–22 September 2004.

(a)

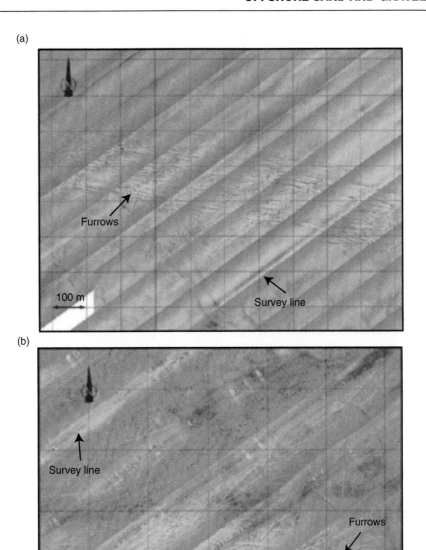

(b)

Figure 5 Side scan sonar mosaics showing sandy furrows, in two adjacent areas, at Graal Müritz, Baltic Sea, in December 2000 (with the survey lines in a NE–SW direction): (a) furrows are observed in the center of the image, with a subequatorial direction (note the reverse in direction of the operating dredger at the western tip of the tracks); (b) the furrows are oriented more randomly, although predominantly NE–SW in the lower part of the image. From Diesing M (2003) *Die Regeneration von Materialentnahmestellen in der südwestlichen Ostsee unter besonderer Berücksichtigung der rezenten Sedimentdynamik*, 158pp. PhD Thesis, Christian-Albrechts-Universität, Kiel.

retain all the dredged material on board, for discharge and processing ashore. However, dredging operations often target only a specific type of aggregate, defined by market demand. Since it is more cost-effective to load and transport only a cargo of the required type, dredgers often have the facility to screen onboard the dredged material, spilling back to the sea the unwanted fines. This screening process spills an important amount of material, up to 3–4 times the retained cargo load, generating sediment plumes that settle down and spread over the extraction zone (making subsequent extractions sometimes more difficult).

Nowadays, some dredgers with cargo capacities of up to 8500 t operate with pumps capable of drawing aggregates at 2600 t of material per hour, in water depths up to 50 m (even 90 m, for Japanese vessels). Discharging on the wharf is performed by a range of machinery, such as bucket wheels, scrapers, or wire-hoisted grabs, which place the aggregates on a

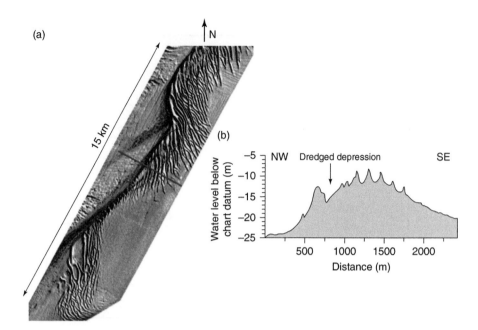

Figure 6 Example of depressions created by repeated trailer suction hopper dredging on a sandbank: the Kwintebank, offshore Belgium. (a) Bathymetric map of the Kwintebank, where the depression is located with the red dotted ellipse; the red straight line indicates the location of the cross section (b); the bank is 15 km in length, 1–2 km in width, and 10–20 m in height. (b) Cross section perpendicular to the bank long-axis, including the dredged depression, 300 m in width, 1 km in length, and up to 5 m deep. (a) Source: Belgian Fund for Sand Extraction.

conveyor belt. Hydraulic discharge is also used, mainly for beach-nourishment schemes.

Environmental Impacts

The environmental effects of sand and gravel extraction include physical effects, such as the modification of the sea bottom topography, the creation of turbidity plumes, substrate alteration, changes in the local wave and current patterns, and impacts on the coast. Biological effects include changes in the density, diversity, biomass, and community structure of the benthos or fish populations, as a consequence of the physical effects. One dredging activity may not have a significant direct or indirect environmental impact, but the cumulative effect of several (adjacent) dredging sites may induce significant changes.

Physical Impacts

The estimation of the regeneration time of dredged features is still very difficult and depends upon the extracted material, the geometry of the excavation, the water depth, and the hydrodynamic regime of the system. Typical timescales for the regeneration of dredged furrows in sandy dynamic substrates lie within the range of months. In very energetic shallow sandy areas, such as those found in estuaries, they may recover after just one (or a few) tidal cycle. In contrast,

in deep and low-energy areas of fine sand (e.g., the German Baltic Sea), recovery could take decades, even for small features such as furrows. In waters of the Netherlands, southern North Sea, dredged tracks in a gravelly substrate exposed to long-period waves were found to disappear completely within 8 months, at 38-m water depth. In contrast, in waters of SE England, dredged tracks in gravelly sediments can take from 3 to more than 7 years to recover. Dredged depressions or pits created by static dredging have also been reported to remain, as recognizable seabed features, for several years and up to decades. In some cases, pits have been observed to migrate slowly in the direction of the dominant current.

Surface turbid plumes are generated by the screening process and by the overflow of material from the hopper, during dredging. A further source of turbidity, with a far lesser quantity of suspended material, results from the mechanical disturbance of the bed sediment, by the head of the pipe. Large increases in suspended solid concentrations tend to be short-lived and localized, close to the operating dredger. However, turbid plumes with low suspended sediment concentrations can affect much larger areas of the seabed, over extended time periods (several days instead of several hours), especially when dredging is occurring simultaneously in adjacent extraction areas.

Changes in sediment composition, as a direct result of dredging, range from minor alterations of superficial granulometry to a large increase in the proportion of fines (sand or silt), or to an increase of gravel through the exposure of coarser sediments. In addition, dredged depressions or furrows, in gravel and coarse sand, can trap gravelly sand remobilized by storms, or finer sediment mobilized by tidal currents. This process may reduce significantly the volume of sediment bypassing in an area. In the case of relatively steep and deep dredged trough, a significant increase in silt and clay material, at the bottom, is sometimes observed following dredging, with long-term consequences (i.e., alteration) upon the *in situ* infaunal assemblage.

Local hydrodynamic changes (i.e., in wave heights and current patterns) may result from the modification of the bathymetry by dredging. In turn, the local erosional and depositional patterns may be affected (near-field effects). In general, the dimensions of dredged features are, in relative terms, so small, that there is little influence of the deepened area on the macroscale hydrodynamics. However, local hydrodynamic changes induced by dredging may result in wider environmental effects (far-field effects). The main issue concerns possible erosion at the coastline, according to various scenarios. Beach erosion may occur, as a result of offshore dredging, due to a reduction of the sediment supply to the coast. This effect is induced either by the direct extraction of material which would normally supply the coast, or by trapping these sediments within the dredged depressions. Any 'beach drawdown' is a particular example of this case: if the area of extraction is not located far enough from the shore, that is, on the subtidal part of a beach, beach sediment may slump down into the excavated area (e.g., during winter storms) and not be able to return subsequently to the beach (in the summer). Coastal erosion may take place also if the protection of the coast is reduced. In particular, dredging may lower significantly the crest heights of (shallow) banks that normally reduce wave breaking and hence, when removed, would lead to larger wave heights at the coast. In addition, if the alterations in the sea bottom topography affect the wave refraction patterns, erosion may result from modification of the wave directions at the coastline.

A substantial list of parameters has to be considered to evaluate whether dredging leads to erosion on the shore: water depth; the distance from the shore; the distribution of the banks; the degree of exposure; the direction of the prevailing waves and tidal currents; the frequency, direction, and severity of storms; the wave reflection and refraction pattern; and seabed type and sediment mobility. Predictions on the changes are performed generally using mathematical models in which input data are provided through a number of sources, including surveys, charts, and offshore buoys. The modeling results have often a wide error range due to the relative uncertainties in the estimation of processes, especially for sediment transport calculations.

Biological Impacts

The most obvious harmful direct effect of marine aggregate extraction results from the direct removal of substrata and associated benthic epifauna and inflora. Available evidence indicates that dredging causes an initial reduction in the abundance, species diversity, and biomass of a benthic community. Other impacts on benthic communities (and referred to as indirect effects) arise from the modification of abiotic factors, which determine the broad-scale benthic community patterns. These modifications concern the nature and stability of the sediment, through the exposition of underlying strata, and the tidal current strength, through the alteration of the local topography. The creation of relatively deep depressions with anoxic conditions is another source of concern. Sediment plumes arising from the dredging activities may also have adverse biological effects, by affecting water quality, increasing the sediment deposition, and modifying the sediment type. In the case of clean mobile sands, increased sedimentation and resuspension, as a consequence of dredging, are considered generally to be of less concern, as the fauna inhabiting such areas tend to be adapted to naturally high levels of suspended sediments, caused by wave and tidal current action. The effects of sediment deposition and resuspension may be more significant in gravelly habitats dominated by encrusting epifaunal taxa, due to the abrasive impacts of suspended sediments. In addition, turbidity plumes may have indirect impacts on fisheries, due to avoidance behavior by some species, the interruption of migratory pathways (e.g., lobsters), and the loss of access to traditional fishing grounds.

The importance of the impacts that affect a benthic community is investigated commonly on the basis of the number of species, abundance, biomass, and the age of the population, compared to an undisturbed similar baseline. The existence of negligible or significant impacts depends largely upon the spatial and temporal intensity of dredging and, likewise, upon the nature of the benthos. While the recolonization of a dredged site by benthic species is generally rapid, the complete recovery (also 'restoration' or 'readjustment') of a benthic community can take many years. The recovery rate, after dredging, depends primarily upon several factors: the community diversity and richness of the area prior to dredging; the life cycle and growth rate of species; the nature, magnitude, and duration of the dredging operations; the nature of the new sediment which is exposed, or subsequently accumulated at the extraction site; the larval and adult pool of potential new colonizers; the hydrodynamics and associated bed load transport processes; and the nature and intensity of the stresses (caused, for example, by storms) which the community normally withstands. Thus, among a number of studies addressing the consequences of long-term dredging operations on the recolonization of biota and the composition of sediment following cessation, some disparity appears in the findings. These results range from minimal effects of disturbance after dredging to significant changes in community structure, persisting over many years.

Supply and Demand: The Future of Offshore Sand and Gravel Mining

Offshore sand and gravel constitute a limited and nonrenewable resource. The notion of supply is linked strongly to the usage of the material. As such, a distinction has to be made between the (reserves of) primary aggregates, suitable for use in construction, and the (resources of) secondary aggregates, which may include contaminants, and used typically as in-filling material. For example, in-filling sand is available in relatively large quantities, whereas sand for construction, whose demand for quality is comparatively higher, is not so abundant. Gravel, which is also considered generally as a high-quality material, is in short supply. In the UK waters, the industry estimates that marine aggregate availability will last for 50 years, at present levels of extraction. However, such estimations are subject to important fluctuations, due to limited knowledge of the seafloor surface and subsurface. Furthermore, the level of accuracy of the identification of the resource varies greatly, from country to country (in general, data are far more complete in countries that have a greater reliance on marine sand and gravel). Another frequent difficulty toward a reliable assessment of the supply comes from the dispersal of the data relevant to the resource among governmental organizations, hydrographic departments, the dredging industry, and other commercial companies. In order to tackle this issue, a number of countries are undertaking seabed mapping programs, dedicated to the recognition of marine aggregates deposits. These programs range from detailed resource assessment, to reconnaissance level of the resource.

The precise estimation of the future demand for aggregates in general, including marine sand and gravel, is not an easy task. Difficulties arise from the highly variable and changing character of a number of parameters, such as the construction market and other economic factors, political strategies, environmental considerations, etc. In addition, the estimates can be distorted by extensive public work projects, involving large amounts of material. Moreover, the predictions should consider the exports and imports of material, and, therefore, take into account the demand at a much broader scale than the national one alone. Few countries have published estimates for the future demand of aggregate (only the UK and the Netherlands, in Europe).

However, in a number of countries, on-land aggregate mining takes place within the context of increasing environmental pressure and the depletion of the resource. At the same time, large parts of the coasts, on a worldwide scale, experience increasing coastal erosion. Therefore, it is expected that the future marine sand and gravel extraction is bound to increase, in order to supply high-quality material and to provide the large quantities of aggregates required for the realization of large-scale infrastructure projects designed to manage coastal retreat and accommodate the development of the coastal zones (i.e., beach replenishment, dune restoration, foreshore nourishment, land reclamation, and other coastal defense schemes). Hence, in order to find these new resources, it is expected that the activity will move farther offshore, within the next decade.

Further Reading

Bellamy AG (1998) The UK marine sand and gravel dredging industry: An application of quaternary geology. In: Latham J-P (ed.) *Engineering Geology Special Publications 13: Advances in Aggregates and Armourstone Evaluation*, pp. 33–45. London: Geological Society.

Boyd SE, Limpenny DS, Rees HL, and Cooper KM (2005) The effects of marine sand and gravel extraction on the macrobenthos at a commercial dredging site (results 6 years post-dredging). *ICES Journal of Marine Science* 62: 145–162.

Byrnes MR, Hammer RM, Thibaut TD, and Snyder DB (2004) Physical and biological effects of sand mining offshore Alabama, USA. *Journal of Coastal Research* 20(1): 6–24.

Desprez M (2000) Physical and biological impact of marine aggregate extraction along the French coast of the Eastern English Channel. Short and long-term post-dredging restoration. *ICES Journal of Marine Science* 57: 1428–1438.

Diesing M (2003) Die Regeneration von Material-entnahmestellen in der Südwestlichen Ostsee unter Besonderer Berücksichtigung der Rezenten Sediment-dynamik, 158pp. PhD Thesis, Christian-Albrechts-Universität, Kiel.

Diesing M, Schwarzer K, Zeiler M, and Kelon H (2006) *Special Issue: Comparison of Marine Sediment Extraction Sites by Mean of Shoreface Zonation. Journal of Coastal Research* 39: 783–788.

ICES (1992) Effects of extraction of marine sediments on fisheries. *Cooperative Research Report, No. 182.* Copenhagen: International Council for the Exploration of the Sea.

ICES (2001) Report of the ICES Working Group on the effects of extraction of marine sediments on the marine ecosystem. *Cooperative Research Report, No. 247.* Copenhagen: International Council for the Exploration of the Sea.

Kenny AJ and Rees HL (1996) The effects of marine gravel extraction on the macrobenthos: Results 2 years post-dredging. *Marine Pollution Bulletin* 32(8/9): 615–622.

Manso F, Diesing M, Schwarzer K, and Garel E (2004) Monitoring of dredged areas by combination of multibeam echosounder and sidescan sonar data. *Conference Littoral 2004*, Aberdeen, UK, 20–22 September 2004.

Newell RC, Seiderer LJ, and Hitchcock DR (1998) The impact of dredging works in coastal waters: A review of the sensitivity to disturbance and subsequent recovery of biological resources on the sea bed. *Oceanography and Marine Biology* 36: 127–178.

Roos PC (2004) *Seabed Pattern Dynamics and Offshore Sand Extraction*, 166pp. PhD Thesis, University of Twente, The Netherlands (ISBN 90-365-2067-3).

van Dalfsen JA, Essink K, Toxvig madsen H, *et al.* (2000) Differential response of macrozoobenthos to marine sand extraction in the North Sea and the Western Mediterranean. *ICES Journal of Marine Science* 57(5): 1439–1445.

van der Veer HW, Bergman MJN, and Beukema JJ (1985) Dredging activities in the Dutch Waddensea: Effects on macrobenthic infauna. *Netherlands Journal of Sea Research* 19: 183–190.

van Moorsel GWNM (1994) The Klaverbank (North Sea), geomorphology, macrobenthic ecology and the effect of gravel extraction. *Report No. 94.24.* Culemborg: Bureau Waardenburg BV.

Van Rijn LC, Soulsby RL, Hoekstra P, and Davies AG (eds.) (2005) *SANDPIT: Sand Transport and Morphology of Offshore Sand Mining Pits Process Knowledge and Guidelines for Coastal Management.* Amsterdam: Aqua Publications.

Relevant Websites

http://www.bmapa.org
– British Marine Aggregate Producers Association (BMAPA).

http://www.dredging.org
– Central Dredging Association (CEDA): Serving Africa, Europe, and the Middle East.

http://www.ciria.org
– Construction Industry Research and Information Association (CIRIA)

http://www.ciria.org
– European Marine Sand and Gravel Group (EMSAGG).

http://www.dvz.be
– ICES Working Group on the Effect of Extraction of Marine Sediments on the Marine Ecosystem (WGEXT).

http://www.iadc-dredging.com
– International Association of Dredging Companies (IADC).

http://www.ices.dk
– International Council for the Exploration of the Sea (ICES).

http://www.sandandgravel.com
– Marine Sand and Gravel Information Service (MAGIS).

http://www.westerndredging.org
– Western Dredging Association (WEDA): Serving the Americas.

http://www.woda.org
– World Organisation of Dredging Associations (WODA).

MINERAL EXTRACTION, AUTHIGENIC MINERALS

J. C. Wiltshire, University of Hawaii, Manoa,
Honolulu, HA, USA

Introduction

The extraction of marine mineral resources represents a worldwide industry of just under two billion dollars per year. There are approximately a dozen general types of marine mineral commodities (depending on how they are classified), about half of which are presently being extracted successfully from the ocean. Those being extracted include sand, coral, gravel and shell for aggregate, cement manufacture and beach replenishment; magnesium for chemicals and metal; salt; sulfur largely for sulfuric acid; placer deposits for diamonds, tin, gold, and heavy minerals. Deposits which have generated continuing interest because of their potential economic interest but which are not presently mined include manganese nodules and crusts, polymetallic sulfides, phosphorites, and methane hydrates.

Authigenic minerals are those formed in place by chemical and biochemical processes. This contrasts with detrital minerals which have been fragmented from an existing rock or geologic formation and accumulated in their present position usually by erosion and sediment transport. The detrital minerals – sand, gravel, clay, shell, diamonds, placer gold, and heavy mineral beach sands – are presently extracted commercially in shallow water. The economically interesting authigenic mineral deposits tend to be found in more than 1000 m of water and have not yet been commercially extracted. Nonetheless, between 1970 and 2000 on the order of one billion US dollars was spent collectively worldwide on studies and tests to recover five authigenic minerals. These minerals are: manganese nodules, manganese crusts, metalliferous sulfide muds, massive consolidated sulfides, and phosphorites. This article will focus on the extraction of these five mineral types.

Descriptions of the formation, geology, geochemistry, and associated microbiology of these deposits are presented elsewhere in this work. As a brief generalization, manganese nodules are black, golf ball to potato sized concretions of ferromanganese oxide sitting on the deep seafloor at depths of 4000–6000 m. They contain potentially economic concentrations of copper, nickel, cobalt, and manganese, and lower concentrations of titanium and molybdenum. Manganese crusts are a flat layered version of manganese oxides found on the tops and sides of seamounts with the highest metal concentrations in water depths of 800–2400 m. They are a potential source of cobalt, nickel, manganese, rare earth elements and perhaps platinum and phosphate. Polymetallic sulfides also come in two forms: metalliferous muds and massive consolidated sulfides. The sulfides contain potentially economic concentrations of gold, silver, copper, lead, zinc, and lesser amounts of cobalt and cadmium. The metalliferous muds are unconsolidated sediments (muds) found at seafloor spreading centers and volcanically active seafloor sites. The metals have been concentrated in the muds by hydrothermal processes operating at and below the seafloor. The best explored site of these metalliferous muds is in the hydrothermally active springs in the central deeps of the Red Sea. By contrast, the massive consolidated sulfides are associated with chimney and mound deposits found at the sites of 'black smokers', hydrothermal vents on the seafloor. The sulfides deposited at these sites have been concentrated by hot seawater percolating through the seafloor and being expelled onto the seabottom at the vents. When in contact with the cold ambient sea water, the hydrothermally heated water drops its mineral riches as sulfide-rich precipitates, forming the sulfide chimneys and mounds. The final authigenic mineral of economic interest is phosphorite, used primarily for phosphate fertilizer. The seabed phosphorites are found as nodules, crusts, irregular masses, pellets, and conglomerates on continental shelves and on the tops of seamounts and rises. Their distribution is widespread throughout the tropical and temperate oceans, although they are preferentially found in areas of oceanic upwelling and high organic productivity.

Whichever of these five mineral types is the target of a mining operation, following regional exploration the process of extracting the commodity of value has seven distinct steps. These steps are: (1) survey the mine site, (2) lease, (3) pick up the mineral, (4) lift and transport, (5) process, (6) refine and sell the metal, and (7) remediate the environmental damage. These steps will be considered in turn (**Figures 1–7**). Naturally, there are differences for each mineral type as well as a range of possible processes that can be used. While many of these

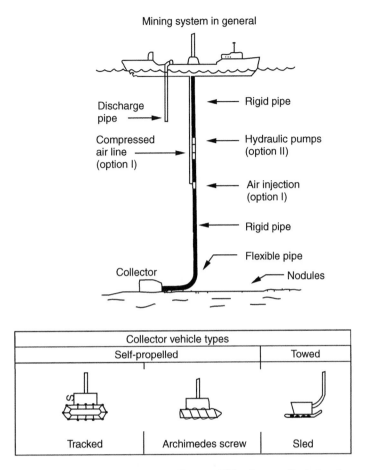

Figure 1 Generalized deep-sea mining system component diagram. This diagram illustrates the proposed collection and lift components which would apply to a wide variety of mineral systems. Note that a considerable variety of collector types are possible on the bottom, including tracked robotic miners, Archimedes screw-driven vehicles, or towed sleds. Both airlift and pumped systems are illustrated as possible lift mechanisms. (Reproduced with permission from Thiel *et al.*, 1998.)

processes have been tested and many are used in traditional terrestrial minerals operations, to date there is no commercially viable full-scale deep-sea authigenic mineral extraction operation.

Survey

The first step in minerals development is to find an economic mine site. This is found by surveying and mineral sampling. A great deal of mineral sampling has already been done over the last 40 years throughout the world's ocean. These data are available for initial planning purposes. Following a detailed literature review the prospective ocean miner would send out a research vessel to sample extensively in the areas under consideration. New acoustical techniques can be calibrated to show certain kinds of bottom cover, including the density of manganese nodule cover. This is one way to rapidly survey the bottom to highlight areas with potentially economic accumulation of authigenic minerals.

Significant advances in marine electronics, navigation, and autonomous underwater vehicles (AUVs) are being brought together. New 'chirp' sonars which transmit a long pulse of sound in which the frequency of the transmitted pulse changes linearly with time give high resolution and long-range seafloor and sub-bottom imagery. Navigation based on the satellite global positioning system (GPS) can now give accurate underwater positions (≤ 1m) when linked to an acoustic relay. This level of survey equipment is now available on underwater autonomous vehicles, meaning that the cost of a ship is not necessarily an impediment.

Sampling for metal concentrations follows the initial surveys. Sampling may be from a ship, a remotely operated vehicle (ROV), or a submersible. Sampling is likely to begin with dredges, progress to some kind of coring, and finish with carefully oriented drilled samples giving a three-dimensional picture of the ore distribution. These data, after chemical analysis of the contained metals, will give

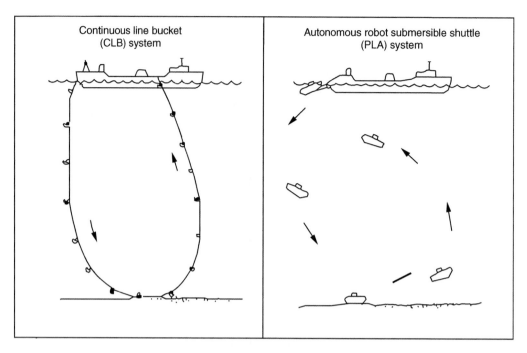

Figure 2 In contrast to the more conventional marine mining systems illustrated in **Figure 1**, two other potential systems are the continuous line bucket system and the autonomous robot submersible shuttle. The continuous line bucket is a series of dredge buckets on a line dragged over the mineral deposit. The submersible shuttle system is a theoretical system using a series of robotic transport submersibles to carry ore from a collection point on the bottom to a waiting surface ship. (Reproduced with permission from Thiel *et al.*, 1998.)

grade and tonnage information. The grade and tonnage estimates of the deposit will be entered into a financial model to determine whether it is economically profitable to mine a given deposit.

In actual fact this process is iterative. A financial model will be used to indicate the type of mineral deposit which must be found. This will narrow the search area. As new data are forthcoming, increasing levels of survey sophistication will follow, assuming that the data continue to indicate an economic mining possibility. In its simplest form the financial model looks at sales of the metals derived from the mine, compared to the total cost of all the operations required to obtain the minerals and contained metals. If the projected sales are greater than the costs by an amount sufficient to allow a profit, typically on the order of a minimum of 20% after taxes, then the mineral deposit is economic. If not, costs must be decreased, sales increased or another more attractive deposit must be found. In the end, a deposit survey will be developed of sufficient detail to allow the production of a mining plan for the development of the property. Normally such a mining plan would need to be approved by the government authority granting mineral leases before actual mining could commence. At some point during the survey and evaluation process, the mining company needs to obtain a lease on the mineral claim in

question. Usually this is done very early in the process, after the first broad area wide surveys.

Lease

Ocean floor minerals are not owned by a mining company. If the minerals are within the exclusive economic zone (EEZ) of a country, they are the property of that country's government. EEZs are normally 200 nautical miles off the coast of a given country, but in the case of a very broad continental shelf may be extended up to 350 miles offshore with recognition of the appropriate United Nations boundary commission. Beyond the EEZ, the ownership of minerals rests with the 'common heritage of mankind' and is administered for that purpose by the United Nations International Seabed Authority in Kingston, Jamaica.

Both national regimes and the International Seabed Authority will lease seabed minerals to *bone fide* mining groups after the payment of fees and the arrangement for filing of mining and exploration plans, environmental impact statements, and remediation plans. The specific details of the requirements vary with the size of the proposed operation, the mineral sought, and the regime under which the application is made. In many countries, the offshore mining laws

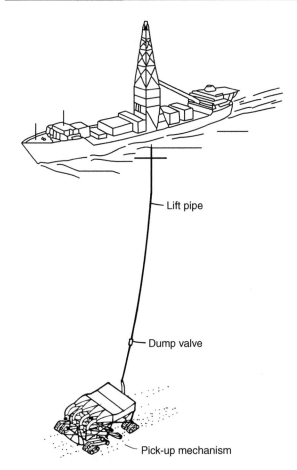

- Lift pipe

- Dump valve

- Pick-up mechanism

Figure 3 An engineering design for a proposed manganese crust mining system. (Reproduced with permission from State of Hawaii (1987). *Mining Development Scenario for Cobalt-Rich Manganese Crusts in the Exclusive Economic Zones of the Hawaiian Archipelago and Johnston Island.* Honolulu, Hawaii, Ocean Resources Branch.)

taken. A typical deep water rental for an oil and gas lease might be $25 per acre per year with a royalty payment equal to 12.5% of the oil extracted. Hard mineral lease rentals and royalties could be expected to be lower, as the commodity is less valuable and the demand for access to offshore mineral resources is much less than for offshore oil.

Mineral Pick-up

Once a deposit has been characterized and leased, a mining plan is drawn up. This plan will depend on the mineral to be mined. In general, several steps must be taken to mine. The first is separating the mineral from the bottom. In the case of polymetallic sulfide, manganese crust or underlying phosphorite, a cutting operation is involved. In the case of phosphorite nodules, manganese nodules, or sulfide muds there is solely a pick-up operation. Cutting requires specialized cutting heads, often these are simply rotating drums with teeth (**Figures 3–5**). Such cutters have been developed for the dredging industry as well as the underground coal industry on land. The size, angle, and spacing of the teeth on a cutter are dependent on the rate of cutting desired and the size to which particles are to be broken. The overall mineral pickup rate is determined by the necessary rate of throughput at the mineral processing plant. This can be worked back to pick-up rate on the bottom. Engineering judgment then dictates whether this is best achieved with larger numbers of smaller cutters or a lower number of larger cutters. This may translate into multiple machines operating on the bottom and feeding one lift system.

When the mineral is broken into sufficiently small pieces, these must then be collected for lifting. This is usually a scooping or vacuuming operation. Scooping is accomplished by blades of various shapes. Vacuuming is usually the result of a powerful airlift or pump farther up the line. One system tested by the Ocean Minerals Co. for picking up manganese nodules involved an Archimedes screw-driven robotic miner, which had two pontoons. A flange in screw-shaped spiral was welded onto each pontoon. The screws both served to drive the vehicle forward as well as pick up the nodules which were sitting in the mud it passed over. There may be a sieving or grinding step between mineral pick-up and lift. This serves two functions; the sieving gets rid of unwanted bottom sediment that may have become entrained in the ore; the grinding ensures that the particle size range going up the lift pipe is in the correct range to get optimum lift without clogging the pipe.

are modeled on the legislation which governs offshore oil development. This is not surprising as the issues are similar and worldwide the annual value of the offshore oil industry is in excess of $100 billion, whereas the offshore mining industry will not be more than a few percent of this value for the foreseeable future.

Leasing arrangements vary from country to country. In the USA, offshore leasing of both hard minerals and oil and gas in the federally controlled waters of the EEZ (normally more than 3 miles offshore) falls under the Outer Continental Shelf Lands Act and is administered by the US Department of the Interior. This act requires a competitive lease sale of offshore tracts of land. The company to whom a lease is awarded is the one which offers the highest payment at a sealed bid auction. In addition to this payment, known as the bonus bid, there is also an annual per acre rental and a royalty payment which would be a percentage of the value of the mineral

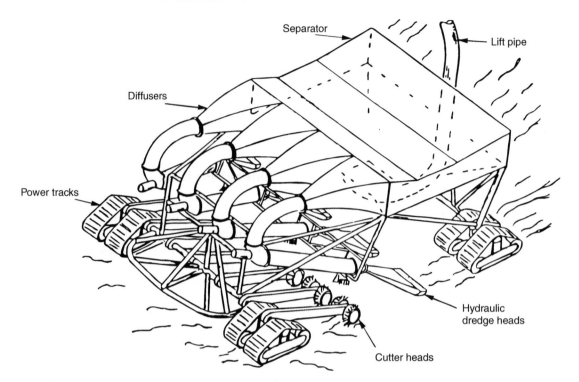

Figure 4   A robotic miner designed to rip up and lift attached flat lying bottom deposits such as manganese crusts. (Reproduced with permission from State of Hawaii (1987) *Mining Development Scenario for Cobalt-Rich Manganese Crusts in the Exclusive Economic Zones of the Hawaiian Archipelago and Johnston Island.* Honolulu, Hawaii, Ocean Resources Branch.)

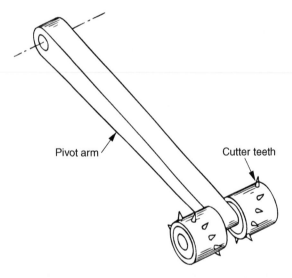

Figure 5   Cutter head design for a bottom miner ripping attached ores such as polymetallic sulfides or manganese crusts. (Reproduced with permission from State of Hawaii (1987) *Mining Development Scenario for Cobalt-Rich Manganese Crusts in the Exclusive Economic Zones of the Hawaiian Archipelago and Johnston Island.* Honolulu, Hawaii, Ocean Resources Branch.)

Lift and Transport

Once the mineral has been cut off or picked up from the bottom it must be lifted to the surface. Over the years a number of systems have been suggested and tested for this purpose. The most successful of these tests have involved bringing the minerals to the surface in a seawater slurry in a steel pipe. This pipe would typically be 30–50 cm in diameter, hence it would be similar if not identical to pipe used in offshore oil drilling operations. Two methods have been tested for bringing the slurry up the pipe, i.e. pumps and airlift. Airlift is a commonly used technique in shallow-water dredging. Compressed air is introduced into the pipe, typically about a third of the way from the top (**Figure 6**). As the air expands moving upward in the pipe it draws the mineral slurry behind it. This works extremely well over lifts of a few hundred meters. It also works over lifts in the deep sea of 6000 m, for example on manganese nodules, except that it is much more difficult to control. Hydraulically driven pump systems along the pipe's length have also been tested to full ocean depth. They also work and although easier to control may ultimately be less efficient lifting systems than an airlift.

Several other lift systems are possible (**Figures 2–5**). The most notable of these is the continuous line bucket, which is a series of buckets on a line which is continuously dragged over the mineral deposit. This system has been successfully tested, although it suffers the disadvantages of lack of control on the bottom, potentially wider spread environmental

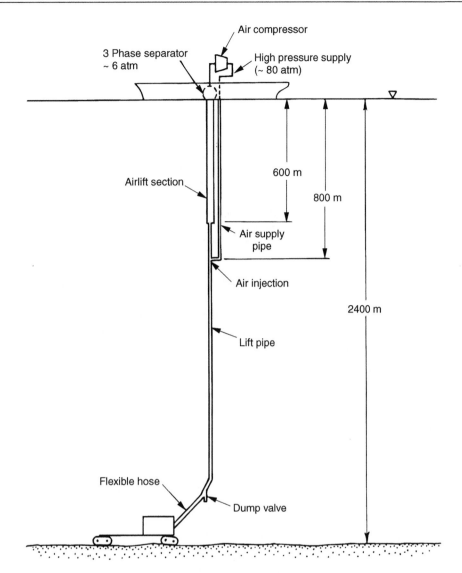

Figure 6 Generalized design of an airlift system to lift ground minerals from the ocean bottom to the surface. (Reproduced with permission from State of Hawaii (1987) *Mining Development Scenario for Cobalt-Rich Manganese Crusts in the Exclusive Economic Zones of the Hawaiian Archipelago and Johnston Island*. Honolulu, Hawaii, Ocean Resources Branch.)

damage, and the possibility of the rope entangling on itself. A system proposed but not tested involves a series of ore-carrying robotic submersible shuttles. These shuttles would use waste mineral tailings as ballast to descend to the bottom, where they would exchange the ballast for a load of ore. Adjusting their buoyancy they would then rise to the surface, off-load the ore onto an ore carrier, reload waste, and return to the bottom. This is potentially a very elegant system in that it handles both tailings waste disposal and mineral lift at the same time, each with the expenditure of very little energy.

Once the ore is on the surface it must be taken to a processing plant. This processing plant can be an existing plant at a site on land, a purpose-built plant near a harbor to process marine minerals, a large

offshore floating platform moored near the mine site on which a plant is built, or a ship converted to mineral processing. The latter two could be right at the mine site. The minerals would be processed as soon as they were lifted to the surface. For a shore-based processing plant a fleet of transport ships or barges would be required to move the lifted ore from the mine site to the plant. This could be a tug and barge operation or a series of dedicated ore carrier vessels.

Processing

Once the minerals have been mined and transported to the processing location they must then be treated to remove the metals of interest. There are two basic

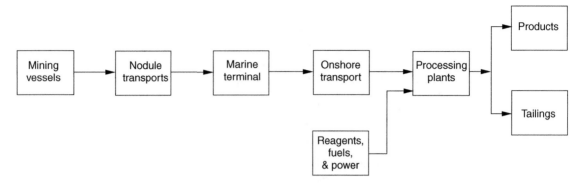

Figure 7 Components involved in a marine minerals extraction operation, each of which has an environmental effect and must be considered in the context of the whole operation. (Reproduced with permission from State of Hawaii (1981) *The Feasibility and Potential Impact of Manganese Nodule Processing in the Puna and Kohala Districts of Hawaii.* Honolulu, Hawaii, Ocean Resources Branch.)

ways to do this: smelting and leaching (pyrometallurgy and hydrometallurgy). Once a basic scenario of either reducing the mineral with acid or melting has been decided upon, there are a number of possible steps to partially separate out unwanted fractions of the mineral. In order to be economically viable, a mineral processing plant needs to process large tonnages of material. In the case of manganese nodules this might be 3 million tonnes y^{-1}. In order to process these large tonnages economically it is important to separate out as much of the waste mineral material as early as possible in the process. This is done most commonly by magnetic separation, froth flotation, or a density separation such as heavy media separation. In an effort to make these processes more efficient the incoming ore would normally be washed to remove salt and ground to decrease the particle size and increase surface area.

Magnetic separation separates magnetic and paramagnetic fractions from nonmagnetic fractions. This technique can be done either wet or dry. Dry magnetic separation involves passing the ground ore through a strong magnetic field underlying a conveyor belt. The magnetic and paramagnetic fraction stay on the conveyor belt while the nonmagnetic fractions drop off as the belt goes over a descending roller. Manganese, cobalt, and nickel are paramagnetic, so this is a good separation technique for manganese nodules and crusts. Wet magnetic separation involves passing a slurry of ore over a series of magnetized steel balls. The magnetic and paramagnetic materials stick to the balls and the nonmagnetic fraction flows through. The level of separation depends on the strength of the magnetic field. Usually this is done with an electromagnet so that the field can be turned on and off.

Froth flotation relies on differences in the surface chemistry of the ore and waste grains. Sulfide ores in particular are more readily wet with oil than water.

Grains of clay, sand, and other seabed detritus have the reverse tendency. Typically something like pine oil might be used to wet the mineral grains. The oil-wet ore is then introduced into a tank with an agent that produces bubbles, a commonly used chemical for this is sodium lauryl sulfate. The bubbles tend to stick to the oil-wet grains, which float them to the surface where they can be taken off in the froth at the top of the tank. The waste material is allowed to settle in the tank and removed at the bottom. The separated ore is then washed to remove the chemicals. The chemicals are recycled if possible.

Another technique commonly used in mineral beneficiation is density separation. This can be as simple as panning for gold. In a simple panning operation or more complex sluice boxes or jigging tables, the heavy mineral (the gold) stays at the bottom, whereas the lighter waste materials are washed out in the swirling motion. A more sophisticated version of the same idea is to use a very dense liquid or colloidal suspension to allow less dense material to float on top and more dense material to sink. The density of the medium is adjusted to the density of the particular ore and waste to be separated. In general, heavy media suspensions, such as extremely finely ground ferro-silicon, are superior to traditional heavy liquids such as tetrabromoethane, which have a high toxicity.

Following mineral beneficiation, which removes the non-ore components, the ore itself must be broken down into its component value metals. Marine minerals can contain up to half a dozen economically desirable metals. In order to separate these metals individually, a reduction process must occur to break the chemical framework of the ore. This will either involve smelting or leaching. In a smelting operation the whole mineral structure is melted and the various metals are taken off in fractions of different densities which float on each other. Normally a series of fluxing agents is used. Clean separation in the case of a multi-metal ore can be a very exacting process. For highly complex polymetallic ores such as manganese nodules or crusts many experts favor a leaching approach over smelting in order to facilitate high purity separation of the metals. However, both leaching and smelting processes have been tried on a variety of marine minerals and both types of process have been proved technically viable.

The most common leach agent is sulfuric acid as it is cheap, efficient, and readily available. Other acids, hydrogen peroxide, and even ammonia are also used as leaching agents. To reduce leaching times, temperature and pressure are raised in the leaching vessel. The leaching process breaks up the mineral structure and dissolves the contained metals into an acidic solution known as a 'pregnant liquor'. The metals become positive ions in solution. These individual metal ions must be separated out of the pregnant liquor and plated out as an elemental metal. This is done primarily in three processes: solvent extraction, ion exchange, and electrowinning. These may initially be done with the primary processing of the ore; however, the final phase of these techniques will be done in a dedicated metal refinery to achieve metal purity >99%.

Solvent extraction relies on an organic solvent which is optimized to select one metal ion. This organic extractant is immiscible with the aqueous pregnant liquor. They are stirred together in a mixer forming an emulsion much like oil stirred into water. This emulsion has a very high contact area. The organic extractant takes the desired metal ion out of the aqueous phase and concentrates it in the organic phase. The aqueous and organic phases are then separated by density (usually the organic phase will float on the aqueous phase). The metal ion is then stripped out of the organic phase by an acidic solution and sent to an electrowinning step.

Instead of doing this extraction in a liquid as in the case of solvent extraction, it is possible to run the pregnant liquor over a series of ion-exchange beads in a column. The ionic beads have the property of collecting or releasing a given metallic ion, e.g. nickel, at a given pH. Therefore by adjusting the pH of the liquid flowing over the column it is possible to remove the nickel, for example, from the pregnant liquor, transferring it to its own tank, then remove copper or cobalt, etc. It is possible to use both solvent extraction and ion-exchange column steps in a particularly complex metal separation process or to achieve very high metal purities.

Following solvent extraction or ion exchange, the various metal ions have been separated from each other and each is in its own acidic solution. The standard concentration technique used to remove these metal ions from solution is electrowinning. The metal ions are positively charged in solution. They will be attracted to the negatively charged plate in a cell. A small current is set up between two or more plates in a tank. The positive metal ions plate out on the negatively charged plate. Depending on the purity of the metal plated out and how strongly it is attached, either the metal is scraped off the plate and sold as powdered metal or the entire plate is sold as metal cathode, e.g. cathode cobalt.

Refining and Metal Sales

Each of the metals produced by an offshore mining operation must be refined to meet highly exacting standards. The metals are sold on various exchanges in bulk lots. Silver, copper, nickel, lead, and zinc are largely sold on the London Metal Exchange. Platinum, silver, and gold are sold on the New York Mercantile Exchange. Both of these exchanges maintain informative internet websites with the current details and requirements for metals transactions (www.lme.co.uk and www.nymex.com). Cobalt, manganese and phosphate are sold through brokers or by direct contract between a mining company and end-users (e.g. a steel mill). Metal sales are by contract. The contracts specify the purity of the metal, its form (e.g. powder, pellets or 2 inch squares), the amount, place, and date that the metal must be delivered. Standard contracts allow for metal delivery as much as 27 months in the future (in the case of the London Metal Exchange). This allows a major futures market in metals and considerable speculation. This speculation allows metal producers to lock in a future price of which they are certain. In reality, only a few percent of the contracts written for future metal deliveries result in actual metal deliveries. The vast majority are traded among speculators over the time between the initial contract settlement and the metal delivery date. The price for a metal is highly dependent on its purity. Often metals are sold at several different grades. For example, cobalt is typically sold at a guaranteed purity of 99.8%. It may also be purchased at a discounted price for 99.3% purity and at a premium for 99.95%.

In order to achieve these grades considerable refining takes place. Often this is at a facility which is removed from the original mine site or processing plant. Metal refineries may be associated with the manufacturing of the final consumer endproduct. For example, a copper refinery will often take lower grade copper metal powder or even scrap and produce copper pipe, wire, cookware, or copper plate. Most modern metal fabricators rely on electric furnaces of some form to cast the final metal product.

One of the techniques commonly used at refineries is known as 'zone refining' whereby a small segment of a piece of metal in a tube is progressively melted while progressing through a slowly moving electric furnace coil. The impurities are driven forward in the liquid phase with the zone of melting. The purified metal resolidifies at the trailing edge of the melting zone. This technique has been used to reduce impurities to the parts per billion level.

Remediation of the Mine Site

Once mining has taken place both the mine site and any processing site must be remediated (**Figure 7**). Considerable scientific work took place in the 1980s and 1990s looking at the rate that the ocean bottom recovers after being scraped in a mining operation. While it is clear that recovery does take place and is slow, it is still unclear how many years are involved. A period of several years to several decades appears likely for natural recolonization of an underwater mined site. It also appears that relatively little can be done to enhance this process. Once all mining equipment is removed from the site, nature is best left to her own processes.

An independent yet perhaps even more important issue is the way the waste products are handled after mineral processing. There are both liquid and solid wastes. The most advanced of a number of clean-up scenarios for the discharged liquids is to use some form of artificial ponds or wetlands, most often involving cattails (*Typha*) and peat moss (*Sphagnum*), the two species shown to be most adept at wastewater clean-up. Typically the wastewater will circulate over several limestone beds and through various artificial wetlands rich with these and related species. At the end of the circulation a certain amount of cleaned water is lost to ground water and the rest is usually sufficiently cleaned to dispose in a natural stream, lake, river, or ocean.

The larger and as yet less satisfactorily engineered problem is with solid waste. In fact, recent environmental work on manganese crusts has shown that 75% of the environmental problems associated with marine ferromanganese operations will be with the processing phase of the operation, particularly tailings disposal. Traditionally, mine tailings are dumped in a tailings pond and left. Current work with manganese tailings has shown them to be a resource of considerable value in their own right. Tailings have applications in a range of building materials as well as in agriculture. Manganese tailings have been shown to be a useful additive as a fine-grained aggregate in concrete, to which they impart higher compressive strength, greater density, and reduced porosity. These tailings serve as an excellent filler for certain classes of resin-cast solid surfaces, tiles, asphalt, rubber, and plastics, as well as having applications in coatings and ceramics. Agricultural experiments extending over 2 years have documented that tailings mixed into the soil can significantly stimulate the growth of commercial hardwood trees and at least half a dozen other plant species. Finding beneficial uses for tailings is an important new direction in the sustainable environmental management of mineral waste.

See also

Manganese Nodules.

Further Reading

Cronan DS (1999) *Handbook of Marine Mineral Deposits*. Boca Raton: CRC Press.

Cronan DS (1980) *Underwater Minerals*. London: Academic Press.

Earney FCF (1990) *Marine Mineral Resources: Ocean Management and Policy*. London: Routledge.

Glasby GP (ed.) (1977) *Marine Manganese Deposits*. Amsterdam: Elsevier.

Nawab Z (1984) Red Sea mining: a new era. *Deep-Sea Research* 31A: 813–822.

Thiel H, Angel M, Foell E, Rice A, and Schriever G (1998) *Environmental Risks from Large-Scale Ecological Research in the Deep Sea – A Desk Study*. Luxembourg: European Commission, Office for Official Publications.

Wiltshire J. (2000) Marine Mineral Resouces – State of Technology Report. *Marine Technology Society Journal* 34: no. 2, p. 56–59

MANGANESE NODULES

D. S. Cronan, Royal School of Mines, London, UK

Introduction

Manganese nodules, together with micronodules and encrustations, are ferromanganese oxide deposits which contain variable amounts of other elements (Table 1). They occur throughout the oceans, although the economically interesting varieties have a much more restricted distribution. Manganese nodules are spherical to oblate in shape and range in size from less than 1 cm in diameter up to 10 cm or more. Most accrete around a nucleus of some sort, usually a volcanic fragment but sometimes biological remains.

The deposits were first described in detail in the Challenger Reports. This work was co-authored by J. Murray and A. Renard, who between them

Table 1 Average abundances of elements in ferromanganese oxide deposits

	Pacific Ocean	Atlantic Ocean	Indian Ocean	Southern Ocean	World Ocean average	Crustal abundance	Enrichment factor	Shallow marine	Lakes
B	0.0277	—	—	—	—	0.0010	27.7		
Na	2.054	1.88	—	—	1.9409	2.36	0.822	0.81	0.22
Mg	1.710	1.89	—	—	1.8234	2.33	0.782	0.55	0.26
Al	3.060	3.27	2.49	—	2.82	8.23	0.342	1.80	1.16
Si	8.320	9.58	11.40	—	8.624	28.15	0.306	8.76	5.38
P	0.235	0.098	—	—	0.2244	0.105	2.13	0.91	0.15
K	0.753	0.567	—	—	0.6427	2.09	0.307	1.30	0.40
Ca	1.960	2.96	2.37	—	2.47	4.15	0.595	2.40	1.14
Sc	0.00097	—	—	—	—	0.0022	0.441		
Ti	0.674	0.421	0.662	0.640	0.647	0.570	1.14	0.212	0.338
V	0.053	0.053	0.044	0.060	0.0558	0.0135	4.13	0.012	0.001
Cr	0.0013	0.007	0.0029	—	0.0035	0.01	0.35	0.002	0.006
Mn	19.78	15.78	15.10	11.69	16.02	0.095	168.6	11.88	12.61
Fe	11.96	20.78	14.74	15.78	15.55	5.63	2.76	21.67	21.59
Co	0.335	0.318	0.230	0.240	0.284	0.0025	113.6	0.008	0.013
Ni	0.634	0.328	0.464	0.450	0.480	0.0075	64.0	0.014	0.022
Cu	0.392	0.116	0.294	0.210	0.259	0.0055	47.01	0.002	0.003
Zn	0.068	0.084	0.069	0.060	0.078	0.007	11.15	0.011	0.051
Ga	0.001	—	—	—	—	0.0015	0.666		
Sr	0.085	0.093	0.086	0.080	0.0825	0.0375	2.20		
Y	0.031	—	—	—	—	0.0033	9.39	0.002	0.002
Zr	0.052	—	—	0.070	0.0648	0.0165	3.92	0.004	0.045
Mo	0.044	0.049	0.029	0.040	0.0412	0.00015	274.66	0.004	0.003
Pd	0.602^{-6}	0.574^{-6}	0.391^{-6}	—	0.553^{-6}	0.665^{-6}	0.832		
Ag	0.0006	—	—	—	—	0.000007	85.71		
Cd	0.0007	0.0011	—	—	0.00079	0.00002	39.50		
Sn	0.00027	—	—	—	—	0.00002	13.50		
Te	0.0050	—	—	—	—	—	—		
Ba	0.276	0.498	0.182	0.100	0.2012	0.0425	4.73	0.287	0.910
La	0.016	—	—	—	—	0.0030	5.33		0.027
Yb	0.0031	—	—	—	—	0.0003	10.33		
W	0.006	—	—	—	—	0.00015	40.00		
Ir	0.939^{-6}	0.932^{-6}	—	—	0.935^{-6}	0.132^{-7}	70.83		
Au	0.266^{-6}	0.302^{-6}	0.811^{-7}	—	0.248^{-6}	0.400^{-6}	0.62		
Hg	0.82^{-4}	0.16^{-4}	0.15^{-6}	—	0.50^{-4}	0.80^{-5}	6.25		
Tl	0.017	0.0077	0.010	—	0.0129	0.000045	286.66		
Pb	0.0846	0.127	0.093	—	0.090	0.00125	72.72	0.002	0.063
Bi	0.0006	0.0005	0.0014	—	0.0008	0.000017	47.05		

Note: Superscript numbers denote powers of ten, e.g. $^{-6} = \times 10^{-6}$.
Reproduced with permission from Cronan (1980).

initiated the first great manganese nodule controversy. Murray believed the deposits to have been formed by submarine volcanic processes whereas Renard believed that they had precipitated from continental runoff products in sea water. This controversy remained unresolved until it was realized that nodules could obtain their metals from either or both sources. The evidence for this included the finding of abundant nodules in the Baltic Sea where there are no volcanic influences, and the finding of rapidly grown ferromanganese oxide crusts associated with submarine hydrothermal activity of volcanic origin on the Mid-Atlantic Ridge. Subsequently, a third source of metals to the deposits was discovered, diagenetic remobilization from underlying sediments. Thus marine ferromanganese oxides can be represented on a triangular diagram (**Figure 1**), the corners being occupied by hydrothermal (volcanically derived), hydrogenous (seawater derived) and diagenetic (sediment interstitial water derived) constituents.

There appears to be a continuous compositional transition between hydrogenous and diagenetic deposits, all of which are formed relatively slowly at normal deep seafloor temperatures. By contrast, although theoretically possible, no continuous compositional gradation has been reported between hydrogenous and hydrothermal deposits, although mixtures of the two do occur. This may be partly because (1) the growth rates of hydrogenous and hydrothermal deposits are very different with the latter accumulating much more rapidly than the former leading to the incorporation of only limited amounts of the more slowly accumulating

hydrogenous material in them, and (2) the temperatures of formation of the deposits are different leading to mineralogical differences between them which can affect their chemical composition. Similarly, a continuous compositional gradation between hydrothermal and diagenetic ferromanganese oxide deposits has not been found, although again this is theoretically possible. However, the depositional conditions with which the respective deposits are associated i.e., high temperature hydrothermal activity in mainly sediment-free elevated volcanic areas on the one hand, and low-temperature accumulation of organic rich sediments in basin areas on the other, would preclude much mixing between the two. Possibly they may occur in sedimented active submarine volcanic areas.

Internal Structure

The main feature of the internal structure of nodules is concentric banding which is developed to a greater or lesser extent in most of them (**Figure 2**). The bands represent thin layers of varying reflectivity in polished section, the more highly reflective layers being generally richer in manganese than the more poorly reflective ones. They are thought to possibly represent varying growth conditions.

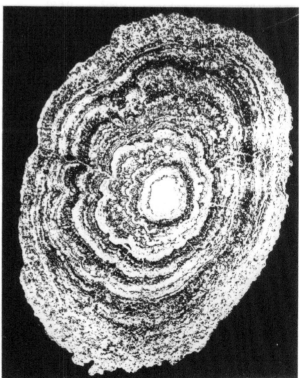

Figure 2 Concentric banding in a manganese nodule. (Reproduced by kind permission of CNEXO, France.)

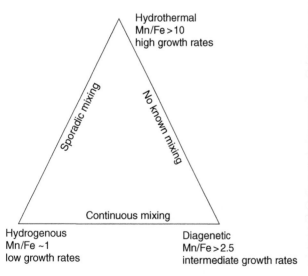

Figure 1 Triangular representation of marine ferromanganese oxide deposits.

On a microscopic scale, a great variety of structures and textures are apparent in nodules, some of them indicative of postdepositional alteration of nodule interiors. One of the most commonly observed and most easily recognizable is that of collomorphic globular segregations of ferromanganese oxides on a scale of tenths of a millimeter or less, which often persist throughout much of the nodule interior. Often the segregations become linked into polygons or cusps elongated radially in the direction of growth of the nodules. Several workers have also recognized organic structures within manganese nodules. Furthermore, cracks and fissures of various sorts are a common feature of nodule interiors. Fracturing of nodules is a process which can lead to their breakup on the seafloor, in some cases as a result of the activity of benthic organisms, or of bottom currents. Fracturing is an important process in limiting the overall size of nodules growing under any particular set of conditions.

Growth Rates

It is possible to assess the rate of growth of nodules either by dating their nuclei, which gives a minimum rate of growth, or by measuring age differences between their different layers. Most radiometric dating techniques indicate a slow growth rate for nodules, from a few to a few tens of millimeters per million years. Existing radiometric and other techniques for nodule dating include uranium series disequilibrium methods utilizing ^{230}Th ^{231}Pa, the ^{10}Be method, the K-Ar method, fission track dating of nodule nuclei, and hydration rind dating.

In spite of the overwhelming evidence for slow growth, data have been accumulating from a number of sources which indicate that the growth of nodules may be variable with periods of rapid accumulation being separated by periods of slower, or little or no growth. In general, the most important factor influencing nodule growth rate is likely to be the rate at which elements are supplied to the deposits, diagenetic sources generally supplying elements at a faster rate than hydrogenous sources (**Figure 1**). Further, the tops, bottoms and sides of nodules do not necessarily accumulate elements at the same rate, leading to the formation of asymmetric nodules in certain circumstances (**Figure 3**). Differences in the surface morphology between the tops, bottoms and sides of nodules *in situ* may also be partly related to growth rate differences. The tops receive slowly accumulating elements hydrogenously supplied from seawater and are smooth, whereas the bottoms receive more rapidly accumulating elements diagenetically supplied

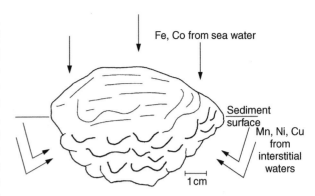

Figure 3 Morphological and compositional differences between the top and bottom of a Pacific nodule. (Reproduced with permission from Cronan, 1980.)

from the interstitial waters of the sediments and are rough (**Figure 3**). The 'equatorial bulges' at the sediment–water interface on some nodules have a greater abundance of organisms on them than elsewhere on the nodule surface, suggesting that the bulges may be due to rapid growth promoted by the organisms.

It is evident therefore that nodule growth cannot be regarded as being continuous or regular. Nodules may accrete material at different rates at different times and on different surfaces. They may also be completely buried for periods of time during which it is possible that they may grow from interstitial waters at rates different from those while on the surface, or possibly not grow at all for some periods. Some even undergo dissolution, as occurs in the Peru Basin where some nodules get buried in suboxic to reducing sediments.

Distribution of Manganese Nodules

The distribution and abundance of manganese nodules is very variable on an oceanwide basis, and can also be highly variable on a scale of a kilometer or less. Nevertheless, there are certain regional regularities in average nodule abundance that permit some broad areas of the oceans to be categorized as containing abundant nodules, and others containing few nodules (**Figure 4**), although it should always be borne in mind that within these regions local variations in nodule abundance do occur.

The distribution of nodules on the seafloor is a function of a variety of factors which include the presence of nucleating agents and/or the nature and age of the substrate, the proximity of sources of elements, sedimentation rates and the influence of organisms. The presence of potential nuclei on the seafloor is of prime importance in determining nodule distribution. As most nodule nuclei are volcanic in origin, patterns of volcanic activity and the

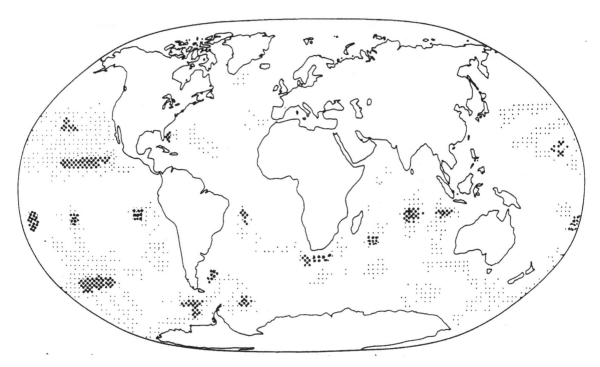

Figure 4 Distribution of mangancese nodules in the oceans (updated from Cronan, 1980 after various sources.), Areas of nodule coverage, areas where nodules are locally abundant.

subsequent dispersal of volcanic materials have an important influence on where and in what amounts nodules occur. Other materials can also be important as nodule nuclei. Biogenic debris such as sharks' teeth, can be locally abundant in areas of slow sedimentation and their distribution will in time influence the abundance of nodules in such areas.

As most nuclei are subject to replacement with time, old nodules have sometimes completely replaced their nuclei and have fractured, thus providing abundant nodule fragments to serve as fresh nuclei for ferromanganese oxide deposition. In this way, given sufficient time, areas which initially contained only limited nuclei may become covered with nodules.

One of the most important factors affecting nodule abundance on the seafloor is the rate of accumulation of their associated sediments, low sedimentation rates favoring high nodule abundances. Areas of the seafloor where sedimentation is rapid are generally only sparsely covered with nodules. For example, most continental margin areas have sedimentation rates that are too rapid for appreciable nodule development, as do turbidite-floored deep-sea abyssal plains. Low rates of sedimentation can result either from a minimal sediment supply to the seafloor or currents inhibiting its deposition. Large areas in the centers of ocean basins receive minimal sediment input. Under these conditions substantial accumulation of nodules at the sediment surface is favored.

Worldwide Nodule Distribution Patterns

Pacific Ocean As shown in, nodules are abundant in the Pacific Ocean in a broad area, called the Clarion–Clipperton Zone, between about 6°N and 20°N, extending from approximately 120°W to 160°W. The limits of the area are largely determined by sedimentation rates. Nodules are also locally abundant further west in the Central Pacific Basin. Sediments in the northern part of the areas of abundant nodules in the North Pacific are red clays with accumulation rates of around 1 mm per thousand years whereas in the south they are siliceous oozes with accumulation rates of 3 mm per thousand years, or more.

Nodule distribution appears to be more irregular in the South Pacific than in the North Pacific, possibly as a result of the greater topographic and sedimentological diversity of the South Pacific. The nodules are most abundant in basin environments such as those of the south-western Pacific Basin, Peru Basin, Tiki Basin, Penrhyn Basin, and the Circum-Antarctic area.

Indian Ocean In the Indian Ocean the most extensive areas of nodule coverage are to the south of the equator. Few nodules have been recorded in the Arabian Sea or the Bay of Bengal, most probably because of the high rates of terrigenous sediment input in these regions from the south Asian rivers. The equatorial zone is also largely devoid of nodules. High nodule concentrations have been recorded in parts of the Crozet Basin, in the Central Indian Ocean Basin and in the Wharton Basin.

Atlantic Ocean   Nodule abundance in the Atlantic Ocean appears to be more limited than in the Pacific or Indian Oceans, probably as a result of its relatively high sedimentation rates. Another feature which inhibits nodule abundance in the Atlantic is that much of the seafloor is above the calcium carbonate compensation depth (CCD). The areas of the Atlantic where nodules do occur in appreciable amounts are those where sedimentation is inhibited. The deep water basins on either side of the Mid-Atlantic Ridge which are below the CCD and which accumulate only limited sediment contain nodules in reasonable abundance, particularly in the western Atlantic. Similarly, there is a widespread occurrence of nodules and encrustations in the Drake Passage–Scotia Sea area probably due to the strong bottom currents under the Circum-Antarctic current inhibiting sediment deposition in this region. Abundant nodule deposits on the Blake Plateau can also be related to high bottom currents.

Buried nodules   Most workers on the subject agree that the preferential concentration of nodules at the sediment surface is due to the activity of benthic organisms which can slightly move the nodules. Buried nodules have, however, been found in all the oceans of the world. Their abundance is highly variable, but it is possible that it may not be entirely random. Buried nodules recovered in large diameter cores are sometimes concentrated in distinct layers. These layers may represent ancient erosion surfaces or surfaces of nondeposition on which manganese nodules were concentrated in the past. By contrast, in the Peru Basin large asymmetrical nodules get buried when their bottoms get stuck in tenacious suboxic sediment just below the surface layer.

Compositional Variability of Manganese Nodules

Manganese nodules exhibit a continuous mixing from diagenetic end members which contain the mineral 10Å manganite (todorokite) and are enriched in Mn, Ni and Cu, to hydrogenous end members which contain the mineral δ MnO_2

(vernadite) and are enriched in Fe and Co. The diagenetic deposits derive their metals at least in part from the recycling through the sediment interstitial waters of elements originally contained in organic phases on their decay and dissolution in the sediments, whereas the hydrogenous deposits receive their metals from normal sea water or diagenetically unenriched interstitial waters. Potentially ore-grade manganese nodules of resource interest fall near the diagenetic end member in composition. These are nodules that are variably enriched in Ni and Cu, up to a maximum of about 3.0% combined.

One of the most striking features shown by chemical data on nodules are enrichments of many elements over and above their normal crustal abundances (**Table 1**). Some elements such as Mn, Co, Mo and Tl are concentrated about 100-fold or more; Ni, Ag, Ir and Pb are concentrated from about 50- to 100-fold, B, Cu, Zn, Cd, Yb, W and Bi from about 10 to 50-fold and P, V, Fe, Sr, Y, Zr, Ba, La and Hg up to about 10-fold above crustal abundances.

Regional Compositional Variability

Pacific Ocean   In the Pacific, potentially ore-grade nodules are generally confined to two zones running roughly east–west in the tropical regions, which are well separated in the eastern Pacific but which converge at about 170°–180°W (**Figure 5**). They follow the isolines of intermediate biological productivity, strongly suggestive of a biological control on their distribution. Within these zones, the nodules preferentially occupy basin areas near or below the CCD. Thus they are found in the Peru Basin, Tiki Basin, Penrhyn Basin, Nova Canton Trough area, Central Pacific Basin and Clarion–Clipperton Zone (**Figure 5**). Nodules in all these areas have features in common and are thought to have attained their distinctive composition by similar processes.

The potentially ore-grade manganese nodule field in the Peru Basin, centered at about 7°–8°S and 90°W (**Figure 5**), is situated under the southern flank of the equatorial zone of high biological productivity on a seafloor composed of pelagic brown mud with variable amounts of siliceous and calcareous remains. Nodules from near the CCD at around 4250 m are characterized by diagenetic growth and are enriched in Mn, Ni and Cu, whereas those from shallower depth are characterized mainly by hydrogenous growth. The Mn/Fe ratio increases from south to north as productivity increases, whereas the Ni and Cu contents reach maximum values in the middle of the area where Mn/Fe ratios are about 5.

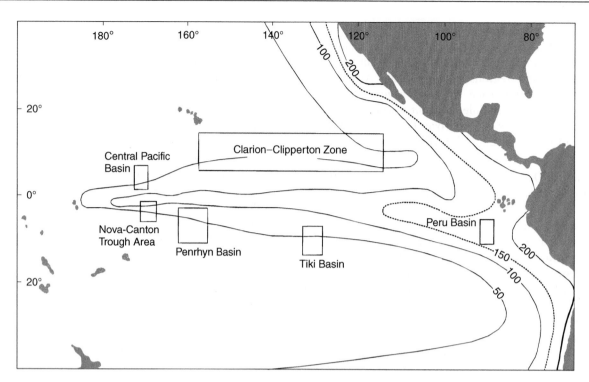

Figure 5 Approximate limits of areas of nickel- and copper-rich nodules in the subequatorial Pacific referred to in the text (productivity isolines if $gCm^{-2}y^{-1}$).

In the Tiki Basin there is also an increase in the Mn/Fe ratio of the nodules from south to north. All Ni + Cu values are above the lower limit expected in diagenetically supplied material.

The Penrhyn Basin nodules fall compositionally within the lower and middle parts of the Mn/Fe range for Pacific nodules as a whole. However, nodules from the northern part of the Basin have the highest Mn/Fe ratios and highest Mn, Ni and Cu concentrations reflecting diagenetic supply of metals to them, although Ni and Cu decrease slightly as the equator is approached. Superimposed on this trend are variations in nodule composition with their distance above or below the CCD. In the Mn-, Ni-, and Cu-rich nodule area, maximum values of these metals in nodules occur within about 200 m above and below the CCD. The latititudinal variation in Mn, Ni and Cu in Penrhyn Basin nodules may be due to there being a hydrogenous source of these metals throughout the Basin, superimposed on which is a diagenetic source of them between about 2° and 6°S at depths near the CCD, but less so in the very north of the Basin (0–2°S) where siliceous sedimentation prevails under highest productivity waters.

In the Nova Canton Trough area, manganese concentrations in the nodules are at a maximum between the equator and 2.5°S, where the Mn/Fe ratio is also highest. Manganese shows a tendency to

decrease towards the south. Nickel and copper show similar trends to Mn, with maximum values of these elements being centered just south of the equator at depths of 5300–5500 m, just below the CCD.

In the central part of the Central Pacific Basin, between the Magellan Trough and the Nova Canton Trough, diagenetic nodules are found associated with siliceous ooze and clay sedimentation below the CCD. Their Ni and Cu contents increase southeastwards reaching a maximum at about 2.5°–3°N and then decrease again towards the equator where productivity is highest.

The Clarion–Clipperton Zone deposits rest largely on slowly accumulated siliceous ooze and pelagic clay below the CCD. The axis of highest average Mn/Fe ratio and Mn, Ni and Cu concentrations runs roughly southwest–northeast with values of these elements decreasing both to the north and south as productivity declines respectively to the north and increases towards the equatorial maximum in the south.

Indian Ocean In the Indian Ocean, Mn-, Ni-, and Cu-rich nodules are present in the Central Indian Ocean Basin between about 5° and 15°S. They are largely diagenetic in origin and rest on siliceous sediments below the CCD under high productivity waters. The deposits show north–south compositional variability with the highest grades occurring in the north.

Atlantic Ocean In the Atlantic Ocean, diagenetic Mn-, Ni-, and Cu-rich nodules occur most notably in the Angola Basin and to a lesser extent in the Cape/Agulhas Basin and the East Georgia Basin. These three areas have in common elevated biological productivity and elevated organic carbon contents in their sediments, which coupled with their depth near or below the CCD would help to explain the composition of their nodules. However, Ni and Cu contents are lower in them than in areas of diagenetic nodules in the Pacific and Indian Oceans.

Economic Potential

Interest in manganese nodules commenced around the mid-1960s and developed during the 1970s, at the same time as the Third United Nations Law of the Sea Conference. However, the outcome of that Conference, in 1982, was widely regarded as unfavorable for the mining industry. This, coupled with a general downturn in metal prices, resulted in a lessening of mining company interest in nodules. About this time, however, several government-backed consortia became interested in them and this work expanded as evaluation of the deposits by mining companies declined. Part 11 of the 1982 Law of the Sea Convention, that part dealing with deep-sea mining, was substantially amended in an agreement on 28 July 1994 which ameliorated some of the provisions relating to deep-sea mining. The Convention entered into force in November 1994.

During the 1980s interest in manganese nodules in exclusive economic zones (EEZs) started to increase. An important result of the Third Law of the Sea Conference, was the acceptance of a 200-nautical-mile EEZ in which the adjacent coastal state could claim any mineral deposits as their own. The nodules found in EEZs are similar to those found in adjacent parts of the International Seabed Area, and are of greatest economic potential in the EEZs of the South Pacific.

At the beginning of the twenty-first century, the out-look for manganese nodule mining remains rather unclear. It is likely to commence some time in this century, although it is not possible to give a precise estimate as to when. The year 2015 has been suggested as the earliest possible date for nodule mining outside of the EEZs. It is possible, however, that EEZ mining for nodules might commence earlier if conditions were favorable. It would depend upon many factors; economic, technological, and political.

Discussion

A model to explain the compositional variability of nodules in the Penrhyn Basin can be summarized as follows. Under the flanks of the high productivity area, reduced sedimentation rates near the CCD due to calcium carbonate dissolution enhance the content of metal-bearing organic carbon rich phases (fecal material, marine snow, etc.) in the sediments, the decay of which drives the diagenetic reactions that in turn promote the enrichment of Mn, Ni, and Cu in the nodules via the sediment interstitial waters. Away from the CCD, organic carbon concentrating processes are less effective. Further south as productivity declines, there is probably insufficient organic carbon supplied to the seafloor to promote the formation of diagenetic nodules at any depth. Under the equator, siliceous ooze replaces pelagic clay as the main sediment builder at and below the CCD, and when its rate of accumulation is high it dilutes the concentrations of organic carbon-bearing material at all depths to levels below that at which diagenetic Mn, Ni, and Cu rich nodules can form.

To a greater or lesser extent, this model can account for much of the variability in nodule composition found in the other South Pacific areas described, although local factors may also apply. In the Peru Basin, as in the Penrhyn Basin, diagenetic Mn-, Ni-, and Cu-rich nodules are concentrated near the CCD and their Ni and Cu contents reach a maximum south of the highest productivity waters. In the Tiki Basin, the greatest diagenetic influences are also found in the north of the Basin. As the South Pacific basins deepen to the west, the areas of diagenetic nodules tend to occur below the CCD as, for example, in the Nova Canton Trough area. This may be because the settling rates of large organic particles are quite fast in the deep ocean. Probably only limited decay of this material takes place between it settling through the CCD and reaching the seafloor, and enough probably gets sedimented to extend the depth of diagenetic nodule formation to well below the CCD under high productivity waters where there is limited siliceous sediment accumulation.

In the North Pacific, the trends in nodule composition in relation to the equatorial zone are the mirror image of those in the south. Thus in both the Central Pacific Basin and the Clarion–Clipperton Zone the highest nodule grades occur in diagenetic nodules on the northern flanks of the high productivity area and decline both to the north and south. The general model erected to explain the Penrhyn Basin nodule variability thus probably applies, at least in part, to these areas also.

The model also has some applicability in the Indian Ocean but less in the Atlantic. In the Indian Ocean, diagenetic nodules associated with sediments containing moderate amounts of organic carbon occur resting on siliceous ooze to the south of the equatorial zone in the Central Indian Ocean Basin. Farther to the south these nodules give way to hydrogenous varieties resting on pelagic clay. However, in the north the changes in nodule composition that might be expected under higher productivity waters do not occur, probably because terrigenous sedimentation becomes important in those areas which in turn reduces the Mn, Ni, and Cu content of nodules. In the Atlantic, the influence of equatorial high productivity on nodule composition that is evident in the Pacific is not seen, mainly because the seafloor in the equatorial area is largely above the CCD. Where diagenetic nodules do occur, as in the Angola, Cape and East Georgia Basins, productivity is also elevated, but the seafloor is near or below the CCD leading to reduced sedimentation rates.

Conclusions

Manganese nodules, although not being mined today, are a considerable resource for the future. They consist of ferromanganese oxides variably enriched in Ni, Cu, and other metals. They generally accumulate around a nucleus and exhibit internal layering on both a macro- and microscale. Growth

rates are generally slow. The most potentially economic varieties of the deposits occur in the subequatorial Pacific under the flanks of the equatorial zone of high biological productivity, at depths near the CCD. Similar nodules occur in the Indian Ocean under similar conditions.

Further Reading

Cronan DS (1980) *Underwater Minerals*. London: Academic Press.

Cronan DS (1992) *Marine Minerals in Exclusive Economic Zones*. London: Chapman and Hall.

Cronan DS (ed.) (2000) *Handbook of Marine Mineral Deposits*. Boca Raton: CRC Press.

Cronan DS (2000) Origin of manganese nodule 'ore provinces'. *Proceedings of the 31st International Geological Congress*, Rio de Janero, Brazil, August 2000.

Earney FC (1990) *Marine Mineral Resources*. London: Routledge.

Glasby GP (ed.) (1977) *Marine Manganese Deposits*. Amsterdam: Elsevier.

Halbach P, Friedrich G, and von Stackelberg U (eds.) (1988) *The Manganese Nodule Belt of the Pacific Ocean*. Stuttgart: Enke.

Nicholson K. Hein J, Buhn B, Dasgupta S (eds.) (1997) *Manganese Mineralisation: Geochemistry and Mineralogy of Terrestrial and Marine Deposits*. Geological Society Special Publication 119, London.

Roy S (1981) *Manganese Deposits*. London: Academic Press.

Teleki PG, Dobson MR, Moore JR, and von Stackelberg U (eds.) (1987) *Marine Minerals: Advances in Research and Resource Assessment*. Dordrecht: D. Riedel.

OCEAN THERMAL ENERGY CONVERSION (OTEC)

S. M. Masutani and P. K. Takahashi, University of
Hawaii at Manoa, Honolulu, HI, USA

Ocean thermal energy conversion (OTEC) generates
electricity indirectly from solar energy by harnessing
the temperature difference between the sun-warmed
surface of tropical oceans and the colder deep
waters. A significant fraction of solar radiation incident on the ocean is retained by seawater in tropical regions, resulting in average year-round surface
temperatures of about 28°C. Deep, cold water,
meanwhile, forms at higher latitudes and descends to
flow along the seafloor toward the equator. The
warm surface layer, which extends to depths of about
100–200 m, is separated from the deep cold water by
a thermocline. The temperature difference, ΔT, between the surface and thousand-meter depth ranges
from 10 to 25°C, with larger differences occurring in
equatorial and tropical waters, as depicted in **Figure 1**. ΔT establishes the limits of the performance of
OTEC power cycles; the rule-of-thumb is that a
differential of about 20°C is necessary to sustain viable operation of an OTEC facility.

Since OTEC exploits renewable solar energy, re-
curring costs to generate electrical power are minimal. However, the fixed or capital costs of OTEC
systems per kilowatt of generating capacity are very
high because large pipelines and heat exchangers are
needed to produce relatively modest amounts of
electricity. These high fixed costs dominate the economics of OTEC to the extent that it currently cannot compete with conventional power systems,
except in limited niche markets. Considerable effort
has been expended over the past two decades to
develop OTEC by-products, such as fresh water, air
conditioning, and mariculture, that could offset the
cost penalty of electricity generation.

State of the Technology

OTEC power systems operate as cyclic heat engines.
They receive thermal energy through heat transfer
from surface sea water warmed by the sun, and
transform a portion of this energy to electrical
power. The Second Law of Thermodynamics precludes the complete conversion of thermal energy in
to electricity. A portion of the heat extracted from
the warm sea water must be rejected to a colder
thermal sink. The thermal sink employed by OTEC
systems is sea water drawn from the ocean depths by
means of a submerged pipeline. A steady-state control volume energy analysis yields the result that net
electrical power produced by the engine must equal
the difference between the rates of heat transfer from
the warm surface water and to the cold deep water.
The limiting (i.e., maximum) theoretical Carnot energy conversion efficiency of a cyclic heat engine
scales with the difference between the temperatures
at which these heat transfers occur. For OTEC, this
difference is determined by ΔT and is very small;

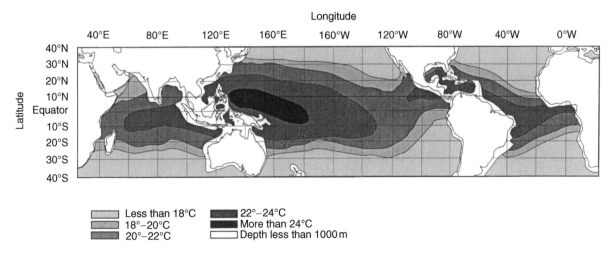

Figure 1 Temperature difference between surface and deep sea water in regions of the world. The darkest areas have the greatest
temperature difference and are the best locations for OTEC systems.

hence, OTEC efficiency is low. Although viable OTEC systems are characterized by Carnot efficiencies in the range of 6–8%, state-of-the-art combustion steam power cycles, which tap much higher temperature energy sources, are theoretically capable of converting more than 60% of the extracted thermal energy into electricity.

The low energy conversion efficiency of OTEC means that more than 90% of the thermal energy extracted from the ocean's surface is 'wasted' and must be rejected to the cold, deep sea water. This necessitates large heat exchangers and seawater flow rates to produce relatively small amounts of electricity.

In spite of its inherent inefficiency, OTEC, unlike conventional fossil energy systems, utilizes a renewable resource and poses minimal threat to the environment. In fact, it has been suggested that widespread adoption of OTEC could yield tangible environmental benefits through avenues such as reduction of greenhouse gas CO_2 emissions; enhanced uptake of atmospheric CO_2 by marine organism populations sustained by the nutrient-rich, deep OTEC sea water; and preservation of corals and hurricane amelioration by limiting temperature rise in the surface ocean through energy extraction and artificial upwelling of deep water.

Carnot efficiency applies only to an ideal heat engine. In real power generation systems, irreversibilities will further degrade performance. Given its low theoretical efficiency, successful implementation of OTEC power generation demands careful engineering to minimize irreversibilities. Although OTEC consumes what is essentially a free resource, poor thermodynamic performance will reduce the quantity of electricity available for sale and, hence, negatively affect the economic feasibility of an OTEC facility.

An OTEC heat engine may be configured following designs by J.A. D'Arsonval, the French engineer who first proposed the OTEC concept in 1881, or G. Claude, D'Arsonval's former student. Their designs are known, respectively, as closed cycle and open cycle OTEC.

Closed Cycle Ocean Thermal Energy Conversion

D'Arsonval's original concept employed a pure working fluid that would evaporate at the temperature of warm sea water. The vapor would subsequently expand and do work before being condensed by the cold sea water. This series of steps would be repeated continuously with the same working fluid, whose flow path and thermodynamic process representation constituted closed loops – hence, the name 'closed cycle.' The specific process adopted for closed cycle OTEC is the Rankine, or vapor power, cycle. **Figure 2** is a simplified schematic diagram of a closed cycle OTEC system. The principal components are the heat exchangers, turbogenerator, and seawater supply system, which, although not shown, accounts for most of the parasitic power consumption and a significant fraction of the capital expense. Also not included are ancillary devices such as separators to remove residual liquid downstream of the evaporator and subsystems to hold and supply working fluid lost through leaks or contamination.

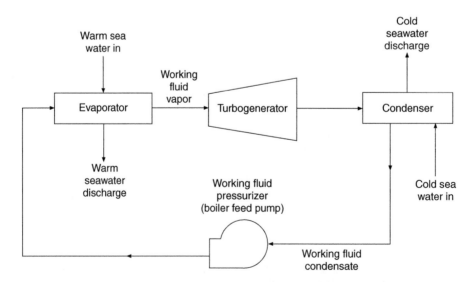

Figure 2 Schematic diagram of a closed-cycle OTEC system. The working fluid is vaporized by heat transfer from the warm sea water in the evaporator. The vapor expands through the turbogenerator and is condensed by heat transfer to cold sea water in the condenser. Closed-cycle OTEC power systems, which operate at elevated pressures, require smaller turbines than open-cycle systems.

In this system, heat transfer from warm surface sea water occurs in the evaporator, producing a saturated vapor from the working fluid. Electricity is generated when this gas expands to lower pressure through the turbine. Latent heat is transferred from the vapor to the cold sea water in the condenser and the resulting liquid is pressurized with a pump to repeat the cycle.

The success of the Rankine cycle is a consequence of more energy being recovered when the vapor expands through the turbine than is consumed in re-pressurizing the liquid. In conventional (e.g., combustion) Rankine systems, this yields net electrical power. For OTEC, however, the remaining balance may be reduced substantially by an amount needed to pump large volumes of sea water through the heat exchangers. (One misconception about OTEC is that tremendous energy must be expended to bring cold sea water up from depths approaching 1000 meters. In reality, the natural hydrostatic pressure gradient provides for most of the increase in the gravitational potential energy of a fluid particle moving with the gradient from the ocean depths to the surface.)

Irreversibilities in the turbomachinery and heat exchangers reduce cycle efficiency below the Carnot value. Irreversibilities in the heat exchangers occur when energy is transferred over a large temperature difference. It is important, therefore, to select a working fluid that will undergo the desired phase changes at temperatures established by the surface and deep sea water. Insofar as a large number of substances can meet this requirement (because pressures and the pressure ratio across the turbine and pump are design parameters), other factors must be considered in the selection of a working fluid including: cost and availability, compatibility with system materials, toxicity, and environmental hazard. Leading candidate working fluids for closed cycle OTEC applications are ammonia and various fluorocarbon refrigerants. Their primary disadvantage is the environmental hazard posed by leakage; ammonia is toxic in moderate concentrations and certain fluorocarbons have been banned by the Montreal Protocol because they deplete stratospheric ozone.

The Kalina, or adjustable proportion fluid mixture (APFM), cycle is a variant of the OTEC closed cycle. Whereas simple closed cycle OTEC systems use a pure working fluid, the Kalina cycle proposes to employ a mixture of ammonia and water with varying proportions at different points in the system. The advantage of a binary mixture is that, at a given pressure, evaporation or condensation occurs over a range of temperatures; a pure fluid, on the other hand, changes phase at constant temperature. This additional degree of freedom allows heat transfer-related irreversibilities in the evaporator and condenser to be reduced.

Although it improves efficiency, the Kalina cycle needs additional capital equipment and may impose severe demands on the evaporator and condenser. The efficiency improvement will require some combination of higher heat transfer coefficients, more heat transfer surface area, and increased seawater flow rates. Each has an associated cost or power penalty. Additional analysis and testing are required to confirm whether the Kalina cycle and assorted variations are viable alternatives.

Open Cycle Ocean Thermal Energy Conversion

Claude's concern about the cost and potential biofouling of closed cycle heat exchangers led him to propose using steam generated directly from the warm sea water as the OTEC working fluid. The steps of the Claude, or open, cycle are: (1) flash evaporation of warm sea water in a partial vacuum; (2) expansion of the steam through a turbine to generate power; (3) condensation of the vapor by direct contact heat transfer to cold sea water; and (4) compression and discharge of the condensate and any residual noncondensable gases. Unless fresh water is a desired by-product, open cycle OTEC eliminates the need for surface heat exchangers. The name 'open cycle' comes from the fact that the working fluid (steam) is discharged after a single pass and has different initial and final thermodynamic states; hence, the flow path and process are 'open.'

The essential features of an open cycle OTEC system are presented in **Figure 3**. The entire system, from evaporator to condenser, operates at partial vacuum, typically at pressures of 1–3% of atmospheric. Initial evacuation of the system and removal of noncondensable gases during operation are performed by the vacuum compressor, which, along with the sea water and discharge pumps, accounts for the bulk of the open cycle OTEC parasitic power consumption.

The low system pressures of open cycle OTEC are necessary to induce boiling of the warm sea water. Flash evaporation is accomplished by exposing the sea water to pressures below the saturation pressure corresponding to its temperature. This is usually accomplished by pumping it into an evacuated chamber through spouts designed to maximize heat and mass transfer surface area. Removal of gases dissolved in the sea water, which will come out of

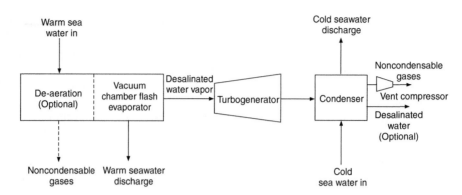

Figure 3 Schematic diagram of an open-cycle OTEC system. In open-cycle OTEC, warm sea water is used directly as the working fluid. Warm sea water is flash evaporated in a partial vacuum in the evaporator. The vapor expands through the turbine and is condensed with cold sea water. The principal disadvantage of open-cycle OTEC is the low system operating pressures, which necessitate large components to accommodate the high volumetric flow rates of steam.

solution in the low-pressure evaporator and compromise operation, may be performed at an intermediate pressure prior to evaporation.

Vapor produced in the flash evaporator is relatively pure steam. The heat of vaporization is extracted from the liquid phase, lowering its temperature and preventing any further boiling. Flash evaporation may be perceived, then, as a transfer of thermal energy from the bulk of the warm sea water of the small fraction of mass that is vaporized. Less than 0.5% of the mass of warm sea water entering the evaporator is converted into steam.

The pressure drop across the turbine is established by the cold seawater temperature. At 4°C, steam condenses at 813 Pa. The turbine (or turbine diffuser) exit pressure cannot fall below this value. Hence, the maximum turbine pressure drop is only about 3000 Pa, corresponding to about a 3:1 pressure ratio. This will be further reduced to account for other pressure drops along the steam path and differences in the temperatures of the steam and seawater streams needed to facilitate heat transfer in the evaporator and condenser.

Condensation of the low-pressure steam leaving the turbine may employ a direct contact condenser (DCC), in which cold sea water is sprayed over the vapor, or a conventional surface condenser that physically separates the coolant and the condensate. DCCs are inexpensive and have good heat transfer characteristics because they lack a solid thermal boundary between the warm and cool fluids. Surface condensers are expensive and more difficult to maintain than DCCs; however, they produce a marketable freshwater by-product.

Effluent from the condenser must be discharged to the environment. Liquids are pressurized to ambient levels at the point of release by means of a pump, or,

if the elevation of the condenser is suitably high, can be compressed hydrostatically. As noted previously, noncondensable gases, which include any residual water vapor, dissolved gases that have come out of solution, and air that may have leaked into the system, are removed by the vacuum compressor.

Open cycle OTEC eliminates expensive heat exchangers at the cost of low system pressures. Partial vacuum operation has the disadvantage of making the system vulnerable to air in-leakage and promotes the evolution of noncondensable gases dissolved in sea water. Power must ultimately be expended to pressurize and remove these gases. Furthermore, as a consequence of the low steam density, volumetric flow rates are very high per unit of electricity generated. Large components are needed to accommodate these flow rates. In particular, only the largest conventional steam turbine stages have the potential for integration into open cycle OTEC systems of a few megawatts gross generating capacity. It is generally acknowledged that higher capacity plants will require a major turbine development effort.

The mist lift and foam lift OTEC systems are variants of the OTEC open cycle. Both employ the sea water directly to produce power. Unlike Claude's open cycle, lift cycles generate electricity with a hydraulic turbine. The energy expended by the liquid to drive the turbine is recovered from the warm sea water. In the lift process, warm seawater is flash evaporated to produce a two-phase, liquid–vapor mixture – either a mist consisting of liquid droplets suspended in a vapor, or a foam, where vapor bubbles are contained in a continuous liquid phase. The mixture rises, doing work against gravity. Here, the thermal energy of the vapor is expended to increase the potential energy of the fluid. The vapor is then condensed with cold sea water and discharged back into the ocean. Flow of the liquid through the hydraulic turbine may occur before or after the lift process. Advocates of the mist and foam lift cycles contend that they are cheaper to implement than closed cycle OTEC because they require no expensive heat exchangers, and are superior to the Claude cycle because they utilize a hydraulic turbine rather than a low pressure steam turbine. These claims await verification.

Hybrid Cycle Ocean Thermal Energy Conversion

Some marketing studies have suggested that OTEC systems that can provide both electricity and water may be able to penetrate the marketplace more readily than plants dedicated solely to power generation. Hybrid cycle OTEC was conceived as a response to these studies. Hybrid cycles combine the potable water production capabilities of open cycle OTEC with the potential for large electricity generation capacities offered by the closed cycle.

Several hybrid cycle variants have been proposed. Typically, as in the Claude cycle, warm surface seawater is flash evaporated in a partial vacuum. This low pressure steam flows into a heat exchanger where it is employed to vaporize a pressurized, low-boiling-point fluid such as ammonia. During this process, most of the steam condenses, yielding desalinated potable water. The ammonia vapor flows through a simple closed-cycle power loop and is condensed using cold sea water. The uncondensed steam and other gases exiting the ammonia evaporator may be further cooled by heat transfer to either the liquid ammonia leaving the ammonia condenser or cold sea water. The noncondensables are then compressed and discharged to the atmosphere.

Steam is used as an intermediary heat transfer medium between the warm sea water and the ammonia; consequently, the potential for biofouling in the ammonia evaporator is reduced significantly. Another advantage of the hybrid cycle related to freshwater production is that condensation occurs at significantly higher pressures than in an open cycle OTEC condenser, due to the elimination of the turbine from the steam flow path. This may, in turn, yield some savings in the amount of power consumed to compress and discharge the noncondensable gases from the system. These savings (relative to a simple Claude cycle producing electricity and water), however, are offset by the additional back-work of the closed-cycle ammonia pump.

One drawback of the hybrid cycle is that water production and power generation are closely coupled. Changes or problems in either the water or power subsystem will compromise performance of the other. Furthermore, there is a risk that the potable water may be contaminated by an ammonia leak. In response to these concerns, an alternative hybrid cycle has been proposed, comprising decoupled power and water production components. The basis for this concept lies in the fact that warm sea water leaving a closed cycle evaporator is still sufficiently warm, and cold seawater exiting the condenser is sufficiently cold, to sustain an independent freshwater production process.

The alternative hybrid cycle consists of a conventional closed-cycle OTEC system that produces electricity and a downstream flash-evaporation-based desalination system. Water production and electricity generation can be adjusted independently, and either can operate should a subsystem fail or require servicing. The primary drawbacks are that the ammonia evaporator uses warm seawater directly and is subject to biofouling; and additional equipment, such as the potable water surface condenser, is required, thus increasing capital expenses.

Environmental Considerations

OTEC systems are, for the most part, environmentally benign. Although accidental leakage of closed cycle working fluids can pose a hazard, under normal conditions, the only effluents are the mixed seawater discharges and dissolved gases that come out of solution when sea water is depressurized. Although the quantities of outgassed species may be significant for large OTEC systems, with the exception of carbon dioxide, these species are benign. Carbon dioxide is a greenhouse gas and can impact global climate; however, OTEC systems release one or two orders of magnitude less carbon dioxide than comparable fossil fuel power plants and those emissions may be sequestered easily in the ocean or used to stimulate marine biomass production.

OTEC mixed seawater discharges will be at lower temperatures than sea water at the ocean surface. The discharges will also contain high concentrations of nutrients brought up with the deep sea water and may have a different salinity. It is important, therefore, that release back into the ocean is conducted in a manner that minimizes unintended changes to the ocean mixed layer biota and avoids inducing long-term surface temperature anomalies. Analyses of OTEC effluent plumes suggest that discharge at depths of 50–100 m should be sufficient to ensure minimal impact on the ocean environment. Conversely, the nutrient-rich OTEC discharges could be exploited to sustain open-ocean mariculture.

Economics of Ocean Thermal Energy Conversion

Studies conducted to date on the economic feasibility of OTEC systems suffer from the lack of reliable cost data. Commercialization of the technology is unlikely until a full-scale plant is constructed and operated continuously over an extended period to provide these data on capital and personnel and maintenance expenses.

Uncertainties in financial analyses notwithstanding, projections suggest very high first costs for OTEC power system components. Small land-based or near-shore floating plants in the 1–10 MW range, which would probably be constructed in rural island communities, may require expenditures of $10 000–$20 000 (in 1995 US dollars) per kW of installed generating capacity. Although there appears to be favorable economies of scale, larger floating (closed cycle) plants in the 50–100 MW range are still anticipated to cost about $5000 kW^{-1}. This is well in excess of the $1000–$2000 kW^{-1} of fossil fuel power stations.

To enhance the economics of OTEC power stations, various initiatives have been proposed based on marketable OTEC by- or co-products. OTEC proponents believe that the first commercial OTEC plants will be shore-based systems designed for use in developing Pacific island nations, where potable water is in short supply. Many of these sites would be receptive to opportunities for economic growth provided by OTEC-related industries.

Fresh Water

The condensate of the open and hybrid cycle OTEC systems is desalinated water, suitable for human consumption and agricultural uses. Analyses have suggested that first-generation OTEC plants, in the 1–10 MW range, would serve the utility power needs of rural Pacific island communities, with the desalinated water by-product helping to offset the high cost of electricity produced by the system.

Refrigeration and Air Conditioning

The cold, deep sea water can be used to maintain cold storage spaces, and to provide air conditioning. The Natural Energy Laboratory of Hawaii Authority (NELHA), which manages the site of Hawaii's OTEC experiments, has air-conditioned its buildings by passing the cold sea water through heat exchangers. A new deep seawater utilization test facility in Okinawa also employs cold seawater air conditioning. Similar small-scale operations would be viable in other locales. Economic studies have been performed for larger metropolitan and resort applications. These studies indicate that air conditioning new developments, such as resort complexes, with cold seawater may be economically attractive even if utility-grid electricity is available.

Mariculture

The cold deep ocean waters are rich in nutrients and low in pathogens, and therefore provide an excellent medium for the cultivation of marine organisms. The 322-acre NELHA facility has been the base for successful mariculture research and development enterprises. The site has an array of cold water pipes, originally installed for the early OTEC research, but since used for mariculture. The cold water is applied to cultivate flounder, *opihi* (limpet; a shellfish delicacy), oysters, lobsters, sea urchins, abalone, kelp, *nori* (a popular edible seaweed used in sushi), and macro- and microalgae. Although many of these ongoing endeavors are profitable, high-value products such as biopharmaceuticals, biopigments, and pearls will need to be advanced to realize the full potential of the deep water.

The cold sea water may have applications for open-ocean mariculture. Artificial upwelling of deep water has been suggested as a method of creating new fisheries and marine biomass plantations. Should development proceed, open-ocean cages can be eliminated and natural feeding would replace expensive feed, with temperature and nutrient differentials being used to keep the fish stock in the kept environment.

Agriculture

An idea initially proposed by University of Hawaii researchers involves the use of cold sea water for agriculture. This involves burying an array of cold water pipes in the ground near to the surface to create cool weather growing conditions not found in tropical environments. In addition to cooling the soil, the system also drip irrigates the crop via condensation of moisture in the air on the cold water pipes. Demonstrations have determined that strawberries and other spring crops and flowers can be grown throughout the year in the tropics using this method.

Energy Carriers

Although the most common scenario is for OTEC energy to be converted into electricity and delivered directly to consumers, energy storage has been considered as an alternative, particularly in applications involving floating plants moored far offshore.

Storage would also allow the export of OTEC energy to industrialized regions outside of the tropics. Long-term proposals have included the production of hydrogen gas via electrolysis, ammonia synthesis, and the development of shore-based mariculture systems or floating OTEC plant-ships as ocean-going farms. Such farms would cultivate marine biomass, for example, in the form of fast-growing kelp which could be converted thermochemically into fuel and chemical co-products.

See also

Carbon Dioxide (CO₂) Cycle.

Further Reading

Avery WH and Wu C (1994) *Renewable Energy from the Ocean: A Guide to OTEC*. New York: Oxford University Press.

Nihous GC, Syed MA, and Vega LA (1989) Conceptual design of an open-cycle OTEC plant for the production of electricity and fresh water in a Pacific island. *Proceedings International Conference on Ocean Energy Recovery.*

Penney TR and Bharathan D (1987) Power from the sea. *Scientific American* 256(1): 86–92.

Sverdrup HV, Johnson MW, and Fleming PH (1942) *The Oceans: Their Physics, Chemistry, and General Biology.* New York: Prentice-Hall.

Takahashi PK and Trenka A (1996) *Ocean Thermal Energy Conversion; UNESCO Energy Engineering Series.* Chichester: John Wiley.

Takahashi PK, McKinley K, Phillips VD, Magaard L, and Koske P (1993) Marine macrobiotechnology systems. *Journal of Marine Biotechnology* 1(1): 9–15.

Takahashi PK (1996) Project blue revolution. *Journal of Energy Engineering* 122(3): 114–124.

Vega LA and Nihous GC (1994) Design of a 5 MW OTEC pre-commercial plant. *Proceedings Oceanology* 94: 5.

Vega LA (1992) Economics of ocean thermal energy conversion. In: Seymour RJ (ed.) *Ocean Energy Recovery: The State of the Art.* New York: American Society of Civil Engineers.

MARINE BIOTECHNOLOGY

H. O. Halvorson, University of Massachusetts Boston, Boston, MA, USA
F. Quezada, Biotechnology Center of Excellence Corporation, Waltham, MA, USA

Introduction

Definitions

Marine biotechnology is the application of biotechnological approaches to marine organisms in the harnessing of commercial products and services. Marine biotechnology is a multidisciplinary activity marrying three traditional marine disciplines – ocean science, marine biology, and marine engineering – with the more modern fields of molecular and cellular biology, genomics, and proteomics.

Marine Biotechnology as an Academic and Research Discipline

The world's oceans, which comprise two-thirds of the surface of planet, possess the largest habitats on Earth and contain the most ancient forms of life. Over time, marine microbes have changed the global climate and structured the atmosphere. The large and unexplored diversity of living species today is found mostly in the oceans. With the tools of molecular biology, these specific adaptations can be understood in detail and applied for the benefit of biomedical and industrial innovation, environmental remediation, food production, and fundamental scientific progress.

Marine organisms have traditionally been good models for the study of communication, defense, adhesion, host interactions, disease, epidemics, nutrition, and adaptation to large, often extreme, variations in their environment. There are numerous examples of marine models contributing to understanding basic concepts within biology and medicine. The possibility to cultivate these organisms, and to study them in the laboratory at a cellular, molecular, or genetic level, has opened new perspectives.

Professional Associations

The formation and establishment of professional associations in marine biotechnology reflects the growing worldwide interest in this area. The Japanese Society for Marine Biotechnology was organized in 1989. Since then, other counterpart associations were formed: the American Society for Molecular Marine Biology and Biotechnology, the Asian-Pacific Marine Biotechnology Society, and the Pan-American Marine Biotechnology Association. Networks of marine laboratories also came together. The National Association of Marine Laboratories, in the United States, and the European Network of Marine Laboratories, in Europe, were formed, with strong interests in marine biotechnology. In 2000, Canada initiated an interdisciplinary program (Aquanet) to support aquaculture in that country.

In 1989, the Japanese Society for Marine Biotechnology organized the first International Marine Biotechnology Conferences (IMBC). The first European meetings on marine biotechnology were held at Montpellier (1992), Willemshaven (Germany) (1998), Edinburgh (1998), and Noordwijkerhout (the Netherlands) (1999). The IMBC meetings focused on new breakthroughs in basic and applied science, industrial applications, environmental science, industrial applications, commerce, and international issues of marine biotechnology. Two new marine journals were established: *Journal of Marine Biotechnology* (Japan) and *Molecular Marine Biology and Biotechnology* (USA). These were merged to form *Marine Biotechnology*, an international journal on the molecular and cellular biology of marine life and its technology applications.

Marine Biotechnology and Marine Conservation

Tools for the Study of Marine Ecology

Oceans are highways of commerce for international trade, providing food, contributing to our energy supply (oil and gas), and serving as recreational areas. The health of our ocean ecosystems is threatened by overfishing and unintended bycatch, pollution, habitat loss, boating traffic, and climate change. This threat comes from a number of human activities: conversion of coastal habitats for other uses, nonpoint pollution, airborne pollution, waterborne diseases, marine debris, ballast water, and invasive species, among others. Some of the tools of marine biotechnology have helped with ongoing monitoring to assess the health of ocean and coastal ecosystems to guide management in decisions on the use of oceans and coastal waters and establishment of standard practices and procedures.

Multiple means are used to collect the desired information: aircraft, ships, moored instruments, drifters, gliders, submersibles, remotely operated vehicle (ROV), and satellites. The satellite communications infrastructure provides global broadband coverage to support ocean observations. Monitoring stations and buoys collect and transmit continuous data streams on climate, weather, air quality, temperature, surface pressure, ocean dynamics, and other selected variables. The Integrated Ocean Observing System (IOOS) and Global Ocean Observing System (GOOS) integrate the collected data.

Additional sensors are being developed to monitor marine ecology. Optical and sonic probes are used to count populations of marine organisms. Quantification of plant and animal biodiversity can be a useful approach to follow the health of the habitat. Molecular techniques used to measure marine microbial diversity should be modified as marine sensors to detect marine pathogens and toxins. Sensors to measure marine pollutants should be expanded. Major challenges are collection, storage, and assimilation of the large volume of data as well as the ability to provide these data rapidly to the user community. Recent developments in molecular biology, ecology, and environmental engineering now offer opportunities to modify organisms so that their basic biological processes are more efficient and can degrade more complex chemicals and higher volumes of waste materials.

Understanding Life Cycles of Marine Organisms

Life cycles in marine organisms are composed of stages that can vary dramatically in size, shape, mobility, and food preferences. Reproductive strategies abound in the ocean and include fission, budding, free-spawned gametes (eggs and sperm), external embryo hatching, internal embryo hatching, and live birth. A significant number of marine organisms are hermaphrodites, containing the reproductive organs of both sexes. Many species of fish develop from a fertilized egg through a juvenile phase and into an adult phase. Deep-sea fish have larval stages distinctly separate from the juvenile and adult stage. The life cycle of some commercial species, such as salmon and trout, have been very well documented.

Marine organisms face challenges to fertilization of free-spawned gametes, nutrient acquisition during larval development, benthic site selection, juvenile survival, and adult reproductive location and timing. Understanding ecological and evolutionary processes at the single organism and population levels is essential.

Major Product Areas of Interest

Medicines from the Sea

Marine molecules often belong to new classes without terrestrial counterparts, for example, halogenated compounds. In addition, marine microorganisms are a source of new genes. Secondary metabolites produced by marine bacteria and invertebrates have yielded pharmaceutical products such as novel anti-inflammatory agents, anticancer agents, and antibiotics. For example, isolation of the compound manoalide from a Pacific sponge has led to the development of more than 300 chemical analogs, with many of these going on to clinical trials as anti-inflammatory agents. Melanins have a range of chromophoric properties that can be exploited for sunscreens, dyes, and coloring. Marine sponges also sequester different kinds of organic compounds, including fungicides and antibiotics, which may allow them to act as slow-release agents.

Other examples include the circulating cells (amoebocytes) of the horseshoe crab that contain molecules that react with the outer coats of Gram-negative bacteria, and thus have found use in the detection of early infection in humans and pyrogens in biotechnological products. Lectins, found in all invertebrates, are being used to target cancer cells, screen bacteria, and immobilize enzymes.

Seaweed produces compounds like laminarin and fucoidans that are known to protect against radiation damage, lower cholesterol levels, and help in wound repair. They also have an anti-inflammatory action, as well as strong immunomodulatory effects. Seaweed-derived products increase resistance to bacterial, viral, and parasitic infections (including infection after surgery) and help with prevention of opportunistic infections in immunocompromised individuals (such as HIV sufferers and geriatric patients). Secondary metabolites, such as halogenated compounds, extracted from macroalgae (seaweeds) are of great promise as antibacterial and antiviral agents or antifouling agents. Extracts from certain red algae are used in bone replacement therapy.

Biomaterials and Nutraceuticals

Historically, macroalgae (seaweeds) have been used as a subsidiary food and have been used extensively in medicine. Macroalgae make use of different photosynthesizing pigments that divide them roughly into three groups – the brown, green, and red algae – which is in contrast to the terrestrial environment with only the green plants. The biochemical diversity present in seaweeds provides the potential for a very large array of products. For example, polysaccharides from certain red algae or brown algae are widely used as thickening agents in the food industry, cosmetics, and even building materials.

Seaweeds are used in aquaculture diets for urchin, abalone, and fish as an alternative food source to fishmeal and fish oil. These seaweeds contain all of the essential amino acids, polyunsaturated fatty acids, vitamins, and minerals, as well as a small amount of protein. Eelgrass produces an effective antifouling agent against bacteria, algal spores, and a variety of hard-fouling barnacles and tubeworms.

Marine diatoms, mollusks, and other marine invertebrates generate elaborate mineralized structures on a nanometer scale that can have unusual and useful properties.

Deep-sea hydrothermal vents now offer a new source of a variety of fascinating microorganisms well adapted to these extreme environments. This newly appreciated bacterial diversity includes strains that are able to produce a number of unusual microbial exopolysaccharides with interesting chemical and physical properties.

Chitin and chitosan are associated with proteins in the exoskeleton of many invertebrate species, such as annelids, shellfish, and insects, and also in the envelope of many fungi, molds, and yeasts. Chitin polymers are natural, nontoxic, and biodegradable and have many applications in food and pharmaceuticals as well as cosmetics.

'Antifreeze glycoproteins' which inhibit ice-crystal formation are found in the tissues of fish in the Arctic and Antarctic (e.g., Arctic char) as well as microorganisms which live at low temperatures. These species may prove useful in industrial and medical cryopreservation processes.

Antioxidant peptides have been isolated from extracts of prawn muscle and seaweeds. These have applications as food additives and in cosmetics.

Industrial Enzymes

Microorganisms provide the basis for development of sophisticated biosensors, and diagnostic devices for medicine, aquaculture, and environmental biomonitoring. Intact cell preparations and isolated enzyme systems for bioluminescence are used as biosensors. The genes encoding these enzymes have been cloned from marine bacteria and have been subsequently transferred successfully to a variety of organisms where they are expressed only under defined environmental conditions. Other bioluminescent proteins from marine organisms are currently under study in order to produce gene probes that can be employed to detect human pathogens in

food, or fish pathogens in aquaculture systems. Green fluorescent protein, a naturally fluorescent protein first found in jellyfish, is now widely employed as a sensitive fluorescent molecular 'tag' to identify and localize individual proteins within a cell or a subset of cells within a tissue and to follow gene expression in various systems.

Aquatic extremophiles produce thermostable DNA-modifying enzymes used in research and industrial applications. Some enzymes produced by marine bacteria are salt resistant: extracellular proteases are used in detergents and industrial cleaning applications. Vibrio species have been found to produce a variety of extracellular proteases. Vibrio alginolyticus produces collagenase, an enzyme used in the dispersion of cells in tissue culture studies.

Novel Marine-derived Bioprocesses

Bioremediation

Marine microorganisms, either as independent strains or as members of microbial consortia, express novel biodegradation pathways for breaking down a wide variety of organic pollutants. Marine microorganisms frequently produce environmentally friendly chemicals such as biopolymers and nontoxic biosurfactants that can also be applied in environmental waste management and treatment. Recent findings into the basis of cell–cell communication have shown that this process is involved in biofilm formation leading to environmental corrosion. This has generated a search for new bioactive molecules active in preventing such communication and controlling subsequent fouling. In addition, further understanding of the interaction of marine microbes with toxic heavy metals has suggested their application in various treatments of contaminated water systems.

Biotechnology Applications to Marine Aquaculture

Reproduction

Over the past two centuries, the practice of agriculture involved classical selective breeding of plants and animals to enhance a desirable characteristic by increasing the expression of a particular gene in the population over subsequent generations. In aquaculture, cells, molecules, and genes can now be modified directly to improve production efficiency through improvements in growth rates, food conversion, disease resistance, product quality, and composition. This same approach can help conserve wild species and genetic resources and provide unique models for biomedical research. Key hormones regulate the processes of molting, development, and reproduction in economically important crustaceans. Foreign genes controlling growth-enhancing hormones have now been introduced into scallops and a number of fish. Hormones have been used to develop monosex lines, agents for smolt enhancement, and enhanced feed-conversion efficiency.

Microbial Diseases and Pathogen Control

Bacteria, viruses, and other pathogens are a natural part of any ecosystem and aquatic organisms have co-evolved with them. Fish diseases do not occur as a single caused event but are the end result of interactions of the disease, the fish, and the environment. In intensive culture, handling, crowding, transporting, drug treatments, undernourishment, fluctuating temperatures, and poor water quality continuously affect fish. These conditions impose considerable stress on fish, rendering them susceptible to a wide variety of pathogens. Bacteria cause the most severe disease problems in aquaculture. Specific bacterial pathogens are responsible for specific disease problems. Gram-negative bacteria are the most frequent causes of disease in finfish, whereas Gram-positive bacteria are most common in crustaceans.

Viral diseases cause serious problems in aquaculture requiring quarantine and destruction of the infected stock. Viral disease in wild stock fish are formidable obstacles to raising fish in net-pen aquaculture in open waters. Fungal infections are frequently encountered when the water quality is poor, or in the presence of stress, inadequate nutrition, and skin trauma which provides a port of entry for molds. Saprolegniosis, one of the most common water molds encountered, occurs with handling, crowding, heavy feeding, and high organic loads.

The use of chemotherapeutics in fish, as in land-based animals, is extremely controversial. Bacterial resistance to antibiotics has developed in animal systems, in which antibiotics are routinely used. Resistant bacteria may be carried on food products, and thus infect the consumer or may infect the handler directly. There are very limited numbers of chemical/antibiotic treatments approved for use in aquacultured fish. Drug development is expensive and pharmaceutical companies are not willing to invest the money necessary to directly approve drugs for use in aquacultured animals. The decline in the use of antibiotics as a defense for man and animals against pathogenic microorganisms provides the biomedical and biotechnology industry with new opportunities to design and rapidly produce vaccines.

Genetic Management and Genetic Engineering

Modern biotechnology approaches to aquaculture of marine species include the use of marker-assisted breeding techniques. These techniques allow hatcheries and breeders to select reproductive pathways on the basis of optimal levels of diversity in the cultivated population and to screen for certain traits.

Genetic engineering, on the other hand, is seen as the more precise and effective method for genetic improvement of specific traits that are controlled by single genes. A transgenic organism is one whose genome contains DNA inserted from another organism. DNA is introduced into embryo stem cells, which are then merged with early-stage embryos. If the gene of interest is present in a germ line, they can be bred to homogeneity by future generations. Genes are activated to produce protein by adjacent genes that produce promoter sequences. The promoter sequences are responsible for switching on genes in specific areas of the body. This technology has been used successfully in numerous marine microalgae and animal systems to alter characteristics in a population by conferring temperature resistance to plants and marine animals, growth hormone, antifreeze protein, or disease resistance.

An effective biological containment of some commercially important species must be developed. Cloning wild genotypes has benefits for stock enhancement, conservation, restoration, and hatchery production.

Improved Feed and Nutrition

Exclusion of traditional raw materials from aquaculture feeds and possible limitation in the use of those remaining impose a great risk to aquaculture sustainability. A certain reduction of production costs can be obtained through adequate larval nutrition, diversification of marine organisms used for the first feeding of larvae, and development of efficient artificial diets for early larval stages. These, together with feed-dispersing techniques, would add to the overall quality of larvae used.

Other considerations include how the plant genome can now be manipulated to produce products economically for use in aquaculture. The use of genetically modified crops to eliminate toxic products and increase specific nutrients is now possible. Further understanding of the fish immune system and the relation of its condition to nutrition and stress would help in improvements in growth and disease resistance.

Oceans and Human Health

Public health authorities, tourism, and the seafood industry have been increasingly concerned over public health risks from the marine sources. The degradation of the oceans makes an impact on how we use the ocean and where we buy our seafood. The dramatic growth of the international seafood market increases the global potential for risk. The survival of human populations may depend on the preservation of healthy, diverse, and sustainable ocean and coastal systems.

Molecular technology has provided an increasing range of diagnostic tools to clinicians and environmental regulators to measure risk assessment. The main effort of such risk assessment should be the protection of individuals and populations from harm. Including biomarkers in the risk assessment process provides a functional measure of exposure to chemicals, which stress an organism. In recent years, an increasing number of biomarkers have been used to measure molecular damage, developmental abnormality, and physiological impairment in invertebrate species.

General biological stress biomarkers include cardiac activity in bivalve mollusks, dye retention in the hemocytes of bivalve mollusks, measurement of immunocompetence in invertebrates, apoptosis assay, and defecation assay with shrimp. Behavior biomarkers include swimming behavior in shrimp, burrowing behavior assay with amphipods, and a starfish-righting assay. Chemical biomarkers include acetylcholinesterase inhibition in crustaceans, florescence assay in urine samples of crabs, assay of endocrine disruptors in gastropod mollusks, and genotoxins.

Developing nations dominate both the production and international trade in seafood. Since it is now recognized that most of the seafood-borne risk comes from the environment, a more global perspective for product protection and coastal waters is needed. A chain of custody/product traceability would assure critical information essential to a comprehensive view of seafood risk mitigation. Such a system would incorporate a set of protocols focusing on simplified determination of associated socioeconomic dynamics. Here again, advanced research in marine biotechnology can help to provide some of the analytical tools to address these seafood safety issues.

Policy Issues

The role of government and public agencies in the development of marine biotechnology and its diverse applications to the environment, commercial endeavors, scientific discovery, new products, and other aspects brings attention to several policy-related issues affecting this field. Among these are issues of access and benefit sharing, intellectual property rights, and ecological concerns. These and other considerations are highlighted below.

Funding

One of these concerns the need for public support for research and human resource development in marine biotechnology. With continuing competition for research funds and dwindling allocations for higher education, those institutions involved in the field of marine biotechnology are faced with the need to highlight the relevance of this field to the high-priority problems of scientific, economic, and ecological nature.

User Conflicts

The growing urbanization of coastal areas along with maritime commercial transport and recreational development of marine zones have created new and competing uses for the marine habitats as a natural environment. Additionally, offshore marine aquaculture itself has environmental impacts, which must be considered along with land-based runoff and related consequences. As groups sustain all of these activities with their respective interests and constituencies, appropriate mechanisms must be found to reach acceptable arrangements for all.

Regulatory Framework

The regulatory provisions for field applications of marine biotechnology need to be in place in a clear and consistent framework at both the national and regional levels. This is a continuing challenge that will require a greater understanding of the delicate marine habitat, the tools of research, and their applications for public benefit through sustainable development of marine resources. Regulations must be science-based, transparent, and enforceable at every level of government. International harmonization of regulations will require continuing attention.

Intellectual Property Protection

Commercialization of knowledge and technological breakthroughs generated by marine biotechnology research requires recognition and protection of the intellectual property involved. Countries that have put intellectual property regimes in place relevant to marine resources are also dealing with issues of assignment of benefits to researchers and to governmental agencies in charge of managing the resources.

In the case of marine-protected areas, the intellectual property issues are widely debated.

Recent National Policy Initiatives

The United States has shown increased interest in marine biotechnology during the past several years. The Oceans Act of 2000 established a 16-member Commission on Ocean Policy to undertake an 18-month study and make recommendations for a national ocean policy for the United States, including stewardship of marine resources, pollution prevention, and commerce and marine science. In 2001, the US Department of Agriculture sponsored a workshop titled Biotechnology–Aquaculture Interface: The Site of Maximum Impact Workshop, which focused on the research and development opportunities of new technologies on genomics, biocomplexity, biocellular technology, biosecurity, and social issues. In 2002, the Ocean Studies Board of the National Academy of Sciences reported on two workshops that highlighted new developments and opportunities in environmental and biomedical applications of marine biotechnology. The report highlighted the need to: (1) develop a fundamental understanding of the genetic, nutritional, and environmental factors that control the production of primary and secondary metabolites in marine organisms, as a basis for developing new and improved products and (2) identify bioactive compounds and determine their mechanisms of action and natural function, to provide models for new lines of selectively active materials for application in medicine and the chemical industry.

Future Perspectives

Marine-related biotechnological tools can now be used to determine the effect of natural and anthropogenic perturbations on commercial stocks, propose management tools, determine predator–prey relationships, and restore habitats. The development of predictive models for analyzing potential global climate changes depends on the acquisition of fundamental information on molecular regulatory mechanisms of plankton growth in the oceans. An increased understanding of marine ecological systems will allow the specification of the 'normal' baseline level of their biological function, the monitoring and prediction of potential changes, or biological impacts on the systems.

To grow and sustain a vital marine biotechnology presence, the essential elements include a supportive government infrastructure, public and private funding for research and development, support for high-risk start-up companies through venture capital funds, a strong intellectual property system, public acceptance through awareness and involvement, and a reliable qualified human resource pipeline. Ideally, all of these areas would be addressed simultaneously, but practical reality and limited resources mean that progress will not be made uniformly in all areas. Having a goal of moving forward in all of these areas is a reasonable expectation for the coming years.

Glossary

anthropogenic Processes, objects, or materials that are derived from human activities.

apoptosis Programmed cell death.

aquatic extremophiles Organisms that live in extreme environments of salt, chemicals, pH, temperature, etc.

benthic Living near the bottom of the sea.

biomarkers Indicators of a particular state of an organism.

bioremediation The use of living organisms to clean up or improve polluted or contaminated environments.

biosurfactants Compounds produced on surfaces of living cells that reduce the surface tension.

bycatch Species caught in a fishery intended to target another species.

chemotherapeutics The branch of therapeutics that deals with the treatment of disease by means of chemical substances or drugs.

cryopreservation A process used to preserve cells or tissues through cooling to subzero temperatures.

genetic management The use of genetic means to control the characteristics of a population.

genotoxins A chemical or other agent that damages cellular DNA, resulting in mutations or cancer.

Gram-negative bacteria Bacteria that do not retain crystal violet dye in the Gram stain.

marine ecology The relationship between marine organisms and their environment.

pyrogens Compounds that produce fever.

user conflicts Competing uses for usage of marine zones, including fishing, aquaculture, recreation, and marine transport.

See also

Coastal Zone Management. Mariculture Diseases and Health. Mariculture Overview. Ocean Zoning.

Further Reading

Attaway DH and Zaborsky OR (1993) *Marine Biotechology, Vol. 1: Pharmaceutical and Bioactive Natural Products*, 524pp. New York: Springer.

Board of Biology, Ocean Studies Board, National Research Council (2000) *Opportunities for Environmental Applications of Marine Biotechnology*, 187pp. Washington, DC: National Academy Press.

Bowen RE, Halvorson H, and Depledge MH (2006) The oceans and human health. *Marine Pollution Bulletin 53*: 539–656.

Fenical W, Greenberg M, Halvorson HO, and Hunter-Cevera JC (1993) *International Marine Biotechnology Conference*, 2 vols. Dubuque, IA: Williams C. Brown Publishers.

Le Gal Y and Halvorson HO (1998) *New Developments in Marine Biotechnology*, 343pp. New York: Plenum.

Le Gal Y and Ulber R (1997) *Marine Biotechnology II*, 297pp. Berlin: Springer.

Matsunaga T (2004) *Marine Biotechnology: Proceedings of Marine Biotechnology Conference 2003*, Chiba, Japan. Marine Biotechnology 6, 554pp.

National Research Council (2007) *Marine Biotechnology in the Twenty-First Century. Problems, Promises and Products*, 132pp. Washington, DC: National Academy Press.

Relevant Websites

http://www.floridabiotech.org
- Center of Excellence in Biomedical and Marine Biotechnology.

http://www.umbi.umd.edu
- Center of Marine Biotechnology, UMBI.

http://www.esmb.org
- European Society for Marine Biotechnology.

http://www.research.noaa.gov
- Marine Biotechnology, Oceans and Coastal Research, NOAA Research.

MARINE CHEMICAL AND MEDICINE RESOURCES

S. Ali and C. Llewellyn, Plymouth Marine Laboratory, Plymouth, UK

Introduction

The marine environment consists of several defined habitats ranging from the sea surface microlayer which encompasses the first few microns of the water column, through the bulk water column itself, down to the ocean floor and the subsurface sediments underneath which can be found hydrothermal vents, cold seeps, hydrocarbon seeps, and saturated brines, as well as a wide range of mineral and geological variation. It has become increasingly apparent that within all these oceanic layers there is a diversity of micro- and macroorganisms capable of generating a plethora of previously undescribed molecules through novel metabolic pathways which could be of value to both industry and the clinic. The biological diversity in some marine ecosystems may exceed that of the tropical rain forests and this is supported by the presence of 34 out of the 36 phyla of life. This biodiversity stems from the wide range of environmental conditions to which marine organisms have adapted for survival, including extremes of pH (acid and alkali), temperature (high and low), salinity, pressure, and chemical toxicity (complex polycyclic hydrocarbons, heavy metals).

Marine organisms currently being exploited for biotechnology include sponges, tunicates, bryozoans, mollusks, bacteria, cyanobacteria, macroalgae (seaweeds), and microalgae. These organisms have produced compounds with good activities for a range of infectious and noninfectious disease with high specificity for the target molecule (usually an enzyme). Targets of marine natural products which may be clinically relevant include ion channels and G-protein-coupled receptors, protein serine-threonine kinases, protein tyrosine kinases, phospholipase A_2, microtubule-interfering agents (of which the largest number identified are of marine origin), and DNA-interactive compounds. In addition to small organic molecules, marine organisms are increasingly being recognized as a potential source of novel enzymes which could be of industrial and pharmaceutical importance. More than 30 000 diseases have been clinically described, yet less than one-third of these can be treated based on symptoms and only a small number can be cured.

Thus, the potential market for novel marine compounds for clinical development is enormous. In addition to providing new molecules for direct clinical intervention, the marine environment is also rich in compounds which are finding uses as natural additives in foods, as nutritional supplements including color additives and antioxidants, and as vitamins, oils, and cofactors which enhance general well-being. Marine organisms are also increasingly providing new solutions to developments in such diverse fields as bioremediation, biocatalysis and chemistry, materials science, nanotechnology, and energy. Some of the potential uses of marine products are summarized in **Figure 1**.

The oceans have long been a source of nutrients, additives, and medicines derived from marine mammals and fish; however, this article focuses on some of the potential which is harbored in predominantly microscopic organisms which are now being increasingly studied for novel bioactive compounds and chemicals and may provide a sustainable alternative source for new compounds and processes.

Novel Metabolites and Drug Discovery

Marine organisms have long been recognized as a source of novel metabolites with applications in human disease therapy. Particular emphasis has been placed on the invertebrates such as sponges, mollusks, tunicates, and bryozoans, but more recently advances in genetics and microbial culture have led to a growing interest in cyanobacteria and marine bacteria. For example, a number of anticancer drugs have been derived from marine sources such as sponges which have proven difficult to cultivate and their metabolites display a structural complexity which often precludes total chemical synthesis as an option for potential drug candidates. In recent years, studies have suggested that many of these complex molecules may in fact be the product of microbes which live in a symbiotic relationship with the sponge and that some of these molecules may be the final product of reactions carried out by different organisms. A major challenge within marine biotechnology will be to ascertain the nature of the organisms present in the symbiotic relationship and to identify the pathways involved in metabolite production. A recent advance in molecular biology with the development of metagenomics has opened up the possibility of organism-independent cultivation of genetic material and subsequent screening and characterization of that

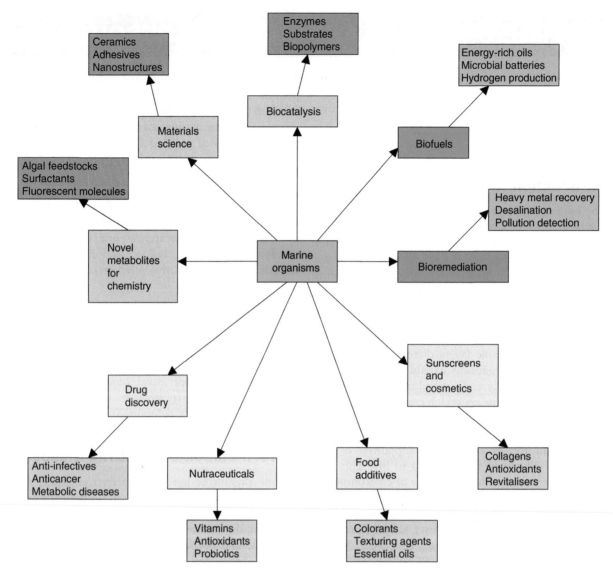

Figure 1 Uses for marine organisms and their products in the chemical, pharmaceutical, energy, and environmental industries.

DNA for novel metabolic pathways and enzymes. This provides a powerful tool for accessing difficult-to-culture microbes which exist in complex symbiotic relationships.

Recent advances in the isolation and culture of marine bacteria using both flow cytometry and microencapsulation-based methods have yielded a vast array of previously unknown bacteria (**Figure 2**). Increasingly, these bacteria are being tested for the presence of bioactive compounds with activities against a diverse range of human and infectious diseases including cancer, HIV, hepatitis C, malaria, and those caused by the increasingly drug-resistant common bacterial pathogens (e.g., *Staphylococcus aureus*, *Enterococcus faecalis*, *Mycobacterium tuberculosis*). For example, the antibiotic abyssomicin C has been isolated from an actinomycete which

was cultured from marine sediment collected in the Sea of Japan. Abyssomicin has been shown to interfere with the synthesis of the essential cofactor folic acid in bacteria and is active against the methicillin-resistant *Staphylococcus aureus* (MRSA) pathogen. Thus, marine bacteria represent a vast untapped source for novel compounds with the potential for development as novel drugs.

Marine-Derived Nutraceuticals and Food Additives

In recent years, there has been an upsurge in the consumption of nutritional supplements such as vitamins and cofactors which are essential to cellular function. One traditional supplement has been

(a)   (b)

Figure 2 Examples of marine microbial cultures. (a) Bacteria isolated from the English Channel. (b) Microalgal cultures of chlorophytes, cryptophytes, and haptophytes.

cod-liver oil, which is rich in the omega-3 and omega-6 long-chain polyunsaturated fatty acids (PUFAs). Fatty acids have long been used as supplements in the aquaculture industry but in recent years their medicinal value to humans has been proposed. PUFAs have been implicated in enhanced blood circulation and brain development, particularly docosahexaenoic acid (omega-3), which plays an important role in early brain development and neurite outgrowth. Although PUFAs have traditionally been extracted from fish oils they do not naturally occur there but rather accumulate from the diet of the fish. The primary source of PUFAs are marine microbes, and in recent years the isolation and characterization of microbes which produce these fatty acids have allowed their production in other organisms. At present considerable effort is being expended on obtaining high-level production of PUFAs in plants using genes from microalgae with the aim of eventually producing PUFAs in plants which have traditionally been a source of natural oils (e.g., linseed oil, rapeseed oil). The technology to grow and extract oils from such plants in large scale already exists and the development of genetically modified plants which can produce PUFAs in a readily accessible and sustainable form for an increasing market is highly desirable.

Consumer-led demand for naturally occurring food colorings and antioxidants has resulted in increased interest in photosynthetic pigments and in particular the carotenoids which occur within marine microalgae (**Figure 3**). There is, for example, widespread use of the carotenoid astaxanthin; this is a pigmented antioxidant produced by many microalgae and is responsible for the red color often associated with crustaceans such as shrimps, crabs, and lobsters. Astaxanthin possesses an unusual antioxidant activity which has been implicated in a wide range of health benefits such as preventing cardiovascular disease, modulating the immune system as well as effects in cancer, diabetes, and ocular health. Its antioxidant activities may also have a neuroprotective effect. It has been used extensively in the feed of farmed fish as a nutritional supplement and is partly responsible for the strong coloration often observed in farmed salmon, a fish which naturally accumulates astaxanthin in the wild resulting in the pink hue of its flesh. Other carotenoids such as beta-carotene and lutein are used widely as food coloring and antioxidants. Inclusion, for example, in the diets of chickens leads to a darkening of the egg yolk resulting in a rich yellow color.

Another group of pigments of commercial importance is that of the phycobiliproteins; these are used as colorings, in cosmetics, and as fluorescent dyes for flow cytometry and in immunological assays. More recent research suggests that phycobiliproteins have anticancer and anti-inflammatory properties. A more unusual and unique pigment which is being used increasingly in personal care products and may also possess anticancer and anti-HIV activity is 'marennine', a blue-green pigment

(a)

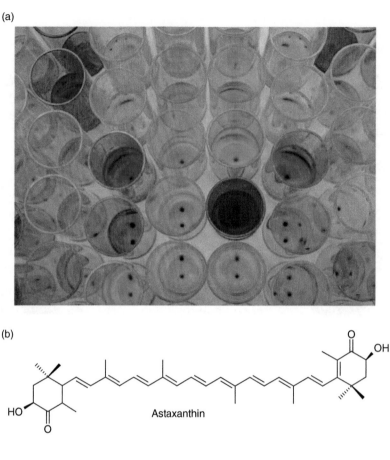

(b)

Astaxanthin

Lutein

B,B-carotene

Figure 3 (a) Any one species of microalgae contains an array of carotenoid pigments, as shown here with the fractionated isolates obtained from the chromatographic analysis of an *Emiliania huxleyi* extract. (b) Chemical structures of cartenoids widely used as color additives and antioxidants.

produced by *Haslea ostrearia*. Several species of microalgae and, in particular, *Haematococcus*, *Dunaliella*, and *Spirulina* are now grown on large commercial scale to accommodate the growing demand for natural pigments.

In addition to being rich in phycobiliproteins, *Spirulina*, a filamentous cyanobacterium, contains a wide variety of nutrients including potentially beneficial proteins, lipids, vitamins, and antioxidants.

Spirulina is also reported to have various beneficial effects including antiviral activity, immunomodulatory effects, and a role in modulating metabolic function in humans which could be of value in managing diseases involving lipids and carbohydrates such as diabetes. Furthermore, studies indicate that pretreatment with *Spirulina* may reduce the toxic side effects observed with some drugs on mammalian organs such as the heart and kidneys.

The marine environment is a rich source for naturally occurring antioxidants and pigments with a diverse range of microorganisms producing a unique and valuable resource. The potential for the discovery of new pigments and other additives which can be used to replace some of the existing artificial additives currently being used in the food industry is significant.

Sunscreens

The continuous exposure of marine organisms to strong sunlight has resulted in some, primarily the macro- and microalgae, evolving compounds which provide a very good screen against ultraviolet (UV) light. These organisms have the ability to synthesize small organic molecules, called mycosporine-like amino acids (MAAs), which are capable of absorbing UV light very efficiently and thus prevent DNA damage. Over 20 different MAAs occur in nature with a wide range of marine organisms utilizing them, including corals, anemones, limpets, shrimp, sea urchins, and some vertebrates including fish and fish eggs. MAAs are widely distributed across the marine environment; however, they can only be synthesized by certain types of bacteria or algae. For example, red seaweeds and some bloom-forming phytoplankton species are a particularly rich source of MAAs. Studies have revealed that in addition to their screening ability, some MAAs, such as mycosporine-glycine, have antioxidant properties. The ability of naturally occurring compounds such as MAAs to act as effective sunscreens has resulted in some interest from the commercial sector as to the value of these compounds in creams and cosmetics.

Biocatalysis

The ability of enzymes to synthesize complex chiral molecules with high efficiency and precision is of considerable interest within the pharmaceutical and chemical industries and marine bacteria present a new source for novel enzymes with not only unusual synthetic properties but also potentially valuable catalytic and structural properties. This arises from the ability of marine bacteria to grow under extreme conditions such as high and low temperature, high pressure (extreme depth), high salinity, and extremes of pH. This has opened up the potential to isolate naturally occurring small molecules which are difficult to synthesize in the laboratory but which could be of value in synthetic organic chemistry as intermediates. The existence of a large number of potentially novel enzymes in these same organisms

which are capable of performing diverse chemical modifications not readily amenable by standard chemical synthesis also opens the route to novel chemical modification of synthetic molecules using biocatalysis and biotransformation.

Growth of microbes under these extreme conditions has led to proteins which possess different temperature optima and improved stability, which has been exploited in the development of new processes and methods such as the polymerase chain reaction, a method for amplifying specific fragments of DNA, and which depends on a thermostable DNA polymerase isolated from a thermophilic microorganism. It has been suggested that enzymes which display high salt tolerance may be of value in the development of enzyme reactions to be performed in organic solvents as they appear to be less prone to denaturing under dehydrating conditions.

Marine bacteria make up the largest potential single source of novelty in the world's oceans and of these the major component are the actinobacteria which includes the actinomycetes. Actinomycetes are readily isolated from the marine environment and consequently are the best studied of the actinobacteria but the other more difficult to culture members are now being identified using advanced culturing and molecular techniques. The actinomycetes in particular hold the promise of tremendous diversity and to date have been underexploited. Terrestrial actinomycetes are responsible for about half of the known bioactive molecules isolated from natural sources to date and include antibiotics, antitumor compounds, immunosuppressants, and novel enzymes. Consequently, the isolation of new organisms from the environment and their analysis for novel metabolites has been a cornerstone in drug discovery. In recent years, however, terrestrial organisms have divulged less novelty than before, and advances in microbiology and genetics have now made the exploitation of marine-derived actinomycetes more attractive. The recognition that the world's oceans are rich in biological diversity and that extreme environmental conditions (e.g., high pressure and temperature at deep-sea hydrothermal vents) have not repressed the development of organisms to form distinct ecological niches suggests that these habitats will be a rich source of chemical novelty.

Although much emphasis has been placed on isolating organisms from extreme environments in the search for novel biocatalysts, the general marine environment should not be ignored. Both micro- and macroalgae have been demonstrated to produce novel enzymes with possible applications in biocatalysis such as the haloperoxidases, enzymes capable of introducing halogen atoms into metabolites. For example, two species of tropical red macroalgae produce halogenated compounds as a defense against predators and such compounds are being tested for medical applications. The availability of haloperoxidases with different catalytic functions would be of use in generating new types of halogenated molecules for the chemical and pharmaceutical industries.

The realization that viruses are the most abundant biological agents in the marine environment and the discovery of highly diverse, ancient, giant viruses with genomes comparable in size to the smallest microbes opens up new sources of genetic diversity. Current indications are that the oceans contain a wide variety of both DNA and RNA viruses with survival strategies which mimic those of terrestrial viruses yet these marine viruses encode a great many proteins of unknown function. Most marine viruses are assumed to be bacteriophages because virus particles are most commonly detected in the vicinity of bacteria, and bacteria are the most abundant organisms in the oceans.

Recent studies have revealed that marine viruses encode unexpected and novel proteins which would not be expected to occur within a virus genome. For example, the giant algal viruses have been shown to encode novel glycosylases, potassium pumps, and a pathway for the synthesis of complex sphingolipids. This biochemical diversity indicates that marine viruses could be a rich source for exploitation in the future for new types of carbohydrate and lipid as well as new proteins and enzymes.

Bioremediation

Pollution of the marine environment is a growing concern particularly with the continuous discharge of both industrial and domestic waste into rivers and estuaries leading to concerns about the impact such pollution could have on long-term human health. The discovery of marine microorganisms capable of detoxifying heavy metals and utilizing complex hydrocarbons as an energy source has provided a new impetus to develop natural solutions to the problems of environmental pollution. However, it should be remembered that toxic substances are not the only causes of marine distress and that the utilization of fertilizers and the disposal of sewage can also result in an imbalance in the marine ecology, resulting in the formation of large, often toxic, algal blooms which although not always a direct threat to human health do lead to widespread ecological damage. Thus, the discovery of microbes capable of growing in the presence of high concentrations of

ammonia could be of value in the treatment of wastewater, and an understanding of the anaerobic oxidation of ammonia could lead to the development of new chemical processes. The same organisms also possess unusual metabolic intermediates such as hydrazine and produce unusual lipids which could also be of value in the search for new chemical intermediates.

Heterocyclic molecules containing sulfur, nitrogen, and oxygen are among the most potent pollutants and inevitably contaminate the marine environment to a considerable extent. The use of microbes to degrade and detoxify such compounds is gaining considerable interest as a process which is environmentally friendly and would represent a long-term solution to removing heterocyclic contaminants. A particularly rich source of such organisms is the marine environment where growth in close proximity to sulfur-rich hydrothermal vents or adjacent to hydrocarbon (oil) seeps on the ocean floor has produced a plethora of microorganisms with metabolisms adapted to the utilization of a wide variety of carbon-, nitrogen-, and sulfur-based chemistries. Again, these organisms are also a very rich source of enzymes with previously unknown characteristics such as unusual substrate specificity, which could be of great value to the chemicals industry where they could be utilized in the production of new or difficult-to-synthesize compounds because they can perform reactions which are difficult to duplicate using traditional synthetic chemistry methods.

Microbial Fuel Cells and Biofuels

The use of marine organisms to produce fuels has also been proposed. The generation of electricity through the degradation of organic matter has recently been demonstrated to occur in marine sediments and may be mediated by complex communities of marine microorganisms. These organisms degrade complex organic matter such as carbohydrates and proteins to simpler molecules such as acetate which are then used by electricity-generating bacteria to reduce metals such as iron and manganese. By replacing the naturally occurring metals with an anode these bacteria, under anoxic conditions, will supply the electrons needed to produce an electric current to a cathode linked to the anode by wires and exposed to the oxygen in the water column. It has been suggested that this type of system could be used to supply the electricity needed to operate equipment in regions where access is difficult and so eliminate the need to replace batteries. Microbial fuel cells would be self-sustaining, would not require the preprocessing of fuels to function efficiently, and would not contribute net CO_2 to the atmosphere nor produce toxic waste as with conventional batteries.

Another area of intense study is the development of renewable biofuels with much focus being given to developing terrestrial plant species to produce the precursors to biodiesel. Microalgae present a potential alternative source of hydrocarbons for the generation of biofuels as some species naturally produce significantly more oil (per year per unit area of land) than terrestrial oil seed crops. Several marine species such as *Porphyridiuim*, *Chlorella*, and *Tetraselmis* are currently under investigation as sources rich in hydrocarbons suitable for biofuel production.

Biomaterials

Another area of interest is the development of novel biomaterials inspired by marine organisms. Areas of particular interest are the mechanism of calcium- and silica-based structure formation which is found in many phytoplankton, formation of hard chitinous shells in many larger marine organisms such as oysters and crabs, as well as the very powerful bioadhesives produced by mussels and barnacles. Proteins form the basis of a number of naturally occurring adhesive molecules which display a number of attractive features, particularly for clinical and other specialist applications. These features include the ability to adhere strongly to both smooth and uneven surfaces with a high degree of bonding strength and the ability to form and maintain bonds in very humid and wet conditions. This bonding ability seems to be strongly linked to the presence of hydroxylated tyrosine residues (L-dopa; L-3,4-dihydroxyphenylalanine) in such proteins and it is thought that adhesion involves interactions between the hydroxyl groups and the target surface. The development of powerful adhesives which can cure rapidly under wet conditions and are nontoxic would be of particular value in the clinic. At present there is considerable interest in using bioadhesives in the field of ophthalmology where the use of alternatives to sutures in, for example, corneal grafts is desired in order to reduce the risks of irritation and scarring to the eye following surgery. Another area where the use of bioadhesives is being actively researched is in drug delivery where the ability to attach naturally occurring polymers which can slowly release a drug over time would be useful. This is particularly relevant for poorly soluble biological drugs based on antibodies and other large proteins which can be difficult to administer. The potential contribution that marine-derived biomaterials and bioadhesives could make to such fields is enormous.

Chitin is the second most abundant natural polysaccharide after cellulose and is found in the exoskeletons of crustaceans such as crabs and shrimp as well as in the cell walls of fungi and cuticles of insects. The deacetylation of chitin produces chitosan, a biopolymer with great potential in medicine. Chitosan and its derivatives possess numerous applications due to their properties which include reactive functional groups, gel-forming capability, low toxicity, and high adsorption capacity, as well as complete biodegradability and antibacterial and antifungal activities. These properties make chitosan particularly attractive in areas of research such as drug delivery and tissue engineering where a nontoxic, biodegradable scaffold with antimicrobial activity would be particularly attractive. Both chitin and chitosan can influence the immune system and are being studied extensively as biomaterials in the development of supports for accelerated wound healing. These chitosan-based materials are also being modified to improve adhesion to wound sites and for the incorporation of antimicrobial agents to minimize the risk of infection. Chitosan and its derivatives are also being used to develop scaffolds for applications in tissue engineering to grow cells to form complex structures which could ultimately be used to replace damaged tissues and organs.

The elaborate silica-based structures (frustules) which are exhibited by many diatoms have been of interest to materials scientists for many years and recent studies have begun to reveal some of the characteristics that are present in these silica shells, including an understanding of the proteins and other molecules involved in structure formation (**Figure 4**). The highly precise nature of the structures has led to suggestions that the silica structures can be used directly as either templates for microfabrication or as materials for use in microprocesses such as filters in microfluidics. By understanding and manipulating the growth environment of any given diatom it may be possible to modify the precise geometry of the natural silica shells it produces and the resultant frustules could then be modified using standard microengineering techniques to create new nanostructures with potential applications in the development of medical devices. The glass-like properties of diatom frustules, the remains of which form diatomite (diatomaceous earth), have over 300 recorded commercial applications. The fine pores present in the frustules make them especially useful in filtration processes and the bulk of diatomite is used for this purpose. It has also been suggested that frustules might have applications in the development of new optical devices.

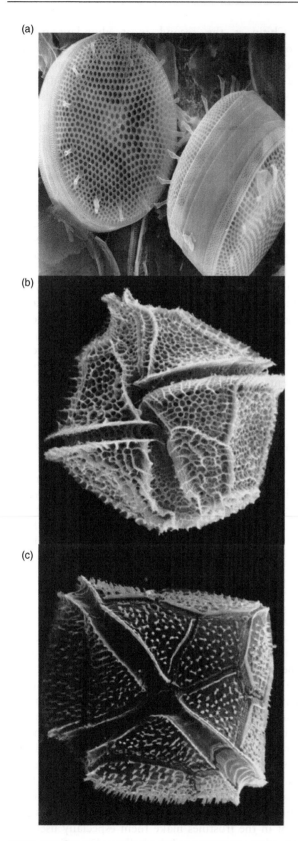

(a)

(b)

(c)

Figure 4 Scanning electron microscope images of (a) the diatom *Thalassiosira*, (b) the dinoflagellate *Gonyualax,* and (c) the dinoflagellate *Peridinium* showing the intricacies of cell walls as inspirations to novel biomaterials.

Future Prospects

The marine environment, representing 70% of the Earth's surface, is a vast untapped resource for new chemicals and enzymes, often with characteristics which are considerably different to anything discovered in the terrestrial environment. A major hurdle to the study and exploitation of this resource has been the inaccessibility of the oceans but advances in science and technology are now providing new approaches to isolating and characterizing the organisms present. This should lead to a considerable increase in the number of novel marine-derived chemicals and medicines available to the market in the near future.

Glossary

actinomycetes   Major group of bacteria commonly isolated from low-nutrient environments. Rich source of unusual small molecules (e.g., antibiotics) and enzymes.

cyanobacteria   Aquatic photosynthetic bacteria widely found throughout nature.

macroalga(e)   Seaweed(s).

microalga(e)   Microscopic single-cell plants.

nutraceuticals   Natural chemicals, usually contained in foods, with potential benefits to human health.

See also

Global Marine Pollution. Marine Biotechnology. Pollution: Approaches to Pollution Control. Pollution, Solids. Radioactive Wastes. Thermal Discharges and Pollution.

Further Reading

Gullo VP, McAlpine J, Lam KS, Baker D, and Petersen F (2006) Drug discovery from natural products. *Journal of Industrial Microbiology and Biotechnology* 33: 523–531.

Higuera-Ciapara I, Félix-Valenzuela L, and Goycoolea FM (2006) Astaxanthin: A review of its chemistry and applications. *Critical Reviews in Food Science and Nutrition* 46: 185–196.

Hussein G, Sankawa U, Goto H, Matsumoto K, and Watanabe H (2006) Astaxanthin, a carotenoid with potential in human health and nutrition. *Journal of Natural Products* 69: 443–449.

Khan Z, Bhadouria P, and Bisen PS (2005) Nutritional and therapeutic potential of *Spirulina*. *Current Pharmaceutical Biotechnology* 6: 373–379.

König GM, Kehraus S, Seibert SF, Abdel-Lateff A, and Müller D (2005) Natural products from marine organisms and their associated microbes. *ChemBioChem* 7: 229–238.

Lovley DR (2006) Microbial fuel cells: Novel microbial physiologies and engineering approaches. *Current Opinion in Biotechnology* 17: 327–332.

Marszalek JR and Lodish HF (2005) Docosahexaenoic acid, fatty acid-interacting proteins, and neuronal function: Breastmilk and fish are good for you. *Annual Review of Cell and Developmental Biology* 21: 633–657.

Napier JA and Sayanova O (2005) The production of very-long-chain PUFA biosynthesis in transgenic plants: Towards a sustainable source of fish oils. *Proceedings of the Nutritional Society* 64: 387–393.

Newman DJ and Hill RT (2005) New drugs from marine microbes: The tide is turning. *Journal of Industrial Microbiology and Biotechnology* 33: 539–544.

Shi C, Zhu Y, Ran X, Wang M, Su Y, and Cheng T (2006) Therapeutic potential of chitosan and its derivatives in regenerative medicine. *Journal of Surgical Research* 133: 185–192.

Shick JM and Dunlap WC (2002) Mycosporine-like amino acids and related gadusols: Biosynthesis, accumulation, and UV-protective functions in aquatic organisms. *Annual Review of Physiology* 64: 223–262.

Suttle CA (2005) Viruses in the sea. *Nature* 437: 356–361.

Wilt FH (2005) Developmental biology meets materials science: Morphogenesis of biomineralized structures. *Developmental Biology* 280: 15–25.

MARITIME ARCHAEOLOGY

R. D. Ballard, Institute for Exploration, Mystic, CT, USA

Introduction

For thousands of years, ancient mariners have traversed the waters of our planet. During this long period of time, many of their ships have been lost along the way, carrying their precious cargo and the history it represents to the bottom of the sea. Although it is difficult to know with any degree of precision, some estimate that there are hundreds of thousands of undiscovered sunken ships littering the floor of the world's oceans.

For hundreds of years, attempts have been made to recover their contents. In *Architettura Militare* by Francesco de Marchi (1490–1574), for example, a device best described as a diving bell was used in a series of attempts to raise a fleet of 'pleasure galleys' from the floor of Lake Nemi, Italy in 1531. In *Treatise on Artillery* by Diego Ufano in the mid-1600s, a diver wearing a crude hood and air-hose of cowhide was shown lifting a cannon from the ocean floor. Clearly, these early attempts at recovering lost cargo were done for economic, not archaeological, reasons and were very crude and destructive.

The field of maritime archaeology, on the other hand, is a relatively young discipline, emerging as a recognizable study in the later 1800s. Not to be confused with nautical archaeology which deals solely with the study of maritime technology, maritime archaeology is much broader in scope, concerning itself with all aspects of marine-related culture including social, religious, political, and economic elements of ancient societies.

Early History

Sunken ships offer an excellent opportunity to learn about those ancient civilizations. Archaeological sites on land can commonly span hundreds to even thousands of years with successive structures being built upon the ruins of older ones. Correlating a find in one area to a similar stratigraphic find in another can introduce errors that potentially represent long periods of time. A shipwreck, on the other hand, represents a 'time capsule', the result of a momentary event where the totality of the artifact assemblage comes from one distinct point in time.

It is important to point out that maritime archaeology's first major field efforts were not conducted underwater but, in fact, were the excavation of boats that are now located in land-locked sites. In 1867, the owners of a farm began to cart away the soil from a large mound some 86 m in length only to discover the timbers of a large Viking ship, the Tune ship, complete with the charred bones of a man and a horse revealing it to be a burial chamber. In 1880, the Gokstad burial ship was discovered in a flat plain on the west side of the Oslo fiord. It was buried in blue clay which resulted in a high state of preservation. Contained within the grave was a Viking king, his weapons, twelve horses, six dogs and various artifacts.

Since the late 1890s, the excavation of boats and harbor installations in terrestrial settings continues to this day, following more or less traditional land excavation protocol. One of the most famous discoveries took place in 1954 near the Great Pyramid of Egypt. While building a new road around the Pyramid, a series of large limestone blocks were encountered beneath which was found an open pit containing the oldest intact ship ever discovered. Dating to 2600 BC, the ship measures 43 m in length, weighs 40 tons, and stands 7.6 m from the keel line.

Although ships found in terrestrial settings provide valuable insight into the culture of their period, many have a more religious significance than one reflecting the economics of the period. Burial ships were commonly modified for this unique purpose and were not engaged in maritime activity at the time of their burial.

Prior to the advent of SCUBA diving technology invented by Captain Jacques-Yves Cousteau and Emile Gagnan in 1942, archaeology conducted underwater by trained archaeologists was extremely rare. In fact, owing to the dangerous aspects of pre-SCUBA diving techniques which included the use of 'hard hats' employed by commercial sponge divers, archaeologists relied upon these commercial divers instead of doing the underwater work themselves.

Even after SCUBA technology became available to the archaeological community it was the commercial or recreational divers by their numbers who tended to discover an ancient shipwreck site. As a result, it was not uncommon for a site to be stripped of its small, unique and easily recovered artifacts before the 'authorities' were notified of a wreck's location.

Even after learning of these sites, many of the early efforts in underwater archaeology were conducted by divers lacking archaeological training.

The first was the famous *Antikythera* wreck off Greece which contained a cargo of bronze and marble sculptures. The site was initially raided in 1900 by sponge divers wearing copper helmets, lead weights, and steel-soled shoes with air supplied to them from the surface by a hand-cranked compressor.

Fortunately, the location of the wreck was soon discovered by the Greek government which with the help of their navy, mounted a follow-up expedition under supervision from the surface by Professor George Byzantinos, their Director of Antiquities, which resulted in the recovery of more of its valuable cargo. The wreck turned out to be a first century Roman ship carrying its Greek treasures back to Rome including the famous bronze statue *Youth*, thought to be done by the last great classical Greek sculptor, Lysippos.

A few years later, another Roman argosy was discovered off Mahdia, Tunisia which was also initially plundered for its Greek statuary. Again, the local government and their Department of Antiquities took control of the salvage operation which continued from 1907 to 1913 enriching the world with its bronze artwork.

Unfortunately, not all of these early shipwreck discoveries were reported to the local government. Their fate was far less fortunate than those just mentioned including a first century BC wreck lost off Albenga, Italy which was torn apart by the Italian salvage ship *Artiglio II* using bucket grabs to penetrate its interior holds.

One of the first ancient shipwrecks to be excavated under some semblance of archaeological control was carried out by Captain Cousteau in 1951 along with Professor Fernand Benoit, Director of Antiquities for Provence off Grand Congloue near Marseilles, France. It, too, was initially discovered by a salvage diver. Although no site plan was ever published after three seasons of diving, the ship was thought to be Roman from 230 BC, 31 m long, at a depth of 35 m, and carrying 10 000 amphora and 15 000 pieces of Campanian pottery bowls, pots, and 40 types of dishes with an estimated 20 tons of lead aboard.

At the same time, a colleague of Cousteau, Commander Philippe Taillez of the French Navy organized a similar excavation of a first century BC wreck on the Titan reef off the French coast but in the absence of an archaeologist failed to actually document the site. Without the presence of trained archaeologists working underwater, there was little hope that acceptable techniques would be developed that met archaeological standards.

In 1958, Peter Throckmorton who was the Assistant Curator of the San Francisco Maritime Museum, went to Turkey hoping to locate an ancient shipwreck site. Like many before him, he learned from local sponge divers the location of a wreck off Cape Gelidonya on the southern coast of Turkey from which a series of broken objects had been recovered. During the following year, Throckmorton's team made a number of dives on the site and recovered a series of bronze tools and ingots revealing the ship to be a late Bronze Age wreck from around 1200 BC but no major effort was mounted.

Throckmorton and veteran diver Frederick Dumas, however, returned in 1960 along with a young archaeologist eager to make his first open-ocean dives, George Bass. With Bass in charge of the archaeological aspects of the diving program, they began to establish for the first time true archaeological mapping and sampling protocol. Although primitive by today's standard, they established a traditional grid system using anchored lines followed by the use of an airlift system to carefully remove the overburden covering the wreck. Working in 28 m of water, the ship proved to be 11 m in length, covered with a coralline concretion some 20 cm thick. Its more than one ton of cargo consisted of four-handled ingots of copper and 'bun' ingots of bronze as well as a large quantity of broken and unfinished tools, including both commercial and personal goods. These assemblages of artifacts suggested the captain was a Syrian scrap-metal dealer who was also a tinker making his way from Cyprus along the coast of Turkey to the Aegean Sea.

Maritime archaeology truly came of age with the excavation of the Yassi Adav wreck from 1961 to 1964 off the south coast of Turkey. Running offshore near the small barren island of Yassi Adav is a reef which is as shallow as 2 m that has claimed countless ships over the years. Its surface is strewn with Ottoman cannon balls from various wrecks of that period. Several of the ships that have run aground on its unforgiving coral outcrops slide off its rampart coming to rest on a sandy bottom. One such ship, lying in 31 m of water was the focus of an intense effort carried out by the University Museum of Pennsylvania under the direction of George Bass.

Prior to this effort, no ancient ship had ever been recovered in its entirety; this became the objective of this project. With the help of fifteen specialists and thousands of hours underwater, the team carefully mapped the site. The techniques developed during this excavation effort became the new emerging standards for this young field of research and continue to be followed today by research teams around the world.

Bass' team first cleared the upper surface of cargo of its encrustation of weed. Its 900 amphora were then mapped, cataloged, and removed. One hundred were taken to the surface for subsequent conservation and preservation whereas the remaining 800 were stored on the bottom at an off-site location. Simple triangulation was first used to initially delineate the wreck site followed by the establishment of a complex series of wire grids. Each object was given a plastic tag and artists hovered over the grid system making numerous sketches of the site before any object was recovered.

Copper and gold coins recovered from the site revealed the ship to be of Byzantine age sinking during the reign of Emperor Heraclius from AD. 610 to 641. Following the recovery of the ship's contents, the now-exposed hull provided marine archaeologists with their first opportunity to develop techniques needed to document and recover the ship's timbers.

This effort was extremely time consuming but the resulting insight proved worth the effort, providing the archaeological community with a transitional method of ship construction between the classical 'shell' technique to the later evolved 'skeleton' technique. Its length to width ratio of 3.6 to 1 further supported earlier suggestions that ships of this period would have to be built with more streamlined hulls to outrun and outmaneuver hostile or piratical adversaries.

The Growth of Maritime Archaeology

Following the excavation of the Yassi Adav wreck, members of Bass' team then conducted the extensive excavation of a fourth century BC wreck near Kyrenia in Cyprus directed by Michael Katzev. This effort mirrored the Yassi Adav project and resulted in the raising and conservation of the ship's preserved hull structure.

Since these early pioneering efforts, numerous maritime archaeology programs have emerged around the world. Off Western Europe and in the Mediterranean maritime archaeology remains a strong focus of activities. Bass and his Institute for Nautical Archaeology (Texas A&M University in College Station, Texas) continue to carry out a growing number of underwater research projects off the southern coast of Turkey centered at their research facility in Bodrum.

His excavation of the late Bronze Age Ulu Burun wreck off the southern Turkish coast led to the recovery of thousands of artifacts that provided valuable insight in a period of time marked by the reign of Egypt's Tutankhamun and the fall of Troy. His Byzantine 'Glass Wreck' found north of the island of Rhodes and dating from the twelfth to thirteenth century AD, continues to generate important information about this particular period of maritime trade.

In addition to the well-known work with the Wasa, Swedish archaeologists have conducted excellent work in the worm-free waters of the Baltic. In Holland, Dutch archaeologists have drained large sections of shallow water areas which contain a rich history of maritime trade dating from the twelfth to nineteenth centuries.

Another major program directed by Margaret Rule took place in England with the recovery and preservation of the *Mary Rose*, a large warship lost in July 1545 during the reign of King Henry VIII. From this project, a great deal was learned about the long-term preservation of wooden timbers which is being incorporated in other similar conservation programs.

Research efforts in America span the length of its human habitation. Recent archaeological research suggests that humans arrived in North America more than 12 000 years ago when a southern route was first thought to have opened in the glacial icesheet covering the continent. Some scientists now suggest that early humans may have circumvented this barrier by way of water or overland surfaces now submerged on the continental shelf. New research programs are now being designed to work on the continental shelf looking for early evidence of Paleoindian settlements.

For years Indian canoes, rafts, dugouts, and reed boats have been discovered in freshwater lakes, and sinkholes in the limestone terrains of North and Central America have attracted researchers for many years in search of human sacrifices and other religious artifacts associated with native American cultures.

Ships associated with early explorers, including Columbus, French explorer Rene La Salle, the British, and countless Spanish explorers, have been the focus of research efforts in the Gulf of Mexico, the Northwest Passage, and the Caribbean while shipwrecks from the Revolutionary War and War of 1812 have been discovered in the Great Lakes and Lake Champlain. Warships associated with the American Civil War have received renewed interest including the *Monitor* lost off Cape Hatteras, the submarine *Huntley* and numerous other recent finds in the coastal waters of the US east coast.

Within the last two decades, deep water search systems developed by the oceanographic community have been used to successfully locate the remains of the RMS *Titanic*, the German Battleship *Bismarck*, fourteen warships lost during the Battle of Guadalcanal, and the US aircraft carrier *Yorktown* lost during the Battle of Midway.

Ships associated with World War II have been carefully documented including the *Arizona* in Pearl Harbor and numerous ships sunk during nuclear bomb testing in the atolls of the Pacific. Maritime archaeology is not limited to European or American investigators. A large number of underwater sites too numerous to mention have been investigated off the coasts of Africa, the Philippines, the Persian Gulf, South America, China, Japan, and elsewhere around the world as this young field begins to experience an explosive growth.

Marine Methodologies

As was previously noted, early underwater archaeological sites were not discovered by professional archaeologists; they were found, instead, by commercial or recreational divers. It wasn't until the mid-1960s that archaeologists, notably George Bass, began to devise their own search strategies. Being divers, their early attempts tended to favor visual techniques from towing divers behind their boats, to towed camera systems, and finally small manned submersibles (**Figure 1**).

It was not until the introduction of side-scan sonars that major new wreck sites were found. Operating at a frequency of 100 kHz, such sonar systems are able to search a swath-width of ocean floor 400 m wide, moving through the water at a typical speed of 3–5 knots (5.5–9 km h^{-1}). Today, numerous companies build side-scan sonars each offering a variety of options ranging from higher frequencies (i.e. 500 kHz) to improved signal processing, recording, and display.

Various magnetic sensors have been used effectively over the years in locating sunken shipwreck sites having a ferrous signature. This is particularly true for warships with large cannons aboard. Magnetic sensors have also proved effective in locating buried objects in extremely shallow water, on beaches, and beneath coastal dunes.

Over the years, a variety of changes have taken place with regard to the actual documentation of a

Figure 1 Archaeological mapping techniques pioneered by Dr George Bass of Texas A&M University for shallow water archaeology. These techniques were heavily dependent on the use of divers and were limited to less than 50 m water depths.

wreck site. Beginning in the early 1960s, various stereophotogrammetry techniques were used. More recently the SHARPS (sonic high accuracy ranging and positioning system) acoustic positioning system has proved extremely rapid and cost effective in accurately mapping submerged sites. This tracking technique coupled with electronic imaging sensors, has produced spectacular photomosaics.

More recently, remotely operated vehicles have begun to enter this field of research. In 1990, the JASON vehicle from the Woods Hole Oceanographic Institution was used to map the *Hamilton* and *Scourge*, two ships lost during the War of 1812 in Lake Ontario. Using a SHARPS tracking system, the vehicle was placed in closed-loop control and made a series of closely spaced survey lines across and along the starboard and port sides of the ships. Mounted on the remotely operated vehicle (ROV) was a pencil-beam sonar and electronic still camera which resulted in volumetric models of the ships as well as electronic mosaics.

Deep-water Archaeology

The shallow waters of the world's oceans where the vast majority of maritime archaeology has been done represent less than 5% of its total surface area. For years, archaeologists have argued that the remaining 95% is unimportant since the ancient mariner stayed close to land and it was there that their ships sank.

This premise was challenged in 1988 when an ancient deep-water trade route was first discovered between the Roman seaport of Ostia and ancient Carthage. The discovery site was situated more than 100 nautical miles (185 km) off Carthage in approximately 1000 m of water. Over a nine-year period from 1988 to 1997, a series of expeditions resulted in the discovery of the largest concentration of ancient ships ever found in the deep sea. In all, eight ships were located in an area of 210 km^2, including five of the Roman era spanning a period of time from 100 BC to AD 400. The project involved the use of highly sophisticated deep submergence technologies including towed acoustic and visual search vehicles, a nuclear research submarine, and an advanced remotely operated vehicle. Precision navigation and control, similar to that first used in Lake Ontario in 1990, permitted rapid yet careful visual and acoustic mapping of each site with a degree of precision never attained before utilizing advanced robotics, the archaeological team recovered selected objects from each site for subsequent analysis ashore without intrusive excavation.

Deep-water wreck sites offer numerous advantages over shallow water sites. Ships lost in shallow water commonly strike an underwater obstacle such as rocks or reefs severely damaging themselves in the process. Each winter more storms continue to damage the site as encrustation begins to form. Commonly, the site is located by nonarchaeologists who frequently retrieve artifacts before reporting the wreck's location.

Ships that sink in deep water, however, tend to be swamped. As a result, they sink intact, falling at a slow speed toward the bottom where they come to rest standing upright in the soft bottom ooze. When they are located, they have not been looted. Sedimentation rates in the deep sea are extremely slow, commonly averaging 1 cm per 1000 years. That coupled with cold bottom temperatures, total darkness, and high pressures result in conditions favoring preservation. Although wood-boring organisms remove exposed wooden surfaces, deep sea muds encase the lower portions of the wreck in an oxygen free environment. When deep-sea excavation techniques are developed in the near future, these wrecks may provide highly preserved organic material normally lost in shallow-water sites.

The Roman shipwrecks located off Carthage were found within a much larger area of isolated artifacts spanning a longer period of time. The isolated artifacts appear to have been discarded from passing ships overhead. Given the slow sedimentation rates in the deep sea, it might be possible to easily delineate ancient trade routes by looking for these debris trails, thus learning a great deal about ancient maritime trading practices.

Since this new field of deep-water archaeology has grown out of the oceanographic community, it brings with it a strong expertise in deep submergence technology. The newly developed ROVs possess the latest in advanced imaging, robotics, and control technologies. Using this technology, archaeologists are able to map underwater sites far faster than their shallow water counterparts (**Figure 2**).

Most recently, a second deep-water archaeological expedition resulted in the discovery of two Phoenician ships lost some 2700 years ago. Located in 450 m of water about 30 nautical miles (55 km) off the coast of Egypt, these two ships are lying upright. Due to local bottom currents, both ships are completely exposed resting in two-meter deep elongated depressions.

Ethics

As pointed out earlier, the salvaging of cargo from lost ships goes back much farther in time than marine archaeological research. As a result, this long

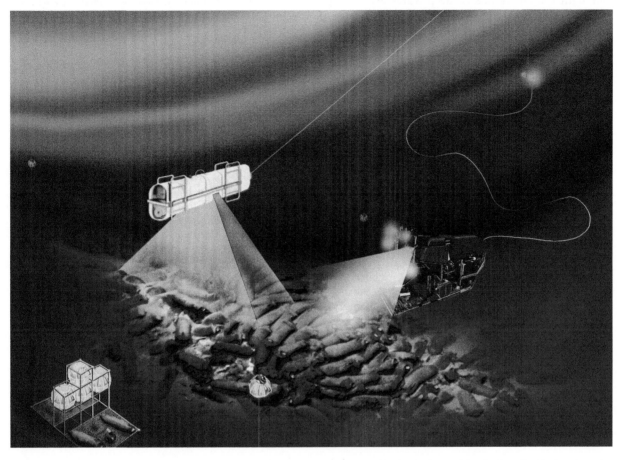

Figure 2 Archaeological techniques pioneered at the Woods Hole Oceanographic Institution and the Institute for Exploration rely exclusively on remotely operated vehicle systems with operating depths down to 6000 m.

history of maritime salvage, rooted in international law, has led to a quasipublic acceptance of salvage operations, making it difficult for the archaeological community to garner moral and legal public support to protect and preserve truly important underwater archaeological sites. 'Finders keepers' remains rooted in the public's mind as a logical policy governing lost ships. Further blurring the boundary between these two extremes in the early years was the fact that marine archaeologists relied upon the very community that was removing artifacts from underwater sites to tell them where they were located.

This uneasy marriage between the diving community and the archaeological community has, in many ways, stifled the growth and acceptance of the field. Its lack of development of systematization which arises from its immaturity and lack of a large database has further hindered its acceptance into mainstream archaeology.

Today's marine salvagers commonly employ individuals with archaeological experience to participate in their operations. In some cases, this results in important documentation of the site as was the case with the salvage of the *Central America*. In other cases, however, they are being used to create a false impression that archaeological standards are being followed when they are not.

American salvagers have, in large part, concentrated their attention on lost ships of the Spanish Main beginning with search efforts off the coast of Florida where a large number of silver- and gold-bearing ships were lost in hurricanes between 1715 and 1733. A famous shipwreck in this area, the *Atocha*, was exploited by salvager Mel Fisher.

On the Silver Bank off the Dominican Republic in the Caribbean the richly laden *Nuestra Senora de la Pura y Limpia Concepcion* sank in October 1641. Salvage efforts seeking to retrieve its valuable cargo began almost immediately including one by the British in 1687. Lost from memory, the *Concepcion* was relocated in 1978 by American treasure hunters who continue their recovery efforts to this day.

Fortunately, more and more countries are beginning to enact laws to protect offshore cultural sites, but with the emergence of deep-water archaeology which is conducted on the high seas, the majority of the world's oceans and the human history contained within them are not protected.

One logical step is to add human history to the present Law of the Sea Convention that governs the exploitation of natural resources. Although this would not protect all future underwater sites, it would serve as an important first step.

Further Reading

Babits LE and Tilburg HV (1998) *Maritime Archaeology.* New York: Plenum Press.

Ballard RD (1993) The MEDEA/JASON remotely operated vehicle system. *Deep-Sea Research* 40(8): 1673–1687.

Ballard RD, McCann AM, Yoerger D, Whitcomb L, Mindell D, Oleson J, Singh H, Foley B, Adams J, Piechota D, and Giangrande C (2000) The discovery of ancient history in the deep sea using advanced deep submergence technology. *Deep Sea Research, Part I* 47: 1591–1620.

Bass GF (1972) *A History of Seafaring Based on Underwater Archaeology.* New York: Walker.

Bass GF (1975) *Archaeology Beneath the Sea.* New York: Walker.

Bass GF (1988) *Ships and Shipwrecks of the Americas.* New York: Thames and Hudson.

Cockrell WA (1981) Some moral, ethical, and legal considerations in archaeology. In: Cockrell WA (ed.) *Realms of Gold.* Proceedings of the Tenth Conference on Underwater Archaeology, Fathom Eight San Marino, California. pp. 215–220.

Dean M and Ferrari B (1992) *Archaeology Underwater: The NAS Guide to Principles and Practice.* London: Nautical Archaeology Society.

Greene J (1990) *Maritime Archaeology.* London: Academic Press.

Muckelroy K (1978) *Maritime Archaeology.* New York: Cambridge University Press.

NESCO (1972) *Underwater Archaeology – A Nascent Discipline.* Paris: UNESCO.

MARINE POLLUTION

GLOBAL MARINE POLLUTION

A. D. McIntyre, University of Aberdeen, Aberdeen, UK

This aritcle is designed to provide an entry to the historical and global context of issues related to pollution of the oceans.

Until recently, the size and mobility of the oceans encouraged the view that they could not be significantly affected by human activities. Freshwater lakes and rivers had been degraded for centuries by effluents, particularly sewage, but, although from the 1920s coastal oil pollution from shipping discharges was widespread, it was felt that in general the open sea could safely dilute and disperse anything added to it. Erosion of this view began in the 1950s, when fallout from the testing of nuclear weapons in the atmosphere resulted in enhanced levels of artificial radionuclides throughout the world's oceans. At about the same time, the effluent from a factory at Minamata in Japan caused illness and deaths in the human population from consumption of mercury-contaminated fish, focusing global attention on the potential dangers of toxic metals. In the early 1960s, buildup in the marine environment of residues from synthetic organic pesticides poisoned top predators such as fish-eating birds, and in 1967 the first wreck of a supertanker, the *Torrey Canyon*, highlighted the threat of oil from shipping accidents, as distinct from operational discharges.

It might therefore be said that the decades of the 1950s and 1960s saw the beginnings of marine pollution as a serious concern, and one that demanded widespread control. It attracted the efforts of national and international agencies, not least those of the United Nations. The fear of effects of radioactivity focused early attention, and initiated the establishment of the International Commission on Radiological Protection (ICRP), which produced a set of radiation protection standards, applicable not just to fallout from weapons testing but also to the increasingly more relevant issues of operational discharges from nuclear reactors and reprocessing plants, from disposal of low-level radioactive material from a variety of sources including research and medicine, and from accidents in industrial installations and nuclear-powered ships. Following Minamata, other metals, in particular cadmium and lead, joined mercury on the list of concerns. Since this was seen, like radioactive wastes, as a public

health problem, immediate action was taken. Metals in seafoods were monitored and import regulations were put in place. As a result, metal toxicity in seafoods is no longer a major issue, and since most marine organisms are resilient to metals, this form of pollution affects ecosystems only when metals are in very high concentrations, such as where mine tailings reach the sea.

Synthetic organics, either as pesticides and antifoulants (notably tributyl tin (TBT)) or as industrial chemicals, are present in seawater, biota, and sediments, and affect the whole spectrum of marine life, from primary producers to mammals and birds. The more persistent and toxic compounds are now banned or restricted, but since many are resistant to degradation and tend to attach to particles, the seabed sediment acts as a sink, from which they may be recirculated into the water column. Other synthetic compounds include plastics, and the increasing use of these has brought new problems to wildlife and amenities.

Oil contaminates the marine environment mainly from shipping and offshore oil production activities. Major incidents can release large quantities of oil over short periods, causing immense local damage; but in the longer term, more oil reaches the sea via operational discharges from ships. These and other threats to the ocean are controlled by the International Convention for the Prevention of Pollution from Ships, administered by the International Maritime Organisation of the UN. Pollution is generated by human activities, and the most ubiquitous item is sewage, which is derived from a variety of sources: as a direct discharge; as a component of urban wastewater; or as sludge to be disposed of after treatment. Sewage in coastal waters is primarily a public health problem, exposing recreational users to pathogens from the local population. The dangers are widely recognized, and many countries have introduced protective legislation. At the global level, the London Dumping Convention controls, among other things, the disposal of sewage sludge. As well as introducing pathogens, sewage also contributes carbon and nutrients to the sea, adding to the substantial quantities of these substances reaching the marine environment from agricultural runoff and industrial effluents. The resulting eutrophication is causing major ecosystem impacts around the world, resulting in excessive, and sometimes harmful, algal blooms.

The need for a global approach to ocean affairs was formally brought to the attention of the United Nations in 1967, and over the next 15 years, while sectoral treaties and agreements were being introduced, negotiations for a comprehensive regime led in 1982 to the adoption of the UN Convention on the Law of the Sea (UNCLOS). This provided a framework for the protection and management of marine resources. In the ensuing years, concepts such as sustainable development and the precautionary approach came to the fore and were endorsed in the Declaration of the Rio Summit, while proposals for Integrated Coastal Zone Management (ICZM) are being widely explored nationally and advanced through the Intergovernmental Oceanographic Commission.

Early ideas of pollution were focused on chemical inputs, but following the Group of Experts on the Scientific Aspects of Marine Pollution (GESAMP) definition pollution is now seen in a much wider context, encompassing any human activity that damages habitats and amenities and interferes with legitimate users of the sea. Thus, manipulation of terrestrial hydrological cycles, and other hinterland activities including alterations in agriculture, or afforestation, can profoundly influence estuarine regimes, and are seen in the context of pollution. In particular, it is now recognized that excessive fishing can do more widespread damage to marine ecosystems than most chemical pollution, and the need for ecosystem-based fisheries management is widely recognized.

Over the years, assessments of pollution effects have altered the priority of concerns, which today are very different from those of the 1950s. Thanks to the rigorous control of radioactivity and metals, these are not now major worries. Also, decades of experience with oil spills have shown that, after the initial damage, oil degrades and the resilience of natural communities leads to their recovery. Today, while the effects of sewage, eutrophication, and harmful algal blooms top the list of pollution concerns, along with the physical destruction of habitats by coastal construction, another item has been added: aquatic invasive species. The particular focus is on the transfer of harmful organisms in ships' ballast water and sediments, which can cause disruption of fisheries, fouling of coastal industry, and reduction of human amenity. In conclusion, most of the impacts referred to above are on the shallow waters and the shelf, associated with continental inputs and activities. The open ocean, although subject to contamination from the atmosphere and from vessels in shipping lanes, is relatively less polluted United Nations Environment Programme (UNEP).

See also

Atmospheric Input of Pollutants. Eutrophication. Exotic Species, Introduction of. International Organizations. Marine Policy Overview. Oil Pollution. Pollution: Approaches to Pollution Control. Pollution, Solids. Radioactive Wastes. Thermal Discharges and Pollution. Viral and Bacterial Contamination of Beaches.

Further Reading

Brackley P (ed.) (1990) *World Guide to Environmental Issues and Organisations*. Harlow: Longman.

Brune D, Chapman DV, and Gwynne DW (1997) Eutrophication. In: *The Global Environment*, ch. 30. Weinheim: VCH.

Coe JM and Rogers DB (1997) *Marine Debris*. Berlin: Springer.

de Mora SJ (ed.) (1996) *Tributyltin Case Study of an Environmental Contaminant*. Cambridge, UK: Cambridge University Press.

GESAMP (1982) *Reports and Studies No. 15: The Review of the Health of the Oceans*. Paris: UNESCO.

Grubb M, Koch M, and Thomson K (1993) *The 'Earth Summit' Agreements: A Guide and Assessment*. London: Earth Scan Publications.

HMSO (1981) *Eighth Report of the Royal Commission on Environmental Pollution, Oil Pollution of the Sea, Cmnd 8358*. London: HMSO.

Hollingworth C (2000) Ecosystem effects of fishing. *ICES Journal of Marine Science* 57(3): 465–465(1).

International Maritime Organization (1991) *IMO MARPOL 73/78 Consolidated Edition MARPOL 73/78*. London: International Maritime Organization.

International Oceanographic Commission (1998) *Annual Report of the International Oceanographic Commission*. Paris: UNESCO.

Kutsuna M (ed.) (1986) *Minamata Disease*. Kunamoto: Kunammoyo University.

Matheickal J and Raaymakers S (eds.) (2004) Second International Ballast Water Treatment R & D Symposium, 21–23 July 2003: Proceedings. *GloBallast Monograph Series No. 15*. London: International Maritime Organization.

National Research Council (1985) *Oil in the Sea*. Washington, DC: National Academies Press.

Park PK, Kester DR, and Duedall IW (eds.) (1983) *Radioactive Wastes and the Ocean*. New York: Wiley.

Pravdic V (1981) *GESAMP the First Dozen Years*. Nairobi: UNEP.

Pritchard SZ (1987) *Oil Pollution Control*. Wolfeboro, NH: Croom Helm.

Sinclair M and Valdimarsson G (eds.) (2003) *Responsible Fisheries in the Marine Ecosystem*. Oxford, UK: CABI.

Tolba MK, El-Kholy OA, and El-Hinnawi E (1992) *The World Environment 1972–1992: Two Decades of Challenge*. London: Chapman and Hall.

UNEP (1990) The State of the Marine Environment. *UNEP Regional Seas Reports and Studies No. 115*. Nairobi: UNEP.

UNO (1983) *The Law of the Sea: Official Text of UNCLOS*. New York: United Nations.

World Commission on Environment and Development (1987) *Our Common Future*. Oxford, UK: Oxford University Press.

EUTROPHICATION

V. N. de Jonge, Department of Marine Biology,
Groningen University, Haren, The Netherlands
M. Elliott, Institute of Estuarine and Coastal Studies,
University of Hull, Hull, UK

Introduction

Eutrophication is the enrichment of the environment
with nutrients and the concomitant production of
undesirable effects, while the presence of excess nutrients *per se* is merely regarded as hypernutrification.
In more detail, eutrophication is

> the process of nutrient enrichment (usually by nitrogen and
> phosphorus) in aquatic ecosystems such that the product-
> ivity of the system ceases to be limited by the availability of
> nutrients. It occurs naturally over geological time, but may
> be accelerated by human activities (e.g. sewage disposal or
> land drainage).
>
> (Oxford English Dictionary)

Anthropogenic nutrient enrichment is important
when naturally productive estuarine and coastal systems receive nutrients from 'point sources', e.g.
as outfall discharges of industrial plants and
sewage treatment works, or human-influenced 'diffuse
sources', such as runoff from an agricultural catchment. Whereas point source discharges are relatively
easy to control, with an appropriate technology, diffuse
and atmospheric sources are more difficult and require
a change in agricultural and technical practice.

Coastal and estuarine areas with tide-associated
accumulation mechanisms for seaborne suspended
matter are organically productive by nature and
represent some of the world's most productive environments. This is the result of the freshwater outflow, and biogeochemical cycling within the
estuarine systems and adjacent shallow coastal seas.
About 28% of the total global primary production
takes place here, while the surface area of these
systems covers only 8% of the Earth's surface (**Figure 1**) and as such the effects of eutrophication are
most manifest in the coastal zone, including estuaries, areas which are the focus of this chapter. As
indicated below, there is generally a good qualitative
understanding of the processes operating, but the
quantitative influence on the ecological processes
and the changes in community structure are still not
well understood. Within the available field studies
attention has been focused on long data series, because time-series with a length of less than 10 years
have to be considered as too narrow a window of
time relative to natural meteorological and climatic
fluctuations influencing the ecosystem.

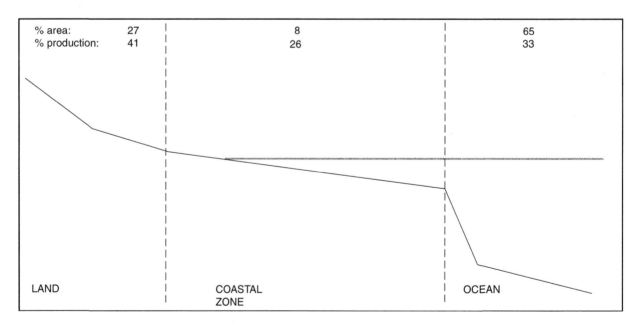

Figure 1 Indication of importance to primary production of different types of area (land, ocean, and shallow coastal seas and its
fringes). (Redrawn after LOICZ, 1993. Report no. 25, Science Plan, Stockholm.)

Nutrient inputs are required for the natural functioning of aquatic systems. Eutrophication merely indicates that the system cannot cope with the available inputs. This chapter focuses on the causes and mechanisms of eutrophication as well as the consequences. It gives examples varying from eutrophication caused by freshwater inflow to that caused mainlyby atmospheric inputs and nutrient import from the sea instead of land and atmosphere. It will be shown that increased organic enrichment may lead to dystrophication, the modification of bacterial activity leading ultimately to anoxia. Despite this, certain areas with low hydrodynamic energy conditions, such as lagoons, part of the estuaries and enclosed seas, can be considered as naturally organically enriched and thus require little additional material to make them eutrophic. In contrast, there are also naturally oligotrophic areas which drain poor upland areas and receive little organic matter.

The symptoms of eutrophication as a response to nutrient enrichment differ greatly, due mainly to differences in the physical characteristics of the different systems receiving this 'excess' organic matter and nutrients. For example, the mixing state (stratification or not) of the receiving body of water and theresidence time (or flushing time) of the fresh water and its nutrients in the system determine the intensity of a particular symptom and thereby the sensitivity of systems to eutrophication events or symptoms. For example, in the UK, the susceptibility of waters to enrichment and adverse effects caused by nutrients is interpreted according to their ability as high natural dispersing areas (HNDA). In general, this reflects the waters' assimilative capacity, i.e. capacity to dilute, degrade, andassimilate nutrients without adverse consequences.

Historical Background

Several attempts have been made to determine either the nutrient loads to coastal zones during the 'pre-development period' (i.e. before widespread human development) or to determine the 'natural background' concentrations of nutrients. These assessments are important, as they may improve our understanding of the way eutrophication in the past may havechanged and in the future will change aquatic systems. The term 'natural' is regarded here as those levels present before the large-scale production and use of artificial fertilizers and detergents started. It also covers the period just prior to the start of chemical monitoring of the aquatic environment. For example, the greatest increase innutrient loads to the Dutch coast and the Wadden Sea occurred after

the early 1950s, due to the introduction of artificial fertilizers and detergents.

Available data for Narragansett Bay (USA) suggest that the system presumably was nitrogen limited in thepast and that the total dissolved inorganic nitrogen (DIN) input has increased five-fold, while that of dissolved inorganic phosphorus (DIP) has increased two-fold (Table 1). During that period the nutrient input from thesea is assumed to have been more important than the supply from the drainage basin, a feature that in the early 1950s has also been postulated for European waters.

Natural background concentrations for someDutch and German rivers as well as the Wadden Sea are given in Table 1. The values represent the situation before the introduction of the artificial fertilizers and detergents. The 4- to 50-fold increase in the values in fresh waters and coastal waters is clear.

Structuring Elements and Processes

In addition to the inputs, the mechanisms and processes which influence the fate and effects of excess nutrient inputs will be considered. In coastal systems, eutrophication will influence not onlythe nutrient-related processes of the system but will also affect structural elements of the ecosystem (Figure 2).

Structuring Elements

The physical and chemical characteristics mainly create the basic habitat conditions and niches of the marine system to be colonized with organisms. These conditions also determine colonization rate, which is dependent on the organisms' tolerances to environmental variables.

Processes

Many nutrient transformation processes occur in estuaries, as they have the appropriate conditions. Estuaries can be considered as reactor vessels with a continuous inflow of components from the sea, the river, and the atmosphere and an outflow of compounds to the sea, the atmosphere, and the bottom sediments after undergoing certain transformations within the estuary (Figure 3). The important process elements are import, transformation, retention, and export of substances related to organic carbon and nutrients. Part of the transformation processes is illustrated in Figure 4. All these processes dictate that estuaries should be considered as both sources and sinks of organic matter and nutrients.

Table 1 Recent and 'prehistoric' loadings of some systemsby nitrogen and phosphorus and the resultant annual primary production of these systems

System: period/year	N influx (mmol m^{-2} a^{-1})	P influx (mmol m^{-2} a^{-1})	Mean annual nitrogen concentration in system (μmol l^{-1})	Mean annual phosphorus concentration in system (μmol l^{-1})	Annual primary production (g C m^{-2} a^{-1})
River Rhine and River Ems 'background' situation			45 ± 25 (tN)	1.8 ± 0.8 (tP)	
English Channel 'background' situation			5.5 ± 0.5 (winter NO$_3$)	0.45 ± 0.05 (winter DIP)	
North Sea 'background' situation			9.1 ± 3.1 (NO$_3$) (near coast)	0.57 ± 0.13 (DIP) (near coast)	
Dutch western Wadden Sea 'background' situation			13 ± 6 (tN) c. 4 (DIN) (for salinity gradient)	0.8 ± 0.3 (tP) c. 0.3 (DIP) (for salinity gradient)	<50
Ems estuary	(rivers)	(rivers)	10–45 (tN)	0.7–1.8 (tP)	
'background' situation	315 (tN)	16 (tP)			
early 1980s	3850 (tN)	90 (tP)			
early 1990s	3850 (tN)	50 (tP)			
Baltic Sea 'background' situation (c. 1900)	57 (tN) (rivers + AD + fix)	0.8 (tP) (rivers + AD)			80–105
early 1980s	230 (tN) (rivers + AD + fix)	6.7 (tP) (rivers + AD)			135
Narragansett Bay 'prehistoric' situation	18–76 (DIN) (rivers + AD) 270–330 (DIN)(+sea input)	~1 (DIP) (rivers+AD) 61 (DIP)(+sea input)			130
'recent' (1990s)	1445 (DIN) (rivers+AD) 1725 (DIN)(+sea input)	73 (DIP) (rivers + AD) 140 (DIP)(+ sea input)			290
Long Island Sound 1952	—	—			
early 1980s	1040 (tN) (rivers + AD)	70 (tP) (rivers + AD)			c. 200 300
Chesapeake Bay mid-1980s	290–2140 (DIN) 1430 (tN)	30 (DIP) 40 (tP)			400–600

tN, total nitrogen; tP, total phosphorus; DIN, dissolved inorganic deposition nitrogen; DIP, dissolved inorganic phosphorus; AD, atmospheric deposition; fix, nitrogen fixation.

Output

Several processes contribute significantly to the prevention of eutrophication symptoms. Active bacterial removal of nitrogen may occur under favorable conditions due to the conversion of nitrogen compounds into nitrogen gas. These conditions are the spatial change from anoxic and hypoxic and oxygenated conditions. Phosphorus may be removed from the system due to either transport to the open sea or the permanent burial of apatites, for example.

There is a wide variation in bacterial processes enhancing the transformation of nutrients. The intensity of the several conversion processes is partly related to the differences in the dimensions of these systems and factors such as tidal range (energy) and

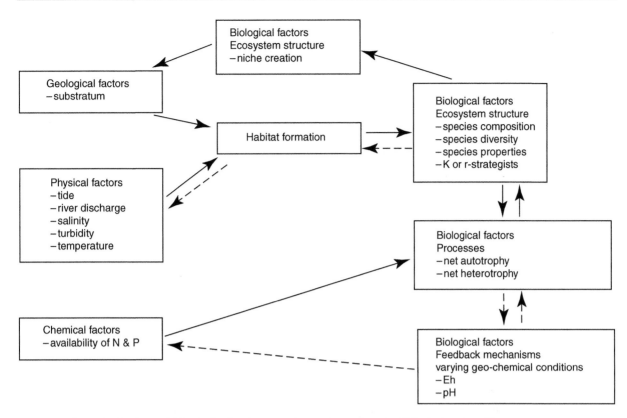

Figure 2 Operating forcing variables in the development of estuarine and marine biological communities.

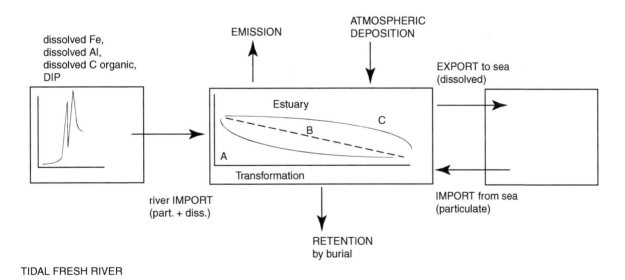

Figure 3 Box model showing fluxes, gradients, and processes in the freshwater tidal river, the estuary, and the coastal zone of the sea; part., particulate; diss., dissolved.

freshwater inflow (flushing time of fresh water, residence time of fresh or sea water and turnover time of the basin water) and related important determinants such as turbidity.

The most important factors in the expression of eutrophication are: flushing time (f_τ), the turbidity (gradient) expressed as light extinction coefficient (k_z), and the input and concentration gradient of the nutrients N, P, and Si. The combination of mainly these three factors determines whether an estuarine system has a low or high risk of producing eutrophication symptoms (**Figure 5**).

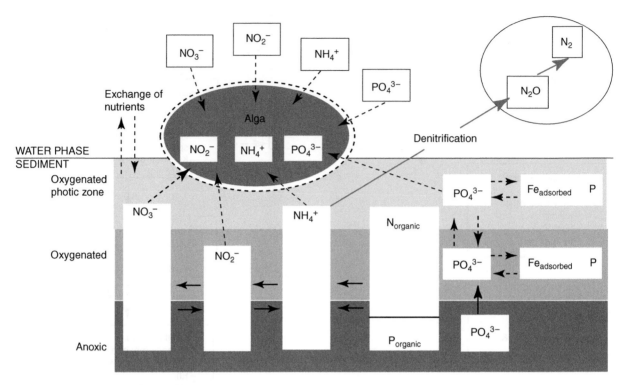

Figure 4 Schematic representation of conversion processes of nitrogen and phosphorus. Dashed lines represent assimilation, full lines represent biotic conversion (mainly microbial), and dotted lines represent geochemical equilibria. (Modified after Wiltshire, 1992 and van Beusekom & de Jonge, 1998 and references therein.)

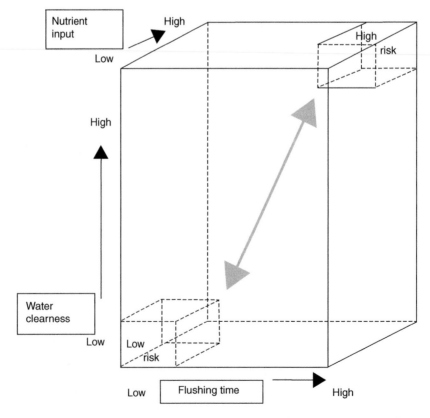

Figure 5 A 3-D classification scheme of eutrophication risk of estuaries based on flushing time, turbidity, and nutrient input.

The longer the flushing time then the more vulnerable the system is to nutrient enrichment, as the primary producers have a greater period available to utilize the excess nutrients. If the flushing time is shorter than the mean growth rate of the algae (areas with high dispersion capacity), flushing of the population will occur and thus prevent the symptoms within the system, although transport to the open sea will increase. Hence, if algal blooming does not occur within the estuary it may develop in the lowest reaches of the system or even just outside the system in the sea.

Case Studies Indicating Trends and Symptoms

In determining any response, it is necessary to describe the natural situation and its spatial and temporal variability, the change from that natural system, and the significance and cause of that change. As described above, it may be difficult to assess anthropogenic nutrient enrichment against a

background of the natural variability in nutrient influxes and turnover and its impact on the productivity of the ecosystem under consideration. The greatest problem is to make a clear distinction between the two and to assess the contribution of natural variation and of human activitiesto the nutrient enrichment and its symptoms in any system. Despite this, there are several case studies which illustrate the main features, as describedbelow.

Forcing Variable 1: Riverine Inputs and Concentrations of Nutrients

There are many scattered data available on input values and concentrations of nutrients, but there are much less consistent long-term data series. Comparisons of historical and present data for the North Sea and Wadden Sea indicate that large-scale variability in inputs (**Figure 6**; inflowing Atlantic water) and large increases in inputs (**Figure 7**; inputs from rivers) resulted in an increased productivity. An analysis of the river, estuarine and coastal dynamics and an understanding of the processes, especially in

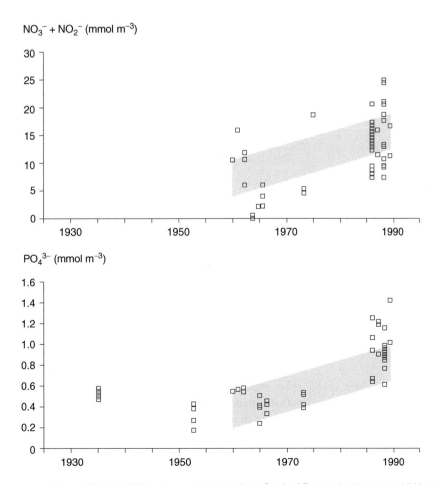

Figure 6 Winter concentrations of DIN and DIP in the central part of the Strait of Dover showing a two-fold increase in DIN and a three-fold increase in DIP. (Reproduced with permission from de Jonge, 1997).

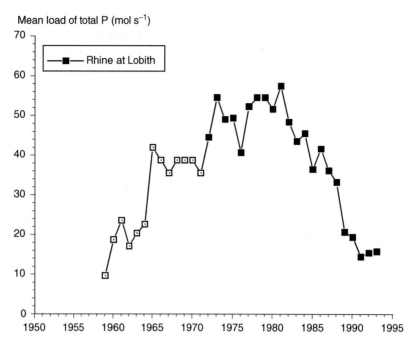

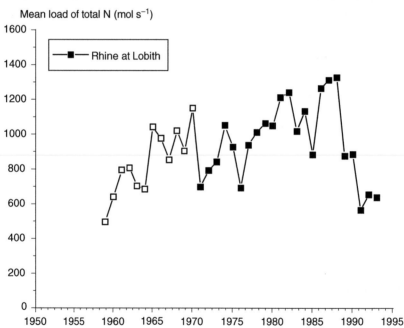

Figure 7　Development of the loads in total phosphorus and total nitrogen as measured on the border between Germany and The Netherlands. (Data from Rijkswaterstaat).

the near-coastal dynamics, showed that there may be a time delay inherent in the system and that the coast may respond later than the estuary. It is of note that detailed and systematic monitoring was required to detect these trends. Consequently, international agreements (the Oslo and Paris Conventions and European Commission Directives) were designed to control these trends. The measures were effective and

thus the remediation may be a model for systems elsewhere (cf. **Figure 2**).

The changes reported include an increase of surface algal blooms (**Figure 8**), a major sudden change in plankton composition from diatoms to small flagellates, and an increase in the area with elevated nitrogen and phosphorus concentrations (**Figure 9**) and oxygen deficiency (**Figure 10**). There are well-defined

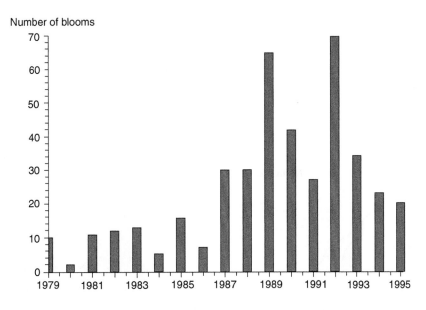

Number of blooms

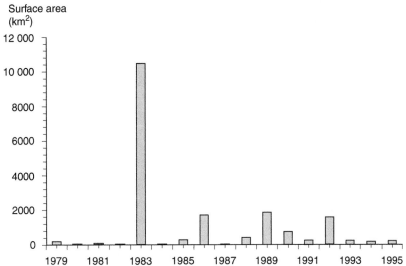

Surface area (km²)

Figure 8 Surface algal blooms observed during Dutch airborne surveys over the period 1979–1995 (after Zevenboom 1993, 1998). (A) Frequency per annum; (B) surface area in km² per annum.

positive relationships between riverborne nutrients and the long-term variation in primary and secondary production in the western Dutch Wadden Sea (**Figure 11**). However, the processes and responses are complicated by inputs from the North Sea, including the inflowing Atlantic water (**Figures 6** and **12**) and meteorological conditions. This is important as it emphasizes the need to consider all aspects when reduction measures have been undertaken.

Forcing Variable 2: Size of Receiving Area

The Baltic Sea is typical of many semi-enclosed seas with nutrient loadings. Although the four- to eightfold increased nutrient loads only changed the primary production by 30–70%, the permanent anoxic layer of the Baltic increased from 19 000 to near 80 000 km² (nearly the entire hypolimnion) as the result of the eutrophication (**Figure 13**). This produced a dramatic structural decline in population size of two important prey species (the large isopod *Saduria entomon* and the snake blenny *Lumpenus lampetraeformis*), which in turn greatly influenced the local cod populations. The increased density of algal mats reduced the development of herring eggs, possibly by exudate production and the large hypoxic areas adversely affected the development of cod eggs. Thus nutrient enrichment in the Baltic greatly damaged some essential parts of the food web and the ecosystem structure.

In additon, more localized eutrophication of the inner Baltic occurs due to fish farming (**Figure 14**),

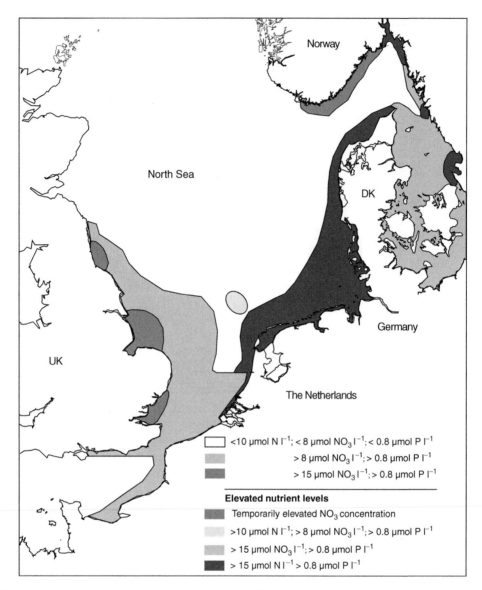

Figure 9 Elevated nutrient levels in North Sea waters (data from OSPAR 1992). The background levels have been agreed to be 10 μmol l^{-1} DIN and 0.6 μmol l^{-1} DIP (from Zevenboom 1988, 1993).

which accounts for over 35% and 55% of the total local nitrogen and phosphorus inputs respectively. It was concluded that phosphorus was the most important determinant, e.g. leading to a strongly reduced N/P ratio. This nutrient enrichment produced an increase in the primary production and an increase in turbidity which negatively affected the macrophyte populations and stimulated the blooming of cyanobacteria. The loss of five benthic crustaceans has been observed, while a gain of four was reported, of which two were polychaetes (*Polydora redeki, Marenzelleria viridis*) new to the area. Furthermore, the macrobenthos showed a structural change from suspension feeders to deposit feeders. Finally, the extensive bloom of the microalga *Chrysochromulina* in the late 1980s in the outer Baltic

apparently developed in response to nutrient build-up on the eastern North Sea and contributed to the hypoxia and eutrophic symptoms.

Forcing Variable 3: Peak Loadings and Changing Ratios in Nutrient Fractions

In Chesapeake Bay (USA; north Atlantic west coast) (**Figure 15**), the total phosphorus concentrations decreased with time, but nitrogen had maximum concentrations in the mid-1980s and the DIN concentrations doubled in the oligohaline zone of the bay. Significant increases were detected in surface chlorophyll-a data between 1950 and 1970 in all the regions of the bay and there has been a significant long-term increase in the DIN/DIP ratios since the

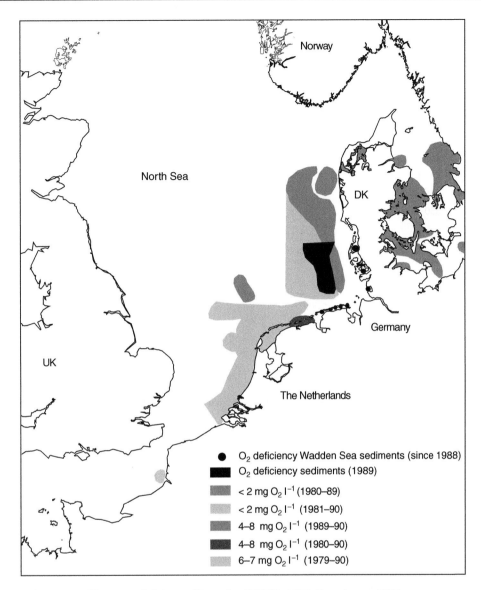

Figure 10 North Sea areas with oxygen deficiency. (Data after OSPAR, 1992; Zevenboom, 1993).

1960s in much of the bay. This ratio was generally above the Redfield ratio of 16 (necessary for optimal plant growth) in all regions in winter and spring, but in summer and autumn the values were below the Redfield ratio in the main part of the bay, suggesting N limitation. Potentially limiting concentrations of reactive silicate and DIP often occurred in the mesohaline to polyhaline bay.

Forcing Variable 4: River Basin Alterations

Alterations in the lower part of the Rhine river basin have led to a decrease in the flushing time of the system and concurrent relative increase in the discharge of nutrient loads to the Dutch Wadden Sea.

Eutrophication in Florida Bay (**Figure 16**) possibly produced a large-scale seagrass die off (4000 ha of

Thalassia testudinum and *Halodule wrightii* disappeared between 1987 and 1988), followed by increased phytoplankton abundance, sponge mortality, and a perceived decline in fisheries. This very large change in the health of the system followed major engineering works to the Everglades area and reflected changes in the nutrient concentrations, the nutrient pool, the chlorophyll levels, and turbidity. The preliminary nutrient budget for the bay assumes a large oceanic and atmospheric input of N and P to the bay, although the denitrification rates are unknown. The cause(s) of the seagrass mortality in 1987 is still unknown. Although not mentioned, synergistic effects (where eutrophication effects are exacerbated by other pollutants) are also expected in urbanized and developed areas. Furthermore, the change in primary producers in this area reflected

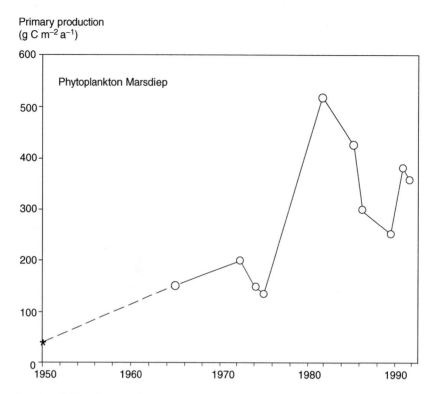

Figure 11 Time-series of available values on primary production in the western Dutch Wadden Sea, as reviewed by de Jonge *et al.* 1996 (with permission).

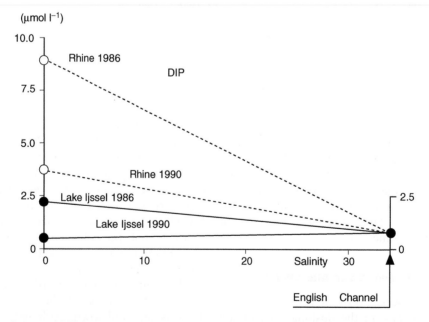

Figure 12 Mixing diagram of DIP for the years 1986 and 1990 when nutrient concentrations in river water declined after having increased for decades. The strong and structural (cf. also main river Rhine values in **Figure 13**) decrease in DIP values over a short time period is remarkable. Also the conclusion that in 1990 the DIP values of fresh water in Lake Ijsselmeer were lower than the values in theStrait of Dover/English Channel (which has consequences for the primary production potential of coastal waters in the southern Bight of the North Sea and coastal policy plans and management plans is striking. (Reproduced with permission from de Jonge, 1997.)

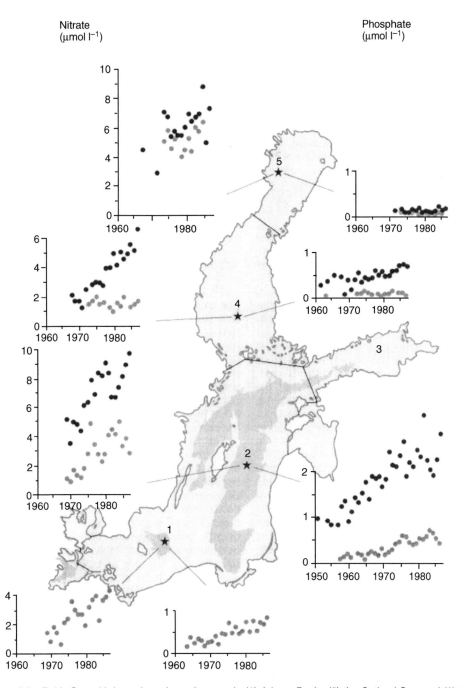

Figure 13 Map of the Baltic Sea with hypoxic and anoxic areas in (1) Arkona Basin, (2) the Gotland Sea, and (3) Gulf of Finland. Further trends in nitrate (left panels) and phosphate (right panels) in surface waters in winter (gray) and at 100 m depth (black). (Reproduced with permission from Elmgren, 1989.)

vascular plants operating as k-strategists under stress and replaced by more opportunistic algae like phytoplankton species.

Forcing Variable 5: Stratification

The Peel-Harvey estuarine system (South Pacific west coast of Australia) receives a high nutrient loading but has a phosphorus release during stratification-induced anoxia from the bottom sediments. This is after a clear loading of the estuary and the subsequent development of dense populations of microphytobenthos which is responsible for the nutrient storage. This release (**Figure 17**) contributed to the changes in macroalgal community structure and increased turbidity due to algal blooms. Remedial measures are now in place to reduce inputs and remove the adverse symptoms.

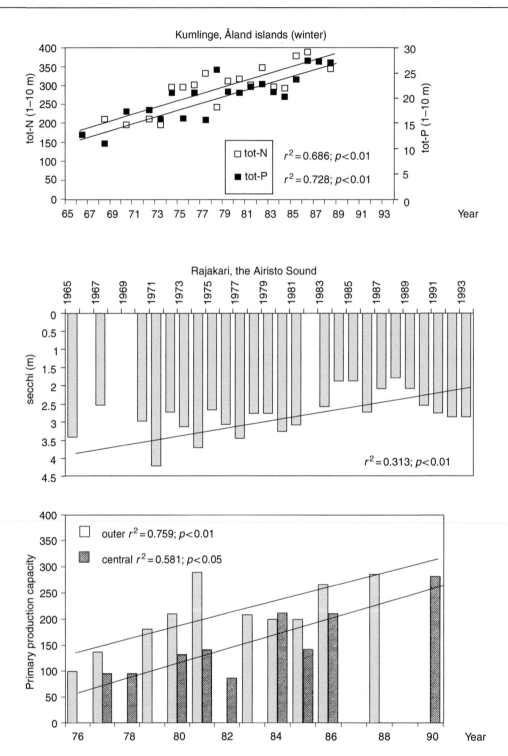

Figure 14 Development of nutrient levels in the open outer archipelago region (Baltic part of Finland) over the period 1963–93, the increase in turbidity over the period 1965–93, and the consequent increase in primary production capacity over the period 1976–90. (Reproduced with permission from Bonsdorff *et al.*, 1997.)

Forcing Variable 6: Hydrographic Regime (Residence Time and Turbity)

The creation of red tides (noxious, toxic, and nuisance microalgal blooms) in Tolo Harbor (Hong Kong) resulted from large urban nutrient inputs, a water residence time of 16–42 days, and a low turbidity which led to the phytoplankton producing dense populations. Diatoms decreased in abundance from 80–90% to 53% in 1982–85, dinoflagellates increased concurrently with red tides (**Figure 18**) and

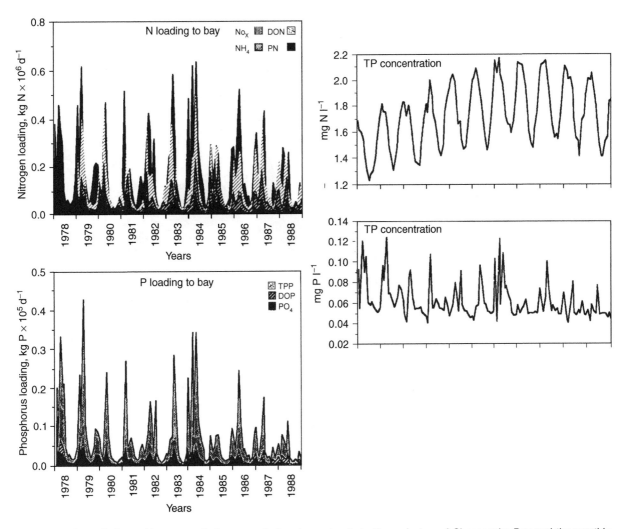

Figure 15 Record of monthly averaged nitrogen and phosphorus inputs to the mainstem of Chesapeake Bay and the monthly average values of total nitrogen and total phosphorus measured at the fall-line of the Susquehanna River (original data from Summers 1989). (Reproduced with permission from Boynton et al., 1995.)

chlorophyll-a levels also increased significantly. Oxygen depletion occurred due to nutrient loadings and the local development of phytoplankton in combination with reduced water exchange (long flushing time), features associated with a low-energy eutrophic environment. In addition, the area is degraded after heavy pollution by heavy metals and organic contaminants. In other systems (e.g. Southampton Water, UK) regular blooms of the nuisance ciliate with a symbiotic red alga, *Mesodinium rubrum*, is the result of increased nutrients, high organic matter, relatively long residence time, and turbid waters.

Conceptual Model of Effects

There have been few studies assessing all possible effects of nutrient enrichment but it is possible through the many different case studies to create a conceptual model of the main effects. **Figure 19** gives a descriptive overview on functional groups of plants and animals and at differing levels of biological organization, thus mainly at the process level.

'Hot Spots' and Remedial Measures

With regard to eutrophication, 'hotspots' may be those being hypernutrified, such as estuaries (e.g.the Ythan, Scotland) or those areas showing regular symptoms ofeutrophication, e.g. the Baltic Sea. Other good examples are the near absence of beaver dams in the USA today, and the absence of large natural wetlands as aresult of reclamation in many low-lying countries. In the past these natural obstacles as beaver dams and large wetlands favored the retention of nutrients resulting in lower more 'near' natural loads of coastal systems. It is clear that

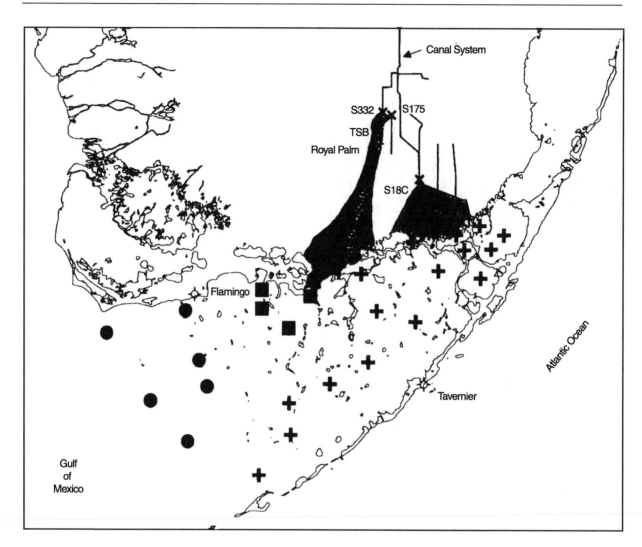

Figure 16 Map of Florida Bay with three zones of similar influence as a result of a cluster analysis on mean and SD of five principal component scores. Stations are labeled as eastern bay ($+$), central bay ($\blacksquare$), and western bay ($\bullet$). (Reproduced with permission from Boyer *et al.*, 1999.)

restoration of river systems or the rehabilitation of the integrity of entire river systems in combination with the application of best possible techniques is the best remedial measure to implement, coupled with river basin and catchment management.

In general 'hot spots' are allclose to intensive land use (agriculture and urbanized areas), with poor waste water treatment and no removal of P and N. Increasing development is usually accompanied by greater waste treatment, for example, European Directives require better treatment depending on the local population and theability of receiving waters to assimilate waste. However, it is axiomatic that sewage treatment removes organic matter but, unless nutrient stripping is installed, which is expensive, it may fail to remove, or hardly remove nutrients. Similarly, the creation of nitrate vulnerable areas requiring fertilizer control, as within the EU Nitrates Directive, will

reduce inputs. However, the fact that ground water may retain nutrients for many years, even decades in the case of aquifers, will dictate that the results of remediation will not be apparent for a while.

Areas requiring attention include populated regions, agricultural lands, and low-energy areas (Baltic Sea with Åland Islands, German Bight in the North Sea, Long Island Sound,Chesapeake Bay), i.e. mainly the large estuarine systems as well as developing countries with no or hardly any wastewater treatment. Anthropogenic eutrophication must be addressed, especially further improvement of wastewater treatment and technical processes to reduce the emissions of nutrients and related (NO_x) compounds to the atmosphere.

Despite increasing knowledge, most countries show the same history when focusing on eutrophication. The fact that the information given above suggests a

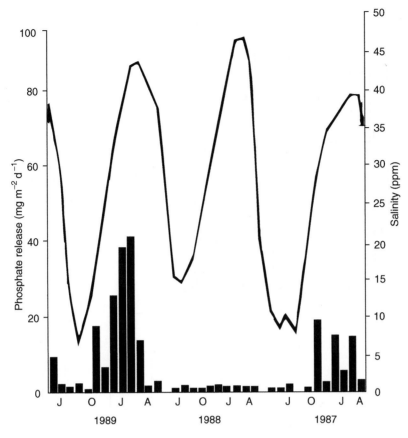

Figure 17 Release of phosphorus from the sediments of Harvey estuary. The data are release rates, recorded under standard conditions in the laboratory, for cores removed from the same site in the estuary at different times of year. The background line shows the estuarine salinity at the time. (Reproduced with permission from McComb, 1995.)

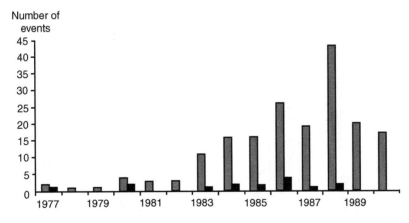

Figure 18 Red tides and associated fish kills in Tolo Harbor. (Reproduced with permission from Hodgkiss and Yim, 1995.)

reduction in the emission of nutrientsshould be interpreted with caution, because differences in nutrient ratios in combination with changes in concentrations may lead to the development of undesirable micro- and macro-algae. For example, Sweden's reduction policy, which focused on phosphorus, failed as phosphorus became depleted along the coasts but not in the central part of the Baltic Sea where it was supplied in excess from anoxic deep water – thus maintaining the near-surface algal blooms. Given the action plans adopted by developed nations to further reduce nutrient loads, it can be argued that in the near future, eutrophication will be caused by sea water that has been enriched with nutrients for decades instead of fresh water. This is due to the expectation that the present nutrient policy on 'diffuse sources' and the

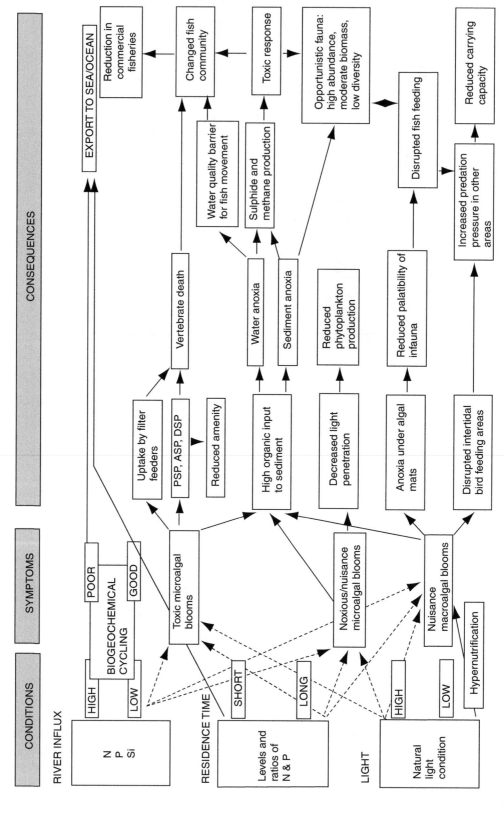

Figure 19 Overview of potential effects of nutrient enrichment in combination with turbidity conditions and residence time of water in an estuarine area.

increasing application of modern, sophisticated wastewater treatment plants will further diminish the freshwater loads. However, the atmospheric deposition of nitrogen as well as phosphorus (in dust) will become increasingly important due to many nutrient sources resulting from land use (burning of fossil carbon, fields, and forests).

The process of nitrogen fixation of increasing future importance as a mechanism during low nutrient conditions to compensate for the remedial measures taken by the different governments. This expectation means a well-balanced reduction in nutrient loads to prevent noxious blooms. It also means continuing to pay attention to eutrophication in all its aspects. At the moment nitrogen fixation is probably a small N-source as is the case in most nutrient-rich estuarine systems. However, some species have developed the ability to cope with very low nitrogen concentrations under conditions where just enough is provided by nitrogen fixation. Further global reduction in nitrogen emissions is required to protectthe environment. It is possible that the problem due to N fixation will be apparent when reduction in phosphorus loads have been taken as far as possible.

The most important 'hot spot' on this planet is the rapidly growing world population. The big question and challenge is how to offer every individual 'sustainable' living conditions while at the same time maintaining the integrity of our aquatic systems. This marked increase in population size is the main cause of the most common and most severe environmental problem of today and tomorrow.

Further Reading

van Beusekom JEE and de Jonge VN (1998) Retention of phosphorus and nitrogen in the Ems estuary. *Estuaries* 21: 527–539.

Boyer JN, Fourqurean JW, and Jones RD (1999) Seasonal and long-term trends in the water quality of Florida Bay (1989–1997). *Estuaries* 22: 417–430.

Boynton WR, Garber JH, Summers R, and Kemp WM (1995) Inputs, transformations, and transport of nitrogen and phosphorus in Chesapeake Bay and selected tributaries. *Estuaries* 18: 285–314.

Daan N and Richardson K (eds.) (1996) Changes in the North Sea ecosystem and their causes: Århus 1975 revisited. *ICES Journal of Marine Science* 53: 879–1226.

de Jonge VN (1997) High remaining productivity in the Dutch western Wadden Sea despite decreasing nutrient inputs from riverine sources. *Marine Pollution Bulletin* 34: 427–436.

de Jonge VN, Bakker JF, and van Stralen M (1996) Recent changes in the contributions of river Rhine and North Sea to the eutrophication of the western Dutch Wadden Sea. *Netherlands Journal of Aquatic Ecology* 30: 27–39.

de Jonge VN and Postma H (1974) Phosphorus compounds in the Dutch Wadden Sea. *Netherlands Journal of Sea Research* 8: 139–153.

Elmgren R (1989) Man's impact on the ecosystem of the Baltic Sea: energy flows today and at the turn of the century. *Ambio* 18: 326–332.

Hodgkiss IJ and Yim WW-S (1995) A case study of Tolo Harbour, Hong Kong. In: McComb AJ (ed.) *Eutrophic Shallow Estuaries and Lagoons*, CRC-Series, pp. 41–57. Boca Raton, FL: CRC Press.

Libes SM (1992) *An Introduction to Marine Biogeochemistry*. New York: John Wiley Sons.

McComb AJ (ed.) (1995) *Eutrophic Shallow Estuaries and Lagoons*. CRC-Series. Boca Raton, FL: CRC Press.

Olausson E and Cato I (eds.) (1980) *Chemistry and Biogeochemistry of Estuaries*. New York: John Wiley & Sons.

Salomons W, Bayne BL, Duursma EK, and Förstner U (eds.) (1998) *Pollution of the North Sea: An Assessment*. Berlin: Springer-Verlag.

Stumm W (1992) *Chemistry of the Solid–Water Interface. Processes at the Mineral–Water and Particle–Water Interface in Natural Systems*. New York: John Wiley Sons.

Summers RM (1989) *Point and Non-point Source Nitrogen and Phosphorus Loading to the Northern Chesapeake Bay*. Maryland Department of the Environment, Water Management Administration, Chesapeake Bay Special Projects Program, Baltimore, MD.

Wiltshire KH (1992) The influence of microphytobenthos on oxygen and nutrient fluxes between eulittoral sediments and associated water phases in the Elbe estuary. In: Colombo G, Ferrari I, Ceccherelli VU, and Rossi R (eds.) *Marine Eutrophication and Population Dynamics*, pp. 63–70. Fredensborg: Olsen & Olsen.

Zevenboom W (1993) Assessment of eutrophication and its effects in marine waters. *German Journal of Hydrography* Suppl 1: 141–170.

HYPOXIA

N. N. Rabalais, Louisiana Universities Marine Consortium, Chauvin, LA, USA

Introduction: Definitions

Hypoxic (low-oxygen) and anoxic (no-oxygen) waters have existed throughout geologic time. Presently, hypoxia occurs in many of the ocean's deeper environs, open-ocean oxygen-minimum zones (OMZs), enclosed seas and basins, below western boundary current upwelling zones, and in fjord. Hypoxia also occurs in shallow coastal seas and estuaries, where their occurrence and severity appear to be increasing, most likely accelerated by human activities (**Figure 1**). A familiar term used in the popular press and literature, 'dead zone', used for coastal and estuarine hypoxia, refers to the fish and shellfish killed by the suffocating conditions or the failure to catch these animals in bottom waters when the oxygen concentration in the water covering the seabed is below a critical level.

Based on laboratory or field observations or both, the level of oxygen stress and related responses of invertebrate and fish faunas vary. The units are often determined by oxygen conditions that are physiologically stressful, but these levels also differ depending on the organisms considered, and the pressure, temperature, and salinity of the ambient waters. The numerical definition of hypoxia varies as do the units used, but hypoxia has mostly been defined as dissolved oxygen levels lower than a range of $3-2\,\mathrm{ml\,l^{-1}}$, with the consensus being in favor of $1.4\,\mathrm{ml\,l^{-1}}\,(=2\,\mathrm{mg\,l^{-1}}$ or ppm). This value is approximately equivalent to 30% oxygen saturation at $25\,^{\circ}\mathrm{C}$ and 35 salinity (psu). Below this concentration, bottom-dragging trawl nets fail to capture fish, shrimp, and swimming crabs. Other fishes, such as rays and sharks, are affected by oxygen levels below $3\,\mathrm{mg\,l^{-1}}$, which prompts a behavioral response to evacuate the area, up into the water column and shoreward. Water-quality standards in the coastal waters of Long Island Sound, New York, and Connecticut, USA, consider that dissolved oxygen conditions below $5\,\mathrm{mg\,l^{-1}}$ result in behavioral effects in marine organisms and fail to support living resources at sustainable levels.

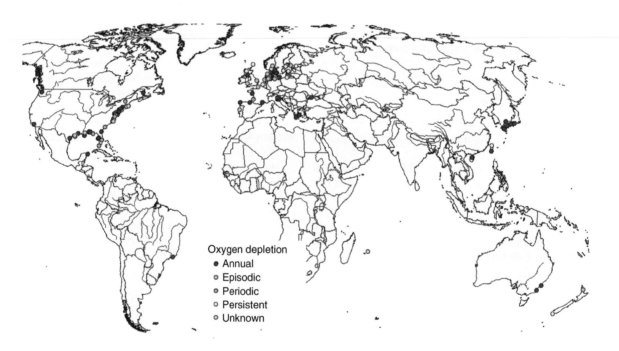

Oxygen depletion
- Annual
- Episodic
- Periodic
- Persistent
- Unknown

Figure 1 Distribution of coastal ocean hypoxic areas; excludes naturally occurring hypoxia, such as upwelling zones and OMZs. Reproduced from Díaz RJ, Nestlerode J, and Díaz ML (2004) A global perspective on the effects of eutrophication and hypoxia on aquatic biota. In: Rupp GL and White MD (eds.) *Proceedings of the 7th International Symposium on Fish Physiology, Toxicology and Water Quality, Tallinn, Estonia, May 12-15, 2003. EPA 600/R-04/049*, pp. 1–33. Athens, GA: Ecosystems Research Division, US Environmental Protection Agency, with permission from Robert J. Díaz.

The most commonly used definition for oceanic waters is dissolved oxygen content less than $1\,ml\,l^{-1}$ (or $0.7\,mg\,l^{-1}$). Disoxyic or disaerobic refers to oxygen levels between 0.1 and $1.0\,ml\,l^{-1}$. OMZs are usually defined as waters less than $0.5\,ml\,l^{-1}$ dissolved oxygen.

Causes

Hypoxia occurs where the consumption of oxygen through respiratory or chemical processes exceeds the rate of supply from oxygen production via photosynthesis, diffusion through the water column, advection, or mixing. The biological and physical water-column characteristics that support the development and maintenance of hypoxia include (1) the production, flux, and accumulation of organic-rich matter from the upper water column; and (2) water-column stability resulting from stratification or long residence time. Dead and senescent algae, zooplankton fecal pellets, and marine aggregates contribute significant amounts of organic detritus to the lower water column and seabed. Aerobic bacteria consume oxygen during the decay of the carbon and deplete the oxygen, particularly when stratification prevents diffusion of oxygen. Stratification is the division of the water column into layers with different densities caused by differences in temperature or salinity or both. Hypoxia will persist as long as oxygen consumption rates exceed those of supply. Oxygen depletion occurs more frequently in estuaries or coastal areas with longer water residence times, with higher nutrient loads and with stratified water columns.

Hypoxia is a natural feature of many oceanic waters, such as OMZs and enclosed seas, or forms in coastal waters as a result of the decomposition of high carbon loading stimulated by upwelled nutrient-rich waters. Hypoxia in many coastal and estuarine waters, however, is but one of the symptoms of eutrophication, an increase in the rate of production and accumulation of carbon in aquatic systems. Eutrophication very often results from an increase in nutrient loading, particularly by forms of nitrogen and phosphorus. Nutrient over-enrichment from anthropogenic sources is one of the major stressors impacting estuarine and coastal ecosystems, and there is increasing concern in many areas around the world that an oversupply of nutrients is having pervasive ecological effects on shallow coastal waters. These effects include reduced light penetration, increased abundance of nuisance macroalgae, loss of aquatic habitat such as seagrass or macroalgal beds, noxious and toxic algal blooms, hypoxia and anoxia, shifts in trophic interactions and food webs, and impacts on living resources.

Hypoxic Systems

Oxygen-Minimum Zones

Persistent hypoxia is evident in mid-water OMZs, which are widespread in the world oceans where the oxygen concentrations are less than $0.5\,ml\,l^{-1}$ (or about 7.5% oxygen saturation, $<22\,\mu M$). They occur at different depths from the continental shelf to upper bathyal zones (down to 1300 m). Many of the OMZs form as a result of high primary production associated with coastal upwelled nutrient-rich waters. Their formation also requires stagnant circulation, long residence times, and the presence of oxygen-depleted source waters. The extensive OMZ development in the eastern Pacific Ocean is attributed to the fact that intermediate depth waters of the region are older and have overall oxygen concentrations lower than other water masses. The largest OMZs are at bathyal depths in the eastern Pacific Ocean, the Arabian Sea, the Bay of Bengal, and off southwest Africa. The upper boundary of an OMZ may come to within 10 or 50 m of the sea surface off Central America, Peru, and Chile. The OMZ is more than 1000-m thick off Mexico and in the Arabian Sea, but off Chile, the OMZ is <400-m thick. Along continental margins, minimum oxygen concentrations occur typically between 200 and 700 m. The area of the ocean floor where oceanic waters permanently less than $0.5\,ml\,l^{-1}$ impinge on continental margins covers $10^6\,km^2$ of shelf and bathyal seafloor, with over half occurring in the northern Indian Ocean. These permanently hypoxic waters account for 2.3% of the ocean's continental margin. These hypoxic areas are not related to eutrophication, but longer-term shifts in meteorological conditions and ocean currents may increase their prevalence in the future with global climate change. Shifts in ocean currents have been implicated in the increased frequency of continental shelf hypoxia along the northwestern US Pacific coast of Oregon.

Deep Basins, Enclosed Seas, and Fjord

Many of the existing permanent or periodic anoxic ocean environments occur in enclosed or semi-enclosed waters where a mass of deep water is bathymetrically isolated from main shelf or oceanic water masses by surrounding landmasses or one or more shallow sills. In conjunction with a pycnocline, the bottom water volume is restricted from exchange with deep open water. Examples of hypoxic and anoxic basins include anoxic deep water fjord, such as Saanich Inlet, the deeper basins of the Baltic, the basin of the Black Sea, the Japanese Seto Inland Sea, deep waters of the Sea of Cortez, Baja California, and Santa Barbara Basin in the southern California borderland.

Coastal Seas and Estuaries

Periodic hypoxia or anoxia also occurs on open continental shelves, for example, the northern Gulf of Mexico and the Namibian and Peruvian shelves where upwelling occurs. More enclosed shelves such as the northern Adriatic and the northwestern shelf of the Black Sea also have periodic hypoxia or anoxia. In these instances, there is minimal exchange of shelf-slope water and/or high oxygen demand on the shallow shelf. Estuaries, embayments, and lagoons are susceptible to the formation of hypoxia and anoxia if the water residence time is sufficiently long, especially where the water column is stratified. Light conditions are also important in these coastal habitats as a limiting factor on phytoplankton growth, which, if excessive, contributes to high organic loading within the confined waters.

Coastal ecosystems that have been substantially changed as a result of eutrophication exhibit a series of identifiable symptoms, such as reduced water clarity, excessive, noxious, and, sometimes, harmful algal blooms, loss of critical macroalgal or seagrass habitat, development or intensification of oxygen depletion in the form of hypoxia or anoxia, and, in some cases, loss of fishery resources. More subtle responses of coastal ecosystems to eutrophication include shifts in phytoplankton and zooplankton communities, shifts in the food webs that they support, loss of biodiversity, changes in trophic interactions, and changes in ecosystem functions and biogeochemical processes.

In a review of anthropogenic hypoxic zones in 1995, Díaz and Rosenberg noted that no other environmental variable of such ecological importance to estuarine and coastal marine ecosystems around the world has changed so drastically, in such a short period of time, as dissolved oxygen. For those reviewed, there was a consistent trend of increasing severity (either in duration, intensity, or size) where hypoxia occurred historically, or hypoxia existed presently when it did not occur before. While hypoxic environments have existed through geologic time and are common features of the deep ocean or adjacent to areas of upwelling, their occurrence in estuarine and coastal areas is increasing, and the trend is consistent with the increase in human activities that result in nutrient over-enrichment.

The largest human-caused hypoxic zone is in the aggregated coastal areas of the Baltic Sea, reaching $84\,000\,km^2$. Hypoxia existed on the northwestern Black Sea shelf historically, but anoxic events became more frequent and widespread in the 1970s and 1980s, reaching over areas of the seafloor up to $40\,000\,km^2$ in depths of 8–40 m. There is also evidence that the suboxic zone of the open Black Sea enlarged toward the surface by about 10 m since 1970. The condition of the northwestern shelf of the Black Sea, in which hypoxia covered up to $40\,000\,km^2$, improved over the period 1990–2000 when nutrient loads from the Danube River decreased, but may be experiencing a worsening of hypoxic conditions more recently.

Similar declines in bottom water dissolved oxygen have occurred elsewhere as a result of increasing nutrient loads and cultural eutrophication, for example, the northern Adriatic Sea, the Kattegat and Skaggerak, Chesapeake Bay, Albemarle-Pamlico Sound, Tampa Bay, Long Island Sound, New York Bight, the German Bight, and the North Sea. In the United States, over half of the estuaries experience hypoxia at some time over an annual period and many experience hypoxia over extensive areas for extended periods on a perennial basis. The number of estuaries with hypoxia or anoxia continues to rise.

Historic data on Secchi disk depth in the northern Adriatic Sea in 1911 through the present, with few interruptions of data collection, provide a measure of water transparency that could be interpreted to depict surface water productivity. These data coupled with surface and bottom water dissolved oxygen content determined by Winkler titrations and nutrient loads outline the sequence of eutrophication in the northern Adriatic Sea. Similar historical data from other coastal areas around the world demonstrate a decrease in water clarity due to phytoplankton production in response to increased nutrient loads that are paralleled by declines in water column oxygen levels.

There are strong relationships between river flow and nutrient flux into the Chesapeake Bay and northern Gulf of Mexico and phytoplankton production and biomass and the subsequent fate of that production in spring deposition of chlorophyll *a*. Further there is a strong relationship between the deposited chlorophyll *a* and the seasonal decline of deep-water dissolved oxygen. Excess nutrients in many watersheds are driven by agricultural activities and atmospheric deposition from burning of fossil fuels. The link with excess nutrients in more urban areas, such as Long Island Sound, is with the flux of nutrients associated from numerous wastewater outfalls.

Swift currents that move materials away from a river delta and that do not permit the development of stratification are not conducive to the accumulation of biomass or depletion of oxygen, for example in the Amazon and Orinoco plumes. Similar processes off the Changjiang (Yantze River) and high turbidity in the plume of the Huanghe (Yellow River) were once thought to be reasons why hypoxia did not develop in those coastal systems. Incipient indications of the beginning of symptoms of cultural eutrophication were becoming evident at the terminus of both these systems as nutrient loads increased. The severely reduced, almost minimal, flow of the Huanghe has prevented the formation of hypoxia, but other coastal ecosystem problems remain. There is, however, now a hypoxic area off the Changjiang Estuary and harmful algal blooms are more frequent in the East China Sea. The likelihood that more and more coastal systems, especially in developing countries, where the physical conditions are appropriate will become eutrophic with accompanying hypoxia is worrisome.

Northern Gulf of Mexico

The hypoxic zone on the continental shelf of the northern Gulf of Mexico is one of the largest hypoxic zones in the world's coastal oceans, and is representative of hypoxia resulting from anthropogenic activities over the last half of the twentieth century (**Figure 2**). Every spring, the dissolved oxygen levels in the coastal waters of the northern Gulf of Mexico decline and result in a vast region of oxygen-starved water that stretches from the Mississippi River westward along the Louisiana shore and onto the Texas coast. The area of bottom covered by hypoxic water can reach $22\,000\,km^2$, and the volume of hypoxic waters may be as much as $10^{11}\,m^3$. Hypoxia

in the Gulf of Mexico results from a combination of natural and human-influenced factors. The Mississippi River, one of the 10 largest in the world, drains 41% of the land area of the lower 48 states of the US and delivers fresh water, sediments, and nutrients to the Gulf of Mexico. The fresh water, when it enters the Gulf, floats over the denser saltier water, resulting in stratification, or a two-layered system. The stratification, driven primarily by salinity, begins in the spring, intensifies in the summer as surface waters warm and winds that normally mix the water subside, and dissipates in the fall with tropical storms or cold fronts.

Hypoxic waters are found at shallow depths near the shore (4–5 m) to as deep as 60 m. The more typical depth distribution is between 5 and 35 m. The hypoxic water is not just located near the seabed, but may rise well up into the water column, often occupying the lower half of a 20-m water column (**Figure 3**). The inshore/offshore distribution of hypoxia on the Louisiana shelf is dictated by winds and currents. During typical winds from the southeast, downwelling favorable conditions force the hypoxic bottom waters farther offshore. When the wind comes from the north, an upwelling favorable current regime promotes the movement of the hypoxic bottom waters close to shore. When the hypoxic waters move onto the shore, fish, shrimp, and crabs are trapped along the beach, resulting sometimes in a 'jubilee' when the stunned animals are easily harvested by beachgoers. A more negative result is a massive fish kill of all the sea life trapped without sufficient oxygen.

Hypoxia occurs on the Louisiana coast west of the Mississippi River delta from February through November, and nearly continuously from mid-May through mid-September. In March and April, hypoxic water masses are patchy and ephemeral. The hypoxic

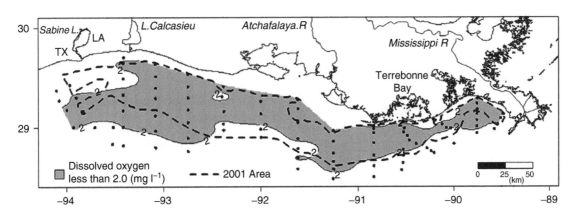

Figure 2 Similar size and expanse of bottom water hypoxia in mid-July 2002 (shaded area) and in mid-July 2001 (outlined with dashed line). Data source: N. N. Rabalais, Louisiana Universities Marine Consortium.

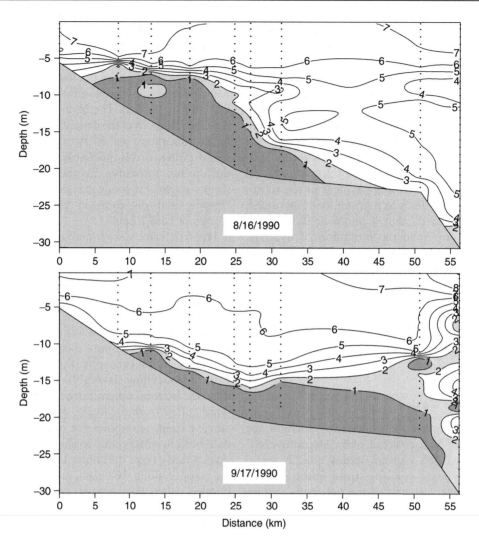

Figure 3 Contours of dissolved oxygen ($mg\,l^{-1}$) across the continental shelf of Louisiana approximately 200 km west of the Mississippi River delta in summer. The distribution across the shelf in August is a response to an upwelling favorable oceanographic regime and that of September to a downwelling favorable oceanographic regime. These contours also illustrate the height above the seabed that hypoxia can reach, i.e., over half the water column. Data source: N. N. Rabalais, Louisiana Universities Marine Consortium.

zone is most widespread, persistent, and severe in June, July, and August, and often well into September, depending on whether tropical storm activity begins to disrupt the stratification and hypoxia. Anoxic waters occur periodically in midsummer.

The midsummer size of the hypoxic zone varies annually, and is most closely related to the nitrate load of the Mississippi River in the 2 months prior to the typically late-July mapping exercise. The load of nitrate is determined by the discharge of the Mississippi River multiplied by the concentration of the nitrate, so that the amount of water coming into the Gulf of Mexico is also a factor. The relationship of the size of hypoxia, however, is stronger with the load of nitrate than with the total river water discharge or any other nutrient or combination of

nutrients. Changes in the severity of hypoxia over time are related mostly to the change in nitrate concentration in the Mississippi River (80%), the remainder to changes in increased discharge (20%).

Historical Change in Oxygen Conditions

Historical dissolved oxygen data such as those for the northern Adriatic Sea beginning in 1911 are not commonly available. A solution is to turn to the sediment record for paleoindicators of long-term transitions related to eutrophication and oxygen deficiency. Biological, mineral, or chemical indicators of plant communities, level of productivity, or conditions of hypoxia preserved in sediments, where sediments accumulate, provide clues to prior hydrographic and biological conditions.

Data from sediment cores taken from the Louisiana Bight adjacent to the Mississippi River where sediments accumulate with their biological and chemical indicators document increased recent eutrophication and increased organic sedimentation in bottom waters, with the changes being more apparent in areas of chronic hypoxia and coincident with the increasing nitrogen loads from the Mississippi River system beginning in the 1950s. This evidence comes as an increased accumulation of diatom remains and marine-origin carbon accumulation in the sediments.

Benthic microfauna and chemical conditions provide several surrogates for oxygen conditions. The mineral glauconite forms under reducing conditions in sediments, and its abundance is an indication of low-oxygen conditions. (Note that glauconite also forms in reducing sediments whose overlying waters are $>2\,mg\,l^{-1}$ dissolved oxygen.) The average glauconite abundance in the coarse fraction of sediments in the Louisiana Bight was $\sim5.8\%$ from 1900 to a transition period between 1940 and 1950, when it increased to $\sim13.4\%$, suggesting that hypoxia 'may' have existed at some level before the 1940–50 time period, but that it worsened since then.

Benthic foraminiferans and ostracods are also useful indicators of reduced oxygen levels because oxygen stress decreases their overall diversity as measured by the Shannon–Wiener diversity index (SWDI) and shifts community composition. Foraminiferan and ostracod diversity decreased since the 1940s and early 1950s, respectively. While present-day foraminiferan diversity is generally low in the Louisiana Bight, comparisons among assemblages from areas of different oxygen depletion indicate that the dominance of *Ammonia parkinsoniana* over *Elphidium* spp. (A–E index) was much more pronounced in oxygen-depleted compared to well-oxygenated waters. The A–E index has also proven to be a strong, consistent oxygen-stress signal in other coastal areas, for example, Chesapeake Bay and Long Island Sound. The A–E index from sediment cores increased significantly after the 1950s, suggesting increased oxygen stress (in intensity or duration) in the last half century. *Buliminella morgani*, a hypoxia-tolerant species, known only from the Gulf of Mexico, dominates the present-day population ($>50\%$) within areas of chronic seasonal hypoxia, and has also increased markedly in recent decades. *Quinqueloculina* sp., a hypoxia-intolerant foraminiferan, was a conspicuous member of the fauna from 1700 to 1900, indicating that oxygen stress was not a problem prior to 1900, but this species is no longer present on northern Gulf of Mexico shelf in the Louisiana Bight.

Multiple lines of evidence from sediment cores indicate an overall increase in phytoplankton productivity and continental shelf oxygen stress (in intensity or duration) in the northern Gulf of Mexico adjacent to the plume of the Mississippi River, especially in the last half of the twentieth century. The changes in these indicators are consistent with the increases in river nitrate-N loading during that same period.

OMZ intensity and distribution vary over geological timescales as a result of shifts in productivity or circulation over a few thousands to 10 ky. These changes affect expansions and contractions of the oxygen-depleted waters both vertically and horizontally. Paleoindicators, including foraminiferans, organic carbon preservation, carbonate dissolution, nitrogen isotopes, and Cd:Ca ratios that reflect productivity maxima and shallow winter mixing of the water column, are used to trace longer-term changes in OMZs, similar to studies of continental shelf sediment indicators.

Consequences

Direct Effects

The obvious effects of hypoxia/anoxia are displacement of pelagic organisms and selective loss of demersal and benthic organisms. These impacts may be aperiodic so that recovery occurs; may occur on a seasonal basis with differing rates of recovery; or may be permanent so that a shift occurs in long-term ecosystem structure and function. As the oxygen concentration falls from saturated or optimal levels toward depletion, a variety of behavioral and physiological impairments affect the animals that reside in the water column or in the sediments or that are attached to hard substrates (**Figure 4**). Hypoxia acts as an endocrine disruptor with adverse effects on reproductive performance of fishes, and loss of secondary production may therefore be a widespread environmental consequence of hypoxia. Mobile animals, such as shrimp, fish, and some crabs, flee waters where the oxygen concentration falls below $3–2\,mg\,l^{-1}$.

As dissolved oxygen concentrations continue to fall, less mobile organisms become stressed and move up out of the sediments, attempt to leave the seabed, and often die (**Figure 5**). As oxygen levels fall from 0.5 toward $0\,mg\,l^{-1}$, there is a fairly linear decrease in benthic infaunal diversity, abundance, and biomass. Losses of entire higher taxa are features of the

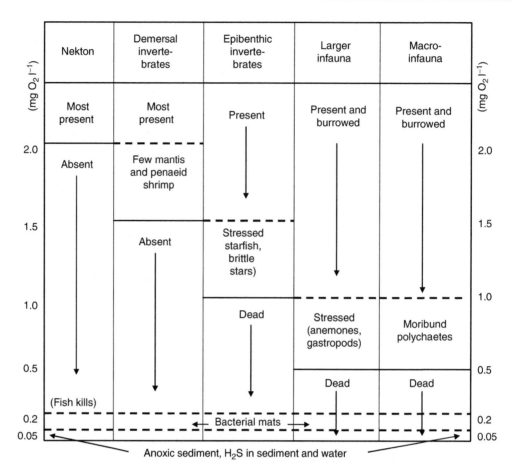

Figure 4 Progressive changes in fish and invertebrate fauna as oxygen concentration decreases from 2 mg l^{-1} to anoxia. From Rabalais NN, Harper DE, Jr., and Tuner RE (2001) Responses of nekton and demersal and benthic fauna to decreasing oxygen concentrations. In: Rabalais NN and Turner RE (eds.) *Coastal and Estuarine Studies 58: Coastal Hypoxia: Consequences for Living Resources and Ecosystems*, pp. 115–128. Washington, DC: American Geophysical Union.

depauperate benthic fauna in the severely stressed seasonal hypoxic/anoxic zone of the Louisiana inner shelf in the northern Gulf of Mexico. Larger, longer-lived burrowing infauna are replaced by short-lived, smaller surface deposit-feeding polychaetes, and certain typical marine invertebrates are absent from the fauna, for example, pericaridean crustaceans, bivalves, gastropods, and ophiuroids. Long-term trends for the Skagerrak coast of western Sweden in semi-enclosed fiordic areas experiencing increased oxygen stress showed declines in the total abundance and biomass of macroinfauna, abundance and biomass of mollusks, and abundance of suspension feeders and carnivores. These changes in benthic communities result in an impoverished diet for bottom-feeding fish and crustaceans and contribute, along with low dissolved oxygen, to altered sediment biogeochemical cycles. In waters of Scandinavia and the Baltic, there was a reduction of 3 million t in benthic macrofaunal biomass during the worst years of hypoxia occurrence. This loss, however, may be partly compensated by the biomass increase that occurred in well-flushed organically enriched coastal areas not subject to hypoxia.

Where oxygen minimum zones impinge on continental margins or sea mounts, they have considerable effects on benthic assemblages. The benthic fauna of OMZs consist mainly of smaller-sized protozoan and meiofaunal organisms, with few or no macrofauna or megafauna. The few eukaryotic organisms are nematodes and foraminiferans. Meiofauna appear to be more broadly tolerant of oxygen depletion than are macrofauna. The numbers of metazoan meiofaunal organisms, primarily nematodes, are not reduced in OMZs, presumably due to abundant particulate food and reduced predation pressure. In hypoxic waters of the northern Gulf of Mexico, harpacticoid copepod meiofauna are reduced at low oxygen levels, but the nematodes maintain their densities. Benthic macrofauna are found in all hypoxic sediments of the northern Gulf of Mexico, although the density is severely reduced

- Direct mortality
- Altered migration
- Reduction in suitable habitat
- Increased susceptibility to predation
- Changes in food resources
- Susceptibility of early life stages

Figure 5 Effects of hypoxia on fishery resources and the benthic communities that support them. Upper right: Dead demersal and bottom-dwelling fishes killed by the encroachment of near-anoxic waters onto a Grand Isle, Louisiana, beach in August 1990. Photo provided by K. M. St. Pé. Lower right: dead spider crab (family Majidae) at sediment surface. Photo provided by Franklin Viola. Lower left: dead polychaete (family Spionidae) and filamentous sulfur bacteria. Photo provided by Franklin Viola.

below $0.5 \, \text{mg} \, \text{l}^{-1}$, and the few remaining organisms are polychaetes of the families Ampharetidae and Magelonidae and some sipunculans.

While permanent deep-water hypoxia that impinges on 2.3% of the ocean's continental margin may be inhospitable to most commercially valuable marine resources, they support the largest, most continuous reducing ecosystems in the world oceans. Large filamentous sulfur bacteria, *Thioploca* and *Beggiatoa*, thrive in hypoxic conditions of $0.1 \, \text{ml} \, \text{l}^{-1}$. OMZ sediments characteristically support large bacteria, both filamentous sulfur bacteria and giant spherical sulfur bacteria with diameters of 100–300 μm. The filamentous sulfur bacteria are also characteristic of severely oxygen depleted waters in the northern Gulf of Mexico.

Secondary Production

An increase in nutrient availability results in an increase of fisheries yield to a maximal point; then there are declines in various compartments of the fishery as further increases in nutrients lead to seasonal hypoxia and permanent anoxia in semi-enclosed seas. Documenting loss of fisheries related

to the secondary effects of eutrophication, such as the loss of seabed vegetation and extensive bottom water oxygen depletion, is complicated by poor fisheries data, inadequate economic indicators, increase in overharvesting that occurred at the time that habitat degradation progressed, natural variability of fish populations, shifts in harvestable populations, and climatic variability.

Eutrophication often leads to the loss of habitat (rooted vegetation or macroalgae) or low dissolved oxygen, both of which may lead to loss of fisheries production. In the deepest bottoms of the Baltic proper, animals have long been scarce or absent because of low oxygen availability. This area was $20\,000 \, \text{km}^2$ until the 1940s. Since then, about a third of the Baltic bottom area has intermittent oxygen depletion. Lowered oxygen concentrations and increased sedimentation have changed the benthic fauna in the deeper parts of the Baltic, resulting in an impoverished diet for bottom fish. Above the halocline in areas not influenced by local pollution, benthic biomass has increased due mostly to an increase in mollusks. On the other hand, many reports document instances where local pollution resulting in severely depressed oxygen levels has greatly impoverished or even annihilated the soft-bottom macrofauna.

Eutrophication of surface waters accompanied by oxygen-deficient bottom waters can lead to a shift in dominance from demersal fishes to pelagic fishes. In the Baltic Sea and Kattegatt where eutrophication-related ecological changes occurred mainly after World War II, changes in fish stocks have been both positive (due to increased food supply; e.g., pike perch in Baltic archipelagos) and negative (e.g., oxygen deficiency reducing Baltic cod recruitment and eventual harvest). Similar shifts are inferred with limited data on the Mississippi River-influenced shelf with the increase in two pelagic species in bycatch from shrimp trawls and a decrease in some demersal species. Commercial fisheries in the Black Sea declined as eutrophication led to the loss of macroalgal habitat and oxygen deficiency, amid the possibility of overfishing. After the mid-1970s, benthic fish populations (e.g., turbot) collapsed, and pelagic fish populations (small pelagic fish, such as anchovy and sprat) started to increase. The commercial fisheries diversity declined from about 25 fished species to about five in 20 years (1960s to 1980s), while anchovy stocks and fisheries increased rapidly. The point on the continuum of increasing nutrients versus fishery yields remains vague as to where benefits are subsumed by environmental problems that lead to decreased landings or reduced quality of production and biomass.

Future Expectations

The continued and accelerated export of nitrogen and phosphorus to the world's coastal ocean is the trajectory to be expected unless societal intervention takes place (in the form of controls or changes in culture). The largest increases are predicted for southern and eastern Asia, associated with predicted large increases in population, increased fertilizer use to grow food to meet the dietary demands of that population, and increased industrialization. The implications for coastal eutrophication and subsequent ecosystem changes such as worsening conditions of oxygen depletion are significant.

See also

Coastal Topography, Human Impact on. Ecosystem Effects of Fishing. Eutrophication. Fishery Management, Human Dimension.

Further Reading

Díaz RJ, Nestlerode J, and Díaz ML (2004) A global perspective on the effects of eutrophication and hypoxia on aquatic biota. In Rupp GL and White MD (eds.) *Proceedings of* the 7th International *Symposium on Fish Physiology, Toxicology and Water Quality*, EPA 600/R-04/049, pp. 1–33. Tallinn, Estonia, 12–15 May 2003. Athens, GA: Ecosystems Research Division, US EPA.

Díaz RJ and Rosenberg R (1995) Marine benthic hypoxia: A review of its ecological effects and the behavioural responses of benthic macrofauna. *Oceanography and Marine Biology Annual Review* 33: 245–303.

Gray JS, Wu RS, and Or YY (2002) Review. Effects of hypoxia and organic enrichment on the coastal marine environment. *Marine Ecology Progress Series* 238: 249–279.

Hagy JD, Boynton WR, and Keefe CW (2004) Hypoxia in Chesapeake Bay, 1950–2001: Long-term change in relation to nutrient loading and river flow. *Estuaries* 27: 634–658.

Helly J and Levin LA (2004) Global distributions of naturally occurring marine hypoxia on continental margins. *Deep-Sea Research* 51: 1159–1168.

Mee LD, Friedrich JJ, and Gomoiu MT (2005) Restoring the Black Sea in times of uncertainty. *Oceanography* 18: 100–111.

Rabalais NN and Turner RE (eds.) (2001) *Coastal and Estuarine Studies 58: Coastal Hypoxia – Consequences for Living Resources and Ecosystems*. Washington, DC: American Geophysical Union.

Rabalais NN, Turner RE, and Scavia D (2002) Beyond science into policy: Gulf of Mexico hypoxia and the Mississippi River. *BioScience* 52: 129–142.

Rabalais NN, Turner RE, Sen Gupta BK, Boesch DF, Chapman P, and Murrell MC (2007) Characterization and long-term trends of hypoxia in the northern Gulf of Mexico: Does the science support the Action Plan? *Estuaries and Coasts* 30(supplement 5): 753–772.

Turner RE, Rabalais NN, and Justić D (2006) Predicting summer hypoxia in the northern Gulf of Mexico: Riverine N, P and Si loading. *Marine Pollution Bulletin* 52: 139–148.

Tyson RV and Pearson TH (eds.) *Geological Society Special Publication No. 58: Modern and Ancient Continental Shelf Anoxia*, 470pp. London: The Geological Society.

Relevant Websites

http://www.gulfhypoxia.net
– Hypoxia in the Northern Gulf of Mexico.

OIL POLLUTION

J. M. Baker, Clock Cottage, Shrewsbury, UK

Introduction

This article describes the sources of oil pollution, composition of oil, fate when spilt, and environmental effects. The initial impact of a spill can vary from minimal to the death of nearly everything in a particular biological community, and recovery times can vary from less than one year to more than 30 years. Information is provided on the range of effects together with the factors which help to determine the course of events. These include oil type and volume, local geography, climate and season, species and biological communities, local economic and amenity considerations, and clean-up methods. With respect to clean-up, decisions sometimes have to be made between different, conflicting environmental concerns. Oil spill response is facilitated by pre-spill contingency planning and assessment including the production of resource sensitivity maps.

Oil: A High Profile Pollutant

Consider some of the worst and most distressing effects of an oil spill: dying wildlife covered with oil; smothered shellfish beds on the shore; unusable amenity beaches. It is not surprising that ever since the *Torrey Canyon* in 1967 (the first major tanker accident) oil spills have been media events. Questions about environmental effects and adequacy of response arise time and time again, but to help answer these it is now possible to draw upon decades of experience from three main types of activity: post-spill case studies and long-term monitoring; field experiments to test clean-up methods; and laboratory tests to investigate toxicities of oils and dispersants.

Oil is a complex substance containing hundreds of different compounds, mainly hydrocarbons (compounds consisting of carbon and hydrogen only). The three main classes of hydrocarbons in oil are alkanes (also known as paraffins), cycloalkanes (cycloparaffins or naphthenes) and arenes (aromatics). Compounds within each class have a wide range of molecular weights. Different refined products ranging from petrol to heavy fuel oil obviously differ in physical properties such as boiling range and

evaporation rates, and this is related to the molecular weights of the compounds they contain. Crude oils are also variable in their chemical composition and physical properties, depending on the field of origin. Chemical and physical properties are factors which partly determine environmental effects of spilt oil.

Tanker Accidents Compared With Chronic Inputs

Tanker accidents represent a small, but highly visible, proportion of the total oil inputs to the world's oceans each year (**Figure 1**). The incidence of large tanker spills has, however, declined since the 1970s. Irrespective of accidental spills, background hydrocarbons are ubiquitous, though at low concentrations. Water in the open ocean typically contains 1–10 parts per billion but higher concentrations are found in nearshore waters. Sediments or organisms may accumulate hydrocarbons in higher concentrations than does water, and it is common for sediments in industrialized bays to contain several hundred parts per million.

Sources of background hydrocarbons include:

- Operational discharges (e.g. bilge water) from ships and boats;
- Land-based discharges (e.g. industrial effluents, rainwater run-off from roads, sewage discharges);
- Natural seeps of petroleum hydrocarbons, such as occur, for example, along the coasts of Baffin Island and California;

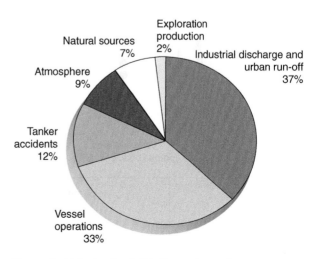

Figure 1 Major inputs of oil to the marine environment.

- Airborne combustion products, either natural (e.g. from forest fires) or artificial (e.g. from the burning of fossil fuels);
- Organisms, e.g. leaf waxes and hydrocarbons synthesized by algae.

Notwithstanding the relatively great total amounts of oil from operational and land-based sources these discharges usually comprise diluted, dissolved and dispersed oil. The sections below focus on accidental spills because it is these which produce visible slicks which may coat wildlife and shores, and which typically require a clean-up response.

Natural Fate of Oil Slicks

When oil is spilt on water, a series of complex interactions of physical, chemical, and biological processes takes place. Collectively these are called 'weathering'; they tend to reduce the toxicity of the oil and in time they lead to natural cleaning. The main physical processes are spreading, evaporation, dispersion as small drops, solution, adsorption onto sediment particles, and sinking of such particles. Degradation occurs through chemical oxidation (especially under the influence of light) and biological action – a large number of different species of bacteria and fungi are hydrocarbon degraders. Case history evidence gives a reasonable indication of natural cleaning timescales in different conditions. For open water sites, half-lives (the time taken for natural removal of 50% of the oil from the water surface) typically range from about half a day for the lightest oils (e.g. kerosene) to seven days or more for heavy oils (e.g. heavy fuel oil). However, for large spills near coastlines, some oil typically is stranded on the shore within a few days; once oil is stranded, the natural cleaning timescale may be prolonged. On the shore observed timescales range from a few days (some very exposed rocky shores) to more than 30 years (some very sheltered shores notably salt marshes). It is estimated that natural shore cleaning may take several decades in extreme cases.

Shore cleaning timescales are affected by factors including:

- exposure of the shore to wave energy (**Figure 2**) from very exposed rocky headlands to sheltered tidal flats, salt marshes and mangroves. This in turn depends on a number of variables that include fetch; speed, direction, duration and frequency of winds; and open angle of the shore;
- localized exposure/shelter – even on an exposed shore, cracks, crevices, and spaces under boulders can provide sheltered conditions where oil may persist;
- steepness/shore profile – extensive, gently sloping shores dissipate wave energy;

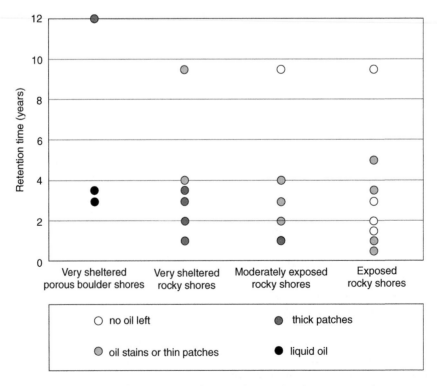

Figure 2 Oil residence times on a variety of rocky shores where no clean up has been attempted.

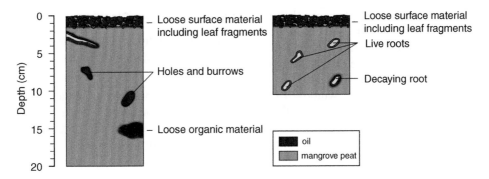

Figure 3 Two core samples taken from a mangrove swamp, showing penetration of oil down biological pathways.

- substratum – oil does not easily penetrate fine sediments (especially if they are waterlogged) unless they have biological pores such as crab and worm burrows or root holes (**Figure 3**). Penetration of shingle, gravel and some sand beaches can take place relatively easily, sometimes to depths of more than one metre;
- clay-oil flocculation – this process (first noticed on some Alaskan shores after the *Exxon Valdez* spill) reduces adherence of oil to shore substrata and facilitates natural cleaning;
- height of the stranded oil on the shore – oil spots taken into the supratidal zone by spray can persist for many years, conversely, oil on the middle and lower shore is more likely to be removed by water action. It is common to have stranded oil concentrated in the high tide area;
- oil type, e.g. viscosity affects movement into and out of sediment shores.

Oil Spill Response

Aims

The aims of oil spill response are to minimize damage and reduce the time for environmental recovery by:

- guiding or re-distributing the oil into less sensitive environmental components (e.g. deflecting oil away from mangroves onto a sandy beach) and/or
- removing an appropriate amount of oil from the area of concern and disposing of it responsibly.

Initiation of a response, or decision to stop cleaning or leave an area for natural clean-up, needs to be focused on agreed definitions of 'how clean is clean', otherwise there is no yardstick for determining whether the response has achieved the desired result.

The main response options are described below.

Booms and Skimmers

Booms and skimmers can be successful if the diversion and containment of the oil starts before it has had too much time to spread over the water surface. Booms tend to work well under calm sea conditions, but they are ineffective in rough seas. When current speeds are greater than 0.7–1.0 knots (1.3–1.85 km h^{-1}), with the boom at right angles to the current, the oil is entrained into the water, passes under the boom and is lost. In some cases it may be possible to angle the boom to prevent this. Booms can also be used for shoreline protection, for example to stop oil from entering sheltered inlets with marshes or mangroves. The time available for protective boom deployment depends on the position and movement of the slick, and can vary from hours to many days. The efficiency of skimmers depends on the oil thickness and viscosity, the sea state and the storage capacity of the skimmer. Skimmers normally work best in sheltered waters. Because of their limitations, recovering 10% of the oil at a large spill in open seas is considered good for these mechanisms.

Dispersants

Dispersants, which contain surfactants (surface active agents), reduce interfacial tension between oil and water. They promote the formation of numerous tiny oil droplets and retard the recoalescence of droplets into slicks. They do not clean oil out of the water, but can improve biodegradation by increasing the oil surface area, thereby increasing exposure to bacteria and oxygen. Information about dispersant effectiveness from accidental spills is limited for various reasons, such as inadequate monitoring and the difficulty of distinguishing between the contributions of different response methods to remove oil from the water surface. However, in at least some situations dispersants appear to remove a greater proportion of oil from the water surface than mechanical methods. Moreover, they can be used relatively quickly and under sea conditions where mechanical collection is impossible. Dispersants, however, do not work well in all circumstances (e.g. on heavy fuel oil) and even for initially dispersable oils, there is only a short 'window of opportunity', typically 1–2 days, after which the oil becomes too weathered for dispersants to be effective. The main environmental concern is that the dispersed oil droplets in the water column may in some cases affect organisms such as corals or fish larvae.

In situ Burning

Burning requires the use of special fireproof containment booms. It is best achieved on relatively fresh oil and is most effective when the sea is fairly calm. It generates a lot of smoke and as with dispersants, the window of opportunity is short, typically a few hours to a day or two, depending on the oil type and the environmental conditions at the time of the spill. When it is safe and logistically feasible, *in situ* burning is highly efficient in removing oil from the water surface.

Nonaggressive Shore Cleaning

Nonaggressive methods of shore cleaning, methods with minimal impact on shore structure and shore organisms, include:

- physical removal of surface oil from sandy beaches using machinery such as front-end loaders (avoiding removal of underlying sediment);
- manual removal of oil, asphalt patches, tar balls etc., by small, trained crews using equipment such as spades and buckets;
- collection of oil using sorbent materials (followed by safe disposal);
- low-pressure flushing with seawater at ambient temperature;
- bioremediation using fertilizers to stimulate indigenous hydrocarbon-degrading bacteria.

In appropriate circumstances these methods can be effective, but they may also be labor-intensive and clean-up crews must be careful to minimize trampling damage.

Nonaggressive methods do not work well in all circumstances. Low-pressure flushing, for example, is ineffective on weathered firmly adhering oil on rocks; and bioremediation is ineffective for subsurface oil in poorly aerated sediments.

Aggressive Shore Cleaning

Aggressive methods of shore cleaning, those that are likely to damage shore structure and/or shore organisms, include:

- removal of shore material such as sand, stones, or oily vegetation together with underlying roots and mud (in some cases the material may be washed and returned to the shore);
- water flushing at high pressure and/or high temperature;
- sand blasting.

In some cases these methods are effective at cleaning oil from the shore, e.g. hot water was more effective than cold water for removing weathered, viscous *Exxon Valdez* oil from rocks. However, heavy machinery, trampling and high-pressure water all can force oil into sediments and make matters worse.

Net Environmental Benefit Analysis for Oil Spill Response

In many cases a possible response to an oil spill is potentially damaging to the environment. The public perception of disaster has sometimes been heightened by headlines such as 'Clean-up makes things worse'. The advantages and disadvantages of different responses need to be weighed up and compared both with each other and with the advantages and disadvantages of entirely natural cleaning. This evaluation process is sometimes known as net environmental benefit analysis. The approach accepts that some response actions cause damage but may be justifiable because they reduce the overall problems resulting from the spill and response.

Example 1   Consider sticky viscous fuel oil adhering to rocks which are an important site for seals. If effective removal of oil could only be achieved by high-pressure hot-water washing or sand blasting, prolonged recovery times of shore organisms such as algae, barnacles and mussels might be accepted because the seals were given a higher priority. A consideration here and in similar cases is that populations of wildlife species are likely to be smaller, more localized, and slower to recover if affected by oil than populations of abundant and widespread shore algae and invertebrates.

Example 2   Consider a slick moving over shallow nearshore water in which there are coral reefs of particular conservation interest. The slick is moving towards sandy beaches important for tourism but of low biological productivity. Dispersant spraying will minimize pollution of the beaches, but some coral reef organisms may be damaged by dispersed oil. From an ecological point of view, it is best not to use dispersants but to allow the oil to strand on the beaches, from where it may be quickly and easily cleaned.

Effects and Recovery

Range of Effects

The range of oil effects after a spill can encompass:

- physical and chemical alteration of habitats, e.g. resulting from oil incorporation into sediments;
- physical smothering effects on flora and fauna;
- lethal or sublethal toxic effects on flora and fauna;
- short- and longer-term changes in biological communities resulting from oil effects on key organisms, e.g. increased abundance of intertidal algae following death of limpets which normally graze the algae;
- tainting of edible species, notably fish and shellfish, such that they are inedible and unmarketable (even though they are alive and capable of self-cleansing in the long term);
- loss of use of amenity areas such as sandy beaches;
- loss of market for fisheries products and tourism because of bad publicity (irrespective of the actual extent of the tainting or beach pollution);
- fouling of boats, fishing gear, slipways and jetties;
- temporary interruption of industrial processes requiring a supply of clean water from sea intakes (e.g. desalination).

The extent of biological damage can vary from minimal (e.g. following some open ocean spills) to the death of nearly everything in a particular biological community. Examples of extreme cases of damage following individual spills include deaths of thousands of sea birds, the death of more than 100 acres of mangrove forest, and damage to fisheries and/or aquaculture with settlements in excess of a million pounds.

Extent of damage, and recovery times, are influenced by the nature of the clean-up operations, and by natural cleaning processes. Other interacting factors include oil type, oil loading (thickness), local geography, climate and season, species and biological communities, and local economic and amenity considerations. With respect to oil type, crude oils and products differ widely in toxicity. Severe toxic effects are associated with hydrocarbons with low boiling points, particularly aromatics, because these hydrocarbons are most likely to penetrate and disrupt cell membranes. The greatest toxic damage has been caused by spills of lighter oil particularly when

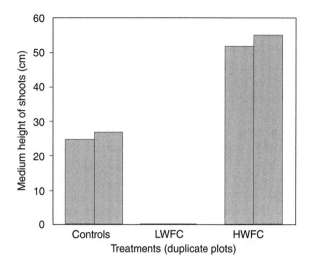

Figure 4 The effects of experimental oil treatments (duplicate plots) on shoot heights of the common salt marsh grass *Spartina anglica*. The measurements shown were taken four months after treatment. Lightly weathered Forties crude (LWFC) killed most of the grass shoots. Heavily weathered Flotta crude (HWFC) stimulated growth.

confined to a small area. Spills of heavy oils, such as some crudes and heavy fuel oil, may blanket areas of shore and kill organisms primarily through smothering (a physical effect) rather than through acute toxic effects. Oil toxicity is reduced as the oil weathers. Thus a crude oil that quickly reaches a shore is more toxic to shore life than oil that weathered at sea for several days before stranding. There have been cases of small quantities of heavy or weathered oils stimulating the growth of salt marsh plants (**Figure 4**).

Geographical factors which have a bearing on the course of events include the characteristics of the water body (e.g. calm shallow sea or deep rough sea), wave energy levels along the shoreline (because these affect natural cleaning) and shoreline sediment characteristics. Temperature and wind speeds influence oil weathering, and according to season, vulnerable groups of birds or mammals may be congregated at breeding colonies, and fish may be spawning in shallow nearshore waters.

Vulnerable Natural Resources

Salt marshes   Salt marshes are sheltered 'oil traps' where oil may persist for many years. In cases where perennial plants are coated with relatively thin oil films, recovery can take place through new growth from underground stems and rootstocks. In extreme cases of thick smothering deposits, recovery times may be decades.

Mangroves   Mangrove forests are one of the most sensitive habitats to oil pollution. The trees are easily killed by crude oil, and with their death comes loss of habitat for the fish, shellfish and wildlife which depend on them. Mangrove estuaries are sheltered 'oil trap' areas into which oil tends to move with the tide and then remain among the prop roots and breathing roots, and in the sediments (**Figure 5**).

Coral reefs   Coral reef species are sensitive to oil if actually coated with it. There is case-history evidence of long-term damage when oil was stranded on a reef flat at low tide. However, the risk of this type of scenario is quite low – oil slicks will float over coral reefs at most stages of the tide, causing little damage. Deep water corals will escape direct oiling at any stage of the tide.

Fish   Eggs, larvae and young fish are comparatively sensitive but there is no definitive evidence

which suggests that oil pollution has significant effects on fish populations in the open sea. This is partly because fish can take avoiding action and partly because oil-induced mortalities of young life stages are often of little significance compared with huge natural losses each year (e.g. through predation). There is an increased risk to some species and life stages of fish if oil enters shallow near-shore waters which are fish breeding and feeding grounds. If oil slicks enter into fish cage areas there may be some fish mortalities, but even if this is not the case there is likely to be tainting. Fishing nets, fish traps and aquaculture cages are all sensitive because adhering oil is difficult to clean and may taint the fish.

Turtles   Turtles are likely to suffer most from oil pollution during the breeding season, when oil at

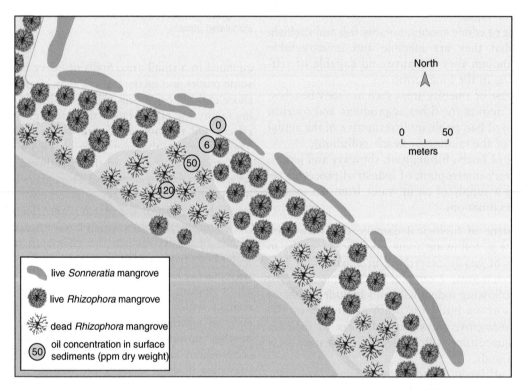

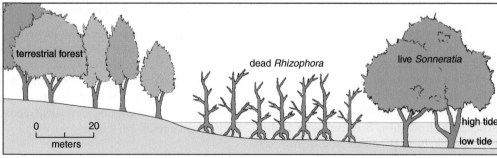

Figure 5   Plan and profile showing mangrove patches killed by small oil slicks.

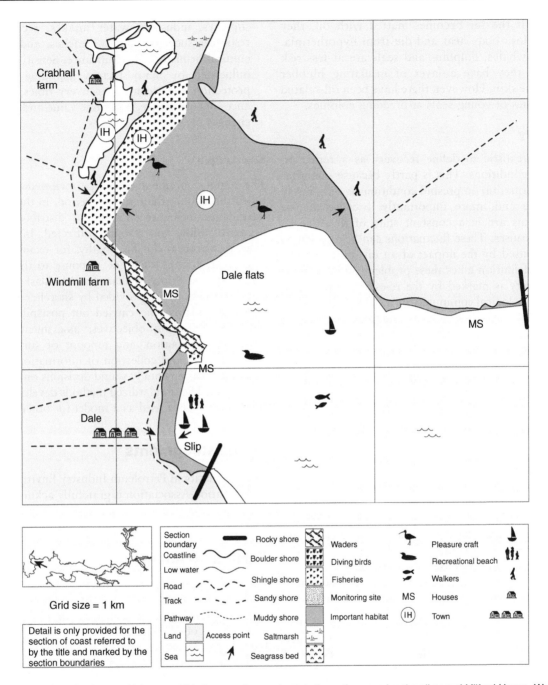

Figure 6 An example of a sensitivity map. This large-scale map is part of an atlas covering the oil port of Milford Haven, Wales. It facilitates response when oil has come near shore or onshore, and the protection or clean-up of specific locations is being planned.

egg-laying sites could have serious effects on eggs or hatchlings. If oiling occurs, the effects from the turtle conservation point of view could be serious, because the various turtle species are endangered.

Birds Seabirds are extremely sensitive to oiling, with high mortality rates of oiled birds. Moreover, there is experimental evidence that small amounts of oil transferred to eggs by sublethally oiled adults can significantly reduce hatching success. Shore birds, notably waders, are also at risk. For them, a worst-case scenario would be oil impacting shore feeding grounds at a time when large numbers of migratory birds were coming into the area.

Mammals Marine mammals with restricted coastal distributions are more likely to encounter oil than wide-ranging species moving quickly through an area. Species at particular risk are those which rely on fur for conservation of body heat (e.g. sea otters). If the fur becomes matted with oil, they rapidly lose body heat and die from hypothermia. At sea, whales, dolphins and seals are at less risk because they have a layer of insulating blubber under the skin. However, there have been oil-related mortalities of young seals at breeding colonies.

Recovery

It is unrealistic to define recovery as a return to prespill conditions. This is partly because quantitative information on prespill conditions is only rarely available and, more importantly, because marine ecosystems are in a constant state of flux due to natural causes. These fluctuations can be as great as those caused by the impact of an oil spill. The following definition takes these problems into account.

Recovery is marked by the re-establishment of a healthy biological community in which the plants and animals characteristic of that community are present and functioning normally. It may not have the same composition or age structure as that which was present before the damage, and will continue to show further change and development. It is impossible to say whether an ecosystem that has recovered from an oil spill is the same as, or different from, that which would have persisted in the absence of the spill.

Assessment

Before a Spill

Before a spill it is important to identify what the particular sensitivities are for the area covered by any particular oil spill contingency plan, and to put the information on a sensitivity map which will be available to response teams. An example is shown in **Figure 6**. Maps should include information on the following.

- Shoreline sensitivity. Shorelines may be ranked using the basic principles that sensitivity to oil increases with increasing shelter of the shore from wave action, penetration of oil into the substratum, natural oil retention times on the shore, and biological productivity of shore organisms. Typically, the least sensitive shorelines are exposed rocky headlands, and the most sensitive are marshes and mangroves forests.
- Other ecological resources such as coral reefs, seagrass and kelp beds, and wildlife such as turtles, birds and mammals.
- Socioeconomic resources, for example fishing areas, shellfish beds, fish and crustacean nursery areas, fish traps and aquaculture facilities. Other features include boat facilities such as harbors and

slipways, industrial water intakes, recreational resources such as amenity beaches, and sites of cultural or historical significance. Sensitivities are influenced by many factors including ease of protection and clean-up, recovery times, importance for subsistence, economic value and seasonal changes in use.

After a Spill

The response options need to be reviewed and fine-tuned throughout the response period, in the light of information being received about distribution and degree of oiling and resources affected. In extreme cases this process can be lengthy, for example over three years for the shoreline response to the *Exxon Valdez* spill in Prince William Sound, Alaska. In this case information was provided by shoreline clean-up assessment teams who carried out postspill surveys with the following objectives: assessment of the presence, distribution, and amount of surface and subsurface oil, and collection of information needed to make environmentally sound decisions on clean-up techniques. The standardized methods developed have subsequently been used as a model for other spills.

Acknowledgments

The International Petroleum Industry Environmental Conservation Association is gratefully acknowledged for permission to use material from its Report series.

Further Reading

American Petroleum Institute, Washington, DC. *Oil Spill Conference Proceedings*, published biennially from 1969 onwards. The primary source detailed papers on all aspects of oil pollution.

IPIECA Report Series, Vol. 1 (1991) *Guidelines on Biological Impacts of Oil Pollution*; vol. 2 (1991) *A Guide to Contingency Planning for Oil Spills on Water*; vol. 3 (1992) *Biological Impacts of Oil Pollution: Coral Reefs*; vol. 4 (1993) *Biological Impacts of Oil Pollution: Mangroves*; vol. 5 (1993) *Dispersants and their Role in Oil Spill Response*; vol. 6 (1994) *Biological Impacts of Oil Pollution: Saltmarshes*; vol. 7 (1995) *Biological Impacts of Oil Pollution: Rocky Shores*; vol. 8 (1997) *Biological Impacts of Oil Pollution: Fisheries*; vol. 9 (1999) *Biological Impacts of Oil Pollution: Sedimentary Shores*; vol. 10 (2000) Choosing Spill Response Options to Minimize Damage. London: International Petroleum Industry Environmental Conservation Association.

ITOPF (1987) *Response to Marine Oil Spills. International Tanker Owners Pollution Federation Ltd, London*. London: Witherby.

POLLUTION, SOLIDS

C. M. G. Vivian and L. A. Murray, The Centre for Environment, Fisheries and Aquaculture Sciences, Lowestoft, UK

Introduction

A very wide variety of solid wastes have been dumped at sea or discharged into the oceans via pipelines or rivers, either deliberately or as a consequence of other activities. The main solid materials involved are dredged material, particulate wastes from sand/gravel extraction and land reclamation, industrial wastes, including mining wastes and munitions, and plastics and litter.

Regulation

The London Convention 1972 has regulated dumping at sea of these materials on a global basis since it came into force in 1974 and currently has 83 contracting parties (mid-2008). In 1996, following a detailed review that began in 1993, the Contracting Parties to the London Convention 1972 agreed a Protocol to the Convention to update and modernize it. This came into force for those states that had ratified the Protocol on 24 March 2006 after the 27th state ratified it and currently (mid-2008) has 34 contracting parties. Regional conventions also regulate the dumping of these materials in some parts of the world, for example, the OSPAR Convention 1992 for the North-East Atlantic, the Helsinki Convention for the Baltic (updated in 2004), the Barcelona Convention 1995 for the Mediterranean, the Cartagena Convention 1983 for the Caribbean, and the Noumea Convention 1986 for the South Pacific. Where materials are discharged into the oceans via pipelines or rivers, this usually falls under national regulation but regional controls or standards may strongly influence this regulation, for example, within the countries of the European Union. When human activities introduce solid materials as a by-product of those activities, this is commonly subject only to national regulation, if any at all.

Impacts

All solid wastes have significant physical impacts at the point of disposal when disposed of in bulk. These impacts can include covering of the seabed, shoaling, and short-term local increases in the levels of water-suspended solids. Physical impacts may also result from the subsequent transport, particularly of the finer fractions, by wave and tidal action and residual current movements. All these impacts can lead to interference with navigation, recreation, fishing, and other uses of the sea. In relatively enclosed waters, the materials with a high chemical or biological oxygen demand (e.g., organic carbon-rich) can adversely affect the oxygen regime of the receiving environment while materials with high levels of nutrients can significantly affect the nutrient flux and thus potentially cause eutrophication.

Biological consequences of these physical impacts can include smothering of benthic organisms, habitat modification, and interference with the migration of fish or shellfish due to turbidity or sediment deposition. The toxicological effects of the constituents of these materials may be significant. In addition, constituents may undergo physical, chemical, and biochemical changes when entering the marine environment and these changes have to be considered in the light of the eventual fate and potential effects of the material.

Dredged Material

Dredging

Dredging is undertaken for a variety of reasons including navigation, environmental remediation, flood control, and the emplacement of structures (e.g., foundations, pipelines, and tunnels). These activities can generate large volumes of waste requiring disposal. Dredging is also undertaken to win materials, for example, for reclamation or beach nourishment. Types of material dredged can include sand, silt, clay, gravel, coral, rock, boulders, and peat. Mining for ores is considered further within the section on industrial wastes.

Most dredging activities, particularly hydraulic dredging, generate an overspill of fine solid material as a consequence of the dredging activity. The particle size characteristics of this material and the hydrodynamics of the site will determine how far it may drift before settling and this will influence the extent and severity of any physical impact outside the immediate dredging site. The chemical characteristics of this material have to be taken into consideration in any risk assessment of the consequences to the environment.

Dredged Material

Waterborne transport is vital to domestic and international commerce. It offers an economical, energy efficient and environmentally friendly transportation for all types of cargo. Dredging is essential to maintain navigable depths in ports and harbors and their approach channels in estuaries and coastal waters (maintenance) as well as for the development of port facilities (capital). Maintenance dredged materials tend generally to consist of sands and silts, whereas capital dredged material may include any of the materials mentioned under dredging. Dredged material can be used for land reclamation or other beneficial uses but most of that generated in marine areas is disposed of in estuaries and coastal waters by dumping from ships or barges. The disposal of dredged material is the largest mass input of wastes deposited directly into the oceans. However, it does not follow that dredged material is the most environmentally significant waste deposited in the ocean, since much dredged material disposal is relocation of sediments already in the marine environment, rather than fresh input.

London Convention data indicate that Contracting Parties dumped around 400 million tonnes dredged material in 2003. However, not all coastal states have signed and ratified the Convention and less than half of the Contracting Parties regularly report on their dumping activities. Data from the International Association of Ports and Harbours indicated that in 1981 around 580 million tonnes of dredged material were being dumped each year and that survey only had data from half the countries contacted. Thus, it would seem likely that actual quantities of dredged material dumped in the world's ocean could be up to 1000 million tonnes per annum.

Regulation

The Specific Guidelines for Assessment of Dredged Material (SGADM) of the global London Convention 1972 is widely accepted as the standard guidance for the management of dredged material disposal at sea. These guidelines were last updated in 2000 and are available on the London Convention website. Regional conventions will often have their own guidance documents that will usually be more specific to take account of local circumstances, for example, The Guidelines for the Management of Dredged Material of the OSPAR Convention last revised in 2004.

Characterization

Dredged material needs to be characterized before disposal as part of the risk assessment procedure if environmental harm is to be minimized. This will usually require consideration of its physical and chemical characteristics and may require assessment of its biological impacts, either inferred from the chemical data, or directly through the conduct of tests for toxicity, bioaccumulation, etc. The physical properties and chemical composition of dredged materials are highly variable depending on the nature of the material concerned and its origin, for example, grain size, mineralogy, bulk properties, organic matter content, and exposure to contamination.

Impacts

In considering the impacts of dredged material disposal, a distinction may be made between natural geological materials not generally exposed to anthropogenic influences, for example, material excavated from beneath the seafloor in deepening a navigation channel and those sediments that have been contaminated by human activities. Although physical, and associated biological, impacts may arise from both types of materials, chemical impacts and their associated biological consequences are usually of particular concern from the latter.

In common with the other materials covered in this article, the physical impact of dredged material disposal is most often the most obvious and significant impact on the aquatic environment. The seabed impacts can be aggravated if the sediment characteristics of the material are very different from that of the sediment at the disposal site or if the material is contaminated with debris such as pieces of cable, scrap metal, and wood.

Most of the material dredged from coastal waters and estuaries is, by its nature, either uncontaminated or only lightly contaminated by human activity (i.e., at, or close to, natural background levels). Sediments in enclosed docks and waterways are often subject to more intense contamination, both because of local practices, and the reduced rate of flushing of such areas. This contamination may be by metals, oil, synthetic organics (e.g., polychlorinated biphenyls, polyaromatic hydrocarbons, and pesticides) or organometallic compounds such as tributyl tin. The latter has been a major concern in recent years due to its previous widespread use in antifouling paints, its consequential occurrence in estuarine sediments in particular, and its wide range of harmful effects to many marine organisms. In recognition of this concern, the international community agreed the International Convention on the Control of Harmful Anti-fouling Systems on Ships, which was adopted on 5 October 2001. This will prohibit the use of harmful organotins in antifouling paints used on ships and will establish a mechanism to prevent the potential future use of other harmful substances in antifouling systems. The convention will enter into force on 17 September 2008.

However, it is generally the case that only a small proportion of dredged material is contaminated to an extent that either major environmental constraints need to be applied when depositing these sediments or the material has to be removed to land for treatment or to specialized confinement facilities.

Beneficial Uses

Dredged material is increasingly being regarded as a resource rather than a waste. Worldwide, around 90% of dredged material is either uncontaminated or only lightly contaminated by human activity so that it can be acceptable for a wide range of uses. The London Convention SGADM recognizes this and requires that possible beneficial uses of dredged material be considered as the first step in examining dredged material management options. Beneficial uses of dredged materials can include beach nourishment, coastal protection, habitat development or enhancement and land reclamation. Operational feasibility is a crucial aspect of many beneficial uses, that is, the availability of suitable material in the required amounts at the right time sufficiently close to the use site. PIANC produced practical guidance on beneficial uses of dredged material in 1992 that is currently (2008) in the process of being revised.

Sand/Gravel Extraction

Introduction

Sand and gravel is dredged from the seabed in various parts of the world for, among other things, land reclamation, concreting aggregate, building sand, beach nourishment, and coastal protection. The largest producer in the world is Japan at around $80–100\,Mm^3$ per year, Hong Kong at $25–30\,Mm^3$ per year, the Netherlands and the UK regularly producing some $20–30\,Mm^3$ per year, and Denmark, the Republic of Korea, and China lesser amounts.

Reclamation with Marine Dredged Sand and Gravel

In recent years, some very large land reclamations have taken place for port and airport developments particularly in Asia. For example, in Hong Kong some $170\,Mm^3$ was placed for the new artificial island airport development with another $80\,Mm^3$ used for port developments over the period 1990–98. Demand projections indicate a need for a further $300\,Mm^3$ by 2010. Also, in Singapore the Jurong

Island reclamation project initiated in 1999 was projected to require $220\,Mm^3$ of material over 3 years. However, in 2003 Singapore estimated that it needed 1.8 billion cubic meters of sand over the following 8 years for reclamation works including Tuas View, Jurong Island, and Changi East. The planned Maasvlakte 2 extension of Rotterdam Port in the Netherlands will reclaim some 2000 ha and require $400\,Mm^3$ of sand with construction due to take place between 2008 and 2014.

Regulation

National authorities generally carry out the regulation of sand and gravel extraction and they may have guidance on the environmental assessment of the practice. There is currently no accepted international guidance other than for the Baltic Sea area under the Helsinki Convention and the North-East Atlantic under the OSPAR Convention that has adopted the ICES guidance.

Impacts

The environmental impacts of sand and gravel dredging depend on the type and particle size of the material being dredged, the dredging technique used, the hydrodynamic situation of the area and the sensitivity of biota to disturbance, turbidity, or sediment deposition. Screening of cargoes as they are loaded is commonly employed when dredging for sand or gravel to ensure specific sand:gravel ratios are retained in the dredging vessel. Exceptionally, this can involve the rejection of up to 5 times as much sediment over the side of the vessel as is kept as cargo. When large volumes of sand or gravel are used in land reclamation, the runoff from placing the material may contain high levels of suspended sediment. The potential effects of this runoff are very similar to the impacts from dredging itself. The *ICES Cooperative Research Report No. 247* published in 2001 deals with the effects of extraction of marine sediments on the marine ecosystem. It summarizes the impacts of dredging for sand and gravel. An update of this report is due for completion in 2007.

Physical impacts   The most obvious and immediate impacts of sand and gravel extraction are physical ones arising from the following.

- *Substrate removal and alteration of bottom topography.* Trailer suction dredgers leave a furrow in the sediment of up to 2 m wide by about 30 cm deep but stationary (or anchor) suction dredgers may leave deep pits of up to 5 m deep or more. Infill of the pits or furrows created depends on the natural stability of the sediment and the rate of sediment movement due to tidal currents or wave action. This can take many years in some instances. A consequence of significant depressions in the seabed is the potential for a localized drop in current strength resulting in the deposition of finer sediments and possibly a localized depletion in dissolved oxygen.

- *Creation of turbidity plumes in the water column.* This results mainly from the overflow of surplus water/sediment from the spillways of dredgers, the rejection of unwanted sediment fractions by screening, and the mechanical disturbance of the seabed by the draghead. It is generally accepted that the latter is of relatively small significance compared to the other two sources. Recent studies in Hong Kong and the UK indicate that the bulk of the discharged material is likely to settle to the seabed within 500 m of the dredger but the very fine material (<0.063 mm) may remain in suspension over greater distances due to the low settling velocity of the fine particles.

- *Redeposition of fines from the turbidity plumes and subsequent sediment transport.* Sediment that settles out from plumes will cover the seabed within and close to the extraction site. It may also be subject to subsequent transport away from the site of deposition due to wave and tidal current action since it is liable to have less cohesion and may be finer (due to screening) than undredged sediments.

Chemical impacts The chemical effects of sand and gravel dredging are likely to be minor due to the very low organic and clay mineral content of commercial sand and gravel deposit from geological deposits not generally exposed to anthropogenic influences and to the generally limited spatial and temporal extent of the dredging operations.

Biological impacts The biological impacts of sand and gravel dredging derive from the physical impacts described above and the most obvious impacts are on the benthic biota. The consequences are as follows.

- *Substrate removal and alteration of bottom topography.* This is the most obvious and immediate impact on the ecosystem. Few organisms are likely to survive intact as a result of passage through a dredger but damaged specimens may provide a food source to scavenging invertebrates and fish. Studies in Europe on dredged areas show very large depletions in species abundance, number of taxa, and biomass of benthos immediately

following dredging. The recolonization process following cessation of dredging depends primarily on the physical stability of the seabed sediments and that is closely related to the hydrodynamic situation. Significant recovery toward the original faunal state of the benthos depends on having seabed sediment of similar characteristics to that occurring before dredging. The biota of mobile sediments tends to be more resilient to disturbance and able to recolonize more quickly than the more long-lived biota of stable environments. Generally, sands and sandy gravels have been found to have recovery times of 2–4 years. However, recovery may take longer where rare slow-growing animals were present prior to dredging. The fauna of coarser deposit are likely to recover more slowly, taking from 5 to 10 years. Recolonizing fauna may move in from adjacent undredged areas or result from larval settlement from the water column.

- *Creation of turbidity plumes in the water column.* As these tend to be limited in space and time, they are not likely to be of great significance for water column fauna unless dredging takes place frequently or continuously.

- *Redeposition of fines from the turbidity plumes and subsequent sediment transport.* Deposition of fines may smother the benthos in and around the area actually dredged but may not cause mortality where the benthos is adapted to dealing with such a situation naturally, for example, in areas with mobile sediments. The deposited fines may be transported away from the site of deposition and affect more distant areas, particularly if the deposited sediment is finer than that naturally on the seabed as a result of screening.

Industrial Solids

Introduction

Industries based around coasts and estuaries have historically used those waters for waste disposal whether by direct discharge or by dumping at sea from vessels. Many of these industries do not generate significant quantities of solid wastes but some, particularly mining and those processing bulk raw materials, may do so.

The sea may be impacted by mining taking place on land, where waste materials are disposed into estuaries and the sea by river inputs, direct tipping, dumping at sea or through discharge pipes. Examples include china clay waste discharged into coastal waters of Cornwall (southwest England) via rivers, colliery waste dumped off the coast of northeast England, and tailings from metalliferous mines in Norway and Canada tipped into fjords and coastal waters. Alternatively, mining may itself be a water-borne activity, such as the mining of diamonds in Namibian waters or tin in Malaysian waters, with waste material discharged directly into the sea.

Fly ash has been dumped at sea off the coast of northeast England since the early 1950s, although this ceased in 1992, and was also carried out off the coast of Israel in the Mediterranean Sea between 1988 and 1998. Fly ash is derived from 'pulverized fuel ash' from the boilers of coal power stations but is mainly 'fly ash' extracted from electrostatic filters and other air pollution control equipment of coal- and oil-fired power stations. This fine material is mainly composed of silicon dioxide with oxides of aluminum and iron but may carry elemental contaminants condensed onto its surfaces following the combustion process.

Surplus munitions, including chemical munitions, were disposed of at sea in various parts of the world in large quantities after both World Wars I and II. In addition, a number of countries have made regular but smaller disposals at sea since 1945. In earlier times much of this material was deposited on continental shelves, but since around 1970 most disposals have taken place in deep ocean waters. OSPAR has collated information from its contracting parties on munitions disposal into a database that is available on its website.

Regulation

The London Convention 1972 has regulated the dumping of industrial wastes at sea since it came into force in 1974. Annex I of the Convention was amended with effect from 1 January 1996 to ban the dumping at sea of industrial wastes by its Contracting Parties. However, not all coastal states have signed and ratified the Convention and less than half of the Contracting Parties regularly report on their dumping activities. The 1996 Protocol to the London Convention restricts the categories of wastes permitted to be considered for disposal at sea to the 7 listed in Annex 2 of the Protocol, which is very similar to the amended Annex I of the Convention referred to above. The protocol was amended in late 2006 to allow the sequestration of carbon dioxide in sub-seabed geological structures as one of a range of climate change mitigation options. In the period 2000–04 Japan dumped some 1.0 million tonnes per annum of bauxite residues into the sea. This activity was controversial as some contracting parties to the Convention regarded the material as industrial waste, and thus banned from disposal, while Japan maintained that it was an Inert Material of Natural Origin and thus allowed to be disposed of at sea. However, Japan announced in 2005 that it would phase out this disposal by 2015.

Characterization

Like dredged material, these materials need to be characterized before disposal can be permitted to ensure that significant environmental harm is not caused as a result of its disposal.

Impacts

The impacts of the disposal of solid industrial wastes into the sea are as described in the introduction, although chemical and biological impacts may be more significant than for some other solid wastes due to the constituents of the materials. However, physical impacts are usually dominant due to the bulk properties of these materials. Fly ash has a special property (pozzolanic activity) that causes the development of strong cohesiveness between the particles on contact with water. This leads to aggregation of the particles and the formation of concretions that may cover all or part of the seabed.

Beneficial Uses

Beneficial uses can often be found for these materials but depend strongly on the characteristics of the material and the opportunities for use in the area of production. Inorganic materials may be used for example for fill, land reclamation, road foundations, or building blocks, and organic materials may be composted, used as animal feed or as a fertilizer.

Plastics and Litter

Introduction

There has been growing concern over the last decade or two about the increasing amounts of persistent plastics and other debris found at sea and on the coastline. Since the 1940s there has been an enormous increase in the use of plastics and other synthetic materials that have replaced many natural and more degradable materials. These materials are generally very durable and cheap but his means that they tend to be both very persistent and readily discarded. They are often buoyant so that they can accumulate in shoreline sinks.

Regulation

A number of international conventions including the London Convention 1972, MARPOL 73/78, and regional seas conventions ban the disposal into the sea of persistent plastics. Annex V of the International Convention for the Prevention of Pollution from Ships 1973, as modified by the 1978 Protocol (known as MARPOL 73/78), states that "the disposal into the sea of all plastics, including but not limited to synthetic ropes, synthetic fishing nets and plastic garbage bags, is prohibited." An exception is made for the accidental loss of synthetic fishing nets provided all reasonable precautions have been taken to prevent such loss. This Annex came into force in 1988. Guidelines for implementing Annex V of MARPOL 73/78 suggest best practice for waste handling on vessels. Many parties to MARPOL 73/78 have built additional waste reception facilities at ports and/or expanded their capacity in order to encourage waste materials to be landed rather than disposed of into the sea. Annex I of the London Convention 1972 prohibits the dumping of "persistent plastics and other persistent synthetic materials, for example netting and ropes, which may float or remain in suspension in the sea in such a manner as to interfere materially with fishing, navigation or other legitimate uses of the sea." Since these materials are not listed in Annex 2 of the 1996 Protocol to the London Convention, this instrument also bans them. All land-based sources of these materials fall under national regulation.

Impacts

Persistent plastics and other synthetic materials present a threat to marine wildlife, as well as being very unsightly and potentially affecting economic interests in recreation areas. They are a visible reminder that the ocean is being used as a dump for plastic and other wastes. These materials also present a threat to navigation through entangling propellers and blocking of cooling water systems of vessels. In 1975 the US Academy of Sciences estimated that 6.4 million tonnes of litter was being discarded each year by the shipping and fishing industries. According to other estimates in a 2005 UNEP report some 8 million items of marine litter are dumped into seas and oceans everyday, approximately 5 million of which (solid waste) are thrown overboard or lost from ships. It has also been estimated that over 13 000 pieces of plastic litter are floating on every square kilometer of ocean today. The UNEP Regional Seas Programme is currently (2006) developing a series of global and regional activities aimed at controlling, reducing, and abating the problem. The debris that is most likely to be disposed of or lost can be divided into three groups.

Fishing gear and equipment Nets that are discarded or lost can continue to trap marine life whether floating on the surface, on the bottom, or at some intermediate level. Marine mammals, fish, seabirds, and turtles are among the animals caught. In 1975 the UN Food and Agriculture Organization estimated that 150 000 tonnes of fishing gear was lost annually worldwide.

Strapping bands and synthetic ropes These are used to secure cargoes, strap boxes, crates, or packing cases and hold materials on pallets. When pulled off rather than cut, the discarded bands may encircle marine mammals or large fish and become progressively tighter as the animal grows. They also entangle limbs, jaws, heads, etc., and affect the animal's ability to move or eat. Plastic straps for four or six packs of cans or bottles can affect smaller animals in a similar way.

Miscellaneous other debris This covers a wide variety of materials including plastic bags or sheeting, packing material, plastic waste materials, and containers for beverages and other liquids. There are numerous studies of turtles, whales, and other marine mammals that were apparently killed by ingesting plastic bags or sheeting. Turtles appear particularly vulnerable to this type of pollution, perhaps by mistaking it for their normal food, for example, jellyfish. A potentially more serious problem may be the increasing quantities of small plastic particles widely found in the ocean, probably from plastics production and insulation, and packing materials. The principal impacts of this material are via ingestion affecting animals' feeding digestion processes. These plastic particles have been found in 25% of the world's sea bird species in one study and up to 90% of the chicks of a single species in another study.

See also

International Organizations. Law of the Sea. Marine Policy Overview. Pollution: Approaches to Pollution Control.

Further Reading

Arnaudo R (1990) The problem of persistent plastics and marine debris in the oceans. In: *Technical Annexes to the GESAMP Report on the State of the Marine*

Environment, UNEP Regional Seas Reports and Studies No. 114/1, Annex I, 20pp. London: UNEP

Duedall IW, Ketchum BH, Park PK, and Kester DR (1983) Global inputs, characteristics and fates of ocean-dumped industrial and sewage wastes: An overview. In: Duedall IW, Ketchum BH, Park PK, and Kester DR (eds.) *Wastes in the Ocean, Industrial and Sewage Wastes in the Ocean*, vol. 1, pp. 3–45. New York, NY: Wiley.

IADC/CEDA (1996–99) *Environmental Aspects of Dredging Guides 1–5*. The Hague: International Association of Dredging Companies.

IADC/IAPH (1997) *Dredging for Development*. The Hague: International Association of Dredging Companies.

ICES (2001) Effects of extraction of marine sediments on the marine ecosystem. *International Council for the Exploration of the Sea, Cooperative Research Report 247*.

Kester DR, Ketchum BH, Duedall IW, and Park PK (1983) The problem of dredged material disposal. In: Kester DR, Ketchum BH, Duedall IW, and Park PK (eds.) *Wastes in the Ocean, Dredged-Material Disposal in the Ocean*, vol. 2, pp. 3–27. New York: Wiley.

Newell RC, Seiderer LJ, and Hitchcock DR (1998) The impact of dredging on biological resources of the seabed. *Oceanography and Marine Biology Annual Review* 36: 127–178.

Thompson RC, Olsen Y, Mitchell RP, *et al.* (2004) Lost at sea: Where is all the plastic? *Science* 304: 838.

United Nations Environment Programme (2005) *Marine Litter: An Analytical overview*. Nairobi: UNEP.

Relevant Websites

http://www.helcom.fi
– Helsinki Commission: Baltic Marine Environment Protection Commission.

http://www.ices.dk
– International Council for the Exploration of the Sea.

http://www.londonconvention.org
– London Convention 1972.

http://www.ospar.org
– OSPAR Commission.

http://www.pianc-aipcn.org
– PIANC.

http://www.unep.org
– United Nations Environment Programme (UNEP).

http://www.woda.org
– World Organisation of Dredging Associations.

RADIOACTIVE WASTES

L. Føyn, Institute of Marine Research, Bergen, Norway

Introduction

The discovery and the history of radioactivity is closely connected to that of modern science. In 1896 Antoine Henri Becquerel observed and described the spontaneous emission of radiation by uranium and its compounds. Two years later, in 1898, the chemical research of Marie and Pierre Curie led to the discovery of polonium and radium.

In 1934 Frédéric Joliot and Irène Curie discovered artificial radioactivity. This discovery was soon followed by the discovery of fission and the enormous amounts of energy released by this process. However, few in the then limited community of scientists working with radioactivity believed that it would be possible within a fairly near future to establish enough resources to develop the fission process for commercial production of energy or even think about the development of mass-destruction weapons.

World War II made a dramatic change to this. The race that began in order to be the first to develop mass-destruction weapons based on nuclear energy is well known. Following this came the development of nuclear reactors for commercial production of electricity. From the rapidly growing nuclear industry, both military and commercial, radioactive waste was produced and became a problem. As with many other waste problems, discharges to the sea or ocean dumping were looked upon as the simplest and thereby the best and final solution.

The Sources

Anthropogenic radioactive contamination of the marine environment has several sources: disposal at sea, discharges to the sea, accidental releases and fallout from nuclear weapon tests and nuclear accidents. In addition, discharge of naturally occurring radioactive materials (NORM) from offshore oil and gas production is a considerable source for contamination.

The marine environment receives in addition various forms of radioactive components from medical, scientific and industrial use. These contributions are mostly short-lived radionuclides and enter the local marine environment through diffuse outlets like muncipal sewage systems and rivers.

Disposal at Sea

The first ocean dumping of radioactive waste was conducted by the USA in 1946 some 80 km off the coast of California. The International Atomic Energy Agency (IAEA) published in August 1999 an 'Inventory of radioactive waste disposal at sea' according to which the disposal areas and the radioactivity can be listed as shown in **Table 1**.

Figure 1 shows the worldwide distribution of disposal-points for radioactive waste.

The majority of the waste disposed consists of solid waste in various forms and origin, only 1.44% of the total activity is contributed by low-level liquid waste. The disposal areas in the north-east Atlantic and the Arctic contain about 95% of the total radioactive waste disposed at sea.

Most disposal of radioactive waste was performed in accordance with national or international regulations. Since 1967 the disposals in the north-east Atlantic were for the most part conducted in accordance with a consultative mechanism of the Organization for Economic Co-operation and Development/Nuclear Energy Agency (OECD/NEA).

The majority of the north-east Atlantic disposals were of low-level solid waste at depths of 1500–5000 m, but the Arctic Sea disposals consist of various types of waste from reactors with spent fuel to containers with low-level solid waste dumped in fairly shallow waters ranging from about 300 m depth in the Kara Sea to less than 20 m depth in some fiords on the east coast of Novaya Zemlya.

Most of the disposals in the Arctic were carried out by the former Soviet Union and were not reported internationally. An inventory of the USSR disposals was presented by the Russian government in 1993. Already before this, the good collaboration between Russian and Norwegian authorities had led

Table 1 Worldwide disposal at sea of radioactive waste[a]

North-west Atlantic Ocean	2.94 PBq
North-west Atlantic Ocean	2.94 PBq
Arctic Ocean	38.37 PBq
North-east Pacific Ocean	0.55 PBq
West Pacific Ocean	0.89 PBq

[a]PBq (petaBq) = 10^{15} Bq (1 Bq1 = disintegration s^{-1}). The old unit for radioactivity was Curie (Ci); 1 Ci3.7 × 10^{10} Bq.

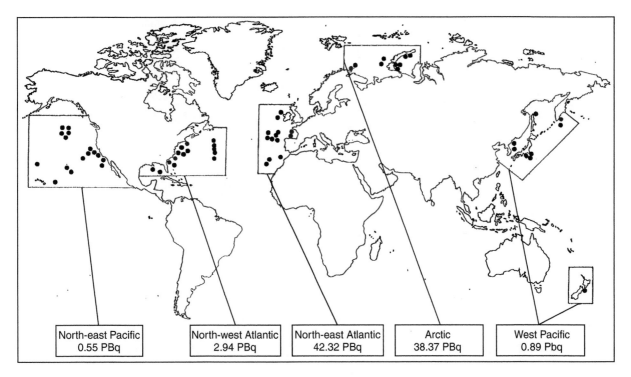

Figure 1 The worldwide location for disposal of radioactive waste at sea.

to a joint Norwegian–Russian expedition to the Kara Sea in 1992 followed by two more joint expeditions, in 1993 and 1994. The main purpose of these expeditions was to locate and inspect the most important dumped objects and to collect samples for assessing the present environmental impact and to assess the possibility for potential future leakage and environmental impacts.

These Arctic disposals differ significantly from the rest of the reported sea disposals in other parts of the world oceans as most of the dumped objects are found in shallow waters and some must be characterized as high-level radioactive waste, i.e. nuclear reactors with fuel. Possible releases from these sources may be expected to enter the surface circulation of the Kara Sea and from there be transported to important fisheries areas in the Barents and Norwegian Seas. **Figures 2** and **3** give examples of some of the radioactive waste dumped in the Arctic. Pictures were taken with a video camera mounted on a ROV (remote operated vehicle). The ROV was also equipped with a NaI-detector for gamma-radiation measurements and a device for sediment sampling close to the actual objects.

The Global Convention on the Prevention of Marine Pollution by Dumping of Wastes and Other Matter was adopted by an Intergovernmental Conference in London in 1972. The convention named the London Convention 1972, formerly the London Dumping Convention (LDC), addressed from the very beginning the problem of radioactive waste. But it was not until 20 February 1994 that a total prohibition on radioactive waste disposal at sea came into force.

Discharges to the Sea

Of the total world production of electricity about 16% is produced in nuclear power plants. In some countries nuclear energy counts for the majority of the electricity produced, France 75% and Lithuania 77%, and in the USA with the largest production of nuclear energy of more than 96 000 MWh this accounts for about 18% of the total energy production.

Routine operations of nuclear reactors and other installations in the nuclear fuel cycle release small amounts of radioactive material to the air and as liquid effluents. However, the estimated total releases of ^{90}Sr, ^{131}I and ^{137}Cs over the entire periods of operation are negligible compared to the amounts released to the environment due to nuclear weapon tests.

Some of the first reactors that were constructed used a single-pass cooling system. The eight reactors constructed for plutonium production at Hanford, USA, between 1943 and 1956, pumped water from Columbia River through the reactor cores then delayed it in cooling ponds before returning it to the

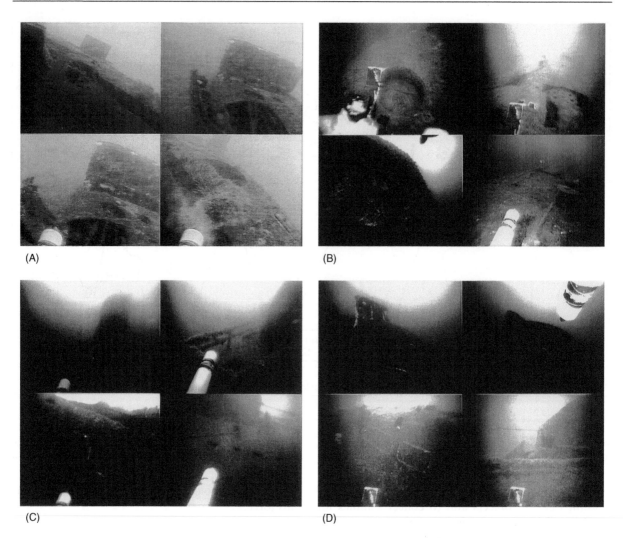

Figure 2   (A)–(D) Pictures of a disposed submarine at a depth of *c.* 30 m in the Stepovogo Fiord, east coast of Novaya Zemyla. The submarine contains a sodium-cooled reactor with spent fuel. Some of the hatches of the submarine are open which allows for 'free' circulation of water inside the vessel.

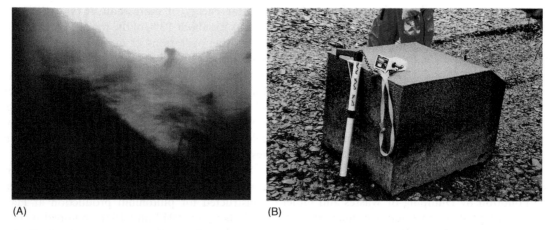

Figure 3   (A) Containers of low level solid waste at the bottom of the Abrosimov Fiord at a depth of *c.* 15 m and (B) a similar container found washed ashore.

river. The river water and its contents of particles and components were thereby exposed to a great neutron flux and various radioactive isotopes were created. In addition corrosion of neutron-activated metal within the reactor structure contributed to the radioactive contamination of the cooling water. Only a limited number of these radionuclides reached the river mouth and only ^{32}P, ^{51}Cr, ^{54}Mn and ^{65}Zn were detected regularly in water, sediments and marine organisms in the near-shore coastal waters of the US Pacific Northwest.

Reactors operating today all have closed primary cooling systems that do not allow for this type of contamination. Therefore, under normal conditions production of electricity from nuclear reactors does not create significant amounts of operational discharges of radionuclides. However, the 434 energy-producing nuclear plants of the world in 1998 created radioactive waste in the form of utilized fuel. Utilized fuel is either stored or reprocessed.

Only 4–5% of the utilized nuclear fuel worldwide is reprocessed. Commercial, nonmilitary, reprocessing of nuclear fuel takes place in France, Japan, India and the United Kingdom. Other reprocessing plants defined as defense-related are in operation and producing waste but without discharges. For example in the USA, at the Savannah River Plant and the Hanford complex, about $83\,000\,m^3$ and $190\,000\,m^3$, respectively, of high-level liquid waste was in storage in 1985.

Reprocessing plants and the nuclear industry in the former Soviet Union have discharged to the Ob and Yenisey river systems ending up in the Arctic ocean. In 1950–51 about $77 \times 10^6\,m^3$ liquid waste of 100 PBq was discharged to the River Techa. The Techa River is connected to the River Ob as is the Tomsk River where the Tomsk-7, a major production site for nuclear weapons plutonium, is situated. Other nuclear plants, such as the Krasnoyarsk industrial complex, have discharged to the Yenisey river. Large amounts of radioactive waste are also stored at the sites.

Radioactive waste stored close to rivers has the potential of contaminating the oceans should an accident happen to the various storage facilities.

The commercial reprocessing plants in France at Cap de la Hague and in the UK at Sellafield have for many years, and still do, contributed to the radioactive contamination of the marine environment. They both discharge low-level liquid radioactive effluents to the sea. Most important, however, these discharges and their behavior in the marine environment have been and are still thoroughly studied and the results are published in the open literature. The importance of these discharges is

Table 2 Total discharges of some radionuclides from Sellafield 1952–92

^3H	39 PBq
^{90}Sr	6.3 PBq
^{134}Cs	5.8 PBq
^{137}Cs	41.2 PBq
^{238}Pu	0.12 PBq
^{239}Pu	0.6 PBq
^{241}Pu	21.5 PBq
^{241}Am	0.5 PBq

extensive as radionuclides from Sellafield and la Hague are traced throughout the whole North Atlantic. Most important is Sellafield; **Table 2** summarizes the reported discharge of some important radionuclides.

In addition a range of other radionuclides have been discharged from Sellafield, but prior to 1978 the determination of radionuclides was, for many components, not specific. Technetium (^{99}Tc) for instance was included in the 'total beta' determinations with an estimated annual discharge from 1952 to 1970 below 5 TBq and from 1970 to 1977 below 50 TBq. Specific determination of ^{99}Tc in the effluents became part of the routine in 1978 when about 180 TBq was discharged followed by about 50 TBq in 1979 and 1980 and then an almost negligible amount until 1994.

The reason for mentioning ^{99}Tc is that this radionuclide, in an oceanographic context, represents an almost ideal tracer in the oceans. Technetium is most likely to be present as pertechnetate, TcO_4^-, totally dissolved in seawater; it acts conservatively and moves as a part of the water masses. In addition the main discharges of technetium originate from point sources with good documentation of time for and amount of the release. The discharges from Sellafield are a good example of this. From 1994 the UK authorities have allowed for a yearly ^{99}Tc discharge of up to 200 TBq.

Based on surveys before and after the discharges in 1994, 30 TBq (March–April) and 32 TBq (September–October), the transit time for technetium from the Irish Sea to the North Sea was calculated to be considerably faster than previous estimations of transit times for released radionuclides. This faster transport is demonstrated by measurements indicating that the first discharge plume of ^{99}Tc had reached the south coast of Norway before November 1996 in about 2.5 years compared to the previously estimated transit time of 3–4 years.

Other reprocessing plants may have discharges to the sea, but without a particular impact in the world oceans. The reprocessing plant at Trombay,

India, may, for example, be a source for marine contamination.

Accidental Releases

Accidents resulting in direct radioactive releases to the sea are not well known as most of them are connected to wreckage of submarines. Eight nuclear submarines with nuclear weapons have been reported lost at sea, two US and six former USSR. The last known USSR wreck was the submarine *Komsomolets* which sank in the Norwegian Sea southwest of Bear Island, on 7 April 1989. The activity content in the wreck is estimated by Russian authorities to be 1.55–2.8 PBq ^{90}Sr and 2.03–3 PBq ^{137}Cs and the two nuclear warheads on board may contain about 16 TBq ^{239,240}Pu equivalent to 6–7 kg plutonium. Other estimates indicate that each warhead may contain 10 kg of highly enriched uranium or 4–5 kg plutonium.

On August 12th 2000, the Russian nuclear submarine *Kursk* sank at a depth of 108 meters in the Barents Sea north of the Kola peninsula. Vigorous explosions in the submarine's torpedo-chambers caused the wreckage where 118 crew-members were entrapped and lost their lives. *Kursk*, and Oscar II attack submarine, was commissioned in 1995 and was powered by two pressurized water reactors. *Kursk* had no nuclear weapons on board. Measurements close to the wreck in the weeks after the wreckage showed no radioactive contamination indicating that the primary cooling-systems were not damaged in the accident. A rough inventory calculation estimates that the reactors at present contain about 56 000 TBq. Russian authorities are planning for a salvage operation where the submarine or part of the submarine will be lifted from the water and transported to land. Both a possible salvage operation or to leave the wreck where it is will be create a demand for monitoring as the location of the wreck is within important fishing grounds.

The wreckage of *Komsomolets* in 1989 and the attempts to raise money for an internationally financed Russian led salvage operation became very public. The Russian explanation for the intensive attempts of financing the salvage was said to be the potential for radioactive pollution. The wreck of the submarine is, however, located at a depth of 1658 m and possible leaching of radionuclides from the wreck will, due to the hydrography of the area, hardly have any vertical migration and radioactive components will spread along the isopycnic surfaces gradually dispersing the released radioactivity in the deep water masses of the Nordic Seas. An explanation for the extensive work laid down for a salvage operation and for what became the final solution, coverage of the torpedo-part of the hull, may be that this submarine was said to be able to fire its torpedo missiles with nuclear warheads from a depth of 1000 m.

In 1990, the Institute of Marine Research, Bergen, Norway, started regular sampling of sediments and water close to the wreck of *Komsomolets*. Values of ^{137}Cs were in the range 1–10 Bq per kg dry weight sediment and 1–30 Bq per m^3 water. No trends were found in the contamination as the variation between samples taken at the same date were equal to the variation observed from year to year. Detectable amounts of ^{134}Cs in the sediment samples indicate that there is some leaching of radioactivity from the reactor.

Accidents with submarines and their possible impact on the marine environment are seldom noticed in the open literature and there is therefore little common knowledge available. An accident, however, that is well known is the crash of a US B-52 aircraft, carrying four nuclear bombs, on the ice off Thule air base on the northwest coast of Greenland in January 1968. Approximately 0.4 kg plutonium ended up on the sea floor at a depth of 100–300 m. The marine environment became contaminated by about 1 TBq ^{239,240}Pu which led to enhanced levels of plutonium in benthic animals, such as bivalves, sea-stars and shrimps after the accident. This contamination has decreased rapidly to the present level of one order of magnitude below the initial levels.

Fallout from Nuclear Weapon Tests and Nuclear Accidents

Nuclear weapon tests in the atmosphere from 1945 to 1980 have caused the greatest man-made release of radioactive material to the environment. The most intensive nuclear weapon tests took place before 1963 when a test-ban treaty signed by the UK, USA and USSR came into force. France and China did not sign the treaty and continued some atmospheric tests, but after 1980 no atmospheric tests have taken place.

It is estimated that 60% of the total fallout has initially entered the oceans, i.e. 370 PBq ^{90}Sr, 600 PBq ^{137}Cs and 12 PBq ^{239,240}Pu. Runoff from land will slightly increase this number. As the majority of the weapon tests took place in the northern hemisphere the deposition there was about three times as high as in the southern hemisphere.

Results from the GEOSECS expeditions, 1972–74, show a considerable discrepancy between the measured inventories in the ocean of 900 PBq ^{137}Cs, 600 PBq ^{90}Sr and 16 PBq ^{239,240}Pu and the estimated input from fallout. The measured values are far higher than would be expected from the assumed fallout data. Thus the exact input of anthropogenic radionuclides may be partly unknown or the geographical coverage of the measurements in the oceans were for some areas not dense enough for accurate calculations.

Another known accident contributing to marine contamination was the burn-up of a US satellite (SNAP 9A) above the Mozambique channel in 1964 which released 0.63 PBq ^{238}Pu and 0.48 TBq ^{239}Pu; 73% was eventually deposited in the southern hemisphere.

The Chernobyl accident in 1986 in the former USSR is the latest major event creating fallout to the oceans. Two-thirds of the c.100 PBq ^{137}Cs released was deposited outside the Soviet Union. The total input to the world oceans of ^{137}Cs from Chernobyl is estimated to be from 15–20 Pbq, i.e. 4.5 PBq in the Baltic Sea; 3–5 PBq in the Mediterranean Sea, 1.2 PBq in the North Sea and about 5 PBq in the northeast Atlantic.

Natural Occurring Radioactive Material

Oil and gas production mobilize naturally occurring radioactive material (NORM) from the deep underground reservoir rock. The radionuclides are primarily ^{226}Ra, ^{228}Ra and ^{210}Pb and appear in sludge and scales and in the produced water. Scales and sludge containing NORM represent an increasing amount of waste. There are different national regulations for handling this type of waste. In Norway, for example, waste containing radioactivity above 10 Bq g^{-1} is stored on land in a place specially designed for this purpose. However, there are reasons to believe that a major part of radioactive contaminated scales and sludge from the worldwide offshore oil and gas production are discharged to the sea.

Reported NORM values in scales are in the ranges of 0.6–57.2 Bq g^{-1} ^{226}Ra + ^{228}Ra (Norway), 0.4–3700 Bq g^{-1} (USA) and 1–1000 Bq g^{-1} ^{226}Ra (UK).

Scales are an operational hindrance in oil and gas production. Frequent use of scale-inhibitors reduce the scaling process but radioactive components are released to the production water adding to its already elevated radioactivity. More than 90% of the radioactivity in produced water is due to ^{226}Ra and ^{228}Ra having a concentration 100–1000 times higher than normal for seawater.

The discharge of produced water is a continuous process and the amount of water discharged is considerable and increases with the age of the production wells. As an example, the estimated amount of produced water discharged to the North Sea in 1998 was 340 million m^3 and multiplying by an average value of 5 Bq^{-1} of ^{226}Ra in produced water, the total input of ^{226}Ra to the North Sea in 1998 was 1.7 TBq.

Discussion

The total input of anthropogenic radioactivity to the world's oceans is not known exactly, but a very rough estimate gives the following amounts: 85 PBq dumped, 100 PBq discharged from reprocessing and 1500 PBq from fallout. Some of the radionuclides have very long half-lives and will persist in the ocean, for example ^{99}Tc has a half-life of 2.1×10^5 years, ^{239,240}Pu, 2.4×10^4 years and ^{226}Ra, 1600 years. ^{137}Cs, ^{90}Sr and ^{228}Ra with half-lives of 30 years, 29 years and 5.75 years, respectively, will slowly decrease depending on the amount of new releases.

In an oceanographic context it is worth mentioning the differences in denomination between radioactivity and other elements in the ocean. The old denomination for radioactivity was named after Curie (Ci) and 1 g radium was defined to have a radioactivity of 1 Ci; 1 Ci 3.7×10^{10} Bq and 1 PBq 27 000 Ci. Therefore released radioactivity of 1 PBq can be compared to the radioactivity of 27 kg radium.

The common denominations for major and minor elements in seawater are given in weight per volume. For comparison if 1 PBq or 27 kg radium were diluted in 1 km^3 of seawater, this would give a radium concentration of 0.027 µg l^{-1} or 1000 Bq l^{-1}. Calculations like this clearly visualize the sensitivity of the analytical methods used for measuring radioactivity. In the Atlantic Ocean for example radium (^{228}Ra) has a concentration of 0.017–3.40 mBq l^{-1}, whereas ^{99}Tc measured in surface waters off the southwest coast of Norway is in the range of 0.9–6.5 mBq l^{-1}.

Measured in weight the total amount of radionuclides do not represent a huge amount compared to the presence of nonradioactive components in seawater. The radioisotopes of cesium and strontium are both important in a radioecological context since they have chemical behavior resembling potassium and calcium, respectively. Cesium follows potassium in and out of the soft tissue cells whereas strontium follows calcium into bone cells and stays. Since uptake and release in organisms is due to the chemical characteristics and rarely if the element is radioactive or not, radionuclides such as ^{137}Cs and ^{90}Sr have to compete with the nonradioactive isotopes of cesium and strontium.

Oceanic water has a cesium content of about 0.5 l µg^{-1} and a strontium content of about 8000 µg l^{-1}. Uptake in a marine organism is most likely to be in proportion to the abundance of the radioactive and the nonradioactive isotopes of the actual element. This can be illustrated by the following example. The sunken nuclear submarine *Komsomolets* contained an estimated (lowest) amount of 1.55 PBq ^{90}Sr (about 300 g) and 2.03 PBq ^{137}Cs (about 630 g). If all this was released at once and diluted in the immediate surrounding 1 km^3 of water the radioactive concentration would have been 1550 Bq l^{-1} for ^{90}Sr and 2030 Bq l^{-1} for ^{137}Cs, the concentration in weight per volume would have been 0.000 3 µg l^{-1} ^{90}Sr and 0.000 63 µg l^{-1} ^{137}Cs. This means that even if the radioactive material was kept in the extremely small volume of 1 km^3, compared to the volume of the deep water of the Norwegian Sea available for a primary dilution, the proportion of radioactive to nonradioactive isotopes of strontium and cesium, available for uptake in marine organisms, would have been about 2.7×10^6 and 7.9×10^{-5}, respectively.

From the examples above it can be seen that if uptake, and thereby impact, in marine organisms follows regular chemical–physiological rules there is a 'competition' in seawater in favor of the nonradioactive isotopes for elements normally present in seawater. Measurable amount of radionuclides of cesium and strontium are detected in marine organisms but at levels far below the concentrations in freshwater fish. Average concentrations of ^{137}Cs in fish from the Barents Sea during the period with the most intensive nuclear weapon tests in that area, 1962–63, never exceeded 90 Bq kg^{-1} fresh weight, whereas fallout from Chernobyl resulted in concentrations in freshwater fish in some mountain lakes in Norway far exceeding 10 000 Bq kg^{-1}.

For radionuclides like technetium and plutonium, which will persist in the marine environment, uptake will be based only on the actual concentrations in seawater of radionuclide. The levels of ^{99}Tc, for example, increased in seaweed (*Fucus vesiculosus*) from 70 Bq per kg dry weight (December 1997) to 124 Bq kg^{-1} in January 1998 in northern Norway which reflected the increased concentration in the water as the peak of the technetium plume from Sellafield reached this area.

Previously the effects of anthropogenic radioactivity have been based on the possible dose effect to humans. Most of the modeling work has been concentrated on assessing the dose to critical population groups eating fish and other marine organisms. But even if the radiation from anthropogenic radionuclides to marine organisms is small compared to natural radiation from radionuclides like potassium, ^{40}K, the presence of additional radiation may give a chronic exposure with possible effects, at least on individual marine organisms.

The input of radioactivity, NORM, from the offshore oil and gas production may also give reason for concern. The input will increase as it is a continuous part of the production. Even if radium as the main radionuclide is not likely to be taken up by marine organisms the use of chemicals like scale inhibitors may change this making radium more available for marine organisms.

Conclusion

The sea began receiving radioactive waste from anthropogenic sources in 1946, in a rather unregulated way in the first decades. Both national and international regulations controlling disposals have now slowly come into force. Considerable amounts are still discharged regularly from nuclear industries and the practice of using the sea as a suitable wastebasket is likely to continue for ever. In 1994 an international total prohibition on radioactive waste disposal at sea came into force, but the approximately 85 PBq of solid radioactive waste that has already been dumped will sooner or later be gradually released to the water masses.

Compared to other wastes disposed of at sea the amount of radioactive waste by weight is rather diminutive. However, contrary to most of the 'ordinary' wastes in the sea, detectable amounts of anthropogenic radioactivity are found in all parts of the world oceans and will continue to contaminate the sea for many thousands of years to come. This means that anthropogenic radioactive material has become an extra chronic radiation burden for marine organisms. In addition, the release of natural occurring radionuclides from offshore oil and gas production will gradually increase the levels of radium, in particular, with a possible, at present unknown, effect.

However, marine food is not, and probably never will be, contaminated at a level that represents any danger to consumers. The ocean has always received debris from human activities and has a potential for receiving much more and thereby help to solve the waste disposal problems of humans. But as soon as a waste product is released and diluted in the sea it is almost impossible to retrieve. Therefore, in principal, no waste should be disposed of in the sea without clear documentation that it will never create any damage to the marine environment and its living resources. This means that with present knowledge no radioactive wastes should be allowed to be released into the sea.

See also

International Organizations.

Further Reading

Guary JC, Guegueniat P, and Pentreath RJ (eds.) (1988) *Radionuclides: A Tool for Oceanography.* London, New York: Elsevier Applied Science.

Hunt GJ, Kershaw PJ, and Swift DJ (eds.) (1998) Radionuclides in the oceans (RADOC 96–97). Distribution, Models and Impacts. *Radiation Protection Dosimetry* 75: 1–4.

IAEA (1995) *Environmental impact of radioactive releases*; Proceedings of an International Symposium on Environmental Impact of Radioactive Releases. Vienna: International Atomic Energy Agency.

IAEA (1999) *Inventory of Radioactive Waste Disposals at Sea.* IAEA-TECDOC-1105 Vienna: International Atomic Energy Agency. pp. 24 A.1–A.22.

THERMAL DISCHARGES AND POLLUTION

T. E. L. Langford, University of Southampton, Southampton, UK

Sources of Thermal Discharges

The largest single source of heat to most water bodies, including the sea, is the sun. Natural thermal springs also occur in many parts of the world, almost all as fresh water, some of which discharge to the sea. In the deep oceans hydrothermal vents discharge mineral-rich hot water at temperatures greatly exceeding any natural temperatures either at depth or at the surface. To add to these natural sources of heat, industrial processes have discharged heated effluents into coastal waters in many parts of the world for at least 150 years. By far the largest volumes of these heated effluents reaching the sea in the past 60 years have originated from the electricity generation industry (power industry). Indeed more than 80% of the volume of heated effluents to the sea originate from the power industry compared with 3–5% from the petroleum industries and up to 7% (in the USA) from chemical and steel industries.

The process known as 'thermal' power generation, in which a fuel such as oil or coal or the process of nuclear fission is used to heat water to steam to drive turbines, requires large volumes of cooling water to remove the waste heat produced in the process. Where power stations are sited on or near the coast all of this waste heat, representing some 60–65% of that used in the process, is discharged to the sea. The heat is then dissipated through dilution, conduction, or convection. In a few, atypical coastal situations, where the receiving water does not have the capacity to dissipate the heat, artificial means of cooling the effluent such as ponds or cooling-towers are used. Here the effluent is cooled prior to discharge and much of the heat dissipated to the air.

The waste heat is related to the theoretical thermal efficiency of the Rankine cycle, which is the modification of the Carnot thermodynamic cycle on which the process is based. This has a maximum theoretical efficiency of about 60% but because of environmental temperatures and material properties the practical efficiency is around 40%. Given this level of efficiency and the normal operating conditions of a modern coal- or oil-fired power station, namely steam at 550°C and a pressure of 10.3×10^6 kg cm^{-2}

with corresponding heat rates of 2200 kg cal kWh^{-1} of electricity, some 1400 kg cal kWh^{-1} of heat is discharged to the environment, usually in cooling water at coastal sites. This assumes a natural water temperature of 10°C. Nuclear power stations usually reject about 50% more heat per unit of electricity generated because they operate at lower temperatures and pressures. Since the 1920s efficiencies have increased from about 20% to 38–40% today with a corresponding reduction by up to 50% of the rate of heat loss. The massive expansion of the industry since the 1920s has, however, increased the total amounts of heat discharged to the sea.

Thus for each conventional modern power station of 2000 MW capacity some 63 m^3 s^{-1} of cooling water is required to remove the heat. Modern developments such as the combined cycle gas turbine (CCGT) power stations with increased thermal efficiencies have reduced water requirements and heat loss further so that a 500 MW power station may require about 9–10 m^3 s^{-1} to remove the heat, a reduction of over 30% on conventional thermal stations.

Water Temperatures

Natural sea surface temperatures vary widely both spatially and temporally throughout the world with overall ranges recorded from −2°C to 30°C in open oceans and −2°C to 43°C in coastal waters. Diurnal fluctuations at the sea surface are rarely more than 1°C though records of up to 1.9°C have been made in shallow seas. The highest temperatures have been recorded in sheltered tropical embayments where there is little exchange with open waters. Most thermal effluents are discharged into coastal waters and these are therefore most exposed to both physical and biological effects. In deeper waters vertical thermal stratification can often exceed 10°C and a natural maximum difference of over 23°C between surface and bottom has been recorded in some tropical waters.

The temperatures of thermal discharges from power stations are typically 8–12°C higher than the natural ambient water temperature though at some sites, particularly nuclear power stations, temperature rises can exceed 15°C (**Figure 1**). Maximum discharge temperatures in some tropical coastal waters have reached 42°C though 35–38°C is more typical. There are seasonal and diurnal fluctuations at many sites related to the natural seasonal

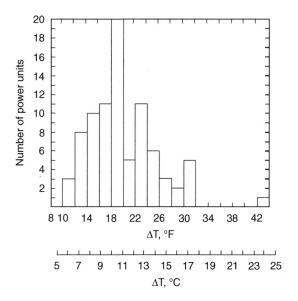

Figure 1 Frequency distribution of designed temperature rises for thermal discharges in US power station cooling water systems. (Reproduced with permission from Langford, 1990.)

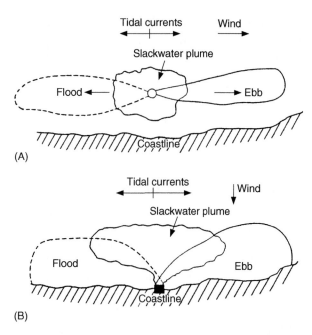

Figure 2 Movements of thermal plumes in tidal waters. (A) Offshore outfall; (B) onshore outfall. (Reproduced with permission from Langford, 1990.)

temperature cycles and to the operating cycles of the power station.

Thermal Plumes and Mixing Zones

Once discharged into the sea a typical thermal effluent will spread and form a three-dimensional layer with the temperature decreasing with distance from the outfall. The behavior and size of the plume will depend on the design and siting of the outfall, the tidal currents, the degree of exposure and the volume and temperature of the effluent itself. Very few effluents are discharged more than 1 km from the shore. Effects on the shore are, however, maximized by shoreline discharge (**Figure 2**).

The concept of the mixing zone, usually in three components, near field, mid-field and far-field, related to the distance from the outfall, has mainly been used in determining legislative limits on the effluents. Most ecological studies have dealt with near and mid-field effects. The boundary of the mixing zone is, for most ecological limits, set where the water is at 0.5°C above natural ambients though this tends to be an arbitrary limit and not based on ecological data. Mixing zones for coastal discharges can be highly variable in both temperature and area of effect (**Figure 3**).

In addition to heat effects, thermal discharges also contain chemicals, mainly those used for the control of marine fouling in pipework and culverts. Chlorine compounds are the most common and because of its

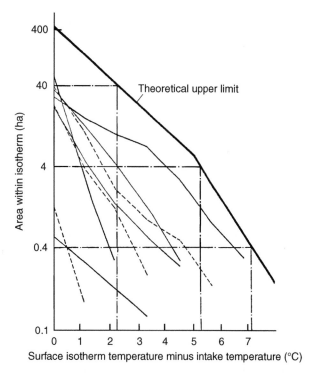

Figure 3 Relationship between surface-temperature elevation and surface area affected for nine different surveys at Moro Bay Power Plant, California. (Reproduced with permission from Langford, 1990.)

strong biocidal properties, the effects of chlorine are of primary concern in many coastal discharges, irrespective of the temperature. Measured chlorine residuals immediately after application vary between

0.5 and 10 mg l^{-1} throughout the world but most are within the 0.5–1.0 mg l^{-1} range. Because of the complex chemical reactions in sea water free chlorine residuals in discharges are usually a factor of 10 lower than the initial dosing rate. Even so, concentrations of chlorine compounds can exceed lethal limits for organisms at some sites.

Biological Effects of Temperature

The biological effects of temperature on marine and coastal organisms have been reviewed by a number of authors. Most animals and plants can survive ranges of temperature which are essentially genetically determined. The ultimate lethal temperature varies within poikilothermic groups but there is a trend of tolerance which is inversely related to the structural complexity of the organism (**Table 1**). Thus groups of microorganisms tend to contain species which have much higher tolerances than invertebrates, fish, or vascular plants. Because the life processes and survival of many organisms is so dependent on water temperature many species have developed physiological or behavioral strategies for optimizing temperature exposure and for survival in extremes of heat or cold. Examples are to be found among intertidal species and in polar fish.

The effects of temperature on organisms can be classified mainly from experimental data as lethal, controlling, direct, and indirect and all are relevant to the effects of thermal discharges in the marine environment. These effects can be defined briefly as follows.

- *Lethal*: high or low temperatures which will kill an organism within a finite time, usually less than its normal life span. The lethal temperature for any one organism depends on many factors within genetic limits. These include acclimatization, rate of change of temperature, physiological state (health) of the organism and any adaptive mechanisms.
- *Controlling*: temperatures below lethal temperatures which affect life processes, i.e., growth, oxygen consumption, digestive rates, or reproduction. There is a general trend for most organisms to show increases in metabolic activity with increasing temperature up to a threshold after which it declines sharply.
- *Direct*: temperatures causing behavioral responses such as avoidance or selection, movements, or migrations. Such effects have been amply demonstrated in experiments but for some work *in situ* the effects are not always clear.
- *Indirect*: where temperatures do not act directly but through another agent, for example poisons or oxygen levels or through effects on prey or predators. Temperature acts synergistically with toxic substances which can be important to its effects on chlorine toxicity *in situ* in thermal plumes. Where temperature immobilizes or kills prey animals they can become much more vulnerable to predation.

Biological Effects of Thermal Discharges

The translation of data obtained from experimental studies to field sites is often not simple. The complexity of the natural environment can mask or exacerbate effects so that they bear little relation to experimental conditions and this has occurred in many studies of thermal discharges *in situ*. Further, factors other than that being studied may be responsible for the observed effects. Examples relevant to thermal discharges are discussed later in this article.

Entrainment

The biological effects of any thermal discharge on marine organisms begin before the effluent is discharged. Cooling water abstracted from the sea usually contains many planktonic organisms notably bacteria, algae, small crustacea, and fish larvae. Within the power station cooling system these

Table 1 Upper temperature limits for aquatic organisms. Data from studies of geothermal waters[a]

Group	Temperature (°C)
Animals	
Fish and other aquatic vertebrates	38
Insects	45–50
Ostracods (crustaceans)	49–50
Protozoa	50
Plants	
Vascular plants	45
Mosses	50
Eukaryotic algae	56
Fungi	60
Prokaryotic microorganisms	
Blue–green algae	70–73
Photosynthetic bacteria	70–73
Nonphotosynthetic bacteria	>99

[a]Reproduced with permission from Langford (1990).

organisms experience a sudden increase in temperature (10–20°C, depending on the station) as they pass through the cooling condensers. They will also experience changes in pressure (1–2 atm) and be dosed with chlorine (0.5–5 mg l^{-1}) during this entrainment, with the effect that many organisms may be killed before they are discharged to the receiving water. Estimates for power stations in various countries have shown that if the ultimate temperatures are less than 23°C the photosynthesis of entrained planktonic marine algae may increase, but at 27–28°C a decrease of 20% was recorded. At 29–34°C the rates decreased by 61–84% at one US power station. Only at temperatures exceeding 40°C has total mortality been recorded. Concentrations of chlorine (total residual) of 1.0 mg l^{-1} have been found to depress carbon fixation in entrained algae by over 90% irrespective of temperature (**Figure 4**). Diatoms were less affected than other groups and the effects in open coastal waters were less marked than in estuarine waters. Unless the dose was high enough to cause complete mortality many algae showed evidence of recovery.

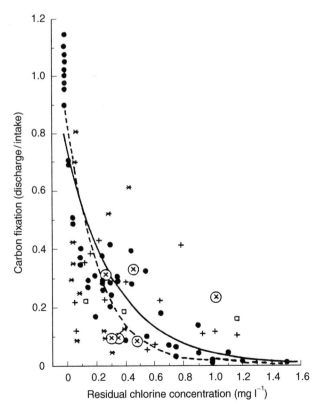

Figure 4 The effect of cooling water chlorination on carbon fixation in marine phytoplankton. ⊗, San Onofre; +, Morgantown; *, Hudson River; ●, Fawley; □, Dunkerque. (Reproduced with permission from Langford, 1990.)

The mean mortality rates for marine zooplankton entrained through cooling-water systems were shown to be less than 30% except in unusual cases where extreme temperatures and high chlorine doses caused 100% mortality. High mortalities can also occur where the entrained organisms are exposed to high temperatures in cooling ponds after discharge from the cooling system. In general, zooplankton do not suffer percentage mortalities as high as those of phytoplankton under typical cooling-water conditions. After passing through the cooling system at a US power station the dead or dying zooplankton were observed being eaten by large numbers of fish gathered at the outfall and hence passed into the food chain.

There are few published observations of marine macro-invertebrates entrained in cooling-water systems though at a site in the UK, field experiments showed that specimens of the prawn *Palaemonetes varians* were killed by mechanical damage as they passed through a cooling system. Larval fish are probably most vulnerable to the effects of entrainment, mostly killed by the combination of mechanical, chemical, and temperature effects. Mortalities of ichthyoplankton have varied from 27 to 100% at sites in the US and UK. Many of these were, however, on estuaries or tidal reaches of rivers. At a coastal site in California the mortality rate increased from 10 to 100% as temperatures rose from 31 to 38°C. Some 13% mortalities occurred with no heat, mainly as a result of pressure and abrasion in the system. The significance of 20% larval mortalities to the natural populations of flounders (*Platichthyes americanus*) calculated for a site on Long Island Sound indicated that the annual mortality was estimated at a factor of 0.01 which might cause a reduction of 9% of the adult population over 35 years provided the fish showed no compensatory reproductive or survival mechanism, or no immigration occurred.

Effects of Discharges in Receiving Waters

Algae

At some US power stations the metabolism of phytoplankton was found to be inhibited by chlorine as far as 200 m from the outfall. Also, intermittent chlorination caused reductions of 80–90% of the photosynthetic activity some 50 m from the outfall at a site on the Californian coast. From an assessment of the total entrainment and discharge effects, however, it was reported that where dilution was 300 times per second the effect of even a 100% kill of phytoplankton would not be detectable in the receiving water. In Southampton Water in the UK, mortality rates of 60% were estimated as causing about 1.2–3% reduction in the productivity of the tidal exchange volume where a power station used 6% of this for cooling. The main problem with most assessments is that replication of samples was typically low and estimates suggest that 88 samples would be needed from control and affected areas to detect a difference of 5% in productivity at a site, 22 samples for a 10% change and at least six for a 20% change. Such replication is rarely recorded in site studies.

Temperatures of 35–36°C killed shore algae at a coastal site in Florida, particularly *Halimeda* sp. and *Penicillus* spp. but factors other than temperature, most likely chlorine and scour, removed macro-algae at another tropical site. Blue–green algae were found where temperatures reached 40°C intermittently and *Enteromorpha* sp. occurred where temperatures of 39°C were recorded. In more temperate waters the algae *Ascophyllum* and *Fucus* were eliminated where temperatures reached 27–30°C at an outfall but no data on chlorination were shown. Replacing the shoreline outfall with an offshore diffuser outfall (which increased dilution and cooling rates) allowed algae to recolonize and recover at a coastal power station in Maine (US). On the Californian coast one of the potentially most vulnerable algal systems, the kelps *Macrocystis*, were predicted to be badly affected by power station effluents, but data suggest that at one site only about 0.7 ha was affected near the outfall.

The seagrass systems (*Thalassia* spp.) of the Florida coastal bays appeared to be affected markedly by the effluents from the Turkey Point power stations and a long series of studies indicated that within the +3 to +4°C isotherm in the plume, seagrass cover declined by 50% over an area of 10–12 ha. However, the results from two sets of studies were unequivocal as to the effects of temperature. The data are outlined briefly in the following summary. First, the effluent was chlorinated. Second it contained high levels of copper and iron. Third, the main bare patch denuded of seagrass, according to some observations, may have been caused by the digging of the effluent canal. Although one set of data concluded that the threshold temperature for adverse effects on seagrass was +1.5°C (summer 33–35°C) a second series of observations noted that *Thalassia* persisted apparently unharmed in areas affected by the thermal discharge, though temperatures rarely exceed 35°C. From an objective analysis, it would appear that the effects were caused mainly by a combination of thermal and chemical stresses.

Zooplankton and Microcrustacea

In Southampton Water in the UK, the barnacle *Elminius modestus* formed large colonies in culverts at the Marchwood power station and discharged large numbers of nauplii into the effluent stream increasing total zooplankton densities. Similar increases occurred where fouling mussels (*Mytilus* sp.) released veliger larvae into effluent streams. Data from 10 US coastal power stations were inconclusive about the effects of thermal discharges on zooplankton in receiving waters with some showing increases and others the reverse. Changes in community composition in some areas receiving thermal discharges were a result of the transport of species from littoral zones to offshore outfalls or vice versa. At Tampa Bay in Florida no living specimens of the benthic ostracod *Haplocytheridea setipunctata* were found when the temperature in the thermal plume exceeded 35°C. Similarly the benthic ostracod *Sarsiella zostericola* was absent from the area of a power effluent channel in the UK experiencing the highest temperature range.

Macro-invertebrates

As with other organisms there is no general pattern of change in invertebrate communities associated with thermal discharges to the sea which can be solely related to temperature. Some of the earliest studies in enclosed temperate saline waters in the UK showed that no species was consistently absent from areas affected by thermal plumes and the studies at Bradwell power station on the east coast showed no evidence of a decline in species richness over some 20 years though changes in methodology could have obscured changes in the fauna. No changes in bottom fauna were recorded at other sites affected by thermal plumes in both Europe and the US. The polychaete *Heteromastus filiformis* was found to be common to many of the thermal plume zones in several countries. In these temperate waters temperatures rarely exceeded 33–35°C.

In contrast, in tropical coastal waters data suggest that species of invertebrates are excluded by thermal plumes. For example in Florida, surveys showed that some 60 ha of the area affected by the Turkey Point thermal plume showed reduced diversity and abundance of benthos in summer, but there was marked recovery in winter. The 60 ha represented 0.0023% of the total bay area. A rise of 4–5°C resulted in a dramatic reduction in the benthic community. Similarly, at Tampa Bay, 35 of the 104 indigenous invertebrate species were excluded from the thermal plume area. Removal of the vegetation was considered to be the primary cause of the loss of benthic invertebrates. In an extreme tropical case few species of macro-invertebrates survived in a thermal effluent where temperatures reached 40°C, 10 species occurred at the 37°C isotherm and the number at the control site was 87. Scour may have caused the absence of species from some areas nearer the outfall (**Figure 5**). The effluent was chlorinated but no data on chlorine concentrations were published. Corals suffered severe mortalities at the Kahe power station in Hawaii but the bleaching of the colonies suggested that again chlorine was the primary cause of deaths, despite temperatures of up to 35°C. It has been suggested that temperature increases of as little as 1–2°C could cause damage to tropical ecosystems but detailed scrutiny of the data indicate that it would be difficult to come to that conclusion from field data, especially where chlorination was used for antifouling.

The changes in the invertebrate faunas of rocky shores in thermal effluents have been less well studied. Minimal changes were found on breakwaters in the paths of thermal plumes at two Californian sites. Any measurable effects were within 200–300 m of the outfalls. In contrast in southern France a chlorinated thermal discharge reduced the numbers of species on rocks near the outfall though 11 species of Hydroida were found in the path of the effluent. In most of the studies, chlorine would appear to be the primary cause of reductions in species and abundance though where temperatures exceeded 37°C both factors were significant. There is some evidence that species showed advanced reproduction and growth in the thermal plumes areas of some power stations where neither temperature nor chlorine were sufficient to cause mortality. Also behavioral effects were demonstrated for invertebrates at a Texas coastal power station. Here, blue crabs (*Callinectes sapidus*) and shrimps (*Penaeus aztecus* and *P. setiferus*) avoided the highest temperatures (exceeding 38°C) in the discharge canal but recolonized as temperatures fell below 35°C. At another site in tropical waters, crabs (*Pachygrapsus transversus*) avoided the highest temperatures (and possibly chlorine) by climbing out of the water on to mangrove roots.

Fish

There are few records of marine fish mortalities caused by temperature in thermal discharges except where fish are trapped in effluent canals. For example mortalities of Gulf menhaden (*Brevoortia petronus*), sea catfish (*Arius felis*) occurred in the canal of a Texas power station when temperatures reached 39°C. Also a rapid rise of 15°C killed menhaden (*Alosa* sp.) in the effluent canal of the Northport power station in the US. Avoidance behavior prevents mortalities where fish can move freely.

The apparent attraction of many fish species to thermal discharges, widely reported, was originally associated with behavioral thermoregulation and the selection of preferred temperatures. Perhaps the best recorded example is the European seabass (*Dicentrarchus labrax*) found associated with cooling-water outfalls in Europe. Temperature selection is, however, not now believed to be the cause of the aggregations. Careful analysis and observations indicate clearly that the main cause of aggregation is the large amounts of food organisms discharged either dead or alive in the discharge. Millions of shrimps and small fish can be passed through into effluents and are readily consumed by the aggregated fish. Further where fish have gathered, usually in cooler weather, they remain active in the warmer water and are readily caught by anglers unlike the individuals outside the plume. This also gives the impression that there are more fish in the warmer water. Active tracking of fish has shown mainly short-term association with outfalls though some species have been

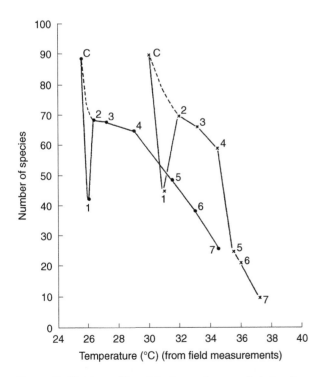

Figure 5 Numbers of invertebrate species recorded at various temperatures, taken from two separate surveys (×, October; ●, winter) at Guyanilla Bay. C, control sampling; 1–7, effluent sampling sites. (Reproduced with permission from Langford, 1990.)

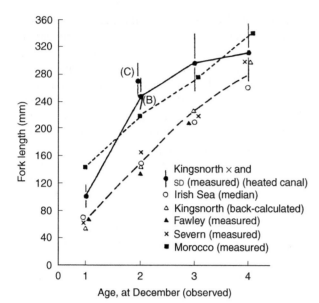

Figure 6 Growth of bass (*Dicentrarchus labrax*) in a thermal discharge canal in comparison to other locations. Note that back calculated lengths are smaller because they probably relate to cold water growth. (Reproduced with permission from Langford, 1990.)

shown as entering water at temperatures above their lethal maximum for very short periods to collect food. There is clear evidence, however, that fish avoid adversely high temperatures for most of the time and will return to a discharge area once the temperatures cool. Avoidance behavior is also apparent at high chlorine concentrations.

Where water temperatures and chemical conditions allow consistent residence, fish in thermal discharge canals show increased growth (**Figure 6**). At the Kingsnorth power station in the UK seabass (*D. labrax*) grew at twice the rate as in cold water, particularly in the first year. Fish showed varying residence times and sequential colonization of the canal at various ages. Winter growth occurred and the scales of older fish with long-term association with the discharge showed no evidence of annual winter growth checks. The fish were able to move into and out of the canal freely.

Occurrence of Exotic Species

Exotic or introduced marine invertebrate species have been recorded from thermal discharges in various parts of the world. Some of the earliest were from the enclosed docks heated by power station effluents near Swansea and Shoreham in the UK. The exotic barnacle *Balanus amphitrite* var. *denticulata*

and the woodborer *Limnoria tripunctata* replaced the indigenous species in the heated areas but declined in abundance when the effluent ceased. The polyzoan *Bugula neritina* originally a favored immigrant species in the heated water disappeared completely as the waters cooled. In New Jersey (USA) subtropical species of *Teredo* bred in a heated effluent and both the ascidian *Styela clavata* and the copepod *Acartia tonsa* have both been regarded as immigrant species favored by heated effluents. However many immigrant species have survived in temperate waters without being associated with heated waters. In UK waters, only the crab *Brachynotus sexdentatus* and the barnacle *B. amphitrite* may be regarded as the only species truly associated with thermal discharges despite records of various other species.

The decline of the American hard shell clam fishery (*Mercenaria mercenaria*) introduced to Southampton Water in the UK was reportedly caused by the closure of the Marchwood power station combined with overfishing. Recruitment of young clams and their early growth were maximized in the heated water and reproduction was advanced but all declined as the thermal discharge ceased.

Aquaculture in Marine Thermal Discharges

The use of marine thermal discharges from power stations for aquaculture has not been highly successful in most parts of the world. Although it is clear that some species will grow faster in warmer water, the presence and unpredictability of chlorination has been a major obstacle. It is not generally economically viable to allow a large power station to become fouled such that efficiency declines merely to allow fish to grow. In Japan some farming ventures are regarded as profitable at power stations but in Europe and the USA such schemes are rarely profitable. At the Hunterston power station in Scotland plaice (*Pleuronectes platessa*) and halibut (*Hippoglossus hippoglossus*) grew almost twice as fast in warm water as in natural ambient but the costs of pump maintenance and capital equipment caused the system to be uneconomic irrespective of chlorination problems. Optimization of temperature is also a problem especially where temperatures fluctuate diurnally or where they exceed optimal growth temperatures. The general conclusion is that commercial uses of heated effluents in marine systems are not yet proven and are unlikely to become large-scale global ventures.

Thermal Discharges and Future Developments

The closure of many older, less-efficient power stations, has led to an increase in the efficiency of the use of water and a decline in the discharge of heat to the sea per unit of electricity generated. However, the increasing industrialization of the developing countries, China, Malaysia, India and the African countries is leading to the construction of new, large power stations in areas not previously developed. The widespread use of CCGT stations can reduce the localized problems of heat loss and water use further as well as reducing emissions of carbon dioxide and sulphur dioxide to the air but the overall increase in power demand and generation will lead to an increase in the total aerial and aquatic emissions in some regions.

In some tropical countries the delicate coastal ecosystems will be vulnerable not only to heat and higher temperature but much more importantly to the biocides used for antifouling. There is as yet no practical alternative that is as economic as chlorine though different methods have been tried with varying success in some parts of the world. There is little doubt that the same problems will be recognized in the areas of new development but as in the past after they have occurred.

Conclusions

It is clear that the problems of the discharge of heated effluents are essentially local and depend on many factors. Although temperatures of over 37°C are lethal to many species which cannot avoid exposure, there are species which can tolerate such temperatures for short periods. Indeed it can be concluded for open coastal waters that discharge temperatures may exceed the lethal limits of mobile species at least for short periods. This, of course, would not apply if vulnerable sessile species were involved, though again some provisos may be acceptable. For example, an effluent which stratified at the surface in deep water would be unlikely to affect the benthos. On the other hand an effluent which impinges on the shore may need strict controls to protect the benthic community. From all the data it is clear that blanket temperature criteria intended to cover all situations would not protect the most vulnerable ecosystems and may be too harsh for those that are more tolerant or less exposed. Constraints should therefore be tailored to each specific site and ecosystem.

Irrespective of temperature it is also very clear that chlorination or other biocidal treatment has been responsible for many of the adverse ecological effects originally associated with temperature. The solution to fouling control and the reduction of chlorination of other antifouling chemicals is therefore probably more important than reducing heat loss and discharge temperatures particularly where vulnerable marine ecosystems are at risk.

See also

Demersal Species Fisheries. Ocean Thermal Energy Conversion (OTEC).

Further Reading

Barnett PRO and Hardy BLS (1984) Thermal deformations. In: Kinne O (ed.) *Marine Ecology*, vol. V. *Ocean Management*, part 4, *Pollution and Protection of the Seas, Pesticides, Domestic Wastes and Thermal Deformations*, pp. 1769–1926. New York: Wiley.

Jenner HA, Whitehouse JW, Taylor CJL and Khalanski M (1998) *Cooling Water Management in European Power Stations: Biology and Control of Fouling*. Hydroecologie Appliquee. Electricité de France.

Kinne O (1970) *Marine Ecology, vol. 1, Environmental Factors, part 1*. New York: Wiley-Interscience.

Langford TE (1983) *Electricity Generation and the Ecology of Natural Waters*. Liverpool: Liverpool University Press.

Langford TE (1990) *Ecological Effects of Thermal Discharges*. London: Elsevier Applied Science.

Newell RC (1970) *The Biology of Intertidal Animals*. London: Logos Press.

ATMOSPHERIC INPUT OF POLLUTANTS

R. A. Duce, Texas A&M University, College Station, TX, USA

Introduction

For about a century oceanographers have tried to understand the budgets and processes associated with both natural and human-derived substances entering the ocean. Much of the early work focused on the most obvious inputs – those carried by rivers and streams. Later studies investigated sewage outfalls, dumping, and other direct input pathways for pollutants. Over the past decade or two, however, it has become apparent that the atmosphere is also not only a significant, but in some cases dominant, pathway by which both natural materials and contaminants are transported from the continents to both the coastal and open oceans. These substances include mineral dust and plant residues, metals, nitrogen compounds from combustion processes and fertilizers, and pesticides and a wide range of other synthetic organic compounds from industrial and domestic sources. Some of these substances carried into the ocean by the atmosphere, such as lead and some chlorinated hydrocarbons, are potentially harmful to marine biological systems. Other substances, such as nitrogen compounds, phosphorus, and iron, are nutrients and may enhance marine productivity. For some substances, such as aluminum and some rare earth elements, the atmospheric input has an important impact on their natural chemical cycle in the sea.

In subsequent sections there will be discussions of the input of specific chemicals via the atmosphere to estuarine and coastal waters. This will be followed by considerations of the atmospheric input to open ocean regions and its potential importance. The atmospheric estimates will be compared with the input via other pathways when possible. Note that there are still very large uncertainties in all of the fluxes presented, both those from the atmosphere and those from other sources. Unless otherwise indicated, it should be assumed that the atmospheric input rates have uncertainties ranging from a factor of 2 to 4, sometimes even larger.

Estimating Atmospheric Contaminant Deposition

Contaminants present as gases in the atmosphere can exchange directly across the air/sea boundary or they may be scavenged by rain and snow. Pollutants present on particles (aerosols) may deposit on the ocean either by direct (dry) deposition or they may also be scavenged by precipitation. The removal of gases and/or particles by rain and snow is termed wet deposition.

Direct Deposition of Gases

Actual measurement of the fluxes of gases to a water surface is possible for only a very few chemicals at the present time, although extensive research is underway in this area, and analytical capabilities for fast response measurements of some trace gases are becoming available. Modeling the flux of gaseous compounds to the sea surface or to rain droplets requires a knowledge of the Henry's law constants and air/sea exchange coefficients as well as atmospheric and oceanic concentrations of the chemicals of interest. For many chemicals this information is not available. Discussions of the details of these processes of air/sea gas exchange can be found in other articles in this volume.

Particle Dry Deposition

Reliable methods do not currently exist to measure directly the dry deposition of the full size range of aerosol particles to a water surface. Thus, dry deposition of aerosols is often estimated using the dry deposition velocity, v_d. For dry deposition, the flux is then given by:

$$F_d = v_d \cdot C_a \quad [1]$$

where F_d is the dry deposition flux (e.g., in $g\,m^{-2}\,s^{-1}$), v_d is the dry deposition velocity (e.g., in $m\,s^{-1}$), and C_a is the concentration of the substance on the aerosol particles in the atmosphere (e.g., in $g\,m^{-3}$). In this formulation v_d incorporates all the processes of dry deposition, including diffusion, impaction, and gravitational settling of the particles to a water surface. It is very difficult to parameterize accurately the dry deposition velocity since each of these processes is acting on a particle population, and they are each dependent upon a number of factors, including wind speed, particle size, relative humidity, etc. The following are dry deposition velocities that have been used in some studies of atmospheric deposition of particles to the ocean:

- Submicrometer aerosol particles, $0.001 \text{ m s}^{-1} \pm$ a factor of three
- Supermicrometer crustal particles not associated with sea salt, $0.01 \text{ m s}^{-1} \pm$ a factor of three
- Giant sea-salt particles and materials carried by them, $0.03 \text{ m s}^{-1} \pm$ a factor of two

Proper use of eqn [1] requires that information be available on the size distribution of the aerosol particles and the material present in them.

Particle and Gas Wet Deposition

The direct measurement of contaminants in precipitation samples is certainly the best approach for determining wet deposition, but problems with rain sampling, contamination, and the natural variability of the concentration of trace substances in precipitation often make representative flux estimates difficult using this approach. Studies have shown that the concentration of a substance in rain is related to the concentration of that substance in the atmosphere. This relationship can be expressed in terms of a scavenging ratio, S:

$$S = C_r \cdot \rho \cdot C_{a/g}^{-1} \qquad [2]$$

where C_r is the concentration of the substance in rain (e.g., in g kg^{-1}), ρ is the density of air ($\sim 1.2 \text{ kg m}^{-3}$), $C_{a/g}$ is the aerosol or gas phase concentration in the atmosphere (e.g., in g m^{-3}), and S is dimensionless. Values of S for substances present in aerosol particles range from a few hundred to a few thousand, which roughly means that 1 g (or 1 ml) of rain scavenges $\leqq 1 \text{ m}^3$ of air. For aerosols, S is dependent upon such factors as particle size and chemical composition. For gases, S can vary over many orders of magnitude depending on the specific gas, its Henry's law constant, and its gas/water exchange coefficient. For both aerosols and gases, S is also dependent upon the vertical concentration distribution and vertical extent of the precipitating cloud, so the use of scavenging ratios requires great care, and the results have significant uncertainties. However, if the concentration of an atmospheric substance and its scavenging ratio are known, the scavenging ratio approach can be used to estimate wet deposition fluxes as follows:

$$F_r = P \cdot C_r = P \cdot S \cdot C_{a/g} \cdot \rho^{-1} \qquad [3]$$

where F_r is the wet deposition flux (e.g., in $\text{g m}^{-2} \text{year}^{-1}$) and P is the precipitation rate (e.g., in m year^{-1}), with appropriate conversion factors to translate rainfall

depth to mass of water per unit area. Note that $P \cdot S \cdot \rho^{-1}$ is equivalent to a wet deposition velocity.

Atmospheric Deposition to Estuaries and the Coastal Ocean

Metals

The atmospheric deposition of certain metals to coastal and estuarine regions has been studied more than that for any other chemicals. These metals are generally present on particles in the atmosphere. Chesapeake Bay is among the most thoroughly studied regions in North America in this regard. Table 1 provides a comparison of the atmospheric and riverine deposition of a number of metals to Chesapeake Bay. The atmospheric numbers represent a combination of wet plus dry deposition directly onto the Bay surface. Note that the atmospheric input ranges from as low as 1% of the total input for manganese to as high as 82% for aluminum. With the exception of Al and Fe, which are largely derived from natural weathering processes (e.g., mineral matter or soil), most of the input of the other metals is from human-derived sources. For metals with anthropogenic sources the atmosphere is most important for lead (32%).

There have also been a number of investigations of the input of metals to the North Sea, Baltic Sea, and Mediterranean Sea. Some modeling studies of the North Sea considered not only the direct input pathway represented by the figures in Table 1, but also considered Baltic Sea inflow, Atlantic Ocean inflow and outflow, and exchange of metals with the

Table 1 Estimates of the riverine and atmospheric input of some metals to Chesapeake Bay

Metal	Riverine input (10^6 g year^{-1})	Atmospheric input (10^6 g year^{-1})	% Atmospheric input
Aluminum	160	700	81
Iron	600	400	40
Manganese	1300	13	1
Zinc	50	18	26
Copper	59	3.5	6
Nickel	100	4	4
Lead	15	7	32
Chromium	15	1.5	10
Arsenic	5	0.8	14
Cadmium	2.6	0.4	13

Data reproduced with permission from Scudlark JR, Conko KM and Church TM (1994) Atmospheric wet desposition of trace elements to Chesapezke Bay: (CBAD) study year 1 results. *Atmospheric Environment* 28: 1487–1498.

sediments, as well as the atmospheric contribution to all of these inputs. **Figure 1** shows schematically some modeling results for lead, copper, and cadmium. Note that for copper, atmospheric input is relatively unimportant in this larger context, while atmospheric input is somewhat more important for cadmium, and it is quite important for lead, being approximately equal to the inflow from the Atlantic Ocean, although still less than that entering the North Sea from dumping. As regards lead, note that approximately 20% of the inflow from the Atlantic to the North Sea is also derived from the atmosphere. This type of approach gives perhaps the most accurate and in-depth analysis of the importance of atmospheric input relative to all other sources of a chemical in a water mass.

Nitrogen Species

The input of nitrogen species from the atmosphere is of particular interest because nitrogen is a necessary nutrient for biological production and growth in the ocean. There has been an increasing number of studies of the atmospheric input of nitrogen to estuaries and the coastal ocean. Perhaps the area most intensively studied is once again Chesapeake Bay. **Table 2** shows that approximately 40% of all the nitrogen contributed by human activity to Chesapeake Bay enters via

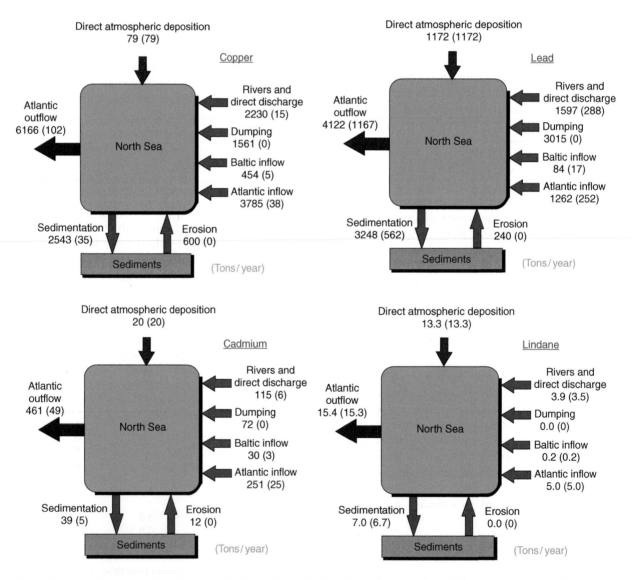

Figure 1   Input of copper, lead, cadmium, and lindane to the North Sea. Values in parentheses denote atmospheric contribution. For example, for copper the atmospheric contribution to rivers and direct discharges is 15 tons per year. (Figure reproduced with permission from Duce, 1998. Data adapted with permission from van den Hout, 1994.)

Table 2 Estimates of the input of nitrogen to Chesapeake Bay

Source	Total input (10^9g year^{-1})	Areal input rate (g m^{-2}year^{-1})	% of the total
Animal waste	5	0.4	3
Fertilizers	48	4.2	34
Point sources	33	2.9	24
Atmospheric precipitation			
nitrate	35	3.1	25
ammonium	19	1.7	14
Total	140	12.3	100

Data reproduced with permission from Fisher D, Ceroso T, Mathew T and Oppenheimer M (1988) *Polluted Coastal Waters: The Role of Acid Rain*. New York: Environmental Defense Fund.

precipitation falling directly on the Bay or its watershed. These studies were different from most earlier studies because the atmospheric contributions were considered not only to be direct deposition on the water surface, but also to include that coming in via the atmosphere but falling on the watershed and then entering the Bay. Note from **Table 2** that atmospheric input of nitrogen exceeded that from animal waste, fertilizers, and point sources. In the case of nitrate, about 23% falls directly on the Bay, with the remaining 77% falling on the watershed. These results suggest that studies that consider only the direct deposition on a water surface (e.g., the results shown in **Table 1**) may significantly underestimate the true contribution of atmospheric input. The total nitrogen fertilizer applied to croplands in the Chesapeake Bay region is ~ 5.4 g m^{-2} year^{-1}, while the atmospheric nitrate and ammonium nitrogen entering the Bay is ~ 4.8 g m^{-2} year^{-1}. Chesapeake Bay is almost as heavily fertilized from atmospheric nitrogen, largely anthropogenic, as the croplands are by fertilizer in that watershed!

Results from studies investigating nitrogen input to some other estuarine and coastal regions are summarized in **Table 3**. In this table atmospheric sources for nitrogen are compared with all other sources, where possible. The atmospheric input ranges from 10% to almost 70% of the total. Note that some estimates compare only direct atmospheric deposition with all other sources and some include as part of the atmospheric input the portion of the deposition to the watershed that reaches the estuary or coast.

Synthetic Organic Compounds

Concern is growing about the input of a wide range of synthetic organic compounds to the coastal ocean. To date there have been relatively few estimates of the atmospheric fluxes of synthetic organic compounds to the ocean, and these estimates have significant uncertainties. These compounds are often both persistent and toxic pollutants, and many have relatively high molecular weights. The calculation of the atmospheric input of these compounds to the

Table 3 Estimates of the input of nitrogen to some coastal areas[a]

Region	Total atmospheric input[b] (10^9g year^{-1})	Total input all sources (10^9g year^{-1})	% Atmospheric input
North Sea	400[c]	1500	27[c]
Western Mediterranean Sea	400[c]	577[d]	69[c]
Baltic Sea	500	$\sim$1200	42
Chesapeake Bay	54	140	39
New York Bight	–	–	13[c]
Long Island Sound	11	49	22
Neuse River Estuary, NC	1.7	7.5	23

[a]Data from several sources in the literature.
[b]Total from direct atmospheric deposition and runoff of atmospheric material from the watershed.
[c]Direct atmospheric deposition to the water only.
[d]Total from atmospheric and riverine input only.

Table 4 Estimates of the input of synthetic organic compounds to the North Sea

Organic compound	Atmospheric input (10^6 g year^{-1})	Input from other sources (10^6 g year^{-1})	% Atmospheric input
PCB	40	3	93
Lindane	36	3	92
Polycyclic aromatic hydrocarbons	80	90	47
Benzene	400	500	44
Trichloroethene	300	80	80
Trichloroethane	90	60	94
Tetrachloroethene	100	10	91
Carbon tetrachloride	6	40	13

Data reproduced with permission from Warmerhoven JP, Duiser JA, de Leu LT and Veldt C (1989) *The Contribution of the Input from the Atmosphere to the Contamination of the North Sea and the Dutch Wadden Sea*. Delft, The Netherlands: TNO Institute of Environmental Sciences.

coastal ocean is complicated by the fact that many of them are found primarily in the gas phase in the atmosphere, and most of the deposition is related to the wet and dry removal of that phase. The atmospheric residence times of most of these compounds are long compared with those of metals and nitrogen species. Thus the potential source regions for these compounds entering coastal waters can be distant and widely dispersed.

Figure 1 shows the input of the pesticide lindane to the North Sea. Note that the atmospheric input of lindane dominates that from all other sources. **Table 4** compares the atmospheric input to the North Sea with that of other transport paths for a number of other synthetic organic compounds. In almost every case atmospheric input dominates the other sources combined.

Atmospheric Deposition to the Open Ocean

Studies of the atmospheric input of chemicals to the open ocean have also been increasing lately. For many substances a relatively small fraction of the material delivered to estuaries and the coastal zone by rivers and streams makes its way through the near shore environment to open ocean regions. Most of this material is lost via flocculation and sedimentation to the sediments as it passes from the freshwater environment to open sea water. Since aerosol particles in the size range of a few micrometers or less have atmospheric residence times of one to several days, depending upon their size distribution and local precipitation patterns, and most substances of interest in the gas phase have similar or even longer atmospheric residence times, there is ample opportunity

for these atmospheric materials to be carried hundreds to thousands of kilometers before being deposited on the ocean surface.

Metals

Table 5 presents estimates of the natural and anthropogenic emission of several metals to the global atmosphere. Note that ranges of estimates and the best estimate are given. It appears from **Table 5** that anthropogenic sources dominate for lead, cadmium, and zinc, with essentially equal contributions for copper, nickel, and arsenic. Clearly a significant fraction of the input of these metals from the atmosphere to the ocean could be derived largely from anthropogenic sources.

Table 6 provides an estimate of the global input of several metals from the atmosphere to the ocean and compares these fluxes with those from rivers. Estimates are given for both the dissolved and particulate forms of the metals. These estimates suggest that rivers are generally the primary source of particulate

Table 5 Emissions of some metals to the global atmosphere

Metal	Anthropogenic emissions (10^9 g year^{-1})		Natural emissions (10^9 g year^{-1})	
	Range	Best estimate	Range	Best estimate
Lead	289–376	332	1–23	12
Cadmium	3.1–12	7.6	0.15–2.6	1.3
Zinc	70–194	132	4–86	45
Copper	20–51	35	2.3–54	28
Arsenic	12–26	18	0.9–23	12
Nickel	24–87	56	3–57	30

Data reproduced with permission from Duce *et al.*, 1991.

Table 6 Estimates of the input of some metals to the global ocean

Metal	Atmospheric input		Riverine input	
	Dissolved (10^9 g year^{-1})	Particulate (10^9 g year^{-1})	Dissolved (10^9 g year^{-1})	Particulate (10^9 g year^{-1})
Iron	1600–4800	14 000–42 000	1100	110 000
Copper	14–45	2–7	10	1 500
Nickel	8–11	14–17	11	1 400
Zinc	33–170	11–55	6	3 900
Arsenic	2.3–5	1.3–3	10	80
Cadmium	1.9–3.3	0.4–0.7	0.3	15
Lead	50–100	6–12	2	1 600

Data reproduced with permission from Duce et al., 1991.

metals in the ocean, although again a significant fraction of this material may not get past the coastal zone. For the dissolved phase atmospheric and riverine inputs are roughly equal for metals such as iron, copper, and nickel; while for zinc, cadmium, and particularly lead atmospheric inputs appear to dominate. These estimates were made based on data collected in the mid-1980s. Extensive efforts to control the release of atmospheric lead, which has been primarily from the combustion of leaded gasoline, are now resulting in considerably lower concentrations of lead in many areas of the open ocean. For example, **Figure 2** shows that the concentration of dissolved lead in surface sea water near Bermuda has been decreasing regularly over the past 15–20 years, as has the atmospheric lead concentration in that region. This indicates clearly that at least for very particle-reactive metals such as lead, which has a short lifetime in the ocean (several years), even the open ocean can recover rather rapidly when the anthropogenic input of such metals is reduced or ended. Unfortunately, many of the other metals of most concern have much longer residence times in the ocean (thousands to tens of thousands of years).

Figure 3 presents the calculated fluxes of several metals from the atmosphere to the ocean surface and from the ocean to the seafloor in the 1980s in the tropical central North Pacific. Note that for most metals the two fluxes are quite similar, suggesting the potential importance of atmospheric input to the

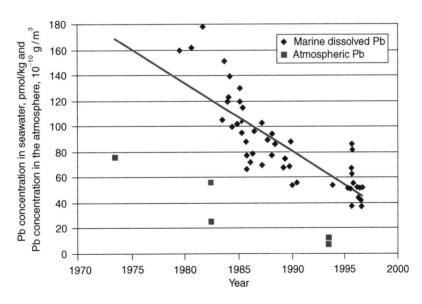

Figure 2 Changes in concentration of atmospheric lead at Bermuda and dissolved surface oceanic lead near Bermuda from the mid-1970s to the mid-1990s. (Data reproduced with permission from Wu J and Boyle EA (1997) Lead in the western North Atlantic Ocean: Completed response to leaded gasoline phaseout. *Geochimica et Cosmochimica Acta* 61: 3279–3283; and from Huang S, Arimoto R and Rahn KA (1996) Changes in atmospheric lead and other pollution elements at Bermuda. *Journal of Geophysical Research* 101: 21 033–21 040.)

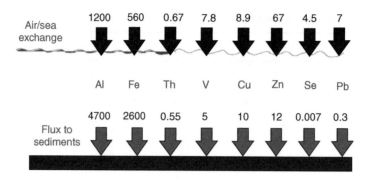

Figure 3 A comparison of the calculated fluxes of aluminum (Al), iron (Fe), thorium (Th), vanadium (V), copper (Cu), zinc (Zn), selenium (Se), and lead (Pb) (in 10^{-9} g cm^{-2} year^{-1}) from the atmosphere to the ocean and from the ocean to the sediments in the central tropical North Pacific. For each metal note the relative similarity in the two fluxes, except for lead and selenium. (Reproduced with permission from Duce, 1998.)

marine sedimentation of these metals in this region. Lead and selenium are exceptions, however, as the atmospheric flux is much greater than the flux to the seafloor. The fluxes to the seafloor represent average fluxes over the past several thousand years, whereas the atmospheric fluxes are roughly for the present time. The atmospheric lead flux is apparently much larger than the flux of lead to the sediments, primarily because of the high flux of anthropogenic lead from the atmosphere to the ocean since the introduction of tetraethyllead in gasoline in the 1920s. (The atmospheric flux is much lower now than in the 1980s, as discussed above.) However, in the case of selenium the apparently higher atmospheric flux is an artifact, because most of the flux of selenium from the atmosphere to the ocean is simply marine-derived selenium that has been emitted from the ocean to the atmosphere as gases, such as dimethyl selenide (DMSe). DMSe is oxidized in the atmosphere and returned to the ocean, i.e., the selenium input is simply a recycled marine flux. Thus, care must be taken when making comparisons of this type.

Nitrogen Species

There is growing concern about the input of anthropogenic nitrogen species to the global ocean. This issue is of particular importance in regions where nitrogen is the limiting nutrient, e.g., the oligotrophic waters of the central oceanic gyres. Estimates to date suggest that in such regions atmospheric nitrogen will in general account for only a few percent of the total 'new' nitrogen delivered to the photic zone, with most of the 'new' nutrient nitrogen derived from the upwelling of nutrient-rich deeper waters and from nitrogen fixation in the sea. It is recognized, however, that the atmospheric input is highly episodic, and at times it may play a much more important role as a source for nitrogen in

surface waters. **Table 7** presents a recent estimate of the current input of fixed nitrogen to the global ocean from rivers, the atmosphere, and nitrogen fixation. From the numbers given it is apparent that all three sources are likely important, and within the uncertainties of the estimates they are roughly equal. In the case of rivers, about half of the nitrogen input is anthropogenic for atmospheric input perhaps the most important information in **Table 7** is that the organic nitrogen flux appears to be equal to or perhaps significantly greater than the inorganic (i.e., ammonium and nitrate) nitrogen flux. The source of the organic nitrogen is not known, but there are indications that a large fraction of it is anthropogenic in origin. This is a form of atmospheric nitrogen input to the ocean that had not been considered until very recently, as there had been few measurements of organic nitrogen input to the ocean before the mid-1990s. The chemical forms of this organic nitrogen are still largely unknown.

Of particular concern are potential changes to the input of atmospheric nitrogen to the open ocean in

Table 7 Estimates of the current input of reactive nitrogen to the global ocean

Source	Nitrogen input (10^{12} g year^{-1})
From the atmosphere	
Dissolved inorganic nitrogen	28–70
Dissolved organic nitrogen	28–84
From rivers (dissolved inorganic + organic nitrogen)	
Natural	14–35
Anthropogenic	7–35
From nitrogen fixation within the ocean	14–42

Data reproduced with permission from Cornell S, Rendell A and Jickells T (1995) Atmospheric inputs of dissolved organic nitrogen to the oceans. *Nature* 376: 243–246.

Table 8   Estimates of anthropogenic reactive nitrogen production, 1990 and 2020

Region	Energy (NOx)					Fertilizer				
	1990	2020	Δ	Factor	% of total increase	1990	2020	Δ	Factor	% of total increase
	(10^{12}g N year^{-1})					(10^{12}g N year^{-1})				
USA/Canada	7.6	10.1	2.5	1.3	10	13.3	14.2	0.9	1.1	1.6
Europe	4.9	5.2	0.3	1.1	1	15.4	15.4	0	1.0	0
Australia	0.3	0.4	0.1	1.3	0.4	—	—	—	—	—
Japan	0.8	0.8	0	1.0	0	—	—	—	—	—
Asia	3.5	13.2	9.7	3.8	39	36	85	49	2.4	88
Central/South America	1.5	5.9	4.4	3.9	18	1.8	4.5	2.7	2.5	5
Africa	0.7	4.2	3.5	6.0	15	2.1	5.2	3.1	2.5	6
Former Soviet Union	2.2	5.7	3.5	2.5	15	10	10	0	1.0	0
Total	21	45	24	2.1	100	79	134	55	1.7	100

Data adapted with permission from Galloway *et al.*, 1995.

the future as a result of increasing human activities. The amount of nitrogen fixation (formation of reactive nitrogen) produced from energy sources (primarily as NOx, nitrogen oxides), fertilizers, and legumes in 1990 and in 2020 as a result of human activities as well as the current and predicted future geographic distribution of the atmospheric deposition of reactive nitrogen to the continents and ocean have been evaluated recently. **Table 8** presents estimates of the formation of fixed nitrogen from energy use and production and from fertilizers, the two processes which would lead to the most important fluxes of reactive nitrogen to the atmosphere. Note that the most highly developed regions in the world, represented by the first four regions in the table, are predicted to show relatively little increase in the formation of fixed nitrogen, with none of these areas having a predicted increase by 2020 of more than a factor of 1.3 nor a contribution to the overall global increase in reactive nitrogen exceeding 10%. However, the regions in the lower part of **Table 8** will probably contribute very significantly to increased anthropogenic reactive nitrogen formation in 2020. For example, it is predicted that the production of reactive nitrogen in Asia from energy sources will increase ≤ fourfold, and that Asia will account for almost 40% of the global increase, while Africa will have a sixfold increase and will account for 15% of the global increase in energy-derived fixed nitrogen. It is predicted that production of reactive nitrogen from the use of fertilizers in Asia will increase by a factor of 2.4, and Asia will account for ~88% of the global increase from this source. Since both energy sources (NOx, and ultimately nitrate) and fertilizer (ammonia and nitrate) result in the extensive release of reactive nitrogen to the atmosphere, the predictions above indicate that there

should be very significant increases in the atmospheric deposition to the ocean of nutrient nitrogen species downwind of such regions as Asia, Central and South America, Africa, and the former Soviet Union.

This prediction has been supported by numerical modeling studies. These studies have resulted in the generation of maps of the 1980 and expected 2020 annual deposition of reactive nitrogen to the global ocean. **Figure 4** shows the expected significant increase in reactive nitrogen deposition from fossil fuel combustion to the ocean to the east of all of Asia, from Southeast Asia to the Asian portion of the former Soviet Union; to the east of South Africa, northeast Africa and the Mideast and Central America and southern South America; and to the west of northwest Africa. This increased reactive nitrogen transport and deposition to the ocean will provide new sources of nutrient nitrogen to some regions of the ocean where biological production is currently nitrogen-limited. There is thus the possibility of significant impacts on regional biological primary production, at least episodically, in these regions of the open ocean.

Synthetic Organic Compounds

The atmospheric residence times of many synthetic organic compounds are relatively long compared with those of the metals and nitrogen species, as mentioned previously. Many of these substances are found primarily in the gas phase in the atmosphere, and they are thus very effectively mobilized into the atmosphere during their production and use. Their long atmospheric residence times of weeks to months leads to atmospheric transport that can often be hemispheric or near hemispheric in scale. Thus

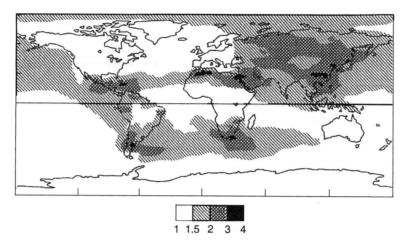

Figure 4 The ratio of the estimated deposition of reactive nitrogen to ocean and land surfaces in 2020 relative to 1980. (Figure reproduced with permission from Watson AJ (1997) Surface Ocean–Lower Atmosphere Study (SOLAS). *Global Change Newsletter, IGBP* no. 31, 9–12; data in figure adapted with permission from Galloway JN, Levy H and Kasibhatla PS (1994) Year 2020: Consequences of population growth and development on deposition of oxidized nitrogen. *Ambio* 23: 120–123.).

Table 9 Estimates of the atmospheric input of organochlorine compounds to the global ocean

Ocean	ΣHCH $(10^6 g\ year^{-1})$	ΣDDT $(10^6 g\ year^{-1})$	ΣPCB $(10^6 g\ year^{-1})$	HCB $(10^6 g\ year^{-1})$	$Dieldrin$ $(10^6 g\ year^{-1})$	$Chlordane$ $(10^6 g\ year^{-1})$
North Atlantic	850	16	100	17	17	8.7
South Atlantic	97	14	14	10	2.0	1.0
North Pacific	2600	66	36	20	8.9	8.3
South Pacific	470	26	29	19	9.5	1.9
Indian	700	43	52	11	6.0	2.4
Global input via the atmosphere	~4700	~170	~230	~80	~40	~22
Global input via rivers	~60	~4	~60	~4	~4	~4
% Atmospheric input	~99%	~98%	~80%	~95%	~91%	~85%

(Data reproduced with permission from Duce *et al.*, 1991.)

atmospheric transport and deposition in general dominates all other sources for these chemicals in sea water in open ocean regions.

Table 9 compares the atmospheric and riverine inputs to the world ocean for a number of synthetic organochlorine compounds. Note that the atmosphere in most cases accounts for 90% or more of the input of these compounds to the ocean. **Table 9** also presents estimates of the input of these same organochlorine compounds to the major ocean basins. Since most of these synthetic organic compounds are produced and used in the northern hemisphere, it is not surprising that the flux into the northern hemisphere ocean is greater than that to southern hemisphere marine regions. There are some differences for specific compounds in different ocean basins. For example, HCH (hexachlorocyclohexane) and DDT have a higher input rate to the North Pacific than the North Atlantic, largely because of the greater use of these compounds in Asia than in North America or

Europe. On the other hand, the input of PCBs (polychlorinated biphenyls) and dieldrin is higher to the North Atlantic than the North Pacific, primarily because of their greater use in the continental regions adjacent to the North Atlantic.

Conclusions

The atmosphere transports materials to the ocean that are both harmful to marine life and that are essential for marine biological productivity. It is now apparent that atmospheric transport and deposition of some metals, nitrogen species, and synthetic organic compounds can be a significant and in some cases dominant pathway for these substances entering both estuarine and coastal waters as well as some open ocean regions. Atmospheric input clearly must be considered in any evaluation of material fluxes to marine ecosystems. However, the uncertainties in the atmospheric fluxes of these materials to the ocean are large. The primary reasons for these large uncertainties are:

- The lack of atmospheric concentration data over vast regions of the coastal and open ocean, particularly over extended periods of time and under varying meteorological conditions;
- The episodic nature of the atmospheric deposition to the ocean;
- The lack of accurate models of air/sea exchange, particularly for gases;
- The inability to measure accurately the dry deposition of particles; and
- The inability to measure accurately the air/sea exchange of gases.

Further Reading

Duce RA (1998) *Atmospheric Input of Pollution to the Oceans*, pp. 9–26. Proceedings of the Commission for Marine Meteorology Technical Conference on Marine Pollution, World Meteorological Organization TD-No. 890, Geneva, Switzerland.

Duce RA, Liss PS, Merrill JT, *et al.* (1991) The atmospheric input of trace species to the world ocean. *Global Biogeochemical Cycles* 5: 193–259.

Galloway JN, Schlesinger WH, Levy H, Michaels A, and Schnoor JL (1995) Nitrogen fixation: anthropogenic enhancement – environmental reponse. *Global Biogeochemical Cycles* 9: 235–252.

Jickells TD (1995) Atmospheric inputs of metals and nutrients to the oceans: their magnitude and effects. *Marine Chemistry* 48: 199–214.

Liss PS and Duce RA (1997) *The Sea Surface and Global Change*. Cambridge: Cambridge University Press.

Paerl HW and Whitall DR (1999) Anthropogenically-derived atmospheric nitrogen deposition, marine eutrophication and harmful algal bloom expansion: Is there a link? *Ambio* 28: 307–311.

Prospero JM, Barrett K, Church T, *et al.* (1996) Nitrogen dynamics of the North Atlantic Ocean – Atmospheric deposition of nutrients to the North Atlantic Ocean. *Biogeochemistry* 35: 27–73.

van den Hout KD (ed.) (1994) *The Impact of Atmospheric Deposition of Non-Acidifying Pollutants on the Quality of European Forest Soils and the North Sea*. Report of the ESQUAD Project, IMW-TNO Report No. R 93/329.

CORALS AND HUMAN DISTURBANCE

N. J. Pilcher, Universiti Malaysia Sarawak, Sarawak, Malaysia

Introduction

Coral reefs are the centers of marine biodiversity on the planet. Reefs are constructed by a host of hermatypic (reef-building) coral species, but also are home to ahermatypic (non-calcium-carbonate depositing) corals, such as soft corals, black corals and gorgonians. The major structural components of reefs are the scleractinian corals. Much like their terrestrial rivals the tropical rainforests, reefs combine a host of microhabitats and a diverse array of life forms that is still being discovered and described. Coral reefs are mostly distributed throughout the tropical belt, and a large fraction are located in developing countries.

To understand how human activities affect coral reefs, it is necessary to briefly review their basic life history. Coral reefs are mostly made up of numerous smaller coral colonies; these colonies are in turn made up of thousands of minute polyps, which secrete a calcium carbonate skeleton. The deposition rate for individual coral species varies, but generally ranges between 0.1 mm and 10.0 cm per year. The accumulation of these skeletons over a long period of time results in massive, three-dimensional geological structures. The actual living tissue, however, is only the thin layer of living coral polyps on the surface. Corals are particularly susceptible to contaminants in sea water because the layer of tissue covering the coral skeleton is thin ($\sim 100 \, \mu m$) and rich in lipids, facilitating direct uptake of chemicals. Coral polyps feed by filtering plankton using nematocyst (stinging cell)-tipped tentacles, and also receive organic matter through their symbiotic relationship with minute dinoflagellates called zooxanthellae. Zooxanthellae live within the gastrodermal tissues, and chemical communication (exchange) occurs via the translocation of metabolites. These small algal cells use sunlight to photosynthesize carbonates and water into organic matter and oxygen, both of which are used by the polyp.

Coral reefs support complex food and energy webs that are interlinked with nutrient inputs from outside sources (such as those brought with ocean currents and runoff from nearby rivers) and from the reef itself (where natural predation and die-off recirculate organic matter). These complex webs mean that any effect on one group of individuals will ultimately impact another, and single disturbances can have multiple effects on reef inhabitants. For example, the complete eradication of the giant Triton *Charonia trinis* through overfishing can result in outbreaks of the crown-of-thorns starfish *Acanthaster planci*. This can lead to massive coral mortality as the starfish reproduce and feed on the coral polyps. The mortality in turn may reduce habitats and food sources for reef fishes, which again, in turn, could lead to declines of larger predatory fishes. Similarly, the introduction of an invasive species either by accident or through ignorance (e.g., dumping of a personal aquarium contents into local habitats) might disrupt feeding processes and kill resident fishes. The death of key organisms on the reef (which then shifts from an autotrophic to a heterotrophic, suspension/detritus-feeding community) changes the dominant ecological process from calcium carbonate deposition to erosion, and ultimate loss of coral reef. Reef ecosystems may respond to environmental change by altering their physical and ecological structure, and through changes in rates of accretion and biogeochemical cycling. However, the potential for adaptation in reef organisms may be overwhelmed by today's anthropogenic stresses. The following sections provide a review of human disturbances and their general effects on coral reefs.

Collection of Corals

Corals have been mined for construction purposes in numerous Pacific Ocean islands and in South-East Asia. Usually the large massive life forms such as *Porites*, *Platygyra*, *Favia*, and *Favites* are collected and broken into manageable sizes or crushed for cement and lime manufacture. Similarly, the shells of the giant clam *Tridacna* and the conch shell *Strombus* are collected. Coral blocks and shells are used for construction of houses, roads, and numerous other projects. Corals are also collected for use in the ornamental trade, either as curios and souvenirs or as jewelry. Entire, small colonies of branching species such as *Acropora* and *Seriatophora* are used in the souvenir trade and for decoration, while black corals *Antipathes* and blue coral *Heliopora* are used for jewelry. The aquarium industry is also responsible for coral collection either for direct sale as live colonies or through the process of fish collecting. In many cases entire colonies full of fish are brought to the surface and are then smashed and discarded.

The removal of coral colonies decreases the shelter and niche areas available to numerous other reef inhabitants. Juvenile stages of fish that seek shelter among the branching species of corals, and worms and ascidians that take up residence on massive life forms, are deprived of protection and may become prey to other reef organisms. Removal of adult colonies also results in a reduction of overall reproductive output, as the corals no longer serve as a source of replenishment larvae. Further, removal of entire colonies reduces the overall structural stability of the reef, and increases rates of erosion through wave and surge damage.

Destructive Fishing

Destructive fishing pressures are taking their toll on coral reefs, particularly in developing countries in South-East Asia. The use of military explosives and dynamite was common shortly after the Second World War, but today this has shifted to the use of home-made explosives of fertilizer, fuel, and fuse caps inserted into empty beer bottles. Bombs weigh approximately 1 kg and have a destructive diameter of 4–5 m. Blast fishers hunt for schooling fish such as sweetlips and fusiliers, which aggregate in groups in the open or hide under large coral heads. Parrot fish and surgeon fish schools grazing on the reef crest are also actively sought. The bombs are usually set on five-second fuses and are dropped into the center of an area judged to have many fish. After the bomb has exploded, the fishers use dip nets, either from the boats or from underwater, to collect the stunned and dying fish. Many larger boats collect the fish using 'hookah' compressors and long air hoses to support divers working underwater.

The pressure wave from the explosion kills or stuns fish, but also damages corals. Natural disturbances may also fragment stony reef corals, and there are few quantitative data on the impacts of skeletal fragmentation on the biology of these corals. Lightly bombed reefs are usually pockmarked with blast craters, while many reefs in developing countries comprise a continuous band of coral rubble instead of a reef crest and upper reef slope. The lower reef slope is a mix of rubble, sand, and overturned coral heads. Typically at the base of the reef slope is a mound of coral boulders that have been dislodged by a blast and then rolled down the slope in an underwater avalanche. The reef slopes are mostly dead coral, loose sand, rubble, or rock and occasionally have overturned clams or coral heads with small patches of living tissue protruding from the rubble. The blasts also change the three-dimensional structure of reefs, and blasted areas no longer provide food or shelter to reef inhabitants. Further, once the reef structure has been weakened or destroyed by blast fishing, it is much more susceptible to wave action and the reef is unable to maintain its role in coastline protection. Larvae do not settle on rubble and thus replenishment and rehabilitation is minimal. Additionally, the destruction of adult colonies also results in a reduction of overall reproductive output, and reefs no longer serve as a source of replenishment larvae. Experimental findings, for instance, indicate that fragmentation reduces sexual reproductive output in the reef-building coral *Pocillopora damicornis*. The recovery of such areas has been measured in decades, and only then with complete protection and cessation of fishery pressure of any kind.

Another type of destructive fishing is 'Muro Ami', in which a large semicircular net is placed around a reef. Fish are driven into the net by a long line of fishermen armed with weighted lines. The weights are repeatedly crashed onto the reef to scare fish in the direction of the net, reducing coral colonies to rubble. The resulting effects are similar to the effects of blast fishing, spread over a larger area.

Cyanide fishing is also among the most destructive fishing methods, in which an aqueous solution of sodium cyanide is squirted at fish to stun them, after which they are collected and sold to the live-fish trade. Other chemicals are also used, including rotenone, plant extracts, fertilizers, and quinaldine. These chemicals all narcotize fish, rendering them inactive enough for collection. The fish are then held in clean water for a short period to allow them to recover, before being hauled aboard boats with live fish holds. In the process of stunning fish, the cyanide affects corals and small fish and invertebrates. The narcotizing solution for large fish is often lethal to smaller ones. Cyanide has been shown to limit coral growth and cause diseases and bleaching, and ultimately death in many coral species.

Among other destructive aspects of fishing are lost fishing gear, and normal trawl and purse fishing operations, when these take place near and over reefs. Trawlers operate close to reefs to take advantage of the higher levels of fish aggregated around them, only to have the trawls caught on the reefs. Many of these have to be cut away and discarded, becoming further entangled on the reefs, breaking corals and smothering others. Similarly, fishing with fine-mesh nets that get entangled in coral structures also results in coral breakage and loss. In South-East Asia, fishing companies have been reported to pull a chain across the bottom using two boats to clear off corals, making it accessible to trawlers.

Spearfishing also damages corals as fishermen trample and break coral to get at fish that disappear into crevices, and crowbars are frequently used to break coral. The collection of reef invertebrates along the reef crest results in breakage of corals that have particular erosion control functions, reducing the reef's potential to act as a coastal barrier.

Discharges

Mankind also effects corals through the uncontrolled and often unregulated discharge of a number of industrial and domestic effluents. Many of these are 'point-source' discharges that affect local reef areas, rather than causing broad-scale reef mortality. Sources of chemical contamination include terrestrial runoff from rivers and streams, urban and agricultural areas, sewage outfalls near coral reefs, desalination plants, and chemical inputs from recreational uses and industries (boat manufacturing, boating, fueling, etc.). Landfills can also leach directly or indirectly into shallow water tables. Industrial inputs from coastal mining and smelting operations are sources of heavy metals. Untreated and partially treated sewage is discharged over reefs in areas where fringing reefs are located close to shore, such as the reefs that fringe the entire length of the Red Sea. Raw sewage can result in tumors on fish, and erosion of fins as a result of high concentrations of bacteria. The resulting smothering sludge produces anaerobic conditions under which all benthic organisms perish, including corals. In enclosed and semi-enclosed areas the sewage causes eutrophication of the coral habitats. For instance, in Kaneohe bay in Hawaii, which had luxuriant reefs, sewage was dumped straight into the bay and green algae grew in plague quantities, smothering and killing the reefs. Evidence indicates that branching species might be more susceptible to some chemical contaminants than are massive corals.

Abattoir refuse is another localized source of excessive nutrients and other wastes that can lead to large grease mats smothering the seabed, local eutrophication, red tides, jellyfish outbreaks, an increase in biological oxygen demand (BOD), and algal blooms. Similarly, pumping/dumping of organic compounds such as sugar cane wastes also results in oxygen depletion.

The oil industry is a major source of polluting discharges. Petroleum hydrocarbons and their derivatives and associated compounds have caused widespread damage to coastal ecosystems, many of which include coral reefs. The effects of these discharges are often more noticeable onshore than offshore where reefs are generally located, but nonetheless have resulted in the loss of reef areas, particularly near major exploration and drilling areas, and along major shipping routes. Although buoyant eggs and developing larvae are sometimes affected, reef flats are more vulnerable to direct contamination by oil. Oiling can lead to the increased incidence of mortality of coral colonies.

In the narrow Red Sea, where many millions of tonnes per annum pass through the region, there have been more than 20 oil spills along the Egyptian coast since 1982, which have smothered and poisoned corals and other organisms. Medium spills from ballast and bilgewater discharges, and leakages from terminals, cause localized damage and smothering of intertidal habitats. Oil leakage is a regular occurrence from the oil terminal and tankers in Port Sudan harbor. Seismic blasts during oil exploration are also a threat to coral reefs. Refineries discharge oil and petroleum-related compounds, resulting in an increase in diatoms and a decrease in marine fauna closer to the refineries. Throughout many parts of the world there is inadequate control and monitoring of procedures, equipment, and training of personnel at refineries and shipping operations.

Drilling activities frequently take place near reef areas, such as the Saudi Arabian shoreline in the Arabian Gulf. Drilling muds smother reefs and contain compounds that disrupt growth and cause diseases in coral colonies. Field assessment of a reef several years after drilling indicated a 70–90% reduction in abundance of foliose, branching, and platelike corals within 85–115 m of a drilling site. Research indicates that exposure to ferrochrome lignosulfate (FCLS) can decrease growth rates in *Montastrea annularis*, and growth rates and extension of calices (skeleton supporting the polyps) decrease in response to exposure to $100 \, mg \, l^{-1}$ of drilling mud.

Oil spills affect coral reefs through smothering, resulting in a lack of further colonization, such as occurred in the Gulf of Aqaba in 1970 when the coral *Stylophora pistillata* did not recolonize oil-contaminated areas after a large spill. Effects of oil on individual coral colonies range from tissue death to impaired reproduction to loss of symbiotic algae (bleaching). Larvae of many broadcast spawners pass through sensitive early stages of development at the sea surface, where they can be exposed to contaminants and surface slicks. Oiling affects not only coral growth and tissue maintenance but also reproduction. Other effects from oil pollution include degeneration of tissues, impairment of growth and reproduction (there can be impaired gonadal development in both brooding and broadcasting species, decreased egg size and decreased fecundity), and decreased photosynthetic rates in zooxanthellae.

In developing countries, virtually no ports have reception facilities to collect these wastes and the problem will continue mostly through a lack of enforcement of existing regulations. The potential exists for large oil spills and disasters from oil tank ruptures and collisions at sea, and there are no mechanisms to contain and clean such spills. The levels of oil and its derivatives (persistent carcinogens) were correlated with coral disease in the Red Sea, where there were significant levels of diseases, especially Black Band Disease. In addition to the impacts of oils themselves are the impacts of dispersants used to combat spills. These chemicals are also toxic and promote the breakup of heavier molecules, allowing toxic fractions of the oil to reach the benthos. They also promote erosion through limiting adhesion among sand particles. The full effect of oil on corals is not fully understood or studied, and much more work is needed to understand the full impact. Although natural degradation by bacteria occurs, it is slow and, by the time bacteria consume the heavy, sinking components, these have already smothered coral colonies.

Industrial effluents, from a variety of sources, also impact coral reefs and their associated fauna and habitats. Heavy metal discharges lead to elevated levels of lead, mercury, and copper in bivalves and fish, and to elevated levels of cadmium, vanadium, and zinc in sediments. Larval stages of crustaceans and fish are particularly affected, and effluents often inhibit growth in phytoplankton, resulting in a lack of zooplankton, a major food source for corals. Industrial discharges can increase the susceptibility of fish to diseases, and many coral colonies end up with swollen tissues, excessive production of mucus, or areas without tissue. Reproduction and feeding in surviving polyps is affected, and such coral colonies rarely contribute to recolonization of reef areas.

Organisms in low-nutrient tropical waters are particularly sensitive to pollutants that can be metabolically substituted for essential elements (such as manganese). Metals enter coral tissues or skeleton by several pathways. Exposed skeletal spines (in response to environmental stress), can take up metals directly from the surrounding sea water. In Thailand, massive species such as *Porites* tended to be smaller in areas exposed to copper, zinc, and tin, there was a reduced growth rate in branching corals, and calcium carbonate accretion was significantly reduced.

Symbiotic algae have been shown to accumulate higher concentrations of metals than do host tissues in corals. Such sequestering in the algae might diminish possible toxic effects to the host. In addition, the symbiotic algae of corals can influence the skeletal concentrations of metals through enhancement of calcification rates. There is evidence, however, that corals might be able to regulate the concentrations of metals in their own tissues. For example, elevated iron in Thai waters resulted in loss of symbiotic algae in corals from pristine areas, but this response was lower in corals that has been exposed to daily runoff from an enriched iron effluent, suggesting that the corals could develop a tolerance to the metal.

Cooling brine is another industrial effluent that affects shoreline-fringing reefs, often originating from industrial installations or as the outflow from desalination plants. These effluents are typically up to 5–10°C higher in temperature and up to 3–10 ppt higher in salinity. Discharges into the marine environment from desalination plants in Jeddah include chlorine and antiscalant chemicals and 1.73 billion $m^3 d^{-1}$ of brine at a salinity of 51 ppt and 41°C. The higher temperatures decrease the water's ability to dissolve oxygen, slowing reef processes. Increases in temperature are particularly threatening to coral reefs distributed throughout the tropics, where reef-building species generally survive just below their natural thermal thresholds. Higher-temperature effluents usually result in localized bleaching of coral colonies. The higher-salinity discharges increase coral mucus production and result in the expulsion of zooxanthellae and eventual bleaching and algal overgrowth in coral colonies. Often these waters are chlorinated to limit growth of fouling organisms, which increases the effects of the effluents on reef areas. The chlorinated effluents contain compounds that are not biodegradable and can circulate in the environment for years, bringing about a reduction in photosynthesis, with blooms of blue/green and red algae. Chlorinated hydrocarbon compounds include aldrin, lindrane, dieldrin, and even the banned DDT. These oxidating compounds are absorbed by phytoplankton and in turn by filter-feeding corals. Through the complex reef food webs these compounds concentrate in carnivorous fishes, which are often poisonous to humans.

Many airborne particles are also deposited over coral reefs, such as fertilizer dust, dust from construction activities and cement dust. At Ras Baridi, on the Red Sea coast of Saudi Arabia, a cement plant that operates without filtered chimneys discharges over $100 t d^{-1}$ of partially processed cement over the nearby coral reefs, which are now smothered by over 10 cm of fine silt.

Solid Waste Dumping

The widespread dumping of waste into the seas has continued for decades, if not centuries. Plastics, metal, wood, rubber, and glass can all be found littering coral reefs. These wastes are often not biodegradable, and those that are can persist over long periods. Damage to reefs through solid waste dumping is primarily physical. Solid wastes damage coral colonies at the time of dumping, and thereafter through natural tidal and surge action. Sometimes the well-intentioned practice of developing artificial reefs backfires and the artificial materials are thrown around by violent storms, wrecking nearby reefs in the process.

Construction

Construction activities have had a major effect on reef habitats. Such activity includes coastal reclamation works, port development, dredging, and urban and industrial development. A causeway across Abu Ali bay in the northern Arabian Gulf was developed right over coral reefs, which today no longer exist. Commercial and residential property developments in Jeddah, on the Red Sea, have filled in reef lagoon areas out to reef crest and bulldozed rocks over reef crest for protection against erosion and wave action. Activities of this type result in increased levels of sedimentation as soils are nearly always dumped without the benefit of screens or silt barriers.

Siltation is invariably the consequence of poorly planned and poorly implemented construction and coastal development, which can result in removal of shoreline vegetation and sedimentation. Coral polyps, although able to withstand moderate sediment loading, cannot displace the heavier loads and perish through suffocation. Partial smothering also limits photosynthesis by zooxanthellae in corals, reducing feeding, growth, and reproductive rates.

The development of ports and marinas involves dredging deep channels through reef areas for safe navigation and berthing. Damage to reefs comes through the direct removal of coral colonies, sediment fallout, churning of water by dredger propellers, which increases sediment loads, and disruption of normal current patterns on which reefs depend for nutrients.

Landfilling is one of the most disruptive activities for coastal and marine resources, and has caused severe and permanent destruction of coastal habitats and changed sedimentation patterns that damage adjacent coral resources. Changes in water circulation caused by landfilling can alter the distribution of coral communities through redistribution of nutrients or increased sediment loads.

Recreation

The recreation industry can cause significant damage to coral reefs. Flipper damage by scuba divers is widespread. Some will argue that today's divers are more environmentally conscious and avoid damaging reefs, but certain activities, such as irresponsible underwater photography finds divers breaking corals to get at subjects and trampling reef habitats in order to get the 'perfect shot'. In areas where divers walk over a reef lagoon and crest to reach the deeper waters, there is a degree of reef trampling, heightened in cases where entry and exit points are limited.

Anchor damage from boats is a common problem at tourist destinations. In South-East Asia many diving operations are switching to nonanchored boat operations, but many others continue the practice unabated. Large tracts of reef can be found in Malaysia that have been scoured by dragging anchors, breaking corals and reducing reef crests to rubble. Experiments have shown that repeated break-age of corals, such as is caused by intensive diving tourism, may lead to substantially reduced sexual reproduction in corals, and eventually to lower rates of recolonization. In the northern Red Sea, another popular diving destination, and in the Caribbean, efforts are underway to install permanent moorings to minimize the damage to reefs from anchors.

Shipping and Port Activities

Congested and high-use maritime areas such as narrow straits, ports, and anchorage zones often lead to physical damage and/or pollution of coral reef areas. Ship groundings and collisions with reefs occur in areas where major shipping routes traverse coral reef areas, such as the Spratley Island complex in the South China Sea, the Red Sea, the Straits of Bab al Mandab and Hormuz, and the Gulf of Suez, to name only a few. Major groundings have occurred off the coast of Florida in the United States, such as the one off Key Largo in State park waters in the 1980s, causing extensive damage to coral reefs. Often these physical blows are severe and destroy decades, if not centuries, of growth. Fish and other invertebrates lose their refuges and foraging habitats, while settlement of new colonies is restricted by the broken-up nature of the substrate. Seismic cables towed during seabed surveys and exploration activities may damage the seabed. Cable damage from towing of vessels (e.g., a tug and barge) has been reported snagging on shallow reefs in the Gulf of Mexico, causing acute damage to sensitive reefs.

Discharges from vessels include untreated sewage, solid wastes, oily bilge, and ballast water. On the high seas these do not have a major noticeable effect on marine ecosystems, but close to shore, particularly at anchorages and near ports, the effects become more obvious. At low tides, oily residues may coat exposed coral colonies, and sewage may cause localized eutrophication and algal blooms. Algal blooms in turn deplete dissolved oxygen levels and prevent penetration of sunlight.

Port activities can have adverse effects on nearby reefs through spills of bulk cargoes and petrochemicals. Fertilizers, phosphates, manganese, and bauxite, for instance, are often shipped in bulk, granular form. These are loaded and offloaded using massive mechanical grabs that spill a little of their contents on each haul. In Jordan, the death of corals was up to four times higher near a port that suffered frequent phosphate spills when compared to control sites. The input of these nutrients often reduces light penetration, inhibits calcification, and increases sedimentation, resulting in slower feeding and growth rates, and limited settlement of new larva.

War-related Activities

The effects of war-related activities on coral reef health and development are often overlooked. Nuclear testing by the United States in Bimini in the early 1960s obliterated complete atolls, which only in recent years have returned to anything like their original form. This redevelopment is nothing like the original geologic structure that had been built by the reefs over millenia. The effects of the nuclear fallout at such sites is poorly understood, and possibly has long-term effects that are not appreciable on a human timescale. The slow growth rate of coral reefs means that those blast areas are still on the path to recovery.

Target practice is another destructive impact on reefs, such as occurred in 1999 in Puerto Rico, where reefs were threatened by aerial bombing practice operations. In Saudi Arabia, offshore islands were used for target practice prior to the Gulf war in 1991. The bombs do not always impact reefs, but those that do cause acute damage that takes long periods to recover.

In the Spratley islands, the development of military structures to support and defend overlapping claims to reefs and islands has brought about the destruction of large tracts of coral reefs. Man-made islands, aircraft landing strips, military bases, and housing units have all used landfilling to one extent or another, smothering complete reefs and resulting in high sediment loads over nearby reefs. Dredging to create channels into reef atolls has also wiped out extensive reef areas.

Indirect Effects

Most anthropogenic effects and disturbances to coral reefs are easily identifiable. Blast fishing debris and discarded fishing nets can be seen. Pollutant levels and sediment loads can be measured. However, many other man-made changes can have indirect impacts on coral reefs that are more difficult to link directly to coral mortality. Global warming is generally accepted as an ongoing phenomenon, resulting from the greenhouse effect and the buildup of carbon dioxide in the atmosphere. Temperatures generally have risen by $1-2\,^{\circ}C$ across the planet, bringing about secondary effects that have had noticeable consequences for coral reefs. The extensive coral beaching event that took place in 1998, which was particularly severe in the Indian Ocean region, is accepted as having been the result of surface sea temperature rise. Bleaching of coral colonies occurs through the expulsion of zooxanthellae, or reductions in chlorophyll content of the zooxanthellae, as coral polyps become stressed by adverse thermal gradients. Some corals are able to survive the bleaching event if nutrients are still available, or if the period of warm water is short.

Coupled with global warming is change of sea level, which is predicted to rise by 25 cm by the year 2050. This sea level rise, if not matched by coral growth, will mean corals will be submerged deeper and will not receive the levels of sunlight required for zooxanthellae photosynthesis. Additionally, the present control of erosion by coral reefs will be lost if waves are able to wash over submerged reefs.

Coral reef calcification depends on the saturation state of carbonate minerals in surface waters, and this rate of calcification may decrease significantly in the future as a result of the decrease in the saturation level due to anthropogenic release of CO_2 into the atmosphere. The concentration of CO_2 in the atmosphere is projected to reach twice the pre-industrial level by the middle of the twenty-first century, which will reduce the calcium carbonate saturation state of the surface ocean by 30%. Carbonate saturation, through changes in calcium concentration, has a highly significant short-term effect on coral calcification. Coral reef organisms do not seem to be able to acclimate to the changing saturation state, and, as calcification rates drop, coral reefs will be less able to cope with rising sea level and other anthropogenic stresses.

The Future

Mankind has contributed to the widespread destruction of corals, reef areas, and their associated fauna through a number of acute and chronic pollutant discharges, through destructive processes, and through uncontrolled and unregulated development. These effects are more noticeable in developing countries, where social and traditional practices have changed without development of infrastructure, finances, and educational resources. Destructive fishing pressures are destroying large tracts of reefs in South-East Asia, while the development of industry affects reefs throughout their range. If mankind is to be the keeper of coral reefs into the coming millennium, there is going to have to be a shift in fishing practices, and adherence to development and shipping guidelines and regulations, along with integrated coastal management programs that take into account the socioeconomic status of people, the environment, and developmental needs.

Glossary

Ahermatypic Non-reef-building corals that do not secrete a calcium carbonate skeletal structure.

DDT Dichlorodiphenyltrichloroethane.

Dinoflagellates One of the most important groups of unicellular plankton organisms, characterized by the possession of two unequal flagella and a set of brownish photosynthetic pigments.

Eutrophication Pollution by excessive nutrient enrichment.

Gastrodemal The epithelial (skin) lining of the gastric cavity.

Hermatypic Reef-building corals that secrete a calcium carbonate skeletal structure.

Quinaldine A registered trademark fish narcotizing agent.

Scleractinians Anthozoa that secrete a calcareous skeleton and are true or stony corals (Order Scleractinia).

Zooxanthellae Symbiotic algae living within coral polyps.

See also

Eutrophication. Oil Pollution.

Further Reading

Birkland C (1997) *Life and Death of Coral Reefs.* New York: Chapman and Hall.

Connel DW and Hawker DW (1991) *Pollution in Tropical Aquatic Systems.* Boca Raton, FL: CRC Press.

Ginsburg RN (ed.) (1994) *Global Aspects of Coral Reefs: Health, Hazards and History,* 7–11 June 1993, p. 420. Miami: University of Miami.

Hatziolos ME, Hooten AJ, and Fodor F (1998) Coral reefs: challenges and opportunities for sustainable management. In: *Proceedings of an Associated Event of the Fifth Annual World Bank Conference on Environmentally and Socially Sustainable Development.* Washington, DC: World Bank.

Peters EC, Glassman NJ, Firman JC, Richmonds RH, and Power EA (1997) Ecotoxicology of tropical marine ecosystems. *Environmental Toxicology and Chemistry* 16(1): 12–40.

Salvat B (ed.) (1987) *Human Impacts on Coral Reefs: Facts and Recommendations,* p. 253. French Polynesia: Antenne Museum E.P.H.E.

Wachenfeld D, Oliver J, and Morrisey JI (1998) *State of the Great Barrier Reef World Heritage Area 1998.* Townsville: Great Barrier Reef Marine Park Authority.

Wilkinson CR (1993) Coral reefs of the world are facing widespread devastation: can we prevent this through sustainable management practices. In: *Proceedings of the Seventh International Coral Reef Symposium* Guam, Micronesia. Mangilao: University of Guam Marine Laboratory.

Wilkinson CR and Buddemeier RW (1994) *Global Climate Change and Coral Reefs: Implications for People and Reefs.* Report of the UNEO-IOC-ASPEI-IUCN Task Team on Coral Reefs. Gland: IUCN.

Wilkinson CR, Sudara S, and Chou LM (1994) Living coastal resources of Southeast Asia: Status and review. In: *Proceedings of the Third ASEAN-Australia Symposium on Living Coastal Resources,* vol. 1. Townsville: ASEAN-Australia Marine Science Project, Living Coastal Resources.

EXOTIC SPECIES, INTRODUCTION OF

D. Minchin, Marine Organism Investigations, Killaloe, Republic of Ireland

Introduction

Exotic species, often referred to as alien, nonnative, nonindigenous, or introduced species, are those that occur in areas outside of their natural geographic range. Vagrant species are those that appear from time to time beyond their normal range and are often confused with exotic species. Since marine science evolved following periods of human exploration and worldwide trade, there are species that may have become introduced at an early time but cannot clearly be ascribed as native or exotic; these are known as cryptogenic species. The full contribution of exotic species among native assemblages remains, and probably will continue to remain, unknown, but these add to the diversity of an area. There are no documented accounts of an introduced species resulting in the extinction of native species in marine habitats as has occurred in freshwater systems. Nevertheless, exotic species can result in habitat modifications that may reduce native species abundance and restructure communities. The greatest numbers of exotic species are inadvertently distributed by shipping either attached to the hull or carried in the large volumes of ballast water used for ship stability. Introductions may also be deliberate. The dependence for food in developing countries and expansion of luxury food products in the developed world have led to increases in food production by cultivation of aquatic plants, invertebrates, and fishes. Many native species do not perform as well as the desired features of some introduced organisms now in widespread cultivation, for example, the Pacific oyster *Crassostrea gigas* and Atlantic salmon *Salmo salar*. Unfortunately, unwanted organisms that have been unintentionally introduced with stock movements can reduce production. Consequently, care is needed when introductions are made. An accidentally introduced exotic species may remain unnoticed until such time as it either becomes abundant or causes harmful effects, whereas larger organisms are normally recognized sooner. Little is known about the movement of the smallest organisms, yet these must be in transit in great numbers every day.

History

Species have been moved for several hundreds of years and some have almost certainly been carried with the earliest human expeditions. Evidence for early movement is scant; this can normally be determined only from hard remains, from a well-understood biogeography of a taxonomic group or perhaps from genetic comparisons. Although the soft-shell clam *Mya arenaria* was present in northern Europe during the late Pliocene, it disappeared during the last glacial period. In the thirteenth century their shells were found in north European middens; this was considered evidence of its introduction by returning Vikings from North America. It may have been used as fresh food during long sea journeys, as it is easily collected and perhaps was maintained in the bilges of their vessels, or may have unintentionally been carried with sediment used as ballast discharged on return to Europe. Certainly it became sufficiently abundant in the Baltic at about this time to be referred to by quaternary geologists as the Mya Sea. This species was also introduced to the Black Sea in the 1960s and rapidly became abundant, resembling its sudden apparent Baltic expansion.

The South American coral *Oculina patagonica*, established in southern Europe, was most probably introduced on sailing ship hulls returning from South America during the sixteenth or seventeenth centuries. Indeed these sailing vessels probably carried a wide range of organisms attached to their hulls. Predictions of the most likely fouling species during these times are based on settlements on panels attached to the hull of a reconstructed sailing ship. Hulls of wooden vessels were fouled with complex communities and excavated by boring organisms whose vacant galleries could provide refugia for a wide range of species including mobile animals. The drag imposed by fouling and the structural damage caused by the boring organisms played its role in the outcome of naval engagements and in the duration of the working life of these vessels, particularly in tropical regions where boring activities and fouling communities evolve rapidly. The periods of stay by sailing ships in ports could provide opportunities for creating new populations arising from drop-off or from reproduction while remaining attached. Some vessels never returned but would have provided instant reefs by sinking, wreckage, or abandonment, thereby distributing a wide assemblage of exotic species. Wooden vessels are still in widespread use and continue to endure problems of drag and structural damage. Many preparations have been used for controlling this, with the most effective being developed over the last 150 years. Wooden vessels became sheathed with thin copper plates and were vulnerable only where these became displaced or damaged. Ironclad sailing vessels also evolved and normal hull life became prolonged because damage was considerably reduced. The building of iron and then steel ship hulls eliminated opportunities for boring and cryptic fauna, yet hull surfaces needed protection from rusting and fouling. The development of protective and also toxic coatings then evolved. The incorporation of copper and other salts in these coatings considerably reduced fouling, and therefore drag, and protected the underlying surfaces from corrosion. Antifouling coatings, in particular, organotins, such as tributyltin compounds, have been very effective in fouling control, and have enabled most vessels to remain in operation before reentering dry dock for servicing, about every 5 years. Unfortunately, tributyltin is a powerful biocide harmful to aquatic organisms in port regions. Its use during the 1980s and 1990s became restricted or banned in many countries on vessels <25 m. And between 2003 and 2008 all merchant shipping are expected to phase out its use. New products are being developed with an emphasis on less toxic and nontoxic applications.

Most exotic species, used in mariculture or for developing fisheries, were spread following the development of steam transport. An early example is the transport from the eastern coast of North America of the striped bass *Morone saxatilis* fingerlings to California in the 1880s by train. Within 10 years, it became regularly sold in the fish markets of San Francisco. Attempts at culturing different species developed slowly due to a poor knowledge of the full life cycle. Planktonic stages were especially misunderstood and pond systems for rearing oyster larvae, for example, were not fully successful until the larval biology became fully known. There were strong economic pressures to develop this knowledge at an early stage. The progressive understanding of behavior and the physical requirements of organisms, and development of algae, brine shrimp, and other cultures and the use of synthetic materials, such as plastics, enabled a rapid expansion of marine culture products ranging from pearls, chemical compounds, to food.

The Vectors of Exotic Species

Organisms are deliberately introduced for culture, to create fisheries, or as ornamental species. Future

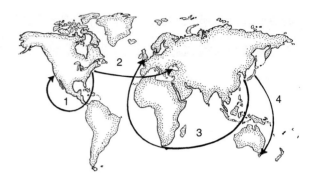

Figure 1 Some examples of primary inoculations: 1, *Carcinus maenas*, the shore crab, a predator of mollusks (vector, probably with shore macroalgae used for transporting fishing baits from the North Atlantic); 2, *Mnemiopsis leidyi* a comb jelly, planktonic predator of larval fishes and crustaceans (vector, ballast water); 3, *Styela clava* an Asian tunicate that causes trophic competition by filtering the water in docks and estuaries (vector, hull fouling); 4, *Asterias amurensis*, an Asian sea star, avid predator of mollusks (vector, ballast water or hull fouling).

introductions may involve species producing products used in pharmacy, as food additives, in the management of diseases, for biological control, or for water quality management.

A primary inoculation is where the first population of an exotic becomes established in a new biogeographic region (**Figure 1**). A secondary inoculation arises by means of further established populations from the first population to other nearby regions by a wide range of activities and so the risk of spread is increased. In many cases, these range extensions are inevitable, either because of human behavior and marketing patterns or because of their mode of life. For example, the Indo-Pacific tubeworm *Ficopomatus enigmaticus* can extensively foul brackish areas and once transported by shipping is easily carried to other estuaries by leisure craft or aquaculture activities, or both.

Species are introduced by a wide range of vectors that can include attachment to seaweeds used for packing material or releases of imported angling baits. The principal vectors are discussed below.

Aquaculture

Useful species have a high tolerance of handling, can endure a wide range of physical conditions, are easily induced to reproduce, and have high survival throughout their production phase. Because of capital costs to retain stock in cages and/or trays, the species must, most usually, be cultivated at densities higher than occurs naturally. The species most likely to adapt to these conditions will be those living under variable conditions at high densities, such as mussels and oysters and some shoaling fishes. Species that do not naturally thrive under these conditions may require more attention and will tend to depend on speciality markets, such as some scallops, tuna, and grouper. Further, aquaculture efforts target those species which grow fast to marketable size.

Mollusks High-value species that are tolerant to handling as well as a wide range of physical conditions are preferred. Some have a long shelf life that allows for live sales. Mollusks, such as clams (e.g., *Ruditapes philippinarum, Mercenaria mercenaria*), scallops (e.g., *Patinopecten yessoensis, Argopecten irradians*), oysters (e.g., *Ostrea edulis, C. gigas*), and abalone (e.g., *Haliotis discus hannai, Haliotis rufescens*), have been introduced for culture (**Figure 2**). Because oysters survive under cool damp conditions for several days, large consignments are easily transported long distances and so have become widely distributed since the advent of steam transportation. With the exception of the east coast of North America, the Pacific oyster *C. gigas*, for example, is now widespread in the Northern Hemisphere and accounts for 80% of world oyster production. Its wide tolerance of salinity, temperature, and turbid conditions and rapid growth make it a desirable candidate for culture. In many cases, it has replaced the native oyster production where this declined either because of overexploitation, diseases of former stocks, or to develop new culture areas. However, pests, parasites, and diseases can be carried within the tissues, mantle cavity, and both in and on the shells of mollusks when they are moved, and can compromise molluskan culture and wild fisheries and may also modify ecosystems. Some examples of pests moved with oysters include the slipper limpet *Crepidula fornicata* and the sea squirt *Styela clava* (**Figure 2**).

Crustaceans The worldwide expansion of penaeid shrimp culture has led to a series of unregulated movements resulting in serious declines of shrimp production caused by pathogenic viruses, bacteria, protozoa, and fungi. Viruses have caused the most serious mortalities of farmed shrimps' broodstock and pose a deterrent to developing culture projects because of casual broodstock movements rather than using those certified as specific-pathogen-free. One parvovirus is the infectious hypodermal and hematopoietic necrosis virus found in wild juvenile and adult prawn stages. This virus was endemic to Southeast Asia, Indonesia, and the Philippines and is now widespread at farms in the tropical regions of the Indo-Pacific and the Americas. About six serious viruses of penaeid shrimp are known. Production may be limited unless disease-resistant

stocks for viruses can be developed together with vaccination against bacterial diseases.

Fishes Of the thousands of fish species only a small number generate a market price high enough to cover production costs. Such fish need to have a high flesh to body weight and high acceptance. While in culture, they should be tolerant of handling and to the wide range of seasonal and diurnal conditions and should not be competitive. The North Atlantic salmon *S. salar* is one of the very few exotics in culture in both hemispheres in the cool to warm temperate climates of North America, Japan, Chile, New Zealand, and Tasmania (**Figure 2**). Fertilized eggs of improved stock are in constant movement between these countries. However, various diseases such as infectious salmon anemia (ISA) may limit the extent of future movements.

It is likely that intensive shore culture facilities for marine species will become more common as the physiological and behavioral requirements of promising species become better understood. For example, in Japan, the bastard sole *Paralichthys olivaceus* is extensively cultivated in shore tanks (and has been introduced to Hawaii) and the sole *Solea senegalensis* in managed lagoons in Portugal. Cultivation under these circumstances is likely to lead to better control of pests, parasites, and diseases, whereas cage culture under more exposed conditions off shore may lead to unexpected mortalities from siphonophores, medusae, algal blooms, and epizootic infestations from parasites, some perhaps introduced.

Shipping

Much of the world trade depends on shipping; the scale and magnitude of these vessels are seldom appreciated. To travel safely, they must either carry cargo or water (as ballast) so that the vessel is correctly immersed to provide more responsive steerage, by allowing better propulsion (without cavitation), rudder bite, and greater stability. The amount of ballast water carried can amount to 30% or more of the overall weight of the ship. Ships also have a large immersed surface area to which organisms attach and result in increased drag that results in higher fuel costs. Nevertheless, despite the best efforts of management, ships carry a great diversity and large numbers of species throughout the world. Unfortunately the risk of introducing further species increases because more ships are in transit and many travel faster than before and operate a wide trading network with new evolving commercial links.

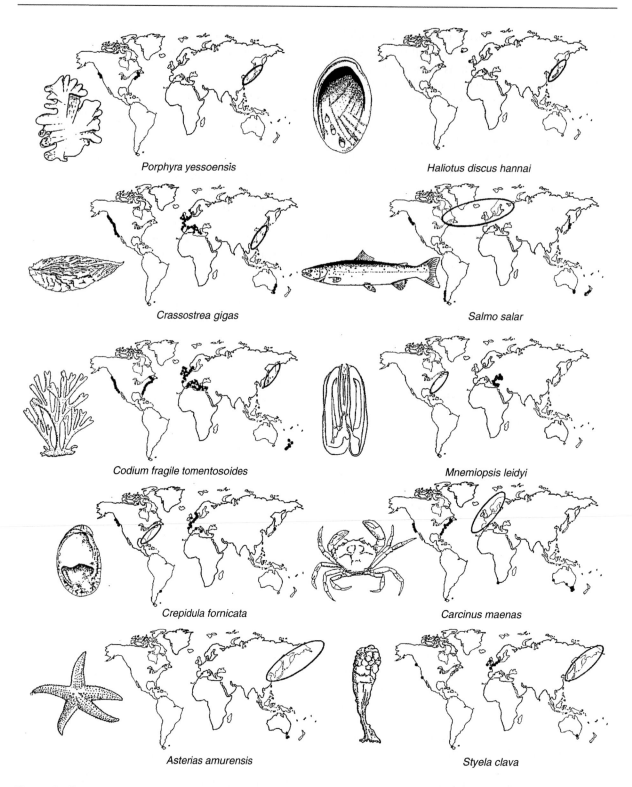

Figure 2 Examples of exotic species. Useful species in culture: *Porphyra yessoensis*, Japanese laver to 30 cm; *H. discus hannai*, Japanese abalone to 18 cm; *C. gigas*, Pacific oyster to 16 cm; *S. salar*, Atlantic salmon to 30 kg. Unwanted invasive species: *Codium fragile tomentosoides*, a green alga to 45 cm; *Mnemiopsis leidyi*, a comb jelly to 11 cm; *Crepidula fornicata*, the slipper limpet to 5 cm; *Carcinus maenas*, the shore or green crab to 9 cm; *Asterias amurensis*, the Asian sea star to 25 cm; *Styela clava*, an Asian tunicate to 18 cm. Open rings, native range; solid dots, established in culture or in the environment.

Ships' ballast water   Ballast water is held on board in specially constructed tanks used only for holding water. The design and size of these tanks depend on the type and size of vessel (**Figure 3**). In some vessels that carry bulk products, a cargo hold may also be flooded with ballast water, but these vessels will also have ballast tanks. Ballast tanks are designed to add structural support to the ship, and will have access ladders, frames, and perforated platforms and baffles to reduce water slopping, making them complex engineering structures. All vessels carry ballast for trim; some of this may be permanent ballast (in which case it is not exchanged at any time) or it may be 'all' released (there is always a small portion of unpumpable water that remains) before taking on cargo. There are several tanks on a ship, and because there may be partial discharges of water, the water within nonpermanent ballast tanks can contain a mixture of water from different ports.

Pumping large volumes of water when ballasting is time-consuming and costly. Should a vessel be ballasted incorrectly, the structural forces produced could compromise the hull. Ships have broken up in port because of incorrect deballasting procedures in relation to the loading/unloading of cargo. Other vessels are known to have capsized due to incorrect ballast water operations. For these reasons, there are special guidelines for the correct ballasting of tanks in relation to cargo load and weather conditions.

Ballasting of water normally occurs close to a port, often in an estuary or shallow harbor; this can include turbid water, sewage discharges, and pollutants. The particulates settle inside the tanks to form sediment accumulations that may be 30 cm in depth. Worms, crustaceans, mollusks, protozoa, and the resting stages of dinoflagellates, as well as other species, can live in these muds. Resting stages may remain dormant for months or years before 'hatching'. Ballast sediments and biota are thus an important component of ballast water uptake adding to the diversity of carried organisms released in new regions through larval releases or from sediment resuspension following journeys with strong winds.

Fresh ballast water normally includes a wide range of different animal and plant groups. However, most of these expire over time, so that on long journeys fewer species survive, and those that do survive, do so in reduced numbers. Because ballast water is held in darkness the contained phytoplankton are unable to photosynthesize. Animals, dependent on these microscopic plants, during vulnerable stages in their development may expire because of insufficient food, unless the tanks are frequently flushed. Scavengers and predators may have better opportunities to survive. To date, observations of organisms surviving in ballast water are incomplete because of the inability or poor efficiency in obtaining adequate samples. Some ballast tanks may only be sampled through a narrow sounding pipe to provide a limited sample of the less active species from one region near the tank floor. Nevertheless, all studies to date point to a wide range of organisms surviving ballast transport, ranging from bacteria to adult fishes. The invasions of the comb jelly *Mnemiopsis leidyi* (from the eastern coast of North America to the Black Sea), the Asian sea star *Asterias amurensis* (from Japan to Tasmania), and the shore crab *Carcinus maenas* (to Australia, the Pacific and Atlantic coasts of North America, South Africa, and the Patagonian coast from Europe) probably arose from ballast water transport (**Figure 2**).

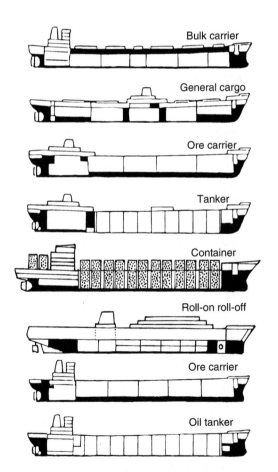

Figure 3   Ballast tanks in ships are dedicated for carrying water for the necessary stabilization of ships. The ballast tanks are outlined in black and show the variation of ballast tank position and size in relation to the type of vessel. Wing tanks (ballast tanks situated along the sides) are not shown in this diagram.

Ships' hulls   Ships are dry-docked for inspection, structural repairs, and recoating the hull with antifouling paint about every 3–5 years. The interdocking time varies according to the type of vessel and its age. While in dry dock, the ship is supported on wooden blocks. The areas beneath the blocks, which may in total cover a ship surface of >100 m^2, are not painted and here fouling may freely develop once the ship is returned to service. The hull is cleaned by shot blasting or using powerful water jets, then recoated. Many coatings have a toxic surface that over time becomes leached or worn from the hull. Unfortunately, some toxic compounds, once released, have caused unexpected and unwanted environmental effects. The use of organotin paints such as tributyltin effectively reduced hull fouling since the 1970s, but it has caused sexual distortion in snails and reproductive and respiratory impairment in other taxa, especially in coastal waters along highly frequented shipping routes. Vessels trading in tropical waters have a higher rate of loss of the active coat and on these vessels a greater fouling can occur at an earlier time.

Most organisms carried on ships' hulls are not harmful but may act as the potential carriers of a wide range of pests, parasites, and diseases. Oysters and mussels frequently attach to hulls and may transmit their diseases to aquaculture sites in the vicinity of shipping ports. Small numbers of disease organisms, once released, may be sufficient to become established if a suitable host is found. Should the host be a cultivated species, it may rapidly spread to other areas in the course of normal trade before the disease becomes recognized.

Many marine organisms spawn profusely in response to sea temperature changes. Spawning could arise following entry of a ship to a port, during unloading of cargo or ballast in stratified water, or from diurnal temperature changes. A ship may turn around in port in a few hours to leave behind larvae that may ultimately settle to form a new population.

Vessels moored for long periods normally acquire a dense and complex fouling community that includes barnacles. Once of sufficient size, these can provide toxic-free surfaces for further attachment, in particular for other barnacles (e.g., the Australasian barnacle *Elminius modestus*, present in many Northern European ports), mollusks, hydroids, bryozoa, and tunicates (e.g., the Asian sea squirt *S. clava*, widely distributed in North America and northern Europe (**Figure 2**)). Vacant shells of barnacles and oysters cemented to the hull can provide shelter for mobile species such as crustaceans and nematodes, and some mobile species such as anemones and flatworms can attach directly to the hull.

Small vessels such as yachts and motorboats also undertake long journeys and may become fouled and thereby extend the range of attached biota. Normally small vessels would be involved in secondary inoculations from port regions to small inlets and lagoons (areas where shipping does not normally have access) and spread species such as the barnacle *Balanus improvisus* to other brackish areas.

Aquarium Species

Aquarium releases seldom become established in the sea whereas in fresh water this is common. The attractive green feather-like alga *Caulerpa taxifolia* was probably released into the Mediterranean Sea from an aquarium in Monaco and by 2000 was present on the French Mediterranean coast, Italy, the Balearic Islands, and the Adriatic Sea. It has a 'root' system that grows over rock, gravel, or sand to form extensive meadows in shallows and may occur to depths of 80 m. It excludes seagrasses and most encrusting fauna and so changes community structure. Although it possesses mild toxins that deter some grazing invertebrates, several fishes will browse on it. This invasive plant is still expanding its range and may extend throughout much of the Mediterranean coastline and perhaps to some Atlantic coasts.

Trade Agreements and Guaranteed Product Production

Trade agreements do not normally take account of the biogeographical regions among trading partners, so introductions of unwanted species are almost inevitable. Often, a population of a harmful species is further spread by trading and thereby has the ability to compromise the production of useful species. Trade in the same species from regions where quarantine was not undertaken may compromise cultured exotic species introduced through the expensive quarantine process which are disease-free. The risk of movement of serious diseases from one country to another is usually recognized, whereas pests are not normally regulated by veterinary regulations, and so are more likely to become freely distributed.

Harvesting of cultured species may be prohibited in areas following the occurrence of toxin-producing phytoplankton. Toxins naturally occur in certain phytoplankton species (most usually dinoflagellates) and are filtered by the mollusk, and these become concentrated within molluskan tissues, with some organs storing greater amounts. Some of these toxins may subsequently accumulate within the tissues of molluskan-feeding crustaceans, such as crabs. If contaminated shellfish is consumed by humans, symptoms such as diahorritic shellfish poisoning, paralytic shellfish poisoning, neurological shellfish poisoning, and amnesic shellfish poisoning may occur. These conditions are caused by different toxins, and new toxins continue to be described. As they all have different breakdown rates in the shellfish tissues, there are varying periods when the products are prohibited from sale. There are monitoring programs for evaluating the levels of these contaminants in most countries. On occasions when harvesting of mollusks is suspended, consignments from abroad may be imported to maintain production levels. If these consignments arrive in poor condition, or are unsuitable because of heavy fouling, they may be relaid or dumped in the sea or on the shore. Such actions can extend the range of unwanted exotic species.

The Increasing Emergence of Unexplained Events

Natural eruptions of endemic or introduced pathogens may be responsible for chronic mortalities or unwanted phenomena. The die-off of the sea urchin *Diadema antillarum* in the western Atlantic may be due to a virus. The unsightly fibropapillomae of the green turtle *Chelonia mydas*, previously known only in the Atlantic Ocean, are now found in the Indo-Pacific, where it was previously unknown. The great mortalities of pilchard *Sardinops sagax neopilchardus* off Australia in the 1990s and mortalities of the bay scallop *A. irradians* in China may be a result of introduced microorganisms. In the case of introduced culture species, mortalities may ensue from a lack of resistance to local pathogens. *Bonamia ostreae*, a protozoan parasite in the blood of some oyster species, was unknown until found in the European flat oyster *O. edulis* following its importation to France from the American Pacific coast, where it had been introduced several years earlier. Very often harmful species first become noticed when aquaculture species are cultivated at high densities, for example, the rhizocephalan-like *Pectinophilus inornata* found in the scallop *Pa. yessoensis* in Japanese waters. Often careful studies of native biota reveal new species to science, such as the generally harmful protozoans, *Perkensis* species found in clams, oysters, and scallops that can be transmitted to each generation by adhering to eggs. Pests, parasites, and diseases in living products used in trade are likely sources of transmission. However, viruses or resting stages of some species may also be transmitted in frozen and dried products. Changes in global climate may be contributing to some of these appearances aided by high levels of human mobility.

Vulnerable Regions

Some areas favor exotic species with opportunities for establishment and so enable their subsequent spread to nearby regions. These areas are normally shipping ports within partly enclosed harbors with low tidal amplitudes and/or with good water retention and a large number of arriving vessels. Although it may be possible to predict which ports are the main sites for primary introductions, the factors involved are not clearly understood and information on the exotic species component is presently only available for a small number of ports. Nevertheless, ports with many exotic species are areas where further exotics will be found. Such regions are likely sites for introductions because ships carry a very large number of a wide range of species from different taxonomic groups. Port regions with known concentrations of exotic species include San Francisco Bay, Prince William Sound, Chesapeake Bay, Port Phillip Bay, Derwent Estuary, Brest Harbor, Cork Harbour, and The Solent (**Figure 4**). In the Baltic Sea, there are several ports that receive ballast water from ships operated via canals from the Black and Caspian Seas as well as arising from direct overseas trade. The component of the exotic species biomass in this region is high. In some areas such as the Curonian Lagoon, Lithuania, exotic species comprise the main biomass. The Black Sea has similar conditions to the Baltic Sea where established species have modified the economy of the region. The effects vary from dying clams *Mya arenaria* creating a stench on tourist beaches, poor recruitment of pilchard and anchovy due to a combination of high exploitation of the fisheries, changes in water quality and an increased predation of their larvae by an introduced comb jelly *M. leidyi*, to the development

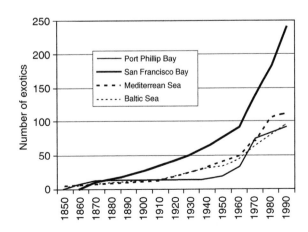

Figure 4 Accumulative numbers of known exotic species in vulnerable areas. The real numbers of exotic species are probably much greater than shown.

of an industry based on harvesting and export of the large introduced predatory snail *Rapana venosa*. The Mediterranean Sea has increasing numbers of unintended introductions arising from trade, mainly by shipping, and from expansion of their range from the Red Sea via the Suez Canal (**Figure 4**).

The majority of exotic species is found in temperate regions of the world. Whether this is a surmise because of a lack of a full understanding, or whether this is due to a real effect, is not known. It may be that tropical species are well dispersed because of natural vectors and that further transmission by shipping is of little consequence to the overall biota present. However, it may be that there is such a daunting diversity of species present in tropical regions, with many still to be described, that it is difficult to grasp the complexity and so be able to understand exotic species' movements in this zone. In the contrasting colder climates, the likely slower development of species may inhibit an introduction from being successful. Temperate ports on either side of the same ocean appear to share several species in common, whereas species from temperate regions from other hemispheres or oceans are less common. This suggests that species that do not undergo undue physiological stresses and those present in shorter voyages have a greater probability of becoming established.

For many species, although opportunities for their dispersal in the past may have taken place, their successful establishment has not succeeded. If an inoculation is to succeed, it must pass through a series of challenges. On most occasions, the populations are unable to be maintained at the point of release in sufficient numbers capable of establishment, even though some may survive. Any sightings of exotic individuals may often pose as a warning that this same species, under different conditions, could become established. The increasing volume and speed of shipping, and of air transport (in the case of foods for human consumption and aquarium species) will provide new opportunities. A successful transfer for a species will depend on: appropriate season, life-history stage in transit, survival time without food, re-immersion, temperature and salinity tolerance, rate of dispersal at the point of release, inoculation size and frequency, presence/absence of predators in the receiving environment, and many unknown factors.

Mode of Life

The dispersal of a species from its point of introduction, and the speed at which it expands its range,

will depend on the behavior of its life-history stages and hydrodynamic conditions in the recipient habitat. Species with very short or no planktonic stages, and with limited mobility, are likely to remain close to the site of introduction, unless carried elsewhere by other means. Tunicates have short larval stages and a sessile adult life and so are normally confined to inlets. Buoyant and planktonic species, which include seaweeds with air bladders and many invertebrates, may be carried by combinations of wind and current and become rapidly dispersed in a directional pattern ruled by the principal vectors. Most species have dispersal potentials that lie between these extremes and some, such as active crustaceans and fishes, may become distributed over a wide range as a result of their own activities. The attempt to establish the pink salmon *Onchorhynchus gorbuscha* on the northern coast of Russia resulted in its capture as far south in the North Atlantic as Ireland. A recent and successful introduction of the king crab *Paralithodes camtschaticus* to the Barents Sea has resulted in its rapid expansion to Norway aided by its planktonic larval stages carried by currents and an ability to travel great distances by walking. With a better knowledge of the behavior of organisms during their life-history stages and of the prominent physical vectors from a point of release, theoretical models of dispersal should be possible. Such models would also be valuable tools for predicting the plumes arising from discharges of ships' ballast water and for evaluating relative risk scenarios of transmitting or receiving exotic species.

The reproductive capability of a species is also important. Those that release broods that are confined to the benthos such as the Chinese hat snail *Calyptraea chinensis*, are likely to remain in one region, and once mature will have a good opportunity for effective reproduction. Species such as *Littorina saxatilis* have the theoretical capability of establishing themselves from the release of a single female by producing miniature crawlers without a planktonic life stage. In contrast, species with long planktonic stages requiring stable conditions are unlikely to succeed. It is doubtful whether spiny lobsters will become inadvertently transferred in ballast water.

Impacts on Society

Exotic species have a wide range of effects: some provide economic opportunities whereas others impose unwanted consequences that can result in serious financial loss and unemployment. In agriculture, the main species utilized in temperate environments for food production were introduced and have taken some thousands of years to develop. In the marine environment, the cultivation of species is relatively new and comparatively few exotic species are utilized. This suggests that in future years an assemblage of exotic species, some presently in cultivation, and others yet to be developed, will form a basis for significant food production. Few exotics in marine cultivation form the basis for subsistence, whereas this is common in freshwater systems. The cultivation of marine species is normally for specialist markets where food quality is an important criterion. Expanding ranges of harmful exotics could erode opportunities by impairing the quality in some way or interfering with production targets.

Unfortunately, many shipping port regions are in areas where conditions for cultivation are suitable, either for the practical reason of lower capital costs for management, or for the ease of operation and/or shelter, or since the conditions favor optimal growth and/or the nearby market. The proximity of shipping to aquaculture activities poses the unquantifiable threat that some imported organisms will impair survival, compromise growth, or render a product unmarketable.

Diseases of organisms and humans are spreading throughout the world. In the marine environment, a large bulk of biota is in transit in ballast water. Ballasting by ships in port may result in loading untreated discharges of human sewage containing bacteria and viruses that may have consequences for human health once discharged elsewhere. In 1991, at the time of the South American cholera epidemic, caused by *Vibrio cholerae*, oysters and fish in Mobile Bay, Alabama (USA), were found with the same strain of this infectious bacterium. Ballast water was considered as a possible source of this event. Subsequently five of 19 ships sampled in Gulf of Mexico ports arriving from Latin America were found with this same strain. The epidemic in South America may have been originally sourced from Asia, and may also have been transmitted by ships.

Of grave concern is the discovery of new algal toxins and the apparent spread of amnesic shellfish poisoning, diahorritic shellfish poisoning, paralytic shellfish poisoning, and neurological shellfish poisoning throughout the world. The associated algae can form dense blooms that, with onshore winds, can form aerosols that may be carried ashore to influence human health. The apparent increase in the frequency of these events may be due to poor historical knowledge of previous occurrences or to a real expansion of the phenomena. There is good evidence that ballast water may be distributing some of these harmful species. Some of the 'bloom-forming' species, such as the naked dinoflagellate *Karenia*

mikimotoi, are almost certainly introduced and cause sufficiently dense blooms to impair respiration in fishes by congestion of the gills, and can also purge the water column of many zooplankton species and cause mortalities of the benthos.

Some introduced invertebrates may act as an intermediate host for human and livestock diseases. The Chinese mitten crab *Eriocheir sinensis*, apart from being a nuisance species, acts as the second intermediate host for the lung fluke *Paragonimus westermanii*. The first intermediate stage appears in snails. The Chinese mitten crab has been introduced to the Mediterranean and Black Seas, North America, and northern Europe. Should the lung fluke be introduced, the ability for it to become established now exists where it may cause health problems for mammals, including humans.

Management of Exotic Species

Aquaculture

There is an expanding interest in aquaculture as an industry to provide employment, revenue, and food. Already several species in production worldwide contribute to these aims. However, any introduction may be responsible for unwanted and harmful introductions of pests, parasites, and diseases. Those involved in future species introductions should consider the International Council for the Exploration of the Sea's (ICES) Code of Practice on Introductions and Transfers of Marine Organisms (**Table 1**). This code takes into account precautionary measures so that unwanted species are unlikely to become unintentionally released. The code also includes provisions for the release of genetically modified organisms (GMOs). These are treated in the same way as if they are exotic species introductions intended for culture. The code is updated from time to time in the light of recent scientific findings. Should this code be ignored the involved parties could be accused of acting inappropriately. By using the code, introductions may take several years before significant production can be achieved. This is because the original broodstock are not released to the wild, only a generation arising from them that is disease-free. This generation must be examined closely for ecological interactions before the species can be freely cultivated. In developing countries it is important that all reasonable precautions are taken to reduce obvious risks. It has been shown historically that direct introductions, even when some precautions have been taken, may lead to problems that can compromise the intended industry or influence other industries, activities, and the environment.

Table 1   Main features of the ICES Code of Practice in relation to an introduction

Conduct a desk evaluation well in advance of the introduction, to include:
  previous known introductions of the species elsewhere;
  review the known diseases, parasites, and pests in the native environment;
  understand its physical tolerances and ecological interactions in its native environment;
  develop a knowledge of its genetics; and
  provide a justification for the introduction
Determine the likely consequences of the introduction and undertake a hazard assessment
Introduce the organisms to a secure quarantine facility and treat all wastewater and waste materials effectively
Cultivate F1 generation in isolation in quarantine and destroy broodstock
Disease-free filial generation may be used in a limited pilot project with a contingency withdrawal plan
Development of the species for culture

At all stages the advice of the ICES Working Group on the Introductions and Transfers of Marine Organisms is sought. Organisms with deliberately modified heritable traits, such as genetically modified native organisms, are considered as exotic species and are required to follow the ICES Code of Practice.

Table 2   Treatment measures of ballast water

Disinfection: Tank wall coatings, biocides, ozone, raised temperature, electrical charges and microwaves, deoxygenation, filtration, ultraviolet light, ultrasonification, mechanical agitation, exchanges with different salinity
By management: Special shore facilities or lighters (transfer vessels) for discharges, provision of clean water (fresh water) by port authorities, no ballasting when organisms are abundant (i.e., during algal blooms, at night) or of turbid water (i.e., during dredging, in shallows), removal of sediments and disposal ashore. Specific port management plans taking account of local port conditions and seasonality of the port as a donor area
Passive effects: Increase time to deballasting, long voyages, reballast at sea.

Experience has shown that good water quality, moderate stocking densities, and meteorological and oceanographic conditions, within the normal limits of species, or the culture system, are of importance for successful cultivation. Maintenance of production in deteriorating conditions, following high sedimentation or pollution, can rarely be achieved by using other introduced species.

Exotic species generally used in aquaculture may not always prove to be beneficial. Although the Pacific oyster C. *gigas* is generally accepted as a useful species, in some parts of the Adriatic Sea it fouls metal ladders, rocks, and stones causing cuts to bathers' skin, this in a region where revenue from tourism exceeds that from aquaculture. This same species is unwelcome in New South Wales because of competition with the Sydney rock oyster *Saccostrea commercialis* and there have been attempts to eradicate it.

Ecomorphology   Organisms can respond to changes in their environment by adapting specific characteristics that provide them with advantage. Sometimes these changes can be noted within a single lifetime, but more usually this takes place over many generations and may ultimately lead to species separation. When considering a species for introduction its morphology may provide clues as to whether it will compete with native species or whether its feeding capabilities or range is likely to

be distinct and separate, overlapping or coinciding. However, it is not possible to evaluate the overall impacts of a species for introduction in advance of the introduction using morphological features alone.

Ships' Ballast Water

Sterilization techniques of ballast water are difficult, and most ideas are not cost-effective or practical, either because the great volumes of water require large amounts of chemicals or because of the added corrosion to tanks or because the treated water, when discharged, has now become an environmental hazard. Reballasting at sea is the current requirement by the International Maritime Organization (IMO), the United Nations body which deals with shipping. Ballast tanks cannot be completely drained and so three exchanges are required to remove >95% of the original ballast water. However, it is not possible for ships to reballast in mid-ocean in every case. Ships that deballast in bad weather can be structurally compromised or become unstable; this could lead to the loss of the vessel and its crew. A further method under consideration that does not compromise the safety of the vessel is the continuous flushing of water while in passage. Several further techniques have either been considered, researched, or are in development (**Table 2**). Exotic species management in ballast water is likely to become a major research area into the twenty-first century.

Ships' Hulls

The use of the toxic yet effective organotins as antifouling agents is likely to become phased out over the first decade of the twenty-first century. Replacement coatings will need to be as, or more, effective, if ships are going to manage fuel costs at current levels. Nontoxic coatings or paint coatings containing deterrents to settling organisms are likely to evolve, rather than coatings containing biocides. However, the effectiveness of these coatings will need to take account the normal interdocking times for ships. In some cases robots may be required for re-activating coat surfaces and removing undue fouling. Locations and/or special management procedures where these activities take place need to be carefully planned to avoid establishment of species from the 'rain' of detritus from cleaning operations.

Biocontrol

Biological control is the release of an organism that will consume or attack a pest species resulting in a population decrease to a level where it is no longer considered a pest. Although there are many effective examples of biological control in terrestrial systems, this has not been practiced in the marine environment using exotic species. However, biological control has been considered in a number of cases.

1. The green alga C. taxifolia has become invasive in the Mediterranean following its likely release from an aquarium; the species presently ranges from the Adriatic Sea to the Balearic Islands and forms meadows over rock, gravels, and sands, displacing many local communities. The introduction of a Caribbean saccoglossan sea slug that does not have a planktonic stage and avidly feeds on this alga has been under consideration for release. These sea slugs were cultured in southern France, but their release to the wild has not been approved.
2. The comb jelly M. leidyi became abundant in the Black Sea in the mid-1980s. It readily feeds on larval fishes and stocks of anchovy and pilchard declined in concert with its expansion. The introduction of either cod Gadus morhua from the Baltic Sea or chum salmon Onchorhynchus gorbusha from North America was considered. Also considered for control was a related predatory comb jelly. However, the predatory comb jelly Beroe became introduced to the Black Sea, possibly in ballast water, and the M. leidyi abundance has since declined. More recently, M. leidyi has entered the Caspian Sea and is repeating the same pattern.
3. The European green crab C. maenas has been introduced to South Africa, Western Australia, and Tasmania as well as to the Pacific and Atlantic coasts of North America. In Tasmania, it avidly feeds on shellfish in culture and on wild mollusks. Here the introduction of a rhizocephalan Sacculina carcini, commonly found within the crab's home range and which reduces reproductive output in populations, was considered. However, because there was evidence that this parasite was not host-specific and may infect other crab species, the project did not proceed.
4. The North Pacific sea star A. amurensis has become abundant in eastern Tasmania and may have been introduced there as a result of either ships' fouling or ballast water releases taken from Japan. This species feeds on a wide range of benthic organisms, including cultured shellfish. A Japanese ciliate Orchitophyra sp. that castrates its host was considered.

Exotic species used in biocontrol need to be species-specific. Generalist predators, parasites, and diseases should be avoided. Complete eradication of a pest species may not be possible or cost-effective except under very special circumstances. In order to evaluate the potential input of the control organism, a good knowledge of the biological system is needed in order to avoid predictable effects that may result in a cascade of changes through the trophic system. Native equivalent species of the biological control organism should be sought for first when introducing a biocontrol species. The ICES Code of Practice should be considered and an external panel of consultants should be involved in all discussions. In some cases, control may be possible by developing a fishery for the pest species as has happened in Turkey following the introduction of the rapa whelk R. venosa to the Black Sea and in Norway to control the population of the red king crab P. camtschaticus.

Management

Shipping is seen as the most widespread means of disseminating species worldwide and the IMO has held conventions that relate to improved management of ships' ballast water and sediments and hull-fouling management. Aquaculture and the trade in living products also has been responsible for many species transmissions. To this end, ICES has developed a Code of Practice, providing wise advice on deliberate movements of marine species that should be consulted. Species may be released in ignorance to the wild with potential consequences for native communities, such as what may have happened with the introduction of lionfish Pterois volitans to the western North Atlantic. Elimination of an impacting species can rarely be achieved unless found soon after arrival. Regular surveys as well as public participation and awareness may reveal an early arrival. However, this involves up-to-date information. The development of rapid assessment surveys for known target species may greatly reduce the duration of surveys and provide early information. Such approaches work. The byssate bivalve *Mytilopsis sallei* was found during a survey in Darwin docks, Australia, and the Mediterranean form of the marine alga *C. taxifolia* was found in a small bay in California, USA; it was first recognized by a member of the public as the result of an awareness campaign. Such management requires up-to-date dissemination of information to convey current affairs to all stakeholders. Rapid dissemination by free-access online journals, such as *Aquatic Invaders*, as well as regular scientific meetings, for example, the International Aquatic Invasive Species Conferences, greatly aid in distributing current knowledge. Management of impacting species may take place in different ways according to their mode of life; it will also depend on the vectors involved in their spread. Where natural vectors disperse a species rapidly, controls may be futile. Programs such as the European Union's specialist projects Assessing Large Scale Risks for Biodiversity with Tested Methods (ALARM) and Delivery Alien Invasive Species Inventories for Europe (DAISIE) to map and manage invading species are likely to lead to new management advances in understanding vector processes and through the development of risk assessments. It is certain that further exotic species will spread and that some will have unpredictable consequences.

Conclusions

Exotic species are an important component of human economic affairs and when used in culture or for sport fisheries, etc., require careful management at the time of introduction. They should pass through a quarantine procedure to reduce transmission of any pests, parasites, and diseases. Aquaculture activities, where practicable, should be sited away from port regions, because there is a risk that shipping may introduce unwanted organisms that may compromise aquaculture production.

Movements of aquarium species also need careful attention and should be sold together with advice not to release these to the wild. Transported fish are normally stressed and many of these have been shown to carry pathogenic bacteria and parasites. There is a risk that serious diseases of fishes could be transmitted outside of the Indo-Pacific region by the aquarium trade.

Trading networks need to consider ways in which organisms transferred alive or organisms and disease agents that may be transferred in or with products,

are not released to the wild in areas that lie beyond their normal range.

More effective and less toxic antifouling agents are needed to replace the effective but highly toxic organotin paint applications on ships. This may lead to less toxic port regions which in turn may now become more suitable for invasive species to become established. Despite widespread usage of new antifouling agents on ships' hulls, unpainted or worn or damaged regions of the hull are areas where fouling organisms will continue to colonize.

Ships' ballast water and its sediments pose a serious threat of transmitting harmful organisms. Future designs of ballast tanks that facilitate complete exchanges, with reduced sediment loading and options for sterilization, could greatly reduce the volume of distributed biota.

Exotic species are becoming established at an apparently increasing rate. Some of these will have serious implications for human health, industry, and the environment.

See also

International Organizations. Large Marine Ecosystems. Mariculture of Aquarium Fishes.

Further Reading

Cohen AN and Carlton JT (1995) *Nonindigenous Aquatic Species in a United States Estuary: A Case Study of the Biological Invasions of the San Francisco Bay and Delta*, 246pp. Washington, DC: United States Fish and Wildlife Service.

Davenport J and Davenport JL (2006) *Environmental Pollution, Vol. 10: The Ecology of Transportation: Managing Mobility for the Environment*, 392pp. Dordrecht: Springer (ISBN 1-4020-4503-4).

Drake LA, Choi KH, Ruiz GM, and Dobbs FC (2001) Global redistribution of bacterioplankton and virioplankton communities. *Biological Invasions* 3: 193–199.

Hewitt CL (2002) The distribution and diversity of tropical Australian marine bio-invasions. *Pacific Science* 56(2): 213–222.

ICES (2005) Vector pathways and the spread of exotic species in the sea. *ICES Co-Operative Report No. 271*, 25pp. Copenhagen: ICES.

Leppäkoski E, Gollasch S, and Olenin S (2002) *Invasive Aquatic Species of Europe: Distribution, Impacts and Management*, 583pp. Dordrecht: Kluwer Academic Publishers (ISBN 1-4020-0837-6).

Minchin D and Gollasch S (2003) Fouling and ships' hulls: How changing circumstances and spawning events may result in the spread of exotic species. *Biofouling* 19(supplement): 111–122.

Padilla DK and Williams SL (2004) Beyond ballast water: Aquarium and ornamental trades as sources of invasive species in aquatic systems. *Frontiers in Ecology and Environment* 2(3): 131–138.

Ruiz GM, Carlton JT, Grozholtz ED, and Hunes AH (1997) Global invasions of marine and estuarine habitats by non-indigenous species: Mechanisms, extent and consequences. *American Zoologist* 37: 621–632.

Williamson AT, Bax NJ, Gonzalez E, and Geeves W (eds.) (2002) Development of a regional risk management framework for APEC economies for use in the control and prevention of introduced pests. *APEC MRC-WG Final Report: Control and Prevention of Introduced Marine Pests*. Singapore: Asia-Pacific Economic Cooperation Secretariat.

Relevant Websites

http://www.alarmproject.net
  – ALARM (Assessing Large Scale Risks for Biodiversity with Tested Methods).
http://www.aquaticinvasions.ru
  – *Aquatic Invasions* (online journal).
http://www.daisie.se
  – Delivering Alien Invasive Species Inventories for Europe (DAISIE).
http://www.ices.dk
  – International Council for the Exploration of the Sea.
http://www.imo.org
  – International Maritime Organization.

VIRAL AND BACTERIAL CONTAMINATION OF BEACHES

J. Bartram, World Health Organization, Geneva, Switzerland
H. Salas, CEPIS/HEP/Pan American Health Organization, Lima, Peru
A. Dufour, United States Environmental Protection Agency, OH, USA

Introduction

Interest in the contamination of beaches by microbes is driven by concern for human health. The agents of concern are human pathogens, microorganisms capable of causing disease. Most are derived from human feces; therefore disposal of excreta and water-borne sewage are of particular importance in their control. Pathogens derived from animal feces may also be significant in some circumstances. The human population of concern constitutes primarily the recreational users, whether local residents, visitors, or tourists. Recreational use of natural waters (including coastal waters) is common worldwide and the associated tourism may be an important component of local and/or national economy.

Scientific underpinning and insight into public health concern for fecal pollution of beaches developed rapidly from around 1980. Approaches to regulation and control (including monitoring) have yet to respond to the increased body of knowledge, although some insights into potential approaches are available.

This article draws heavily on two recent substantial publications: the World Health Organization *Guidelines for Safe Recreational Water Environment*, released as a 'draft for consultation' in 1998, and *Monitoring Bathing Waters* by Bartram and Rees, published in 2000.

Public Health Basis for Concern

Recreational waters typically contain a mixture of pathogenic (i.e., disease-causing) and nonpathogenic microbes derived from multiple sources. These sources include sewage effluents, non-sewage excreta disposals (such as septic tanks), the recreational user population (through defecation and/or shedding), industrial processes (including food processing, for example), farming activities (especially feed lots and animal husbandry), and wildlife, in addition to the indigenous aquatic microflora. Exposure to pathogens in recreational waters may lead to adverse health effects if a suitable quantity (infectious dose) of a pathogen is ingested and colonizes (infects) a suitable site in the body and leads to disease.

What constitutes an infectious dose varies with the agent (pathogen) concerned, the form in which it is encountered, the conditions (route) of exposure, and host susceptibility and immune status. For some viruses and protozoa, this may be very few viable infectious particles (conceptually one). The infectious dose for bacteria varies widely from few particles (e.g., some *Shigella* spp., the cause of bacillary dysentery) to large numbers (e.g. 10^8 for *Vibrio cholerae*, the cause of cholera). In all cases it is important to recall that microorganisms rarely exist as homogeneous dispersions in water and are often aggregated on particles, where they may be partially protected from environmental stresses and as a result of which the probability of ingestion of an infectious dose is increased.

Transmission of disease through recreational water use is biologically plausible and is supported by a generalized dose–response model and the overall body of evidence. For infectious disease acquired through recreational water use, most attention has been paid to diseases transmitted by the fecal–oral route, in which pathogens are excreted in feces, are ingested by mouth, and establish infection in the alimentary canal.

Other routes of infectious disease transmission may also be significant as a result of exposure though recreational water use. Surface exposure can lead to ear infections and inhalation exposure may result in respiratory infections.

Sewage-polluted waters typically contain a range of pathogens and both individuals and recreational user populations are rarely limited to exposure to a single encounter with a single pathogen. The effects of multiple and simultaneous or consecutive exposure to pathogens remain poorly understood.

Water is not a natural ambient medium for the human body, and use of water (whether contaminated or not) for recreational purposes may compromise the body's natural defenses. The most obvious example of this concerns the eye. Epidemiological studies support the logical inference that recreational water use involving repeated immersion will increase the likelihood of eye infection through compromising natural resistance mechanisms, regardless of the quality of the water.

On the basis of a review of all identified and accessible publications concerning epidemiological studies on health outcomes associated with recreational water exposure, the WHO has recently concluded the following:

- The rate of occurrence of certain symptoms or symptom groups is significantly related to the count of fecal indicator bacteria. An increase in outcome rate with increasing indicator count is reported in most studies.
- Mainly gastrointestinal symptoms (including 'highlycredible' or 'objective' gastrointestinal symptoms) are associated with fecal indicator bacteria such as enterococci, fecal streptococci, thermotolerant coliforms and *Escherichia coli*.
- Overall relative risks for gastroenteric symptoms of exposure to relatively clean water lie between 1.0 and 2.5.
- Overall relative risks of swimming in relatively polluted water versus swimming in clean water vary between 0.4 and 3.

- Many studies suggest continuously increasing risk models with thresholds for various indicator organisms and health outcomes. Most of the suggested threshold values are low in comparison with the water qualities often encountered in coastal waters used for recreation.
- The indicator organisms that correlate best with health outcome are enterococci/fecal streptococci for marine and freshwater, and *E.coli* for freshwater. Other indicators showing correlation are fecal coliforms and staphylococci. The latter may correlate with density of bathers and were reported to be significantly associated with ear, skin, respiratory, and enteric diseases.

In assessing the adequacy of the overall body of evidence for the association of bathing water quality and gastrointestinal symptoms, WHO referred to Bradford Hill's criteria forcausation in environmental studies (**Table 1**). Seven of the nine criteria were fulfilled. The criterion on specificity of association was considered inapplicable because the etiological agents were suspected to be numerous and relatively outcome-nonspecific. Results of experiments on the impact of preventive actions on health outcome frequency have not been reported.

Table 1   Criteria for causation in environmental studies (according to Bradford Hill, 1965). Application to bathing water quality and gastrointestinal symptoms

Criterion	Explanation	Fulfillment
1. Strength of association	Difference in illness rate between exposed and nonexposed groups, measured as a ratio	Yes Significant associations have been found; the ratios are relatively low (usually < 3)
2. Consistency	Has it been observed by different people at different places?	Yes In several countries and by various authors
3. Specificity of association	A particular type of exposure is linked with a particular site of infection or a particular disease	No
4. Temporality	Does the exposure precede the disease rather than following it?	Yes Most studies indicate temporal relationship
5. Biological gradient	A dose–response curve can be detected	Yes Most of the selected studies show significant exposure–response relationships
6. Plausibility	Does the present relationship seem likely in terms of present knowledge?	Yes For example, the results are in line with findings on ingestion of infective doses of pathogens
7. Coherence	Cause-and-effect interpretation of the data should not conflict with knowledge of natural history and biology of the disease	Yes
8. Experiment	Did preventive actions change the disease frequency?	Preventive actions have not yet been described in the studies
9. Analogy	Are similar agents known to cause similar diseases in similar circumstances?	Yes Similar to ingestion of recreational water, gastrointestinal symptoms are known to be caused by fecally polluted drinking water

This degree of fulfillment suggests that the association is causal.

Because of the study areas used, especially for the available randomized controlled trials, the results are primarily indicated for adult populations in temperate climates. Greater susceptibility among younger age groups has been shown and the overall roles of endemicity and immunity in relation to exposure and response are inadequately understood.

The overall conclusions of the work of WHO-concerning fecal contamination of recreational waters and the different potential adverse health outcomes among user groups were as follows:

- The overall body of evidence suggests a casual relationship between increasing exposure to fecal contamination and frequency of gastroenteritis. Limited information concerning the dose–response relationships narrows the ability to apply cost–benefit approaches to control. Misclassification of exposure is likely to produce artificially lowthreshold values in observational studies. The one randomized trial indicated a higher threshold of 33 fecal streptococci per 100 ml for gastrointestinal symptoms.
- A cause–effect relationship between fecal pollution or bather-derived pollution and acute febrile respiratory illness is biologically plausible since associations have been reported and a significant exposure–response relationship with a threshold of 59 fecal streptococci per 100 ml was reported.
- Associations between ear infections and microbiological indicators of fecal pollution and bather load have been reported. A significant dose–response effect has been reported in one study. A cause–effect relationship between fecal or bather derived pollution and ear infection is biologically plausible.
- Increased rates of eye symptoms have been reported among bathers and evidence suggests that bathing, regardless of water quality, compromises the eye's immune defenses. Despite biological plausibility, no credible evidence is available for increased rates of eye ailments associated with water pollution.
- No credible evidence is available for an association of skin disease with either water exposure or microbiological water quality.
- Most investigations have either not addressed severe health outcomes such as hepatitis, enteric fever, or poliomyelitis or have not been undertaken in areas of low or zero endemicity. By inference, transmission of enteric hepatitis viruses and of poliomyelitis – should exposure of susceptible persons occur – is biologically plausible, and one study reported enteric fever (typhoid) causation.

The WHO work of 1998 led to the derivation of draft guideline values as summarized in **Table 2**.

Sources and Control

The principal sources of fecal pollution are sewage (and industrial) discharges, combined sewer overflows, urban runoff, and agriculture. These may lead to pollution remote from their source or point of discharge because of transport in rivers or through currents in coastal areas or lakes. The public health significance of any of these sources may be modified by a number of factors, some of which provide management opportunities for controlling human health risk.

With regard to public health, most attention has, logically, been paid to sewage as the source of fecal pollution. Pollution abatement measures for sewage may be grouped into three disposal alternatives, although there is some variation within and overlap between these: treatment, dispersion through sea outfalls, and discharge not to surface water bodies(e.g., to agriculture or ground water injection).

Where significant attention has been paid to sewage management, it has often been found that other sources of fecal contamination are also significant. Most important among these are combined sewer overflows (and 'sanitary sewer' overflows) and riverine discharges to coastal areas and lakes. Combined sewer overflows(CSOs) generally operate as a result of rainfall. Their effect is rapid and discharge may be directly to areas used for recreation. Riverine discharge may derive from agriculture, from upstream sewage discharges (treated or otherwise), and from upstream CSOs. The effect may be continuous (e.g., from upstream sewage treatment) or rainfall-related (agricultural runoff, urban runoff, CSOs). Where it is rainfall-related, the effect on downstream recreational water use areas may persist for several days. In river systems the decrease in microbiological concentrations downstream of a source (conventionally termed 'die-off') largely reflects sedimentation. After settlement in riverbed sediments, survival times are significantly increased and re-suspension will occur when river flow increases. Because of this and the increased inputs from sources such as CSOs and urban and agricultural runoff during rainfall events, rivers may demonstrate a close correlation between flow and bacterial indicator concentration.

Table 2 Draft guideline values for microbiological quality of marine recreational waters (fecal streptococci per 100 ml)

95th centile value of fecal streptococci per 100 ml	Basis of derivation	Estimated disease burden
10	This value is below the no-observed-adverse-effect level (NOAEL) in most epidemiological studies that have attempted to define a NOAEL	Using the indicator level/burden of disease relationship it corresponds to the 95th centile value that is associated with less than a single excess incidence of enteric symptoms for a family of four healthy adult bathers having 80 exposures per bathing season (rounded value), over a 5-year period, making a total of 400 exposures
50	This value is above the threshold and lowest-observed-adverse-effect level (LOAEL) for gastroenteritis in most epidemiological studies that have attempted to define a LOAEL	Using the indicator level/burden of disease relationship it corresponds to the 95th centile value that is associated with a single excess incidence of enteric symptoms for a family of four healthy adult bathers having 80 exposures per bathing season (rounded value)
200	This value is above the threshold and lowest-observed-adverse-effect level for all adverse health outcomes in most epidemiological studies	Using the indicator level/burden of disease relationship it corresponds to the 95th centile value that is associated with a single excess incidence of enteric symptoms for a healthy adult bather having 20 exposures per bathing season (rounded value)
1000	Derived from limited evidence regarding transmission of typhoid fever in areas of low-level typhoid endemicity and of paratyphoid. These are used in this context as indicators of severe health outcome	The exceedence of this level should be considered a public health risk leading to immediate investigation by the competent authorities. Such an interpretation should generally be supported by evidence of human fecal contamination (e.g., a sewage outfall)

Notes
1. This table would produce protection of 'healthy adult bathers' exposed to marine waters in temperate north European waters.
2. It does not relate to children, the elderly, or the immunocompromised who would have lower immunity and might require a greater degree of protection. There are no available data with which to quantify this and no correction factors are therefore applied.
3. Epidemiological data on fresh waters or exposure other than bathing (e.g., high exposures activities such as surfing or whitewater canoeing) are currently inadequate to present a parallel analysis for defined reference risks. Thus a single guideline value is proposed, at this time, for all recreational uses of water because insufficient evidence exists at present to do otherwise. However, it is recommended that the severity and frequency of exposure encountered by special-interest groups (such as body-, board-, and wind-surfers, subaqua divers, canoeists, and dinghy sailors) be taken into account.
4. Where disinfection is used to reduce the density of indicator bacteria in effluents and discharges, the presumed relationship between fecal streptococci (as indicators of fecal contamination) and pathogen presence may be altered. This alteration is, at present, poorly understood. In water receiving such effluents and discharges, fecal streptococci counts may not provide an accurate estimate of the risk of suffering from mild gastrointestinal symptoms.
5. The values calculated here assume that the probability on each exposure is additive.
Reproduced with permission from WHO (1998).

The efficiency of removal of major groups of microorganisms of concern in various types of treatment processes is described in **Table 3**.

Advanced sewage treatment (for instance based upon ultrafiltration or nanofiltration) can also be effective in removal of viruses and other pathogens. The role and efficiency of ultraviolet light, ozone, and other disinfectants are being critically re-evaluated. Treatment in oxidation ponds may remove significant numbers of pathogens, especially the larger protozoan cysts and helminth ova. However, short-circuiting due to poor design, thermal gradients, or hydraulic overload may reduce residence time from the typical design range of 30–90 days.

During detention in oxidation ponds, pathogens are removed or inactivated by sedimentation, sunlight, temperature, predation, and time.

Disposal of sewage through properly designed-long-sea outfalls provides a high degree of protection for human health, minimizing the risk that bathers will come into contact with sewage. In addition, long-sea outfalls reduce demand on land area in comparison with treatment systems, but they may be considered to have unacceptable environmental impacts (for instance, nutrient discharge into areas wheredilution or flushing is limited). They tend to have high capital costs, although these are comparable to those of land-based treatment systems

Table 3 Pathogen removal during sewage treatment

Treatment	Enterococci (cfu l^{-1})[a]	Enteric viruses	Salmonella	C. perfringens[b]
Raw sewage (l^{-1})	2 800 000	100 000–1 000 000	5000–80 000	100 000
Primary treatment[c]				
Percentage removal	32	50–98.3	95.5–99.8	30
Number remaining (l^{-1})	1 900 000	1700–500 000	10–3600	70 000
Secondary treatment[d]				
Percentage removal	96	53–99.92	98.65–99.996	98
Number remaining (l^{-1})	110 000	80–470 000	< 1–1080	2000
Tertiary treatment[e]				
Percentage removal	99.6	99.983–99.999 9998	99.99–99.999 995	99.9
Numbers remaining (l^{-1})	11 000	< 1–170	< 1–8	100

[a]Miescier JJ and Cabelli VJ (1982) Enterococci and other microbial indicators in municipal wastewater effluents, *Journal of Water Pollution Control Federation* 54: 1599–1606.
[b]Long and Ashbolt (1994) *Microbiological Quality of Sewage Treatment Plant Effluents*, AWT-Science and Environmental Report No. 94/123, Water Board, Sydney.
[c]Secondary = primary sedimentation, trickling filter/activated sludge and disinfection.
[d]Tertiary = primary sedimentation, trickling filter/activated sludge, disinfection, coagulation–sand filtration, and disinfection; note that tertiary does not involve coagulation–sand filtration and second disinfection steps for *C. perfringens*.
[e]Primary = physical sedimentation.
Adapted from Yates and Gerba (1998) *Microbial Considerations in Wastewater Reclamation and Reuse*. Vol. 10, Water Quality Management Library, Technomic Publishing Co., Inc. Lancaster, PA, pp. 437–488.

depending on the degree of treatment, whereas recurrent costs are relatively much lower. Ludwig (1988) has presented a comparison of costs and ecological impacts of long-sea outfalls versus treatment levels. Diffuser length, depth, and orientation, as well as the area and spacing of ports are key design considerations. Pathogens are diluted and dispersed and suffer die-off in the marine environment. These are major considerationsin length of outfall and outfall locations. Pretreatment by screening removes large particulates and 'floatables'. Grease and oil removal are also often undertaken.

Re-use of wastewater and groundwater recharge are two methods of sewage disposal that have minimal impact upon recreational waters. Especially in arid areas, sewage can be a safe and important resource (of water and nutrients) used for agricultural purposes such as crop irrigation. Direct injection or infiltration of sewagefor ground water recharge generally presents very low risk for human health-through recreational water use.

Control of human health hazards associated with recreational use of the water environment may be achieved through control of the hazard itself (that is, pollution control) or through control of exposure. Fecal pollution of recreational waters may be subject to substantial variability whether temporally (e.g., time-limited changes in response to rainfall) or spatially (e.g., because, as aresult of the effects of discharge and currents, one part of a beach may behighly contaminated while another part is of good quality). This temporaland spatial variability

provides opportunities to reduce human exposure while pollution control is planned or implemented or in areas where pollution control cannot or will not be implemented for reasons such as cost. The measures used may include public education, control/limitation of access, or posting of advisory notices; they are often relatively affordable and can be implemented relatively rapidly.

Monitoring, Assessment and Regulation

Present regulatory schemes for the microbiological quality of recreational water are primarily or exclusively based upon percentage compliance with fecal indicator counts(**Table 4**).

These regulations and standards have had some success in driving cleanup, increasing public awareness, and contributing to improved personal choice. Not withstanding these successes, a number of constraints are evident in established approaches to regulation and standardsetting:

● Management actions are retrospective and can only be deployed after human exposure to the hazard.
● The risk to human health is primarily from human feces, the traditional indicators of which may also derive from other sources.
● There is poor interlaboratory and international comparability of microbiological analytical data.

Table 4 Microbiological quality of water guidelines/standards per100 ml[i]

Country	Primary contact recreation			References
	TC[a]	FC[b]	Other	
Brazil	80% < 5000[c]	80% < 1000[c]		Brazil, Ministerio del Interior (1976)
Colombia	1000	200		Colombia, Ministerio de Salud (1979)
Cuba	1000[d]	200[d] 90% < 400		Cuba, Ministerio de Salud (1986)
EEC[e], Europe	80% < 500[f] 95% < 10 000[g]	80% < 100[f] 95% < 2000[g]	Fecal streptococci 100[f] Salmonella 0 l^{-1g} Enteroviruses 0 pfu l^{-1} Enterococci 90% < 100	EEC (1976) CEPPOL (1991)
Ecuador	1000	200		Ecuador, Ministerio de Salud Publica (1987)
France	< 2000	< 500	Fecal streptococci < 100	WHO (1977)
Israel	80% < 1000[h]			Argentina, INCYTH (1984)
Japan	1000			Japan, Environmental Agency (1981)
Mexico	80% < 1000[j] 100% < 10 000[k]			Mexico, SEDUE (1983)
Peru	80% < 5000[j]	80% < 1000[j]		Peru, Ministerio de Salud (1983)
Poland			E. coli < 1000	WHO (1975)
Puerto Rico		200[l] 80% < 400		Puerto Rico, JCA (1983)
California	80% < 1000[m,n] 100% < 10 000[k]	200[d,n] 90% < 400[o]		California State Water Resources Board (no date)
United States, USEPA			Enterococci 35[d] (marine), 33[d] (fresh) E. coli 126[d] (fresh)	USEPA (1986)
Former USSR			E. coli < 100	WHO (1977)
UNEP/WHO		50% < 100[p] 90% < 1000[p]		WHO/UNEP (1978)
Uruguay		< 500[l] < 1000[q]		Uruguay, DINAMA (1998)
Venezuela	90% < 1000 100% < 5000	90% < 200 100% < 400		Venezuela (1978)
Yugoslavia	2000			Argentina, INCYTH (1984)

[a]Total coliforms.
[b]Fecal or thermotolerant coliforms.
[c]'Satisfactory' waters, samples obtained in each of the preceding 5 weeks.
[d]Logarithmic average for a period of 30 days of at least five samples.
[e]Minimum sampling frequency – fortnightly.
[f]Guide.
[g]Mandatory.
[h]Minimum 10 samples per month.
[i]Monthly average.
[j]At least 5 samples per month.
[k]Not a sample taken during the verification period of 48 hours should exceed 10 000/100 ml.
[l]At least 5 samples taken sequentially from the waters in a given instance.
[m]Period of 30 days.
[n]Within a zone bounded by the shoreline and a distance of 1000 feet from the shoreline or the 30-foot depth contour, whichever is further from the shoreline.
[o]Period of 60 days.
[p]Geometric mean of at least 5 samples.
[q]Not to be exceeded in at least 5 samples.
Reproduced with permission from Bartram and Rees (2000).
Salas H (1998) History and application of microbiological water quality standards in the marine environment. Pan-American Center for Sanitary Engineering and Environmental Sciences (CEPIS)/Pan-American Health Organization, Lima, Peru.

Table 5 Risk potential to human health through exposure tosewage

Treatment	Discharge type		
	Directly on beach	Short outfall[a]	Effective outfall[b]
None[c]	Very high	High	NA
Preliminary	Very high	High	Low
Primary (including septic tanks)	Very high	High	Low
Secondary	High	High	Low
Secondary plus disinfection	Medium	Medium	Very Low
Tertiary	Medium	Medium	Very Low
Tertiary plus disinfection	Very Low	Very Low	Very Low
Lagoons	High	High	Low

[a]The relative risk is modified by population size. Relative risk is increased for discharges from large populations and decreased for discharges from small populations.
[b]This assumes that the design capacity has not been exceeded and that climatic and oceanic extreme conditions are considered in the design objective (i.e., no sewage on the beach zone).
[c]Includes combined sewer overflows.
NA, not applicable.
Reproduced with permission from Bartram and Rees (2000).

Table 6 Risk potential to human health through exposure to sewage through riverine flow and discharge

Dilution effect[a, b]	Treatment level				
	None	Primary	Secondary	Secondary plus disinfection	Lagoon
High population with low river flow	Very high	Very high	High	Low	Medium
Low population with low river flow	Very high	High	Medium	Very low	Medium
Medium population with medium river flow	High	Medium	Low	Very low	Low
High population with high river flow	High	Medium	Low	Very low	Low
Low population with high river flow	High	Medium	Very low	Very low	Very low

[a]The population factor includes all the population upstream from the beach to be classified and assumes no instream reduction in hazard factor used to classify the beach.
[b]Stream flow is the 10% flow during the period of active beach use. Stream flow assumes no dispersion plug flow conditions to the beach.
Reproduced with permission from Bartram and Rees (2000).

- While beaches are classified as 'safe' or 'unsafe', there is a gradient of increasing frequency and variety of adverse health effects with increasing fecal pollution and it is desirable to promote incremental improvements by prioritizing 'worst failures'.

The present form of regulation also tends to focus attention upon sewage, treatment, and outfall management as the principal or only effective solutions. Owing to high costs of these measures, local authorities may be effectively disenfranchised and few options for effective local intervention in securing bather safety appear to be available.

A modified approach to regulation of recreational water quality could provide for improved protection of public health, possibly with reduced monitoring effort and greater scope for interventions, especially within the scope for local authority intervention. This was discussed in detail at an international meeting of experts in 1998 leading to the development of the 'Annapolis Protocol'.

Table 7 Risk potential to human health through exposure to sewage from bathers

Bather shedding	Category
High bather density, high dilution[a]	Low
Low bather density, high dilution	Very low
High bather density, low dilution[a, b]	Medium
Low bather density, low dilution[b]	Low

[a]Move to next higher category if no sanitary facilities are available at beach site.
[b]If no water movement. Reproduced with permission from Bartram and Rees (2000).

Table 8 Possible sewage contamination indicators and their functions

Indicator/use	Function	
	Pros	Cons
Fecal streptococci/ enterococci	Marine and potentially freshwater human health indicator More persistent in water and sediments than coliforms Fecal streptococci may be cheaper than enterococci to assay	May not be valid for tropical waters, due to potential growth in soils
Thermotolerant coliforms	Indicator of recent fecal contamination	Possibly not suitable for tropical waters owing to growth in soils and waters Confounded by non-sewage sources (e.g., *Klebsiella* spp. in pulp and paper wastewaters)
E. coli	Potentially a freshwater human health indicator. Indicator of recent fecal contamination. Potential for typing *E. coli* to aid identifying sources of fecal contamination. Rapid identification possible if defined as β-glucuronidase-producing bacteria	Possibly not suitable for tropical waters owing to growth in soils and waters
Sanitary plastics	Immediate assessment can be made for each bathing day Can be categorized Little training of staff required	May reflect old sewage contamination and be of little health significance Subjective and prone to variable description
Rainfall in preceding 12, 24, 48 or 72 h	Simple regressions may account for 30–60% of the variation in microbial indicators for a particular beach	Each beach catchment may need to have its rainfall response assessed Response may depend on the period before the event
Sulfite-reducing clostridia/*Clostridium perfringens*	Always in sewage impacted waters Possibly correlated with enteric viruses and parasitic protozoa Inexpensive assay with H_2S production	May also come from dog feces May be too conservative an indicator Enumeration requires anaerobic culture
Somatic coliphages	Standard method well established Similar physical behavior to human enteric viruses	Not specific to sewage May not be as persistent as human enteric viruses Host may grow in the environment
F-specific RNA phages	Standard ISO method available More persistent than some coliphages Host does not grow in environmental waters below 30°C	Not specific to sewage WG49 host may lose plasmid (although F-amp is more stable) Not as persistent in marine waters
Bacteroides fragilis phages	Appear to be specific to sewage ISO method recently published More resistant than other phages in the environment and similar to hardy human enteric viruses	Requires anaerobic culture Numbers in sewage are lower than other phages, and many humans do not excrete this phage (hence no value for small populations)
Fecals sterols	Coprostanol largely specific to sewage Coprostanol degradation in water similar to die-off of thermotolerant coliforms Ratio of $5\beta/5\alpha$ stanols > 0.5 is indicative of fecal contamination; i.e., coprostanol/5α-cholestanol > 0.5 indicates human fecal contamination; while C_{29} 5β (24-ethylcoprostanol)/5α stanol ratio > 0.5 indicates herbivore feces Ratio of coprostanol/24-ethylcoprostanol can be used to indicate the proportion of human fecal contamination, which can be further supported by ratios with fecal indicator bacteria	Requires gas-chromatographic analysis and is expensive (about \$100/sample) Requires up to 10 litres of sample to be filtered through a glass fiber filter (Whatman) to concentrate particulate stanols

(Continued)

Table 8 *Continued*

Indicator/use	Function	
	Pros	*Cons*
Caffeine	May be specific to sewage, but unproven to date Could be developed into a dipstick assay	Yet to be proven as a reliable method
Detergents (calcufluors)	Relatively routine methods available	May not be related to sewage (e.g., industrial pollution)
Turbidity	Simple, direct, and inexpensive assay available in the field	May not be related to sewage; correlation must be shown for each site type
Cryptosporidium (animal source pathogens)	Required for potential zoonoses, such as *Cryptosporidium* spp., where fecal indicator bacteria may have died out, or are not present	Expensive and specialized assay (e.g., Method 1622, USEPA) Human/animal speciation of serotypes not currently defined

Reproduced with permission from Bartram and Rees (2000).

The 'Annapolis Protocol' requires field-testing and improvement based upon the experience gained before application. Its application leads to a classification scheme through which a beach may be assigned to a class related to health risk. By enabling local management to respond to sporadic or limited areas of pollution (and thereby to upgrade the classification of a beach), it provides significant incentive for local management action as well as for pollution abatement. The protocol recognizes that a large number of factors can influence the safety of a given beach. In order to better reflect risk to public health, the classification scheme takes account of three aspects:

1. Counts of fecal indicator bacteria in samples collected from the water adjoining the beach.
2. An inspection-based assessment of the susceptibility of the area to direct influence from human fecal contamination.
3. Assessment of the effectiveness of management interventions if they are deployed to reduce human exposure at times or in places of increased risk.

The process of beach classification is undertaken in two phases:

1. Initial classification based upon the combination of inspection-based assessment and the results of microbiological monitoring.
2. Taking account of the management interventions.

Inspection-based assessment takes account of the three most important sources of human fecal contamination for public health: sewage (including CSO and storm water discharges); riverine discharges where the river is receiving water from sewage discharges and is used either directly for recreation or discharges near a coastal or lake area used for recreation; and bather-derived contamination. The result of assessment is an estimate of relative risk potential in bands as outlined in **Tables 5, 6** and **7** Use of microbial and nonmicrobial indicators of fecal pollution requires an understanding of their characteristics and properties and their applicability for different purposes. Some very basic indicators such as sanitary plastics and grease in marine environments may be used for some purposes under some circumstances. Some newer indicators are under extensive study, but conventional fecal indicator bacteria remain those of greatest importance. Indicators of fecal contamination and their principal uses are summarized in **Table 8**.

By combining the results of microbiological testing with those of inspection, it is possible to derive a primary beach classification using a simple lookup table of the type outlined in **Table 9**. This primary classification may be modified to take account of management interventions that reduce or prevent exposure at times when or in areas where pollution is unusually high. Such 'reclassification' requires a database adequate to describe the times or locations of elevated contamination and demonstration that management action is effective. Since this 'reclassification' may have significant economic importance, independent audit and verification may be appropriate.

Implementation of a monitoring and assessment scheme of the type envisaged in the Annapolis Protocol would be likely to have a significant impact upon the nature and cost of monitoring activities. In comparison with established practice, it would typically involve a greater emphasis on inspection and relatively less on sampling and analysis than

Table 9 Primary classification matrix

Sanitary inspection category (susceptibility to fecal influence)	Microbiological assessment category (indicator counts)				
	A	B	C	D	E
Very low	Excellent	Excellent	Good	Good (+)	Fair (+)
Low	Excellent	Good	Good	Fair	Fair (+)
Moderate	Good[a]	Good	Fair	Fair	Poor
High	Good[a]	Fair[a]	Fair	Poor	Very poor
Very high	Fair[a]	Fair[a]	Poor[a]	Very poor	Very poor

[a]Unexpected result requiring verification.
(+) implies non-sewage sources of fecal indicators (e.g., livestock) and this should be verified.
Reproduced with permission from Bartram and Rees (2000).

is presently common place. At the level of an administrative area with a number of diverse beaches, it would imply an increased short-term monitoring effort when beginning monitoring, but a decreased overall workloadin the medium to longterm.

Recreational use of the water environment provides benefits as well as potential dangers for human health andwell-being. It may also create economic benefits but can add tocompeting local demands upon a finite and sometimes already over-exploited local environment. Regulation, monitoring, and assessmentof areas of coastal recreational water use should be seen or undertaken not in isolation but within this broader context. Integrated approaches to management that take account of overlapping, competing, and sometimes incompatible uses ofthe coastal environment have been increasingly developed and applied in recent years. Extensive guidance concerning integrated coastal management is now available. However, recreational use of coastal areas is also significantly affected by river discharge and therefore upstream discharge and land use practice. While the need to integrate management around the water cycle is recognized, no substantial experience has yet accrued and tools for its implementation remain unavailable.

See also

Pollution: Approaches to Pollution Control.

Further Reading

Bartram J and Rees G (eds.) (2000) *Monitoring Bathing Waters*. London: EFN Spon.

Bradford-Hill A (1965) The environment and disease: association or causation? *Proceedings of the Royal Society of Medicine* 58: 295–300.

Esrey S, Feachem R, and Hughes J (1985) Interventions for the control of diarrhoeal diseases among young children: improving water supplies and excreta disposal facilities. *Bulletin of the World Health Organization* 63(4): 757–772.

Ludwig RG (1988) *Environmental Impact Assessment. Siting and Design of Submarine Outfalls. An EIA Guidance Document.* MARC Report No. 43. Geneva: Monitoring and Assessment Research Centre/World Health Organization.

Mara D and Cairncross S (1989) *Guidelines for the Safe Use of Wastewater and Excreta in Agriculture and Aquaculture.* Geneva: WHO.

WHO (1998) *Guidelines for Safe Recreational-water Environments: Coastal and Freshwaters.* Draft for Consultation. Document EOS/DRAFT/98.14 Geneva: World Health Organization.

COASTAL TOPOGRAPHY, HUMAN IMPACT ON

D. M. Bush, State University of West Georgia, Carrollton, GA, USA

O. H. Pilkey, Duke University, Durham, NC, USA

W. J. Neal, Grand Valley State University, Allendale, MI, USA

Introduction

The trademark of humans throughout time is the modification of the natural landscape. Topography has been modified from the earliest farming to the modern modifications of nature for transportation and commerce (e.g., roads, utilities, mining), and often for recreation, pleasure, and esthetics. While human modifications of the environment have affected vast areas of the continents, and small portions of the ocean floor, nowhere have human intentions met headlong with nature's forces as in the coastal zone.

A most significant change in human behavior since the 1950s has been the dramatic, rapid increase in population and nonessential development in the coastal zone (**Figure 1**). The associated density of development is in an area that is far more vulnerable and likely to be impacted by natural processes (e.g., wind, waves, storm-surge flooding, and coastal erosion) than most inland areas. Not only are more

Figure 1 The coastal population explosion has resulted in too many people and buildings crowded too close to the shoreline. As sea level rises, the shoreline naturally moves back and encounters the immovable structures of human development. In this example from San Juan, Puerto Rico, erosion in front of buildings has necessitated engineering of the shoreline.

people and development in harm's way, but the human modifications of the coastal zone (e.g., dune removal) have increased the frequency and severity of the hazards. Finally, coastal engineering as a means to combat coastal erosion and management of waterways, ports, and harbors has had profound and often deleterious effects on coastal environments. The endproduct is a total interruption of sediment interchange between land and sea, and a heavily modified topography. Natural hazard mitigation is now moving with a more positive, albeit small, approach by restoring natural features, such as beaches and dunes, and their associated interchangeable sediment supply.

The Scope of Human Impact on the Coast

The natural coastal zone is highly dynamic, with geomorphic changes occurring over several time scales. Equally significant changes are made by humans. On Ocean Isle, NC, USA, an interior dune ridge, the only one on the island, was removed to make way for development. The lowered elevation put the entire development in a higher hazard zone, with a corresponding greater risk for property damage from flooding and other storm processes.

Another example of change, impacting on property damage risk, can be seen in Kitty Hawk, North Carolina. A large shorefront dune once extended in front of the entire community. The dune was constructed in the 1930s by the Civilian Conservation Corps to halt shoreline erosion, and provide a 'protected' area along which to build a road. The modification was done before barrier island migration was understood. Erosion was assumed to be permanent land loss. The artificial dune actually increased erosion here by acting as a seawall in a long-term sense, blocking overwash sand which would have raised island elevation and brought sand to the backside of the island, although the dune did afford some protection for development. As a consequence, buildings by the hundreds were built in the lee of the dune. Fifty years on, however, the price is being paid. During the 1980s, the dune began to deteriorate due to storm penetration, and the 1991 Halloween northeaster finished the job by creating large gaps in the dune, resulting in flooding of portions of the community. The dune cannot be rebuilt in place because the old dune location is now occupied by the beach, backed up against the frontal road. Between the time of dune construction and 1991, the community had only experienced major flooding once, in the great 1962 Ash Wednesday storm. Between 1991 and 2000, the community was flooded four times.

The effect of shoreline engineering on a whole-island system is starkly portrayed by the contrast between Ocean City, MD and the next island to the south, Assateague Island, MD. It has taken several decades to be fully realized, but the impact of the jetties is now apparent. Assateague Island has moved back one entire island width due to sand trapping by an updrift jetty. Similar stories abound along the coast. The Charleston lighthouse, once on the

backside of Morris Island, SC, now stands some 650 m at sea; a sentinel that watched Morris Island rapidly migrate away after the Charleston Harbor jetties, built in 1898, halted the supply of sand to the island (**Figure 2**).

Human alterations of the natural environment have direct and indirect effects. Some types of human modifications to the coastal environment include: (1) construction site modification, (2) building and infrastructure construction, (3) hard shoreline stabilization, (4) soft shoreline stabilization, and (5) major coastal engineering construction projects for waterway, port, and harbor management and inlet channel alteration. Each of the modification types impacts the coastal environment in a variety of ways and also

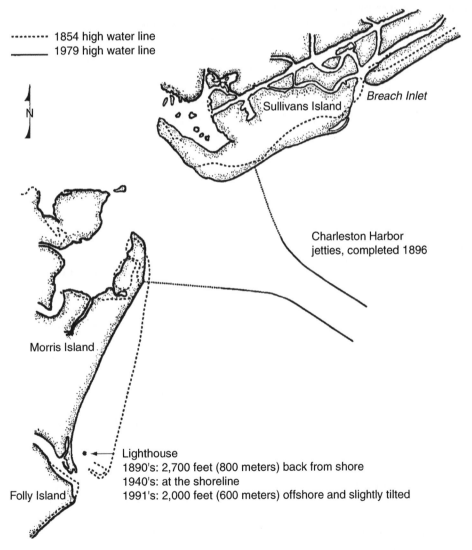

Figure 2 The jetties that stabilize Charleston Harbor, South Carolina, were completed in 1896. The jetties block the southward transport sediment along the beach. As a result, the islands to the north of the jetties have grown seaward slightly, but the islands to the south of the inlet have eroded back more than 1400 m. The Morris Island Lighthouse, once on the back side of the island, is now 650 m offshore.

has several direct and indirect effects. Some of the effects are obvious and intuitive, but many are surprising in that there can be a domino effect as one simple modification creates potential for damage and destruction by increasing the frequency and intensity of natural hazards at individual sites.

Construction Site Modification

Building sites are often flattened and vegetation is removed for ease of construction. Activities such as grading of the natural coastal topography include dune and forest removal. Furthermore, paving of large areas is common, as roads, parking lots, and driveways are constructed. Direct effects of building site modification, in addition to changes in the natural landform configuration, include demobilization of sediment in some places by paving and building footpaths, but also sediment mobilization by removal of vegetation. In either case, rates of onshore–offshore sediment transport and storm-recovery capabilities are changed, which can increase or decrease erosion rates as sediment supply changes.

Other common site modifications include excavating through dunes (dune notching) to improve beach access or sea views. This is particularly common at the ends of streets running toward the beach. After Hurricane Hugo in South Carolina in 1989, shore-perpendicular streets where dunes were notched at their ocean termini were seen to have acted as storm-surge ebb conduits, funneling water back to the sea and increasing scour and property damage. The same effect was noted after Hurricane Gilbert along the northern coast of the Yucatán Peninsula of Mexico in 1988.

Building and Infrastructure Construction

A variety of buildings are constructed in the coastal zone, ranging from single-family homes to high-rise hotels and commercial structures. Some of the common direct effects of building construction are alteration of wind patterns as the buildings themselves interact with natural wind flow, obstruction of sediment movement, marking the landward limit of the beach or dune, channelizing storm surge and storm-surge ebb flow, and reflection of wave energy. Indirect effects result from the simple fact that once there is construction in an area, people tend to want to add more construction, and to increase and improve infrastructure and services. As buildings become threatened by shoreline erosion, coastal engineering endeavors begin.

Roads, streets, water lines, and other utilities are often laid out in the standard grid pattern used inland, cutting through interior and frontal dunes instead of over and around coastal topography (**Figure 3**). Buildings block natural sediment flow (e.g., overwash) while the ends of streets and gaps between rigid buildings funnel and concentrate flow, accentuating the erosive power of flood waters. As noted above, during Hurricane Hugo, water, sand, and debris were carried inland along shore-perpendicular roads in several South Carolina communities. Storm-surge ebb along the shore-perpendicular roads caused scour channels, which undermined roadways and damaged adjacent houses and property. Even something as seemingly harmless as buried utilities may cause a problem as the excavation disrupts the substrate, resulting in a less stable topography after post-construction restorations.

Plugging dune gaps can be a part of nourishment and sand conservation projects. Because dunes are critical coastal geomorphic features with respect to property damage mitigation, they are now often protected, right down to vegetation types that are critical to dune growth. Prior to strict coastal-zone management regulations, however, frontal dunes were often excavated for ocean views or building sites, or notched at road termini for beach access. These artificially created dune gaps are exploited by waves and storm-surge, and by storm surge ebb flows. Wherever dune removal for development has occurred, the probability is increased for the likelihood of complete overwash and possible inlet formation.

Hard Shoreline Stabilization

Hard shoreline stabilization includes various fixed, immovable structures designed to hold an eroding shoreline in place. Hard stabilization is one of the most common modifiers of topography in the coastal zone and is discussed in more detail below. Seawalls, jetties, groins, and offshore breakwaters interrupt sediment exchange and reduce shoreline flexibility to respond to wave and tidal actions. Armoring the shoreline changes the location and intensity of erosion and deposition. Indirectly, hard shoreline stabilization gives a false sense of security and encourages increased development landward of the walls, placing more and more people and property at risk from coastal hazards including waves, storm surge, and wind. Eventual loss of the recreational beach as shoreline erosion continues and catches up with the static line of stabilization is almost a certainty. In addition, structures beget more structures as small walls or groins are replaced by larger and larger walls and groins.

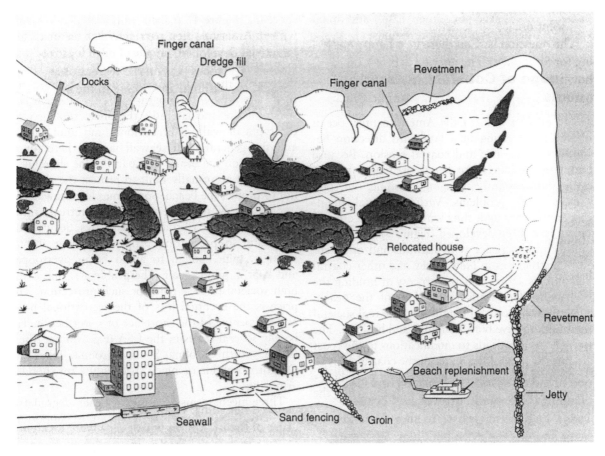

Figure 3 A compilation of many of the impacts humans have on the coastal topography. In this fictional barrier island, roads have been cut through excavated dunes, maritime forest removed for building sites, finger canals dredged, structures built too close to the water, and several types of coastal engineering projects undertaken.

Soft Shoreline Stabilization

The most common forms of soft shoreline stabilization are beach nourishment, dune building, sand fencing, beach bulldozing (beach scraping), and planting of vegetation to grow or stabilize dunes. Direct effects of such manipulations are changes in sedimentation rates and severity of erosion, and interruption of the onshore–offshore sediment transfer, similar in that respect to hard shoreline stabilization. Indirectly, soft shoreline stabilization may make it more difficult to recognize the severity of an erosion problem, i.e., 'masking' the erosion problem. Moreover, as with hard shoreline stabilization, development is actually encouraged in the high-hazard zone behind the beach.

Coastal Engineering Construction Projects

The construction of harbors, port facilities, waterways (e.g., shipping channels, canals) and inlet channel alterations significantly change the coastal outline as well as eliminating land topographic features or erecting artificial shorelines and dredge spoil banks. The Intracoastal Waterway of the Atlantic and Gulf Coasts is one of the longest artificial coastal modifications in the world. Large harbors in many places around the world represent significant alteration of the landscape. Many examples of coastal fill or artificial shorelines exist, but one of the best examples of such a managed shoreline is Chicago's 18 miles of continuous public waterfront. Major canals such as the Suez, Panama, Cape Cod, or Great Dismal Swamp Canal also represent major modifications in the coastal zone. The Houston Ship Channel made the city of Houston, Texas, a major port some 40 miles from the Gulf of Mexico.

Tidal inlets, either on the mainland or between barrier islands, can be altered by dredging, relocation, or artificial closure. Direct effects of dredging tidal inlets are changes in current patterns, which may change the location and degree of erosion and deposition events, and prevention of sand transfer across inlets. In either case, additional shoreline hardening is a common response.

The Scope of Coastal Engineering Impacts

Between 80 and 90% of the American open-ocean shoreline is retreating in a landward direction because of sea-level rise and coastal erosion. Because more static buildings are being sited next to this moving and constantly changing coastline, our society faces major problems. Various coastal engineering approaches to dealing with the coastal erosion problem have been developed (**Figure 3**). More than a century of experience with seawalls and other engineering structures in New Jersey and other coastal developments shows that the process of holding the shoreline in place leads to the loss of the beach, dunes, and other coastal landforms. The real societal issue is how to save both buildings and beaches. The action taken often leads to modifications to the coast that limit the natural flexibility of the coastal zone to respond to storms, that inhibit the natural onshore–offshore exchange of sand, and that interrupt the natural alongshore flow of sand.

Seawalls

Seawalls include a family of coastal engineering structures built either on land at the back of the beach or on the beach, parallel to the shoreline. Strictly defined, seawalls are free-standing structures near the surf-zone edge. The best examples are the giant walls of the northern New Jersey coast, the end result of more than a century of armoring the shoreline (**Figure 4**). If such walls are filled in behind with soil or sand, they are referred to as bulkheads. Revetments, commonly made of piled loose rock, are walls built up against the lower dune-face or land at the back of the beach. For the purpose of considering their alteration of topography both at their construction site and laterally, the distinction between the types of walls is gradational and unimportant, and the general term seawall is used here for all structures on the beach that parallel the shoreline.

Seawalls are usually built to protect the property, not to protect the beach. Sometimes low seawalls are intended only to prevent shoreline retreat, rather than to block wave attack on buildings. Seawalls are successful in preventing property damage if built strongly, high enough to avoid being overtopped, and kept in good repair. The problem is that a very high societal price is paid for such protection. That price is the eventual loss of the recreational beach and steepening of the shoreface or outer beach. This is why several states in the USA (e.g. Maine, Rhode Island, North Carolina, South Carolina, Texas, and Oregon) prohibit or place strict limits on shoreline armoring.

Three mechanisms account for beach degradation by seawalls. Passive loss is the most important. Whatever is causing the shoreline to retreat is unaffected by the wall, and the beach eventually retreats up against the wall. Placement loss refers to the emplacement of walls on the beach seaward of the high-tide line, thus removing part or all of the beach when the wall is constructed (**Figure 5**). Seawall placement was responsible for much of the beach loss in Miami Beach, Florida, necessitating a major beach nourishment project, completed in 1981. Active loss is the least understood of the beach

Figure 4 Cape May, New Jersey has been a popular seaside resort since 1800. Several generations of larger and larger seawalls have been built as coastal erosion caught up with the older structures. Today in many places there is no beach left in front of the seawall.

Figure 5 An example of placement loss in Virginia Beach, Virginia. The seawall was built out on the recreational beach, instantaneously narrowing the beach in front of the wall.

degradation mechanisms. Seawalls are assumed to interact with the surf during storms, which enhances the rate of beach loss. This interaction can occur in a number of ways including seaward reflection of waves, refraction of waves toward the end of the wall, and intensification of surf-zone currents.

By the year 2000, 50% of the developed shoreline on Florida's western (Gulf of Mexico) coast was armored, the same as the New Jersey coast. Similarly 45% of developed shoreline on Florida's eastern (Atlantic Ocean) coast was armored, in contrast to 27% for South Carolina, and only 6% of the developed North Carolina open-ocean shoreline. These figures represent the armored percentage of developed shorelines and do not include protected areas such as parks and National Seashores.

Shoreline stabilization is a difficult political issue because seawalls take as long as five or six decades to destroy beaches, although the usual time range for the beach to be entirely eroded at mid-to-high tide may be only one to three decades. Thus it takes a politician of some foresight to vote for prohibition of armoring. Another issue of political difficulty is that there is no room for compromise. Once a seawall is in place, it is rarely removed. The economic reasoning is that the wall must be maintained and even itself protected, so most walls grow higher and longer.

Groins and Jetties

Groins and jetties are walls or barriers built perpendicular to the shoreline. A jetty, often very long (thousands of feet), is intended to keep sand from flowing into a ship channel within an inlet and to reduce the cost of channel maintenance by dredging. Groins are much shorter structures built on straight stretches of beach away from inlets. Groins are intended to trap sand moving in longshore currents. They can be made of wood, stone, concrete, steel, or fabric bags filled with sand. Some designs are referred to as T-groins because the end of the structure terminates in a short shore-parallel segment.

Both groins and jetties are very successful sand traps. If a groin is working correctly, more sand should be piled up on one side of the groin than on the other. The problem with groins is that they trap sand that is flowing to a neighboring beach. Thus, if a groin is growing the topographic beach updrift, it must be causing downdrift beach loss. Per Bruun, past director of the Coastal Engineering program at the University of Florida, has observed that, on a worldwide basis, groins may be a losing proposition, i.e. more beach may be lost than gained by the use of groins. After one groin is built, the increased rate of

Figure 6 A groin field along Pawleys Island, South Carolina. Trapping of sand on the updrift side of a groin, and erosion of the beach on the downdrift side usually results in a sawtooth pattern to the beach. Note that in this example the beach is the same width on both sides of each groin, indicating little or no longshore transport of sand.

erosion effect on adjacent beaches has to be addressed. So other groins are constructed, in self-defense. The result is a series of groins sometimes extending for miles (**Figure 6**). The resulting groin field is a saw-toothed beach in plan view.

Groins fail when continued erosion at their landward end causes the groin to become detached, allowing water and sand to pass behind the groin. When detachment occurs, beach retreat is renewed and additional alteration of the topography occurs.

Jetties, because of their length, can cause major topographic changes. After jetty emplacement, massive tidal deltas at most barrier island inlents will be dispersed by wave activity. In addition, major build-out of the updrift and retreat of the downdrift shorelines may occur. In the case of the Charleston, SC, jetties noted earlier, beach accretion occurred on the updrift Sullivans Island and Isle of Palms.

Offshore Breakwaters

Offshore breakwaters are walls built parallel to the shoreline but at some distance offshore, typically a few tens of meters seaward of the normal surf zone. These structures dampen the wave energy on the 'protected' shoreline behind the breakwater, interrupting the longshore current and causing sand to be deposited and a beach to form. Sometimes these deposits will accumulate out to the breakwater, creating a feature like a natural tombolo. As in the case of groins, the sand trapped behind breakwaters causes a shortage of sediment downdrift in the directions of dominant longshore transport, leading to additional shoreline retreat (e.g. beach and dune loss, scarping of the fastland, accelerated mass wasting).

Beach Nourishment

Beach nourishment consists of pumping or trucking sand onto the beach. The goal of most communities is to improve their recreational beach, to halt shoreline erosion, and to afford storm protection for beachfront buildings. Many famous beaches in developed areas, in fact, are now artificial!

The beach or zone of active sand movement actually extends out to a water depth of 9–12 m below the low-tide line. This surface is referred to as the shoreface. With nourishment, only the upper beach is covered with new sand so that a steeper beach is created, i.e. the topographic profile is modified on land and offshore. This new steepened profile often increases the rate of erosion; in general, replenished beaches almost always disappear at a faster rate than their natural predecessors.

Beach scraping (bulldozing) should not be confused with beach nourishment. Beach sand is moved from the low-tide beach to the upper back beach (independent of building artificial dunes) as an erosion-mitigation technique. In effect this is beach erosion! A relatively thin layer of sand (≤ 30 cm) is removed from over the entire lower beach using a variety of heavy machinery (drag, grader, bulldozer, front-end loader) and spread over the upper beach. The objectives are to build a wider, higher, high-tide dry beach; to fill in any trough-like lows that drain across the beach; and to encourage additional sand to accrete to the lower beach.

The newly accreted sand in turn, can be scraped, leading to a net gain of sand on the manicured beach. An enhanced recreational beach may be achieved for the short term, but no new sand has been added to the system. Ideally, scraping is intended to encourage onshore transport of sand, but most of the sand 'trapped' on the lower beach is brought in by the longshore transport. Removal of this lower beach sand deprives downdrift beaches of their natural nourishment, steepens the beach topographic profile, and destroys beach organisms.

Dune building is often an important part of beach nourishment design, or it may be carried out independently of beach nourishment. Coastal dunes are a common landform at the back of the beach and part of the dynamic equilibrium of barrier beach systems. Although extensive literature exists about dunes, their protective role often is unknown or misunderstood. Frontal dunes are the last line of defense against ocean storm wave attack and flooding from overwash, but interior dunes may provide high ground and protection against penetration of overwash, and against the damaging effects of storm-surge ebb scour.

Human Impact on Sand Supply

In most of the preceding discussion the impact of humans on beaches and shoreline shape and position was emphasized. The beach plays a major role in supplying sand to barrier islands and, in fact, is important in supplying sand and gravel to any kind of upland, mainland, or island. In this sense, any topographic modification, however small, that affects the sand supply of the beach will affect the topography. In beach communities, sand is routinely removed from the streets and driveways after storms or when sand deposited by wind has accumulated to an uncomfortable level for the community. This sand would have been part of the island or coastal evolution process. Often, dunes are replaced by flat, well-manicured lawns. Sand-trapping dune vegetation is often removed altogether.

The previously mentioned Civilian Conservation Corps construction of the large dune line along almost the entire length of the Outer Banks of North Carolina is an example of a major topographic modification that had unexpected ramifications, namely the increased rate of erosion on the beach as well as on the backside of the islands. Prior to dune construction, the surf zone, especially during storms, expended its energy across a wide band of island surface which was overwashed several times a year. After construction of the frontal dune, wave energy was expended in a much narrower zone, leading to increased rates of shoreline retreat, and overwash no longer nourished the backside of the island. Now that the frontal dune is deteriorating, North Carolina Highway-12 is buried by overwash sand in a least a dozen places 1–4 times each year. Overwash sand is an important part of the island migration process, because these deposits raise the elevation of islands, and when sediment extends entirely across an island, widening occurs. If not for human activities, much of the Outer Banks would be migrating at this point in time, but because preservation of the highway is deemed essential to connect the eight villages of the southern Outer Banks, the NC Department of Transportation removes sand and places it back on the beach. As a result, the island fails to gain elevation.

Inlet formation also is an important part of barrier island evolution. Each barrier island system is different, but inlets form, evolve, and close in a manner to allow the most efficient means of moving water in and out of estuaries and lagoons. Humans interfere by preventing inlets from forming, by closing them after they open naturally (usually during storms), or by preventing their natural migration by construction of jetties. The net result is clogging of navigation channels by construction of huge tidal deltas and reduced water circulation and exchange between the sea and estuaries.

Globally shoreline change is being affected by human activity that causes subsidence and loss of sand supply. The Mississippi River delta is a classic example. The sediment discharge from the Mississippi River has been substantially reduced by upstream dam construction on the river and its tributaries. Large flood-control levees constructed along the lower Mississippi River prevent sediment from reaching the marshes and barrier islands along the rim of the delta in the Gulf of Mexico. Natural land subsidence caused by compaction of muds has added to the problem by creating a rapid (1–2 m per century) relative sea-level rise. Finally, maintaining the river channel south of New Orleans has extended the river mouth to the edge of the continental shelf, causing most remaining sediment to be deposited in the deep sea rather than on the delta. The end result is an extraordinary loss rate of salt marshes and very rapid island migration. The face of the Mississippi River delta is changing with remarkable rapidity.

Other deltas around the world have similar problems that accelerate changes in the shape of associated marshes and barrier islands. The Niger and Nile deltas have lost a significant part of their sediment supply because of trapping sand behind dams. Land loss on the Nile delta is permanent and not just migration of the outermost barrier islands. On the Niger delta the lost sediment supply is compounded by the subsidence caused by oil, gas, and water extraction. The barrier islands there are rapidly thinning.

Sand mining is a worldwide phenomenon whose quantitative importance is difficult to guage. Mining dunes, beaches, and river mouths for sand has reduced the sand supply to the shoreface, beaches, and barrier islands. In developing countries beach-sand mining is ubiquitous, while in developed countries beach and dune mining often is illegal and certainly less extensive, although still a problem. For example, sand mining has adversely affected the beaches of many West Indies nations going through the growing pains of development. Dune mining has been going on for so long that many current residents cannot remember sand dunes ever being present on the beaches, although they must have been there at one time, given the sand supply and the strong winds. For example, on the dual-island nation of Antigua and Barbuda, beach ridges – evidence of accumulating sand – can be observed on Barbuda, but are missing on Antigua. The beach ridges of Barbuda have survived to date only because it is much less heavily developed and populated. Sadly, Barbuda's beach ridges are being actively mined.

Puerto Rico is a heavily developed Caribbean island, much larger than Antigua or Barbuda, and with a more diversified economy. Many of Puerto Rico's dunes have been trucked away (**Figure 7**). East of the capital city of San Juan, large sand dunes were mined to construct the International Airport at Isla Verde by filling in coastal wetlands. As a result of removing the dunes, the highway was regularly overwashed and flooded during even moderate winter storms. First an attempt was made to rebuild the dune, then a major seawall was built to protect the lone coastal road.

Dredging or pumping sand from offshore seems like a quick and simple solution to replace lost beach sand; however, such operations must be considered with great care. The offshore dredge hole may allow larger waves to attack the adjacent beach. Offshore sand may be finer in grain size, or it may be

Figure 7 The dunes here near Camuy, Puerto Rico, used to be over 20 m high. After mining for construction purposes, all that remains is a thin veneer of sand over a rock outcrop.

composed of calcium carbonate, which breaks up quickly under wave abrasion. In all of these cases, the new beach will erode faster than the original beach. Dredging also may create turbidity that can kill bottom organisms. Offshore, protective reefs may be damaged by increased turbidity. Loss of reefs will mean faster beach erosion, as well as the obvious loss to the fishery habitat.

Sand can also be brought in from land sources by dump truck, but this may prove to be more expensive. Sand is a scarce resource, and beaches/dunes have been regarded as a source for mining rather than areas that need artificial replenishment. Past beach and dune mining may well be a principal cause of present beach erosion. In some cases, gravel may be better for nourishment than sand, but the recreational value of beaches declines when gravel is substituted.

Sand mining of beaches and dunes accounts for many of Puerto Rico's problem erosion areas. Such sand removal is now illegal, but permits are given to remove sand for highway construction and emergency repair purposes. However, the extraction limits of such permits are often exceeded – and illegal removal of sand for construction aggregate continues. In all cases, the sand removal eliminates natural shore protection in the area of mining, and robs from the sand budget of downdrift beaches, accelerating erosion. Even a small removal operation can set off a sequence of major shoreline changes.

The Caribe Playa Seabeach Resort along the southeastern coast illustrates just such a chain reaction. Located west of Cabo Mala Pascua, the resort has lost nearly 15 m of beach in recent years according to the owner. The problem dates to the days before permits and regulation, when an updrift property owner sold beach sand for 50 cents per dump truck load; a bargain by anyone's standards but a swindle to downdrift property owners. Where the sand was removed the beach eroded, resulting in shoreline retreat and tree kills. In an effort to restabilize the shore, and ultimately protect the highway, a rip-rap groin and seawall were constructed. Today, only a narrow gravel beach remains. Undoubtedly much of the aggregate in the concrete making up the buildings lining the shore, and now endangered by beach erosion, was beach sand. What extreme irony: taking sand from the beach to build structures that were subsequently endangered by the loss of beach sand.

Conclusion

The majority of the world's population lives in the coastal zone, and the percentage is growing. As this trend continues, the coastal zone will see increased impact of humans as more loss of habitats, more inlet dredging and jetties, continued sand removal, topography modification for building, sand starvation from groins and jetties, and the increased tourism and industrial use of coasts and estuaries. Our society's history illustrates the impact of humans as geomorphic agents, and nowhere is that fact borne out as it is in the coastal zone. The ultimate irony is that many of the human modifications on coastal topography actually decrease the esthetics of the area or increase the potential hazards.

See also

Coastal Zone Management. Viral and Bacterial Contamination of Beaches.

Further Reading

Bush DM and Pilkey OH (1994) Mitigation of hurricane property damage on barrier islands: a geological view. *Journal of Coastal Research* Special issue no. 12: 311–326.

Bush DM, Pilkey OH, and Neal WJ (1996) *Living by the Rules of the Sea.* Durham, NC: Duke University Press.

Bush DM, Neal WJ, Young RS, and Pilkey OH (1999) Utilization of geoindicators for rapid assessment of coastal-hazard risk and mitigation. *Ocean and Coastal Management* 42: 647–670.

Carter RWG and Woodroffe CD (eds.) (1994) *Coastal Evolution: Late Quaternary Shoreline Morphodynamics.* Cambridge: Cambridge University Press.

Carter RWG (1988) *Coastal Environments: An Introduction to the Physical, Ecological, and Cultural Systems of Coastlines.* London: Academic Press.

Davis RA Jr (1997) *The Evolving Coast.* New York: Scientific American Library.

French PW (1997) *Coastal and Estuarine Management, Routledge Environmental Management Series.* London: Routledge Press.

Kaufmann W and Pilkey OH Jr (1983) *The Beaches are Moving: The Drowning of America's Shoreline.* Durham, NC: Duke University Press.

Klee GA (1999) *The Coastal Environment: Toward Integrated Coastal and Marine Sanctuary Management.* Upper Saddle River, NJ: Prentice Hall.

Nordstrom KF (1987) Shoreline changes on developed coastal barriers. In: Platt RH, Pelczarski SG, and Burbank BKR (eds.) *Cities on the Beach: Management Issues of Developed Coastal Barriers*, pp. 65–79. University of Chicago, Department of Geography, Research Paper no. 224.

Nordstrom KF (1994) Developed coasts. In: Carter RWG and Woodroffe CD (eds.) *Coastal Evolution: Late Quaternary Shoreline Morphodynamics*, pp. 477–509. Cambridge: Cambridge University Press.

Nordstrom KF (2000) *Beaches and Dunes of Developed Coasts*. Cambridge: Cambridge University Press.

Pilkey OH and Dixon KL (1996) *The Corps and the Shore*. Washington, DC: Island Press.

Platt RH, Pelczarski SG and Burbank BKR (eds.) (1987) *Cities on the Beach: Management Issues of Developed Coastal Barriers*. University of Chicago, Department of Geography, Research Paper no. 224.

Viles H and Spencer T (1995) *Coastal Problems: Geomorphology, Ecology, and Society at the Coast*. New York: Oxford University Press.

POLLUTION: APPROACHES TO POLLUTION CONTROL

J. S. Gray[†], University of Oslo, Oslo, Norway
J. M. Bewers, Bedford Institute of Oceanography, Dartmouth, NS, Canada

Approaches applied to the control of pollution have altered substantially over the last four decades. Historically, emphasis was given to management initiatives to ensure that damage to the marine environment was avoided by limiting the introduction of substances to the sea. This was typified by the attention given to contaminants such as mercury and oil in early agreements for the prevention of marine pollution. Marine environmental protection was achieved through prior scientific evaluations of the transport and effects of substances proposed for disposal at sea and defining allowable amounts that were not thought to result in significant or unacceptable effects. This reflects largely a management and control philosophy. In the closing stages of the twentieth century, however, the philosophy underlying pollution control has undergone substantial revision. Recent policy initiatives rely less on scientific assessments and place greater emphasis on policy and regulatory controls to restrict human activities potentially affecting the marine environment. During this period of change, practical pollution control and avoidance procedures have been adapted to improve their alignment with these new policy perspectives. Simultaneously, it has been widely recognized that pollutants represent only part of the problem. Other human activities such as overexploitation of fisheries, coastal development, land clearance, and the physical destruction of marine habitat are equally important, and often more serious threats to the marine environment. In recent years, the concept of marine pollution has been broadened to consider the adverse effects on the marine environment of all human activities rather than merely those associated with the release of substances. This is a most positive development, partly influenced by improved scientific understanding that has led to an improved balance of attention among the sources of environmental damage and threats.

Background

In this article, the term 'pollution' implies adverse effects on the environment resulting from human activities. This is consistent with, but broader than, the definition of pollution formulated by the United Nations Joint Group of Experts on Marine Environmental Protection (GESAMP) in 1969 that is restricted to adverse effects associated with the introduction of substances to the marine environment from human activities. The term 'contamination' infers augmentation of natural levels of substances in the environment but without any presumption of associated adverse effects. Indeed, early approaches to marine pollution prevention reflected the distinction between these terms while more recent approaches are based on more or less identical interpretations of these expressions with both implying adverse effects.

Early Agreements on Marine Pollution Prevention

The earliest international marine pollution prevention agreements of the modern era were the Oslo and London conventions of 1972. These conventions were developed at the same time as the heightened awareness of marine pollution issues led to the first major international conference on the topic, the United Nations Conference on the Human Environment, that took place in Stockholm in the same year. Both the original formulation of the two conventions and the results of this conference reflect a commitment to management actions toward the prevention of marine pollution caused primarily by the release of contaminants from human activities. The Oslo and London conventions adopted 'black' and 'gray' lists of substances and a set of measures to prevent pollution resulting from the dumping of wastes and other matter into the sea. Black list substances are essentially prohibited substances that may not be dumped in the ocean except in trace amounts. Gray list substances are those requiring specific special care measures to be considered in judging their suitability for disposal at sea. In addition, these conventions require that all candidate materials for dumping at sea require a prior assessment to ensure that they do not cause significant adverse effects on the marine environment or pose unacceptable risks to human health.

[†] Deceased

In light of the rate of introduction of new chemicals into the market economy, flexibility is required in the assignment of substances to 'black' and 'gray' lists. Thus there is need for an international mechanism for reviewing and updating the list of substances based on an evaluation of their properties. GESAMP provides such a service for the assessment of hazards posed by chemicals transported by ships. There are no similar mechanisms for substances either dumped at sea or entering the marine environment from land-based activities. Meanwhile, each year around a thousand new chemicals are being produced in volume. Their toxicity to a sufficiently wide spectrum of marine species cannot be ensured before use and thus there are constant surprises about the effects of chemicals that were originally thought to entail low risk. The black and gray list approach is too simplistic and does not have the necessary supporting mechanisms to make it reasonably effective. The ocean has the ability to assimilate some finite amount of most substances without adverse effect consistent with the concept of contamination as distinct from pollution. Thus, while not representing a wholly scientific approach, the adoption of these conventions was a major step forward in the introduction of management measures to minimize the risks of marine pollution. However, as will be demonstrated, more recently perceived deficiencies in the provisions of these agreements resulted in their later revision.

Early Approaches to Marine Environmental Protection

The oldest strategy for protecting the marine environment from the adverse effects of the disposal of waste in the sea is that based on the application of 'water quality standards'. This concept was borrowed from practices in freshwater environments, such as rivers, to which it had been applied successfully. The use of water quality standards is based on an assumption that the levels at which contaminants become damaging are well established. Even for chemicals having known effects, for which the severity of effect is proportional to exposure with an assumed threshold for the induction of adverse effects, this assumption has been shown repeatedly to be erroneous as more is learnt about their properties and interactions (see later discussion of the effects of tributyltin (TBT)). For some contaminants, no 'safe' level can be established from entirely scientific considerations because they are postulated to pose risks of adverse effect at any concentration. Such substances have what are termed 'stochastic effects'

where the probability of an adverse effect is a function of exposure without any assumption of threshold. The regulation and management of radionuclides and the effects of nuclear practices are based on a postulation of stochastic effects at low doses. A further problem is that the concentrations of contaminants in the marine environment are frequently lower than in fresh water, making measurement more difficult. In coastal areas, the concentrations of heavy metals, for example, are low and thus are difficult to monitor. Chemical contaminants of most concern are generally particle-reactive with a strong tendency to attach to particles that end up in sediments, especially in depositional areas. Accordingly, marine sediments are usually a more appropriate focus for assessing the quality of the environment and for monitoring than seawater. However, sediments accumulate relatively slowly and, even where deposition rates are high, biological activity tends to mix sediment layers in the vertical, thus smearing out the record of particle accumulation and of the contaminants co-deposited with particles.

Scientific Perspectives

There has long been recognition by scientists that protection of the marine environment *per se* is inappropriate. The overall approach to environmental protection should be holistic and take account of all human activities and their effects on all compartments of the environment – land, sea, and air. It has similarly been noted that contemporary government and intergovernmental arrangements and structures are inappropriately designed for such a task because they are segregated by development sector, and, often, approaches to environmental protection are considered compartment by compartment rather than comprehensively. The development of protection measures for specific environmental compartments is entirely appropriate but it should follow, rather than precede, the formulation of a holistic framework for environmental management and protection.

One of the longest-standing management systems is the radiological protection system, largely developed by the International Commission on Radiological Protection (ICRP), primarily for the protection of human health from the effects of ionizing radiation. This is a scientifically based system that has pioneered concepts such as justification and optimization that require demonstration of the net benefits to society of new practices and minimization of the additional exposures to radiation resulting from all practices. Yet, this same system is predicated, on weakly supported arguments, that if human health is adequately protected, protection of the environment is ensured. While this assumption has been shown to be invalid in some rather special instances, only recently has there been significant professional scientific pressure to extend the protective focus of the system of radiological protection specifically to the environment.

Recent Changes in Policy Perspectives

There have also been some significant shifts in policy and management perspectives regarding marine pollution during the last three decades. In the 1960s and 1970s, primary concerns about marine pollution were expressed as concerns about chemical substances disseminated by human activities. With growing evidence of the physical effects of human activities on the land environment and increased public desires to protect the environment, especially areas of natural beauty and wildlife abundance, policy perspectives regarding the range of human activities that could cause adverse effects on the marine environment broadened considerably. Indeed, the term 'marine pollution' became perceived as a much broader topic than that defined by the Joint Group of Experts on the Scientific Aspects of Marine Environmental Protection (GESAMP) in 1969. This led to greater consideration of the physical effects on the marine environment of coastal development and watershed activities and resource exploitation, for example. However, this broadening in perspective came about relatively slowly and preoccupation with chemical contaminants was, and remains, evident in international agreements of recent years such as the Global Programme of Action on the Protection of the Marine Environment from Land-Based Activities concluded in 1995. Indeed, the process leading up to this agreement resulted in the adoption of the term 'activities' as the final word as a replacement for 'sources' at a relatively late stage. Nevertheless, increasingly concerns became extended toward manifestations of human activities on the marine environment other than those solely associated with chemical contaminants.

Despite the foregoing, there has been little policy change regarding the adoption of holistic environmental management. Indeed, most of the international and national instruments for the protection of the marine environment are just that – they do not specifically consider the effects of human activities on other environments. It is this fact that creates the impression that the marine environment is being accorded preferential protection, at least in the context of international agreements. This may well be due to the largely international nature of ocean space, whereas other environments, particularly land and fresh water, lie within national jurisdiction making governments less inclined to reaching international agreements potentially infringing on their sovereignty. Nevertheless, the fact that the marine environment has been the prime subject for the initial advancement and adoption of new policy initiatives, especially the precautionary approach, supports the impression that the sea is being accorded a greater degree of protection than other compartments of the environment.

There was also a growing perception among the public, which soon became part of the policy perspective, that in some way previous regulatory approaches to environmental protection had been a failure. Whether this was associated with some disillusionment about the benefits and efficacy of science is a matter of conjecture, but it was abundantly clear that scientists working on behalf of governments or industrial proponents were regarded with skepticism if not outright suspicion when giving professional judgments on environmental matters. The growth in membership and advocacy of the various green lobbies throughout the world is a reflection of this distrust. This, in turn, led to demands for more stringent measures, including extreme policy decisions, to reduce the effects of anthropogenic activities on the environment. More recently advocated approaches to the control of marine pollution reflect these altered perspectives.

Recent Approaches to Marine Environmental Protection

A more recent approach to pollution prevention was the so-called 'best available technology' strategy that requires the best technology to be applied in human practices in order to minimize associated effects on the environment. At first sight, this approach appears to be logical and appropriate, but it is flawed fundamentally. It provides neither a guarantee of environmental protection nor of the effective use of resources. This approach may result in ineffective environmental protection because the substances being regulated in this way still have adverse effects on the environment, even at the lower release rates achieved. This results in a waste or misdirection of resources. On the other hand, the approach may result in far greater expenditure of effort than that required to prevent environmental damage, thereby also being wasteful of resources that could be used to better purpose. The point is that, in itself, such an approach achieves little without the knowledge base that allows the tailoring of technology to the needs of society while providing appropriate protection of the environment. A good example is insisting on nitrogen removal for all sewage discharged into the coastal areas of Europe. Primary treatment and phosphorus removal are relatively cheap. Nitrogen treatment is expensive and the amounts of nitrogen discharged in human sewage are small compared to the huge amounts in agricultural wastes that are washed down rivers. In areas with poor water exchange that suffer from eutrophication (enhanced biological production associated with adverse effects such as increased light attenuation, toxic effects on organisms, and increased oxygen demand), such as the inner Oslofjord, there is no doubt that nitrogen treatment of sewage is warranted. However, it is a waste of money to apply the best available sewage technology in other areas where there is no significant risk of eutrophication because there exists sufficient dispersion to ensure that the nutrients are assimilated with little change in the rates and distribution of primary production. There may not be appropriate technology to solve some waste problems, so that even the best contemporarily available technology remains wholly inadequate. For example, some organic chemicals are known to result in widespread effects even at very low concentrations so that contemporary technology, no matter how effective and expensive, is unable to prevent these effects. Various attempts have been made to improve the best technology approach by referring to terms such as 'best practical technology', but all these derivatives suffer from similar difficulties in achieving environmental protection at optimal cost.

One of the more recent approaches to pollution prevention that has been widely advocated and adopted in agreements during the last decade is the so-called 'precautionary approach'. Initially, it had been promoted in the form of 'the precautionary principle'. This appears to have been an outgrowth of a policy development in Germany called the *Vorsorgeprinzip* (or literally the 'principle of foresight'). In its original form, the explanation adopted by the German Ministry of the Environment avoided many of the pitfalls that later became intrinsic components of its application in other arenas. The German version, for example, states that not all effects should be treated as representing significant damage and that all risks cannot be avoided. It also placed considerable emphasis on science as a basis for defining risks whose acceptability could be judged in management and policy contexts.

There have been many subsequent definitions of the precautionary approach. Such revised versions were later adopted as legal instruments in: the North Sea Declaration of 1987; the declaration adopted at the Rio Conference on Environment and Development (UNCED) in 1992; and several other international agreements, most notably the UN Law of the Sea Convention and the Global Plan of Action for the Prevention of Marine Pollution from Land-Based Activities in 1995. The general form of these new formulations has been given as:

> When an activity raises threats of serious or irreversible harm to human health or the environment, precautionary measures that prevent the possibility of harm shall be taken even if the causal link between the activity and the possible harm has not been proven or the causal link is weak and the harm is unlikely to occur. (Holm and Harris, 1999)

Essentially, as later developed from its German origin, the precautionary approach implies that any lack of knowledge about the environmental hazards associated with a practice justify the adoption of special precautionary measures. Subsequently, such precaution has been deemed to warrant the adoption of extreme preventive measures including the banning of certain practices or chemicals. Several of the agreements that have adopted the precautionary approach emphasize its application to contaminants that are described as "persistent, toxic and liable to bioaccumulate." However, threshold values for these three properties at which greater precaution is warranted are not specified, either individually or in combination. This is a fundamental flaw in the application of precaution expressed in this way as all substances have the properties of persistence, toxicity, and liability to bioaccumulate to some degree. Clearly, more precision is needed if the precautionary approach is to be of any practical value.

The precautionary approach has been debated and criticized from both practical and philosophical perspectives and is not the panacea for the prevention of environmental problems its exponents claim it is. Its biggest danger is that it makes science essentially redundant; mere suspicion of an effect or lack of complete scientific knowledge is a good-enough argument to initiate bans on the use and dissemination of a substance. It has resulted in the replacement of the black and gray list approaches with so-called 'reverse lists' of materials that can be considered for dumping at sea under the Oslo and London conventions. These severely curtail management flexibility in the options available to management with little probability of increased environmental protection as a whole. Following pressure from 'green' lobbies, for example, the Swedish government is considering banning the discharge of chemicals unless they are proven to be safe.

At the 1996 Esbjerg Ministerial Conference on the North Sea, the participants went one step further, requiring that chemicals be proven not to cause effects before they are allowed to be discharged to the sea. The problem here is that safety is the opposite of risk and nothing is devoid of risk. Accordingly, complete safety can neither be proven nor guaranteed no matter how good the associated science. Are we to stop all human activities because of the inherent risk (lack of absolute safety) they entail? Surely not.

Another element of recent approaches to pollution prevention generally is 'environmental impact assessment' (EIA). It is a procedure for prior evaluation of the environmental consequences of some proposed human activity such as the construction and operation of a new industrial facility. It is a requirement of organizations of all kinds under European and most national legislation where impacts on the environment may occur. As such, EIA is a useful component of the arsenal of environmental protection measures. Yet, far too frequently, EIAs are simply paper exercises incorporating inadequate accountability if their predictions are wrong and the environment is destroyed. This is deplorable and yet need not be the rule.

This concept can be applied fairly widely not only to specific industrial installations but also to potential investments in entire new industries and other human activities such as coastal development and tourism, for example. In this sense, it is somewhat analogous to the justification of practices in radiological protection. New ideas regarding EIAs provide environmental authorities improved means of assessing and controlling potentially damaging activities. The process needs detailed and careful science to:

- make quantitative and realistic predictions of effects;
- suggest criteria for testing such predictions; and
- design proper and effective monitoring programs with regulatory feedback.

In this context, it should be noted that the process of preparing EIAs needs to involve all parties, including the green movement. It should not be solely the prerogative of the company concerned and its experts. It must also act as a basis for designing scientifically based assessment and monitoring programs.

Predicting the Effects of Chemicals on the Marine Environment

The classical way of predicting effects of chemicals on the marine environment is by first conducting toxicity tests. These usually involve testing a variety of organisms from bacteria through algae to small animals and then fish. Likely effects are predicted, assuming by extension from freshwater models that a concentration of 10% or 1% of the LC_{50} (50% lethality concentration) is safe. This does not always work as can be demonstrated by data pertaining to the effects of TBT used as an antifoulant for vessel hulls and marine structures.

Typical toxicity data show that concentrations of TBT of the order of $1 \, \mu g \, l^{-1}$ induce toxic effects. A level of 1/100th of this value should not lead to negative effects. However, misshapen oysters, round as golf balls, were found in the Blackwater Estuary in the United Kingdom and elsewhere. Through excellent detective work it was later shown that TBT was responsible for such effects even at concentrations below $2 \, ng \, l^{-1}$. These effects had not been predicted because toxicity tests are generally short-term-lasting, at best, for 48 h. Growth effects occur over much longer periods of time. Other scientists discovered that TBT caused the female gastropod snail *Nucella lapillus* to develop a penis. The most extreme effect was the appearance of a penis having the same size as the male, a condition known as imposex, with the affected females being infertile. Imposex in the field has since been used as an excellent marker for long-term exposures to TBT (due to the use of TBT-containing antifouling paints). France was the first country to introduce a ban on the use of such paints on vessels less than 25 m in length because of the value of its shellfish industry. Many other countries subsequently followed suit.

These lessons teach us that laboratory toxicity tests are unable to predict the long-term effects of some chemicals. TBT, with its effects on reproductive organs, now falls into a general category known collectively as hormone-disrupting chemicals. There exist a large range of chemicals having little commonality in chemical structure, other than that they are all organic, that produce similar effects. Organic chemicals are now justifiably the clear focus of toxicity research rather than the 'heavy metals', which were regarded as the key contaminants in the 1970s and 1980s.

Recently, a range of new techniques for assessing effects on individuals have been produced. They are referred to as 'biomarkers' that reflect stress in marine organisms. One example of such techniques involves the measurement of an enzyme Ethoxyresorufin-o-deethylase (EROD) in flatfish that is induced by exposures to polycyclic aromatic hydrocarbons (PAHs). Another is the measurement of acetylcholinesterase production that is inhibited by chemical stressors, primarily organophosphates in fertilizers used in agriculture. The successful development of biomarker techniques suggests that strategies for monitoring the condition of the marine environment should emphasize the application of biological rather than chemical measurements. A suite of biomarkers covering the range of responses from genetic to whole organism, combined with the use of multivariate statistical techniques now in widespread use for the analysis of marine community data, now offer both the sensitivity and efficiency required for impact detection in the environment. These can be followed up by chemical measurements when indicative biological response signals are detected. In this way, science is providing methods needed for an effective environmental management and protection framework.

Uncertainty, Risk Assessment, and Power Analyses

Generalized frameworks for the management of activities potentially affecting the marine environment have been devised. These involve initial desk studies of the sources and amounts of chemicals released and physical disturbance planned. These can then be combined with specifications of the physical and chemical properties of substances (i.e., hazards, including the properties of toxicity, persistence, and liability to bioaccumulate) and biological effects information. Such information can then be considered in the context of understanding of the physical, chemical, and biological characteristics of the potentially affected area to yield potential changes in these conditions caused by the human activity proposed including the exposures of marine organisms and humans to chemicals. This will provide a basis for assessing the consequences of the proposed activity including the risks to human health, the effects on marine organisms posed by chemical exposures, and any effects on other marine resources and amenities caused by changed sedimentation rates, altered physical dynamics, etc. This constitutes a risk-assessment process.

The risk assessment will indicate what effects are likely and the uncertainties that require to be considered in judging the associated risk to the marine environment, its resources and amenities, and to human health. The risk assessment should incorporate appropriate degrees of pessimism that is the equivalent of conservatism to allow for uncertainties in scientific terms and should correspond to precaution in policy terms. Ultimately, the acceptability of effects and risks is not a scientific matter – it lies within the policy and management spheres (although scientists may be consulted). If certain risks are deemed unacceptable, the regulatory authority will legitimately require additional mitigation measures and the risk assessment incorporating the new measures iterated. If this reveals no significant problem from management and protection perspectives, the conditions and predictions should be used as a basis for compliance and effects monitoring of the activity. In this context, the purpose of environmental monitoring is to ensure that the predicted changes are within expectations and not exceeded. If they are, feedback from monitoring to management should ensure that regulatory constraints are revised to reduce the impact further. Equally important is that the results of monitoring can be used to reveal deficiencies or invalid assumptions in previous risk assessments, thereby offering the benefits of future improvements in predictive ability.

Previous environmental-monitoring activities have been widely criticized as being ill-conceived, ill-conducted, and the results inadequately evaluated. In too few cases have monitoring programs been designed around testable hypotheses that enable rigorous scientific evaluation. Furthermore, there has been an unwillingness to evaluate results periodically to ensure that a program is meeting its objectives and yielding useful information. Far too often, the results of monitoring are archived without benefit of human analysis, thereby constituting a waste of resources. These tend to be more general criticisms but there are also improvements that could be made at a more detailed level. Power analysis to determine the ability of a measurement sequence to detect a change of a given amount is not used sufficiently and greater attention to type II, rather than type I, statistical errors is warranted. A type I error occurs when it is accepted that a harmful effect occurs when it does not. We guard against making this error by accepting at least 95% probability ($p < 0.05$) and thereby allowing a 5% error due to pure chance. A type II error, on the other hand, is where it is accepted that a harmful effect has not occurred when in fact it has. Several scientists have argued that this is a far more serious error in the context of marine environmental protection than a type I error and yet is rarely considered. Commonly, the criterion adopted for the probability of type II errors is not 95% but 80%.

Improving Marine Environmental Protection

Some of the improvements needed in scientific endeavours supporting marine environmental protection have been outlined. Use of pessimistic approaches to take account of uncertainties, improvements in the design and objectives of monitoring programs, and greater consideration of type II statistical errors are among the more important of these. Scientists have been poor at explaining the benefits and limitations of science not only to environmental managers but more crucially to the public. Indeed, previous failures of environmental protection have often been attributed to scientific deficiencies rather than to those of management faced with compromises between political pressures and environmental protection. This is somewhat analogous to similar conflicts and failures in fisheries management. The green movement has been far better at getting its message across not only to managers and legislators, but more importantly to the general public. Unfortunately, this has led to perception being used as a measure of the severity of environmental damage or threat and to the adoption of unwarranted and extreme policy measures for the resolution of perceived problems. The public, and ultimately governmental, reaction to the proposal to dispose of the used field storage tank, *Brent Spar*, was out of all proportion to the scale of possible damage to the marine environment. A couple of years have passed and a further $30 million spent, but the platform still lies in a Norwegian fiord demanding surveillance. Herein lies the danger: that extreme measures may not solve existing problems or prevent future problems while placing unnecessarily severe constraints or disincentives on economic and technological development as argued by Holm and Harris. Attention is being diverted from the adoption of rational, considered, and scientifically based approaches to the protection of the environment that will permit the greatest opportunity for social and economic development.

Society should be demanding the increased use of prior environmental impact assessments to provide quantitative predictions of what effects are to be expected and of their spatial distribution. These should be backed up by properly designed monitoring programs that have adequate power to detect the changes expected and provide routine feedback to regulatory controls. The necessary science already exists – what is needed is the will to apply it effectively to ensure that the best possible environmental protection is achieved in the face of the demands for human development.

See also

Atmospheric Input of Pollutants. Global Marine Pollution. Oil Pollution. Pollution, Solids. Thermal Discharges and Pollution.

Further Reading

Bewers JM (1995) The declining influence of science on marine environmental policy. *Chemistry and Ecology* 10: 9–23.

Freestone D and Hay E (1996) *The Precautionary Principle and International Law: The Challenge of Implementation*, 274pp. The Hague: Kluwer Law International.

FRG (1986) *Umweltpolitik: Guidelines on Anticipatory Environmental Protection*, 43pp. Berlin: Federal Ministry for the Environment, Nature Conservation and Nuclear Safety.

GESAMP (1991) Scientific strategies for marine environmental protection. *GESAMP Reports and Studies No. 45.* London: International Maritime Organization.

Gray JS (1998) Risk assessment and management in the exploitation of the sea. In: Calow P (ed.) *Handbook of Environmental Risk Assessment and Management*, pp. 452–473. Oxford, UK: Blackwell.

Holm S and Harris J (1999) Pecautionary principle stifles discovery. *Nature* 400: 398.

Milne A (1993) The perils of green pessimism. *New Scientist* 1877: 34–37.

NRC (1990) *Managing Troubled Waters: The Role of Environmental Monitoring.* Washington, DC: United States National Research Council, National Academies Press.

USEPA (1992) *Framework for Ecological Risk Assessment.* Washington, DC: Office of Research and Development, United States Environmental Protection Agency.

Wynne BG and Mayer S (1993) Science and the environment. *New Scientist* 1876: 33–35.

NATURAL HAZARDS

TSUNAMI

P. L.-F. Liu, Cornell University, Ithaca, NY, USA

Introduction

Tsunami is a Japanese word that is made of two characters: *tsu* and *nami*. The character *tsu* means harbor, while the character *nami* means wave. Therefore, the original word *tsunami* describes large wave oscillations inside a harbor during a 'tsunami' event. In the past, tsunami is often referred to as 'tidal wave', which is a misnomer. Tides, featuring the rising and falling of water level in the ocean in a daily, monthly, and yearly cycle, are caused by gravitational influences of the moon, sun, and planets. Tsunamis are not generated by this kind of gravitational forces and are unrelated to the tides, although the tidal level does influence a tsunami striking a coastal area.

The phenomenon we call a tsunami is a series of water waves of extremely long wavelength and long period, generated in an ocean by a geophysical disturbance that displaces the water within a short period of time. Waves are formed as the displaced water mass, which acts under the influence of gravity, attempts to regain its equilibrium. Tsunamis are primarily associated with submarine earthquakes in oceanic and coastal regions. However, landslides, volcanic eruptions, and even impacts of objects from outer space (such as meteorites, asteroids, and comets) can also trigger tsunamis.

Tsunamis are usually characterized as shallow-water waves or long waves, which are different from wind-generated waves, the waves many of us have observed on a beach. Wind waves of 5–20-s period (T = time interval between two successive wave crests or troughs) have wavelengths ($\lambda = T^2(g/2\pi)$ distance between two successive wave crests or troughs) of c. 40–620 m. On the other hand, a tsunami can have a wave period in the range of 10 min to 1 h and a wavelength in excess of 200 km in a deep ocean basin. A wave is characterized as a shallow-water wave when the water depth is less than 5% of the wavelength. The forward and backward water motion under the shallow-water wave is felted throughout the entire water column. The shallow water wave is also sensitive to the change of water depth. For instance, the speed (celerity) of a shallow-water wave is equal to the square root of the product of the gravitational acceleration ($9.81\,\mathrm{m\,s^{-2}}$) and the water depth. Since the average water depth in the Pacific Ocean is 5 km, a tsunami can travel at a speed of about $800\,\mathrm{km\,h^{-1}}$ ($500\,\mathrm{mi\,h^{-1}}$), which is almost the same as the speed of a jet airplane. A tsunami can move from the West Coast of South America to the East Coast of Japan in less than 1 day.

The initial amplitude of a tsunami in the vicinity of a source region is usually quite small, typically only a meter or less, in comparison with the wavelength. In general, as the tsunami propagates into the open ocean, the amplitude of tsunami will decrease for the wave energy is spread over a much larger area. In the open ocean, it is very difficult to detect a tsunami from aboard a ship because the water level will rise only slightly over a period of 10 min to hours. Since the rate at which a wave loses its energy is inversely proportional to its wavelength, a tsunami will lose little energy as it propagates. Hence in the open ocean, a tsunami will travel at high speeds and over great transoceanic distances with little energy loss.

As a tsunami propagates into shallower waters near the coast, it undergoes a rapid transformation. Because the energy loss remains insignificant, the total energy flux of the tsunami, which is proportional to the product of the square of the wave amplitude and the speed of the tsunami, remains constant. Therefore, the speed of the tsunami decreases as it enters shallower water and the height of the tsunami grows. Because of this 'shoaling' effect, a tsunami that was imperceptible in the open ocean may grow to be several meters or more in height.

When a tsunami finally reaches the shore, it may appear as a rapid rising or falling water, a series of breaking waves, or even a bore. Reefs, bays, entrances to rivers, undersea features, including vegetations, and the slope of the beach all play a role modifying the tsunami as it approaches the shore. Tsunamis rarely become great, towering breaking waves. Sometimes the tsunami may break far offshore. Or it may form into a bore, which is a step-like wave with a steep breaking front, as the tsunami moves into a shallow bay or river. **Figure 1** shows the incoming 1946 tsunami at Hilo, Hawaii.

The water level on shore can rise by several meters. In extreme cases, water level can rise to more than 20 m for tsunamis of distant origin and over 30 m for tsunami close to the earthquake's epicenter. The first wave may not always be the largest in the series of waves. In some cases, the water level will fall significantly first, exposing the bottom of a bay

Figure 1   1946 tsunami at Hilo, Hawaii (Pacific Tsunami Museum). Wave height may be judged from the height of the trees.

or a beach, and then a large positive wave follows. The destructive pattern of a tsunami is also difficult to predict. One coastal area may see no damaging wave activity, while in a neighboring area destructive waves can be large and violent. The flooding of an area can extend inland by 500 m or more, covering large expanses of land with water and debris. Tsunamis may reach a maximum vertical height onshore above sea level, called a runup height, of 30 m.

Since scientists still cannot predict accurately when earthquakes, landslides, or volcano eruptions will occur, they cannot determine exactly when a tsunami will be generated. But, with the aid of historical records of tsunamis and numerical models, scientists can get an idea as to where they are most likely to be generated. Past tsunami height measurements and computer modeling can also help to forecast future tsunami impact and flooding limits at specific coastal areas.

Historical and Recent Tsunamis

Tsunamis have been observed and recorded since ancient times, especially in Japan and the Mediterranean areas. The earliest recorded tsunami occurred in 2000 BC off the coast of Syria. The oldest reference of tsunami record can be traced back to the sixteenth century in the United States.

During the last century, more than 100 tsunamis have been observed in the United States alone. Among them, the 1946 Alaskan tsunami, the 1960 Chilean tsunami, and the 1964 Alaskan tsunami were the three most destructive tsunamis in the US history. The 1946 Aleutian earthquake $(M_w = 7.3)$ generated catastrophic tsunamis that attacked the Hawaiian Islands after traveling about 5 h and killed 159 people.

(The magnitude of an earthquake is defined by the seismic moment, M_0 (dyn cm), which is determined from the seismic data recorded worldwide. Converting the seismic moment into a logarithmic scale, we define $M_w = (1/1.5)\log_{10} M_0 - 10.7$.) The reported property damage reached \$26 million. The 1960 Chilean tsunami waves struck the Hawaiian Islands after 14 h, traveling across the Pacific Ocean from the Chilean coast. They caused devastating damage not only along the Chilean coast (more than 1000 people were killed and the total property damage from the combined effects of the earthquake and tsunami was estimated as \$417 million) but also at Hilo, Hawaii, where 61 deaths and \$23.5 million in property damage occurred (see **Figure 2**). The 1964 Alaskan tsunami triggered by the Prince William Sound earthquake $(M_w = 8.4)$, which was recorded as one of the largest earthquakes in the North American continent, caused the most destructive damage in Alaska's history. The tsunami killed 106 people and the total damage amounted to \$84 million in Alaska.

Within less than a year between September 1992 and July 1993, three large undersea earthquakes strike the Pacific Ocean area, causing devastating tsunamis. On 2 September 1992, an earthquake of magnitude 7.0 occurred c. 100 km off the Nicaraguan coast. The maximum runup height was recorded as 10 m and 168 people died in this event. A few months later, another strong earthquake $(M_w = 7.5)$ attacked the Flores Island and surrounding area in Indonesia on 12 December 1992. It was reported that more than 1000 people were killed in the town of Maumere alone and two-thirds of the population of Babi Island were swept away by the tsunami. The maximum runup was estimated as 26 m. The final toll of this Flores earthquake stood at 1712 deaths and more than 2000 injures. Exactly

Figure 2 The tsunami of 1960 killed 61 people in Hilo, destroyed 537 buildings, and damages totaled over $23 million.

7 months later, on 12 July 1993, the third strong earthquake ($M_w = 7.8$) occurred near the Hokkaido Island in Japan (Hokkaido Tsunami Survey Group 1993). Within 3–5 min, a large tsunami engulfed the Okushiri coastline and the central west of Hokkaido, impinging extensive property damages, especially on the southern tip of Okushiri Island in the town of Aonae. The runup heights on the Okushiri Island were thoroughly surveyed and they varied between 15 and 30 m over a 20-km stretch of the southern part of the island, with several 10-m spots on the northern part of the island. It was also reported that although the runup heights on the west coast of Hokkaido are not large (less than 10 m), damage was extensive in several towns. The epicenters of these three earthquakes were all located near residential coastal areas. Therefore, the damage caused by subsequent tsunamis was unusually large.

On 17 July 1998, an earthquake occurred in the Sandaun Province of northwestern Papua New Guinea, about 65 km northwest of the port city of Aitape. The earthquake magnitude was estimated as $M_w = 7.0$. About 20 min after the first shock, Warapo and Arop villages were completely destroyed by tsunamis. The death toll was at over 2000 and many of them drowned in the Sissano Lagoon behind the Arop villages. The surveyed maximum runup height was 15 m, which is much higher than the predicted value based on the seismic information. It has been suggested that the Papua New Guinea tsunami could be caused by a submarine landslide.

The most devastating tsunamis in recent history occurred in the Indian Ocean on 26 December 2004.

An earthquake of $M_w = 9.0$ occurred off the west coast of northern Sumatra. Large tsunamis were generated, severely damaging coastal communities in countries around the Indian Ocean, including Indonesia, Thailand, Sri Lanka, and India. The estimated tsunami death toll ranged from 156 000 to 178 000 across 11 nations, with additional 26 500–142 000 missing, most of them presumed dead.

Tsunami Generation Mechanisms

Tsunamigentic Earthquakes

Most tsunamis are the results of submarine earthquakes. The majority of earthquakes can be explained in terms of plate tectonics. The basic concept is that the outermost part the Earth consists of several large and fairly stable slabs of solid and relatively rigid rock, called plates (see **Figure 3**). These plates are constantly moving (very slowly), and rub against one another along the plate boundaries, which are also called faults. Consequently, stress and strain build up along these faults, and eventually they become too great to bear and the plates move abruptly so as to release the stress and strain, creating an earthquake. Most of tsunamigentic earthquakes occur in subduction zones around the Pacific Ocean rim, where the dense crust of the ocean floor dives beneath the edge of the lighter continental crust and sinks down into Earth's mantle. These subduction zones include the west coasts of North and South America, the coasts of East Asia (especially Japan), and many Pacific island chains (**Figure 3**). There are different types of faults along subduction margins. The interplate fault usually accommodates a large relative motion between two tectonic plates and the overlying plate is typically pushed upward. This upward push is impulsive; it occurs very quickly, in a few seconds. The ocean water surface responds immediately to the upward movement of the seafloor and the ocean surface profile usually mimics the seafloor displacement (see **Figure 4**). The interplate fault in a subduction zone has been responsible for

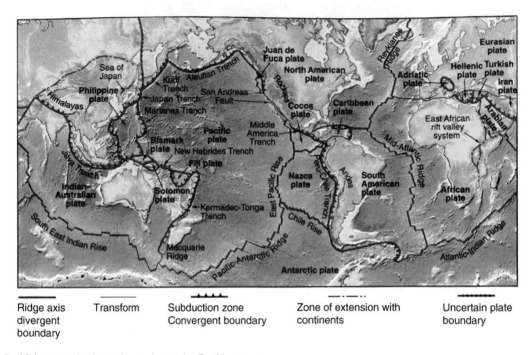

Figure 3   Major tectonic plates that make up the Earth's crust.

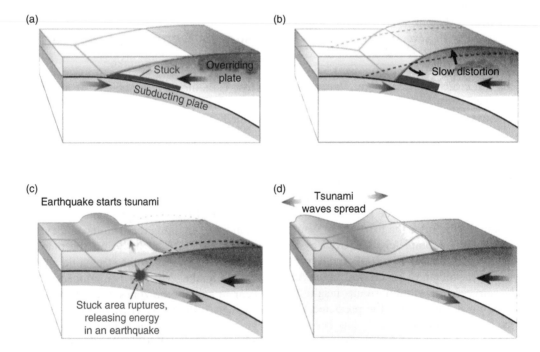

Figure 4   Sketches of the tsunami generation mechanism caused by a submarine earthquake. An oceanic plate subducts under an overriding plate (a). The overriding plate deforms due to the relative motion and the friction between two tectonic plates (b). The stuck area ruptures, releasing energy in an earthquake (c). Tsunami waves are generated due to the vertical seafloor displacement (d).

most of the largest tsunamis in the twentieth century. For example, the 1952 Kamchatka, 1957 Aleutian, 1960 Chile, 1964 Alaska, and 2004 Sumatra earthquakes all generated damaging tsunamis not only in the region near the earthquake epicenter, but also on faraway shores.

For most of the interplate fault ruptures, the resulting seafloor displacement can be estimated based on the dislocation theory. Using the linear elastic theory, analytical solutions can be derived from the mean dislocation field on the fault. Several parameters defining the geometry and strength of the fault rupture need to be specified. First of all, the mean fault slip, D, is calculated from the seismic moment M_0 as follows:

$$M_0 = \mu DS \qquad [1]$$

where S is the rupture area and μ is the rigidity of the Earth at the source, which has a range of $6–7 \times 10^{11}$ dyn cm^{-2} for interplate earthquakes. The seismic moment, M_0, is determined from the seismic data recorded worldwide and is usually reported as the Harvard Centroid-Moment-Tensor (CMT) solution within a few minutes of the first earthquake tremor. The rupture area is usually estimated from the aftershock data. However, for a rough estimation, the fault plane can be approximated as a rectangle with length L and width W. The aspect ratio L/W could vary from 2 to 8. To find the static displacement of the seafloor, we need to assign the focal depth d, measuring the depth of the upper rim of the fault plane, the dip angle δ, and the slip angle λ of the dislocation on the fault plane measured from the horizontal axis (see **Figure 5**). For an oblique slip on a dipping fault, the slip vector can be decomposed into dip-slip and strike-slip components. In general, the magnitude of the vertical displacement is less for the strike-slip component than for the dip-slip component. The closed form expressions for vertical seafloor displacement caused by a slip along a rectangular fault are given by Mansinha and Smylie.

For more realistic fault models, nonuniform stress-strength fields (i.e., faults with various kinds of barriers, asperities, etc.) are expected, so that the actual seafloor displacement may be very complicated compared with the smooth seafloor displacement computed from the mean dislocation field on the fault. As an example, the vertical seafloor displacement caused by the 1964 Alaska earthquake is sketched in **Figure 6**. Although several numerical models have considered geometrically complex faults, complex slip distributions, and elastic layers of variable thickness, they are not yet disseminated in

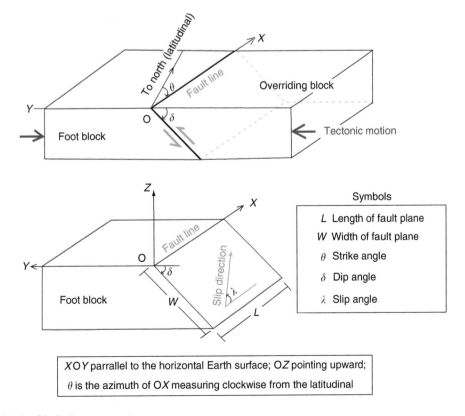

Figure 5 A sketch of fault plane parameters.

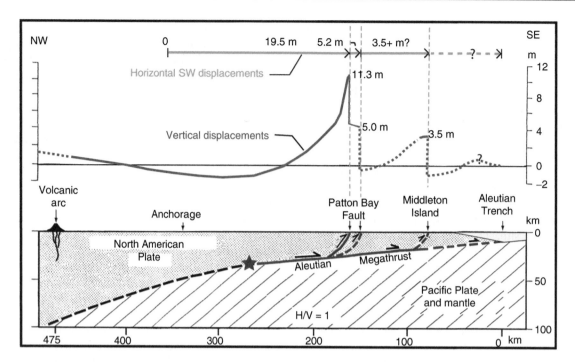

Figure 6 A sketch of 1964 Alaska earthquake generated vertical seafloor displacement (G. Plafker, 2006).

tsunami research. One of the reasons is that our knowledge in source parameters, inhomogeneity, and nonuniform slip distribution is too incomplete to justify using such a complex model.

Certain earthquakes referred to as tsunami earthquakes have slow faulting motion and very long rupture duration (more than several minutes). These earthquakes occur along the shallow part of the interplate thrust or décollement near the trench (the wedge portion of the thin crust above the interface of the continental crust and the ocean plate). The wedge portion consisting of thick deformable sediments with low rigidity, and the steepening of rupture surface in shallow depth all favor the large displacement of the crust and possibility of generating a large tsunami. Because of the extreme heterogeneity, accurate modeling is difficult, resulting in large uncertainty in estimated seafloor displacement.

Landslides and Other Generation Mechanisms

There are occasions when the secondary effects of earthquakes, such as landslide and submarine slump, may be responsible for the generation of tsunamis. These tsunamis are sometimes disastrous and have gained increasing attentions in recent years. Landslides are generated when slopes or sediment deposits become too steep and they fail to remain in equilibrium and motionless. Once the unstable conditions are present, slope failure can be triggered by storm, earthquakes, rains, or merely continued

deposition of materials on the slope. Alternative mechanisms of sediment instability range between soft sediment deformations in turbidities, to rotational slumps in cohesive sediments. Certain environments are particularly susceptible to the production of landslides. River delta and steep underwater slopes above submarine canyons are likely sites for landslide-generated tsunamis. At the time of the 1964 Alaska earthquake, numerous locally landslide-generated tsunamis with devastating effects were observed. On 29 November 1975, a landslide was triggered by a 7.2 magnitude earthquake along the southeast coast of Hawaii. A 60-km stretch of Kilauea's south coast subsided 3.5 m and moved seaward 8 m. This landslide generated a local tsunami with a maximum runup height of 16 m at Keauhou. Historically, there have been several tsunamis whose magnitudes were simply too large to be attributed to the coseismic seafloor movement and landslides have been suggested as an alternative cause. The 1946 Aleutian tsunami and the 1998 Papua New Guinea tsunami are two significant examples.

In terms of tsunami generation mechanisms, two significant differences exist between submarine landslide and coseismic seafloor deformation. First, the duration of a landslide is much longer and is in the order of magnitude of several minutes or longer. Hence the time history of the seafloor movement will affect the characteristics of the generated wave and needs to be included in the model. Second, the effective size of the landslide region is usually much smaller than the coseismic seafloor deformation zone. Consequently, the typical wavelength of the tsunamis generated by a submarine landslide is also shorter, that is, c. 1–10 km. Therefore, in some cases, the shallow-water (long-wave) assumption might not be valid for landslide-generated tsunamis.

Although they are rare, the violent geological activities associated with volcanic eruptions can also generate tsunamis. There are three types of tsunami-generation mechanism associated with a volcanic eruption. First, the pyroclastic flows, which are mixtures of gas, rocks, and lava, can move rapidly off an island and into an ocean, their impact displacing seawater and producing a tsunami. The second mechanism is the submarine volcanic explosion, which occurs when cool seawater encounters hot volcanic magma. The third mechanism is due to the collapse of a submarine volcanic caldera. The collapse may happen when the magma beneath a volcano is withdrawn back deeper into the Earth, and the sudden subsidence of the volcanic edifice displaces water and produces a tsunami. Furthermore, the large masses of rock that accumulate on the sides of volcanoes may suddenly slide down the slope into the sea, producing tsunamis. For example, in 1792, a large mass of the mountain slided into Ariake Bay in Shimabara on Kyushu Island, Japan, and generated tsunamis that reached a height of 10 m in some places, killing a large number of people.

In the following sections, our discussions will focus on submarine earthquake-generated tsunamis and their coastal effects.

Modeling of Tsunami Generation, Propagation, and Coastal Inundation

To mitigate tsunami hazards, the highest priority is to identify the high-tsunami-risk zone and to educate the citizen, living in and near the risk zone, about the proper behaviors in the event of an earthquake and tsunami attack. For a distant tsunami, a reliable warning system, which predicts the arrival time as well as the inundation area accurately, can save many lives. On the other hand, in the event of a near-field tsunami, the emergency evacuation plan must be activated as soon as the earth shaking is felt. This is only possible, if a predetermined evacuation/inundation map is available. These maps should be produced based on the historical tsunami events and the estimated 'worst scenarios' or the 'design tsunamis'. To produce realistic and reliable inundation maps, it is essential to use a numerical model that calculates accurately tsunami propagation from a source region to the coastal areas of concern and the subsequent tsunami runup and inundation.

Numerical simulations of tsunami have made great progress in the last 50 years. This progress is made possible by the advancement of seismology and by the development of the high-speed computer. Several tsunami models are being used in the National Tsunami Hazard Mitigation Program, sponsored by the National Oceanic and Atmospheric Administration (NOAA), in partnership with the US Geological Survey (USGS), the Federal Emergency Management Agency (FEMA), to produce tsunami inundation and evacuation maps for the states of Alaska, California, Hawaii, Oregon, and Washington.

Tsunami Generation and Propagation in an Open Ocean

The rupture speed of fault plane during earthquake is usually much faster than that of the tsunami. For instance, the fault line of the 2004 Sumatra earthquake was estimated as 1200-km long and the rupture process lasted for about 10 min. Therefore, the rupture speed was c. 2–3 km s^{-1}, which is considered as a relatively slow rupture speed and is still about 1 order of magnitude faster than the speed of tsunami (0.17 km s^{-1} in a typical water depth of 3 km). Since the compressibility of water is negligible, the initial free surface response to the seafloor deformation due to fault plane rupture is instantaneous. In other words, in terms of the tsunami propagation timescale, the initial free surface profile can be approximated as having the same shape as the seafloor deformation at the end of rupture, which can be obtained by the methods described in the previous section. As illustrated in **Figure 6**, the typical cross-sectional free surface profile, perpendicular to the fault line, has an N shape with a depression on the landward side and an elevation on the ocean side. If the fault plane is elongated, that is, $L \gg W$, the free surface profile is almost uniform in the longitudinal (fault line) direction and the generated tsunamis will propagate primarily in the direction perpendicular to the fault line. The wavelength is generally characterized by the width of the fault plane, W.

The measure of tsunami wave dispersion is represented by the depth-to-wavelength ratio, that is, $\mu^2 = h/\lambda$, while the nonlinearity is characterized by the amplitude-to-depth ratio, that is, $\varepsilon = A/h$. A tsunami generated in an open ocean or on a continental shelf could have an initial wavelength of several tens to hundreds of kilometers. The initial tsunami wave height may be on the order of magnitude of several meters. For example, the 2004 Indian Ocean tsunami had a typical wavelength of 200 km in the Indian Ocean basin with an amplitude of 1 m. The water depth varies from several hundreds of meters on the continental shelf to several kilometers in the open ocean. It is quite obvious that during the early stage of tsunami propagation both the nonlinear and frequency dispersion effects are small and can be ignored. This is particularly true for the 2004 Indian tsunami. The bottom frictional force and Coriolis force have even smaller effects and can be also neglected in the generation area. Therefore, the linear shallow water (LSW) equations are adequate equations describing the initial stage of tsunami generation and propagation.

As a tsunami propagates over an open ocean, wave energy is spread out into a larger area. In general, the tsunami wave height decreases and the nonlinearity remains weak. However, the importance of the frequency dispersion begins to accumulate as the tsunami travels a long distance. Theoretically, one can estimate that the frequency dispersion becomes important when a tsunami propagates for a long time:

$$t \gg t_d = \sqrt{\frac{h}{g}\left(\frac{\lambda}{h}\right)^3} \qquad [2]$$

or over a long distance:

$$x \gg x_d = t_d \sqrt{gh} = \frac{\lambda^3}{h^2} \qquad [3]$$

In the case of the 2004 Indian Ocean tsunami, $t_d \approx 700$ h and $x_d \approx 5 \times 10^5$ km. In other words, the frequency dispersion effect will only become important when tsunamis have gone around the Earth several times. Obviously, for a tsunami with much shorter wavelength, for example, $\lambda \approx 20$ km, this distance becomes relatively short, that is, $x_d \approx 5 \times 10^2$ km, and can be reached quite easily. Therefore, in modeling transoceanic tsunami propagation, frequency dispersion might need to be considered if the initial wavelength is short. However, nonlinearity is seldom a factor in the deep ocean and only becomes significant when the tsunami enters coastal region.

The LSW equations can be written in terms of a spherical coordinate system as:

$$\frac{\partial \zeta}{\partial t} + \frac{1}{R\cos\varphi}\left[\frac{\partial P}{\partial \psi} + \frac{\partial}{\partial \varphi}(\cos\varphi Q)\right] = -\frac{\partial h}{\partial t} \qquad [4]$$

$$\frac{\partial P}{\partial t} + \frac{gh}{R\cos\varphi}\frac{\partial \zeta}{\partial \psi} = 0 \qquad [5]$$

$$\frac{\partial Q}{\partial t} + \frac{gh}{R}\frac{\partial \zeta}{\partial \varphi} = 0 \qquad [6]$$

where (ψ,φ) denote the longitude and latitude of the Earth, R is the Earth's radius, ζ is free surface elevation, P and Q the volume fluxes ($P = hu$ and $Q = hv$, with u and v being the depth-averaged velocities in longitude and latitude direction, respectively), and h the water depth. Equation [4] represents the depth-integrated continuity equation, and the time rate of change of water depth has been included. When the fault plane rupture is approximated as an instantaneous process and the initial free surface profile is prescribed, the water depth remains time-invariant during tsunami propagation and the right-hand side becomes zero in eqn [4].

The 2004 Indian Ocean tsunami provided an opportunity to verify the validity of LSW equations for modeling tsunami propagation in an open ocean. For the first time in history, satellite altimetry measurements of sea surface elevation captured the Indian Ocean tsunami. About 2 h after the earthquake occurred, two NASA/French Space Agency joint mission satellites, *Jason-1* and *TOPEX/Poseidon*, passed over the Indian Ocean from southwest to northeast (*Jason-1* passed the equator at 02:55:24UTC on 26 December 2004 and *TOPEX/Poseidon* passed the equator at 03:01:57UTC on 26 December 2004) (see **Figure 7**). These two altimetry satellites measured sea surface elevation with accuracy better than 4.2 cm.

Using the numerical model COMCOT (Cornell Multi-grid Coupled Tsunami Model), numerical simulations of tsunami propagation over the Indian Ocean with various fault plane models, including a transient seafloor movement model, have been carried out. The LSW equation model predicts accurately the arrival time of the leading wave and is insensitive of the fault plane models used. However, to predict the trailing waves, the spatial variation of seafloor deformation needs to be taken into consideration. In **Figure 8**, comparisons between LSW results with an optimized fault plane model and *Jason-1/TOPEX* measurements are shown. The excellent agreement between the numerical results and satellite data provides a direct evidence for the validity of the LSW modeling of tsunami propagation in deep ocean.

Coastal Effects – Inundation and Tsunami Forces

Nonlinearity and bottom friction become significant as a tsunami enters the coastal zone, especially during the runup phase. The nonlinear shallow water (NLSW) equations can be used to model certain aspects of coastal effects of a tsunami attack. Using the same notations as those in eqns [4]–[6], the NLSW

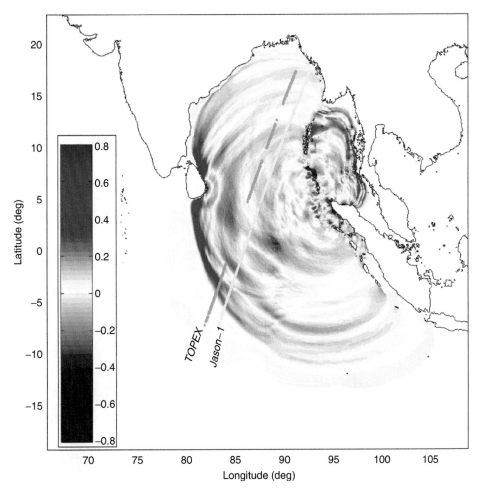

Figure 7 Satellite tracks for *TOPEX* and *Jason-1*. The colors indicate the numerically simulated free surface elevation in meter at 2 h after the earthquake struck.

equations in the Cartesian coordinates are

$$\frac{\partial \zeta}{\partial t} + \frac{\partial P}{\partial x} + \frac{\partial Q}{\partial y} = 0 \qquad [7]$$

$$\frac{\partial P}{\partial t} + \frac{\partial}{\partial x}\left(\frac{P^2}{H}\right) + \frac{\partial}{\partial y}\left(\frac{PQ}{H}\right) + gH\frac{\partial \zeta}{\partial x} + \tau_x H = 0 \qquad [8]$$

$$\frac{\partial Q}{\partial t} + \frac{\partial}{\partial x}\left(\frac{PQ}{H}\right) + \frac{\partial}{\partial y}\left(\frac{Q^2}{H}\right) + gH\frac{\partial \zeta}{\partial y} + \tau_y H = 0 \qquad [9]$$

The bottom frictional stresses are expressed as

$$\tau_x = \frac{gn^2}{H^{10/3}}P(P^2 + Q^2)^{1/2} \qquad [10]$$

$$\tau_y = \frac{gn^2}{H^{10/3}}Q(P^2 + Q^2)^{1/2} \qquad [11]$$

where n is the Manning's relative roughness co-efficient. For flows over a sandy beach, the typical value for the Manning's n is 0.02.

Using a modified leapfrog finite difference scheme in a nested grid system, COMCOT is capable of solving both LSW and NLSW equations simultaneously in different regions. For the nested grid system, the inner (finer) grid adopts a smaller grid size and time step compared to its adjacent outer (larger) grid. At the beginning of a time step, along the interface of two different grids, the volume flux, P and Q, which is product of water depth and depth-averaged velocity, is interpolated from the outer (larger) grids into its inner (finer) grids. And at the end of this time step, the calculated water surface elevations, ζ, at the inner finer grids are averaged to update those values of the larger grids overlapping the finer grids, which are used to compute the volume fluxes at next time step in the outer grids. With this procedure, COMCOT can capture near-shore

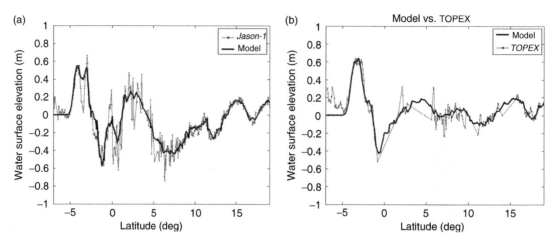

Figure 8 Comparisons between optimized fault model results and *Jason-1* measurements (a)/*TOPEX* measurements (b).

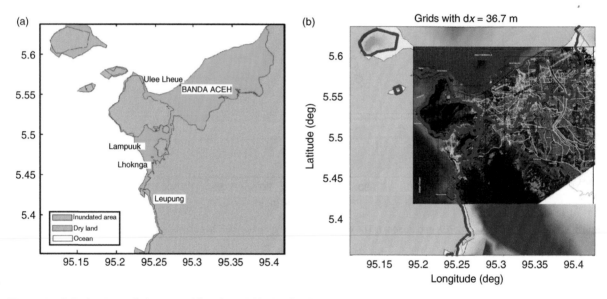

Figure 9 Calculated inundation areas (a) and overlaid with *QUICKBIRD* image (b) in Banda Aceh, Indonesia.

features of a tsunami with a higher spatial and temporal resolution and at the same time can still keep a high computational efficiency.

To estimate the inundation area caused by a tsunami, COMCOT adopts a simple moving boundary scheme. The shoreline is defined as the interface between a wet grid and its adjacent dry grids. Along the shoreline, the volume flux is assigned to be zero. Once the water surface elevation at the wet grid is higher than the land elevation in its adjacent dry grid, the shoreline is moved by one grid toward the dry grid and the volume flux is no longer zero and need to be calculated by the governing equations.

COMCOT, coupled up to three levels of grids, has been used to calculate the runup and inundation areas at Trincomalee Bay (Sri Lanka) and Banda Aceh (Indonesia). Some of the numerical results for Banda Aceh are shown here.

The calculated inundation area in Banda Aceh is shown in **Figure 9**. The flooded area is marked in blue, the dry land region is rendered in green, and the white area is ocean region. The calculated inundation area is also overlaid with a satellite image taken by *QUICKBIRD* in **Figure 9(b)**.

In the overlaid image, the thick red line indicates the inundation line based on the numerical simulation. In the satellite image, the dark green color (vegetation) indicates areas not affected by the tsunami and the area shaded by semitransparent red color shows flooded regions by this tsunami. Obviously, the calculated inundation area matches reasonably well with the satellite image in the neighborhood of Lhoknga and the western part of Banda Aceh. However, in the region of eastern Banda Aceh, the simulations significantly underestimate the inundation area. However, in general, the agreement between the numerical simulation and the satellite observation is surprisingly good.

In **Figure 10**, the tsunami wave heights in Banda Aceh are also compared with the field measurements by two Japan survey teams. On the coast between Lhoknga and Leupung, where the maximum height is measured more than 30 m, the numerical results match very well with the field measurements. However, beyond Lhoknga to the north, the numerical results, in general, are only half of the measurements, except in middle regions between Lhoknga and Lampuuk.

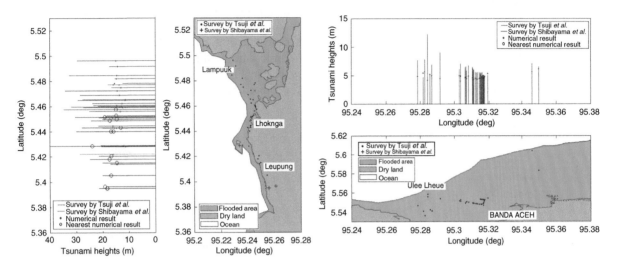

Figure 10 Tsunami heights on eastern and northern coast of Banda Aceh, Indonesia. The field survey measurements are from Tsuji *et al.* (2005) and Shibayama *et al.* (2005).

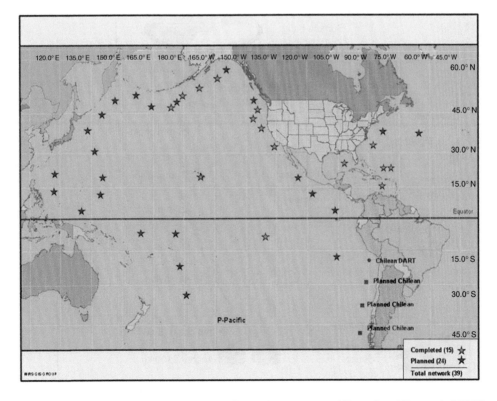

Figure 11 The locations of the existing and planned Deep-Ocean Assessment and Reporting of Tsunamis (DART) system in the Pacific Ocean (NOAA magazine, 17 Apr. 2006).

Tsunami Hazard Mitigation

The ultimate goal of the tsunami hazard mitigation effort is to minimize casualties and property damages. This goal can be met, only if an effective tsunami early warning system is established and a proper coastal management policy is practiced.

Tsunami Early Warning System

The great historical tsunamis, such as the 1960 Chilean tsunami and the 1964 tsunami generated near Prince William Sound in Alaska, prompted the US government to develop an early warning system in the Pacific Ocean. The Japanese government has also developed a tsunami early warning system for the entire coastal community around Japan. The essential information needed for an effective early warning system is the accurate prediction of arrival time and wave height of a forecasted tsunami at a specific location. Obviously, the accuracy of these predictions relies on the information of the initial water surface displacement near the source region, which is primarily determined by the seismic data. In many historical events, including the 2004 Indian Ocean tsunami, evidences have shown that accurate seismic data could not be verified until those events were over. To delineate the source region problem, in the United States, several federal agencies and states

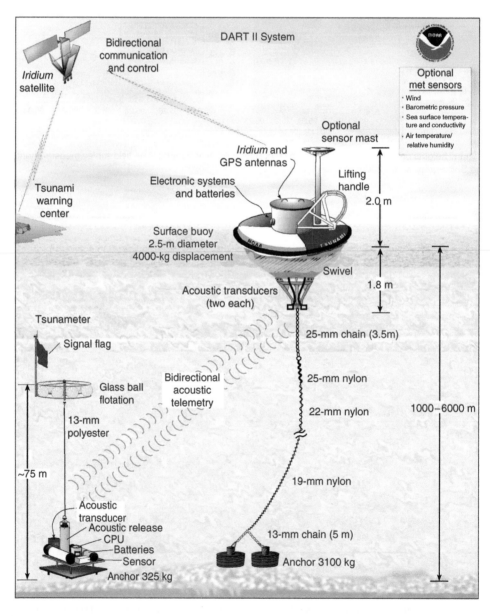

Figure 12   A sketch of the second-generation DART (II) system.

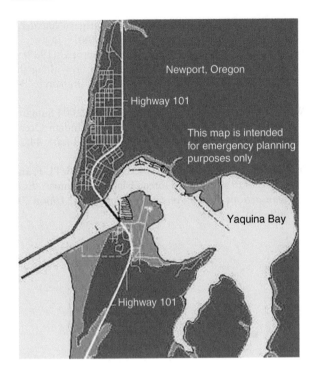

Figure 13 Tsunami inundation map for the coastal city of Newport, Oregon.

These models can simulate a 'design tsunami' approaching a coastline, and they can predict which areas are most at risk to being flooded. The tsunami inundation maps are an integral part of the overall strategy to reduce future loss of life and property. Emergency managers and local governments of the threatened communities use these and similar maps to guide evacuation planning.

As an example, the tsunami inundation map (**Figure 13**) for the coastal city of Newport (Oregon) was created using the results from a numerical simulation using a design tsunami. The areas shown in orange are locations that were flooded in the numerical simulation

Acknowledgment

The work reported here has been supported by National Science Foundation with grants to Cornell University.

Further Reading

Geist EL (1998) Local tsunami and earthquake source parameters. *Advances in Geophysics* 39: 117–209.

Hokkaido Tsunami Survey Group (1993) Tsunami devastates Japanese coastal region. *EOS Transactions of the American Geophysical Union* 74: 417–432.

Kajiura K (1981) Tsunami energy in relation to parameters of the earthquake fault model. *Bulletin of the Earthquake Research Institute, University of Tokyo* 56: 415–440.

Kajiura K and Shuto N (1990) Tsunamis. In: Le Méhauté B and Hanes DM (eds.) *The Sea: Ocean Engineering Science*, pp. 395–420. New York: Wiley.

Kanamori H (1972) Mechanism of tsunami earthquakes. *Physics and Earth Planetary Interactions* 6: 346–359.

Kawata Y, Benson BC, Borrero J, *et al.* (1999) Tsunami in Papua New Guinea was as intense as first thought. *EOS Transactions of the American Geophysical Union* 80: 101, 104–105.

Keating BH and Mcguire WJ (2000) Island edifice failures and associated hazards. *Special Issue: Landslides and Tsunamis. Pure and Applied Geophysics* 157: 899–955.

Liu PL-F, Lynett P, Fernando H, *et al.* (2005) Observations by the International Tsunami Survey Team in Sri Lanka. *Science* 308: 1595.

Lynett PJ, Borrero J, Liu PL-F, and Synolakis CE (2003) Field survey and numerical simulations: A review of the 1998 Papua New Guinea tsunami. *Pure and Applied Geophysics* 160: 2119–2146.

Mansinha L and Smylie DE (1971) The displacement fields of inclined faults. *Bulletin of Seismological Society of America* 61: 1433–1440.

Satake K, Bourgeois J, Abe K, *et al.* (1993) Tsunami field survey of the 1992 Nicaragua earthquake. *EOS*

have joined together to create a warning system that involves the use of deep-ocean tsunami sensors to detect the presence of a tsunami.

These deep-ocean sensors have been deployed at different locations in the Pacific Ocean before the 2004 Indian Ocean tsunami. After the 2004 Indian Ocean tsunami, several additional sensors have been installed and many more are being planned (see **Figure 11**). The sensor system includes a pressure gauge that records and transmits the surface wave signals instantaneously to the surface buoy, which sends the information to a warning center via *Iridium* satellite (**Figure 12**). In the event of a tsunami, the information obtained by the pressure gauge array can be used as input data for modeling the propagation and evolution of a tsunami.

Although there have been no large Pacific-wide tsunamis since the inception of the warning system, warnings have been issued for smaller tsunamis, a few of which were hardly noticeable. This tends to give citizens a lazy attitude toward a tsunami warning, which would be fatal if the wave was large. Therefore, it is very important to keep people in a danger areas educated of tsunami hazards.

Coastal Inundation Map

Using numerical modeling, hazards in areas vulnerable to tsunamis can be assessed, without the area ever having experienced a devastating tsunami.

Transactions of the American Geophysical Union 74: 156–157.

Shibayama T, Okayasu A, Sasaki J, *et al.* (2005) The December 26, 2004 Sumatra Earthquake Tsunami, Tsunami Field Survey in Banda Aceh of Indonesia. http://www.drs.dpri.kyoto-u.ac.jp/sumatra/indonesia-ynu/indonesia_survey_ynu_e.html (accessed Feb. 2008).

Synolakis CE, Bardet J-P, Borrero JC, *et al.* (2002) The slump origin of the 1998 Papua New Guinea tsunami. *Proceedings of Royal Society of London, Series A* 458: 763–789.

Tsuji Y, Matsutomi H, Tanioka Y, *et al.* (2005) Distribution of the Tsunami Heights of the 2004 Sumatra Tsunami in Banda Aceh measured by the Tsunami Survey Team. http://www.eri.u-tokyo.ac.jp/namegaya/sumatera/surveylog/eindex.htm (accessed Feb. 2008).

von Huene R, Bourgois J, Miller J, and Pautot G (1989) A large tsunamigetic landslide and debris flow along the Peru trench. *Journal of Geophysical Research* 94: 1703–1714.

Wang X and Liu PL-F (2006) An analysis of 2004 Sumatra earthquake fault plane mechanisms and Indian Ocean tsunami. *Journal of Hydraulics Research* 44(2): 147–154.

Yeh HH, Imamura F, Synolakis CE, Tsuji Y, Liu PL-F, and Shi S (1993) The Flores Island tsunamis. *EOS Transactions of the American Geophysical Union* 74: 369–373.

STORM SURGES

R. A. Flather, Bidston Observatory, Proudman
Oceanographic Laboratory, Bidston Hill, Prenton, UK

Introduction and Definitions

Storm surges are changes in water level generated by
atmospheric forcing; specifically by the drag of the
wind on the sea surface and by variations in the
surface atmospheric pressure associated with storms.
They last for periods ranging from a few hours to 2
or 3 days and have large spatial scales compared
with the water depth. They can raise or lower the
water level in extreme cases by several meters; a
raising of level being referred to as a 'positive' surge,
and a lowering as a 'negative' surge. Storm surges are
superimposed on the normal astronomical tides
generated by variations in the gravitational attraction of the moon and sun. The storm surge
component can be derived from a time-series of sea
levels recorded by a tide gauge using:

$$\text{surgeresidual} = (\text{observedsealevel})$$
$$-(\text{predictedtidelevel}) \qquad [1]$$

producing a time-series of surge elevations. **Figure 1**
shows an example.

Sometimes, the term 'storm surge' is used for the
sea level (including the tidal component) during a
storm event. It is important to be clear about the
usage of the term and its significance to avoid con-
fusion. Storms also generate surface wind waves that
have periods of order seconds and wavelengths,
away from the coast, comparable with or less than
the water depth.

Positive storm surges combined with high tides
and wind waves can cause coastal floods, which, in
terms of the loss of life and damage, are probably the
most destructive natural hazards of geophysical origin. Where the tidal range is large, the timing of the
surge relative to high water is critical and a large
surge at low tide may go unnoticed. Negative surges
reduce water depth and can be a threat to navigation.
Associated storm surge currents, superimposed on
tidal and wave-generated flows, can also contribute
to extremes of current and bed stress responsible for
coastal erosion. A proper understanding of storm
surges, the ability to predict them and measures to
mitigate their destructive effects are therefore of vital
concern.

Storm Surge Equations

Most storm surge theory and modeling is based on
depth-averaged hydrodynamic equations applicable
to both tides and storm surges and including nonlinear terms responsible for their interaction. In
vector form, these can be written:

$$\frac{\partial \zeta}{\partial t} + \nabla \cdot (D\mathbf{q}) = 0 \qquad [2]$$

$$\frac{\partial \mathbf{q}}{\partial t} + \mathbf{q} \cdot \nabla \mathbf{q} - f\mathbf{k} \times \mathbf{q} = -g\nabla(\zeta - \bar{\zeta}) - \frac{1}{\rho}\nabla p_a$$
$$+ \frac{1}{\rho D}(\tau_s - \tau_b) + A\nabla^2 \mathbf{q} \qquad [3]$$

where t is time; ζ the sea surface elevation; $\bar{\zeta}$ the
equilibrium tide; $\mathbf{q}$ the depth-mean current; τ_s the
wind stress on the sea surface; τ_b the bottom stress;
p_a atmospheric pressure on the sea surface; D the
total water depth ($D = h + \zeta$, where h is the undisturbed depth); ρ the density of sea water, assumed
to be uniform; g the acceleration due to gravity; f the
Coriolis parameter ($= 2\omega \sin\varphi$, where ω is the an-
gular speed of rotation of the Earth and φ is the
latitude); $\mathbf{k}$ a unit vector in the vertical; and A the
coefficient of horizontal viscosity. Eqn [2] is the
continuity equation expressing conservation of volume. Eqn [3] equates the accelerations (left-hand
side) to the force per unit mass (right-hand side).

In this formulation, bottom stress, τ_b is related to
the current, $\mathbf{q}$, using a quadratic law:

$$\tau_b = k\rho\mathbf{q}|\mathbf{q}| \qquad [4]$$

where k is a friction parameter (~ 0.002). Similarly,
the wind stress, τ_s, is related to $\mathbf{W}$, the wind velocity
at a height of $10\,\text{m}$ above the surface, also using a
quadratic law:

$$\tau_s = c_D\rho_a\mathbf{W}|\mathbf{W}| \qquad [5]$$

where ρ_a is the density of air and c_D a drag coefficient. Measurements in the atmospheric boundary
layer suggest that c_D increases with wind speed, W,
accounting for changes in surface roughness associated with wind waves. A typical form due to J. Wu is:

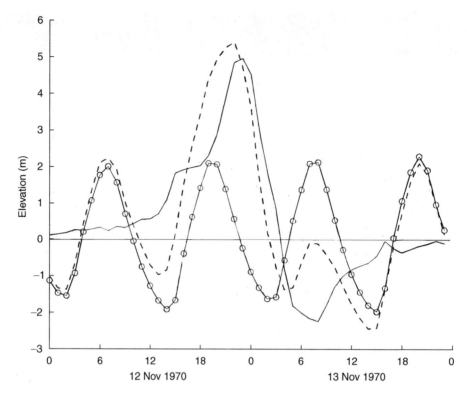

Figure 1 Water level (dashed line), predicted tide (line with ○), and the surge residual (continuous line) at Sandwip Island, Bangladesh, during the catastrophic storm surge of 12–13 November 1970 (times are GMT).

$$10^3 c_D = 0.8 + 0.065\,W \qquad [6]$$

Alternatively, from dimensional analysis, H. Charnock obtained $gz_0/u_*^2 = \alpha$, where z_0 is the aerodynamic roughness length associated with the surface wavefield, u_* is the friction velocity ($u_*^2 = \tau_s/\rho_a$), and α is the Charnock constant. So, the roughness varies linearly with surface wind stress. Assuming a logarithmic variation of wind speed with height z above the surface, $W(z) = (u_*/\kappa)\,\ln(z/z_0)$, where κ is von Kármán's constant. It follows that for $z = 10$ m:

$$c_D = \left[(1/\kappa)\ln\{gz/(\alpha c_D W^2)\}\right]^{-2} \qquad [7]$$

Estimates of α range from 0.012 to 0.035.

Generation and Dynamics of Storm Surges

The forcing terms in eqn [3] which give rise to storm surges are those representing wind stress and the horizontal gradient of surface atmospheric pressure. Very simple solutions describe the basic mechanisms. The sea responds to atmospheric pressure variations by adjusting sea level such that, at depth, pressure in the water is uniform, the hydrostatic approximation. Assuming in eqn [3] that $\mathbf{q} = 0$ and $\tau_s = 0$, then

$g\nabla\zeta + (1/\rho)\nabla p_a = 0$, so $\rho g\zeta + p_a = $ constant . This gives the 'inverse barometer effect' whereby a decrease in atmospheric pressure of 1 hPa produces an increase in sea level of approximately 1 cm. Wind stress produces water level variations on the scale of the storm. Eqns [5] and [6] imply that the strongest winds are most important since effectively $\tau_s \propto W^3$. Both pressure and wind effects are present in all storm surges, but their relative importance varies with location. Since wind stress is divided by D whereas ∇p_a is not, it follows that wind forcing increases in importance in shallower water. Consequently, pressure forcing dominates in the deep ocean whereas wind forcing dominates in shallow coastal seas. Major destructive storm surges occur when extreme storm winds act over extensive areas of shallow water.

As well as the obvious wind set-up, with the component of wind stress directed towards the coast balanced by a surface elevation gradient, winds parallel to shore can also generate surges at higher latitudes. Wind stress parallel to the coast with the coast on its right will drive a longshore current, limited by bottom friction. Geostrophic balance gives (in the Northern Hemisphere) a surface gradient raising levels at the coast (see).

Amplification of surges may be caused by the funneling effect of a converging coastline or estuary and by a resonant response; e.g. if the wind forcing travels at the same velocity as the storm wave, or matches the natural period of oscillation of a gulf, producing a seiche.

As a storm moves away, surges generated in one area may propagate as free waves, contributing as externally generated components to surges in another area. Generally, away from the forcing center, the response of the ocean consists of longshore propagating coastally trapped waves. Examples include external surges in the North Sea (see **Figure 2**), which are generated west and north of Scotland and propagate anticlockwise round the basin like the diurnal tide; approximately as a Kelvin wave. Low mode continental shelf waves have been identified in surges on the west coast of Norway, in the Middle Atlantic Bight of the US, in the East China Sea, and on the north-west shelf of Australia. Currents associated with low mode continental shelf waves generated by a tropical cyclone crossing the north-west shelf of Australia have been observed and explained by numerical modeling. Edge waves can also be generated by cyclones travelling parallel to the coast in the opposite direction to shelf waves.

From eqns [3] and [4], bottom stress (which dissipates surges) also depends on water depth and is non-linear; the current including contributions from tide and surge, $q = q_T + q_s$. Consequently, dissipation of surges is stronger in shallow water and where tidal currents are also strong. In deeper water and where tides are weak, free motions can persist for long times or propagate long distances. For example, the Adriatic has relatively small tides and seiches excited by storms can persist for many days.

In areas with substantial tides and shallow water, non-linear dynamical processes are important, resulting in interactions between the tide and storm surge such that both components are modified. The main contribution arises from bottom stress, but time-dependent water depth, $D = h + \zeta(t)$, can also be significant (e.g. $\tau_s/(\rho D)$ will be smaller at high tide than at low tide). An important consequence is that the linear superposition of surge and tide without accounting for their interaction gives substantial errors in estimating water level. For example, for surges propagating southwards in the North Sea into the Thames Estuary, surge maxima tend to occur on the rising tide rather than at high water (**Figure 3**).

Interactions also occur between the tide–surge motion and surface wind waves (see below).

Areas Affected by Storm Surges

Major storm surges are created by mid-latitude storms and by tropical cyclones (also called hurricanes and typhoons) which generally occur in geographically separated areas and differ in their scale. Mid-latitude storms are relatively large and evolve slowly enough to allow accurate predictions of their wind and pressure fields from atmospheric forecast models. In tropical cyclones, the strongest winds occur within a few tens of kilometers of the storm center and so are poorly resolved by routine weather prediction models. Their evolution is also rapid and much more difficult to predict. Consequently prediction and mitigation of the effects of storm surges is further advanced for mid-latitude storms than for tropical cyclones.

Tropical cyclones derive energy from the warm surface waters of the ocean and develop only where the sea surface temperature (SST) exceeds 26.5°C. Since their generation is dependent on the effect of the local vertical component of the Earth's rotation, they do not develop within 5° of the equator. **Figure 4** shows the main cyclone tracks. Areas affected include: the continental shelf surrounding the Gulf of Mexico and on the east coast of the US (by hurricanes); much of east Asia including Vietnam, China, the Philippines and Japan (by typhoons); the Bay of Bengal, in particular its shallow north-east corner, and northern coasts of Australia (by tropical cyclones). Areas affected by mid-latitude storms include the North Sea, the Adriatic, and the Patagonian Shelf. Inland seas and large lakes, including the Great Lakes, Lake Okeechobee (Florida), and Lake Biwa (Japan) also experience surges.

The greatest loss of life due to storm surges has occurred in the northern Bay of Bengal and Meghna Estuary of Bangladesh. A wide and shallow continental shelf bounded by extensive areas of low-lying poorly protected land is impacted by tropical cyclones. Cyclone-generated storm surges on 12–13 November 1970 and 29–30 April 1991 (**Figure 5**) killed approximately 250 000 and 140 000 people, respectively, in Bangladesh.

A severe storm in the North Sea on 31 January–1 February 1953 generated a large storm surge, which coincided with a spring tide to cause catastrophic floods in the Netherlands (**Figure 6**) and south-east England, killing approximately 2000 people. Subsequent government enquiries resulted in the 'Delta Plan' to improve coastal defences in Holland, led to the setting up of coastal flood warning authorities, and accelerated research into storm surge dynamics.

The city of Venice, in Italy, suffers frequent 'acqua alta' which flood the city, disrupting its life and accelerating the disintegration of the unique historic buildings.

In 1969 Hurricane Camille created a surge in the Gulf of Mexico which rose to 7 m above mean sea level, causing more than 100 deaths and about 1

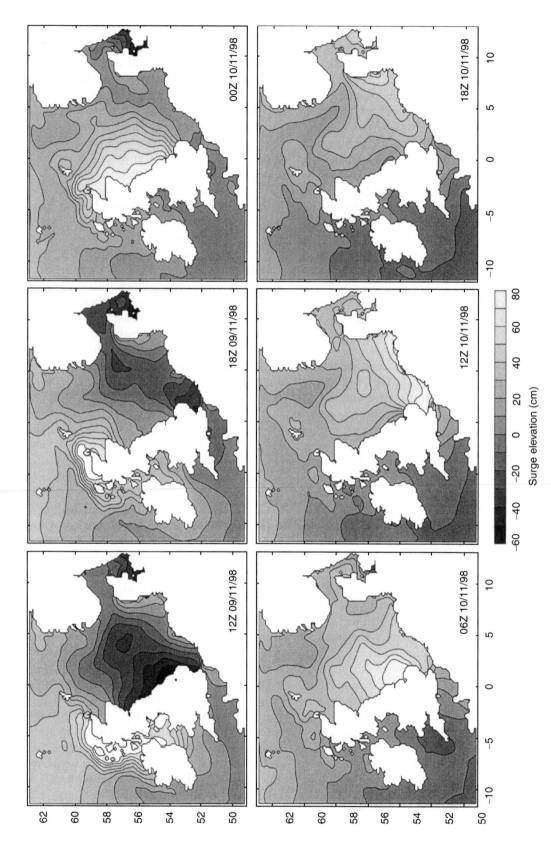

Figure 2 Propagation of an external surge in the North Sea from a numerical model simulation.

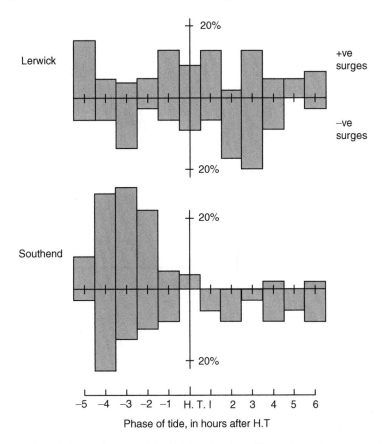

Figure 3 Frequency distribution relative to the time of tidal high water of positive and negative surges at Lerwick (northern North Sea) and Southend (Thames Estuary). The phase distribution at Lerwick is random, whereas due to tide–surge interaction most surge peaks at Southend occur on the rising tide (re-plotted from Prandle D and Wolf J, 1978, The interaction of surge and tide in the North Sea and River Thames. *Geophys. J.R. Astr. Soc.*, 55: 203–216, by permission of the Royal Astronomical Society).

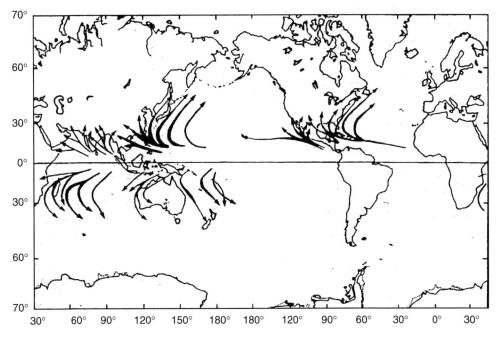

Figure 4 Tropical cyclone tracks (from Murty, 1984, reproduced by permission of the Department of Fisheries and Oceans, Canada).

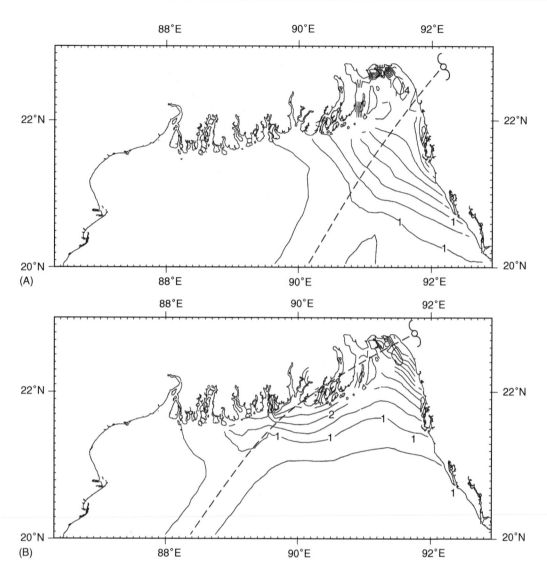

Figure 5 Cyclone tracks (dashed lines) and maximum computed surge elevation measured in meters in the northern Bay of Bengal during the cyclones of (A) 1970 and (B) 1991 from numerical model simulations (contour interval 0.5 m). (Re-plotted from Flather, 1994, by permission of the American Meteorological Society.)

billion dollars worth of damage. An earlier cyclone in the region, in 1900, flooded the island of Galveston, Texas, with the loss of 6000 lives.

Storm Surge Prediction

Early research on storm surges was based on analysis of observations and solution of simplified – usually linearized – equations for surges in idealized channels, rectangular gulfs, and basins with uniform depth. The first self-recording tide gauge was installed in 1832 at Sheerness in the Thames Estuary, England, so datasets for analysis were available from an early stage. Interest was stimulated by events such as the 1953 storm surge in the North Sea, which highlighted the need for forecasts.

First prediction methods were based on empirical formulae derived by correlating storm surge elevation with atmospheric pressure, wind speed and direction and, where appropriate, observed storm surges from a location 'upstream'. Long time-series of observations are required to establish reliable correlations. Where such observations existed, e.g. in the North Sea, the methods were quite successful.

From the 1960s, developments in computing and numerical techniques made it possible to simulate and predict storm surges by solving discrete approximations to the governing equations (eqns [1] and [2]). The earliest and simplest methods, pioneered in Europe by W. Hansen and N.S. Heaps and in the USA by R.O. Reid and C.P. Jelesnianski, used a time-stepping approach based on finite difference approximations

Figure 6 Breached dyke in the Netherlands after the 1953 North Sea storm surge. (Reproduced by permission of RIKZ, Ministry of Public Works, The Netherlands.)

on a regular grid. Surge–tide interaction could be accounted for by solving the non-linear equations and including tide. Effects of inundation could also be included by allowing for moving boundaries; water levels computed with a fixed coast can be O(10%) higher than those with flooding of the land allowed.

Recent developments have revolutionized surge modeling and prediction. Among these, coordinate transformations, curvilinear coordinates and grid nesting allow better fitting of coastal boundaries and enhanced resolution in critical areas. A simple example is the use of polar coordinates in the SLOSH (Sea, Lake and Overland Surges from Hurricanes) model focusing on vulnerable sections of the US east coast. Finite element methods with even greater flexibility in resolution (e.g. **Figure 7**) have also been used in surge computations in recent years.

There has also been increasing use of three-dimensional (3-D) models in storm surge studies. Their main advantage is that they provide information on the vertical structure of currents and, in particular, allow the bottom stress to be related to flow near the seabed. This means that in a 3-D formulation the bottom stress need not oppose the direction of the depth mean flow and hence of the water transport. Higher surge estimates result in some cases.

In the last decades many countries have established and now operate model-based flood warning systems. Although finite element methods and 3-D models have been developed and are used extensively for research, most operational models are still based on depth-averaged finite-difference formulations.

A key requirement for accurate surge forecasts is accurate specification of the surface wind stress. Surface wind and pressure fields from numerical weather prediction (NWP) models are generally used for mid-latitude storms. Even here, resolution of small atmospheric features can be important, so preferably NWP data at a resolution comparable with that of the surge model should be used. For tropical cyclones, the position of maximum winds at landfall is critical, but prediction of track and evolution (change in intensity, etc.) is problematic. Presently, simple models are often used based on basic parameters: p_c, the central pressure; W_m, the maximum sustained 10 m wind speed; R, the radius to maximum winds; and the velocity, V, of movement of the cyclone's eye. Assuming a pressure profile, e.g. that due to G.J. Holland:

$$p_a(r) = p_c + \Delta p \exp\left[-(R/r)^B\right] \qquad [8]$$

where r is the radial distance from the cyclone center, Δp the pressure deficit (difference between the ambient and central pressures), and B is a 'peakedness' factor typically $1.0 < B < 2.5$. Wind fields can then be estimated using further assumptions and approximations. First, the gradient or cyclostrophic wind can be calculated as a function of r. An empirical factor (~ 0.8) reduces this to W, the 10 m wind. A contribution from the motion of the storm (maybe 50% of V) can be added, introducing asymmetry to the wind field, and finally to account for frictional effects in the atmospheric boundary layer, wind vectors may be turned inwards by a cross-isobar angle of $10°$–$25°$. Such procedures are rather crude, so that cyclone surges computed using the resulting winds are unlikely to be very accurate. Simple vertically integrated models of the atmospheric boundary layer have been used to compute winds from a pressure

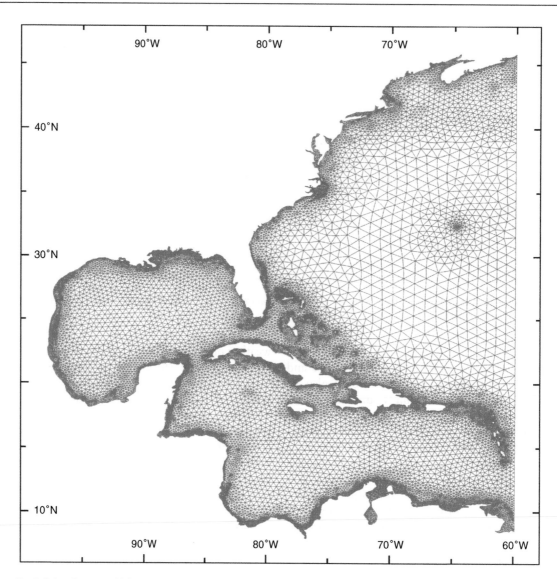

Figure 7 A finite element grid for storm surge calculations on the east coast of the USA, the Gulf of Mexico, and Caribbean (from Blain CA, Westerink JJ and Leuttich RL (1994), The influence of domain size on the response characteristics of a hurricane storm surge model. *J. Geophys. Res.*, 99(C9) 18467–18479. Reproduced by permission of the American Geophysical Union.).

distribution such as eqn [8], providing a more consistent approach. In reality, cyclones interact with the ocean. They generate wind waves, which modify the sea surface roughness, z_0, and hence the wind stress generating the surge. Wind- and wave-generated turbulence mixes the surface water changing its temperature and so modifies the flux of heat from which the cyclone derives its energy. Progress requires improved understanding of air–sea exchanges at extreme wind speeds and high resolution coupled atmosphere–ocean models.

Interactions with Wind Waves

As mentioned above, observations suggested that Charnock's α was not constant but depended on

water depth and 'wave age', a measure of the state of development of waves. Young waves are steeper and propagate more slowly, relative to the wind speed, than fully developed waves and so are aerodynamically rougher enhancing the surface stress. These effects can be incorporated in a drag coefficient which is a function of wave age, wave height, and water depth and agrees well with published datasets over the whole range of wave ages. Further research, considering the effects of waves on airflow in the atmospheric boundary layer, led P.A.E.M. Janssen to propose a wave-induced stress enhancing the effective roughness. Application of this theory requires the dynamical coupling of surge and wave models such that friction velocity and roughness determine and are determined by the waves. Mastenbroek *et al.* obtained improved agreement with observed surges on the Dutch coast by including the wave-induced stress in a model experiment (**Figure 8**). However, they also found that the same improvement could be obtained by a small increase in the standard drag coefficient.

In shallow water, wave orbital velocities also reach the seabed. The bottom stress acting on surge and tide is therefore affected by turbulence introduced at the seabed in the wave boundary layer. With simplifying assumptions, models describing these effects have been developed and can be used in storm surge modeling. Experiments using both 2-D and 3-D surge models have been carried out. 3-D modeling of surges in the Irish Sea using representative waves shows significant effects on surge peaks and improved agreement with observations. Bed stresses are much enhanced in shallow water. Because the processes depend on the nature of the bed, a more complete treatment should take account of details of bed types.

Non-linear interactions give rise to a wave-induced mean flow and a change in mean water depth (wave set-up and set-down). The former has contributions from a mean momentum density produced by a non-zero mean flow in the surface layer (above the trough level of the waves), and from wave breaking. Set-up and set-down arise from the 'radiation stress', which is defined as the excess momentum flux due to the waves (*see* **Waves on Beaches**). Mastenbroek *et al.* showed that the radiation stress has a relatively small influence on the calculated water levels in the North Sea but cannot be neglected in all cases. It is important where depth-induced changes in the waves, as shoaling or breaking, dominate over propagation and generation, i.e. in coastal areas. The effects should be included in the momentum equations of the surge model.

Although ultimately coupled models with a consistent treatment of exchanges between atmosphere and ocean and at the seabed remain a goal, it appears that with the present state of understanding the benefits may be small compared with other inherent uncertainties. In particular, accurate definition of the wind field itself and details of bed types (rippled or smooth, etc.) are not readily available.

Data Assimilation

Data assimilation plays an increasing role, making optimum use of real-time observations to improve the accuracy of initial data in forecast models. Bode and Hardy (see Further Reading section) reviewed two approaches, involving solution of adjoint equations and Kalman filtering. The Dutch operational system has used Kalman filtering since 1992,

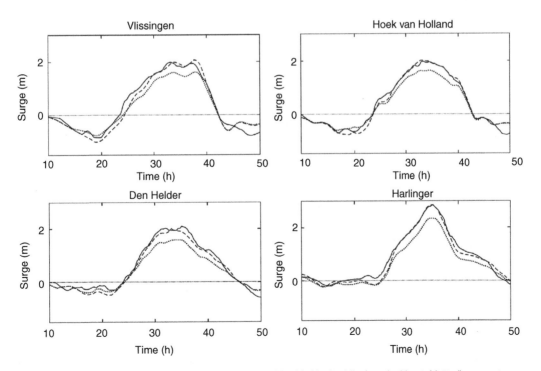

Figure 8 Computed surge elevations during 13–16 February 1989 with (dashed line) and without (dotted) wave stress compared with observations (continuous line). (Re-plotted from Mastenbroek *et al.*, 1994 by permission of the American Geophysical Union.)

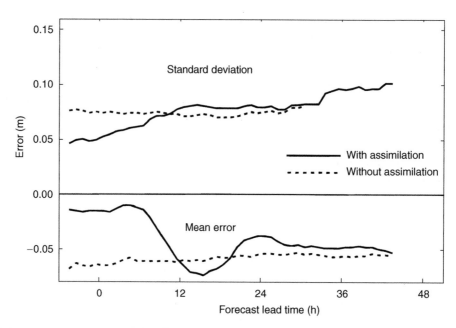

Figure 9 Variation of forecast errors in surge elevation with forecast lead-time from the Dutch operational model with (continuous line) and without (dashed line) assimilation of tide gauge data. (Replotted from Flather, 2000, by permission of Elsevier Science BV.)

incorporating real-time tide gauge data from the east coast of Britain. Accuracy of predictions (**Figure 9**) is improved for the first 10–12 h of the forecast.

Related Issues

Extremes

Statistical analysis of storm surges to derive estimates of extremes is important for the design of coastal defenses and safety of offshore structures. This requires long time- series of surge elevation derived from observations of sea level where available or, increasingly, from model hindcasts covering O(50 years) forced by meteorological analyses.

Climate Change Effects

Climate change will result in a rise in sea level and possible changes in storm tracks, storm intensity and frequency, collectively referred to as 'storminess'. Changes in water depth with rising mean sea level (MSL) will modify the dynamics of tides and surges, increasing wavelengths and modifying the generation, propagation, and dissipation of storm surges. Increased water depth implies a small reduction in the effective wind stress forcing, suggesting smaller surges. However, effects of increased storminess may offset this. It has been suggested, for example, that increased temperatures in some regions could raise

sea surface temperatures resulting in more intense and more frequent tropical cyclones. Regions susceptible to tropical cyclones could also be extended. Research is in progress to assess and quantify some of these effects, e.g. with tide–surge models forced by outputs from climate GCMs. An important issue is that of distinguishing climate-induced change from the natural inter-annual and decadal variability in storminess and hence surge extremes.

Further Reading

Bode L and Hardy TA (1997) Progress and recent developments in storm surge modelling. *Journal of Hydraulic Engineering* 123(4): 315–331.

Flather RA (1994) A storm surge prediction model for the northern Bay of Bengal with application to the cyclone disaster in April 1991. *Journal of Physical Oceanography* 24: 172–190.

Flather RA (2000) Existing operational oceanography. *Coastal Engineering* 41: 13–40.

Heaps NS (1967) Storm surges. In: Barnes H (ed.) *Oceanography and Marine Biology Annual Review 5*, pp. 11–47. London: Allen and Unwin.

Mastenbroek C, Burgers G, and Janssen PAEM (1993) The dynamical coupling of a wave model and a storm surge model through the atmospheric boundary layer. *Journal of Physical Oceanography* 23: 1856–1866.

Murty TS (1984) Storm surges – meteorological ocean tides. *Canadian Bulletin of Fisheries and Aquatic Sciences* 212: 1–897.

Murty TS, Flather RA, and Henry RF (1986) The storm surge problem in the Bay of Bengal. *Progress in Oceanography* 16: 195–233.

Pugh DT (1987) *Tides, Surges, and Mean Sea Level.* Chichester: John Wiley Sons.

World Meteorological Organisation (1978) *Present Techniques of Tropical Storm Surge Prediction.* Report Report 13. Marine science affairs, WMO No. 500. Geneva, Switzerland.

MONSOONS, HISTORY OF

N. Niitsuma, Shizuoka University, Shizuoka, Japan
P. D. Naidu, National Institute of Oceanography, Dona Paula, India

Introduction

The difference in specific heat capacity between continents and oceans (and specifically between the large Asian continent and the Indian Ocean) induces the monsoons, strong seasonal fluctuations in wind direction, and precipitation over oceans and continents. Over the Indian Ocean, strong winds blow from the southwest during boreal summer, whereas weaker winds from the northeast blow during boreal winter. The monsoons are strongest over the western part of the Indian Ocean and the Arabian Sea (**Figure 1**). The high seasonal variability affects various fluctuations in the environment and its biota which are reflected in marine sediments. Marine sedimentary sequences from the continental margins thus contain a record of the history of monsoonal occurrence and intensity, which may be deciphered to obtain insight in the history of the monsoon system. Such insight is important not only to understand the mechanisms that cause monsoons, but also to understand the influence of monsoons on the global climate system including the Walker circulation, the large-scale, west–east circulation over the tropical ocean associated with convection. The summer monsoons may influence the Southern Oscillation because of interactions between the monsoons and the Pacific trade wind systems.

Geologic Records of the Monsoons

Sedimentary sequences record the effects of the monsoons. One such effect is the upwelling of deeper waters to the surface, induced by the strong southwesterly winds in boreal summer in the Arabian Sea, and the associated high productivity of planktonic organisms. Continents supply sediments to the continental margin through the discharge of rivers, as the result of coastal erosion, and carried by winds (eolian sediments). Monsoon-driven wind direction and strength and precipitation therefore control the sediment supply to the continental margin.

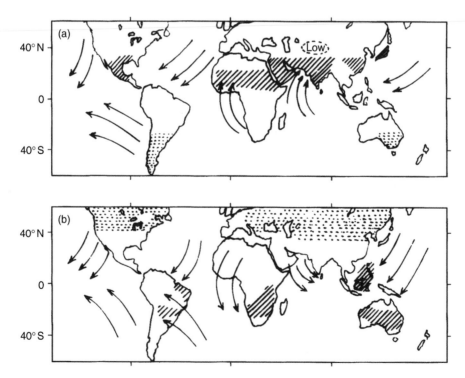

Figure 1 The domains of the monsoon system of the atmosphere during Northern Hemisphere (a) summer and (b) winter. The hatching shows the land areas with maximum surface temperature and stippling indicates the coldest land surface.

Monsoonal precipitation supplies a large volume of fresh water, discharged by rivers from the continents into the oceans, with the flow directed by the topography in the regions of the river mouths. The density of fresh water is less than that of seawater, and the seasonal freshwater flow dilutes surface ocean water which then has a lower density than average seawater. The existence of strong vertical density gradients causes the development of stratification in the water column.

The elevation of the continent governs the circulation of the atmosphere, and influences vegetation patterns. The vegetation cover on the continent directly affects the albedo and heat capacity of the land surface, both of which are important factors in the generation of monsoonal circulation patterns. The high Himalayan mountain range acts as a barrier for air circulation; east–west-oriented mountain ranges particularly affect the course of the jet streams. Continental topography, vegetation coverage, and elevation thus affect the monsoonal circulation and therefore the rate of sediment supply to the continental margin, as well as the composition of continental margin sedimentary sequences.

Sedimentary Indicators of Monsoons

Monsoon-induced seasonal contrasts in wind direction and precipitation are recorded in the sediments by many different proxies. Most of these monsoonal indicators, however, record qualitative and/or quantitative changes of the monsoons in one, but not in both, season. For example, upwelling indicators in the Arabian Sea represent the strength of the southwest (summer) monsoon.

In interpreting the sedimentary record, one must note that bottom-dwelling fauna bioturbates the sediment to depths of *c.* 10 cm, causing the co-occurrence of sedimentary material deposited at different times within a single sediment sample. Laminated sediments without bioturbation are only deposited in oxygen-minimum zones, where anoxia prevents activity of burrowing metazoa. The time resolution of studies of various monsoonal indicators thus is limited by this depth of bioturbation, but each proxy by itself does record information on the environment in which it formed at the time that it was produced. Single-shell measurements of oxygen and carbon isotopes in foraminiferal tests (see below) thus may show the variability within each sample, thus within the zone of sediment mixing.

Oxygen isotopes. The oxygen isotopic ratio (δ^{18}O) of the calcareous tests of fossil organisms (such as foraminifera) provides information on the oxygen isotopic ratio in seawater and on the temperature of formation of the test. The volume of the polar ice sheets and the local influx of fresh water controls the oxygen isotopic ratio of seawater. The δ^{18}O record of the volume of the polar ice sheets can be used for global correlation of oxygen isotopic stages, corresponding to glacial and interglacial intervals. Freshwater discharge caused by monsoonal precipitation lowers the δ^{18}O value of seawater. The temperature record of various species with different depth habits, such as benthic species at the bottom, and planktonic species at various depths below the surface, can provide a temperature profile of the water column. Temperature is the most basic physical parameter, which provides information on the stratification or mixing of the water column, as well as on changes in thermocline depth. Different species of planktonic foraminifera grow in different seasons, and temperatures derived from the oxygen isotopic composition of their tests thus delineates the seasonal monsoonal variation in sea surface temperatures.

Carbon isotopes. The carbon isotopic ratio in the calcareous test of foraminifera provides information on the carbon isotopic ratio of dissolved inorganic carbon (DIC) in seawater and on biofractionation during test formation. Carbon isotopic ratios of various species of foraminifera which live at different depths provide information on the carbon isotopic profile of DIC in the water column. The carbon isotope ratio of DIC in seawater is well correlated with the nutrient concentration in the water, because algae preferentially extract both the lighter carbon isotope (^{12}C) and nutrients to form organic matter by photosynthesis. Decomposition of organic matter releases lighter carbon as well as nutrients, and lowers the carbon isotopic composition of DIC. The carbon isotopic profile with depth thus provides information on the balance of photosynthesis and decomposition.

Rate of sedimentation. Marine sediments are composed of biogenic material produced in the water column (dominantly calcium carbonate and opal), and terrigenous material supplied from the continents. The sediments are transported laterally on the seafloor and down the continental slope, and eventually settle in topographic depressions in the seafloor. Both seafloor topography and the supply of biogenic and lithogenic material control the apparent rate of sedimentation.

Organic carbon content. Organisms produce organic carbon in the euphotic zone, which is strongly recycled by organisms in the upper waters, but a small percentage of the organic material eventually settles on the seafloor. Organisms (including bacteria) decompose the organic carbon on the surface of the seafloor as well as within the sediment, using dissolved oxygen in the process. The flux of organic matter and the availability of oxygen thus control the organic carbon content in the sediments. In high-productivity areas, such as regions where monsoon-induced upwelling occurs, a large amount of organic matter sinks from the sea surface through the water column. The sinking organic matter decomposes and consumes dissolved oxygen, and at mid-water depths an oxygen minimum zone may develop in such high productivity zones. Oxygen concentrations may fall to zero in high-productivity environments, and in these anoxic environments, eukaryotic benthic life becomes impossible, so that there is no bioturbation. The organic carbon content in the sediment deposited below oxygen-minimum zones may become very high, and values of up to 7% organic carbon have been recorded in areas with intense upwelling, such as the Oman Margin.

Calcium carbonate. Three factors control the calcium carbonate content of pelagic sediments: productivity, dilution, and dissolution. Calcium carbonate tests are produced by pelagic organisms, including photosynthesizing calcareous nannoplankton and heterotrophic planktonic foraminifera. At shallow water depths (above the calcite compensation depth, CCD) dissolution is negligible; therefore, the calcium carbonate content in continental margin sediments is regulated by a combination of biotic productivity and dilution by terrigenous sediment.

Magnetic susceptibility. Magnetic susceptibility provides information about the terrigenous material supply and its source. The part of the sediment provided by biotic productivity (calcium carbonate and opaline silica) has no carriers of magnetic material. Magnetic susceptibility is thus used as an indicator of terrigenous supply to the ocean.

Clay mineral composition. Weathering and erosion processes on land lead to the formation of various clay minerals. The clay mineral composition in the sediments is an indicator for the intensity of weathering processes and thus temperature and humidity in the region of origin. The clay mineral composition thus can be used as a proxy for the aridity and vegetation coverage in the continents from which the material derived. In addition, the crystallinity of the clay mineral illite depends on the moisture content of soils in the area of origin. At high moisture, illite decomposes and dehydrates, so that its crystallinity decreases. Therefore, the crystallinity of illite can be used as a proxy for the humidity in its source area.

Eolian dust. Eolian dust consists of fine-grained quartz and clay minerals, such as illite. The content and grain size of eolian dust in marine sediment, therefore, is an indicator of wind strength and direction, as well as of the aridity in the source area.

Fossil abundance and diversity. The abundance and diversity of foraminifera, calcareous nannoplankton, diatoms, and radiolarians depend on the chemical and physical environmental conditions in the oceans. The diversity of nannofossil assemblages decreases with sea surface temperature and is thus generally correlated with latitude. In upwelling areas, low sea surface temperatures caused by the monsoon-driven upwelling disturb the zonal diversity patterns of nannoflora. Therefore, the diversity of nannofossils can be used as an upwelling indicator. In addition, the faunal and floral assemblages can be used to estimate sea surface temperature and salinity as well as productivity.

UK_{37} *ratio.* The biomarker UK_{37} (an alkenone produced by calcareous nannoplankton) is used to estimate sea surface temperatures within the photic zone where the photosynthesizing algae dwell. The records are not always easy to calibrate, especially in the Tropics and at high latitudes, but global calibrations are now available.

Indicators of Present Monsoons

The specific effect of the monsoons on the supply of lithogenic and biogenic material to the seafloor varies in different regions, so that specific tracers cannot be efficiently applied in all oceans.

Arabian Sea

During the boreal summer, strong southwest monsoonal winds produce intense upwelling in the Arabian Sea (**Figure 2**). These upwelling waters are characterized by low temperatures and are highly enriched in nutrients. The process of upwelling fuels the biological productivity in June through August in the Arabian Sea. The weaker, dry, northeast winds which prevail during the boreal winter do not produce upwelling, and productivity is thus lower in the winter months. Thus, the southwest and northeast monsoonal winds produce a strong seasonal contrast in primary productivity in the Arabian Sea. Sediment trap mooring experiments have demonstrated that up to 70% of the biogenic and lithogenic flux to the seafloor occurs during the summer monsoon. Biological and terrestrial particles thus strongly reflect summer conditions, and eventually settle on the seafloor to contribute to a distinct biogeochemical record of monsoonal upwelling. Therefore, regional sediments beneath the areas affected by monsoon-driven upwelling record long-term variations in the

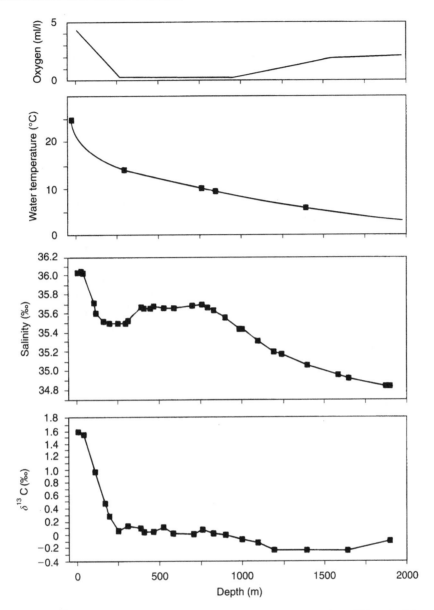

Figure 2 Vertical profiles of $\delta^{13}C$, salinity, temperature, and oxygen values in the Arabian Sea. $\delta^{13}C$ and salinity are from GEOSECS Station 413, while water temperature and oxygen values are from ODP Site 723 in the Oman Margin.

strength and timing of the monsoonal circulation. The following proxies were used to study the upwelling strength in the Arabian Sea.

Globigerina bulloides abundance. Seasonal plankton tows and sedimentary trap data document that the planktonic foraminifer species *G. bulloides* is abundant during the summer upwelling season in the Arabian Sea. Core top data from the upwelling zones of the Arabian Sea show that the dominance of *G. bulloides* in the living assemblage is preserved in the sediment. Changes in the abundance of *G. bulloides* in the sediments thus have been used to infer the history of upwelling intensity in the western Indian Ocean.

Oxygen and carbon isotopes. More recently, it was proposed that oxygen and carbon isotopic difference between the surface and subsurface dwelling planktonic foraminifera can be used as proxy records to reconstruct intensity of upwelling and monsoon.

Lithogenic material. Lithogenic material deposited in the northwest Arabian Sea is dominantly eolian (diameter up to $18.5\,\mu m$), and is transported exclusively during the summer southwest monsoon. The lithogenic grain size in sediment cores thus provides information about the strength of the southwest monsoonal winds and associated upwelling in the Arabian Sea.

China Sea

The surface circulation patterns in the China Sea are also closely associated with the large-scale seasonal reversal of the atmospheric circulation over the Asian continent. High precipitation over Asia during the summer leads to increased input of fresh water into the China Sea, lowering sea surface salinity. Therefore, the $\delta^{18}O$ record of planktonic foraminifera in sediment cores documents the magnitude of fresh-water discharge, sea surface salinity, and summer monsoonal rainfall in the past. During the winter, the westerly winds lower the sea surface temperature and deliver a large amount of eolian dust to the South China Sea. The rate of eolian dust supply and sea surface temperature changes in this region thus reflect the strength of the winter monsoon.

Japan Sea

The winter monsoon's westerlies transport eolian dust from the Asian desert regions to the Japan Sea and the Japanese Islands 3–5 days after sandstorms in the source area. The thickness of the dust layer is larger in the western Japan Sea. The concentration and grain size of eolian particles in sediments of Japan Sea thus represent the strength of the winter monsoon.

Continental Records

Lake levels in the monsoon-influenced regions are highly dependent on the monsoon rainfall; therefore researchers have been using lake levels to trace the monsoon history during the late Quaternary period. The layers of stalagmite (carbonate mineral) preserves the oxygen isotopic composition of monsoon rains that were falling when the stalagmite got precipitated. The oxygen isotope composition of stalagmite is proportional to the amount of rainfall: the more the rainfall, the lighter the $\delta^{18}O$ of stalagmite, and vice versa. Thus, the oxygen isotopic ratios of stalagmites from the caves of monsoon-influenced regions have been used to infer the ultra-high-resolution variability of monsoon precipitation.

Variability of Monsoons during Glacial and Interglacials

Arabian Sea

Along the Oman Margin of the Arabian Sea, strong southwest summer monsoon winds induce upwelling. Detailed analyses of various monsoon tracers such as abundance of *G. bulloides* and *Actinomma* spp. (a radiolarian) and pollen reveal that the southwest monsoon winds were more intense during interglacials (warm periods) and weaker during glacials (cold periods), recognized by the oxygen isotopic stratigraphy in the same samples.

Carbon isotope differences between planktonic and benthic foraminifera show lower gradients during interglacials, and higher gradients during glacials. The lower gradients indicate that upwelling was strong (and pelagic productivity high) during interglacials due to a strong summer southwest monsoon. Similarly, the oxygen isotope difference between planktonic and benthic foraminifera along the Oman Margin reflects changes in thermocline depth, associated with summer monsoon-driven upwelling. A larger oxygen isotope difference between planktonic and benthic foraminifera during interglacials reflects the presence of a shallow thermocline, as a result of the strong summer monsoons. Oxygen and carbon isotope records from various planktonic and benthic foraminifera in several cores located within and away from the axis of the Somali Jet along the Oman Margin indicate that sea surface temperatures were lower and varied randomly during interglacials, reflecting strong upwelling induced by a strong summer monsoon (**Figure 3**).

Studies of the oxygen and carbon isotopic composition of individual tests of planktonic foraminifera enable us to understand the seasonal temperature variability induced by monsoons in the Arabian Sea. Such studies show that the seasonality was stronger during glacials and weaker during interglacials, because during glacials the southwest summer monsoon was weaker and the northeast winter monsoon stronger. The variability in seasonal contrast during glacials and interglacials suggests that interannual and interdecadal changes in monsoonal strength were also greater during glacial periods than during interglacials (**Figure 4**).

High-resolution monsoon records from the Arabian Sea show millennial timescale variability of SW monsoon over last 20 ky. Synchronous changes between SW monsoon intensity and variations of temperatures in North Atlantic and Greenland suggest some kind link between monsoons to the high-latitude temperature.

South China Sea

High-resolution studies in the South China Sea document a high rate of delivery of eolian dust during glacial periods, as well as lower sea surface temperatures, documented by the UK_{37} records. These data indicate that during glacial periods the winter monsoon was more intense. During interglacials the sea surface salinity (derived from oxygen isotope records) was much lower than during

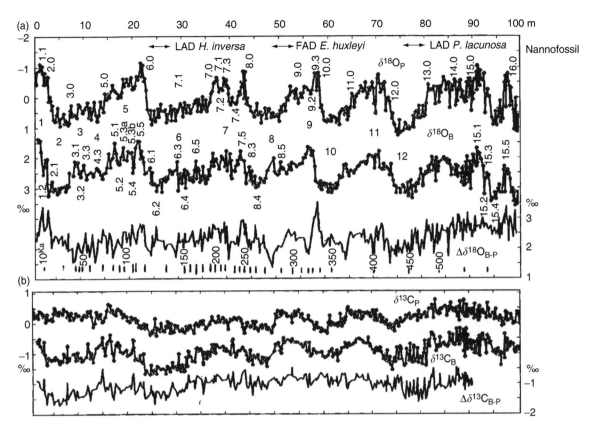

Figure 3 (a) Oxygen isotope profiles of planktonic foraminifera (*Pulleniatina obliquiloculata*, $\delta^{18}O_P$) and benthic foraminifera (*Uvigerina excellens*, $\delta^{18}O_B$), and the difference between oxygen isotopes of planktonic and benthic foraminifera ($\Delta\delta^{18}O_{B-P}$). (b) Carbon isotope profiles of planktonic foraminifera ($\delta^{13}C_P$) and benthic foraminifera ($\delta^{13}C_B$), and the difference between planktonic and benthic records ($\Delta\delta^{13}C_{B-P}$) over last 800 000 years at ODP Site 723 in the Arabian Sea. Large planktonic–benthic differences indicate a more vigorous monsoonal circulation during the summer monsoon.

glacials, probably as a result of increased freshwater discharge from rivers caused by the high summer monsoon precipitation. Over the South China Sea, glacials were thus characterized by an intense winter monsoon, and interglacials by a strong summer monsoon circulation.

Japan Sea

The oceanographic conditions of the Japan Sea are strongly influenced by eustatic sea level changes, because shallow straits connect this sea to the Pacific Ocean. In glacial times, at low sea level, most of the straits were above sea level and the Japan Sea was connected to the ocean only by a narrow channel located on the present continental shelf. River discharge of fresh water into the semi-isolated Japan Sea caused the development of strong stratification, and the development of anoxic conditions, as documented by the occurrence of annually laminated sediments.

The sedimentary sequence in the Japan Sea thus consists of mud, with laminated sections alternating with bioturbated, homogeneous muds. During interglacials the waters at the bottom were oxygenated, although the organic carbon content of the sediment can be high (up to 5%) and diatoms abundant (up to 30% volume). During glacials, conditions on the seafloor alternated between euxinic (anoxic) and noneuxinic. Glacial and interglacial parts of the sediment section can thus be easily recognized.

The sedimentary sequences contain eolian dust transported from the deserts of west China and the Chinese Loess Plateau by the westerly winter monsoon. The eolian dust content of the sediments is thus an indicator of the strength of the winter monsoon. The crystallinity of illite, a main component of the eolian dust, indicates that the source region on the Asian continent was more humid during interglacials. A high content of eolian dust and a high crystallinity of illite during glacial stages indicate that during these intervals the winter monsoon was strong and the summer monsoon weak. Summer monsoons were strong during interglacials, but winter monsoons were strong during glacials.

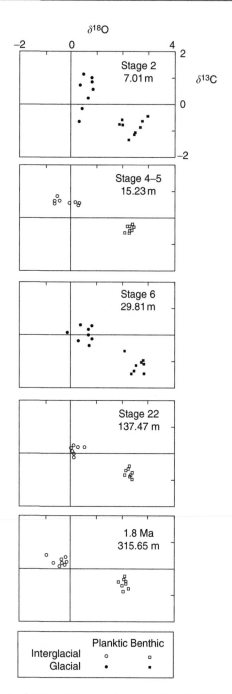

Figure 4 Carbon and oxygen isotope ratios of individual tests of planktonic foraminifera (*Pulleniatina obliquiloculata*) and benthic foraminifera (*Uvigerina excellens*) at ODP Site 723 in the Arabian Sea. The variations in difference between planktonic and benthic values reflect the magnitude of the seasonal changes in surface and bottom waters through time.

Long-Term Evolution of the Asian Monsoon

Arabian Sea

Long-term variations of proxies of monsoonal intensity tracers, especially the species diversity of

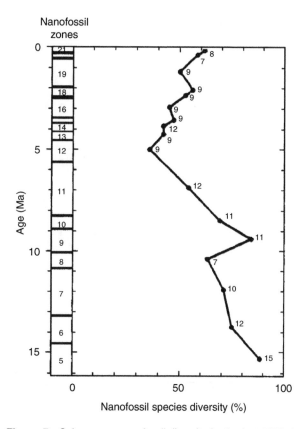

Figure 5 Calcareous nannofossil diversity for the last 15 My in the Indian Ocean. High diversity indicates weak upwelling and low diversity represents strong upwelling. The upwelling intensity is controlled by the strength of the summer monsoon winds over the Indian Ocean. The numbers on the species diversity profile represent number of species.

calcareous nannofossil species, and the abundance of the planktonic foraminifera G. *bulloides* and the radiolarian *Actinomma* spp. (upwelling indicators) show that the evolution of the Asian monsoon started in the late Miocene, at about 9.5 Ma. Between 9.5 and 5 Ma, the monsoon increased noticeably in strength, with smaller fluctuations in monsoonal intensity from 5 to 2 Ma (**Figure 5**).

Upwelling indicators such as the differences in oxygen and carbon isotope between planktonic foraminifera, and the organic carbon content in sediments from the Oman Margin indicate that until about 0.8 Ma the summer monsoon was intense, as during interglacials. Oxygen and carbon isotope ratios in individual tests of planktonic and benthic foraminifera show that from 0.8 Ma onward, the strength of the summer monsoon changed with glacial and interglacial cycles, as described above.

China Sea

In sediments from the South China Sea, magnetic susceptibility and calcium carbonate percentages started to decrease at about 7 Ma, reflecting increased deposition of terrigenous, eolian dust. This increase in dust supply indicates that monsoons started to become noticeable at that time. Both color reflectance and magnetic susceptibility show cyclic fluctuations in monsoonal intensity from 0.8 Ma on, in response to glacial and interglacial cycles (as described above).

Japan Sea

Cyclic changes with large amplitude of magnetic susceptibility and illite crystallinity in sediment cores indicate that the imprint of the monsoon signature in the Japan Sea sediment record started at about 0.8 Ma. The crystallinity of illite was much higher until 0.8 Ma, reflecting that the summer monsoon was weak until that time, as it was afterward during glacials.

Asian Continent

Loess and paleosol sequences on the Chinese Plateau show cyclic fluctuations in magnetic susceptibility and illite crystallinity for the last 0.6 My. Loess and paleosol sequences in the Kathamandu Basin show cyclic fluctuations in magnetic susceptibility and illite crystallinity for the last 1.1 My. The cyclic sequences of low crystallinity, representing a humid climate in the interglacial periods in both areas, is consistent with the marine sediment records of the Japan Sea.

Palearctic elements first became represented in the molluskan faunas in the Siwalik group sediments in the Himalayas after 8 Ma, suggesting the beginning of seasonal migrations of water birds crossing the Himalayas at this time. The diversity of this molluskan fauna increased around 5 Ma, a time when the summer monsoon was strong. The timing of developments in the molluskan faunas in the Himalayas is thus consistent with the long-term evolution of the Asian summer monsoon as derived from marine records.

The Asian Monsoon and the Global Climate System

Uplift of the Himalayas and the Tibetan Plateau occurred coeval with the increase in strength of the Asian monsoon between 9.5 and 5 Ma, as documented by the heavy mineral composition of deposits in the Bengal Fan, derived from the weathering and erosion of the rising Himalayas. Cyclic fluctuations in the strength of the summer monsoon started at about 1.1 Ma in the southern Himalayas and Tibet, whereas in the Arabian Sea, South China Sea, and Japan Sea such changes started at about 0.8 Ma.

Peru Margin

As a result of strong southeasterly trade winds, nutrient-rich water upwells along the Peru coast and reaches the photic zone in a belt that is c. 10 km wide, and parallels the coastline. In this region, upwelling-induced productivity is very high. The organic carbon concentration in the sediments below this high-productivity zone has been used to trace the upwelling strength in the past. Upwelling is absent or less intense during El Niño/Southern Oscillations (ENSO) events, and the Southern Oscillation is linked to the Asian monsoons in the tropical Walker circulation.

Upwelling along the Peru Margin started at around 3.5 Ma, as indicated by an increase in the organic carbon content, and the decrease of sea surface temperatures (derived from UK_{37} records). Upwelling along the Peru Margin thus started after the Asian monsoons reached their full strength at about 5 Ma.

Equatorial Upwelling

The intensity of the southeast Asian monsoon controls the easterly trade winds associated with the north and south equatorial currents, and the strength of the easterly trade winds controls the intensity of equatorial upwelling. In the equatorial Pacific, the intensity of trade winds and equatorial upwelling increased at about 5 Ma (as indicated by a high abundance of siliceous and calcareous pelagic microfossils), at the time that the Asian monsoons developed their full intensity.

Teleconnections

Numerous data sets show synchroneity between changes of monsoon intensity and abrupt climate shifts in Greenland during the deglaciation (between 15 and 16 ky). In addition, monsoon proxy records from the Arabian Sea reveal that the intervals of weak summer monsoon coincide with cold periods in the North Atlantic region and vice versa. All these evidences suggest some kind of teleconnection between monsoon and global climate. However, the exact physical mechanism underlying the link between high-latitude temperature changes and SW monsoon has not been addressed yet.

Cyclicity of Monsoon

Paleomonsoon records show periodicity of SW monsoon at 100, 41, and 23 ky corresponding to Earth's orbital changes of eccentricity, obliquity, and precession, respectively. Subsequently, high-resolution monsoon variability exhibits suborbital periodicities of 2200, 1700, 1500, and 775 years.

Conclusions

The evolution of the Asian monsoon started at around 9.5 Ma, in response to the uplift of the Himalayas. The monsoonal intensity reached its maximum at around 5 Ma, and from that time the associated easterly trade winds caused intense upwelling in the equatorial Pacific. Before 1.1 Ma, the summer monsoon was strong over the Arabian Sea, whereas the winter monsoon was strong over the Japan Sea. The glacial and interglacial cycles in intensity of the monsoons in the Arabian Sea, the South China Sea, and the Japan Sea started around 0.8 Ma, coinciding with the uplift of the Himalayas to their present-day elevation. Therefore, the chronological sequence of monsoonal events and the strength of trade winds and equatorial upwelling suggest that the Asian monsoons (linked to the development of the Himalayan mountains) were an important control on global climate and oceanic productivity.

The Tropics receive by far the most radiative energy from the sun, and the energy received in these regions and the ways in which it is transported to higher latitudes controls the global climate. In the Tropics, the atmospheric circulation over the Asian continent is dominated by the area of highest elevation: the Himalayas and the Tibetan Plateau. The high heat capacity of this region causes the strong seasonality in wind directions, temperature, and rainfall, involving extensive transport of moisture and thus also latent heat from sea to land during summer. The Himalayas–Tibetan Plateau thus influence the transport of sensible and latent heat from low-latitude oceanic areas to mid- and high-latitude land areas. These high mountains act as a mechanical barrier to the air currents, and the north–south contrast across these mountains varied in magnitude with the glacial–interglacial cycles from 0.8 Ma onward, at which time the glacial–interglacial climatic fluctuations reached their largest amplitude. The Asian monsoons thus control the atmospheric heat budget in the Northern Hemisphere, and changes in monsoonal intensity trigger global climate change.

See also

Holocene Climate Variability.

Further Reading

Clemens SE, Prell W, Murray D, Shimmield G, and Weedon G (1991) Forcing mechanisms of the Indian Ocean monsoon. *Nature* 353: 720–725.

Fein JS and Stephenes PL (eds.) (1987) *Monsoons*. Chichester, UK: Wiley.

Kutzbach JE and Guetter PJ (1986) The influence of changing orbital parameters and surface boundary conditions on climate simulations for the past 18,000 years. *Journal of Atmospheric Science* 43: 1726–1759.

Naidu PD and Malmgren BA (1995) A 2,200 years periodicity in the Asian Monsoon system. *Geophysical Research Letters* 22: 2361–2364.

Niitsuma N, Oba T, and Okada M (1991) Oxygen and carbon isotope stratigraphy at site 723, Oman Margin. *Proceedings of Ocean Drilling Program Scientific Results* 117: 321–341.

Prell WL, Murray DW, Clemens SC, and Anderson DM (1992) Evolution and variability of the Indian Ocean summer monsoon: Evidence from western Arabian Sea Drilling Program. *Geophysical Monographs* 70: 447–469.

Prell WL and Niitsuma N, *et al.* (eds.) (1991) *Proceedings of the Ocean Drilling Program Scientific Results, Vol. 117*. College Station, TX: Ocean Drilling Program.

Street FA and Grove AT (1979) Global maps of lake-level fluctuations since 30,000 years BP. *Quaternary Research* 12: 83–118.

Takahashi K and Okada H (1997) Monsoon and quaternary paleoceanography in the Indian Ocean. *Journal of the Geological Society of Japan* 103: 304–312.

Wang L, Sarnthein M, and Erlenkeuser H (1999) East Asian monsoon climate during the late Pleistocene: High-resolution sediment records from the South China Sea. *Marine Geology* 156: 245–284.

Wang P, Prell WL, and Blum MP (2000) Leg 184 summary: Exploring the Asian monsoon through drilling in the South China Sea. *Proceedings of the Ocean Drilling Program, Initial Reports* 184: 1–77.

CLIMATE CHANGE AND IMPACTS

HOLOCENE CLIMATE VARIABILITY

M. Maslin, C. Stickley, and V. Ettwein, University
College London, London, UK

Introduction

Until a few decades ago it was generally thought that
significant large-scale global and regional climate
changes occurred at a gradual pace within a timescale of many centuries or millennia. Climate change
was assumed to be scarcely perceptible during a
human lifetime. The tendency for climate to change
abruptly has been one of the most surprising outcomes of the study of Earth history. In particular,
paleoceanographic records demonstrate that our
present interglacial, the Holocene (the last ~10 000
years), has not been as climatically stable as first
thought. It has been suggested that Holocene climate
is dominated by millennial-scale variability, with
some authorities suggesting that this is a 1500 year
cyclicity. These pronounced Holocene climate changes can occur extremely rapidly, within a few centuries or even within a few decades, and involve
regional-scale changes in mean annual temperature
of several degrees Celsius. In addition, many of these
Holocene climate changes are stepwise in nature and
may be due to thresholds in the climate system.

Holocene decadal-scale transitions would presumably have been quite noticeable to ancient civilizations. For instance, the emergence of crop
agriculture in the Middle East corresponds very
closely with a sudden warming event marking the
beginning of the Holocene, and the widespread collapse of the first urban civilizations, such as the Old
Kingdom in Egypt and the Akkadian Empire, coincided with a cooling event at around 4300 BP. In
addition, paleo-records from the late Holocene
demonstrate the possible influence of climate change
on the collapse of the Mayan civilization (Classic
Period), while Andean ice core records suggests that
alternating wet and dry periods influenced the rise
and fall of coastal and highland cultures of Ecuador
and Peru.

It would be foolhardy not to bear in mind such
sudden stepwise climate transitions when considering the effects that humans might have upon the
present climate system, via the rapid generation of
greenhouse gases for instance. Judging by what we

have already learnt from Holocene records, it is not
improbable that the system may gradually build up
over hundreds of years to a 'breaking point' or
threshold, after which some dramatic change in the
system occurs over just a decade or two. At the
threshold point, the climate system is in a delicate
and somewhat critical state. It may take only a
relatively minor 'adjustment' to trigger the transition
and tip the system into abrupt change.

This article summarizes the current paleoceanographic records of Holocene climate variability and
the current theories for their causes. Concentrating on records that cover a significant portion of the
Holocene. The discussion is limited to centennial–
millennial-scale variations. **Figure 1** illustrates the
Holocene and its climate variability in context of the
major global climatic changes that have occurred
during the last 2.5 million years. Short-term variations such as the North Atlantic Oscillation and the
El Niño-Southern Oscillation will not be discussed.

The Importance of the Oceans and Holocene Paleoceanography

Climate is created from the effects of differential
latitudinal solar heating. Energy is constantly transferred from the equator (relatively hot) toward the
poles (relatively cold). There are two transporters of
such energy – the atmosphere and the oceans. The
atmosphere responds to an internal or external
change in a matter of days, months, or may be a few
years. The oceans, however, have a longer response
time. The surface ocean can change over months to a
few years, but the deep ocean takes decades to centuries. From a physical point of view, in terms of
volume, heat capacity and inertia, the deep ocean is
the only viable contender for driving and sustaining
long-term climate change on centennial to millennial
timescales.

Since the process of oceanic heat transfer largely
regulates climate change on longer time-scales and
historic records are too short to provide any record
of the ocean system prior to human intervention, we
turn to marine sediment archives to provide information about ocean-driven climate change. Such
archives can often provide a continuous record on a
variety of timescales. They are the primary means for
the study and reconstruction of the stability and
natural variability of the ocean system prior to anthropogenic influences.

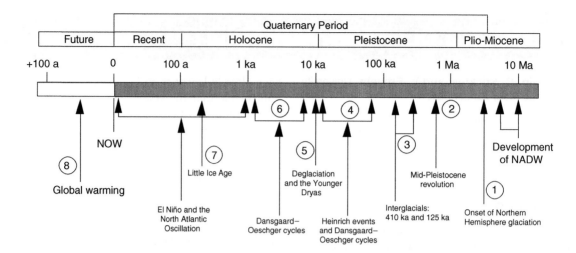

Figure 1 Log time scale cartoon, illustrating the most important climate events in the Quaternary Period. (a, ka, Ma, refer to years ago.) (1) Onset of Northern Hemisphere Glaciation (3.2–2.5 Ma), ushering in the strong glacial–interglacial cycles which are characteristic of the Quaternary Period. (2) Mid-Pleistocene Revolution when the dominant periodicity of glacial–interglacial cycles switched from 41 000 y, to every 100 000 y. The external forcing of the climate did not change; thus, the internal climate feedback's must have altered. (3) The two closest analogues to the present climate are the interglacial periods at 420 000 to 390 000 years ago (oxygen isotope stage 11) and 130 000 to 115 000 years ago (oxygen isotope stage 5e, also known as the Eemian). (4) Heinrich events and Dansgaard–Oeschger cycles (see text). (5) Deglaciation and the Younger Dryas events. (6) Holocene Dansgaard–Oeschger cycles (see text). (7) Little Ice Age (AD 1700), the most recent climate event that seems to have occurred throughout the Northern Hemisphere. (8) Anthropogenic global warming.

One advantage of marine sediments is that they can provide long, continuous records of Holocene climate at annual (sometimes intraannual) to centennial time-resolutions. However, there is commonly a trade-off between temporal and spatial resolution. Deep-ocean sediments usually represent a large spatial area, but sedimentation rates in the deep-ocean are on average between 0.002 and 0.005 cm y^{-1}, with very productive areas producing a maximum of 0.02 cm y^{-1}. This limits the temporal resolution to a maximum of 200 years per cm (50 y cm^{-1} for productive areas). Mixing by the process of bioturbation will reduce the resolution further.

On continental shelves and in bays and other specialized sediment traps such as anoxic basins and fiords, sedimentation rates can exceed 1 cm y^{-1} providing temporal resolution of over 1 y cm^{-1}. More local conditions are recorded in laminated marine sediments formed in anoxic environments, where biological activity can not disturb the sediments. For example, Pike and Kemp (1997) analysed annual and intraannual variability within the Gulf of California from laminated sediments containing a record of diatom-mat accumulation. Time series analysis highlighted a decadal-scale variability in mat-deposition associated with Pacific-wide changes in surface water circulation, suggested to be influenced by solar-cycles. In addition, anoxic sediments from the Mediterranean Ridge (ODP Site 971) reveal

seasonal-scale variability during the late Quaternary from a laminated diatom-ooze sapropel. Pearce *et al.*, (1998) inferred changes in the monsoon-related nutrient input to the Mediterranean Basin via the Nile River as the main cause of the variations in the laminated sediments, which suggests a wide influence of changes in seasonality. Other potentially extremely high-resolution studies will come from Saanich Inlet, a Canadian fiord and Prydz Bay in Antarctica, sites recently drilled by the Ocean Drilling Program.

However, the main drawback to such high-resolution locations is that they contain highly localized environmental and climate information. An additional problem associated mainly with continental margins is reworking, erosion, and redistribution of the sediment by mass density flows such as turbidities and slumps. Hence we concentrate on wider-scale records of Holocene climate change.

Holocene Climatic Variability

Initial studies of the Greenland ice core records concluded the absence of major climate variation within the Holocene. This view is being progressively eroded, particular in the light of new information being obtained from marine sediments (**Figure 2**). Long-term trends indicate an early to mid-Holocene climatic optimum with a cooling trend in the late

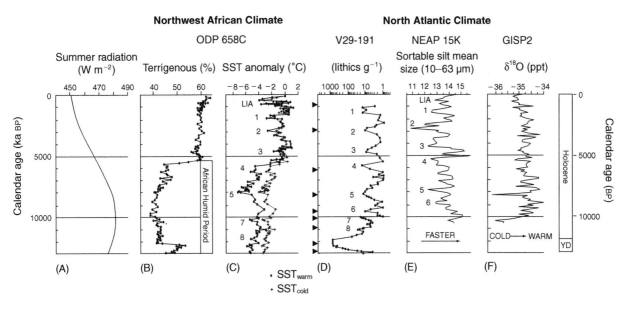

Figure 2 Comparison of summer insolation for 65°N with north-west African climate (deMenocal *et al.*, 2000) and North Atlantic climate (V29-191, Bond *et al.*, 1997; NEAP 15 K, Bianchi and McCave, 1999; GISP2, O'Brien *et al.*, 1996). Note the similarity of events labeled 1 to 8 and the Little Ice Age (LIA).

Holocene. Superimposed on this trend are several distinct oscillations or climatic cooling steps that appear to be of widespread significance (see **Figure 2**), the most dramatic of which occurred 8200, 5500, and 4400 years ago and between AD 1200 and AD 1650.

The event 8200 years ago is the most striking and abrupt, leading to widespread cool and dry conditions lasting perhaps 200 years, before a rapid return to climates warmer and generally moister than at present. This event is noticeably present in the GISP2 Greenland ice cores, from which it appears to have been about half as severe as the Younger Dryas to Holocene transition. Marine records of North African to Southern Asian climate suggest more arid conditions involving a failure of the summer monsoon rains. Cold and/or arid conditions also seem to have occurred in northernmost South America, eastern North America and parts of north-west Europe.

In the middle Holocene approximately 5500–5300 years ago there was a sudden and widespread shift in precipitation, causing many regions to become either noticeably drier or moister. The dust and sea surface temperature records off north-west Africa show that the African Humid Period, when much of subtropical West Africa was vegetated, lasted from 14 800 to 5500 years ago and was followed by a 300 year transition to much drier conditions (de Menocal *et al.*, 2000). This shift also corresponds to the decline of the elm (*Ulmus*) in Europe about 5700, and of hemlock (*Tsuga*) in North America about 5300

years ago. Both vegetation changes were initially attributed to specific pathogen attacks, but it is now thought they may have been related to climate deterioration. The step to colder and drier conditions in the middle of an interglacial period is analogous to a similar change that is observed in records of the last interglacial period referred to as Marine Oxygen Isotope Stage 5e (Eemian).

There is also evidence for a strong cold and arid event occurring about 4400 years ago across the North Atlantic, northern Africa, and southern Asia. This cold, and arid event coincides with the collapse of a large number of major urban civilizations, including the Old Kingdom in Egypt, the Akkadian Empire in Mesopotamia, the Early Bronze Age societies of Anatolia, Greece, Israel, the Indus Valley civilization in India, the Hilmand civilization in Afganistan, and the Hongshan culture of China.

Little Ice Age (LIA)

The most recent Holocene cold event is the Little Ice Age (see **Figures 2** and **3**). This event really consists of two cold periods, the first of which followed the Medieval Warm Period (MWP) that ended ~1000 years ago. This first cold period is often referred to as the Medieval Cold Period (MCP) or LIAb. The MCP played a role in extinguishing Norse colonies on Greenland and caused famine and mass migration in Europe. It started gradually before AD 1200 and ended at about AD 1650. This second cold period,

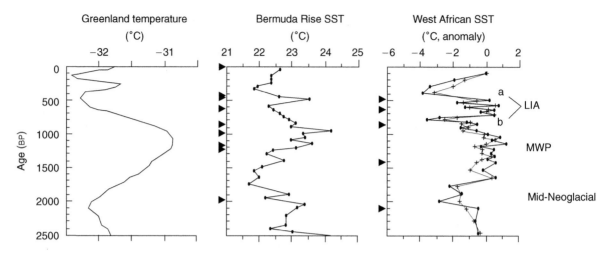

Figure 3 Comparison of Greenland temperatures, the Bermuda Rise sea surface temperatures (SST) (Keigwin, 1996), and west African and a sea surface temperature (deMenocal *et al.*, 2000) for the last 2500 years. LIALittle Ice Age; MWPMedieval Warm Period. Solid triangles indicate radiocarbon dates.

may have been the most rapid and the largest change in the North Atlantic during the Holocene, as suggested from ice-core and deep-sea sediment records. The Little Ice Age events are characterized by a drop in temperature of 0.5–1°C in Greenland and a sea surface temperature falls of 4°C off the coast of west Africa and 2°C off the Bermuda Rise (see **Figure 3**).

Holocene Dansgaard-Oeschger cycles

The above events are now regarded as part of the millennial-scale quasiperiodic climate changes characteristic of the Holocene (see **Figure 2**) and are thought to be similar to glacial Dansgaard–Oeschger (D/O) cycles. The periodicity of these Holocene D/O cycles is a subject of much debate. Initial analysis of the GISP2 Greenland ice core and North Atlantic sediment records revealed cycles at approximately the same 1500 ($\pm$500)-year rhythm as that found within the last glacial period. Subsequent analyses have also found a strong 1000-year cycle and a 550-year cycle. These shorter cycles have also been recorded in the residual $\delta^{14}C$ data derived from dendrochronologically calibrated bidecadal tree-ring measurements spanning the last 11 500 years. In general, during the coldest point of each of the millennial-scale cycles shown in **Figure 2**, surface water temperatures of the North Atlantic were about 2–4°C cooler than during the warmest part.

One cautionary note is that Wunsch has suggested a more radical explanation for the pervasive 1500-year cycle seen in both deep-sea and ice core, glacial and interglacial records. Wunsch suggests that the

extremely narrow spectral lines (less than two bandwidths) that have been found at about 1500 years in many paleo-records may be due to aliasing. The 1500-year peak appears precisely at the period predicted for a simple alias of the seasonal cycle sampled inadequately (under the Nyquist criterion) at integer multiples of the common year. When Wunsch removes this peak from the Greenland ice core data and deep-sea spectral records, the climate variability appears as expected to be a continuum process in the millennial band. This work suggests that finding a cyclicity of 1500 years in a dataset may not represent the true periodicity of the millennial-scale events. The Holocene Dansgaard–Oeschger events are quasi periodic, with different and possibly stochastic influences.

Causes of Millennial Climate Fluctuation during the Holocene

As we have already suggested, deep water circulation plays a key role in the regulation of global climate. In the North Atlantic, the north-east-trending Gulf Stream carries warm and relatively salty surface water from the Gulf of Mexico up to the Nordic seas. Upon reaching this region, the surface water has cooled sufficiently that it becomes dense enough to sink, forming the North Atlantic Deep Water (NADW). The 'pull' exerted by this dense sinking maintains the strength of the warm Gulf Stream, ensuring a current of warm tropical water into the North Atlantic that sends mild air masses across to the European continent. Formation of the NADW can be weakened by two processes. (1) The presence of huge ice sheets over North America and Europe changes the position of the atmospheric polar front, preventing the Gulf Stream from traveling so far north. This reduces the amount of cooling and the capacity of the surface water to sink. Such a reduction of formation occurred during the last glacial period. (2) The input of fresh water forms a lens of less-dense water, preventing sinking. If NADW formation is reduced, the weakening of the warm Gulf Stream causes colder conditions within the entire North Atlantic region and has a major impact on global climate. Bianchi and McCave, using deep-sea sediments from the North Atlantic, have shown that during the Holocene there have been regular reductions in the intensity of NADW (**Figure 2E**), which they link to the 1500-year D/O cycles identified by O'Brien and by Bond (1997). There are two possible causes for the millennial-scale changes observed in the intensity of the NADW: (1) instability in the North Atlantic region caused by varying

freshwater input into the surface waters; and (2) the 'bipolar seesaw'.

There are a number of possible reasons for the instability in the North Atlantic region caused by varying fresh water input into the surface waters:

● Internal instability of the Greenland ice sheet, causing increased meltwater in the Nordic Seas that reduces deep water formation.
● Cyclic changes in sea ice formation forced by solar variations.
● Increased precipitation in the Nordic Seas due to more northerly penetration of North Atlantic storm tracks.
● Changes in surface currents, allowing a larger import of fresher water from the Pacific, possibly due to reduction in sea ice in the Arctic Ocean.

The other possible cause for the millennial-scale changes is an extension of the suggested glacial intrinsic millennial-scale 'bipolar seesaw' to the

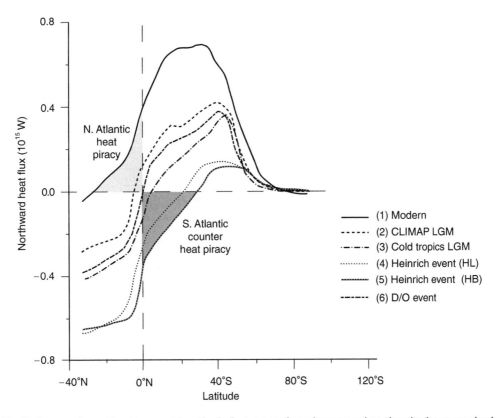

Figure 4 Atlantic Ocean poleward heat transport (positive indicates a northward movement) as given by the ocean circulation model (Seidov and Maslin, 1999) for the following scenarios: (1) present-day (warm interglacial) climate; (2) last glacial maximum (LGM) with generic CLIMAP data; (3) 'Cold tropics' LGM scenario; (4) a Heinrich-type event driven by the meltwater delivered by icebergs from decaying Laurentide ice sheet; (5) a Heinrich-type event driven by meltwater delivered by icebergs from decaying Barents Shelf ice sheet or Scandinavian ice sheet; (6) a general Holocene or glacial Dansgaard–Oeschger (D/O) meltwater confined to the Nordic Seas. Note that the total meridional heat transport can only be correctly mathematically computed in the cases of cyclic boundary conditions (as in Drake Passage for the global ocean) or between meridional boundaries, as in the Atlantic Ocean to the north of the tip of Africa. Therefore the northward heat transport in the Atlantic ocean is shown to the north of 30°S only.

Holocene. One of the most important finds in the study of glacial millennial-scale events is the apparent out-of-phase climate response of the two hemispheres seen in the ice core climate records from Greenland and Antarctica. It has been suggested that this bipolar seesaw can be explained by variations in the relative amount of deep water formation in the two hemispheres and heat piracy (**Figure 4**). This mechanism of altering dominance of the NADW and the Antarctic Bottom Water (AABW) can also be applied to the Holocene. The important difference with this theory is that the trigger for a sudden 'switching off' or a strong decrease in rate of deep water formation could occur either in the North Atlantic or in the Southern Ocean. AABW forms in a different way than NADW, in two general areas around the Antarctic continent: (1) near-shore at the shelf–ice, sea-ice interface and (2) in open ocean areas. In near-shore areas, coastal polynya are formed where katabatic winds push sea ice away from the shelf edge, creating further opportunity for sea ice formation. As ice forms, the surface water becomes saltier (owing to salt rejection by the ice) and colder (owing to loss of heat via latent heat of freezing). This density instability causes sinking of surface waters to form AABW, the coldest and

saltiest water in the world. AABW can also form in open-ocean Antarctic waters; particularly in the Weddell and Ross seas; AABW flows around Antarctica and penetrates the North Atlantic, flowing under the less dense NADW. It also flows into the Indian and Pacific Oceans, but the most significant gateway to deep ocean flow is in the south-west Pacific, where 40% of the world's deep water enters the Pacific. Interestingly, Seidov and colleagues have shown that the Southern Ocean is twice as sensitive to meltwater input as is the North Atlantic, and that the Southern Ocean can no longer be seen as a passive player in global climate change. The bipolar seesaw model may also be self sustaining, with meltwater events in either hemisphere, triggering a train of climate changes that causes a meltwater event in the opposite hemisphere, thus switching the direction of heat piracy (**Figure 5**).

Conclusion

The Holocene, or the last 10 000 years, was once thought to be climatically stable. Recent evidence, including that from marine sediments, have altered this view, showing that there are millennial-scale

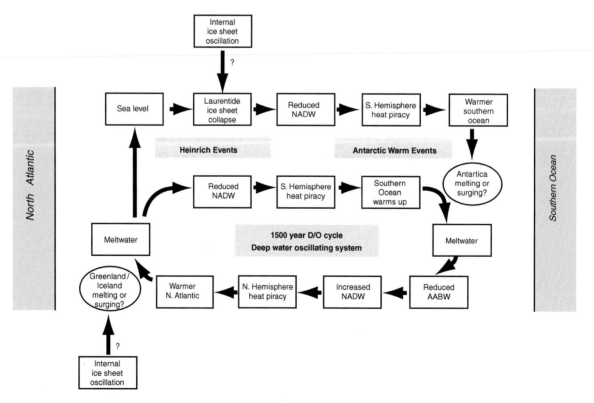

Figure 5 Possible deep water oscillatory system explaining the glacial and interglacial Dansgaard–Oeschger cycles. Additional loop demonstrates the possible link between interglacial Dansgaard–Oeschger cycles and Heinrich events.

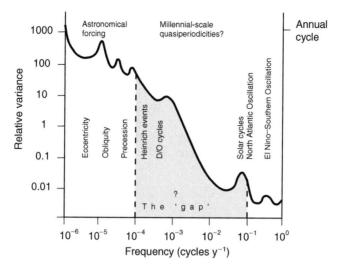

Figure 6 Spectrum of climate variance showing the climatic cycles for which we have good understanding and the 'gap' between hundreds and thousands of years for which we still do not have adequate understanding of the causes.

climate cycles throughout the Holocene. In fact we are still in a period of recovery from the last of these cycles, the Little Ice Age. It is still widely debated whether these cycles are quasiperiodic or have a regular cyclicity of 1500 years. It is also still widely debated whether these Holocene Dansgaard–Oeschger cycles are similar in time and characteristic to those observed during the last glacial period. A number of different theories have been put forward for the causes of these Holocene climate cycles, most suggesting variations in the deep water circulation system. One suggestion is that these cycles are caused by the oscillating relative dominance of North Atlantic Deep Water and Antarctic Bottom Water. Holocene climate variability still has no adequate explanation and falls in the 'gap' of our knowledge between Milankovitch forcing of ice ages and rapid variations such as El Niño and the North Atlantic Oscillation (**Figure 6**). Future research is essential for understanding these climate cycles so that we can better predict the climate response to anthropogenic 'global warming.'

Further Reading

Adams J, Maslin MA, and Thomas E (1999) Sudden climate transitions during the Quaternary. *Progress in Physical Geography* 23(1): 1–36.

Alley RB and Clark PU (1999) The deglaciation of the Northern hemisphere: A global perspective. *Annual Review of Earth and Planetary Science* 27: 149–182.

Bianchi GG and McCave IN (1999) Holocene periodicity in North Atlantic climate and deep-ocean flow south of Iceland. *Nature* 397: 515–523.

Broecker W (1998) Paleocean circulation during the last deglaciation: a bipolar seesaw? *Paleoceanography* 13: 119–121.

Bond G, Showers W, Cheseby M, *et al.* (1997) A pervasive millenial-scale cycle in North Atlantic Holocene and glacial climates. *Science* 278: 1257–1265.

Chapman MR and Shackleton NJ (2000) Evidence of 550 year and 1000 years cyclicities in North Atlantic pattern during the Holocene. *Holocene* 10: 287–291.

Cullen HM, *et al.* (2000) Climate change and the collapse of the Akkadian Empire: evidence from the deep sea. *Geology* 28: 379–382.

Dansgaard W, Johnson SJ, and Clausen HB (1993) Evidence for general instability of past climate from a 250-kyr ice-core record. *Nature* 364: 218–220.

deMenocal P, Ortiz J, Guilderson T, and Sarnthein M (2000) Coherent high- and low-latitude climate variability during the Holocene warm period. *Science* 288: 2198–2202.

Keigwin LD (1996) The Little Ice Age and Medieval warm period in the Sargasso sea. *Science* 274: 1504–1507.

Maslin MA, Seidov D, and Lowe J (2001) *Synthesis of the nature and causes of sudden climate transitions during the Quaternary. AGU Monograph: Oceans and Rapid Past and Future Climate Changes: North–South Connections.* Washington DC: American Geophysical. Union No. 119.

O'Brien SR, Mayewski A, and Meeker LD (1996) Complexity of Holocene climate as reconstructed from a Greenland ice core. *Science* 270: 1962–1964.

Pearce RB, Kemp AES, Koizumi I, Pike J, Cramp A, and Rowland SJ (1998) A lamina-scale, SEM-based study of a late quaternary diatom-ooze sapropel from the Mediterranean ridge, Site 971. In: Robertson AHF, Emeis K-C, Richter C, and Camerlenghi A (eds.) *Proceedings of the Ocean Drilling Program*, Scientific Results 160, 349–363.

Peiser BJ (1998) Comparative analysis of late Holocene environmental and social upheaval: evidence for a disaster around 4000 BP. In: Peiser BJ, Palmer T, and Bailey M (eds.) *Natural Catastrophes during Bronze Age Civilisations*, 117–139.

Pike J and Kemp AES (1997) Early Holocene decadal-scale ocean variability recorded in Gulf of California laminated sediments. *Paleoceanography* 12: 227–238.

Seidov D and Maslin M (1999) North Atlantic Deep Water circulation collapse during the Heinrich events. *Geology* 27: 23–26.

Seidov D, Barron E, Haupt BJ, and Maslin MA (2001) *Meltwater and the ocean conveyor: past, present and future of the ocean bi-polar seesaw. AGU Monograph: Oceans and Rapid Past and Future Climate Changes: North–South Connections*. Washington DC: American Geophysical. Union No. 119..

Wunsch C (2000) On sharp spectral lines in the climate record and millennial peak. *Paleoceanography* 15: 417–424.

CARBON DIOXIDE (CO_2) CYCLE

T. Takahashi, Lamont Doherty Earth Observatory, Columbia University, Palisades, NY, USA

Introduction

The oceans, the terrestrial biosphere, and the atmosphere are the three major dynamic reservoirs for carbon on the earth. Through the exchange of CO_2 between them, the atmospheric concentration of CO_2 that affects the heat balance of the earth, and hence the climate, is regulated. Since carbon is one of the fundamental constituents of living matter, how it cycles through these natural reservoirs has been one of the fundamental questions in environmental sciences. The oceans contain about 50 times as much carbon (about 40 000 Pg-C or10^{15} g as carbon) as the atmosphere (about 750Pg-C). The terrestrial biosphere contains about three times as much carbon (610 Pg-C in living vegetation and 1580 Pg-C in soil organic matter) as the atmosphere. The air–sea exchange of CO_2 occurs via gas exchange processes across the sea surface; the natural air-to-sea and sea-to-air fluxes have been estimated to be about 90 Pg-C y^{-1} each. The unperturbed uptake flux of CO_2 by global terrestrial photosynthesis is roughly balanced with the release flux by respiration, and both have been estimated to be about 60 Pg-C y^{-1}. Accordingly, atmospheric CO_2 is cycled through the ocean and terrestrial biosphere with a time scale of about 7 years.

The lithosphere contains a huge amount of carbon (about 100 000 000 Pg-C) in the form of limestones ((Ca, Mg) CO_3), coal, petroleum, and other forms of organic matter, and exchanges carbon slowly with the other carbon reservoirs via such natural processes as chemical weathering and burial of carbonate and organic carbon. The rate of removal of atmospheric CO_2 by chemical weathering has been estimated to be of the order of 1 Pg-Cy^{-1}. Since the industrial revolution in the nineteenth century, the combustion of fossil fuels and the manufacturing of cement have transferred the lithospheric carbon into the atmosphere at rates comparable to the natural CO_2 exchange fluxes between the major carbon reservoirs, and thus have perturbed the natural balance significantly (6 Pg-Cy^{-1} is about an order of magnitude less than the natural exchanges with the oceans (90 Pg-C y^{-1} and land (60 Pg-C y^{-1})). The industrial carbon emission rate has been about 6 Pg-C y^{-1} for the 1990s, and the cumulative industrial emissions since the nineteenth century to the end of the twentieth century have been estimated to be about 250 Pg-C. Presently, the atmospheric CO_2 content is increasing at a rate ofabout 3.5 Pg-C y^{-1} (equivalent to about 50% of the annual emission) and the remainder of the CO_2 emitted into the atmosphere is absorbed by the oceans and terrestrial biosphere in approximately equal proportions. These industrial CO_2 emissions have caused the atmospheric CO_2 concentration to increase by as much as 30% from about 280 ppm (parts per million mole fraction in dry air) in the pre-industrial year 1850 to about 362 ppm in the year 2000. The atmospheric CO_2 concentration may reach 580 ppm, double the pre-industrial value, by the mid-twenty first century. This represents a significant change that is wholly attributable to human activities on the Earth.

It is well known that the oceans play an important role in regulating our living environment by providing water vapor into the atmosphere and transporting heat from the tropics to high latitude areas. In addition to these physical influences, the oceans partially ameliorate the potential CO_2-induced climate changes by absorbing industrial CO_2 in the atmosphere.

Therefore, it is important to understand how the oceans take up CO_2 from the atmosphere and how they store CO_2 in circulating ocean water. Furthermore, in order to predict the future course of the atmospheric CO_2 changes, we need to understand how the capacity of the ocean carbon reservoir might be changed in response to the Earth's climate changes, that may, in turn, alter the circulation of ocean water. Since the capacity of the ocean carbon reservoir is governed by complex interactions of physical, biological, and chemical processes, it is presently not possible to identify and predict reliably various climate feedback mechanisms that affect the ocean CO_2 storage capacity.

Units

In scientific and technical literature, the amount of carbon has often been expressed in three different units: giga tons of carbon (Gt-C), petagrams of carbon (Pg-C) and moles of carbon or CO_2. Their relationships are: 1Gt-C = 1 Pg-C = 1×10^{15} g of carbon = 1000 million metric tonnes of carbon = $(1/12) \times 10^{15}$ moles of carbon. The equivalent quantity as CO_2 may be obtained by multiplying the above numbers by 3.67 (= 44/12 = the molecular weight of CO_2 divided by the atomic weight of carbon).

The magnitude of CO_2 disequilibrium between the atmosphere and ocean water is expressed by the difference between the partial pressure of CO_2 of ocean water, $(pCO_2)sw$, and that in the overlying air, $(pCO_2)air$. This difference represents the thermodynamic driving potential for CO_2 gas transfer across the sea surface. The pCO_2 in the air may be estimated using the concentration of CO_2 in air, that is commonly expressed in terms of ppm (parts per million) in mole fraction of CO_2 in dry air, in the relationship:

$$p(CO_2)air = (CO_2 \text{ conc.})air \times (Pb - pH_2O) \quad [1]$$

where Pb is the barometric pressure and pH_2O is the vapor pressure of water at the sea water temperature. The partial pressure of CO_2 in sea water, $(pCO_2)sw$, may be measured by equilibration methods or computed using thermodynamic relationships. The unit of microatmospheres (μatm) or 10^{-6} atm is commonly used in the oceanographic literature.

History

The air–sea exchange of CO_2 was first investigated in the 1910s through the 1930s by a group of scientists including K. Buch, H. Wattenberg, and G.E.R. Deacon. Buch and his collaborators determined in land-based laboratories CO_2 solubility, the dissociation constants for carbonic and boric acids in sea water, and their dependence on temperature and chlorinity (the chloride ion concentration in sea water). Based upon these dissociation constants along with the shipboard measurements of pH and titration alkalinity, they computed the partial pressure of CO_2 in surface ocean waters. The Atlantic Ocean was investigated from the Arctic to Antarctic regions during the period 1917–1935, especially during the METEOR Expedition 1925–27, in the North and South Atlantic. They discovered that temperate and cold oceans had lower pCO_2 than air (hence the sea water was a sink for atmospheric CO_2), especially during spring and summer seasons, due to the assimilation of CO_2 by plants. They also observed that the upwelling areas of deep water (such as African coastal areas) had greater pCO_2 than the air (hence the sea water was a CO_2 source) due to the presence of respired CO_2 in deep waters.

With the advent of the high-precision infrared CO_2 gas analyzer, a new method for shipboard measurements of pCO_2 in sea water and in air was introduced during the International Geophysical Year, 1956–59. The precision of measurements was improved by more than an order of magnitude. The global oceans were investigated by this new method, which rapidly yielded high precision data. The

equatorial Pacific was identified as a major CO_2 source area. The GEOSECS Program of the International Decade of Ocean Exploration, 1970–80, produced a global data set that began to show systematic patterns for the distribution of CO_2 sink and source areas over the global oceans.

Methods

The net flux of CO_2 across these a surface, Fs-a, may be estimated by:

$$\begin{aligned}
\text{Fs-a} &= E \times [(pCO_2)sw - (pCO_2)air] \\
&= k \times \alpha \times [(pCO_2)sw - (pCO_2)air]
\end{aligned} \quad [2]$$

where E is the CO_2 gas transfer coefficient expressed commonly in (moles $CO_2/m^2/y/uatm$); k is the gas transfer piston velocity (e.g. in (cmh^{-1})) and α is the solubility of CO_2 in sea water at a given temperature and salinity (e.g. (moles $CO_2\,kg\text{-}sw^{-1}\,atm^{-1}$)). If $(pCO_2)sw < (pCO_2)$ air, the net flux of CO_2 is from the sea to the air and the ocean is a source of CO_2; if $(pCO_2)sw < (pCO_2)$ air, the ocean water is a sink for atmospheric CO_2. The sea–air pCO_2 difference may be measured at sea and α has been determined experimentally as a function of temperature and salinity. However, the values of E and k that depend on the magnitude of turbulence near the air–water interface cannot be simply characterized over complex ocean surface conditions. Nevertheless, these two variables have been commonly parameterized in terms of wind speed over the ocean. A number of experiments have been performed to determine the wind speed dependence under various wind tunnel conditions as well as ocean and lake environments using different nonreactive tracer gases such as SF_6 and ^{222}Rn. However, the published results differ by as much as 50% over the wind speed range of oceanographic interests.

Since ^{14}C is in the form of CO_2 in the atmosphere and enters into the surface ocean water as CO_2 in a timescale of decades, its partition between the atmosphere and the oceans yields a reliable estimate for the mean CO_2 gas transfer rate over the global oceans. This yields a CO_2 gas exchange rate of $20 \pm 3\ mol\ CO_2\ m^{-2}\ y^{-1}$ that corresponds to a sea–air CO_2 transfer coefficient of $0.067\ mol\ CO_2\ m^{-2}\ y^{-1}\ uatm^{-1}$. Wanninkhof in 1992 presented an expression that satisfies the mean global CO_2 transfer coefficient based on ^{14}C and takes other field and wind tunnel results into consideration. His equation for variable wind speed conditions is:

$$k\left(cm\ h^{-1}\right) = 0.39 \times (u_{av})^2 \times (Sc/660)^{-0.5} \quad [3]$$

where u_{av} is the average wind speed in ms^{-1} corrected to 10 m above sea surface; Sc(dimensionless) is the Schmidt number (kinematic viscosity ofwater)/ (diffusion coefficient of CO_2 gas inwater); and 660 represents the Schmidt number for CO_2 in seawater at 20°C.

In view of the difficulties in determining gas transfer coefficients accurately, direct methods for CO_2 flux measurements aboard the ship are desirable. Sea–air CO_2 flux was measured directly by means of the shipboard eddy-covariance method over the North Atlantic Ocean by Wanninkhof and McGillis in 1999. The net flux of CO_2 across the sea surface was determined by a covariance analysis of the tri-axial motion of air with CO_2 concentrations in the moving air measured in short time intervals ($\sim$ms) as a ship moved over the ocean. The results obtained over a wind speed range of 2–13.5 m s^{-1} are consistent with eqn [3] within about $\pm 20\%$. If the data obtained in wind speeds up to 15 m s^{-1} are taken into consideration, they indicate that the gas transfer piston velocity tends to increase as a cube of wind speed. However, because of a large scatter ($\pm 35\%$) of the flux values at high wind speeds, further work is needed to confirm the cubic dependence.

In addition to the uncertainties in the gas transfer coefficient (or piston velocity), the CO_2 flux estimated with eqn [2] is subject to errors in $(pCO_2)sw$ caused by the difference between the bulk water temperature and the temperature of the thin skin of ocean water at the sea–air interface. Ordinarily the $(pCO_2)sw$ is obtained at the bulk seawater temperature, whereas the relevant value for the flux calculation is (pCO_2) sw at the 'skin' temperature, that depends on the rate of evaporation, the incoming solar radiation, the wind speed, and the degree of turbulence near the interface. The 'skin' temperature is often cooler than the bulk water temperatureby as much as 0.5°C if the water evaporates rapidly to a dry air mass, but is not always so if a warm humid air mass covers over the ocean. Presently, the time–space distribution of the 'skin' temperature is not well known. This, therefore, could introduce errors in $(pCO_2)sw$ up to about 6 µatm or 2%.

CO₂ Sink/Source Areas of the Global Ocean

The oceanic sink and source areas for atmospheric CO_2 and the magnitude of the sea–air CO_2 flux over the global ocean vary seasonally and annually as well as geographically. These changes are the manifestation of changes in the partial pressure of sea water,

$(pCO_2)sw$, which are caused primarily by changes in the water temperature, in the biological utilization of CO_2, and in the lateral/vertical circulation of ocean waters including the upwelling of deep water rich in CO_2. Over the global oceans, sea water temperatures change from the pole to the equator by about 32°C. Since the pCO_2 in sea water doubles with each 16°C of warming, temperature changes should cause a factor of 4 change in pCO_2. Biological utilization of CO_2 over the global oceans is about 200 µmol $CO_2 kg^{-1}$, which should reduce pCO_2 in sea water by a factor of 3. If this is accompanied with growths of $CaCO_3$-secreting organisms, the reduction of pCO_2 could be somewhat smaller. While these effects are similar in magnitude, they tend to counteract each other seasonally, since the biological utilization tends to be large when waters are warm. In subpolar and polar areas, winter cooling of surface waters induces deep convective mixing that brings high pCO_2 deep waters to the surface. The lowering effect on CO_2 by winter cooling is often compensated for or some times over compensated for by the increasing effect of the upwelling of high CO_2 deep waters. Thus, in high latitude oceans, surface waters may become a source for atmospheric CO_2 during the winter time when the water is coldest.

In **Figure 1**, the global distribution map of the sea–air pCO_2 differences for February and August 1995, are shown. These maps were constructed on the basis of about a half million pairs of atmospheric and seawater pCO_2 measurements made at sea over the 40-year period, 1958–98, by many investigators. Since the measurements were made in different years, during which the atmospheric pCO_2 was increasing, they were corrected to a single reference year (arbitrarily chosen to be 1995) on the basis of the following observations. Warm surface waters in subtropical gyres communicate slowly with the underlying subsurface waters due to the presence of a strong stratification at the base of the mixed layer. This allows a long time for the surface mixed-layer-waters ($\sim$75 m thick) to exchange CO_2 with the atmosphere. Therefore, their CO_2 chemistry tends to follow the atmospheric CO_2 increase. Accordingly, the pCO_2 in the warm water follows the increasing trend of atmospheric CO_2, and the sea–air pCO_2 difference tends to be independent of the year of measurements. On the other hand, since surface waters in high latitude regions are replaced partially with subsurface waters by deep convection during the winter, the effect of increased atmospheric CO_2 is diluted to undetectable levels and their CO_2 properties tend to remain unchanged from year to year. Accordingly, the sea–air pCO_2 difference measured in a given year increases as the atmospheric CO_2

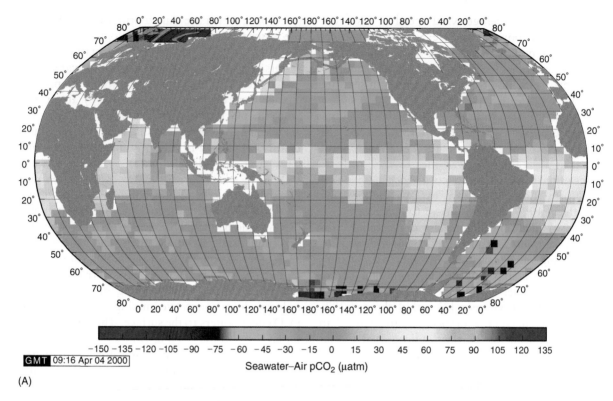

(A)

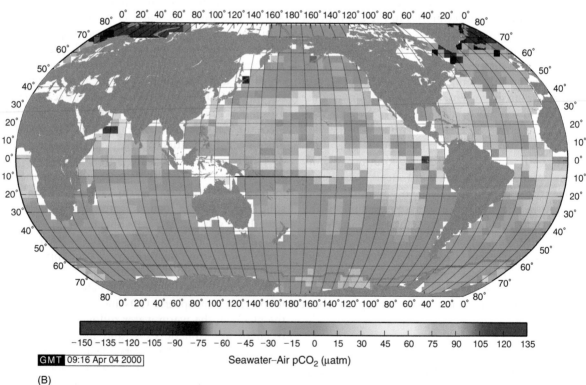

(B)

Figure 1 The sea–air pCO₂ difference in μatm (ΔpCO₂) for (A) February and (B) August for the reference year 1995. The purple-blue areas indicate that the ocean is a sink for atmospheric CO₂, and the red-yellow areas indicate that the ocean is source. The pink lines in the polar regions indicate the edges of ice fields.

concentration increases with time. This effect was corrected to the reference year using the observed increase in the atmospheric CO_2 concentration. During El Niño periods, sea–air pCO_2 differences over the equatorial belt of the Pacific Ocean, which are large in normal years, are reduced significantly and observations are scarce. Therefore, observations made between 10°N and 10°S in the equatorial Pacific for these periods were excluded from the maps. Accordingly, these maps represent the climatological means for non-El Niño period oceans for the past 40 years. The purple-blue areas indicate that the ocean is a sink for atmospheric CO_2, and the red-yellow areas indicate that the ocean is a source.

Strong CO_2 sinks (blue and purple areas) are present during the winter months in the Northern (**Figure 1A**) and Southern (**Figure 1B**) Hemispheres along the poleward edges of the subtropical gyres, where major warm currents are located. The Gulf Stream in the North Atlantic and the Kuroshio Current in the North Pacific are both major CO_2 sinks (**Figure 1A**) due primarily to cooling as they flow from warm tropical oceans to subpolar zones.

Similarly, in the Southern Hemisphere, CO_2 sink areas are formed by the cooling of poleward-flowing currents such as the Brazil Current located along eastern South America, the Agulhus Current located south of South Africa, and the East Australian Current located along south-eastern Australia. These warm water currents meet with cold currents flowing equator ward from the Antarctic zone along the northern border of the Southern (or Antarctic) Ocean. As the sub Antarctic waters rich in nutrients flow northward to more sunlit regions, CO_2 is drawn down by photosynthesis, thus creating strong CO_2 sink conditions, as exemplified by the Falkland Current in the western South Atlantic (**Figure 1A**). Confluence of subtropical waters with polar waters forms broad and strong CO_2 sink zones as a result of the juxta position of the lowering effects on pCO_2 of the cooling of warm waters and the photosynthetic drawdown of CO_2 in nutrient-rich subpolar waters. This feature is clearly depicted in a zone between 40°S and 60°S in **Figure 1A** representing the austral summer, and between 20°S and 40°S in **Figure 1B** representing the austral winter.

During the summer months, the high latitude areas of the North Atlantic Ocean (**Figure 1A**) and the Weddell and Ross Seas, Antarctica (**Figure 1B**), are intense sink areas for CO_2. This is attributed to the intense biological utilization of CO_2 within the strongly stratified surface layer caused by solar warming and ice melting during the summer. The winter convective mixing of deep waters rich in CO_2 and nutrient seliminates the strong CO_2 sink and

replenishes the depleted nutrients in the surface waters.

The Pacific equatorial belt is a strong CO_2 source which is caused by the warming of upwelled deep waters along the coast of South America as well as by the upward entrainment of the equatorial under current water. The source strengths are most intense in the eastern equatorial Pacific due to the strong upwelling, and decrease to the west as a result of the biological utilization of CO_2 and nutrients during the westward flow of the surface water.

Small but strong source areas in the north-western subArctic Pacific Ocean are due to the winter convective mixing of deep waters (**Figure 1A**). The lowering effect on pCO_2 of cooling in the winter is surpassed by the increasing effect of high CO_2 concentration in the upwelled deep waters. During the summer (**Figure 1B**), however, these source areas become a sink for atmospheric CO_2 due to the intense biological utilization that overwhelms the increasing effect on pCO_2 of warming. A similar area is found in the Arabian Sea, where upwelling of deepwaters is induced by the south-west monsoon during July–August (**Figure 1B**), causing the area to become a strong CO_2 source. This source area is eliminated by the photosynthetic utilization of CO_2 following the end of the upwelling period (**Figure 1A**).

As illustrated in **Figure 1A** and **B**, the distribution of oceanic sink and source areas for atmospheric CO_2 varies over a wide range in space and time. Surface ocean waters are out of equilibrium with respect to atmospheric CO_2 by as much as $\pm 200\,\mu atm$ (or by $\pm 60\%$). The large magnitudes of CO_2 disequilibrium between the sea and the air is in contrast with the behavior of oxygen, another biologically mediated gas, that shows only up to $\pm 10\%$ sea–air disequilibrium. The large CO_2 disequilibrium may be attributed to the fact that the internal ocean processes that control pCO_2 in sea water, such as the temperature of water, the photosynthesis, and the upwelling of deep waters, occur at much faster rates than the sea–air CO_2 transfer rates. The slow rate of CO_2 transfer across the sea surface is due to the slow hydration rates of CO_2 as well as to the large solubility of CO_2 in sea water attributable to the formation of bicarbonate and carbonate ions. The latter effect does not exist at all for oxygen.

Net CO$_2$ Flux Across the Sea Surface

The net sea–air CO_2 flux over the global oceans may be computed using eqns [2] and [3]. **Figure 2** shows the climatological mean distribution of the annual sea–air CO_2 flux for the reference year 1995 using the following set of information. (1) The monthly mean ΔpCO_2 values in $4° \times 5°$ pixel areas for the reference year 1995 (**Figure 1A** and **B** for all other months); (2) the Wanninkhof formulation, eqn [3], for the effect of wind speed on the CO_2 gas transfer coefficient; and (3) the climatological mean wind speeds for each month compiled by Esbensen and Kushnir in 1981. This set yields a mean global gas transfer rate of $0.063 \, mole \, CO_2 \, m^{-2} \, \mu atm^{-1} \, y^{-1}$, that is consistent with $20 \, moles \, CO_2 \, m^{-2} \, y^{-1}$ estimated on the basis of carbon-14 distribution in the atmosphere and the oceans.

Figure 2 shows that the equatorial Pacific is a strong CO_2 source. On the other hand, the areas along the poleward edges of the temperate gyres in both hemispheres are strong sinks for atmospheric CO_2. This feature is particularly prominent in the southern Indian and Atlantic Oceans between 40°S and 60°S, and is attributable to the combined effects of negative sea–air pCO_2 differences with strong winds ('the roaring 40 s') that accelerate sea–air gas transfer rates. Similarly strong sink zones are formed in the North Pacific and North Atlantic between 45°N and 60°N. In the high latitude Atlantic, strong

sink areas extend into the Norwegian and Greenland Seas. Over the high latitude Southern Ocean areas, the sea–air gas transfer is impeded by the field of ice that covers the sea surface for ≥ 6 months in a year.

The net sea–air CO_2 fluxes computed for each ocean basin for the reference year of 1995, representing non-El Niño conditions, are summarized in **Table 1**. The annual net CO_2 uptake by the global ocean is estimated to be about 2.0 Pg-C y^{-1}. This is consistent with estimates obtained on the basis of a number of different ocean–atmosphere models including multi-box diffusion advection models and three-dimensional general circulation models.

The uptake flux for the Northern Hemisphere ocean (north of 14°N) is 1.2 Pg-C y^{-1}, whereas that for the Southern Hemisphere ocean (south of 14°S) is 1.7 Pg-C y^{-1}. Thus, the Southern Hemisphere ocean is a stronger CO_2 sink by about 0.5 Pg-C y^{-1}. This is due partially to the much greater oceanic areas in the Southern Hemisphere. In addition, the Southern Ocean south of 50°S is an efficient CO_2 sink, for it takes up about 26% of the global ocean CO_2 uptake, while it has only 10% of the global ocean area. Cold temperature and moderate photosynthesis are both

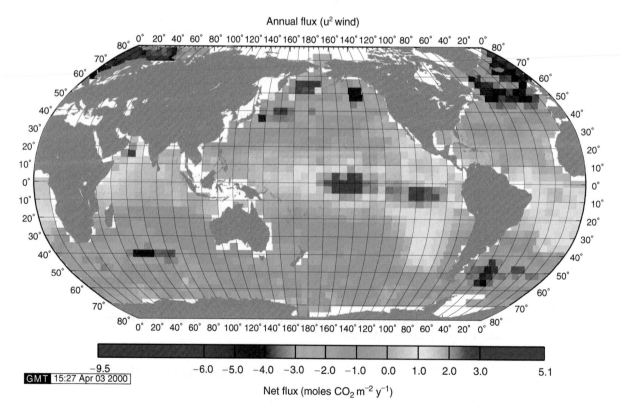

Figure 2 The mean annual sea–air flux of CO_2 in the reference year 1995. The red-yellow areas indicate that the flux is from sea to air, whereas blue-purple areas indicate that the flux is from air to sea. The flux is given in moles of $CO_2 \, m^{-2} y^{-1}$. The map gives a total annual air-to-sea flux of 2.0 Pg-C y^{-1}.

Table 1 The net sea–air flux of CO_2 estimated for a reference year of 1995 using the effect of wind speed on the CO_2 gas transfer coefficient, eqn [3], of Wanninkhof and the monthly wind field of Esbensen and Kushnir

Latitudes	Pacific Ocean	Atlantic Ocean	Indian Ocean	Southern Ocean	Global Oceans
	Sea–air flux in 10^{15}g Carbon y^{-1}				
North of 50°N	− 0.02	− 0.44	—	—	− 0.47
50°N–14°N	− 0.47	− 0.27	+ 0.03	—	− 0.73
14°N–14°S	+ 0.64	+ 0.13	+ 0.09	—	+ 0.86
14°S–50°S	− 0.37	− 0.20	− 0.60	—	− 1.17
South of 50°S	—	—	—	− 0.52	− 0.52
Total	− 0.23	− 0.78	− 0.47	− 0.52	− 2.00
%Uptake	11%	39%	24%	26%	100%
Area (10^6 km^2)	151.6	72.7	53.2	31.7	309.1
Area (%)	49.0%	23.5%	17.2%	10.2%	100%

Positive values indicate sea-to-air fluxes, and negative values indicate air-to-sea fluxes.

responsible for the large uptake by the Southern Ocean.

The Atlantic Ocean is the largest net sink for atmospheric CO_2 (39%); the Southern Ocean (26%) and the Indian Ocean (24%) are next; and the Pacific Ocean (11%) is the smallest. The intense biological drawdown of CO_2 in the high latitude areas of the North Atlantic and Arctic seas during the summer months is responsible for the Atlantic being a major sink. This is also due to the fact that the upwelling deep waters in the North Atlantic contain low CO_2 concentrations, which are in turn caused primarily by the short residence time (~ 80y) of the North Atlantic Deep Waters. The small uptake flux of the Pacific can be attributed to the fact that the combined sink flux of the northern and southern subtropical gyres is roughly balanced by the source flux from the equatorial Pacific during non-El Niño periods. On the other hand, the equatorial Pacific CO_2 source flux is significantly reduced or eliminated during El Niño events. As a result the equatorial zone is covered with the eastward spreading of the warm, low pCO_2 western Pacific waters in response to the relaxation of the trade wind. Although the effects of El Niño and Southern Ocean Oscillation may be far reaching beyond the equatorial zone as far as to the polar areas, the El Niño effects on the equatorial Pacific alone could reduce the equatorial CO_2 source. Hence, this could increase the global ocean uptake flux by up to 0.6 Pg-C y^{-1} during an El Niño year.

The sea–air CO_2 flux estimated above is subject to three sources of error: (1) biases in sea–air ΔpCO_2 values interpolated from relatively sparse observations, (2) the 'skin' temperature effect, and (3) uncertainties in the gas transfer coefficient estimated on the basis of the wind speed dependence. Possible biases in ΔpCO_2 differences have been tested using sea surface temperatures (SST) as a proxy. The systematic error in the global sea–air CO_2 flux resulting from sampling and interpolation has been estimated to be about $\pm 30\%$ or ± 0.6 Pg-C y^{-1}. The 'skin' temperature of ocean water may affect ΔpCO_2 by as much as ± 6 μatm depending upon time and place, as discussed earlier.

Although the distribution of the 'skin' temperature over the global ocean is not known, it may be cooler than the bulk water temperature by a few tenths of a degree on the global average. This may result in an under estimation of the ocean uptake by 0.4 Pg-C y^{-1}. The estimated global sea–air flux depends on the wind speed data used. Since the gas transfer rate increases nonlinearly with wind speed, the estimated CO_2 fluxes tend to be smaller when mean monthly wind speeds are used instead of high frequency wind data.

Furthermore, the wind speed dependence on the CO_2 gas transfer coefficient in high wind speed regimes is still questionable. If the gas transfer rate is taken to be a cubic function of wind speed instead of the square dependence as shown above, the global ocean uptake would be increased by about 1 Pg-C y^{-1}. The effect is particularly significant over the high latitude oceans where the winds are strong. Considering various uncertainties discussed above, the global ocean CO_2 uptake presented in **Table 1** is uncertain by about 1 Pg-C y^{-1}.

Further Reading

Broecker WS and Peng TH (1982) *Tracers in the Sea.* Palisades, NY: Eldigio Press.

Broecker WS, Ledwell JR, Takahashi, *et al.* (1986) Isotopic versus micrometeorologic ocean CO_2 fluxes a: serious conflict. *Journal of Geophysical Research* 91: 10517–10527.

Keeling R, Piper SC, and Heinmann M (1996) Global and hemispheric CO_2 sinks deduced from changes in atmospheric O_2 concentration. *Nature* 381: 218–221.

Sarmiento JL, Murnane R, and Le Quere C (1995) Air–sea CO_2 transfer and the carbon budget of the North Atlantic. *Philosophical Transactions of the Royal Society of London, series B* 343: 211–219.

Sundquist ET (1985) Geological perspectives on carbon dioxide and carbon cycle. In: Sundquist ET and Broecker WS (eds.) *The Carbon Cycle and Atmospheric CO_2 N:atural Variations, Archean to Present, Geophysical Monograph 32*, pp. 5–59. Washington, DC: American Geophysical Union.

Takashahi T, Olafsson J, Goddard J, Chipman DW, and Sutherland SC (1993) Seasonal variation of CO_2 and nutrients in the high-latitude surface oceans a: comparative study. *Global Biogeochemical Cycles* 7: 843–878.

Takahashi T, Feely RA, Weiss R, *et al.* (1997) Global air–sea flux of CO_2 a:n estimate based on measurements of sea–air pCO_2 difference. *Proceedings of the National Academy of science USA* 94: 8292–8299.

Tans PP, Fung IY, and Takahashi T (1990) Observational constraints on the global atmospheric CO_2 budget. *Sciece* 247: 1431–1438.

Wanninkhof R (1992) Relationship between wind speed and gas exchange. *Journal of Geophysical Research* 97: 7373–7382.

Wanninkhof R and McGillis WM (1999) A cubic relationship between gas transfer and wind speed. *Geophysical Research Letters* 26: 1889–1893.

ABRUPT CLIMATE CHANGE

S. Rahmstorf, Potsdam Institute for Climate Impact Research, Potsdam, Germany

Introduction

High-resolution paleoclimatic records from ice and sediment cores and other sources have revealed a number of dramatic climatic changes that occurred over surprisingly short times – a few decades or in some cases a few years. In Greenland, for example, temperature rose by 5–10 °C, snowfall rates doubled, and windblown dust decreased by an order of magnitude within 40 years at the end of the last glacial period. In the Sahara, an abrupt transition occurred around 5500 years ago from a relatively green shrubland supporting significant populations of animals and humans to the dry desert we know today.

One could define an abrupt climate change simply as a large and rapid one – occurring faster than in a given time (say 30 years). The change from winter to summer, a very large change (in many places larger than the glacial–interglacial transition) occurring within 6 months, is, however, not an abrupt change in climate (or weather), it is rather a gradual transition following the solar forcing in its near-sinusoidal path. The term 'abrupt' implies not just rapidity but also reaching a breaking point, a threshold – it implies a change that does not smoothly follow the forcing but is rapid in comparison to it. This physical definition thus equates abrupt climate change with a strongly nonlinear response to the forcing. In this definition, the quaternary transitions from glacial to interglacial conditions and back, taking a few hundred or thousand years, are a prime example of abrupt climate change, as the underlying cause, the Earth's orbital variations (Milankovich cycles), have timescales of tens of thousands of years. On the other hand, anthropogenic global warming occurring within a hundred years is not as such an abrupt climate change as long as it smoothly follows the increase in atmospheric carbon dioxide. Only if global warming triggered a nonlinear response, like a rapid ocean circulation change or decay of the West Antarctic Ice Sheet (WAIS), would one speak of an abrupt climate change.

Paleoclimatic Data

A wealth of paleoclimatic data has been recovered from ice cores, sediment cores, corals, tree rings, and other sources, and there have been significant advances in analysis and dating techniques. These advances allow a description of the characteristics of past climatic changes, including many abrupt ones, in terms of geographical patterns, timing, and affected climatic variables. For example, the ratio of oxygen isotopes in ice cores yields information about the temperature in the cloud from which the snow fell. Another way to determine temperature is to measure the isotopic composition of the nitrogen gas trapped in the ice, and it is also possible to directly measure the temperature in the borehole with a thermometer. Each method has advantages and drawbacks in terms of time resolution and reliability of the temperature calibration. Dust, carbon dioxide, and methane content of the prehistoric atmosphere can also be determined from ice cores.

On long timescales, climatic variability throughout the past 2 My at least has been dominated by the Milankovich cycles in the Earth's orbit around the sun – the cycles of precession, obliquity, and eccentricity with periods of roughly 23 ky, 41 ky, and 100 ky, respectively. Since the middle Pleistocene transition 1.2 Ma, the regular glaciations of our planet follow the 100-ky eccentricity cycle; even though this has only a rather weak direct influence on the solar radiation reaching the Earth, it modulates the much stronger other two cycles. The prevalence of the 100-ky cycle in climate is thus apparently a highly nonlinear response to the forcing that is likely linked to the nonlinear continental ice sheet and/or carbon cycle dynamics. The terminations of glaciations occur rather abruptly (**Figure 1**). Greenland ice cores show that the transition from the last Ice Age to the warm Holocene climate took about 1470 years, with much of the change occurring in only 40 years. The local Greenland response is not typical for the global response; however, since Greenland temperatures can be strongly affected by Atlantic Ocean circulation, which went through rapid changes during deglaciation. Globally, the transition from full Ice Age to Holocene conditions took around 5 ky.

The ice ages were not just generally colder than the present climate but were also punctuated by abrupt climatic transitions. The best evidence for these transitions, known as Dansgaard–Oeschger (D/O) events, comes from the last ice age (**Figure 2**). D/O events typically start with an abrupt warming by up to 12 °C within a few decades or less, followed by gradual cooling over several hundred or thousand

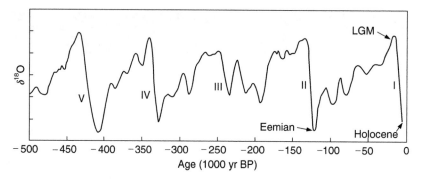

Figure 1 Record of δ^{18}O from marine sediments (arbitrary units), reflecting mainly the changes in global ice volume during the past 50 ky. Note the rapid terminations (labeled with roman numbers) of glacial periods.

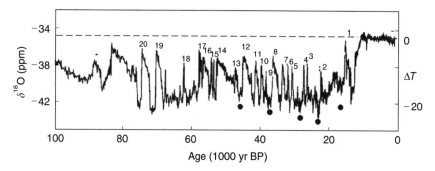

Figure 2 Record δ^{18}O from the GRIP ice core, a proxy for atmospheric temperature over Greenland (approximate temperature range ΔT (°C) is given on the right). Note the relatively stable Holocene climate during the past 10 ky and before that the much colder glacial climate punctuated by Dansgaard–Oeschger warm events (numbered). The timing of Heinrich events 1 to 6 is marked by black dots.

years. The cooling phase often ends with an abrupt final temperature drop back to cold ('stadial') conditions. Although first seen in the Greenland ice cores, the D/O events are not a local feature of Greenland climate. **Figure 3** shows that subtropical sea surface temperatures in the Atlantic closely mirror the sequence of events in Greenland. Similar records have been found near Santa Barbara, California, in the Cariaco Basin off Venezuela, and off the coast of India. D/O climate change is centered on the North Atlantic and regions with strong atmospheric response to changes in the North Atlantic, with little response in the Southern Ocean or Antarctica. The 'waiting time' between successive D/O events is most often around 1470 years or, with decreasing probability, multiples of this period. This suggests the existence of an as yet unexplained 1470-year cycle that often (but not always) triggers a D/O event. The second major type of abrupt event in glacial times is the Heinrich (H) event. H events involve surging of the Laurentide Ice Sheet through Hudson Strait, occurring in the cold stadial phase of some D/O cycles. They have a variable spacing of several thousand years. The icebergs released to the

North Atlantic during H events leave telltale dropstones in the ocean sediments when they melt, the so-called Heinrich layers. Sediment data suggest that H events shut down or at least drastically reduce the formation of North Atlantic Deep Water (NADW). Records from the South Atlantic and parts of Antarctica show that the cold H events in the North Atlantic were associated with unusual warming there (a fact sometimes referred to as 'bipolar seesaw').

At the end of the last glacial, a particularly interesting abrupt climatic change took place, the so-called Younger Dryas event (12 800–11 500 years ago). Conditions had already warmed to near-interglacial conditions and continental ice sheets were retreating, when within decades the climate in the North Atlantic region switched back to glacial conditions for more than a thousand years. It has been speculated that the cooling resulted from a sudden influx of fresh water into the North Atlantic through St. Lawrence River, when an ice barrier holding back a huge meltwater lake on the North American continent broke. This could have shut down the Atlantic thermohaline circulation (i.e., the circulation driven by temperature and salinity differences), but evidence is controversial.

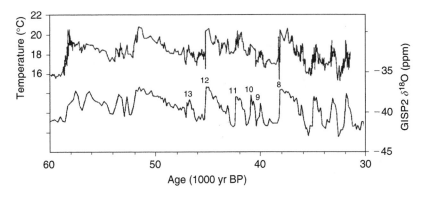

Figure 3 Sea-surface temperatures derived from alkenones in marine sediments from the subtropical Atlantic (Bermuda Rise, upper curve) compared to $\delta^{18}O$ values from the GISP2 ice core in Greenland (lower curve).

Alternatively, the Younger Dryas may simply have been the last cold stadial period of the glacial following a temporary D/O warming event.

Does abrupt climate change occur only during glacial times? Early evidence for the last interglacial, the Eemian, suggested abrupt changes there, but has since been refuted. During the present interglacial, the Holocene, climate was much more stable than during the last glacial. However, two abrupt events stand out. One is the 8200-year event that shows up as a cold spike in Arctic ice cores and affected the North Atlantic region. The second major change is the abrupt desertification of the Sahara 5500 years ago. There is much evidence from cave paintings, fire remains, bones, ancient lake sediments, and the like that the Sahara was a partly swampy savannah before this time. The best evidence for the abrupt ending of this benign climate comes from Atlantic sediments off northeastern Africa, which show a sudden and dramatic step-function increase in windblown dust, witnessing a drying of the adjacent continent.

Mechanisms of Abrupt Climate Change

The increased spatial coverage, quality, and time resolution of paleoclimatic data as well as advances in computer modeling have led to a greater understanding of the mechanisms of abrupt climate change, although many aspects are still in dispute and not fully understood.

The simplest concept for a mechanism causing abrupt climatic change is that of a threshold. A gradual change in external forcing (e.g., the change in insolation due to the Milankovich cycles) or in an internal climatic parameter (e.g., the slow buildup or melting of continental ice) continues until a specific

threshold value is reached where some qualitative change in climate is triggered. Various such critical thresholds are known to exist in the climate system. Continental ice sheets may have a stability threshold where they start to surge; the thermohaline ocean circulation has thresholds where deep-water formation shuts down or shifts location; methane hydrates in the seafloor have a temperature threshold where they change into the gas phase and bubble up into the atmosphere; and the atmosphere itself may have thresholds where large-scale circulation regimes (such as the monsoon) switch.

For the D/O events, H events, and the Younger Dryas event discussed above, the paleoclimatic data clearly point to a crucial role of Atlantic Ocean circulation changes. Modeling and analytical studies of the Atlantic thermohaline circulation (sometimes called the 'conveyor belt') show that there are two positive feedback mechanisms leading to threshold behavior. The first, called advective feedback, is caused by the large-scale northward transport of salt by the Atlantic currents, which in turn strengthens the circulation by increasing density in the northern latitudes. The second, called convective feedback, is caused by the fact that oceanic convection creates conditions favorable for further convection. These (interconnected) feedbacks make convection and the large-scale thermohaline circulation self-sustaining within certain limits, with well-defined thresholds where the circulation changes to a qualitatively different mode.

Three main circulation modes have been identified both in sediment data and in models (**Figure 4**): (1) a warm or interglacial mode with deep-water formation in the Nordic Seas and large oceanic heat transport to northern high latitudes (**Figure 4(a)**); (2) a cold or stadial mode with deep-water formation south of the shallow sill between Greenland, Iceland, and Scotland and with greatly reduced heat transport

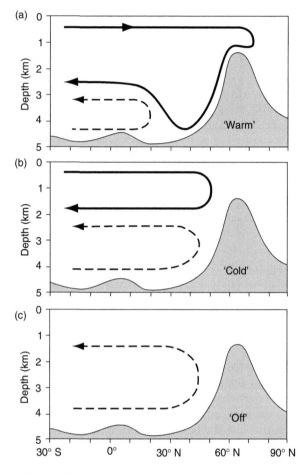

Figure 4 Schematic of three major modes of Atlantic Ocean circulation. (a) 'Warm' or interglacial mode; (b) 'cold' or stadial mode; (c) 'off' or Heinrich mode. In the warm mode the Atlantic thermohaline circulation reaches north over the Greenland–Iceland–Scotland ridge into the Nordic Seas, while in the cold mode it stops south of Iceland. Switches between circulation modes at certain thresholds can pace and amplify climatic changes.

circulation starts to gradually weaken and temperatures start to decline again immediately after the incursion, until the threshold is reached where convection in the Nordic Seas stops and the system reverts to the stable stadial mode. H events can be interpreted as a switch from the stadial mode to the H mode, that is, a shutdown of North Atlantic deepwater formation. As this mode is probably also unstable in glacial conditions, the system spontaneously reverts to the stadial or to the warm mode after a waiting time of centuries, the timescale being determined partly by slow oceanic mixing processes.

This interpretation is consistent with the observed patterns of surface temperature change. The warming during D/O events is centered on the North Atlantic because this is where the change in oceanic heat transport occurs; the warm mode delivers heat to much higher latitudes than does the cold mode. A switch to the H mode, on the other hand, strongly reduces the interhemispheric heat transport from the South Atlantic to the North Atlantic. This cools the Northern Hemisphere while warming the Southern Hemisphere, explaining the 'bipolar seesaw' response in climate. It should also be noted that the initial transient response can differ from the equilibrium response as the oceanic heat storage capacity is large. The patterns of these abrupt changes differ from the longer-timescale (many thousands of years) response to the Milankovich cycles because, for the latter, the slow changes in atmospheric greenhouse gases (e.g., CO_2) and continental ice cover act to globally synchronize and amplify climatic change.

While the threshold behavior of the Atlantic ocean can dramatically shape and amplify climatic change, the question remains what triggers the mode switches. As mentioned above, D/O switches appear to be paced by an underlying 1470-year cyclicity that is as yet unexplained. This could either be an external (astronomical or solar) cycle or an internal oscillation of the climate system, perhaps also involving the Atlantic thermohaline circulation. A superposition of two major shorter solar cycles can, in climate model simulations, trigger events spaced 1470 years apart. The irregularity in D/O event timing is probably the result of the presence of stochastic variability in the climate system as well as the presence of longer-term trends such as the slow buildup of large continental ice sheets.

The ocean circulation change during H events, on the other hand, can be explained by the large amounts of fresh water entering the North Atlantic at these times in the form of icebergs. Simulations show that the observed amounts of fresh water are sufficient to shut down deep-water formation in the North Atlantic. The nonlinear dynamics of ice sheets

to high latitudes (**Figure 4(b)**); and (3) a 'switched off' or 'Heinrich' mode with practically no deep-water formation in the North Atlantic (**Figure 4(c)**). In the last mode, the Atlantic deep circulation is dominated by inflow of Antarctic Bottom Water (AABW) from the south.

Many features of abrupt glacial climate can be explained by switches between these three circulation modes. Model simulations suggest that the cold stadial mode is the only stable mode in a glacial climate; it prevails during the cold stadial periods of the last glacial. D/O events can be interpreted as temporary incursions of warm Atlantic waters into the Nordic Seas and deep-water formation there, that is, a switch to the warm mode causing abrupt climatic warming in the North Atlantic region. As this mode is not stable in glacial conditions, the provide a plausible trigger mechanism. Ice sheets may grow for many thousands of years until their base melts owing to geothermal heating, when the ice sheet becomes unstable and surges.

Thresholds in ocean circulation and continental or sea ice dynamics are not the only mechanisms that can cause abrupt climatic changes. In the desertification of the Sahara in the mid-Holocene, probably neither of these mechanisms were involved. Rather, an unstable positive feedback between vegetation cover (affecting albedo and evapotranspiration) and monsoon circulation in the atmosphere appears to have been responsible.

It is almost certain that there are further nonlinearities in the climate system that could have caused abrupt climatic changes in the past or may do so in the future. We are only beginning to understand abrupt climate change, and the interpretations presented here – while consistent with data and model results – are not the only possible interpretations. Reflecting the state of this science, they are current working hypotheses rather than established and well-tested theory.

Risk of Future Abrupt Changes

The prevalence of abrupt nonlinear (rather than smooth and gradual) climatic change in the past naturally leads to the question whether such changes can be expected in the future, either by natural causes or by human interference. The main outside driving forces of past climatic changes are the Milankovich cycles. Close inspection of these cycles as well as modeling results indicate that we are presently enjoying an unusually quiet period in the climatic effect of these cycles, owing to the present minimum in eccentricity of the Earth's orbit. The next large change in solar radiation that could trigger a new ice age is probably at least 30 ky away. If this is correct, it makes the Holocene an unusually long interglacial, comparable to the Holstein interglacial that occurred around 400 ka when the Earth's orbit went through a similar pattern. This stable orbital situation leaves unpredictable events (such as meteorite impacts or a series of extremely large volcanic eruptions) and anthropogenic interference as possible causes for abrupt climatic changes in the lifetime of the next few generations of humans.

Significant anthropogenic warming of the lower atmosphere and ocean surface will almost certainly occur in this century, raising concerns that nonlinear thresholds in the climate system could be exceeded and abrupt changes could be triggered at some point. Processes that have been (rather speculatively) mentioned in this context include a collapse of the WAIS, a strongly enhanced greenhouse effect due to melting of permafrost or triggering of methane hydrate deposits at the seafloor, a large-scale wilting of forests when drought-tolerance thresholds are exceeded, nonlinear changes in monsoon regimes, and abrupt changes in ocean circulation.

Of those possibilities, the risk of a change in ocean circulation is probably the best understood and perhaps also the least unlikely. Two factors could weaken the circulation and bring it closer to a threshold: the warming of the surface and a dilution of high-latitude waters with fresh water. The latter could result from an enhanced atmospheric water cycle and precipitation as well as meltwater runoff from Greenland and other glaciers. Both warming and fresh water input reduce surface density and thereby inhibit deep-water formation. Model simulations of global-warming scenarios so far suggest three possible responses: a shutdown of convection in the Labrador Sea, one of the two main NADW formation sites; a complete shutdown of NADW formation (i.e., similar to a switch to the H mode); and a shutdown of AABW formation. A transition to the stadial circulation mode has so far not been simulated, perhaps because convection in the Nordic Seas is strongly wind driven and is more effectively switched off by increased sea ice cover than by warming.

A shutdown of Labrador Sea convection would be a significant qualitative change in the Atlantic Ocean circulation, but would probably affect only the surface climate of a smaller region surrounding the Labrador Sea. Effects on ecosystems and fisheries have not been investigated but could be severe. A complete shutdown of NADW formation would have wider climatic repercussions. Temperatures in northwestern Europe could initially rise several degrees in step with global warming, then abruptly drop back to near present values (the competing effects of raised atmospheric CO_2 and reduced oceanic heat transport almost balancing). If CO_2 levels decline again in future centuries as expected, European temperatures could remain several degrees below present as the Atlantic thermohaline circulation is not expected to recover perhaps for millennia. Further effects of a shutdown of deep-water renewal include reduced oceanic uptake of CO_2 (enhancing the greenhouse effect), shifts in tropical rainfall belts, accelerated global sea level rise (due to a faster warming of the deep oceans), and rapid regional sea level rise in the northern Atlantic.

The probability of major climatic thresholds being crossed in the coming centuries is difficult to establish and largely unknown. Currently, this possibility lies within the (still rather large) uncertainty range for future climate projections, so the risk cannot be ruled out. The IPCC 4th assessment report assigns a probability of up to 10% to a shutdown of the Atlantic overturning circulation within this century.

Further Reading

Abrantes F and Mix A (eds.) (1999) *Reconstructing Ocean History – A Window Into the Future*. New York: Plenum.

Broecker W (1987) Unpleasant surprises in the greenhouse? *Nature* 328: 123.

Clark PU, Webb RS, and Keigwin LD (eds.) (1999) *Mechanisms of Global Climate Change at Millennial Time Scales*. Washington, DC: American Geophysical Union.

Clark PU, Alley RB, and Pollard D (1999) Northern Hemisphere ice sheet influences on global climate change. *Science* 286: 1104–1111.

Houghton JT, Meira Filho LG, and Callander BA (1995) *Climate Change 1995*. Cambridge, UK: Cambridge University Press.

Sachs JP and Lehman SJ (1999) Subtropical North Atlantic temperatures 60,000 to 30,000 years ago. *Science* 286: 756–759.

Stocker T (2000) Past and future reorganisations in the climate system. *Quarterly Science Review (PAGES Special Issue)* 19: 301–319.

Taylor K (1999) Rapid climate change. *American Scientist* 87: 320.

SEA LEVEL CHANGE

J. A. Church, Antarctic CRC and CSIRO Marine Research, TAS, Australia
J. M. Gregory, Hadley Centre, Berkshire, UK

Introduction

Sea-level changes on a wide range of time and space scales. Here we consider changes in mean sea level, that is, sea level averaged over a sufficient period of time to remove fluctuations associated with surface waves, tides, and individual storm surge events. We focus principally on changes in sea level over the last hundred years or so and on how it might change over the next one hundred years. However, to understand these changes we need to consider what has happened since the last glacial maximum 20 000 years ago. We also consider the longer-term implications of changes in the earth's climate arising from changes in atmospheric greenhouse gas concentrations.

Changes in mean sea level can be measured with respect to the nearby land (relative sea level) or a fixed reference frame. Relative sea level, which changes as either the height of the ocean surface or the height of the land changes, can be measured by a coastal tide gauge.

The world ocean, which has an average depth of about 3800 m, contains over 97% of the earth's water. The Antarctic ice sheet, the Greenland ice sheet, and the hundred thousand nonpolar glaciers/ ice caps, presently contain water sufficient to raise sea level by 61 m, 7 m, and 0.5 m respectively if they were entirely melted. Ground water stored shallower than 4000 m depth is equivalent to about 25 m (12 m stored shallower than 750 m) of sea-level change. Lakes and rivers hold the equivalent of less than 1 m, while the atmosphere accounts for only about 0.04 m.

On the time-scales of millions of years, continental drift and sedimentation change the volume of the ocean basins, and hence affect sea level. A major influence is the volume of mid-ocean ridges, which is related to the arrangement of the continental plates and the rate of sea floor spreading.

Sea level also changes when mass is exchanged between any of the terrestrial, ice, or atmospheric reservoirs and the ocean. During glacial times (ice ages), water is removed from the ocean and stored in large ice sheets in high-latitude regions. Variations in the surface loading of the earth's crust by water and ice change the shape of the earth as a result of the elastic response of the lithosphere and viscous flow of material in the earth's mantle and thus change the level of the land and relative sea level. These changes in the distribution of mass alter the gravitational field of the earth, thus changing sea level. Relative sea level can also be affected by local tectonic activities as well as by the land sinking when ground water is extracted or sedimentation increases. Sea water density is a function of temperature. As a result, sea level will change if the ocean's temperature varies (as a result of thermal expansion) without any change in mass.

Sea-Level Changes Since the Last Glacial Maximum

On timescales of thousands to hundreds of thousands of years, the most important processes affecting sea-level are those associated with the growth and decay of the ice sheets through glacial–interglacial cycles. These are also relevant to current and future sea level rise because they are the cause of ongoing land movements (as a result of changing surface loads and the resultant small changes in the shape of the earth – postglacial rebound) and ongoing changes in the ice sheets.

Sea-level variations during a glacial cycle exceed 100 m in amplitude, with rates of up to tens of millimetres per year during periods of rapid decay of the ice sheets (**Figure 1**). At the last glacial maximum (about 21 000 years ago), sea level was more than 120 m below current levels. The largest contribution to this sea-level lowering was the additional ice that formed the North American (Laurentide) and European (Fennoscandian) ice sheets. In addition, the Antarctic ice sheet was larger than at present and there were smaller ice sheets in presently ice-free areas.

Observed Recent Sea-Level Change

Long-term relative sea-level changes have been inferred from the geological records, such as radiocarbon dates of shorelines displaced from present day sea level, and information from corals and sediment cores. Today, the most common method of measuring sea level relative to a local datum is by tide gauges at coastal and island sites. A global data set is maintained by the Permanent Service for Mean Sea Level (PSMSL). During the 1990s, sea level has been measured globally with satellites.

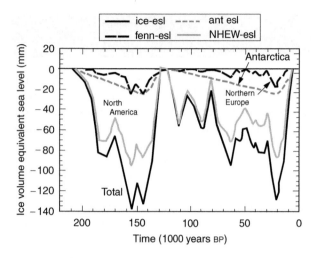

Figure 1 Change in ice sheet volume over the last 200 000 years. Fenn, Fennoscandian; ant, Antarctic; NHEW; North America; esl, equivalent sea level. (Reproduced from Lambeck, 1998.)

Tide-gauge Observations

Unfortunately, determination of global-averaged sea-level rise is severely limited by the small number of gauges (mostly in Europe and North America) with long records (up to several hundred years, **Figure 2**). To correct for vertical land motions, some sea-level change estimates have used geological data, whereas others have used rates of present-day vertical land movement calculated from models of postglacial rebound.

A widely accepted estimate of the current rate of global-average sea-level rise is about $1.8\,\text{mm}\,\text{y}^{-1}$. This estimate is based on a set of 24 long tide-gauge

records, corrected for land movements resulting from deglaciation. However, other analyses produce different results. For example, recent analyses suggest that sea-level change in the British Isles, the North Sea region and Fennoscandia has been about $1\,\text{mm}\,\text{y}^{-1}$ during the past century. The various assessments of the global-average rate of sea-level change over the past century are not all consistent within stated uncertainties, indicating further sources of error. The treatment of vertical land movements remains a source of potential inconsistency, perhaps amounting to $0.5\,\text{mm}\,\text{y}^{-1}$. Other sources of error include variability over periods of years and longer and any spatial distribution in regional sea level rise (perhaps several tenths of a millimeter per year).

Comparison of the rates of sea-level rise over the last 100 years (1.0–$2.0\,\text{mm}\,\text{y}^{-1}$) and over the last two millennia (0.1–$0.0\,\text{mm}\,\text{y}^{-1}$) suggests the rate has accelerated fairly recently. From the few very long tide-gauge records (**Figure 2**), it appears that an acceleration of about 0.3–$0.9\,\text{mm}\,\text{y}^{-1}$ per century occurred over the nineteenth and twentieth century. However, there is little indication that sea-level rise accelerated during the twentieth century.

Altimeter Observations

Following the advent of high-quality satellite radar altimeter missions in the 1990s, near-global and homogeneous measurement of sea level is possible, thereby overcoming the inhomogeneous spatial sampling from coastal and island tide gauges. However, clarifying rates of global sea-level change requires continuous satellite operations over many

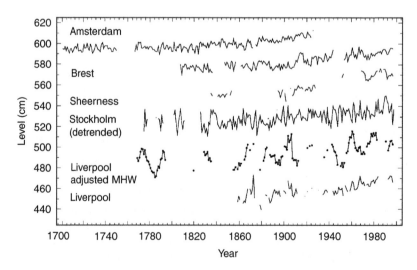

Figure 2 Time series of relative sea level over the last 300 years from several European coastal locations. For the Stockholm record, the trend over the period 1774 to 1873 has been removed from the entire data set. For Liverpool two series are given. These are mean sea level and, for a longer period, the mean high water (MHW) level. (Reproduced with permission from Woodworth, 1999.)

years and careful control of biases within and between missions.

To date, the TOPEX/POSEIDON satellite-altimeter mission, with its (near) global coverage from 66°N to 66°S (almost all of the ice-free oceans) from late 1992 to the present, has proved to be of most value in producing direct estimates of sea-level change. The present data allow global-average sea level to be estimated to a precision of several millimeters every 10 days, with the absolute accuracy limited by systematic errors. The most recent estimates of global-average sea level rise based on the short (since 1992) TOPEX/POSEIDON time series range from 2.1 mm y^{-1} to 3.1 mm y^{-1}.

The alimeter record for the 1990s indicates a rate of sea-level rise above the average for the twentieth century. It is not yet clear if this is a result of an increase in the rate of sea-level rise, systematic differences between the tide-gauge and altimeter data sets or the shortness of the record.

Processes Determining Present Rates of Sea-Level Change

The major factors determining sea-level change during the twentieth and twenty-first century are ocean thermal expansion, the melting of nonpolar glaciers and ice caps, variation in the mass of the Antarctic and Greenland ice sheets, and changes in terrestrial storage.

Projections of climate change caused by human activity rely principally on detailed computer models referred to as atmosphere–ocean general circulation models (AOGCMs). These simulate the global three-dimensional behavior of the ocean and atmosphere by numerical solution of equations representing the underlying physics. For simulations of the next hundred years, future atmospheric concentrations of gases that may affect the climate (especially carbon dioxide from combustion of fossil fuels) are estimated on the basis of assumptions about future population growth, economic growth, and technological change. AOGCM experiments indicate that the global-average temperature may rise by 1.4–5.8°C between 1990 and 2100, but there is a great deal of regional and seasonal variation in the predicted changes in temperature, sea level, precipitation, winds, and other parameters.

Ocean Thermal Expansion

The broad pattern of sea level is maintained by surface winds, air–sea fluxes of heat and fresh water (precipitation, evaporation, and fresh water runoff from the land), and internal ocean dynamics. Mean sea level varies on seasonal and longer timescales. A particularly striking example of local sea-level variations occurs in the Pacific Ocean during El Niño events. When the trade winds abate, warm water moves eastward along the equator, rapidly raising sea level in the east and lowering it in the west by about 20 cm.

As the ocean warms, its density decreases. Thus, even at constant mass, the volume of the ocean increases. This thermal expansion is larger at higher temperatures and is one of the main contributors to recent and future sea-level change. Salinity changes within the ocean also have a significant impact on the local density, and thus on local sea level, but have little effect on the global-average sea level.

The rate of global temperature rise depends strongly on the rate at which heat is moved from the ocean surface layers into the deep ocean; if the ocean absorbs heat more readily, climate change is retarded but sea level rises more rapidly. Therefore, time-dependent climate change simulation requires a model that represents the sequestration of heat in the ocean and the evolution of temperature as a function of depth. The large heat capacity of the ocean means that there will be considerable delay before the full effects of surface warming are felt throughout the depth of the ocean. As a result, the ocean will not be in equilibrium and global-average sea level will continue to rise for centuries after atmospheric greenhouse gas concentrations have stabilized. The geographical distribution of sea-level change may take many decades to arrive at its final state.

While the evidence is still somewhat fragmentary, and in some cases contradictory, observations indicate ocean warming and thus thermal expansion, particularly in the subtropical gyres, at rates resulting in sea-level rise of order 1 mm y^{-1}. The observations are mostly over the last few decades, but some observations date back to early in the twentieth century. The evidence is most convincing for the subtropical gyre of the North Atlantic, for which the longest temperature records (up to 73 years) and most complete oceanographic data sets exist. However, the pattern also extends into the South Atlantic and the Pacific and Indian oceans. The only areas of substantial ocean cooling are the subpolar gyres of the North Atlantic and perhaps the North Pacific. To date, the only estimate of a global average rate of sea-level rise from thermal expansion is 0.55 mm y^{-1}.

The warming in the Pacific and Indian Oceans is confined to the main thermocline (mostly the upper 1 km) of the subtropical gyres. This contrasts with the North Atlantic, where the warming is also seen at greater depths.

AOGCM simulations of sea level suggest that during the twentieth century the average rate

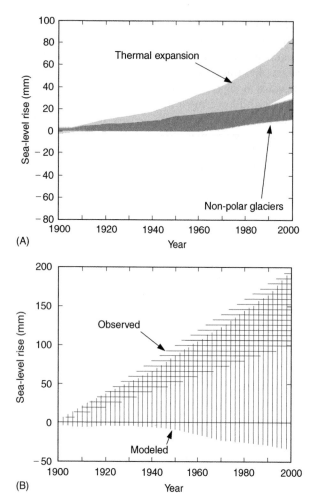

Figure 3 Computed sea-level rise from 1900 to 2000 AD. (A) The estimated thermal expansion is shown by the light stippling; the estimated nonpolar glacial contribution is shown by the medium-density stippling. (B) The computed total sea level change during the twentieth century is shown by the vertical hatching and the observed sea level change is shown by the horizontal hatching.

of change due to thermal expansion was of the order of 0.3–0.8 mm y^{-1} (**Figure 3**). The rate rises to 0.6–1.1 mm y^{-1} in recent decades, similar to the observational estimates of ocean thermal expansion.

Nonpolar Glaciers and Ice Caps

Nonpolar glaciers and ice caps are rather sensitive to climate change, and rapid changes in their mass contribute significantly to sea-level change. Glaciers gain mass by accumulating snow, and lose mass (ablation) by melting at the surface or base. Net accumulation occurs at higher altitude, net ablation at lower altitude. Ice may also be removed by discharge into a floating ice shelf and/or by direct calving of icebergs into the sea.

In the past decade, estimates of the regional totals of the area and volume of glaciers have been improved. However, there are continuous mass balance records longer than 20 years for only about 40 glaciers worldwide. Owing to the paucity of measurements, the changes in mass balance are estimated as a function of climate.

On the global average, increased precipitation during the twenty-first century is estimated to offset only 5% of the increased ablation resulting from warmer temperatures, although it might be significant in particular localities. (For instance, while glaciers in most parts of the world have had negative mass balance in the past 20 years, southern Scandinavian glaciers have been advancing, largely because of increases in precipitation.) A detailed computation of transient response also requires allowance for the contracting area of glaciers.

Recent estimates of glacier mass balance, based on both observations and model studies, indicate a contribution to global-average sea level of 0.2 to 0.4 mm y^{-1} during the twentieth century. The model results shown in **Figure 3** indicate an average rate of 0.1 to 0.3 mm y^{-1}.

Greenland and Antarctic Ice Sheets

A small fractional change in the volume of the Greenland and Antarctic ice sheets would have a significant effect on sea level. The average annual solid precipitation falling onto the ice sheets is equivalent to 6.5 mm of sea level, but this input is approximately balanced by loss from melting and iceberg calving. In the Antarctic, temperatures are so low that surface melting is negligible, and the ice sheet loses mass mainly by ice discharge into floating ice shelves, which melt at their underside and eventually break up to form icebergs. In Greenland, summer temperatures are high enough to cause widespread surface melting, which accounts for about half of the ice loss, the remainder being discharged as icebergs or into small ice shelves.

The surface mass balance plays the dominant role in sea-level changes on a century timescale, because changes in ice discharge generally involve response times of the order of 10^2 to 10^4 years. In view of these long timescales, it is unlikely that the ice sheets have completely adjusted to the transition from the previous glacial conditions. Their present contribution to sea-level change may therefore include a term related to this ongoing adjustment, in addition to the effects of climate change over the last hundred years. The current rate of change of volume of the polar ice sheets can be assessed by estimating the individual mass balance terms or by monitoring surface elevation changes directly (such as by airborne and satellite altimetry during the 1990s). However, these techniques give results with large uncertainties. Indirect methods (including numerical modeling of ice-sheets, observed sea-level changes over the last few millennia, and changes in the earth's rotation parameters) give narrower bounds, suggesting that the present contribution of the ice sheets to sea level is a few tenths of a millimeter per year at most.

Calculations suggest that, over the next hundred years, surface melting is likely to remain negligible in Antarctica. However, projected increases in precipitation would result in a net negative sea-level contribution from the Antarctic ice sheet. On the other hand, in Greenland, surface melting is projected to increase at a rate more than enough to offset changes in precipitation, resulting in a positive contribution to sea-level rise.

Changes in Terrestrial Storage

Changes in terrestrial storage include reductions in the volumes of some of the world's lakes (e.g., the Caspian and Aral seas), ground water extraction in excess of natural recharge, more water being impounded in reservoirs (with some seeping into aquifers), and possibly changes in surface runoff. Order-of-magnitude evaluations of these terms are uncertain but suggest that each of the contributions could be several tenths of millimeter per year, with a small net effect (**Figure 3**). If dam building continues at the same rate as in the last 50 years of the twentieth century, there may be a tendency to reduce sea-level rise. Changes in volumes of lakes and rivers will make only a negligible contribution.

Permafrost currently occupies about 25% of land area in the northern hemisphere. Climate warming leads to some thawing of permafrost, with partial runoff into the ocean. The contribution to sea level in the twentieth century is probably less than 5 mm.

Projected Sea-Level Changes for the Twenty-first Century

Detailed projections of changes in sea level derived from AOCGM results are given in material listed as Further Reading. The major components are thermal expansion of the ocean (a few tens of centimeters), melting of nonpolar glaciers (about 10–20 cm), melting of Greenland ice sheet (several centimeters), and increased storage in the Antarctic (several centimeters).

After allowance for the continuing changes in the ice sheets since the last glacial maximum and the

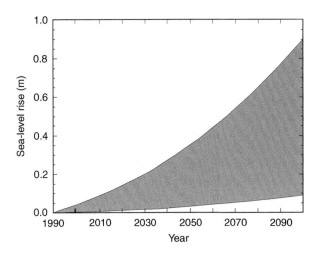

Figure 4 The estimated range of future global-average sea-level rise from 1990 to 2100 AD for a range of plausible projections in atmospheric greenhouse gas concentrations.

melting of permafrost (but not including changes in terrestrial storage), total projected sea-level rise during the twenty-first century is currently estimated to be between about 9 and 88 cm (**Figure 4**).

Regional Sea-Level Change

Estimates of the regional distribution of sea-level rise are available from several AOGCMs. Our confidence in these distributions is low because there is little similarity between model results. However, models agree on the qualitative conclusion that the range of regional variation is substantial compared with the global-average sea-level rise. One common feature is that nearly all models predict less than average sea-level rise in the Southern Ocean.

The most serious impacts of sea-level change on coastal communities and ecosystems will occur during the exceptionally high water levels known as storm surges produced by low air pressure or driving winds. As well as changing mean sea level, climate change could also affect the frequency and severity of these meteorological conditions, making storm surges more or less severe at particular locations.

Longer-term Changes

Even if greenhouse gas concentrations were to be stabilized, sea level would continue to rise for several hundred years. After 500 years, sea-level rise from thermal expansion could be about 0.3–2 m but may be only half of its eventual level. Glaciers presently contain the equivalent of 0.5 m of sea level. If the CO_2 levels projected for 2100 AD were sustained, there would be further reductions in glacier mass.

Ice sheets will continue to react to climatic change during the present millennium. The Greenland ice sheet is particularly vulnerable. Some models suggest that with a warming of several degrees no ice sheet could be sustained on Greenland. Complete melting of the ice sheet would take at least a thousand years and probably longer.

Most of the ice in Antarctica forms the East Antarctic Ice Sheet, which would disintegrate only if extreme warming took place, beyond what is currently thought possible. The West Antarctic Ice Sheet (WAIS) has attracted special attention because it contains enough ice to raise sea level by 6 m and because of suggestions that instabilities associated with its being grounded below sea level may result in rapid ice discharge when the surrounding ice shelves are weakened. However, there is now general agreement that major loss of grounded ice, and accelerated sea-level rise, is very unlikely during the twenty-first century. The contribution of this ice sheet to sea level change will probably not exceed 3 m over the next millennium.

Summary

On timescales of decades to tens of thousands of years, sea-level change results from exchanges of mass between the polar ice sheets, the nonpolar glaciers, terrestrial water storage, and the ocean. Sea level also changes if the density of the ocean changes (as a result of changing temperature) even though there is no change in mass. During the last century, sea level is estimated to have risen by 10–20 cm, as a result of combination of thermal expansion of the ocean as its temperature rose and increased mass of the ocean from melting glaciers and ice sheets. Over the twenty-first century, sea level is expected to rise as a result of anthropogenic climate change. The main contributors to this rise are expected to be thermal expansion of the ocean and the partial melting of nonpolar glaciers and the Greenland ice sheet. Increased precipitation in Antarctica is expected to offset some of the rise from other contributions. Changes in terrestrial storage are uncertain but may also partially offset rises from other contributions. After allowance for the continuing changes in the ice sheets since the last glacial maximum, the total projected sea-level rise over the twenty-first century is currently estimated to be between about 9 and 88 cm.

Further Reading

Church JA, Gregory JM, Huybrechts P, et al. (2001) Changes in sea level. In: Houghton JT (ed.) Climate Change 2001; The Scientific Basis. Cambridge: Cambridge University Press.

Douglas BC, Keaney M, and Leatherman SP (eds.) (2000) Sea Level Rise: History and Consequences, 232 pp. San Diego: Academic Press.

Fleming K, Johnston P, Zwartz D, et al. Refining the eustatic sea-level curve since the Last Glacial Maximum using far- and intermediate-field sites. Earth and Planetary Science Letters 163: 327–342.

Lambeck K (1998) Northern European Stage 3 ice sheet and shoreline reconstructions: Preliminary results. News 5, Stage 3 Project, Godwin Institute for Quaternary Research, 9 pp.

Peltier WR (1998) Postglacial variations in the level of the sea: implications for climate dynamics and solid-earth geophysics. Review of Geophysics 36: 603–689.

Summerfield MA (1991) Global Geomorphology. Harlowe: Longman.

Warrick RA, Barrow EM, and Wigley TML (1993) Climate and Sea Level Change: Observations, Projections and Implications. Cambridge: Cambridge University Press.

WWW pages of the Permanent Service for Mean Sea Level, http://www.pol.ac.uk/psmsl/

PLANKTON AND CLIMATE

A. J. Richardson, University of Queensland,
St. Lucia, QLD, Australia

Introduction: The Global Importance of Plankton

Unlike habitats on land that are dominated by massive immobile vegetation, the bulk of the ocean environment is far from the seafloor and replete with microscopic drifting primary producers. These are the phytoplankton, and they are grazed by microscopic animals known as zooplankton. The word 'plankton' derives from the Greek *planktos* meaning 'to drift' and although many of the phytoplankton (with the aid of flagella or cilia) and zooplankton swim, none can progress against currents. Most plankton are microscopic in size, but some such as jellyfish are up to 2 m in bell diameter and can weigh up to 200 kg. Plankton communities are highly diverse, containing organisms from almost all kingdoms and phyla.

Similar to terrestrial plants, phytoplankton photosynthesize in the presence of sunlight, fixing CO_2 and producing O_2. This means that phytoplankton must live in the upper sunlit layer of the ocean and obtain sufficient nutrients in the form of nitrogen and phosphorus for growth. Each and every day, phytoplankton perform nearly half of the photosynthesis on Earth, fixing more than 100 million tons of carbon in the form of CO_2 and producing half of the oxygen we breathe as a byproduct.

Photosynthesis by phytoplankton directly and indirectly supports almost all marine life. Phytoplankton are a major food source for fish larvae, some small surface-dwelling fish such as sardine, and shoreline filter-feeders such as mussels and oysters. However, the major energy pathway to higher trophic levels is through zooplankton, the major grazers in the oceans. One zooplankton group, the copepods, is so numerous that they are the most abundant multicellular animals on Earth, outnumbering even insects by possibly 3 orders of magnitude. Zooplankton support the teeming multitudes higher up the food web: fish, seabirds, penguins, marine mammals, and turtles. Carcasses and fecal pellets of zooplankton and uneaten phytoplankton slowly yet consistently rain down on the cold dark seafloor, keeping alive the benthic (bottom-dwelling) communities of sponges, anemones, crabs, and fish.

Phytoplankton impact human health. Some species may become a problem for natural ecosystems and humans when they bloom in large numbers and produce toxins. Such blooms are known as harmful algal blooms (HABs) or red tides. Many species of zooplankton and shellfish that feed by filtering seawater to ingest phytoplankton may incorporate these toxins into their tissues during red-tide events. Fish, seabirds, and whales that consume affected zooplankton and shellfish can exhibit a variety of responses detrimental to survival. These toxins can also cause amnesic, diarrhetic, or paralytic shellfish poisoning in humans and may require the closure of aquaculture operations or even wild fisheries.

Despite their generally small size, plankton even play a major role in the pace and extent of climate change itself through their contribution to the carbon cycle. The ability of the oceans to act as a sink for CO_2 relies largely on plankton functioning as a 'biological pump'. By reducing the concentration of CO_2 at the ocean surface through photosynthetic uptake, phytoplankton allow more CO_2 to diffuse into surface waters from the atmosphere. This process continually draws CO_2 into the oceans and has helped to remove half of the CO_2 produced by humans from the atmosphere and distributed it into the oceans. Plankton play a further role in the biological pump because much of the CO_2 that is fixed by phytoplankton and then eaten by zooplankton sinks to the ocean floor in the bodies of uneaten and dead phytoplankton, and zooplankton fecal pellets. This carbon may then be locked up within sediments.

Phytoplankton also help to shape climate by changing the amount of solar radiation reflected back to space (the Earth's albedo). Some phytoplankton produce dimethylsulfonium propionate, a precursor of dimethyl sulfide (DMS). DMS evaporates from the ocean, is oxidized into sulfate in the atmosphere, and then forms cloud condensation nuclei. This leads to more clouds, increasing the Earth's albedo and cooling the climate.

Without these diverse roles performed by plankton, our oceans would be desolate, polluted, virtually lifeless, and the Earth would be far less resilient to the large quantities of CO_2 produced by humans.

Beacons of Climate Change

Plankton are ideal beacons of climate change for a host of reasons. First, plankton are ecthothermic (their body temperature varies with the surroundings), so their physiological processes such as nutrient uptake, photosynthesis, respiration, and reproductive development are highly sensitive to temperature, with their speed doubling or tripling with a 10 °C temperature rise. Global warming is thus likely to directly impact the pace of life in the plankton. Second, warming of surface waters lowers its density, making the water column more stable. This increases the stratification, so that more energy is required to mix deep nutrient-rich water into surface layers. It is these nutrients that drive surface biological production in the sunlit upper layers of the ocean. Thus global warming is likely to increase the stability of the ocean and diminish nutrient enrichment and reduce primary productivity in large areas of the tropical ocean. There is no such direct link between temperature and nutrient enrichment in terrestrial systems. Third, most plankton species are short-lived, so there is tight coupling between environmental change and plankton dynamics. Phytoplankton have lifespans of days to weeks, whereas land plants have lifespans of years. Plankton systems will therefore respond rapidly, whereas it takes longer before terrestrial plants exhibit changes in abundance attributable to climate change. Fourth, plankton integrate ocean climate, the physical oceanic and atmospheric conditions that drive plankton productivity. There is a direct link between climate and plankton abundance and timing. Fifth, plankton can show dramatic changes in distribution because they are free floating and most remain so their entire life. They thus respond rapidly to changes in temperature and oceanic currents by expanding and contracting their ranges. Further, as plankton are distributed by currents and not by vectors or pollinators, their dispersal is less dependent on other species and more dependent on physical processes. By contrast, terrestrial plants are rooted to their substrate and are often dependent upon vectors or pollinators for dispersal. Sixth, unlike other marine groups such as fish and many intertidal organisms, few plankton species are commercially exploited so any long-term changes can more easily be attributed to climate change. Last, almost all marine life has a planktonic stage in their life cycle because ocean currents provide an ideal mechanism for dispersal over large distances. Evidence suggests that these mobile life stages known as meroplankton are even more sensitive to climate change than the holoplankton, their neighbors that live permanently in the plankton.

All of these attributes make plankton ideal beacons of climate change. Impacts of climate change on plankton are manifest as predictable changes in the distribution of individual species and communities, in the timing of important life cycle events or phenology, in abundance and community structure, through the impacts of ocean acidification, and through their regulation by climate indices. Because of this sensitivity and their global importance, climate impacts on plankton are felt throughout the ecosystems they support.

Changes in Distribution

Plankton have exhibited some of the fastest and largest range shifts in response to global warming of any marine or terrestrial group. The general trend, as on land, is for plants and animals to expand their ranges poleward as temperatures warm. Probably the clearest examples are from the Northeast Atlantic. Members of a warm temperate assemblage have moved more than 1000-km poleward over the last 50 years (**Figure 1**). Concurrently, species of a subarctic (cold-water) assemblage have retracted to higher latitudes. Although these translocations have been associated with warming in the region by up to 1 °C, they may also be a consequence of the stronger northward flowing currents on the European shelf edge. These shifts in distribution have had dramatic impacts on the food web of the North Sea. The cool water assemblage has high biomass and is dominated by large species such as *Calanus finmarchicus*. Because this cool water assemblage retracts north as waters warm, *C. finmarchicus* is replaced by *Calanus helgolandicus*, a dominant member of the warm-water assemblage. This assemblage typically has lower biomass and contains relatively small species. Despite these *Calanus* species being indistinguishable to all but the most trained eye, the two species contrast starkly in their seasonal cycles: *C. finmarchicus* peaks in spring whereas *C. helgolandicus* peaks in autumn. This is critical as cod, which are traditionally the most important fishery of the North Sea, spawn in spring. As cod eggs hatch into larvae and continue to grow, they require good food conditons, consisting of large copepods such as *C. finmarchicus*, otherwise mortality is high and recruitment is poor. In recent warm years, however, *C. finmarchicus* is rare, there is very low copepod biomass during spring, and cod recruitment has crashed.

Changes in Phenology

Phenology, or the timing of repeated seasonal activities such as migrations or flowering, is very sensitive to global warming. On land, many events in spring are happening earlier in the year, such as the arrival of swallows in the UK, emergence of butterflies in the US, or blossoming of cherry trees in Japan. Recent

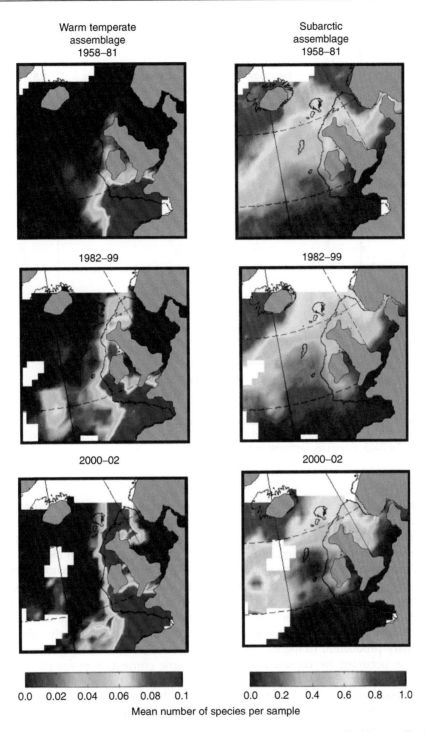

Warm temperate
assemblage
1958–81

Subarctic
assemblage
1958–81

1982–99

1982–99

2000–02

2000–02

0.0 0.02 0.04 0.06 0.08 0.1

0.0 0.2 0.4 0.6 0.8 1.0

Mean number of species per sample

Figure 1 The northerly shift of the warm temperate assemblage (including *Calanus helgolandicus*) into the North Sea and retraction of the subarctic assemblage (including *Calanus finmarchicus*) to higher latitudes. Reproduced by permission from Gregory Beaugrand.

evidence suggests that phenological changes in plankton are greater than those observed on land. Larvae of benthic echinoderms in the North Sea are now appearing in the plankton 6 weeks earlier than they did 50 years ago, and this is in response to warmer temperatures of less than 1 °C. In echinoderms, temperature stimulates physiological developments and larval release. Other meroplankton such as larvae of fish, cirrepedes, and decapods have also responded similarly to warming (**Figure 2**).

Timing of peak abundance of plankton can have effects that resonate to higher trophic levels. In the

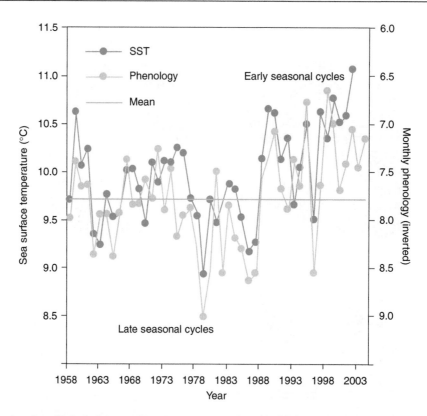

Figure 2 Monthly phenology (timing) of decapod larval abundance and sea surface temperature in the central North Sea from 1958 to 2004. Reproduced from Edwards M, Johns DG, Licandro P, John AWG, and Stevens DP (2006) Ecological status report: Results from the CPR Survey 2004/2005. *SAHFOS Technical Report* 3: 1–8.

North Sea, the timing each year of plankton blooms in summer over the last 50 years has advanced, with phytoplankton appearing 23 days earlier and copepods 10 days earlier. The different magnitude of response between phytoplankton and zooplankton may lead to a mismatch between successive trophic levels and a change in the synchrony of timing between primary and secondary production. In temperate marine systems, efficient transfer of marine primary and secondary production to higher trophic levels, such as those occupied by commercial fish species, is largely dependent on the temporal synchrony between successive trophic production peaks. This type of mismatch, where warming has disturbed the temporal synchrony between herbivores and their plant food, has been noted in other biological systems, most notably between freshwater zooplankton and diatoms, great tits and caterpillar biomass, flycatchers and caterpillar biomass, winter moth and oak bud burst, and the red admiral butterfly and stinging nettle. Such mismatches compromise herbivore survival.

Dramatic ecosystem repercussions of climate-driven changes in phenology are also evident in the subarctic North Pacific Ocean. Here a single copepod species, *Neocalanus plumchrus*, dominates the

zooplankton biomass. Its vertical distribution and development are both strongly seasonal and result in an ephemeral (2-month duration) annual peak in upper ocean zooplankton biomass in late spring. The timing of this annual maximum has shifted dramatically over the last 50 years, with peak biomass about 60 days earlier in warm than cold years. The change in timing is a consequence of faster growth and enhanced survivorship of early cohorts in warm years. The timing of the zooplankton biomass peak has dramatic consequences for the growth performance of chicks of the planktivorous seabird, Cassin's auklet. Individuals from the world's largest colony of this species, off British Columbia, prey heavily on *Neocalanus*. During cold years, there is synchrony between food availability and the timing of breeding. During warm years, however, spring is early and the duration of overlap of seabird breeding and *Neocalanus* availability in surface waters is small, causing a mismatch between prey and predator populations. This compromises the reproductive performance of Cassin's auklet in warm years compared to cold years. If Cassin's auklet does not adapt to the changing food conditions, then global warming will place severe strain on its long-term survival.

Changes in Abundance

The most striking example of changes in abundance in response to long-term warming is from foraminifera in the California Current. This plankton group is valuable for long-term climate studies because it is more sensitive to hydrographic conditions than to predation from higher trophic levels. As a result, its temporal dynamics can be relatively easily linked to changes in climate. Foraminifera are also well preserved in sediments, so a consistent time series of observations can be extended back hundreds of years. Records in the California Current show increasing numbers of tropical/subtropical species throughout the twentieth century reflecting a warming trend, which is most dramatic after the 1960s (**Figure 3**). Changes in the foraminifera record echo not only increase in many other tropical and subtropical taxa in the California Current over the last few decades, but also decrease in temperate species of algae, zooplankton, fish, and seabirds.

Changes in abundance through alteration of enrichment patterns in response to enhanced stratification is often more difficult to attribute to climate change than are shifts in distribution or phenology, but may have greater ecosystem consequences. An illustration from the Northeast Atlantic highlights the role that global warming can have on stratification and thus plankton abundances. In this region, phytoplankton become more abundant when cooler regions warm, probably because warmer temperatures boost metabolic rates and enhance stratification in these often windy, cold, and well-mixed regions.

But phytoplankton become less common when already warm regions get even warmer, probably because warm water blocks nutrient-rich deep water from rising to upper layers where phytoplankton live. This regional response of phytoplankton in the North Atlantic is transmitted up the plankton food web. When phytoplankton bloom, both herbivorous and carnivorous zooplankton become more abundant, indicating that the plankton food web is controlled from the 'bottom up' by primary producers, rather than from the 'top down' by predators. This regional response to climate change suggests that the distribution of fish biomass will change in the future, as the amount of plankton in a region is likely to influence its carrying capacity of fish. Climate change will thus have regional impacts on fisheries.

There is some evidence that the frequency of HABs is increasing globally, although the causes are uncertain. The key suspect is eutrophication, particularly elevated concentrations of the nutrients nitrogen and phosphorus, which are of human origin and discharged into our oceans. However, recent evidence from the North Sea over the second half of the twentieth century suggests that global warming may also have a key role to play. Most areas of the North Sea have shown no increase in HABs, except off southern Norway where there have been more blooms. This is primarily a consequence of the enhanced stratification in the area caused by warmer temperatures and lower salinity from meltwater. In the southern North Sea, the abundance of two key HAB species over the last 45 years is positively related to warmer ocean temperatures. This work

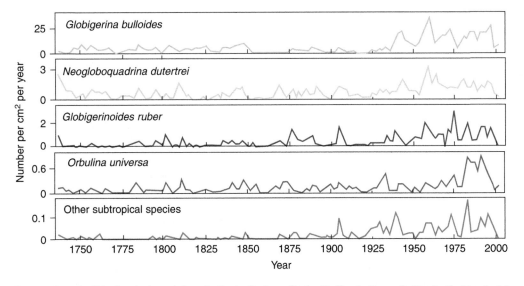

Figure 3 Fluxes of tropical/planktonic foraminifera in Santa Barbara Basin (California Current). Tropical/subtropical foraminifera showing increased abundance in the twentieth century. Reproduced from Field DB, Baumgartner TR, Charles CD, Ferreira-Bartrina V, and Ohman M (2006) Planktonic foraminifera of the California Current reflect 20th century warming. *Science* 311: 63–66.

supports the notion that the warmer temperatures and increased meltwater runoff anticipated under projected climate change scenarios are likely to increase the frequency of HABs.

Although most evidence for changes in abundance in response to climate change are from the Northern Hemisphere because this is where most (plankton) science has concentrated, there is a striking example from waters around Antarctica. Over the last 30 years, there has been a decline in the biomass of krill *Euphausia superba* in the Southern Ocean that is a consequence of warmer sea and air temperatures. In many areas, krill has been replaced by small gelatinous filter-feeding sacs known as salps, which occupy the less-productive, warmer regions of the Southern Ocean. The decline in krill is likely to be a consequence of warmer ocean temperatures impacting sea ice. It is not only that sea ice protects krill from predation, but also the algae living beneath the sea ice and photosynthesizing from the dim light seeping through are a critical food source for krill. As waters have warmed, the extent of winter sea ice and its duration have declined, and this has led to a deterioration in krill density since the 1970s. As krill are major food items for baleen whales, penguins, seabirds, fish, and seals, their declining population may have severe ramifications for the Southern Ocean food web.

Impact of Acidification

A direct consequence of enhanced CO_2 levels in the ocean is a lowering of ocean pH. This is a consequence of elevated dissolved CO_2 in seawater altering the carbonate balance in the ocean, releasing more hydrogen ions into the water and lowering pH. There has been a drop of 0.1 pH units since the Industrial Revolution, representing a 30% increase in hydrogen ions.

Impacts of ocean acidification will be greatest for plankton species with calcified (containing calcium carbonate) shells, plates, or scales. For organisms to build these structures, seawater has to be supersaturated in calcium carbonate. Acidification reduces the carbonate saturation of the seawater, making calcification by organisms more difficult and promoting dissolution of structures already formed.

Calcium carbonate structures are present in a variety of important plankton groups including coccolithophores, mollusks, echinoderms, and some crustaceans. But even among marine organisms with calcium carbonate shells, susceptibility to acidification varies depending on whether the crystalline form of their calcium carbonate is aragonite or

calcite. Aragonite is more soluble under acidic conditions than calcite, making it more susceptible to dissolution. As oceans absorb more CO_2, undersaturation of aragonite and calcite in seawater will be initially most acute in the Southern Ocean and then move northward.

Winged snails known as pteropods are probably the plankton group most vulnerable to ocean acidification because of their aragonite shell. In the Southern Ocean and subarctic Pacific Ocean, pteropods are prominent components of the food web, contributing to the diet of carnivorous zooplankton, myctophids, and other fish and baleen whales, besides forming the entire diet of gymnosome mollusks. Pteropods in the Southern Ocean also account for the majority of the annual flux of both carbonate and organic carbon exported to ocean depths. Because these animals are extremely delicate and difficult to keep alive experimentally, precise pH thresholds where deleterious effects commence are not known. However, even experiments over as little as 48 h show shell deterioration in the pteropod *Clio pyrimidata* at CO_2 levels approximating those likely around 2100 under a business-as-usual emissions scenario. If pteropods cannot grow and maintain their protective shell, their populations are likely to decline and their range will contract toward lower-latitude surface waters that remain supersaturated in aragonite, if they can adapt to the warmer temperature of the waters. This would have obvious repercussions throughout the food web of the Southern Ocean.

Other plankton that produce calcite such as foraminifera (protist plankton), mollusks other than pteropods (e.g., squid and mussel larvae), coccolithophores, and some crustaceans are also vulnerable to ocean acidification, but less so than their cousins with aragonite shells. Particularly important are coccolithophorid phytoplankton, which are encased within calcite shells known as liths. Coccolithophores export substantial quantities of carbon to the seafloor when blooms decay. Calcification rates in these organisms diminish as water becomes more acidic (**Figure 4**).

A myriad of other key processes in phytoplankton are also influenced by seawater pH. For example, pH is an important determinant of phytoplankton growth, with some species being catholic in their preferences, whereas growth of other species varies considerably between pH of 7.5 and 8.5. Changes in ocean pH also affect chemical reactions within organisms that underpin their intracellular physiological processes. pH will influence nutrient uptake kinetics of phytoplankton. These effects will have repercussions for phytoplankton community

Emiliania huxleyi   _Gephyrocapsa oceanica_

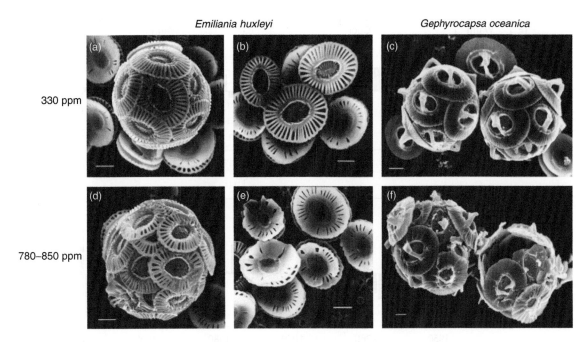

Figure 4 Scanning electron microscopy photographs of the coccolithophores _Emiliania huxleyi_ and _Gephyrocapsa oceanica_ collected from cultures incubated at CO_2 levels of about 300 and 780–850 ppm. Note the difference in the coccolith structure (including distinct malformations) and in the degree of calcification of cells grown at normal and elevated CO_2 levels. Scale bar = 1 mm. Reprinted by permission from Macmillan Publishers Ltd., _Nature_, Riebesell U, Zondervan I, Rost B, Tortell PD, Zeebe RE, and Morel FMM, Reduced calcification of marine plankton in response to increased atmospheric CO_2, 407: 364–376, Copyright (2000).

composition and productivity, with flow-on effects to higher trophic levels.

Climate Variability

Many impacts of climate change are likely to act through existing modes of variability in the Earth's climate system, including the well-known El Niño/ Southern Oscillation (ENSO) and the North Atlantic Oscillation (NAO). Such large synoptic pressure fields alter regional winds, currents, nutrient dynamics, and water temperatures. Relationships between integrative climate indices and plankton composition, abundance, or productivity provide an insight into how climate change may affect ocean biology in the future.

ENSO is the strongest climate signal globally, and has its clearest impact on the biology of the tropical Pacific Ocean. Observations from satellite over the past decade have shown a dramatic global decline in primary productivity. This trend is caused by enhanced stratification in the low-latitude oceans in response to more frequent El Niño events. During an El Niño, upper ocean temperatures warm, thereby enhancing stratification and reducing the availability of nutrients for phytoplankton

growth. Severe El Niño events lead to alarming declines in phytoplankton, fisheries, marine birds and mammals in the tropical Pacific Ocean. Of concern is the potential transition to more frequent El Niño-like conditions predicted by some climate models. In such circumstances, enhanced stratification across vast areas of the tropical ocean may reduce primary productivity, decimating fish, mammal, and bird populations. Although it is unknown whether the recent decline in primary productivity is already a consequence of climate change, the findings and underlying understanding of climate variability are likely to provide a window to the future.

Further north in the Pacific, the Pacific Decadal Oscillation (PDO) has a strong multi-decadal signal, longer than the ENSO period of a few years. When the PDO is negative, upwelling winds strengthen over the California Current, cool ocean conditions prevail in the Northeast Pacific, copepod biomass in the region is high and is dominated by large cool-water species, and fish stocks such as coho salmon are abundant (**Figure 5**). By contrast, when the PDO is positive, upwelling diminishes and warm conditions exist, the copepod biomass declines and is dominated by small less-nutritious species, and the abundance of coho salmon plunges.

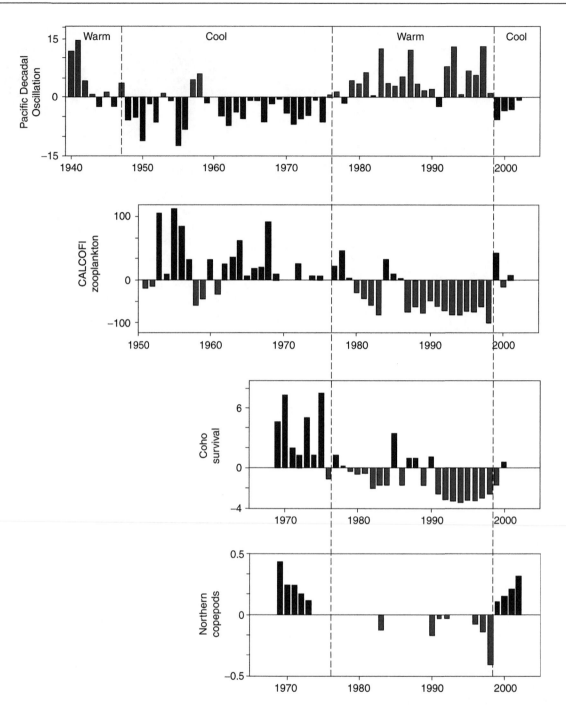

Figure 5 Annual time series in the Northeast Pacific: the PDO index from May to September; anomalies of zooplankton biomass (displacement volumes) from the California Current region (CALCOFI zooplankton); anomalies of coho salmon survival; and biomass anomalies of cold-water copepod species (northern copepods). Positive (negative) PDO index indicates warmer (cooler) than normal temperatures in coastal waters off North America. Reproduced from Peterson WT and Schwing FB (2003) A new climate regime in Northeast Pacific ecosystems. *Geophysical Research Letters* 30(17): 1896 (doi:10.1029/2003GL017528).

These transitions between alternate states have been termed regime shifts. It is possible that if climate change exceeds some critical threshold, some marine systems will switch permanently to a new state that is less favorable than present.

In Hot Water: Consequences for the Future

With plankton having relatively simple behavior, occurring in vast numbers, and amenable to experimental manipulation and automated measurements, their dynamics are far more easily studied and modeled than higher trophic levels. These attributes make it easier to model potential impacts of climate change on plankton communities. Many of our insights gained from such models confirm those already observed from field studies.

The basic dynamics of plankton communities have been captured by nutrient–phytoplankton–zooplankton (NPZ) models. Such models are based on a functional group representation of plankton communities, where species with similar ecological function are grouped into guilds to form the basic biological units in the model. Typical functional groups represented include diatoms, dinoflagellates, coccolithophores, microzooplankton, and mesozooplankton. There are many global NPZ models constructed by different research teams around the world. These are coupled to global climate models (GCMs) to provide future projections of the Earth's climate system. In this way, alternative carbon dioxide emission scenarios can be used to investigate possible future states of the ocean and the impact on plankton communities.

One of the most striking and worrisome results from these models is that they agree with fieldwork that has shown general declines in lower trophic levels globally as a result of large areas of the surface tropical ocean becoming more stratified and nutrient-poor as the oceans heat up (see sections titled 'Changes in abundance' and 'Climate variability'). One such NPZ model projects that under a middle-of-the-road emissions scenario, global primary productivity will decline by 5–10% (**Figure 6**). This trend will not be uniform, with increases in productivity by 20–30% toward the Poles, and marked declines in the warm stratified tropical ocean basins. This and other models show that warmer, more-stratified conditions in the Tropics will reduce nutrients in surface waters and lead to smaller phytoplankton cells dominating over larger diatoms. This will lengthen food webs and ultimately support fewer fish, marine mammals, and seabirds, as more trophic linkages are needed to transfer energy from small phytoplankton to higher trophic levels and 90% of the energy is lost within each trophic level through respiration. It also reduces the oceanic uptake of CO_2 by lowering the efficiency of the

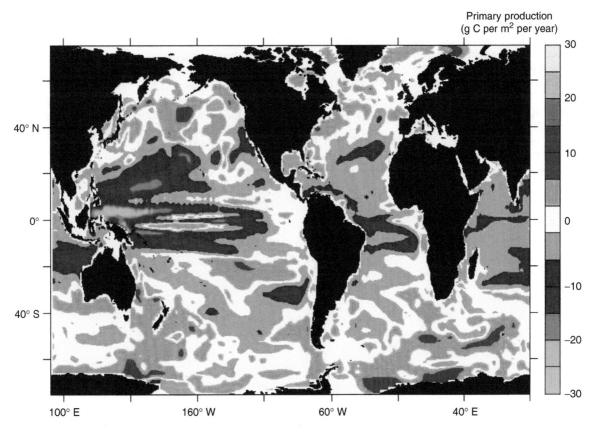

Figure 6 Change in primary productivity of phytoplankton between 2100 and 1990 estimated from an NPZ model. There is a global decline in primary productivity by 5–10%, with an increase at the Poles of 20–30%. Reproduced from Bopp L, Monfray P, Aumont O, et al. (2001) Potential impact of climate change on marine export production. *Global Biogeochemical Cycles* 15: 81–99.

biological pump. This could cause a positive feedback between climate change and the ocean carbon cycle: more CO_2 in the atmosphere leads to a warmer and more stratified ocean, which supports less and smaller plankton, and results in less carbon being drawn from surface ocean layers to deep waters. With less carbon removed from the surface ocean, less CO_2 would diffuse into the ocean and more CO_2 would accumulate in the atmosphere.

It is clear that plankton are beacons of climate change, being extremely sensitive barometers of physical conditions. We also know that climate impacts on plankton reverberate throughout marine ecosystems. More than any other group, they also influence the pace and extent of climate change. The impact of climate change on plankton communities will not only determine the future trajectory of marine ecosystems, but the planet.

Further Reading

Atkinson A, Siegel V, Pakhomov E, and Rothery P (2004) Long-term decline in krill stock and increase in salps within the Southern Ocean. *Nature* 432: 100–103.

Beaugrand G, Reid PC, Ibanez F, Lindley JA, and Edwards M (2002) Reorganisation of North Atlantic marine copepod biodiversity and climate. *Science* 296: 1692–1694.

Behrenfield MJ, O'Malley RT, Siegel DA, *et al.* (2006) Climate-driven trends in contemporary ocean productivity. *Nature* 444: 752–755.

Bertram DF, Mackas DL, and McKinnell SM (2001) The seasonal cycle revisited: Interannual variation and ecosystem consequences. *Progress in Oceanography* 49: 283–307.

Bopp L, Aumont O, Cadule P, Alvain S, and Gehlen M (2005) Response of diatoms distribution to global warming and potential implications: A global model study. *Geophysical Research Letters* 32: L19606 (doi:10.1029/2005GL023653).

Bopp L, Monfray P, Aumont O, *et al.* (2001) Potential impact of climate change on marine export production. *Global Biogeochemical Cycles* 15: 81–99.

Edwards M, Johns DG, Licandro P, John AWG, and Stevens DP (2006) Ecological status report: Results from the CPR Survey 2004/2005. *SAHFOS Technical Report* 3: 1–8.

Edwards M and Richardson AJ (2004) The impact of climate change on the phenology of the plankton community and trophic mismatch. *Nature* 430: 881–884.

Field DB, Baumgartner TR, Charles CD, Ferreira-Bartrina V, and Ohman M (2006) Planktonic forminifera of the California Current reflect 20th century warming. *Science* 311: 63–66.

Hays GC, Richardson AJ, and Robinson C (2005) Climate change and plankton. *Trends in Ecology and Evolution* 20: 337–344.

Peterson WT and Schwing FB (2003) A new climate regime in Northeast Pacific ecosystems. *Geophysical Research Letters* 30(17): 1896 (doi:10.1029/2003GL017528).

Raven J, Caldeira K, Elderfield H, *et al.* (2005) *Royal Society Special Report: Ocean Acidification Due to Increasing Atmospheric Carbon Dioxide*. London: The Royal Society.

Richardson AJ (2008) In hot water: Zooplankton and Climate change. *ICES Journal of Marine Science* 65: 279–295.

Richardson AJ and Schoeman DS (2004) Climate impact on plankton ecosystems in the Northeast Atlantic. *Science* 305: 1609–1612.

Riebesell U, Zondervan I, Rost B, Tortell PD, Zeebe RE, and Morel FMM (2000) Reduced calcification of marine plankton in response to increased atmospheric CO_2. *Nature* 407: 364–367.

FISHERIES AND CLIMATE

K. M. Brander, DTU Aqua, Charlottenlund, Denmark

Introduction

Concern over the effects of climate has increased in recent years as concentrations of greenhouse gases have risen in the atmosphere. We all observe changes which are attributed, with more or less justification, to the influence of climate. Anglers and fishermen in many parts of the world have noticed a steady increase in warm-water species. The varieties of locally caught fish which are sold in markets and shops are changing. For example, fish shops in Scotland now sell locally caught bass (*Dicentrarchus labrax*) and red mullet (*Mullus surmuletus*) – species which only occurred in commercial quantities south of the British Isles (600 km to the south) until the turn of the millennium. The northward spread of two non-commercial, subtropical species is shown in **Figure 1**.

They were first recorded off Portugal in the 1960s and were then progressively found further north, until by the mid-1990s they occurred over 1000 km north of the Iberian Peninsula. *Zenopsis conchifer* was recorded at Iceland for the first time in 2002.

In addition to distribution changes, the growth, reproduction, migration, and seasonality of fish are affected by climate. The productivity and composition of the ecosystems on which fish depend are altered, as is the incidence of pathogens. The changes are not only due to temperature; but winds, ocean currents, vertical mixing, sea ice, freshwater runoff, cloud cover, oxygen, salinity, and pH are also part of climate change, with effects on fish. The processes by which these act will be explored in the examples cited later.

The effect of climate on fisheries is not a new phenomenon or a new area of scientific investigation. Like the effects of climate on agriculture or on hunted and harvested animal populations, it has probably been systematically observed and studied since humans began fishing. However, the anthropogenic component of climate change is a new phenomenon, which is gradually pushing the ranges of atmospheric and oceanic conditions outside the envelope experienced during human history. This article begins by describing some of the past effects of climate on fisheries and then goes on to review expected impacts of anthropogenic climate change.

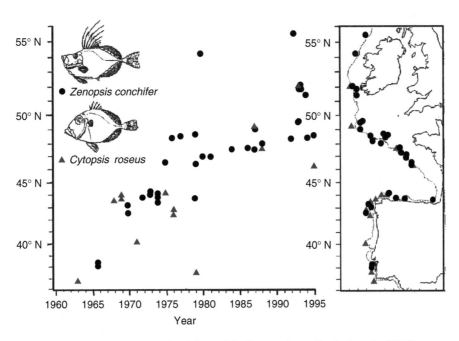

Figure 1 First records of two subtropical fish species (silvery John Dory and rosy Dory) along the NW European continental shelf. From Quéro JC, Buit HD, and Vayne JJ (1998) Les observations de poissons tropicaux et le réchauffement des eaux dans l'Atlantique européen. *Oceanologica Acta* 21: 345–351.

Climate Timescales and Terminology

A wide range of timescales of change in the physical and chemical environment may be included in the term 'climate'. In this article, 'climate variability' refers to changes in temperature, wind fields, hydrological cycles, etc., at annual to decadal timescales and 'climate change' denotes longer-term shifts in the mean values. There are both natural and anthropogenic causes of climate variability and it is not always easy to distinguish the underlying causes of a particular observed effect. Changes in the physical and chemical environment occur naturally on daily, seasonal, and longer-term (e.g., 18.6-year nodal tide) cycles, which can be related to planetary motion. Natural variability in the environment overlies these cycles, so that one can, for example, speak of a windy month or a wet year. Underlying such statements is the idea of a 'normal' month or year, which is generally defined in relation to a climatology, that is, by using a long-term mean and distribution of the variable in question. Volcanic activity and solar fluctuations are other nonanthropogenic factors which affect climate.

History

The Norwegian spring spawning herring (*Clupea harengus* L.) has been a major part of the livelihood of coastal communities in western and northern Norway for over 1000 years.

> It comes up to the shore here from the great fish pond which is the Icelandic Sea, towards the winter when the great part of other fish have left the land. And the herring does not seek the shore along the whole, but at special points which God in his Good Grace has found fitting, and here in my days there have been two large and wonderful herring fisheries at different places in Norway. The first was between Stavanger and Bergen and much further north, and this fishery did begin to diminish and fall away in the year 1560. And I do not believe there is any man to know how far the herring travelled. For the Norwegian Books of Law show that the herring fishery in most of the northern part of Norway has continued for many hundreds of years, although it may well be that in punishment for the unthankfulness of men it has moved from place to place, or has been taken away for a long period. (Clergyman Peder Claussøn Friis (1545–1614))

The changes in distribution of herring which Clergyman Friis wrote about in the sixteenth century have occurred many times since. We now know that herring, which spawn along the west coast of Norway in spring, migrate out to feed in the Norwegian Sea in summer, mainly on the copepod *Calanus finmarchicus*. The distribution shifts in response to decadal changes in oceanic conditions.

Beginning in the 1920s much of the North Atlantic became warmer, the polar front moved north, and the summer feeding migration of this herring stock expanded along the north coast of Iceland. For the next four decades, the resulting fishery provided economic prosperity for northern Iceland and for the country as a whole. When the polar front, which separates cold, nutrient-poor polar water from warmer, nutrient-rich Atlantic water, shifted south and east again, in 1964–65, the herring stopped migrating along the north coast of Iceland. At the same time, the stock collapsed due to overfishing and the herring processing business in northern Iceland died out (**Figure 2**).

There are many similar examples of fisheries on pelagic and demersal fish stocks, which have changed their distribution or declined and recovered as oceanic conditions switched between favorable and unfavorable periods. The Japanese Far Eastern sardine has undergone a series of boom and bust cycles lasting several decades as has the Californian sardine and anchovy. Along the west coast of South America the great Peruvian anchoveta fishery is subject to great fluctuations in abundance, which are not only driven by El Niño/Southern Oscillation (ENSO), but also by decadal-scale variability in the Pacific circulation. In the North Atlantic, the economy of fishing communities in Newfoundland, Greenland, and Faroe has been severely affected by climate-induced fluctuations in the cod stocks (*Gadus morhua* L.) on which they depend.

A number of lessons can be learned from history:

- Fish stocks have always been subject to climate-induced changes.
- Stocks can recover and recolonize areas from which they had disappeared, but such recoveries may take a long time.
- The risk of collapse increases when unfavorable environmental changes overwhelm the resilience of heavily fished stocks.

The present situation is different from the past in at least two important respects: (1) the current rate of climate change is very rapid; and (2) fish stocks are currently subjected to extra mortality and stress due to overfishing and other anthropogenic impacts. History may therefore not be a reliable guide to the future changes in fisheries.

Effects on Individuals, Populations, and Ecosystems

Climatic factors act directly on the growth, survival, reproduction, and movement of individuals and

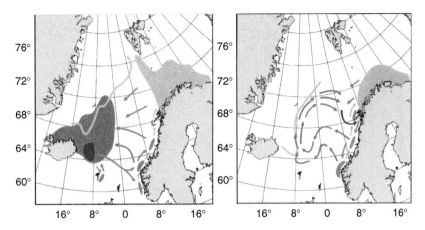

Figure 2 Left panel shows the distribution of Norwegian spring spawning herring between the 1920s and 1965, with areas of spawning (dark blue), nursery (light blue), feeding (green), and overwintering (red). Right panel is the distribution after 1990, when the stock had recovered, but had not reoccupied the area north of Iceland. The polar front separating warm, nutrient-rich Atlantic water from cooler, nutrient-depleted polar water is shown as a light blue line. From Vilhjalmsson H (1997) Climatic variations and some examples of their effects on the marine ecology of Icelandic and Greenland waters, in particular during the present century. *Rit Fiskideildar* XV(1): 9–27.

some of these processes can be studied experimentally, as well as by sampling in the ocean. The integrated effects of the individual processes, are observed at the level of populations, communities, or ecosystems.

In order to convincingly attribute a specific change to climate, one needs to be able to identify and describe the processes involved. Even if the processes are known, and have been studied experimentally, such as the direct effects of temperature on growth rate, the outcome at the population level, over periods of years, can be complex and uncertain. Temperature affects frequency of feeding, meal size, rate of digestion, and rate of absorption of food. Large-scale experiments in which a range of sizes of fish were fed to satiation (Atlantic cod, *G. morhua* L. in this case; see **Figure 3**) show that there is an optimum temperature for growth, but that this depends on the size of the fish, with small fish showing a higher optimum. The optimum temperature for growth also depends on how much food is available, since the energy required for basic maintenance metabolism increases at higher temperature. This means that if food is in short supply then fish will grow faster at cooler temperature, but if food is plentiful then they will grow faster at higher temperatures. A further complication is that temperature typically has a seasonal cycle so the growth rate may decline due to higher summer temperature, but increase due to higher winter temperature.

Recruitment of young fish to an exploited population is variable from year to year for a number of reasons, including interannual and long-term climatic variability. The effects of environmental

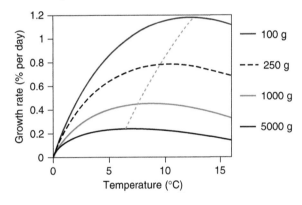

Figure 3 Growth rate of four sizes of Atlantic cod (*Gadus morhua* L.) in rearing experiments at different temperatures in which they were provided unlimited food. The steep dashed line intersects the growth curves at their maximum values, to show how the temperature for maximum growth rate declines as fish get bigger.

variability on survival during early life, when mortality rates in the plankton are very high, is thought to be critical in determining the number of recruits. When recruitment is compared across a number of cod populations, a consistent domed pattern emerges (**Figure 4**).

The relationship between temperature and recruitment for cod has an ascending limb from *c.* 0 to 4 °C and a descending limb above *c.* 7 °C.

In addition to temperature, the distribution of fish depends on salinity, oxygen, and depth, which is only affected by climate at very long timescales. A 'bioclimate envelope', defining the limits of the range, can be constructed by taking all climate-related variables together. Such 'envelopes' are in reality

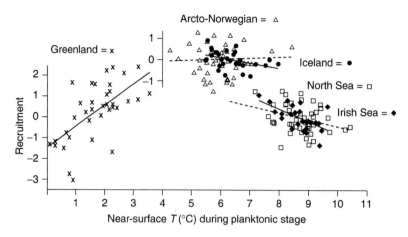

Figure 4 Composite pattern of recruitment for five of the North Atlantic cod stocks to illustrate the effect of temperature during the planktonic stage of early life on the number of recruiting fish. The scales are log$_e$ (number of 1-year-old fish) with the means adjusted to zero. The axes for the Arcto-Norwegian and Iceland stocks have been displaced vertically.

quite complex because the tolerance and response of fish may vary with size (as shown for growth in **Figure 3**); there may be interactions between the factors (e.g., tolerance of low oxygen level depends on salinity) and different subpopulations may have different tolerances because they have developed local adaptations (e.g., cold-adapted populations may produce antifreeze molecules).

Distribution and abundance of a population are the outcome of their rates of growth, reproductive output, survival to maturity, mortality (due to fishing and natural causes), migration, and location of spawning. Unfavorable changes in any of these, due to climate and other factors, will cause changes in distribution and abundance, with the result that a species may no longer be able to maintain a population in areas which are affected. Favorable changes allow a species to increase its population or to colonize previously unsuitable areas.

Climate and fishing can both be considered as additional stresses on fish populations. In order to manage fisheries in a sustainable way the effects of both (and other factors, such as pollution or loss of essential habitat) need to be recognized, as well as the interactions between them. When the age structure and geographic extent of fish populations are reduced by fishing pressure, then they become less resilient to climate change. Conversely, if climate change reduces the surplus production of a population (by altering growth, survival, or reproductive success), then that population will decline at a level of fishing which had previously been sustainable. Because of these interactions it is not possible to deal with adaption to climate change and fisheries management as separate issues. The most effective strategy to assist fish stocks in adapting to climate change is to reduce the mortality due to fishing. Sustainable

fisheries require continuous monitoring of the consequences of climate change.

Regional Effects of Climate

Tropical Pacific

The tuna of the Pacific provide one of the very few examples in which the consequences of climate change have been modeled to include geographic and trophic detail all the way through from primary and secondary production to top fish predators. The tuna species skipjack (*Katsuwonus pelamis*) and yellowfin (*Thunnus albacares*) are among the top predators of the tropical pelagic ecosystem and produced a catch of 3.6 million t in 2003, which represents *c.* 5.5% of total world capture fisheries in weight and a great deal more in value. Their forage species include the Japanese sardine *Engraulis japonicus*, which itself provided a catch of over 2 million t in 2003. The catches and distribution of these species and other tuna species (e.g., albacore *Thunnus alalunga*) are governed by variability in primary and secondary production and location of suitable habitat for spawning and for adults, which in turn are linked to varying regimes of the principal climate indices, such as the El Niño–La Niña Southern Oscillation index (SOI) and the related Pacific Decadal Oscillation (PDOI). Statistical and coupled biogeochemical models have been developed to explore the causes of regional variability in catches and their connection with climate. The model area includes the Pacific from 40° S to 60° N and timescales range from short-term to decadal regime shifts. The model captures the slowdown of Pacific meridional overturning circulation and decrease of equatorial upwelling, which has caused primary production and biomass to decrease by about 10% since 1976–77 in the equatorial Pacific. Further climate change will affect the distribution and production of tuna fisheries in rather complex ways. Warmer surface waters and lower primary production in the central and eastern Pacific may result in a redistribution of tuna to higher latitudes (such as Japan) and toward the western equatorial Pacific.

North Pacific

Investigations into the effects of climate change in the North Pacific have focused strongly on regime shifts. The physical characteristics of these regime shifts and the biological consequences differ between the major regions within the North Pacific.

The PDO tracks the dominant spatial pattern of sea surface temperature (SST). The alternate phases of the PDO represent cooling/warming in the central subarctic Pacific and warming/cooling along the North American continental shelf. This 'classic' pattern represents change along an east–west axis, but since 1989 a north–south pattern has also emerged. Other commonly used indices track the intensity of the winter Aleutian low-pressure system and the sea level pressure over the Arctic. North Pacific regime shifts are reported to have occurred in 1925, 1947, 1977, 1989, and 1998, and paleo-ecological records show many earlier ones. The duration of these regimes appears to have shortened from 50–100 years to c. 10 years for the two most recent regimes, although whether this apparent shortening of regimes is real and whether it is related to other aspects of climate change is a matter of current debate and concern. The SOI also has a large impact on the North Pacific, adding an episodic overlay with a duration of 1 or 2 years to the decadal-scale regime behavior.

Regime shifts, such as the one in 1998, have well-documented effects on ocean climate and biological systems. Sea surface height (SSH) in the central North Pacific increased, indicating a gain in thickness of the upper mixed layer, while at the same time SSH on the eastern and northern boundaries of the North Pacific dropped. The position of the transition-zone chlorophyll front, which separates subarctic from subtropical waters and is a major migration and forage habitat for large pelagic species, such as albacore tuna, shifted northward. In addition to its effects on pelagic fish species, shifts in the winter position of the chlorophyll front affect other species, such as Hawaiian monk seals, whose pup survival rate is lower when the front, with its associated production, is far north of the islands. Spiny lobsters (*Panulirus*

marginatus) recruitment in the Northwestern Hawaiian Islands is also affected.

In the California Current System (CCS), zooplankton species characteristic of shelf waters have since 1999 replaced the southerly, oceanic species which had been abundant since 1989 and northern fish species (Pacific salmon, cod, and rockfish species) have increased, while the southern migratory pelagics such as Pacific sardines, have declined. The distribution of Pacific hake (*Merluccius productus*), which range from Baja California to the Gulf of Alaska, is closely linked to hydrographic conditions. During the 1990s, the species occurred as far north as the Gulf of Alaska, but following a contraction of range by several hundred kilometers in 2000 and 2001, its northern limit has reverted to northern Vancouver Island, a return to the distribution observed in the 1980s.

The biological response to the 1998 regime shift was weaker in the Gulf of Alaska and the Bering Sea than in the central North Pacific and CCS. The northern regions of the western North Pacific resembled the southern regions of the eastern North Pacific in showing an increase in biological production. Zooplankton biomass increased in the Sea of Okhotsk and the previously dominant Japanese sardine was replaced by herring, capelin, and Japanese anchovy.

North Atlantic

Some of the consequences of the warming of the North Atlantic from the 1920s to the 1960s have already been described. There are numerous excellent publications on the subject, dating back to the 1930s and 1940s, which show how much scientific interest there was in the effects of climate change 65 years ago.

The history of cod stocks at Greenland since the early 1900s is particularly well documented, showing how rapidly a species can extend its range (at a rate of $50 \, \mathrm{km \, yr^{-1}}$) and then decline again. This rate of range extension is matched by the examples shown in **Figure 1** and also by the plankton species which have been collected systematically since the 1930s by ships of opportunity towing continuous plankton recorder on many routes in the North Atlantic. This and other evidence indicate that the rates of change in distribution in response to climate are much more rapid in the sea than they are on land. There are fewer physical barriers in the sea; many marine species have dispersive planktonic life stages; the diverse, immobile habitats, which are created on land by large, long-lived plants, do not occur in the

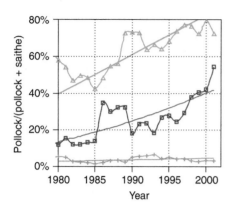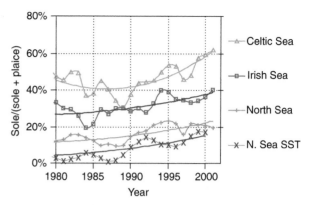

Figure 5 Increasing abundance of warm-adapted species (pollock: Pollachius pollachius and sole: *Solea solea*) relative to similar cold-adapted species (saithe: *Pollachius virens* and plaice: *Pleuronectes platessa*) as shown by catch ratios in the Celtic Sea, Irish Sea, and North Sea. The SST for the North Sea is also shown.

sea, with the exception of kelp forests in coastal areas. Coral reefs are another exception.

One of the difficulties in ascribing observed changes in fish stocks to climate is that fishing now has such a pervasive effect and exerts such high mortalities on most species. Where similar warm and cold-adapted species co-occur in a particular fishing area, the ratio of their catches gives an indication of which of them is being favored by climate trends (see **Figure 5**).

Baltic Sea

The Baltic Sea is the largest brackish water sea in the world, with much lower species diversity than the adjacent North Sea. The fisheries depend on just three marine species (cod, herring, and sprat), which are at the extreme limits of their tolerance of low salinity and oxygen. Over time, the Baltic has become fresher and less oxygenated, but there are periodic inflows of high-salinity, oxygen-rich water from the Skagerrak (North Sea) which flush through the deep basins and restore more favorable, saline, oxygenated conditions for the fish to reproduce. A specific, short-term weather pattern is required to generate an inflow: a period of easterly wind lowers the sea level in the Baltic, then several weeks of westerly wind force water from the Skagerrak through the shallow Danish Straits and into the deeper areas to the east.

On average, there has been approximately one major inflow per year since 1897 and the benefit to the fish stocks lasts for more than 1 year, but the frequency with which the required short-term weather pattern occurs depends on longer-term climatic factors. Inflows have been much less frequent over the past two decades, the last two major inflows happening in 1993 and 2003. The changes in

windfields associated with climate change may continue to reduce the frequency of inflows to the Baltic.

The reproductive potential of cod in particular has been badly affected by the reduced volume of suitable water for development of its eggs. Cod spawn in the deep basins and if the salinity and density are too low, then their eggs sink into the anoxic layers near the bottom, where they die. This is one of the very few examples in which laboratory studies on individual fish can be applied almost directly to make inferences about the effects of climate. The buoyancy of cod eggs can be measured in a density gradient and their tolerance of low oxygen and salinity studied in incubation chambers. With this information, it is possible to determine what depth they will sink to under the conditions of temperature and salinity found in the deep basins and hence whether they remain in sufficiently oxygenated water for survival.

Coral Reef Fisheries

Coral reefs have suffered an increasing frequency of bleaching events due to loss of symbiotic algae. This occurs when SST remains $\sim 1\,^{\circ}$C above the current seasonal maxima. Mortality of corals occurs when the increment in SST is $> 2\,^{\circ}$C. Extensive and extreme bleaching occurred in 1998, associated with the strong El Niño and the hottest year on record.

The mass coral bleaching in the Indian Ocean in 1998 apparently did not result in major short-term impacts on coastal reef fisheries. However, in the longer term, the loss of coral communities and reduced structural complexity of the reefs are expected to have serious consequences for fisheries production with reduced fish species richness, local extinctions, and loss of species within key functional groups of reef fish.

Indirect Effects and Interactions

Spread of Pathogens

Pathogens have been implicated in mass mortalities of many aquatic species, including plants, fish, corals, and mammals, but lack of standard epidemiological data and information on pathogens generally make it difficult to attribute causes. An exception is the northward spread of two protozoan parasites (*Perkinsus marinus* and *Haplosporidium nelsoni*) from the Gulf of Mexico to Delaware Bay and further north, where they have caused mass mortalities of Eastern oysters (*Crassostrea virginica*). Winter temperatures consistently lower than 3 °C limit the development of the MSX disease caused by *Perkinsus* and the poleward spread of this and other pathogens can be expected to continue as such winter temperatures become rarer. This example also illustrates the relevance of seasonal information when considering the effects of climate change, since in this case it is winter temperature which controls the spread of the pathogen.

Effects of Changing Primary Production

Changes in primary production and in food-chain processes due to climate will probably be the major cause of future changes in fisheries production. Reduced nutrient supply to the upper ocean due to slower vertical mixing and changes in the balance between primary production and respiration at higher temperature will result in less energy passing to higher trophic levels. Many other processes will also have an effect, which makes prediction uncertain. For example, nutrient-depleted conditions favor small phytoplankton and longer food chains at the expense of diatoms and short food chains. Altered seasonality of primary production will cause mismatches with zooplankton whose phenology is adapted to the timing of the spring bloom. As a result, a greater proportion of the pelagic production may settle out of the water column to the benefit of benthic production.

There is evidence from satellite observations and from *in situ* studies that primary production in some parts of the ocean has begun to decline, but the results of modeling studies suggest that primary production will change very little up to 2050. Within this global result, there are large regional differences. For example, the highly productive sea–ice margin in the Arctic is retreating, but as a result a greater area of ocean is exposed to direct light and therefore becomes more productive.

Impacts of Changes in Fisheries on Human Societies

Fluctuations in fish stocks have had major economic consequences for human societies throughout history. Fishing communities which were dependent on local resources of just a few species have always been vulnerable to fluctuations in stocks, whether due to overfishing, climate, or other causes. The increase in distant water fleets during the last century reduced the dependence of that sector of the fishing industry on a particular area or species, but the resulting increase in rates of exploitation also reduced stock levels and increased their variability.

Many examples can be cited to show the effects of fish-stock fluctuations. The history of herring in European waters over the past 1000 years influenced the economic fortunes of the Hanseatic League and had a major impact on the economy of northern Europe. Climate-dependent fluctuations in the Far Eastern sardine population influenced their fisheries and human societies dependent on them. Variability in cod stocks at Newfoundland, Greenland, and the Faroe Islands due to a combination of climate-related effects and overfishing had major impacts on the economies and societies, resulting in changes in migration and human demography. The investigation of economic effects of climate change on fisheries is a rapidly developing field, which can be expected to help considerably when planning strategies for adaptation or, in some cases, mitigation of future impacts.

Given the uncertainties over future marine production and consequences for fish stocks, it is not surprising that projections of impacts on human societies and economies are also uncertain. Global aquaculture production increased by c. 50% between 1997 and 2003, while capture production decreased by c. 5% and the likelihood that these trends will continue also affects the way in which climate change will affect fisheries production.

Some areas, such as Greenland, which are strongly affected by climate variability and which have been undergoing a relatively cold period since the 1960s, can be expected to benefit from warmer oceanic conditions and changes in the marine ecosystem are occuring there quite rapidly. In other areas, such as Iceland, the positive and negative impacts are more finely balanced.

It is very difficult to judge at a global level who the main losers and winners will be from changes in fisheries as a result of climate change, aside from the obvious advantages of being well informed, well capitalized, and able to shift to alternative areas or kinds of fishing activity (or other nonfishery activities) as circumstances change. Some of the most vulnerable systems may be in the mega-deltas of rivers in Asia, such as the Mekong, where 60 million people are active in fisheries in some way or other. These are mainly seasonal floodplain fisheries, which, in addition to overfishing, are increasingly threatened by changes in the hydrological cycle and in land use, damming, irrigation, and channel alteration. Thus the impact of climate change is just one of a number of pressures which require integrated international solutions if the fisheries are to be maintained.

See also

Dynamics of Exploited Marine Fish Populations. Ecosystem Effects of Fishing. Fisheries Economics. Fisheries Overview. Fishery Management. Fishery Management, Human Dimension. Marine Fishery Resources, Global State of.

Further Reading

ACIA (2005) *Arctic Climate Impact Assessment Scientific Report*. Cambridge, UK: Cambridge University Press.

Cushing DH (1982) *Climate and Fisheries*. London: Academic Press.

Drinkwater KF, Loeng H, Megrey BA, Bailey N, and Cook RM (eds.) (2005) The influence of climate change on North Atlantic fish stocks. *ICES Journal of Marine Science* 62(7): 1203–1542.

German Advisory Council on Global Change (2006) *The Future Oceans – Warming up, Rising High, Turning Sour*. Special Report. Berlin: German Advisory Council on Global Change (WBGU). http://www.wbgu.de/wbgu_sn2006_en.pdf (accessed Mar. 2008).

King JR (ed.) (2005) *Report of the Study Group on Fisheries and Ecosystem Responses to Recent Regime Shifts*. PICES Scientific Report 28, 162pp.

Lehodey P, Chai F, and Hampton J (2003) Modelling climate-related variability of tuna populations from a coupled ocean biogeochemical-populations dynamics model. *Fisheries Oceanography* 12: 483–494.

Quéro JC, Buit HD, and Vayne JJ (1998) Les observations de poissons tropicaux et le réchauffement des eaux dans l'Atlantique européen. *Oceanologica Acta* 21: 345–351.

Stenseth Nils C, Ottersen G, Hurrell JW, and Belgrano A (eds.) (2004) *Marine Ecosystems and Climate Variation: The North Atlantic. A Comparative Perspective*. Oxford, UK: Oxford University Press.

Vilhjalmsson H (1997) Climatic variations and some examples of their effects on the marine ecology of Icelandic and Greenland waters, in particular during the present century. *Rit Fiskideildar* XV(1): 9–27.

Wood CM and McDonald DG (eds.) (1997) *Global Warming: Implications for Freshwater and Marine Fish*. Cambridge, UK: Cambridge University Press.

Relevant Websites

http://www.ipcc.ch
  – IPCC Fourth Assessment Report.

EFFECTS OF CLIMATE CHANGE ON MARINE MAMMALS

I. Boyd and N. Hanson, University of St. Andrews, St. Andrews, UK

Introduction

The eff‘ects of climate change on marine mammals will be caused by changes in the interactions between the physiological state of these animals and the physical changes in their environment caused by climate change. In this article, climate change is defined as a long-term (millennial) trend in the physical climate. This distinguishes it from short-term, regional fluctuations in the physical climate. Marine mammals are warm-blooded vertebrates living in a highly conductive medium often with a steep temperature gradient across the body surface. They also have complex behavioral repertoires that adapt rapidly to changes in the conditions of the external environment. In general, we would expect the changes in the physical environment at the scales envisaged under climate change scenarios to be well within the homeostatic capacity of these species. Effects of climate upon the prey species normally eaten by marine mammals, most of which do not have the same level of homeostatic control to stresses in their physical environment, may be the most likely mechanism of interaction between marine mammals and climate change. However, we should not assume that effects of climate change on marine mammals should necessarily be negative.

Responses to Normal Environmental Variation

Marine mammals normally experience variation in their environment that is very large compared with most variance predicted due to climate change. Examples include the temperature gradients that many marine mammals experience while diving through the water column and the extreme patchiness of the prey resources for marine mammals. Marine mammals in the Pacific have life-histories adapted to transient climatic phenomena such as El Niño, which oscillate every 4 years or so. Consequently, the morphologies, physiologies, behaviors, and life histories of marine mammals will have evolved to cope with this high level of variance. However, it is generally accepted that climate change is occurring too rapidly for the life histories of marine mammals to adapt to longer periods of adverse conditions than are experienced in examples like El Niño.

Marine mammals appear to cope with other longer wavelength oscillations including the North Pacific and North Atlantic Oscillations and the Antarctic Circumpolar Wave and it is possible that their life histories have evolved to cope with this type of long wavelength variation. Nevertheless, nonoscillatory climate change could result in nonlinear processes of change in some of the physical and biological features of the environment that are important to some marine mammal species. Although speculative, obvious changes such as the extent of Arctic and Antarctic seasonal ice cover could affect the presence of essential physical habitat for marine mammals as well as food resources, and there may be other changes in the structure of marine mammal habitats that are less obvious and difficult to both identify and quantify. Changes in the trophic structure of the oceans in ice-bound regions, where the ecology is very reliant on sea ice, may lead the trophic pyramid that supports these top predators to alter substantially. The polar bear is particularly an obvious example of this type of effect where both loss of hunting habitat in the form of sea ice and the potential effects on prey abundance are already having measurable effects upon populations. Similarly, changes in coastal habitats resulting from changes in sea level, changes in run-off and salinity, and changes in nutrient and sediment loads are likely to have important effects on some species of small cetaceans with localized distributions. Many sirenians rely upon seagrass communities and anything that affects the sustainability of this food source is likely to have a negative effect on these species.

Many marine mammal species have already experienced range retraction and population depletion because of direct interaction with man. Monk seals appear to be particularly vulnerable because they rely upon small pockets of beach habitat, many of which are threatened directly by man and also by rising sea level.

Marine mammals are known to be vulnerable to the effects of toxic algal blooms. Toxins may lead to sublethal effects, such as reduced rates of reproduction as well as direct mortality. Several mortality events including coastal whale and dolphin species as well as seals have been attributed to these effects.

Climate change could result in increased frequency of the conditions that lead to such effects, perhaps as a result of interactions between temperature and eutrophication of coastal habitats.

Classifying Effects

A common approach to the assessment of the effects of climate change is to divide these into 'direct' and 'indirect' effects. In this case, 'direct' effects are those associated with changes in the physical environment, such as those that affect the availability of suitable habitat. 'Indirect' effects are those that operate through the agency of food availability because of changes in ecology, susceptibility to disease, changed exposure to pollution, or changes in competitive interactions. Würsig et al. in 2002 added a third level of effect which was the result of human activities occurring in response to climate change that tend to increase conflicts between man and marine mammals. This division has little utility in terms of rationalizing the effects of climate change because, in simple terms, the effects will operate ultimately through the availability of suitable habitat. Assessing the effects of climate change rests upon an assessment of whether there is a functional relationship between the availability of suitable habitat and climate and the form of these functional relationships, which will differ between species, has not been determined.

The expansion and contraction of suitable habitat can be affected by a broad range of factors and some of these can operate on their own but others are often closely related and synergistic, such as the combined effect of retraction of sea ice upon the availability of breeding habitat for seals and also for the food chains that support these predators.

Evidence for the Effects of Climate Change on Marine Mammals

There is no strong evidence that current climate change scenarios are affecting marine mammals although there are studies that suggest some typical effects of climate change could affect marine mammal distribution and abundance. There is an increasing body of literature that links apparent variability in marine mammal abundance, productivity, or behavior with climate change processes. However, with the probable exception of those documenting the changes occurring to the extent of breeding habitat for ringed seals within some sections of the Arctic, and the consequences of this also for polar bears, most of these studies simply reflect a trend toward the interpretation of responses of marine mammals to large-scale regional variability in

the physical environment, as has already been well documented in the Pacific for El Niño, in terms of climate trends. Long-term trends in the underlying regional ecosystem structure are sometimes extrapolated as evidence of climate change. In few, if any, of these cases is there strong evidence that the physical environmental variability being observed is derived from irreversible trends in climate. Some of the current literature confounds understanding of the responses of marine mammals to regional variability with that of climate change, albeit that an understanding of one may be useful in the interpretation and prediction of the effects of the other.

Based upon records of species from strandings, MacLeod et al. in 2005 have suggested that the species diversity of cetaceans around the UK has increased recently and that this may be evidence of range expansion in some species. However, the sample sizes involved are small and there are difficulties in these types of studies accounting for observer effort. This is a common story for marine mammals, and many other marine predators including seabirds, in that, there is a great deal of theory about what the effects of climate change might be but little convincing evidence that backs up these suggestions. Even process studies, involving research on the mechanisms underlying how climate change could affect marine mammals, when considered in detail make a tenuous linkage between the physical variables and the biological response of the marine mammals.

Is Climate Change Research on Marine Mammals Scientific?

Although it is beyond dispute that marine mammals respond to physical changes in habitat suitability, the relationship between a particular effect and the response from the marine mammal is seldom clear. Where data from time series are analyzed, as in the case of Forcada et al., they are used to test post hoc for relationships between climate and biological variables. There is a tendency in these circumstances to test for all possible relationships using a range of physical and biological variables. Such post hoc testing is fraught with pitfalls because invariably the final apparently statistically significant relationships are not downweighted in their significance by all the other nonsignificant relationships that were investigated alongside those that proved to be statistically significant. Of course, there may be a priori reasons for accepting that a particular relationship is true, but the approach to examining time series rarely provides an analysis of the relationships that were not statistically significant or the a priori reasons there might be for rejection of these. Consequently, current suggestions from the literature about the potential effects of climate change may be exaggerated because of the strong possibility of the presence of type I and type II statistical error in the assessment process. Moreover, in the great majority of examples, it will be almost impossible to clearly demonstrate effects of climate change, as has been the case with partitioning the variance between a range of causes of the decline of the Steller sea lion (*Eumetopias jubatus*) in the North Pacific and Bering Sea.

Identifying Situations in Which Climate Change is Likely to Have a Negative Effect on Marine Mammals: Future Work

To date, little has been done to build predictive frameworks for assessing the effects of climate change of marine mammals. There have been broad assessments and focused ecological studies but these are a fragile foundation for guiding policy and management, and for identifying populations that are at greatest risk. The resilience of marine mammal populations to climate change will reflect resilience to any other change in habitat quality, that is, it will depend upon the extent of suitable habitat, the degree to which populations currently fill that habitat, the dispersal capacity of the species, and the structure of the current population, including its capacity for increase and demographics. Clearly, populations that are already in a depleted state, or that are dependent upon habitat that is diminishing for reasons other than climate change, will be more vulnerable to the effects of climate change. There are also some, as yet unconvincing, suggestions that habitat degradation may occur through effects of climate upon pollutant burdens.

The general demographic characteristics of marine mammal populations are relatively well known so there are simple ways of assessing the risk to populations under different scenarios of demographic stochasticity, population size, and isolation. An analysis of this type could only provide a very broad guide to the types of effects that could be expected but, whereas no such analysis has been carried out to date, this should be seen as a first step in the risk-assessment process.

The metapopulation structure of many marine mammal populations will affect resilience to climate change and will be reflected in the dispersal capacity of the population. Again, this type of effect could be included within an analysis of the sensitivity of marine mammal populations to climate change under different metapopulation structures. A feature of climate change is that it is likely to have global as well as local effects and the sensitivity to the relative contribution from these would be an important feature of such an analysis.

Further Reading

Atkinson A, Siegel V, Pakhamov E, and Rothery P (2004) Long-term decline in krill stocks and increase in salps within the Southern Ocean. *Nature* 432: 100–103.

Cavalieri DJ, Parkinson CL, and Vinnikov KY (2003) 30-year satellite record reveals contrasting Arctic and Antarctic decadal sea ice variability. *Geophysical Research Letters* 30: 1970 (doi:10.1029/2003GL018031).

Derocher E, Lunn N, and Stirling I (2004) Polar bears in a warming climate. *Integrative and Comparative Biology* 44: 163–176.

Ferguson S, Stirling I, and McLoughlin P (2005) Climate change and ringed seal (*Phoca hispida*) recruitment in western Hudson Bay. *Marine Mammal Science* 21: 121–135.

Forcada J, Trathan P, Reid K, and Murray E (2005) The effects of global climate variability in pup production of Antarctic fur seals. *Ecology* 86: 2408–2417.

Grebmeier J, Overland J, Moore S, *et al.* (2006) A major ecosystem shift in the northern Bering Sea. *Science* 311: 1461–1464.

Green C and Pershing A (2004) Climate and the conservation biology of North Atlantic right whales: The right whale at the wrong time? *Frontiers in Ecology and the Environment* 2: 29–34.

Heide-Jorgensen MP and Lairde KL (2004) Declining extent of open water refugia for top predators in Baffin Bay and adjacent waters. *Ambio* 33: 487–494.

Hunt G, Stabeno P, Walters G, *et al.* (2002) Climate change and control of southeastern Bering Sea pelagic ecosystem. *Deep Sea Research II* 49: 5821–5853.

Laidre K and Heide-Jorgensen M (2005) Artic sea ice trends and narwhal vulnerability. *Biological Conservation* 121: 509–517.

Leaper R, Cooke J, Trathan P, Reid K, Rowntree V, and Payne R (2005) Global climate drives southern right whale (*Eubaena australis*) population dynamics. *Biology Letters* 2 (doi:10.1098/rsbl.2005.0431).

Lusseau RW, Wilson B, Grellier K, Barton TR, Hammond PS, and Thompson PM (2004) Parallel influence of climate on the behaviour of Pacific killer whales and Atlantic bottlenose dolphins. *Ecology Letters* 7: 1068–1076.

MacDonald R, Harner T, and Fyfe J (2005) Recent climate change in the Artic and its impact on contaminant pathways and interpretation of temporal trend data. *Science of the Total Environment* 342: 5–86.

MacLeod C, Bannon S, Pierce G, *et al.* (2005) Climate change and the cetacean community of north-west Scotland. *Biological Conservation* 124: 477–483.

McMahon C and Burton C (2005) Climate change and seal survival: Evidence for environmentally mediated changes in elephant seal *Mirounga leonina* pup survival. *Proceedings of the Royal Society B* 272: 923–928.

Robinson R, Learmouth J, Hutson A, *et al.* (2005) Climate change and migratory species. *BTO Research Report 414*. London: Defra. http://www.bto.org/research/reports/researchrpt_abstracts/2005/RR414%20_summary_report.pdf (accessed Mar. 2008).

Sun L, Liu X, Yin X, Zhu R, Xie Z, and Wang Y (2004) A 1,500-year record of Antarctic seal populations in response to climate change. *Polar Biology* 27: 495–501.

Trillmich F, Ono KA, Costa DP, *et al.* (1991) The effects of El Niño on pinniped populations in the eastern Pacific. In: Trillmich F and Ono KA (eds.) *Pinnipeds and El Niño: Responses to Environmental Stress*, pp. 247–270. Berlin: Springer.

Trites A, Miller A, Maschner H, *et al.* (2006) Bottom up forcing and decline of Stellar Sea Lions in Alaska: Assessing the ocean climate hypothesis. *Fisheries Oceanography* 16: 46–67.

Walther G, Post E, Convey P, *et al.* (2002) Ecological responses to recent climate change. *Nature* 416: 389–395.

Würsig B, Reeves RR, and Ortega-Ortiz JG (2002) Global climate change and marine mammals. In: Evans PGH and Raga JA (eds.) *Marine Mammals – Biology and Conservation*, pp. 589–608. New York: Kluwer Academic/Plenum Publishers.

SEABIRD RESPONSES TO CLIMATE CHANGE

David G. Ainley, H.T. Harvey and Associates, San Jose, CA, USA

G. J. Divoky, University of Alaska, Fairbanks, AK, USA

Introduction

This article reviews examples showing how seabirds have responded to changes in atmospheric and marine climate. Direct and indirect responses take the form of expansions or contractions of range; increases or decreases in populations or densities within existing ranges; and changes in annual cycle, i.e., timing of reproduction. Direct responses are those related to environmental factors that affect the physical suitability of a habitat, e.g., warmer or colder temperatures exceeding the physiological tolerances of a given species. Other factors that can affect seabirds directly include: presence/absence of sea ice, temperature, rain and snowfall rates, wind, and sea level. Indirect responses are those mediated through the availability or abundance of resources such as food or nest sites, both of which are also affected by climate change.

Seabird response to climate change may be most apparent in polar regions and eastern boundary currents, where cooler waters exist in the place of the warm waters that otherwise would be present. In analyses of terrestrial systems, where data are in much greater supply than marine systems, it has been found that range expansion to higher (cooler but warming) latitudes has been far more common than retraction from lower latitudes, leading to speculation that cool margins might be more immediately responsive to thermal variation than warm margins. This pattern is evident among sea birds, too. During periods of changing climate, alteration of air temperatures is most immediate and rapid at high latitudes due to patterns of atmospheric circulation. Additionally, the seasonal ice and snow cover characteristic of polar regions responds with great sensitivity to changes in air temperatures. Changes in atmospheric circulation also affect eastern boundary currents because such currents exist only because of wind-induced upwelling.

Seabird response to climate change, especially in eastern boundary currents but true elsewhere, appears to be mediated often by El Niño or La Niña. In other words, change is expressed stepwise, each step coinciding with one of these major, short-term climatic perturbations. Intensive studies of seabird populations have been conducted, with a few exceptions, only since the 1970s; and studies of seabird responses to El Niño and La Niña, although having a longer history in the Peruvian Current upwelling system, have become commonplace elsewhere only since the 1980s. Therefore, our knowledge of seabird responses to climate change, a long-term process, is in its infancy. The problem is exacerbated by the long generation time of seabirds, which is 15–70 years depending on species.

Evidence of Sea-bird Response to Prehistoric Climate Change

Reviewed here are well-documented cases in which currently extant seabird species have responded to climate change during the Pleistocene and Holocene (last 3 million years, i.e., the period during which humans have existed).

Southern Ocean

Presently, 98% of Antarctica is ice covered, and only 5% of the coastline is ice free. During the Last Glacial Maximum (LGM: 19 000 BP), marking the end of the Pleistocene and beginning of the Holocene, even less ice-free terrain existed as the ice sheets grew outward to the continental shelf break and their mass pushed the continent downward. Most likely, land-nesting penguins (Antarctic genus *Pygoscelis*) could not have nested on the Antarctic continent, or at best at just a few localities (e.g., Cape Adare, northernmost Victoria Land). With warming, loss of mass and subsequent retreat of the ice, the continent emerged.

The marine-based West Antarctic Ice Sheet (WAIS) may have begun to retreat first, followed by the land-based East Antarctic Ice Sheet (EAIS). Many Adélie penguin colonies now exist on the raised beaches remaining from successive periods of rapid ice retreat. Carbon-dated bones from the oldest beaches indicate that Adélie penguins colonized sites soon after they were exposed. In the Ross Sea, the WAIS receded south-eastward from the shelf break to its present position near Ross Island approximately 6200 BP. Penguin remains from Cape Bird, Ross Island (southwestern Ross Sea), date back to

7070 ± 180 BP; those from the adjacent southern Victoria Land coast (Terra Nova Bay) date to 7505 ± 230 BP. Adélie penguin remains at capes Royds and Barne (Ross Island), which are closest to the ice-sheet front, date back to 500 BP and 375 BP, respectively. The near-coast Windmill Islands, Indian Ocean sector of Antarctica, were covered by the EAIS during the LGM. The first islands were deglaciated about 8000 BP, and the last about 5500 BP. Penguin material from the latter was dated back to 4280–4530 BP, with evidence for occupation 500–1000 years earlier. Therefore, as in Victoria Land, soon after the sea and land were free from glaciers, Adélie penguins established colonies.

The study of raised beaches at Terra Nova Bay also investigated colony extinction. In that area several colonies were occupied 3905–4930 BP, but not since. The period of occupancy, called 'the penguin optimum' by geologists, corresponds to one of a warmer climate than at present. Currently, this section of Victoria Land is mostly sea-ice bound and penguins nest only at Terra Nova Bay owing to a small, persistent polynya (open-water area in the sea ice).

A study that investigated four extinct colonies of chinstrap penguin in the northern part of the Antarctic Peninsula confirmed the rapidity with which colonies can be founded or deserted due to fluctuations in environmental conditions. The colonies were dated at about 240–440 BP. The chinstrap penguin, an open-water loving species, occupied these former colonies during infrequent warmer periods indicated in glacial ice cores from the region. Sea ice is now too extensive for this species offshore of these colonies. Likewise, abandoned Adélie penguin nesting areas in this region were occupied during the Little Ice Age (AD 1500–1850), but since have been abandoned as sea ice has dissipated in recent years (see below).

South-east Atlantic

A well-documented avifaunal change from the Pleistocene to Recent times is based on bone deposits at Ascension and St Helena Islands. During glacial maxima, winds were stronger and upwelling of cold water was more pronounced in the region. This pushed the 23°C surface isotherm north of St Helena, thus accounting for the cool-water seabird avifauna that was present then. Included were some extinct species, as well as the still extant sooty shearwater and white-throated storm petrel. As the glacial period passed, the waters around St Helena became warmer, thereby encouraging a warm-water avifauna similar to that which exists today at Ascension Island; the cool-water group died out or decreased

substantially in representation. Now, a tropical avifauna resides at St Helena including boobies, a frigatebird not present earlier, and Audubon's shearwater. Most recently these have been extirpated by introduced mammals.

North-west Atlantic/Gulf of Mexico

Another well-documented change in the marine bird fauna from Plio-Pleistocene to Recent times is available for Florida. The region experienced several major fluctuations in sea level during glacial and interglacial periods. When sea level decreased markedly to expose the Isthmus of Panama, thus, changing circulation, there was a cessation of upwelling and cool, productive conditions. As a result, a resident cool-water avifauna became extinct. Subsequently, during periods of glacial advance and cooler conditions, more northerly species visited the area; and, conversely, during warmer, interglacial periods, these species disappeared.

Direct Responses to Recent Climate Change

A general warming, especially obvious after the mid-1970s, has occurred in ocean temperatures especially west of the American continents. Reviewed here are sea-bird responses to this change.

Chukchi Sea

The Arctic lacks the extensive water–ice boundaries of the Antarctic and as a result fewer seabird species will be directly affected by the climate-related changes in ice edges. Reconstructions of northern Alaska climatology based on tree rings show that temperatures in northern Alaska are now the warmest within the last 400 years with the last century seeing the most rapid rise in temperatures following the end of the Little Ice Age (AD 1500–1850). Decreases in ice cover in the western Arctic in the last 40 years have been documented, but the recent beginnings of regional ornithological research precludes examining the response of birds to these changes.

Changes in distribution and abundance related to snow cover have been found for certain cavity-nesting members of the auk family (Alcidae). Black guillemots and horned puffins require snow-free cavities for a minimum of 80 and 90 days, respectively, to successfully pair, lay and incubate eggs, and raise their chicks(s). Until the mid-1960s the snow-free period in the northern Chukchi Sea was usually shorter than 80 days but with increasing spring air temperatures, the annual date of spring snow-melt has advanced more than 5 days per decade over the last five decades (**Figure 1**). The annual snow-free period now regularly exceeds 90 days, which reduces the likelihood of chicks being trapped in their nest sites before fledging. This has allowed black guillemots and horned puffins to establish colonies (range expansion) and increase population size in the northern Chukchi.

California Current

Avifaunal changes in this region are particularly telling because the central portion of the California Current marks a transitional area where subtropical and subarctic marine faunas meet, and where north–south faunal shifts have been documented at a variety of temporal scales. Of interest here is the invasion of brown pelicans northward from their 'usual' wintering grounds in central California to the Columbia River mouth and Puget Sound, and the invasion of various terns and black skimmers northward from Mexico into California. The pelican and terns are tropical and subtropical species.

During the last 30 years, air and sea temperatures have increased noticeably in the California Current region. The response of seabirds may be mediated by thermoregulation as evidenced by differences in the amount of subcutaneous fat held by polar and tropical seabird species, and by the behavioral responses of seabirds to inclement air temperatures.

The brown pelican story is particularly well documented and may offer clues to the mechanism by which similar invasions come about. Only during the very intense El Niño of 1982–83 did an unusual number of pelicans move northward to the Columbia River mouth. They had done this prior to 1900, but then came a several-decade long period of cooler temperatures. Initially, the recent invasion involved juveniles. In subsequent years, these same birds returned to the area, and young-of-the-year birds followed. Most recently, large numbers of adult pelicans have become a usual feature feeding on anchovies that have been present all along. This is an example of how tradition, or the lack thereof, may facilitate the establishment (or demise) of expanded range, in this case, compatible with climate change.

The ranges of skimmers and terns have also expanded in pulses coinciding with El Niño. This pattern is especially clear in the case of the black skimmer, a species whose summer range on the east coast of North America retracts southward in winter. On the west coast, almost every step in a northward expansion of range from Mexico has coincided with ocean warming and, in most cases, El Niño: first California record, 1962 (not connected to ocean warming); first invasion *en masse*, 1968 (El Niño); first nesting at Salton Sea, 1972 (El Niño); first nesting on coast (San Diego), 1976 (El Niño); first nesting farther north at Newport Bay and inland, 1986 (El Niño). Thereafter, for Southern California as a whole, any tie to El Niño became obscure, as (1) average sea-surface temperatures off California became equivalent to those reached during the intense 1982–83 El Niño, and (2) population increase became propelled not just by birds dispersing north from Mexico, but also by recruits raised locally. By 1995, breeding had expanded north to Central California. In California, with warm temperatures year round, skimmers winter near where they breed.

The invasion northward by tropical/subtropical terns also relates to El Niño or other warm-water incursions. The first US colony (and second colony in the world) of elegant tern, a species usually present off California as post-breeders (July through October), was established at San Diego in 1959, during the strongest El Niño event in modern times. A third colony, farther north, was established in 1987 (warm-water year). The colony grew rapidly, and in 1992–93 (El Niño) numbers increased 300% (to 3000 pairs). The tie to El Niño for elegant terns is confused by the strong correlation, initially, between numbers breeding in San Diego and the biomass of certain prey (anchovies), which had also increased. Recently, however, anchovies have decreased. During the intense 1997–98 El Niño, hundreds of elegant terns were observed in courtship during spring even farther north (central California). No colony formed.

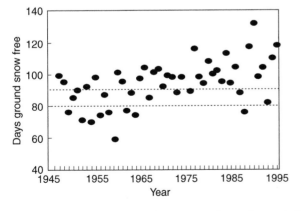

Figure 1 Changes in the length of the annual snow-free period at Barrow, Alaska, 1947–1995. Dashed lines show the number of days that black guillemots and horned puffins require a snow-free cavity (80 and 90 days, respectively). Black guillemots first bred near Barrow in 1966 and horned puffins in 1986. (Redrawn from Divoky, 1998.)

Climate change, and El Niño as well, may be involved in the invasion of Laysan albatross to breed on Isla de Guadalupe, in the California Current off northern Mexico. No historical precedent exists for the breeding by this species anywhere near this region. First nesting occurred in 1983 (El Niño) and by 1988, 35–40 adults were present, including 12 pairs. Ocean temperatures off northern Mexico are now similar to those in the Hawaiian Islands, where nesting of this species was confined until recently. In the California Current, sea temperatures are the warmest during the autumn and winter, which, formerly, were the only seasons when these albatross occurred there. With rising temperatures in the California Current, more and more Laysan albatross have been remaining longer into the summer each year. Related, too, may be the strengthening of winds off the North American west coast to rival more closely the trade winds that buffet the Hawaiian Islands. Albatross depend on persistent winds for efficient flight, and such winds may limit where albatrosses occur, at least at sea.

Several other warm-water species have become more prevalent in the California Current. During recent years, dark-rumped petrel, a species unknown in the California Current region previously, has occurred regularly, and other tropical species, such as Parkinson's petrel and swallow-tailed gull have been observed for the first time in the region.

In response to warmer temperatures coincident with these northward invasions of species from tropical and subtropical regions, a northward retraction of subarctic species appears to be underway, perhaps related indirectly to effects of prey availability. Nowadays, there are markedly fewer black-footed albatross and sooty and pink-footed shearwaters present in the California Current system than occurred just 20 years ago (**Figure 2**). Cassin's auklet is becoming much less common at sea in central California, and its breeding populations have also been declining. Similarly, the southern edge of the breeding range of common murres has retreated north. The species no longer breeds in Southern California (Channel Islands) and numbers have become greatly reduced in Central California (**Figure 3**). Moreover, California Current breeding populations have lost much of their capacity, demonstrated amply as late as the 1970s, to recover

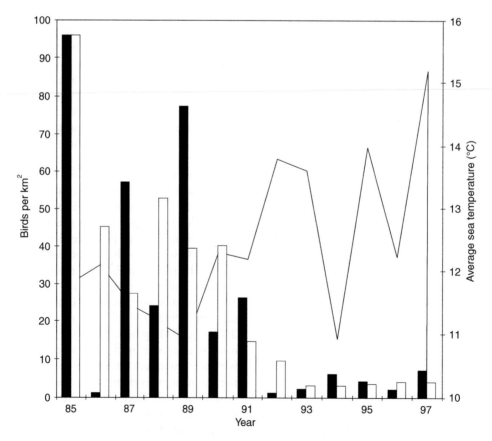

Figure 2 The density (■) (plus 3-point moving average, □) of a cool-water species, the sooty shearwater, in the central portion of the California Current, in conjunction with changes in marine climate (sea surface temperature, (—)), 1985–1997.

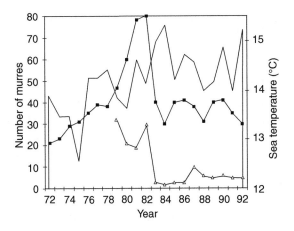

Figure 3 Changes in the number of breeding common murres in central (California, ■) and northern (Washington, △) portions of the California Current during recent decades. Sea surface temperature (—) from central California shown for comparison. During the 1970s, populations had the capacity to recover from reductions due to anthropogenic factors (e.g., oil spills). Since the 1982–83 El Niño event and continued higher temperatures, however, the species' capacity for population recovery has been lost (cf. **Figure 5**)

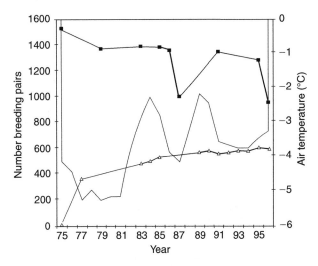

Figure 4 Changes in the number of breeding pairs of two species of penguins at Arthur Harbor, Anvers Island, Antarctica (64°S, 64°W), 1975–1996. The zoogeographic range of Adélie penguins (■) is centered well to the south of this site; the range of chinstrap penguins (△) is centered well to the north. Arthur Harbor is located within a narrow zone of overlap (200 km) between the two species and is at the northern periphery of the Adélie penguins' range.

from catastrophic losses. The latter changes may or may not be more involved with alterations of the food web (see below).

Northern Bellingshausen Sea

Ocean and air temperatures have been increasing and the extent of pack ice has been decreasing for the past few decades in waters west of the Antarctic peninsula. In response, populations of the Adélie penguin, a pack-ice species, have been declining, while those of its congener, the open-water dwelling chinstrap penguin have been increasing (**Figure 4**). The pattern contrasts markedly with the stability of populations on the east side of the Antarctic Peninsula, which is much colder and sea-ice extent has changed little.

The reduction in Adélie penguin populations has been exacerbated by an increase in snowfall coincident with the increased temperatures. Deeper snow drifts do not dissipate early enough for eggs to be laid on dry land (causing a loss of nesting habitat and eggs), thus also delaying the breeding season so that fledging occurs too late in the summer to accommodate this species' normal breeding cycle. This pattern is the reverse of that described above for black guillemots in the Arctic. The penguin reduction has affected mostly the smaller, outlying subcolonies, with major decreases being stepwise and occurring during El Niño.

Similar to the chinstrap penguin, some other species, more typical of the Subantarctic, have been

expanding southward along the west side of the Antarctic Peninsula. These include the brown skua, blue-eyed shag, southern elephant seal and Antarctic fur seal.

Ross Sea

A large portion (32%) of the world's Adélie penguin population occurs in the Ross Sea (South Pacific sector), the southernmost incursion of ocean on the planet (to 78°S). This species is an obligate inhabitant of pack ice, but heavy pack ice negatively affects reproductive success and population growth. Pack-ice extent decreased noticeably in the late 1970s and early 1980s and air temperatures have also been rising. The increasing trends in population size of Adélie penguins in the Ross Sea are opposite to those in the Bellingshausen Sea (see above; **Figure 5**). The patterns, however, are consistent with the direction of climate change: warmer temperatures, less extensive pack ice. As pack ice has become more dispersed in the far south, the penguin has benefited.

As with the Antarctic Peninsula region, subantarctic species are invading southward. The first brown skua was reported in the southern Ross Sea in 1966; the first known breeding attempt occurred in 1982; and the first known successful nesting occurred in 1996. The first elephant seal in the Ross Sea was reported in 1974; at present, several individuals occur there every year.

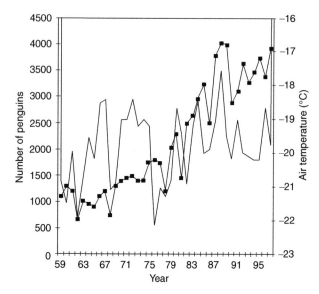

Figure 5 Changes in numbers of breeding pairs of Adélie penguins (■) at Cape Royds, Ross Island, Antarctica (77°S, 166°E), during the past four decades. This is the southernmost breeding site for any penguin species. Although changes in sea ice (less extensive now) is the major direct factor involved, average air temperatures (—) (indirect factor) of the region are shown.

Indirect Responses to Recent Climate Change

California Current

The volume of zooplankton has declined over the past few decades, coincident with reduced upwelling and increased sea-surface temperatures. In response, numbers of sooty shearwaters, formerly, the most abundant avian species in the California Current avifauna, have declined 90% since the 1980s (**Figure 2**). The shearwater feeds heavily on euphausiids in the California Current during spring. The decline, however, has occurred in a stepwise fashion with each El Niño or warm-water event. Sooty shearwaters are now ignoring the Peru and California currents as wintering areas, and favoring instead those waters of the central North Pacific transition zone, which have been cooling and increasing in productivity (see below).

The appearance of the elegant and royal terns as nesting species in California (see above) may in part be linked to the surge in abundance in northern anchovy, which replaced the sardine in the 1960s–1980s. More recently, the sardine has rebounded and the anchovy has declined, but the tern populations continue to grow (see above). Similarly, the former breeding by the brown pelican as far north as Central California was linked to the former presence of sardines. However, the pelicans recently invaded northward (see above) long before the sardine resurgence began. Farthest north the pelicans feed on anchovies.

Central Pacific

In the central North Pacific gyre, the standing crop of chlorophyll-containing organisms increased gradually between 1965 and 1985, especially after the mid-1970s. This was related to an increase in storminess (winds), which in turn caused deeper mixing and the infusion of nutrients into surface waters. The phenomenon reached a maximum during the early 1980s, after which the algal standing crop subsided. As ocean production increased, so did the reproductive success of red-billed tropicbirds and red-footed boobies nesting in the Leeward Hawaiian Islands (southern part of the gyre). When production subsided, so did the breeding success of these and other species (lobsters, seals) in the region. Allowing for lags of several years as dictated by demographic characteristics, the increased breeding success presumably led to larger populations of these seabird species.

Significant changes in the species composition of seabirds in the central Pacific (south of Hawaii) occurred between the mid-1960s and late 1980s. Densities of Juan Fernandez and black-winged petrels and short-tailed shearwaters were lower in the 1980s, but densities of Stejneger's and mottled petrels and sooty shearwaters were higher. In the case of the latter, the apparent shift in migration route (and possibly destination) is consistent with the decrease in sooty shearwaters in the California Current (see above).

Peru Current

The Peruvian guano birds – Peruvian pelican, piquero, and guanay – provide the best-documented example of changes in seabird populations due to changes in prey availability. Since the time of the Incas, the numbers of guano birds have been strongly correlated with biomass of the seabirds' primary prey, the anchoveta. El Niño 1957 (and earlier episodes) caused crashes in anchoveta and guano bird populations, but these were followed by full recovery. Then, with the disappearance of the anchoveta beginning in the 1960s (due to over-fishing and other factors), each subsequent El Niño (1965, 1972, etc.) brought weaker recovery of the seabird populations.

Apparently, the carrying capacity of the guano birds' marine habitat had changed, but population decreases occurred stepwise, coinciding with mortality caused by each El Niño. However, more than just fishing caused anchoveta to decrease; without fishing pressure, the anchoveta recovered quickly (to its lower level) following El Niño 1982–83, and

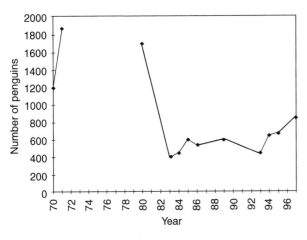

Figure 6 Changes in numbers of breeding Galapágos penguins, 1970–1997. With the 1982–83 El Niño event, the species lost the capacity to recover from periodic events leading to increased mortality. Compare to **Figure 3**, which represents another cool-water species having a similar life-history strategy, but which resides in the other eastern Pacific boundary current (i.e., the two species are ecological complements in the Northern and Southern Hemispheres).

trends in the sardine were contrary to those of the anchoveta beginning in the late 1960s. The seabirds that remain have shifted their breeding sites southward to southern Peru and northern Chile in response to the southward, compensatory increase in sardines. A coincident shift has occurred in the zooplankton and mesopelagic fish fauna. All may be related to an atmospherically driven change in ocean circulation, bringing more subtropical or oceanic water onshore. It is not just breeding seabird species that have been affected, nonbreeding species of the region, such as sooty shearwater, have been wintering elsewhere than the Peru Current (see above).

Trends in penguin populations on the Galapagos confirm that a system-wide change has occurred off western South America (**Figure 6**). Galapagos penguins respond positively to cool water and negatively to warm-water conditions; until recently they recovered after each El Niño, just like the Peruvian guano birds. Then came El Niño 1982–83. The population declined threefold, followed by just a slight recovery. Apparently, the carrying capacity of the habitat of this seabird, too, is much different now than a few decades ago. Like the diving, cool-water species of the California Current (see above), due to climate change, the penguin has lost its capacity for population growth and recovery.

Gulf of Alaska and Bering Sea

A major 'regime' shift involving the physical and biological make-up of the Gulf of Alaska and Bering Sea is illustrated amply by oceanographic and fisheries data. Widespread changes and switches in populations of ecologically competing fish and invertebrate populations have been underway since the mid-1970s. Ironically, seabird populations in the region show few geographically consistent patterns that can be linked to the biological oceanographic trends. There have been no range expansions or retractions, no doubt because this region, in spite of its great size, does not constitute a faunal transition; and from within the region, species to species, some colonies have shown increases, others decreases, and others stability. Unfortunately, the picture has been muddled by the introduction of exotic terrestrial mammals to many seabird nesting islands. Such introductions have caused disappearance and serious declines in sea-bird numbers. In turn, the subsequent eradication of mammals has allowed recolonization and increases in seabird numbers.

In the Bering Sea, changes in the population biology of seabirds have been linked to decadal shifts in the intensity and location of the Aleutian Low Pressure System (the Pacific Decadal Oscillation), which affects sea surface temperatures among other things. For periods of 15–30 years the pressure (North Pacific Index) shifts from values that are above (high pressure state) or below (low pressure state) the long-term average. Kittiwakes in the central Bering Sea (but not necessarily the Gulf of Alaska) do better with warmer sea temperatures; in addition, the relationship of kittiwake productivity to sea surface temperature changes sign with switches from the high to low pressure state. Similarly, the dominance among congeneric species of murres at various sympatric breeding localities may flip-flop depending on pressure state. Although these links to climate have been identified, cause–effect relationships remain obscure. At the Pribilof Islands in the Bering Sea, declines in seabird numbers, particularly of kittiwakes, coincided with the regime shift that began in 1976. Accompanying these declines has been a shift in diets, in which lipid-poor juvenile walleye pollock have been substituted for lipid-rich forage fishes such as sand lance and capelin. Thus, the regime shifts may have altered trophic pathways to forage fishes and in turn the availability of these fish to seabirds. Analogous patterns are seen among the seabirds of Prince William Sound.

Chukchi Sea

Decrease of pack ice extent in response to recent global temperature increases has been more pronounced in the Arctic than the Antarctic. This decrease has resulted in changes in the availability of under-ice fauna, fish, and zooplankton that are important to certain arctic seabirds. The decline of black guillemot populations in northern Alaska in the last decade may be associated with this pack ice decrease.

North Atlantic

The North Atlantic is geographically confined and has been subject to intense human fishery pressure for centuries. Nevertheless, patterns linked to climate change have emerged. In the North Sea, between 1955 and 1987, direct, positive links exist between the frequency of westerly winds and such biological variables as zooplankton volumes, stock size of young herring (a seabird prey), and the following parameters of black-legged kittiwake biology: date of laying, number of eggs laid per nest, and breeding success. As westerly winds subsided through to about 1980, zooplankton and herring decreased, kittiwake laying date retarded and breeding declined. Then, with a switch to increased westerly winds, the biological parameters reversed.

In the western Atlantic, changes in the fish and seabird fauna correlate to warming sea surface temperatures near Newfoundland. Since the late 1800s, mackerel have moved in, as has one of their predators, the Atlantic gannet. Consequently, the latter has been establishing breeding colonies farther and farther north along the coast.

South-east Atlantic, Benguela Current

A record of changes in sea-bird populations relative to the abundance and distribution of prey in the Benguela Current is equivalent to that of the Peruvian upwelling system. As in other eastern boundary currents, Benguela stocks of anchovy and sardine have flip-flopped on 20–30 year cycles for as long as records exist (to the early 1900s). Like other eastern boundary currents, the region of concentration of prey fish, anchovy versus sardine, changes with sardines being relatively more abundant poleward in the region compared to anchovies. Thus, similar to the Peruvian situation, seabird populations have shifted, but patterns also are apparent at smaller timescales depending on interannual changes in spawning areas of the fish. As with all eastern boundary currents, the role of climate in changing pelagic fish populations is being intensively debated.

See also

Sea Level Change.

Further Reading

Aebischer NJ, Coulson JC, and Colebrook JM (1990) Parallel long term trends across four marine trophic levels and weather. *Nature* 347: 753–755.

Crawford JM and Shelton PA (1978) Pelagic fish and seabird interrelationships off the coasts of South West and South Africa. *Biological Conservation* 14: 85–109.

Decker MB, Hunt GL, and Byrd GV (1996) The relationship between sea surface temperature, the abundance of juvenile walleye pollock (*Theragra chalcogramma*), and the reproductive performance and diets of seabirds at the Pribilof islands, southeastern Bering Sea. *Canadian Journal of Fish and Aquatic Science* 121: 425–437.

Divoky GJ (1998) *Factors Affecting Growth of a Black Guillemot Colony in Northern Alaska.* PhD Dissertation, University of Alaska, Fairbanks.

Emslie SD (1998) *Avian Community, Climate, and Sea-level Changes in the Plio-Pleistocene of the Florida Peninsula.* Ornithological Monograph No. 50. Washington, DC: American Ornithological Union.

Furness RW and Greenwood JJD (eds.) (1992) *Birds as Monitors of Environmental Change.* London, New York: Chapman and Hall.

Olson SL (1975) *Paleornithology of St Helena Island, South Atlantic Ocean.* Smithsonian Contributions to Paleobiology, No. 23. Washington, Dc..

Smith RC, Ainley D, Baker K, *et al.* (1999) Marine ecosystem sensitivity to climate change. *BioScience* 49(5): 393–404.

Springer AM (1998) Is it all climate change? Why marine bird and mammal populations fluctuate in the North Pacific. In: Holloway G, Muller P, and Henderson D (eds.) *Biotic Impacts of Extratropical Climate Variability in the Pacific.* Honolulu: University of Hawaii: SOEST Special Publication.

Stuiver M, Denton GH, Hughes T, and Fastook JL (1981) History of the marine ice sheet in West Antarctica during the last glaciation: a working hypothesis. In: Denton GH and Hughes T (eds.) *The Last Great Ice Sheets.* New York: Wiley.

METHANE HYDRATES AND CLIMATIC EFFECTS

B. U. Haq, Vendome Court, Bethesda, MD, USA

Introduction

Natural gas hydrates are crystalline solids that occur widely in marine sediment of the world's continental margins. They are composed largely of methane and water, frozen in place in the sediment under the dual conditions of high pressure and frigid temperatures at the sediment–water interface (**Figure 1**). When the breakdown of the gas hydrate (also known as clathrate) occurs in response to reduced hydrostatic pressure (e.g., sea level fall during glacial periods), or an increase in bottom-water temperature, it causes dissociation of the solid hydrate at its base, creating a zone of reduced sediment strength that is prone to structural faulting and sediment slumping. Such sedimentary failure at hydrate depths could inject large quantities of methane (a potent greenhouse gas) in the water column, and eventually into the atmosphere, leading to enhanced greenhouse warming. Ice core records of the recent geological past show that climatic warming occurs in tandem with rapid increase in atmospheric methane. This suggests that catastrophic release of methane into the atmosphere during periods of lowered sea level may have been a causal factor for abrupt climate change. Massive injection of methane in sea water following hydrate

Figure 1 A piece of natural gas hydrate from the Gulf of Mexico. (Photograph courtesy of I. MacDonald, Texas A & M University.)

dissociation during periods of warm bottom temperatures are also suspected to be responsible for major shifts in carbon-isotopic ratios of sea water and associated changes in benthic assemblages and hydrographic conditions.

Hydrate Stability and Detection

Gas hydrate stability requires high hydrostatic pressure (>5 bars) and low bottom-water temperature (<7°C) on the seafloor. These requirements dictate that hydrates occur mostly on the continental slope and rise, below 530 m water depth in the low latitudes, and below 250 m depth in the high latitudes. Hydrated sediments may extend from these depths to c. 1100 m sub-seafloor. In higher latitudes, hydrates also occur on land, in association with the permafrost.

Rapidly deposited sediments with high biogenic content are amenable to the genesis of large quantities of methane by bacterial alteration of the organic matter. Direct drilling of hydrates on the Blake Ridge, a structural high feature off the US east coast, indicated that the clathrate is only rarely locally concentrated in the otherwise widespread field of thinly dispersed hydrated sediments. The volume of the solid hydrate based on direct measurements on Blake Ridge suggested that it occupies between 0 and 9% of the sediment pore space within the hydrate stability zone (190–450 m sub-seafloor). It has been estimated that a relatively large amount, c. 35 Gt (Gigaton $= 10^{15}$ g), of methane carbon was tied up on Blake Ridge, which is equal to carbon from about 7% of the total terrestrial biota.

Hydrates can be detected remotely through the presence of acoustic reflectors, known as bottom simulating reflectors (BSR), that mimic the seafloor and are caused by acoustic velocity contrast between the solid hydrate above and the free gas below. However, significant quantities of free gas need to be present below the hydrate to provide the velocity contrast for the presence of a BSR. Thus, hydrate may be present at the theoretical hydrate stability depths even when no BSR is observed. Presence of gas hydrate can also be inferred through the sudden reduction in pore-water chlorinity (salinity) of the hydrated sediments during drilling, as well as through gas escape features on land and on the seafloor.

Global estimates of methane trapped in gas hydrate reservoirs (both in the hydrate stability zone and as free gas beneath it) vary widely. For example, the Arctic permafrost is estimated to hold anywhere between 7.6 and 18 000 Gt of methane carbon, while marine sediments are extrapolated to hold between 1700 and 4 100 000 Gt of methane carbon globally. Obtaining more accurate global estimates of methane sequestered in the clathrate reservoirs remains one of the more significant challenges in gas hydrate research.

Another major unknown, especially for climatic implications, is the mode of expulsion of methane from the hydrate. How and how much of the gas escapes from the hydrate zone and how much of it is dissolved in the water column versus escaping into the atmosphere? In a steady state much of the methane diffusing from marine sediments is believed to be oxidized in the surficial sediment and the water column above. However, it is not clear what happens to significant volumes of gas that might be catastrophically released from the hydrates when they disintegrate. How much of the gas makes it to the atmosphere (in the rapid climate change scenarios it is assumed that much of it does), or is dissolved in the water column?

Hydrate dissociation and methane release into the atmosphere from continental margin and permafrost sources and the ensuing accelerated greenhouse heating also have important implications for the models of global warming over the next century. The results of at least one modeling study play down the role of methane release from hydrate sources. When heat transfer and methane destabilization process in oceanic sediments was modeled in a coupled atmosphere–ocean model with various input assumptions and anthropogenic emission scenarios, it was found that the hydrate dissociation effects were smaller than the effects of increased carbon dioxide emissions by human activity. In a worst case scenario global warming increased by 10–25% more with clathrate destabilization than without. However, these models did not take into account the associated free gas beneath the hydrate zone that may play an additional and significant role as well. It is obvious from drilling results on Blake Ridge that large volumes of free methane are readily available for transfer without requiring dissociation.

Hydrate Breakdown and Rapid Climate Change

The temperature and pressure dependency for the stability of hydrates implies that any major change in either of these controlling factors will also modify the zone of hydrate stability. A notable drop in sea level, for example, will reduce the hydrostatic pressure on the slope and rise, altering the temperature–pressure regime and leading to destabilization of the gas hydrates. It has been suggested that a sea level drop of nearly 120 m during the last glacial maximum (c. 30 000 to 18 000 years BP) reduced the hydrostatic pressure sufficiently to raise the lower limit of gas hydrate stability by about 20 m in the low latitudes. When a hydrate dissociates, its consistency changes from a solid to a mixture of sediment, water, and free gas. Experiments on the mechanical strength behavior of hydrates has shown that the hydrated sediment is markedly stronger than water ice (10 times stronger than ice at 260 K). Thus, such conversion would create a zone of weakness where sedimentary failure could take place, encouraging low-angle faulting and slumping on the continental margins. The common occurrence of Pleistocene slumps on the seafloor have been ascribed to this catastrophic mechanism and major slumps have been identified in sediments of this age in widely separated margins of the world.

When slumping occurs it would be accompanied by the liberation of a significant amount of methane trapped below the level of the slump, in addition to the gas emitted from the dissociated hydrate itself. These emissions are envisaged to increase in the low latitudes, along with the frequency of slumps, as glaciation progresses, eventually triggering a negative feedback to advancing glaciation, encouraging the termination of the glacial cycle. If such a scenario is true then there may be a built-in terminator to glaciation, via the gas hydrate connection.

In this scenario, the negative feedback to glaciation, can initially function effectively only in the lower latitudes. At higher latitudes glacially induced freezing would tend to delay the reversal, but once deglaciation begins, even a relatively small increase in atmospheric temperature of the higher latitudes could cause additional release of methane from near-surface sources, leading to further warming. One scenario suggests that a small triggering event and liberation of one or more Arctic gas pools could initiate massive release of methane frozen in the permafrost, leading to accelerated warming. The abrupt nature of the termination of the Younger Dryas glaciation (some 10 000 years ago) has been ascribed to such an event. Modeling results of the effect of a pulse of 'realistic' amount of methane release at the glacial termination as constrained by ice core records indicate that the direct radiative effects of such an emission event may be too small to account for deglaciation alone. However, with certain combinations of methane, CO_2, and heat transport changes, it may be possible to simulate changes of the same magnitude as those indicated by empirical data.

The Climate Feedback Loop

The paleoclimatic records of the recent past, e.g., Vostock ice core records of the past 420 000 years from Antarctica, show the relatively gradual decrease in atmospheric carbon dioxide and methane at the onset of glaciations. Deglaciations, on the other hand, tend to be relatively abrupt and are associated with equally rapid increases in carbon dioxide and methane. Glaciations are thought to be initiated by Milankovitch forcing (a combination of variations in the Earth's orbital eccentricity, obliquity and precession), a mechanism that also can explain the broad variations in glacial cycles, but not the relatively abrupt terminations. Degassing of carbon dioxide from the ocean surface alone cannot explain the relatively rapid switch from glacials to interglacials.

The delayed response to glacially induced sea level fall in the high latitudes (as compared with low latitudes) is a part of a feedback loop that could be an effective mechanism for explaining the rapid warmings at the end of glacial cycles (also known as the Dansgaard-Oeschger events) in the late Quaternary. These transitions often occur only on decadal to centennial time scales. In this scenario it is envisioned that the low-stand-induced slumping and methane emissions in lower latitudes lead to greenhouse warming and trigger a negative feedback to glaciation. This also leads to an increase in carbon dioxide degassing for the ocean. Once the higher latitudes are warmed by these effects, further release of methane from near-surface sources could provide a positive feedback to warming. The former (methane emissions in the low latitudes) would help force a reversal of the glacial cooling, and the latter (additional release of methane from higher latitudes) could reinforce the trend, resulting in apparent rapid warming observed at the end of the glacial cycles (see **Figure 2**).

The record of stable isotopes of carbon from Santa Barbara Basin, off California, has revealed rapid warmings in the late Quaternary that are synchronous with warmings associated with Dansgaard-Oeschger (D-O) events in the ice record from Greenland. The energy needed for these rapid warmings could have come from methane hydrate dissociation. Relatively large excursions of $\delta^{13}C$ (up to 5 ppm) in benthic foraminifera are associated with the D-O events. However, during several brief intervals the planktonics also show large negative shifts in $\delta^{13}C$ (up to 2.5 ppm), implying that the

entire water column may have experienced rapid ^{12}C enrichment. One plausible mechanism for these changes may be the release of methane from the clathrates during the interstadials. Thus, abrupt warmings at the onset of D-O events may have been forced by dissociation of gas hydrates modulated by temperature changes in overlying intermediate waters.

For the optimal functioning of the negative–positive feedback model discussed above, methane would have to be constantly replenished from new and larger sources during the switchover. Although as a greenhouse gas methane is nearly 10 times as effective as carbon dioxide, its residence time in the atmosphere is relatively short (on the order of a decade and a half), after which it reacts with the hydroxyl radical and oxidizes to carbon dioxide and water. The atmospheric retention of carbon dioxide is somewhat more complex than methane because it is readily transferred to other reservoirs, such as the oceans and the biota, from which it can re-enter the atmosphere. Carbon dioxide accounts for up to 80% of the contribution to greenhouse warming in the atmosphere. An effective residence time of about 230 years has been estimated for carbon dioxide. These retention times are short enough that for cumulative impact of methane and carbon dioxide through the negative–positive feedback loop to be effective methane levels would have to be continuously sustained from gas hydrate and permafrost sources. The feedback loop would close when a threshold is reached where sea level is once again high enough that it can stabilize the residual clathrates and encourage the genesis of new ones.

Several unresolved problems remain with the gas hydrate climate feedback model. The negative–positive feedback loop assumes a certain amount of time lag between events as they shift from lower to higher latitudes, but the duration of the lag remains unresolved, although a short duration (on decadal to centennial time scales) is implied by the ice core records. Also, it is not clear whether hydrate dissociation leads to initial warming, or warming caused by other factors leads to increased methane emissions from hydrates. Data gathered imply a time lag of c. 50 ($\pm$10) years between abrupt warming and the peak in methane values at the Blling transition (around 14 500 years Bp), although an increase in methane emissions seems to have begun almost simultaneously with the warming trend ($\pm$5 years). However, this does not detract from the notion that there may be a built-in feedback between increased methane emissions from gas hydrate sources and accelerated warming. If smaller quantities of methane released from hydrated sediments are

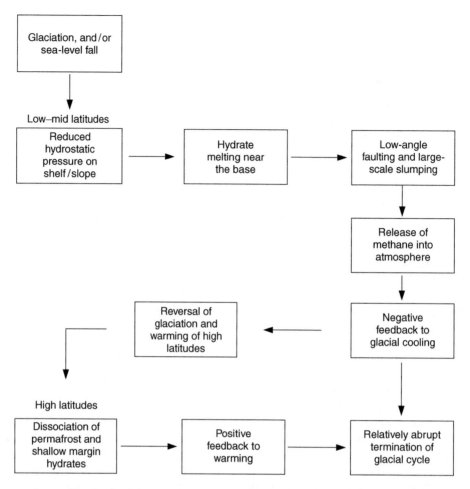

Figure 2 The negative–positive feedback loop model of sea level fall, hydrate decomposition, and climate change (reversal of glaciation and rapid warming) through methane release in the low and high latitudes. (Adapted with permission from Haq, 1993.)

oxidized in the water column, initial releases of methane from dissociated hydrates may not produce a significant positive shift in the atmospheric content of methane. However, as the frequency of catastrophic releases from this source increases, more methane is expected to make it to the atmosphere. And, although the atmospheric residence time of methane itself is relatively short, when oxidized, it adds to the greenhouse forcing of carbon dioxide. This may explain the more gradual increase in methane, and is not inconsistent with the short temporal difference between the initiation of the warming trend and methane increase, as well as the time lag between the height of warming trend and the peak in methane values.

Although there is still no evidence to suggest that the main forcing for the initiation of deglaciation is to be found in hydrate dissociation, once begun, a positive feedback of methane emissions from hydrate sources (and its by-product, carbon dioxide) can only help accelerate the warming trend.

Gas Hydrates and the Long-term Record of Climate Change

Are there any clues in the longer term geological record where cause and/or effect can be ascribed to gas hydrates? One potential clue for the release of significant volumes of methane into the ocean waters is the changes in $\delta^{13}C$ composition of the carbon reservoir. The $\delta^{13}C$ of methane in hydrates averages $c. -60$ ppm; perhaps the lightest (most enriched in ^{12}C) carbon anywhere in the Earth system. It has been argued that massive methane release from gas hydrate sources is the most likely mechanism for the pronounced input of carbon greatly enriched in ^{12}C during a period of rapid bottom-water warming. The dissolution of methane (and its oxidative by-product, CO_2) in the sea water should also coincide with increased dissolution of carbonate on the seafloor. Thus, a major negative shift in $\delta^{13}C$ that occurs together with an increase in benthic temperature (bottom-water warming) or a sea level fall event (reducing hydrostatic pressure) may provide clues to past behavior of gas hydrates.

A prominent excursion in global carbonate and organic matter $\delta^{13}C$ during the latest Paleocene peak warming has been explained as a consequence of such hydrate breakdown due to rapid warming of the bottom waters. The late Paleocene–early Eocene was a period of peak warming, and overall the warmest interval in the Cenozoic when latitudinal thermal gradients were greatly reduced. In the latest Paleocene bottom-water temperature also increased rapidly by as much as $4°C$, with a coincident excursion of about -2 to -3 ppm in $\delta^{13}C$ of all carbon reservoirs in the global carbon cycle. A high resolution study of a sediment core straddling the Paleocene–Eocene boundary concluded that much of this carbon-isotopic shift occurred within no more than a few thousand years and was synchronous in oceans and on land, indicating a catastrophic release of carbon, probably from a methane source. The late Paleocene thermal maximum was also coincident with a major benthic foraminiferal mass extinction and widespread carbonate dissolution and low oxygen conditions on the seafloor. This rapid excursion cannot be explained by conventional mechanisms (increased volcanic emissions of carbon dioxide, changes in oceanic circulation and/or terrestrial and marine productivity, etc.). A rapid warming of bottom waters from 11 to $15°C$ could abruptly alter the sediment thermal gradients leading to methane release from gas hydrates. Increased flux of methane into the ocean–atmosphere system and its subsequent oxidation is considered sufficient to explain the -2.5 ppm excursion in $\delta^{13}C$ in the inorganic carbon reservoir. Explosive volcanism and rapid release of carbon dioxide and changes in the sources of bottom water during this time are considered to be plausible triggering mechanisms for the peak warming leading to hydrate dissociation. Another recent high-resolution study supports the methane hydrate connection to latest Paleocene abrupt climate change. Stable isotopic evidence from two widely separated sites from low- and southern high-latitude Atlantic Ocean indicates multiple injections of methane with global consequences during the relatively short interval at the end of Paleocene. Modeling results, as well as wide empirical data, suggest warm and wet climatic conditions with less vigorous atmospheric circulation during the late Paleocene thermal optimum.

The eustatic record of the late Paleocene–early Eocene could offer further clues for the behavior of the gas hydrates and their contribution to the overall peak warm period of this interval. The longer term trend shows a rising sea level through the latest Paleocene and early Eocene, but there are several shorter term sea level drops throughout this period and one prominent drop straddling the Paleocene–Eocene boundary (which could be an additional forcing component to hydrate dissociation for the terminal Paleocene event). Early Eocene is particularly rich in high-frequency sea level drops of several tens of meters. Could these events have contributed to the instability of gas hydrates, adding significant quantities of methane to the atmosphere and maintaining the general warming of the period? These ideas seem testable if detailed faunal and isotopic data for the interval in question were available with at least the same kind of resolution as that obtained for the latest Paleocene interval.

Timing of the Gas Hydrate Development

When did the gas hydrates first develop in the geological past? The specific low temperature–high pressure requirement for the stability of gas hydrates suggests that they may have existed at least since the latest Eocene, the timing of the first development of the oceanic psychrosphere and cold bottom waters. Theoretically clathrates could exist on the slope and rise when bottom-water temperatures approach those estimated for late Cretaceous and Paleogene (c. 7–15°C), although they would occur deeper within the sedimentary column and the stability zone would be relatively slimmer. A depth of c. 900 m below sea level has been estimated for the hydrate stability zone in the late Palecene. If the bottom waters were to warm up to $22°C$ only then would most margins of the world be free of gas hydrate accumulation. The implied thinner stability zone during warm bottom-water regimes, however, does not necessarily mean an overall reduced methane reservoir, since it also follows that the sub-hydrate free gas zone could be larger, making up to the hydrate deficiency.

Prior to late Eocene there is little evidence of large polar ice caps, and the mechanism for short-term sea level changes remains uncertain. And yet, the Mesozoic–Early Cenozoic eustatic history is replete with major sea level falls of 100 m or more that are comparable in magnitude, if not in frequency, to glacially induced eustatic changes of the late Neogene. If gas hydrates existed in the pre-glacial times, major sea level falls would imply that hydrate dissociation may have contributed significantly to climate change and shallow-seated tectonics along continental margins. However, such massive methane emissions should also be accompanied by prominent $\delta^{13}C$ excursions, as exemplified by the terminal Paleocene climatic optimum.

The role of gas hydrate as a significant source of greenhouse emissions in global change scenarios and as a major contributor of carbon in global carbon cycle remains controversial. It can only be resolved with more detailed studies of hydrated intervals, in conjunction with high-resolution studies of the ice cores, preferably with decadal time resolution. A better understanding of gas hydrates may well show their considerable role in controlling continental margin stratigraphy and shallow structure, as well as in global climatic change, and through it, as agents of biotic evolution.

Further Reading

Dickens GR, O'Neil JR, Rea DK, and Owen RM (1995) Dissociation of oceanic methane hydrate as a cause of the carbon isotope excursion at the end of the Paleocene. *Paleoceanography* 10: 965–971.

Dillon WP Paul CK (1983) Marine gas hydrates II. Geophysical evidence. In: Cox JS (ed.) *Natural Gas Hyrate*. London: Butterswarth, pp. 73–90.

Haq BU (1998) Gas hydrates: Greenhouse nightmare? Energy panacea or pipe dream? *GSA Today, Geological Society of America* 8(11): 1–6.

Henriet J-P and Mienert J (eds.) (1998) *Gas Hydrates: Relevance to World Margin Stability and Climate Change*, vol. 137. London: Geological Society Special Publications.

Kennett JP, Cannariato KG, Hendy IL, and Behl RJ (2000) Carbon isotopic evidence for methane hydrate instability during Quaternary interstadials. *Science* 288: 128–133.

Kvenvolden KA (1998) A primer on the geological occurrence of gas hydrates. In: Henriet J-P and Mienert J (eds.) *Gas Hydrates: Relevance to World Margin Stability and Climate Change*, vol. 137, pp. 9–30. London: Geological Society, Special Publications.

Max MD (ed.) (2000) Natural Gas Hydrates: In: *Oceanic and Permafrost Environments*. Dordrecht: Kluwer Academic Press.

Nisbet EG (1990) The end of ice age. *Canadian Journal of Earth Sciences* 27: 148–157.

Paull CK, Ussler W, and Dillon WP (1991) Is the extent of glaciation limited by marine gas hydrates. *Geophysical Research Letters* 18(3): 432–434.

Sloan ED Jr (1998) *Clathrate Hydrates of Natural Gases*. New York: Marcel Dekker.

Thorpe RB, Pyle JA, and Nisbet EG (1998) What does the ice-core record imply concerning the maximum climatic impact of possible gas hydrate release at Termination 1A? In: Henriet J-P and Mienert J (eds.) *Gas Hydrates: Relevance to World Margin Stability and Climate Change*, vol. 137, pp. 319–326. London: Geological Society, Special Publications.

ECONOMICS OF SEA LEVEL RISE

R. S. J. Tol, Economic and Social Research Institute, Dublin, Republic of Ireland

Introduction

The economics of sea level rise is part of the larger area of the economics of climate change. Climate economics is concerned with five broad areas:

- What are the economic implications of climate change, and how would this be affected by policies to limit climate change?
- How would and should people, companies, and government adapt and at what cost?
- What are the economic implications of greenhouse gas emission reduction?
- How much should emissions be reduced?
- How can and should emissions be reduced?

The economics of sea level rise is limited to the first two questions, which can be rephrased as follows:

- What are the costs of sea level rise?
- What are the costs of adaptation to sea level rise?
- How would and should societies adapt to sea level rise?

These three questions are discussed below.

There are a number of processes that cause the level of the sea to rise. Thermal expansion is the dominant cause for the twenty-first century. Although a few degrees of global warming would expand seawater by only a fraction, the ocean is deep. A 0.01% expansion of a column of 5 km of water would give a sea level rise of 50 cm, which is somewhere in the middle of the projections for 2100. Melting of sea ice does not lead to sea level rise – as the ice currently displaces water. Melting of mountain glaciers does contribute to sea level, but these glaciers are too small to have much of an effect. The same is true for more rapid runoff of surface and groundwater. The large ice sheets on Greenland and Antarctica could contribute to sea level rise, but the science is not yet settled. A complete melting of Greenland and West Antarctica would lead to a sea level rise of some 12 m, and East Antarctica holds some 100 m of sea level. Climate change is unlikely to warm East Antarctica above freezing point, and in fact additional snowfall is likely to store more ice on East Antarctica. Greenland and West Antarctica are much warmer, and climate change may

well imply that these ice caps melt in the next 1000 years or so. It may also be, however, that these ice caps are destabilized – which would mean that sea level would rise by 10–12 m in a matter of centuries rather than millennia.

Concern about sea level rise is only one of many reasons to reduce greenhouse gas emissions. In fact, it is only a minor reason, as sea level rise responds only with a great delay to changes in emissions. Although the costs of sea level rise may be substantial, the benefits of reduced sea level rise are limited – for the simple fact that sea level rise can be slowed only marginally. However, avoiding a collapse of the Greenland and West Antarctic ice sheets would justify emission reduction – but for the fact that it is not known by how much or even whether emission abatement would reduce the probability of such a collapse. Therefore, the focus here is on the costs of sea level rise and the only policy considered is adaptation to sea level rise.

What Are the Costs of Sea Level Rise?

The impacts of sea level rise are manifold, the most prominent being erosion, episodic flooding, permanent flooding, and saltwater intrusion. These impacts occur onshore as well as near shore/on coastal wetlands, and affect natural and human systems. Most of these impacts can be mitigated with adaptation, but some would get worse with adaptation elsewhere. For instance, dikes protect onshore cities, but prevent wetlands from inland migration, leading to greater losses.

The direct costs of erosion and permanent flooding equal the amount of land lost times the value of that land. Ocean front property is often highly valuable, but one should not forget that sea level rise will not result in the loss of the ocean front – the ocean front simply moves inland. Therefore, the average land value may be a better approximation of the true cost than the beach property value.

The amount of land lost depends primarily on the type of coast. Steep and rocky coasts would see little impact, while soft cliffs may retreat up to a few hundred meters. Deltas and alluvial plains are at a much greater risk. For the world as a whole, land loss for 1 m of sea level rise is a fraction of a percent, even without additional coastal protection. However, the distribution of land loss is very skewed. Some islands, particularly atolls, would disappear altogether – and this may lead to the disappearance of entire nations and cultures. The number of people involved is small, though. This is different for deltas, which tend to be densely populated and heavily used because of superb soils and excellent transport. Without additional protection, the deltas of the Ganges–Brahmaputra, Mekong, and Nile could lose a quarter of their area even for relative modest sea level rise. This would force tens of millions of people to migrate, and ruin the economies of the respective countries.

Besides permanent inundation, sea level rise would also cause more frequent and more intense episodic flooding. The costs of this are more difficult to estimate. Conceptually, one would use the difference in the expected annual flood damage. In practice, however, floods are infrequent and the stock-at-risk changes rapidly. This makes estimates particularly uncertain. Furthermore, floods are caused by storms, and climate models cannot yet predict the effect of the enhanced greenhouse effect on storms. That said, tropical cyclones kill only about 10 000 people per year worldwide, and economic damages similarly are only a minor fraction of gross national product (GDP). Even a 10-fold increase because of sea level rise, which is unlikely even without additional protection, would not be a major impact at the global scale. Here, as above, the global average is likely to hide substantial regional differences, but there has been too little research to put numbers on this.

Sea level rise would cause salt water to intrude in surface and groundwater near the coast. Saltwater intrusion would require desalination of drinking water, or moving the water inlet upstream. The latter is not an option on small islands, which may lose all freshwater resources long before they are submerged by sea level rise. The economic costs of desalination are relatively small, particularly near the coast, but desalination is energy-intensive and produces brine, which may cause local ecological problems. Because of saltwater intrusion, coastal agriculture would suffer a loss of productivity, and may become impossible. Halophytes (salt-tolerant plants) are a lucrative nice market for vegetables, but not one that is constrained by a lack of brackish water. Halophytes' defense mechanisms against salt work at the expense of overall plant growth. Saltwater intrusion could induce a shift from agriculture to aquaculture, which may be more profitable. Few studies are available for saltwater intrusion.

Sea level rise would erode coastal wetlands, particularly if hard structures protect human occupations. Current estimates suggest that a third of all coastal wetlands could be lost for less than 40-cm sea level rise, and up to half for 70 cm. Coastal wetlands provide many services. They are habitats for fish (also as nurseries), shellfish, and birds (including migratory birds). Wetlands purify water, and protect coast against storms. Wetlands also provide food and recreation. If wetlands get lost, so do these functions – unless scarce resources are reallocated to provide these services. Estimates of the value of wetlands vary between $100 and $10 000 per hectare, depending on the type of wetland, the services it provides, population density, and per capita income. At present, there are about 70 million ha of coastal wetlands, so the economic loss of sea level rise would be measured in billions of dollars per year.

Besides the direct economic costs, there are also higher-order effects. A loss of agricultural land would restrict production and drive up the price of food. Some estimates have that a 25-cm sea level rise would increase the price of food by 0.5%. These effects would not be limited to the affected regions, but would be spread from international trade. Australia, for instance, has few direct effects from sea level rise – and may therefore benefit from increased exports to make up for the reduced production elsewhere. Other markets would be affected too, as farmers would buy more fertilizer to produce more on the remaining land, and as workers elsewhere would demand higher salaries to compensate for the higher cost of living. Such spillovers are small in developed economies, because agriculture is only a small part of the economy. They are much larger in developing countries.

What Are the Costs of Adaptation to Sea Level Rise?

There are three basic ways to adapt to sea level rise, although in reality mixes and variations will dominate. The two extremes are protection and retreat. In between lies accommodation. Protection entails such things as building dikes and nourishing beaches – essentially, measures are taken to prevent the impact of sea level rise. Retreat implies giving up land and moving people and infrastructure further inland – essentially, the sea is given free range, but people and their things are moved out of harm's way. Accommodation means coping with the consequences of sea level rise. Examples include placing houses on stilts and purchasing flood insurance.

Besides adaptation, there is also failure to (properly) adapt. In most years, this will imply that adaptation costs (see below) are less. In some years, a storm will come to kill people and damage property. The next section has a more extensive discussion on adaptation.

Estimating the costs of protection is straightforward. Dike building and beach nourishment are routine operations practised by many engineering companies around the world. The same holds for accommodation. Estimates have that the total annual cost of coastal protection is less than 0.01% of global GDP. Again, the average hides the extremes. In small island nations, the protection bill may be over 1% of GDP per year and even reach 10%. Note that a sacrifice of 10% is still better than 100% loss without protection.

There may be an issue of scale. Gradual sea level rise would imply a modest extension of the 'wet engineering' sector, if that exists, or a limited expense for hiring foreign consultants. However, rapid sea level rise would imply (local) shortages of qualified engineers – and this would drive up the costs and drive down the effectiveness of protection and accommodation. If materials are not available locally, they will need to be transported to the coast. This may be a problem in densely populated deltas. Few studies have looked into this, but those that did suggest that a sea level rise of less than 2 m per century would not cause logistical problems. Scale is therefore only an issue in case of ice sheet collapse.

Costing retreat is more difficult. The obvious impact is land loss, but this may be partly offset by wetland gains (see above). Retreat also implies relocation. There are no solid estimates of the costs of forced migration. Again, the issue is scale. People and companies move all the time. A well-planned move by a handful of people would barely register. A hasty retreat by a large number would be noticed. If coasts are well protected, the number of forced migrants would be limited to less than 10 000 a year. If coastal protection fails, or is not attempted, over 100 000 people could be displaced by sea level rise each year.

Like impacts, adaptation would have economy-wide implications. Dike building would stimulate the construction sector and crowd out other investments and consumption. In most countries, these effects would be hard to notice, but in small island economies this may lead to stagnation of economic growth.

How Would and Should Societies Adapt to Sea Level Rise?

Decision analysis of coastal protection goes back to Von Dantzig, one of the founding fathers of operations research. This has also been applied to additional protection for sea level rise. Some studies simply compare the best guess of the costs of dike building against the best guess of the value of land lost if no dikes were built, and decide to protect or not on the basis of a simple cost–benefit ratio. Other studies use more advanced methods that, however, conceptually boil down to cost–benefit analysis as well. These studies invariably conclude that it is economically optimal to protect most of the populated coast against sea level rise. Estimates go up to 85% protection, and 15% abandonment. The reason for these high protection levels are that coastal protection is relatively cheap, and that low-lying deltas are often very densely populated so that the value per hectare is high even if people are poor.

However, coastal protection has rarely been based on cost–benefit analysis, and there is no reason to assume that adaptation to sea level rise will be different. One reason is that coastal hazards manifest themselves irregularly and with unpredictable force. Society tends to downplay most risks, and overemphasize a few. The selection of risks shifts over time, in response to events that may or may not be related to the hazards. As a result, coastal protection tends to be neglected for long periods, interspersed with short periods of frantic activity. Decisions are not always rational. There are always 'solutions' waiting for a problem, and under pressure from a public that demands rapid and visible action, politicians may select a 'solution' that is in fact more appropriate for a different problem. The result is a fairly haphazard coastal protection policy.

Some policies make matters worse. Successful past protection creates a sense of safety and hence neglect, and attracts more people and business to the areas deemed safe. Subsidized flood insurance has the same effect. Local authorities may inflate their safety record, relax their building standards and land zoning, and relocate their budget to attract more people.

A fundamental problem is that coastal protection is partly a public good. It is much cheaper to build a dike around a 'community' than around every single property within that community. However, that does require that there is an authority at the appropriate level. Sometimes, there is no such authority and homeowners are left to fend for themselves. In other cases, the authority for coastal protection sits at provincial or national level, with civil servants who are occupied with other matters.

A number of detailed studies have been published on decision making on adaptation to coastal and other hazards. The unfortunate lesson from this work is that every case is unique – extrapolation is not possible except at the bland, conceptual level.

Conclusion

The economics of climate change are still at a formative stage, and so are the economics of sea level rise. First estimates have been developed of the order of magnitude of the problem. Sea level rise is not a substantial economic problem at the global or even the continental scale. It is, however, a substantial problem for a number of countries, and a dominant issue for some local economies. The main issue, therefore, is the distribution of the impacts, rather than the impacts themselves.

Future estimates are unlikely to narrow down the current uncertainties. The uncertainties are not so much due to a paucity of data and studies, but are intrinsic to the problem. The impacts of sea level rise are local, complex, and in a distant future. The priority for economic research should be in developing dynamic models of economies and their interaction with the coast, to replace the current, static assessments.

See also

Abrupt Climate Change. Coastal Zone Management. Monsoons, History of. Sea Level Change.

Further Reading

Burton I, Kates RW, and White GF (1993) *The Environment as Hazard*, 2nd edn. New York: The Guilford Press.

Nicholls RJ, Wong PP, Burkett VR, *et al.* (2007) Coastal systems and low-lying areas. In: Parry ML, Canziani O, Palutikof J, van der Linden P, and Hanson C (eds.) *Climate Change 2007: Impacts, Adaptation and Vulnerability – Contribution of Working Group II to the Fourth Assessment Report of the Intergovernmental Panel on Climate Change*, pp. 315–356. Cambridge, UK: Cambridge University Press.

CARBON SEQUESTRATION VIA DIRECT INJECTION INTO THE OCEAN

E. E. Adams, Massachusetts Institute of Technology, Cambridge, MA, USA
K. Caldeira, Stanford University, Stanford, CA, USA

Introduction

Global climate change, triggered by a buildup of greenhouse gases, is emerging as perhaps the most serious environmental challenge in the twenty-first century. The primary greenhouse gas is CO_2, whose concentration in the atmosphere has climbed from its preindustrial level of *c.* 280 to >380 ppm. Stabilization at no more than 500–550 ppm is a target frequently discussed to avoid major climatic impact.

The primary source of CO_2 is the burning of fossil fuels – specifically gas, oil, and coal – so stabilization of atmospheric CO_2 concentration will clearly require substantial reductions in CO_2 emissions from these sources. For example, one commonly discussed scenario to stabilize at 500 ppm by the mid-twenty-first century suggests that about 640 Gt CO_2 (*c.* 175 Gt C) would need to be avoided over 50 years, with further emission reductions beyond 50 years. As references, a 1000 MW pulverized coal plant produces 6–8 Mt CO_2 (*c.* 2 Mt C) per year, while an oil-fired single-cycle plant produces about two-thirds this amount and a natural gas combined cycle plant produces about half this amount. Thus the above scenario would require that the atmospheric emissions from the equivalent of 2000–4000 large power plants be avoided by approximately the year 2050.

Such changes will require a dramatic reduction in our current dependence on fossil fuels through increased conservation and improved efficiency, as well as the introduction of nonfossil energy sources like solar, wind, and nuclear. While these strategies will slow the buildup of atmospheric CO_2, it is probable that they will not reduce emissions to the required level. In other words, fossil fuels, which currently supply over 85% of the world's energy needs, are likely to remain our primary energy source for the foreseeable future. This has led to increased interest in a new strategy termed carbon capture and storage, or sequestration. The importance of this option for mitigating climate change is highlighted by the recent *Special Report on Carbon Dioxide Capture and Storage* published by the Intergovernmental Panel on Climate Change, to which the reader is referred for more information.

Carbon sequestration is often associated with the planting of trees. As they mature, the trees remove carbon from the atmosphere. As long as the forest remains in place, the carbon is effectively sequestered. Another type of sequestration involves capturing CO_2 from large, stationary sources, such as a power plant or chemical factory, and storing the CO_2 in underground reservoirs or the deep ocean, the latter being the focus of this article. There has been much attention paid recently to underground storage with several large-scale field sites in operation or being planned. Conversely, while there have been many studies regarding use of the deep ocean as a sink for atmospheric carbon, there have been only a few small-scale field studies.

Why is the ocean of interest as a sink for anthropogenic CO_2? The ocean already contains an estimated 40 000 Gt C compared with about 800 Gt C in the atmosphere and 2200 Gt C in the land biosphere. As a result, the amount of carbon that would cause a doubling of the atmospheric concentration would only change the ocean concentration by about 2%. In addition, natural chemical equilibration between the atmosphere and ocean would result in about 80% of present-day emissions ultimately residing in the ocean. Discharging CO_2 directly to the ocean would accelerate this slow, natural process, thus reducing both peak atmospheric CO_2 concentrations and their rate of increase. It is noted that a related strategy for sequestration – not discussed here – would be to enhance the biological sink using nutrients such as iron to fertilize portions of the world's oceans, thus stimulating phytoplankton growth. The phytoplankton would increase the rate of biological uptake of CO_2, and a portion of the CO_2 would be transported to ocean depths when the plankton die.

The indirect flux of CO_2 to the ocean from the atmosphere is already quite apparent: since preindustrial times, the pH of the surface ocean has been reduced by about 0.1 units, from an initial surface pH of about 8.2. **Figure 1** illustrates what could happen to ocean pH under conditions of continued atmospheric release of CO_2. Under the conditions simulated, the pH of the surface would drop by over 0.7 units. Conversely, by injecting some of the CO_2

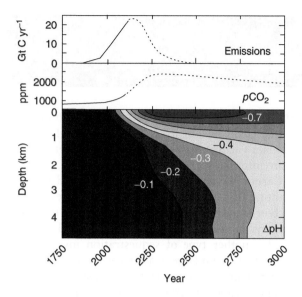

Figure 1 Model simulations of long-term ocean pH changes, averaged horizontally, as a result of atmospheric CO_2 emissions shown in the top panel. Reprinted from Caldeira K and Wickett ME (2003) Anthropogenic carbon and ocean pH. *Nature* 425: 365.

to the deep ocean, the change in pH could be more uniformly distributed.

Ocean sequestration of CO_2 by direct injection assumes that a relatively pure CO_2 stream has been generated at a power plant or chemical factory and transported to an injection point. To better understand the role the ocean can play, we address the capacity of the ocean to sequester CO_2, its effectiveness at reducing atmospheric CO_2 levels, how to inject the CO_2, and possible environmental consequences and issues of public perception.

Capacity

How much carbon can the ocean sequester? At over 70% of the Earth's surface and an average depth of 3800 m, the ocean has enormous storage capacity; based on physical chemistry, the amount of CO_2 that could be dissolved in the deep ocean far exceeds the estimated available fossil energy resources of 5000–10 000 Gt C. However, a more realistic criterion needs to be based on an understanding of ocean biogeochemistry and expected environmental impact.

CO_2 exists in seawater in various forms as part of the carbonate system:

$$CO_2(aq) + H_2O \leftrightarrow H_2CO_3(aq)$$
$$\leftrightarrow H^+ + HCO_3^- \leftrightarrow 2H^+ + CO_3^{2-} \quad [1]$$

Dissolving additional CO_2 increases the hydrogen ion concentration (lowering the pH), but the change

is buffered by the fact that total alkalinity is conserved, which results in carbonate ion being converted into bicarbonate. Thus, the principal reactions occurring when CO_2 is dissolved in seawater are

$$CO_2 + H_2O + CO_3^{2-} \rightarrow 2HCO_3^- \quad [2]$$

$$CO_2 + H_2O + H^+ \rightarrow HCO_3^- \quad [3]$$

which result in a decrease in pH and carbonate ion, and an increase in bicarbonate ion.

Reduced pH is one of the principal environmental impacts threatening marine organisms, the other being the concentration of CO_2 itself. At short travel times from the injection point, the changes in pH and CO_2 concentration will be greatest, which suggests that injection schemes should achieve the maximum dilution possible to minimize potential acute impacts in the vicinity of injection. See further discussion below.

At longer travel times, injected carbon would be distributed widely in the oceans and any far-field impact of the injected CO_2 on the oceans would be similar to the impact of anthropogenic CO_2 absorbed from the atmosphere. As indicated above, such changes are already taking place within the surface ocean, where the pH has been reduced by about 0.1 unit. Adding about 2000 Gt CO_2 to the ocean would reduce the average ocean pH by about 0.1 unit, while adding about 5600 Gt CO_2 (about 200 years of current emissions) would decrease the average ocean pH by about 0.3 units. (It should be noted that with stabilization of atmospheric CO_2 at 550 ppm, natural chemical equilibration between the atmosphere and ocean will result in eventual storage of over 6000 Gt CO_2 in the ocean.)

The impacts of such changes are poorly understood. The deep-ocean environment has been relatively stable and it is unknown to what extent changes in dissolved carbon or pH would affect these ecosystems. However, one can examine measured spatial and temporal variation in ocean pH to understand how much change might be tolerated. The spatial variability within given zoogeographic regions and bathymetric ranges (where similar ecosystems might be expected), and the temporal variability at a particular site, have both been found to vary by about 0.1 pH unit. If it is assumed that a change of 0.1 unit is a threshold tolerance, and that CO_2 should be stored in the bottom half of the ocean's volume (to maximize retention), nearly 1000 Gt CO_2 might be stored, which exceeds the 640 Gt CO_2 over 50 years estimated above. It is important to recognize that the long-term changes in ocean pH would ultimately be much the same whether the CO_2 is released into the atmosphere or the deep ocean. However, in the shorter term, releasing the CO_2 in the deep ocean will diminish the pH change in the near-surface ocean, where marine biota are most plentiful. Thus, direct injection of CO_2 into the deep ocean could reduce adverse impacts presently occurring in the surface ocean. In the long run, however, a sustainable solution to the problem of climate change must ultimately entail a drastic reduction of total CO_2 emissions.

Effectiveness

Carbon dioxide is constantly exchanged between the ocean and atmosphere. Each year the ocean and atmosphere exchange about 350 Gt CO_2, with a net ocean uptake currently of about 8 Gt CO_2. Because of this exchange, questions arise as to how effective ocean sequestration will be at keeping the CO_2 out of the atmosphere. Specifically, is the sequestration permanent, and if not, how fast does the CO_2 leak back to the atmosphere. Because there has been no long-term CO_2 direct-injection experiment in the ocean, the long-term effectiveness of direct CO_2 injection must be predicted based on observations of other oceanic tracers (e.g., radiocarbon) and on computer models of ocean circulation and chemistry.

As implied earlier, because the atmosphere and ocean are currently out of equilibrium, most CO_2 emitted to either media will ultimately enter the ocean. The percentage that is permanently sequestered depends on the atmospheric CO_2 concentration, through the effect of atmospheric CO_2 on surface ocean chemistry (see **Table 1**). At today's concentration of *c.* 380 ppm, nearly 80% of any

carbon emitted to either the atmosphere or the ocean would be permanent, while at a concentration of 550 ppm, 74% would be permanent. Of course, even at equilibrium, CO_2 would continue to be exchanged between the atmosphere and oceans, so the carbon that is currently being injected is not exactly the same carbon that will reside in equilibrium.

For CO_2 injected to the ocean today, the net quantity retained in the ocean ranges from 100% (now) to about 80% as equilibrium between the atmosphere and oceans is approached. (A somewhat greater percentage will ultimately be retained as CO_2 reacts with ocean sediments over a timescale of thousands of years.) The nomenclature surrounding ocean carbon storage can be somewhat confusing. The percentage retained in the ocean shown in **Figure 2** is the fraction of injected CO_2 that has never interacted with the atmosphere. **Table 1** shows the fraction of CO_2 that contacts the atmosphere that remains permanently in the ocean. So, for example, for a 550 ppm atmosphere, even as the 'retained fraction' approaches zero (**Figure 2**), the amount permanently stored in the ocean approaches 74% (see **Table 1**). The exact time course depends on the location and depth of the injection.

Several computer modeling studies have studied the issue of retention. The most comprehensive summary is the Global Ocean Storage of Anthropogenic Carbon (GOSAC) intercomparison study of several ocean general circulation models (OGCMs). In this study a number of OGCMs simulated the fate of CO_2, injected over a period of 100 years at seven locations and three depths, for a period of 500 years. The CO_2 retained as a function of time, averaged over the seven sites, is shown in **Figure 2**. While there is variability among models, they all show that retention increases with injection depth, with most simulations predicting over 70% retention after 500 years for an injection depth of 3000 m.

The time required for injected carbon to mix from the deep ocean to the atmosphere is roughly equal to the time required for carbon to mix from the atmosphere to the deep ocean. This can be estimated through observations of radiocarbon (carbon-14) in the ocean. Correcting for mixing of ocean waters from different sources, the age of North Pacific deep water is in the range of 700–1000 years, while other basins, such as the North Atlantic, have overturning times of 300 years or more. These estimates are consistent with output from OGCMs and, collectively, suggest that outgassing of the 20% of injected carbon would occur on a timescale of 300–1000 years.

It is important to stress that leakage to the atmosphere would take place gradually and over large areas of the ocean. Thus, unlike geological sequestration, it

Table 1 Percent of injected CO_2 permanently sequestered from the atmosphere as a function of atmospheric CO_2 stabilization concentration

Atmospheric carbon dioxide concentration (ppm)	Percentage of carbon dioxide permanently sequestered
350	80
450	77
550	74
650	72
750	70
1000	66

Based on data in IPCC (2005) *Special Report on Carbon Dioxide Capture and Storage*. Prepared by Working Group III of the Intergovernmental Panel on Climate Change. Cambridge, UK: Cambridge University Press. http://arch.rivm.nl/env/int/ipcc/pages_media/SRCCS-final/IPCCSpecialReportonCarbondioxideCaptureandStorage.htm (accessed Mar. 2008) and references therein.

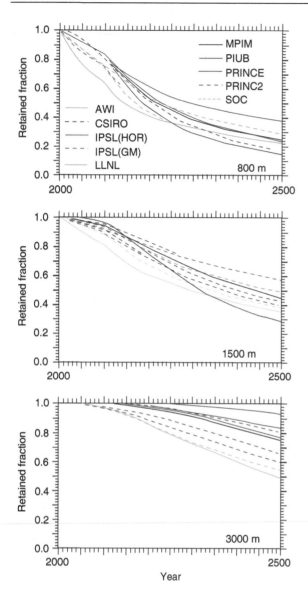

Figure 2 Model-intercomparison study reported by Orr in 2004 showing fraction of CO_2, injected from 2000 through 2100, that remains isolated from the atmosphere as a function of time and injection depth. Results are averaged over seven injection locations. Most of the CO_2 that does interact with the atmosphere remains in the ocean (see **Table 1**), so the amount of CO_2 remaining in the ocean is much greater than shown here. Reprinted with permission from IEA Greenhouse Gas R&D Programme.

would not be possible to produce a sudden release that could lead to harmful CO_2 concentrations at the ocean or land surface.

Injection Methods

The first injection concept was proposed by the Italian physicist Cesare Marchetti, who thought to introduce CO_2 into the outflow of the Mediterranean

Sea, where the relatively dense seawater would cause the CO_2 to sink as it entered the Atlantic Ocean. As illustrated in **Figure 3**, a number of options have been considered since then.

Understanding these methods requires some background information on the CO_2–seawater system. Referring to **Figure 4**, at typical ocean pressures and temperatures, pure CO_2 would be a gas above a depth of 400–500 m and a liquid below that depth. Liquid CO_2 is more compressible than seawater, and would be positively buoyant (i.e., it will rise) down to about 3000 m, but negatively buoyant (i.e., it will sink) below that depth. At about 3700 m, the liquid becomes negatively buoyant compared to seawater saturated with CO_2. In seawater–CO_2 systems, CO_2 hydrate ($CO_2 \cdot nH_2O$, $n \sim 5.75$) can form below c. 400 m depth depending on the relative compositions of CO_2 and H_2O. CO_2 hydrate is a solid with a density about 10% greater than that of seawater.

The rising droplet plume has been the most studied and is probably the easiest scheme to implement. It would rely on commercially available technology to inject the CO_2 as a stream of buoyant droplets from a bottom manifold. Effective sequestration can be achieved by locating the manifold below the thermocline, and dilution can be increased by increasing the manifold length. Even better dilution can be achieved by releasing the CO_2 droplets from a moving ship whose motion provides additional dispersal. Although the means of delivery are different, the plumes resulting from these two options would be similar, each creating a vertical band of CO_2-enriched seawater over a prescribed horizontal region.

Another promising option is to inject liquid CO_2 into a reactor where it can react at a controlled rate with seawater to form hydrates. While it is difficult to achieve 100% reaction efficiency, laboratory and field experiments indicate that negative buoyancy, and hence sinking, can be achieved with as little as about 25% reaction efficiency. The hydrate reactor could be towed from a moving ship to encourage dilution, or attached to a fixed platform, where the large concentration of dense particles, and the increased seawater density caused by hydrate dissolution, would induce a negatively buoyant plume.

The concept of a CO_2 lake is based on a desire to minimize leakage to the atmosphere and exposure to biota. This would require more advanced technology and perhaps higher costs, as the depth of the lake should be at least 3000 m, which exceeds the depths at which the offshore oil industry currently works. The CO_2 in the lake would be partly in the form of solid hydrates. This would limit the CO_2 dissolution into the water column, further slowing leakage to the atmosphere from that shown in **Figure 2**, which

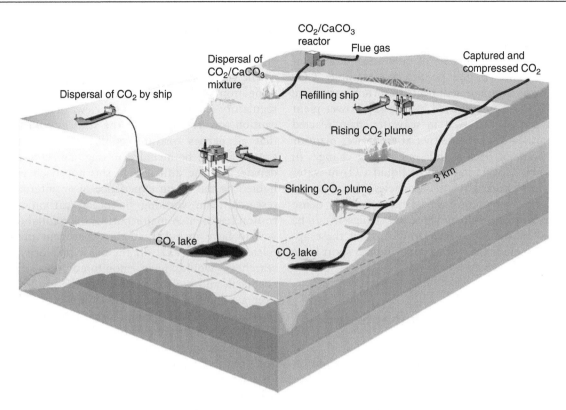

Figure 3 Different strategies for ocean carbon sequestration. Reprinted with permission from IPCC (2005) *Special Report on Carbon Dioxide Capture and Storage*, figure TS-9 (printed as *Special Report on Safeguarding the Ozone Layer and the Global Climate System*, Figure 6.1). Prepared by Working Group III of the Intergovernmental Panel on Climate Change. Cambridge, UK: Cambridge University Press. http://arch.rivm.nl/env/int/ipcc/pages_media/SRCCS-final/IPCCSpecialReportonCarbondioxideCaptureandStorage. htm, with permission from the Intergovernmental Panel on Climate Change.

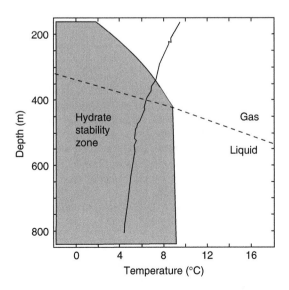

Figure 4 Phase diagram for CO_2 including typical ocean temperature profile (solid line). Reprinted from Brewer PG, Peltzer E, Aya I, *et al.* (2004) Small scale field study of an ocean CO_2 plume. *Journal of Oceanography* 60(4): 751.

assumes that CO_2 is injected into the water column. It is also possible that various approaches could be engineered to physically contain CO_2 on the seafloor and isolate the CO_2 from the overlying water column (and perhaps the sediments); however, this would entail an additional cost.

Another method that has received attention is injecting a dense CO_2–seawater mixture at a depth of 500–1000 m, forming a sinking bottom gravity current. CO_2-enriched seawater is less than 1% heavier than seawater, but this is sufficient to promote a sinking density current, especially if the current were formed along a submarine canyon. However, the environmental impacts would be greater with this option due to the concentrated nature of the plume, and its contact with the seafloor.

As discussed earlier, the deep ocean equilibrates with the surface ocean on the scale of 300–1000 years, and by injecting anthropogenic CO_2 into the deep ocean, the surface-to-deep mixing timescale is effectively bypassed. Anthropogenic CO_2 also equilibrates with carbonate sediments, but over a much longer time, about 6000 years. Technical means could also be used to bypass this timescale, thereby increasing the effectiveness and diminishing the environmental impacts of intentional storage of carbon dioxide in the ocean. For example, CO_2 reacts with carbonate sediments to form bicarbonate ions (HCO_3^-) as indicated by eqn [2]. Power plant CO_2 could be dissolved in seawater, then reacted with crushed limestone, either at the power plant or at the point of release, thus minimizing changes in plume pH. Or an emulsion of liquid CO_2-in-water could be stabilized by fine particles of pulverized limestone; the emulsion would be sufficiently dense to form a sinking plume, whose pH change would be buffered by the limestone. Drawbacks of these approaches include the cost to mine and transport large quantities of carbonate minerals.

Local Environmental Impacts and Public Perception

Environmental impacts may be the most significant factor determining the acceptability of ocean storage, since the strategy is predicated on the notion that impacts to the ocean will be significantly less than the avoided impacts of continued emission to the atmosphere. Earlier, environmental impacts were discussed from the global viewpoint. Here, we examine the environmental impacts near the injection point.

A number of studies have summarized potential impacts to different types of organisms, including adult fish, developmental fish, zooplankton, and benthic fauna. While earlier studies focused mainly on lethal impacts to coastal fauna exposed to strong acids, recent data have focused on deep-water organisms exposed to CO_2, and have included sublethal effects. Impacts include respiratory stress (reduced pH limits oxygen binding and transport of respiratory proteins), acidosis (reduced pH disrupts an organism's acid/basis balance), and metabolic depression (elevated CO_2 causes some animals to reach a state of torpor).

Data generally show that CO_2 causes greater stress than an equivalent change in pH caused by a different acid, that there are strong differences in tolerance among different species and among different life stages of the same species, and that the duration of stress, as well as the level of stress, are important. While some studies imply that deep organisms would be less tolerant than surface organisms, other studies have found the opposite. Likewise, some animals are able to avoid regions of high CO_2 concentration, while others appear less able to. Results generally suggest that lethal effects can be avoided by achieving high near-field dilution. However, more research is needed to resolve impacts, especially at the community level (e.g., reduced lifespan and reproduction effects).

The viability of ocean storage as a greenhouse gas mitigation option hinges on social, political, and regulatory considerations. In view of public precaution toward the ocean, the strategy will require that all parties (private, public, nongovernmental organizations) be included in ongoing research and debate. But the difficulty in this approach is highlighted by the recent experience of an international research team working on ocean carbon sequestration research. A major part of their collaboration was to have included a field experiment involving release of 5 t of CO_2 off the coast of Norway. Researchers would have monitored the physical, chemical, and biological effects of the injected CO_2 over a period of about a week. However, lobbying from environmental groups caused the Norwegian Minister of Environment to rescind the group's permit that had previously been granted. Such actions are unfortunate, because field experiments of this type are what is needed to produce data that would help policymakers decide if full-scale implementation would be prudent.

See also

Abrupt Climate Change. Carbon Dioxide (CO_2) Cycle.

Further Reading

Alendal G and Drange H (2001) Two-phase, near field modeling of purposefully released CO_2 in the ocean. *Journal of Geophysical Research* 106(C1): 1085–1096.

Brewer PG, Peltzer E, Aya I, *et al.* (2004) Small scale field study of an ocean CO_2 plume. *Journal of Oceanography* 60(4): 751–758.

Caldeira K and Rau GH (2000) Accelerating carbonate dissolution to sequester carbon dioxide in the ocean: Geochemical implications. *Geophysical Research Letters* 27(2): 225–228.

Caldeira K and Wickett ME (2003) Anthropogenic carbon and ocean pH. *Nature* 425: 365.

Giles J (2002) Norway sinks ocean carbon study. *Nature* 419: 6.

Golomb D, Pennell S, Ryan D, Barry E, and Swett P (2007) Ocean sequestration of carbon dioxide: Modeling the deep ocean release of a dense emulsion of liquid CO_2-in-water stabilized by pulverized limestone particles. *Environmental Science and Technology* 41(13): 4698–4704.

Haugan H and Drange H (1992) Sequestration of CO_2 in the deep ocean by shallow injection. *Nature* 357(28): 1065–1072.

IPCC (2005) *Special Report on Carbon Dioxide Capture and Storage.* Prepared by Working Group III of the Intergovernmental Panel on Climate Change. Cambridge, UK: Cambridge University Press. http://arch.rivm.nl/env/int/ipcc/pages_media/SRCCS-final/IPCCSpecialReporton CarbondioxideCaptureandStorage.htm (accessed Mar. 2008).

Ishimatsu A, Kikkawa T, Hayashi M, and Lee KS (2004) Effects of CO_2 on marine fish: Larvae and adults. *Journal of Oceanography* 60: 731–741.

Israelsson P and Adams E (2007) Evaluation of the Acute Biological Impacts of Ocean Carbon Sequestration. *Final Report for US Dept. of Energy, under grant DE-FG26-98FT40334.* Cambridge, MA: Massachusetts Institute of Technology.

Kikkawa T, Ishimatsu A, and Kita J (2003) Acute CO_2 tolerance during the early developmental stages of four marine teleosts. *Environmental Toxicology* 18(6): 375–382.

Ohsumi T (1995) CO_2 storage options in the deep-sea. *Marine Technology Society Journal* 29(3): 58–66.

Orr JC (2004) *Modeling of Ocean Storage of CO_2 – The GOSAC Study,* Report PH4/37, 96pp. Paris: Greenhouse Gas R&D Programme, International Energy Agency.

Ozaki M, Minamiura J, Kitajima Y, Mizokami S, Takeuchi K, and Hatakenka K (2001) CO_2 ocean sequestration by moving ships. *Journal of Marine Science and Technology* 6: 51–58.

Pörtner HO, Reipschläger A, and Heisler N (2004) Biological impact of elevated ocean CO_2 concentrations: Lessons from animal physiology and Earth history. *Journal of Oceanography* 60(4): 705–718.

Riestenberg D, Tsouris C, Brewer P, *et al.* (2005) Field studies on the formation of sinking CO_2 particles for ocean carbon sequestration: Effects of injector geometry on particle density and dissolution rate and model simulation of plume behavior. *Environmental Science and Technology* 39: 7287–7293.

Sato T and Sato K (2002) Numerical prediction of the dilution process and its biological impacts in CO_2 ocean sequestration. *Journal of Marine Science and Technology* 6(4): 169–180.

Vetter EW and Smith CR (2005) Insights into the ecological effects of deep-ocean CO_2 enrichment: The impacts of natural CO_2 venting at Loihi seamount on deep sea scavengers. *Journal of Geophysical Research* 110: C09S13 (doi:10.1029/2004JC002617).

Wannamaker E and Adams E (2006) Modeling descending carbon dioxide injections in the ocean. *Journal of Hydraulic Research* 44(3): 324–337.

Watanabe Y, Yamaguchi A, Ishida H, *et al.* (2006) Lethality of increasing CO_2 levels on deep-sea copepods in the western North Pacific. *Journal of Oceanography* 62: 185–196.

IRON FERTILIZATION

K. H. Coale, Moss Landing Marine Laboratories, CA, USA

Introduction

The trace element iron has been shown to play a critical role in nutrient utilization and phytoplankton growth and therefore in the uptake of carbon dioxide from the surface waters of the global ocean. Carbon fixation in the surface waters, via phytoplankton growth, shifts the ocean–atmosphere exchange equilibrium for carbon dioxide. As a result, levels of atmospheric carbon dioxide (a greenhouse gas) and iron flux to the oceans have been linked to climate change (glacial to interglacial transitions). These recent findings have led some to suggest that large-scale iron fertilization of the world's oceans might therefore be a feasible strategy for controlling climate. Others speculate that such a strategy could deleteriously alter the ocean ecosystem, and still others have calculated that such a strategy would be ineffective in removing sufficient carbon dioxide to produce a sizable and rapid result. This article focuses on carbon and the major plant nutrients, nitrate, phosphate, and silicate, and describes how our recent discovery of the role of iron in the oceans has increased our understanding of phytoplankton growth, nutrient cycling, and the flux of carbon from the atmosphere to the deep sea.

Major Nutrients

Phytoplankton growth in the oceans requires many physical, chemical, and biological factors that are distributed inhomogenously in space and time. Because carbon, primarily in the form of the bicarbonate ion, and sulfur, as sulfate, are abundant throughout the water column, the major plant nutrients in the ocean commonly thought to be critical for phytoplankton growth are those that exist at the micromolar level such as nitrate, phosphate, and silicate. These, together with carbon and sulfur, form the major building blocks for biomass in the sea. As fundamental cellular constituents, they are generally thought to be taken up and remineralized in constant ratio to one another. This is known as the Redfield ratio (Redfield, 1934, 1958) and can be expressed on a molar basis relative to carbon as 106C : 16N : 1P.

Significant local variations in this uptake/regeneration relationship can be found and are a function of the phytoplankton community and growth conditions, yet this ratio can serve as a conceptual model for nutrient uptake and export.

The vertical distribution of the major nutrients typically shows surface water depletion and increasing concentrations with depth. The schematic profile in **Figure 1** reflects the processes of phytoplankton uptake within the euphotic zone and remineralization of sinking planktonic debris via microbial degradation, leading to increased concentrations in the deep sea. Given favorable growth conditions, the nutrients at the surface may be depleted to zero. The rate of phytoplankton production of new biomass, and therefore the rate of carbon uptake, is controlled by the resupply of nutrients to the surface waters, usually via the upwelling of deep waters. Upwelling occurs over the entire ocean basin at the rate of approximately 4 m per year but increases in coastal and

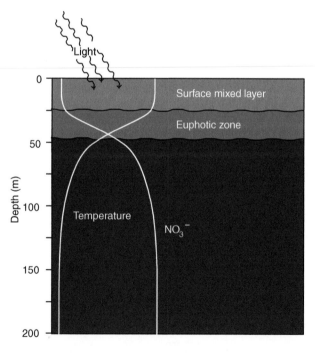

Figure 1 A schematic profile indicating the regions of the upper water column where phytoplankton grow. The surface mixed layer is that region that is actively mixed by wind and wave energy, which is typically depleted in major nutrients. Below this mixed layer temperatures decrease and nutrients increase as material sinking from the mixed layer is regenerated by microbial decomposition.

regions of divergent surface water flow, reaching average values of 15 to 30 or greater. Thus, those regions of high nutrient supply or persistent high nutrient concentrations are thought to be most important in terms of carbon removal.

Nitrogen versus Phosphorus Limitation

Although both nitrogen and phosphorus are required at nearly constant ratios characteristic of deep water, nitrogen has generally been thought to be the limiting nutrient in sea water rather than phosphorus. This idea has been based on two observations: selective enrichment experiments and surface water distributions. When ammonia and phosphate are added to sea water in grow-out experiments, phytoplankton growth increases with the ammonia addition and not with the phosphate addition, thus indicating that reduced nitrogen and not phosphorus is limiting. Also, when surface water concentration of nitrate and phosphate are plotted together (**Figure 2**), it appears that there is still residual phosphate after the nitrate has gone to zero.

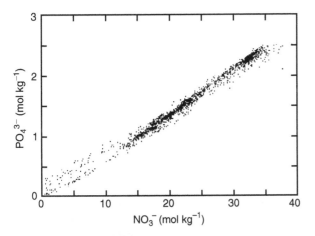

Figure 2 A plot of the global surface water concentrations of phosphate versus nitrate indicating a general positive intercept for phosphorus when nitrate has gone to zero. This is one of the imperical observations favoring the notion of nitrate limitation over phosphate limitation.

The notion of nitrogen limitation seems counter-intuitive when one considers the abundant supply of dinitrogen (N_2) in the atmosphere. Yet this nitrogen gas is kinetically unavailable to most phytoplankton because of the large amount of energy required to break the triple bond that binds the dinitrogen molecule. Only those organisms capable of nitrogen fixation can take advantage of this form of nitrogen and reduce atmospheric N_2 to biologically available nitrogen in the form of urea and ammonia. This is, energetically, a very expensive process requiring specialized enzymes (nitrogenase), an anaerobic microenvironment, and large amounts of reducing power in the form of electrons generated by photosynthesis. Although there is currently the suggestion that nitrogen fixation may have been underestimated as an important geochemical process, the major mode of nitrogen assimilation, giving rise to new plant production in surface waters, is thought to be nitrate uptake.

The uptake of nitrate and subsequent conversion to reduced nitrogen in cells requires a change of five in the oxidation state and proceeds in a stepwise fashion. The initial reduction takes place via the nitrate/nitrite reductase enzyme present in phytoplankton and requires large amounts of the reduced nicotinamide–adenine dinucleotide phosphate (NADPH) and of adenosine triphosphate (ATP) and thus of harvested light energy from photosystem II. Both the nitrogenase enzyme and the nitrate reductase enzyme require iron as a cofactor and are thus sensitive to iron availability.

Ocean Regions

From a nutrient and biotic perspective, the oceans can be generally divided into biogeochemical provinces that reflect differences in the abundance of macronutrients and the standing stocks of phytoplankton. These are the high-nitrate, high-chlorophyll (HNHC); high-nitrate, low-chlorophyll (HNLC); low-nitrate, high-chlorophyll (LNHC); and low-nitrate, low-chlorophyll (LNLC) regimes (**Table 1**). Only the HNLC and LNLC regimes are relatively stable, because the high phytoplankton

Table 1 The relationship between biomass and nitrate as a function of biogeochemical province and the approximate ocean area represented by these regimes

	High-chlorophyll	*Low-chlorophyll*
High-nitrate	Unstable/coastal (5%)	Stable/Subarctic/Antarctic/equatorial Pacific (20%)
Low-nitrate	Unstable/coastal (5%)	Oligotrophic gyres (70%)

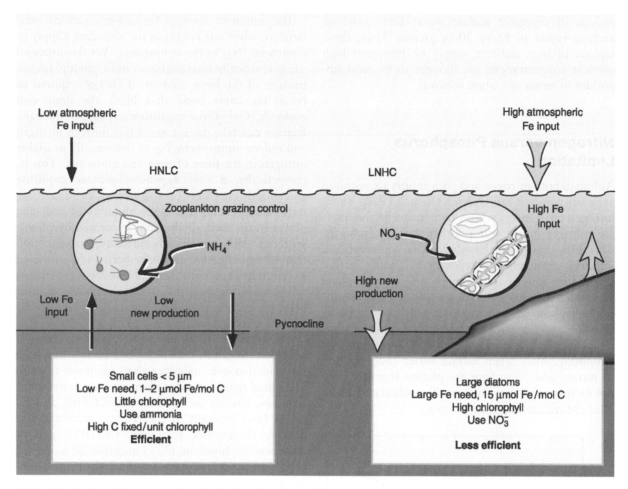

Figure 3 A schematic representation of the 'iron theory' as it functions in offshore HNLC regions and coastal transient LNHC regions. It has been suggested that iron added to the HNLC regions would induce them to function as LNHC regions and promote carbon export.

growth rates in the other two systems will deplete any residual nitrate and sink out of the system. The processes that give rise to these regimes have been the subject of some debate over the last few years and are of fundamental importance relative to carbon export (**Figure 3**).

High-nitrate, Low-chlorophyll Regions

The HNLC regions are thought to represent about 20% of the areal extent of the world's oceans. These are generally regions characterized by more than $2 \, \mu mol \, l^{-1}$ nitrate and less than $0.5 \, \mu g \, l^{-1}$ chlorophyll-a, a proxy for plant biomass. The major HNLC regions are shown in **Figure 4** and represent the Subarctic Pacific, large regions of the eastern equatorial Pacific and the Southern Ocean. These HNLC regions persist in areas that have high macronutrient concentrations, adequate light, and physical characteristics required for phytoplankton growth but have very low plant biomass. Two explanations have been

given to describe the persistence of this condition. (1) The rates of zooplankton grazing of the phytoplankton community may balance or exceed phytoplankton growth rates in these areas, thus cropping plant biomass to very low levels and recycling reduced nitrogen from the plant community, thereby decreasing the uptake of nitrate. (2) Some other micronutrient (possibly iron) physiologically limits the rate of phytoplankton growth. These are known as top-down and bottom-up control, respectively.

Several studies of zooplankton grazing and phytoplankton growth in these HNLC regions, particularly the Subarctic Pacific, confirm the hypothesis that grazers control production in these waters. Recent physiological studies, however, indicate that phytoplankton growth rates in these regions are suboptimal, as is the efficiency with which phytoplankton harvest light energy. These observations indicate that phytoplankton growth may be limited by something other than (or in addition to) grazing. Specifically, these studies implicate the lack of

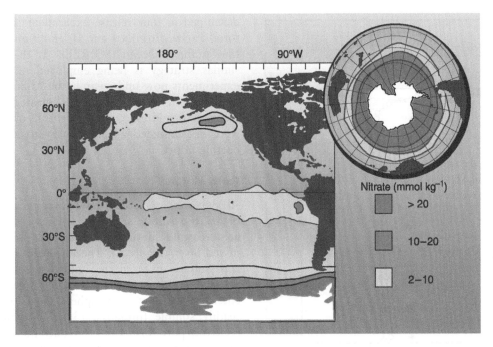

Figure 4   Current HNLC regions of the world's oceans covering an extimated 20% of the ocean surface. These regions include the Subarctic Pacific, equatorial Pacific and Southern Ocean.

sufficient electron transport proteins and the cell's ability to transfer reducing power from the photocenter. These have been shown to be symptomatic of iron deficiency.

The Role of Iron

Iron is a required micronutrient for all living systems. Because of its d-electron configuration, iron readily undergoes redox transitions between Fe(II) and Fe(III) at physiological redox potentials. For this reason, iron is particularly well suited to many enzyme and electron carrier proteins. The genetic sequences coding for many iron-containing electron carriers and enzymes are highly conserved, indicating iron and iron-containing proteins were key features of early biosynthesis. When life evolved, the atmosphere and waters of the planet were reducing and iron was abundant in the form of soluble Fe(II). Readily available and at high concentration, iron was not likely to have been limiting in the primordial biosphere. As photosynthesis evolved, oxygen was produced as a by-product. As the biosphere became more oxidizing, iron precipitated from aquatic systems in vast quantities, leaving phytoplankton and other aquatic life forms in a vastly changed and newly deficient chemical milieu. Evidence of this mass Fe(III) precipitation event is captured in the ancient banded iron formations in many parts of the world. Many primitive aquatic and terrestrial

organisms have subsequently evolved the ability to sequester iron through the elaboration of specific Fe(II)-binding ligands, known as siderophores. Evidence for siderophore production has been found in several marine dinoflagellates and bacteria and some researchers have detected similar compounds in sea water.

Today, iron exists in sea water at vanishingly small concentrations. Owing to both inorganic precipitation and biological uptake, typical surface water values are on the order of 20 pmol l^{-1}, perhaps a billion times less than during the prehistoric past. Iron concentrations in the oceans increase with depth, in much the same manner as the major plant nutrients (**Figure 5**).

The discovery that iron concentrations in surface waters is so low and shows a nutrient-like profile led some to speculate that iron availability limits plant growth in the oceans. This notion has been tested in bottle enrichment experiments throughout the major HNLC regions of the world's oceans. These experiments have demonstrated dramatic phytoplankton growth and nutrient uptake upon the addition of iron relative to control experiments in which no iron was added.

Criticism that such small-scale, enclosed experiments may not accurately reflect the response of the HNLC system at the level of the community has led to several large-scale iron fertilization experiments in the equatorial Pacific and Southern Ocean. These have been some of the most dramatic

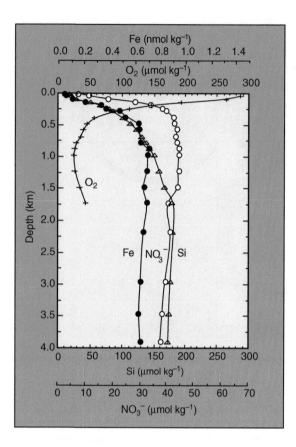

Figure 5 The vertical distributions of iron, nitrate, silicate, and oxygen in sea water. This figure shows how iron is depleted to picomolar levels in surface waters and has a profile that mimics other plant nutrients.

oceanographic experiments of our times and have led to a profound and new understanding of ocean systems.

Open Ocean Iron Enrichment

The question of iron limitation was brought into sharp scientific focus with a series of public lectures, reports by the US National Research Council, papers, special publications, and popular articles between 1988 and 1991. What was resolved was the need to perform an open ocean enrichment experiment in order to definitively test the hypothesis that iron limits phytoplankton growth and nutrient and carbon dioxide uptake in HNLC regions. Such an experiment posed severe logistical challenges and had never been conducted.

Experimental Strategy

The mechanics of producing an iron-enriched experimental patch and following it over time was

developed in four release experiments in the equatorial Pacific (IronEx I and II) and more recently in the Southern Ocean (SOIREE). At this writing, a similar strategy is being employed in the Caruso experiments now underway in the Atlantic sector of the Southern Ocean. All of these strategies were developed to address certain scientific questions and were not designed as preliminary to any geoengineering effort.

Form of Iron

All experiments to date have involved the injection of an iron sulfate solution into the ship's wake to achieve rapid dilution and dispersion throughout the mixed layer (**Figure 6**). The rationale for using ferrous sulfate involved the following considerations: (1) ferrous sulfate is the most likely form of iron to enter the oceans via atmospheric deposition; (2) it is readily soluble (initially); (3) it is available in a relatively pure form so as to reduce the introduction of other potentially bioactive trace metals; and (4) its counterion (sulfate) is ubiquitous in sea water and not likely to produce confounding effects. Although mixing models indicate that Fe(II) carbonate may reach insoluble levels in the ship's wake, rapid dilution reduces this possibility.

New forms of iron are now being considered by those who would seek to reduce the need for subsequent infusions. Such forms could include iron lignosite, which would increase the solubility and residence time of iron in the surface waters. Since this is a chelated form of iron, problems of rapid precipitation are reduced. In addition, iron lignosulfonate is about 15% Fe by weight, making it a space-efficient form of iron to transport. As yet untested is the extent to which such a compound would reduce the need for re-infusion.

Although solid forms of iron have been proposed (slow-release iron pellets; finely milled magnetite or iron ores), the ability to trace the enriched area with an inert tracer has required that the form of iron added and the tracer both be in the dissolved form.

Inert Tracer

Concurrent with the injection of iron is the injection of the inert chemical tracer sulfur hexafluoride (SF_6). By presaturating a tank of sea water with SF_6 and employing an expandable displacement bladder, a constant molar injection ratio of $Fe : SF_6$ can be achieved (**Figure 6**). In this way, both conservative and nonconservative removal of iron can be quantified. Sulfur hexafluoride traces the physical properties of the enriched patch; the relatively rapid shipboard detection of SF_6 can be used to track and

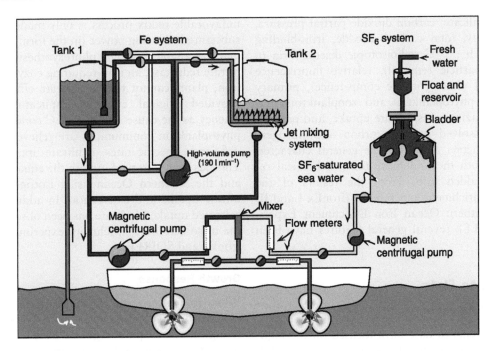

Figure 6 The iron injection system used during the IronEx experiments utilized two polyethylene tanks that could be sequentially filled with sea water and iron sulfate solution while the other was being injected behind the ship's propellers. A steel tank of sea water saturated with 40 g of sulfur hexafluoride (SF_6) was simultaneously mixed with the iron sulfate solution to provide a conservative tracer of mixing.

map the enriched area. The addition of helium-3 to the injected tracer can provide useful information regarding gas transfer.

Fluorometry

The biophysical response of the phytoplankton is rapid and readily detectable. Thus shipboard measurement of relative fluorescence (F_v/F_m) using fast repetition rate fluorometry has been shown to be a useful tactical tool and gives nearly instantaneous mapping and tracking feedback.

Shipboard Iron Analysis

Because iron is rapidly lost from the system (at least initially), the shipboard determination of iron is necessary to determine the timing and amount of subsequent infusions. Several shipboard methods, using both chemiluminescent and catalytic colorimetric detection have proven useful in this regard.

Lagrangian Drifters

A Lagrangian point of reference has proven to be very useful in every experiment to date. Depending upon the advective regime, this is the only practical way to achieve rapid and precise navigation and mapping about the enriched area.

Remote Sensing

A variety of airborne and satellite-borne active and passive optical packages provide rapid, large-scale mapping and tracking of the enriched area. Although SeaWiffs was not operational during IronEx I and II, AVHRR was able to detect the IronEx II bloom and airborne optical LIDAR was very useful during IronEx I. SOIREE has made very good use of the more recent SeaWiffs images, which have markedly extended the observational period and led to new hypotheses regarding iron cycling in polar systems.

Experimental Measurements

In addition to the tactical measurements and remote sensing techniques required to track and ascertain the development of the physical dynamics of the enriched patch, a number of measurements have been made to track the biogeochemical development of the experiment. These have typically involved a series of underway measurements made using the ship's flowing sea water system or towed fish. In addition, discrete measurements are made in the vertical dimension at every station occupied both inside and outside of the fertilized area. These measurements include temperature salinity, fluorescence (a measure of plant biomass), transmissivity (a measure of suspended particles), oxygen, nitrate, phosphate, silicate, carbon dioxide partial pressure, pH, alkalinity, total carbon dioxide, iron-binding ligands, $^{234}Th : ^{238}U$ radioisotopic disequilibria (a proxy for particle removal), relative fluorescence (indicator of photosynthetic competence), primary production, phytoplankton and zooplankton enumeration, grazing rates, nitrate uptake, and particulate and dissolved organic carbon and nitrogen. These parameters allow for the general characterization of both the biological and geochemical response to added iron. From the results of the equatorial enrichment experiments (IronEx I and II) and the Southern Ocean Iron Enrichment Experiment (SOIREE), several general features have been identified.

Findings to Date

Biophysical Response

The experiments to date have focused on the high-nitrate, low-chlorophyll (HNLC) areas of the world's oceans, primarily in the Subarctic, equatorial Pacific and Southern Ocean. In general, when light is abundant many researchers find that HNLC systems are iron-limited. The nature of this limitation is similar between regions but manifests itself at different levels of the trophic structure in some characteristic ways. In general, all members of the HNLC photosynthetic community are physiologically limited by iron availability. This observation is based primarily on the examination of the efficiency of photosystem II, the light-harvesting reaction centers. At ambient levels of iron, light harvesting proceeds at suboptimal rates. This has been attributed to the lack of iron-dependent electron carrier proteins at low iron concentrations. When iron concentrations are increased by subnanomolar amounts, the efficiency of light harvesting rapidly increases to maximum levels. Using fast repetition rate fluorometry and non-heme iron proteins, researchers have described these observations in detail. What is notable about these results is that iron limitation seems to affect the photosynthetic energy conversion efficiency of even the smallest of phytoplankton. This has been a unique finding that stands in contrast to the hypothesis that, because of diffusion, smaller cells are not iron limited but larger cells are.

Nitrate Uptake

As discussed above, iron is also required for the reduction (assimilation) of nitrate. In fact, a change of oxidation state of five is required between nitrate and the reduced forms of nitrogen found in amino acids and proteins. Such a large and energetically unfavorable redox process is only made possible by substantial reducing power (in the form of NADPH) made available through photosynthesis and active nitrate reductase, an iron-requiring enzyme. Without iron, plants cannot take up nitrate efficiently. This provided original evidence implicating iron deficiency as the cause of the HNLC condition. When phytoplankton communities are relieved from iron deficiency, specific rates of nitrate uptake increase. This has been observed in both the equatorial Pacific and the Southern Ocean using isotopic tracers of nitrate uptake and conversion. In addition, the accelerated uptake of nitrate has been observed in both the mesoscale iron enrichment experiments to date, IronEx and SOIREE.

Growth Response

When iron is present, phytoplankton growth rates increase dramatically. Experiments over widely differing oceanographic regimes have demonstrated that, when light and temperature are favorable, phytoplankton growth rates in HNLC environments increase to their maximum at dissolved iron concentrations generally below 0.5 nmol l^{-1}. This observation is significant in that it indicates that phytoplankton are adapted to very low levels of iron and they do not grow faster if given iron at more than 0.5 nmol l^{-1}. Given that there is still some disagreement within the scientific community about the validity of some iron measurements, this phytoplankton response provides a natural, environmental, and biogeochemical benchmark against which to compare results.

The iron-induced transient imbalance between phytoplankton growth and grazing in the equatorial Pacific during IronEx II resulted in a 30-fold increase in plant biomass (**Figure 7**). Similarly, a 6-fold increase was observed during the SOIREE experiment in the Southern Ocean. These are perhaps the most dramatic demonstrations of iron limitation of nutrient cycling, and phytoplankton growth to date and has fortified the notion that iron fertilization may be a useful strategy to sequester carbon in the oceans.

Heterotrophic Community

As the primary trophic levels increase in biomass, growth in the small microflagellate and heterotrophic bacterial communities increase in kind. It appears that these consumers of recently fixed carbon (both particulate and dissolved) respond to the food source and not necessarily the iron (although some have been found to be iron-limited). Because their division rates are fast, heterotrophic bacteria, ciliates, and flagellates can rapidly divide and

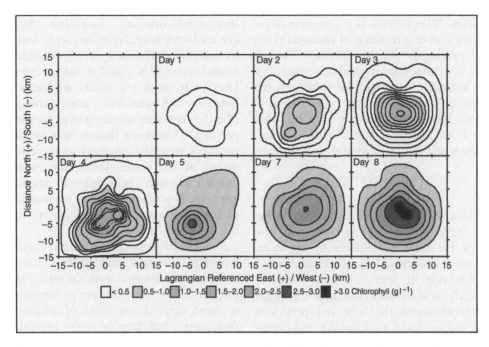

Figure 7 Chlorophyll concentrations during IronEx II were mapped daily. This figure shows the progression of the phytoplankton bloom that reached over 30 times the background concentrations.

respond to increasing food availability to the point where the growth rates of the smaller phytoplankton can be overwhelmed by grazing. Thus there is a much more rapid turnover of fixed carbon and nitrogen in iron replete systems. M. Landry and coworkers have documented this in dilution experiments conducted during IronEx II. These results appear to be consistent with the recent SOIREE experiments as well.

Nutrient Uptake Ratios

An imbalance in production and consumption, however, can arise at the larger trophic levels. Because the reproduction rates of the larger micro- and mesozooplankton are long with respect to diatom division rates, iron-replete diatoms can escape the pressures of grazing on short timescales (weeks). This is thought to be the reason why, in every iron enrichment experiment, diatoms ultimately dominate in biomass. This result is important for a variety of reasons. It suggests that transient additions of iron would be most effective in producing net carbon uptake and it implicates an important role of silicate in carbon flux. The role of iron in silicate uptake has been studied extensively by Franck and colleagues. The results, together with those of Takeda and coworkers, show that iron alters the uptake ratio of nitrate and silicate at very low levels (**Figure 8**). This is thought to be brought about by the increase in nitrate uptake rates relative to silica.

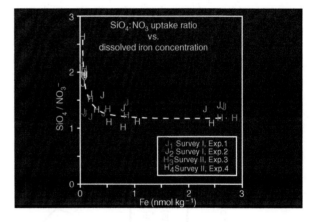

Figure 8 Bottle enrichment experiments show that the silicate : nitrate uptake ratio changes as a function of the iron added. This is thought to be due to the increased rate of iron uptake relative to silicate in these experimental treatments.

Organic Ligands

Consistent with the role of iron as a limiting nutrient in HNLC systems is the notion that organisms may have evolved competitive mechanisms to increase iron solubility and uptake. In terrestrial systems this is accomplished using extracellularly excreted or membrane-bound siderophores. Similar compounds have been shown to exist in sea water where the competition for iron may be as fierce as it is on land. In open ocean systems where it has been measured, iron-binding ligand production increases with the addition of iron. Whether this is a competitive response to added iron or a function of phytoplankton biomass and grazing is not yet well understood. However, this is an important natural mechanism for reducing the inorganic scavenging of iron from the surface waters and increasing iron availability to phytoplankton. More recent studies have considerably advanced our understanding of these ligands, their distribution and their role in ocean ecosystems.

Carbon Flux

It is the imbalance in the community structure that gives rise to the geochemical signal. Whereas iron stimulation of the smaller members of the community may result in chemical signatures such as an increased production of beta-dimethylsulfoniopropionate (DMSP), it is the stimulation of the larger producers that decouples the large cell producers from grazing and results in a net uptake and export of nitrate, carbon dioxide, and silicate.

The extent to which this imbalance results in carbon flux, however, has yet to be adequately described. The inability to quantify carbon export has primarily been a problem of experimental scale. Even though mesoscale experiments have, for the first time, given us the ability to address the effect of iron on communities, the products of surface water processes and the effects on the midwater column have been difficult to track. For instance, in the IronEx II experiment, a time-series of the enriched patch was diluted by 40% per day. The dilution was primarily in a lateral (horizontal/isopycnal) dimension. Although some correction for lateral dilution can be made, our ability to quantify carbon export is dependent upon the measurement of a signal in waters below the mixed layer or from an uneroded enriched patch. Current data from the equatorial Pacific showed that the IronEx II experiment advected over six patch diameters per day. This means that at no time during the experiment were the products of increased export reflected in the waters below the enriched area. A transect through the IronEx II patch is shown in **Figure 9**. This figure indicates the massive production of plant biomass with a concomitant decrease in both nitrate and carbon dioxide.

The results from the equatorial Pacific, when corrected for dilution, suggest that about 2500 t of carbon were exported from the mixed layer over a 7-day period. These results are preliminary and subject to more rigorous estimates of dilution and export production, but they do agree favorably with estimates based upon both carbon and nitrogen budgets. Similarly, thorium export was observed in this experiment, confirming some particle removal.

The results of the SOIREE experiment were similar in many ways but were not as definitive with respect to carbon flux. In this experiment biomass increased 6-fold, nitrate was depleted by $2 \, \mu mol \, l^{-1}$ and carbon dioxide by 35–40 microatmospheres (3.5–4.0 Pa). This was a greatly attenuated signal relative to IronEx II. Colder water temperatures likely led to slower rates of production and bloom evolution and there was no observable carbon flux.

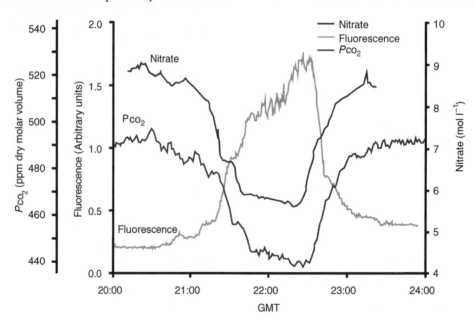

Figure 9 A transect through the IronEx II patch. The x-axis shows GMT as the ship steams from east to west through the center of the patch. Simultaneously plotted are the iron-induced production of chlorophyll, the drawdown of carbon dioxide, and the uptake of nitrate in this bloom.

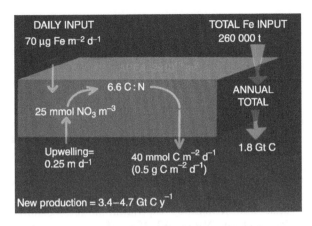

DAILY INPUT
70 µg Fe m^{-2} d^{-1}

TOTAL Fe INPUT
260 000 t

6.6 C:N

25 mmol NO$_3$ m^{-3}

ANNUAL
TOTAL

Upwelling=
0.25 m d^{-1}

40 mmol C m^{-2} d^{-1}
(0.5 g C m^{-2} d^{-1})

1.8 Gt C

New production = 3.4–4.7 Gt C y^{-1}

Figure 10 Simple calculations of the potential for carbon export for the Southern Ocean. These calculations are based on the necessary amount of iron required to efficiently utilize the annual upwelled nitrate and the subsequent incorporation into sinking organic matter. An estimated 1.8×10^9 t (Gt) of carbon export could be realized in this simple model.

Original estimates of carbon export in the Southern Ocean based on the iron-induced efficient utilization of nitrate suggest that as much as 1.8×10^9 t of carbon could be removed annually (**Figure 10**). These estimates of carbon sequestration have been challenged by some modelers yet all models lack important experimental parameters which will be measured in upcoming experiments.

Remaining Questions

A multitude of questions remain regarding the role of iron in shaping the nature of the pelagic community. The most pressing question is whether iron enrichment accelerates the downward transport of carbon from the surface waters to the deep sea? More specifically, how does iron affect the cycling of carbon in HNLC, LNLC, and coastal systems? Recent studies indicate that coastal systems may be iron-limited and the iron requirement for nitrogenase activity is quite large, suggesting that iron may limit nitrogen fixation, but there have been limited studies to test the former and none to test the latter. If iron does stimulate carbon uptake, what are the spatial scales over which this fixed carbon may be remineralized? This is crucial to predicting whether fertilization is an effective carbon sequestration mechanism.

Given these considerations, the most feasible way to understand and quantify carbon export from an enriched water mass is to increase the scale of the experiment such that both lateral dilution and sub-mixed-layer relative advection are small with respect to the size of the enriched patch. For areas such as the equatorial Pacific, this would be very large (hundreds of kilometers on a side). For other areas, it could be much smaller.

The focus of the IronEx and SOIREE experiments has been from the scientific perspective, but this focus is shifting toward the application of iron enrichment as a carbon sequestration strategy. We have come about rapidly from the perspective of trying to understand how the world works to one of trying to make the world work for us. Several basic questions remain regarding the role of natural or anthropogenic iron fertilization on carbon export. Some of the most pressing questions are: What are the best proxies for carbon export? How can carbon export best be verified? What are the long-term ecological consequences of iron enrichment on surface water community structure, midwater processes, and benthic processes? Even with answers to these, there are others that need to be addressed prior to any serious consideration of iron fertilization as an ocean carbon sequestration option.

Simple technology is sufficient to produce a massive bloom. The technology required either for a large-scale enrichment experiment or for purposeful attempts to sequester carbon is readily available. Ships, aircraft (tankers and research platforms), tracer technology, a broad range of new Autonomous Underwater Vehicles (AUVs) and instrument packages, Lagrangian buoy tracking systems, together with aircraft and satellite remote sensing systems and a new suite of chemical sensors/*in situ* detection technologies are all available, or are being developed. Industrial bulk handling equipment is available for large-scale implementation. The big questions, however, are larger than the technology.

With a slow start, the notion of both scientific experimentation through manipulative experiments, as well as the use of iron to purposefully sequester carbon, is gaining momentum. There are now national, international, industrial, and scientific concerns willing to support larger-scale experiments. The materials required for such an experiment are inexpensive and readily available, even as industrial by-products (of paper, mining, and steel processing).

Given the concern over climate change and the rapid modernization of large developing countries such as China and India, there is a pressing need to address the increased emission of greenhouse gases. Through the implementation of the Kyoto accords or other international agreements to curb emissions (Rio), financial incentives will reach into the multibillion dollar level annually. Certainly there will soon be an overwhelming fiscal incentive to investigate, if not implement, purposeful open ocean carbon sequestration trials.

A Societal Challenge

The question is not whether we have the capability of embarking upon such an engineering strategy but whether we have the collective wisdom to responsibly negotiate such a course of action. Posing the question another way: If we do not have the social, political and economic tools or motivation to control our own population and greenhouse gas emissions, what gives us the confidence that we have the wisdom and ability to responsibly manipulate and control large ocean ecosystems without propagating yet another massive environmental calamity? Have we as an international community first tackled the difficult but obvious problem of overpopulation and implemented alternative energy technologies for transportation, industry, and domestic use?

Other social questions arise as well. Is it appropriate to use the ocean commons for such a purpose? What individuals, companies, or countries would derive monetary compensation for such an effort and how would this be decided?

It is clear that there are major scientific investigations and findings that can only benefit from large-scale open ocean enrichment experiments, but certainly a large-scale carbon sequestration effort should not proceed without a clear understanding of both the science and the answers to the questions above.

Glossary

ATP	Adenosine triphosphate
AVHRR	Advanced Very High Resolution Radiometer
HNHC	High-nitrate high-chlorophyll
HNLC	High-nitrate low-chlorophyll
IronEx	Iron Enrichment Experiment
LIDAR	Light detection and ranging
LNHC	Low-nitrate high-chlorophyll
LNLC	Low-nitrate low-chlorophyll
NADPH	Reduced form of nicotinamide–adenine dinucleotide phosphate
SOIREE	Southern Ocean Iron Enrichment Experiment

Further Reading

Abraham ER, Law CS, Boyd PW, et al. (2000) Importance of stirring in the development of an iron-fertilized phytoplankton bloom. Nature 407: 727–730.

Barbeau K, Moffett JW, Caron DA, Croot PL, and Erdner DL (1996) Role of protozoan grazing in relieving iron limitation of phytoplankton. Nature 380: 61–64.

Behrenfeld MJ, Bale AJ, Kobler ZS, Aiken J, and Falkowski PG (1996) Confirmation of iron limitation of phytoplankton photosynthesis in Equatorial Pacific Ocean. Nature 383: 508–511.

Boyd PW, Watson AJ, Law CS, et al. (2000) A mesoscale phytoplankton bloom in the polar Southern Ocean stimulated by iron fertilization. Nature 407: 695–702.

Cavender-Bares KK, Mann EL, Chishom SW, Ondrusek ME, and Bidigare RR (1999) Differential response of equatorial phytoplankton to iron fertilization. Limnology and Oceanography 44: 237–246.

Coale KH, Johnson KS, Fitzwater SE, et al. (1996) A massive phytoplankton bloom induced by an ecosystem-scale iron fertilization experiment in the equatorial Pacific Ocean. Nature 383: 495–501.

Coale KH, Johnson KS, Fitzwater SE, et al. (1998) IronEx-I, an in situ iron-enrichment experiment: experimental design, implementation and results. Deep-Sea Research Part II 45: 919–945.

Elrod VA, Johnson KS, and Coale KH (1991) Determination of subnanomolar levels of iron (II) and total dissolved iron in seawater by flow injection analysis with chemiluminescence dection. Analytical Chemistry 63: 893–898.

Fitzwater SE, Coale KH, Gordon RM, Johnson KS, and Ondrusek ME (1996) Iron deficiency and phytoplankton growth in the equatorial Pacific. Deep-Sea Research Part II 43: 995–1015.

Greene RM, Geider RJ, and Falkowski PG (1991) Effect of iron litiation on photosynthesis in a marine diatom. Limnology Oceanogrography 36: 1772–1782.

Hoge EF, Wright CW, Swift RN, et al. (1998) Fluorescence signatures of an iron-enriched phytoplankton community in the eastern equatorial Pacific Ocean. Deep-Sea Research Part II 45: 1073–1082.

Johnson KS, Coale KH, Elrod VA, and Tinsdale NW (1994) Iron photochemistry in seawater from the Equatorial Pacific. Marine Chemistry 46: 319–334.

Kolber ZS, Barber RT, Coale KH, et al. (1994) Iron limitation of phytoplankton photosynthesis in the Equatorial Pacific Ocean. Nature 371: 145–149.

Landry MR, Ondrusek ME, Tanner SJ, et al. (2000) Biological response to iron fertilization in the eastern equtorial Pacific (Ironex II). I. Microplankton community abundances and biomass. Marine Ecology Progress Series 201: 27–42.

LaRoche J, Boyd PW, McKay RML, and Geider RJ (1996) Flavodoxin as an in situ marker for iron stress in phytoplankton. Nature 382: 802–805.

Law CS, Watson AJ, Liddicoat MI, and Stanton T (1998) Sulfer hexafloride as a tracer of biogeochemical and physical processes in an open-ocean iron fertilization experiment. Deep-Sea Research Part II 45: 977–994.

Martin JH, Coale KH, Johnson KS, et al. (1994) Testing the iron hypothesis in ecosystems of the equatorial Pacific Ocean. Nature 371: 123–129.

Nightingale PD, Liss PS, and Schlosser P (2000) Measurements of air–gas transfer during an open ocean algal bloom. Geophysical Research Letters 27: 2117–2121.

Obata H, Karatani H, and Nakayama E (1993) Automated determination of iron in seawater by chelating resin concentration and chemiluminescence detection. *Analytical Chemistry* 65: 1524–1528.

Redfield AC (1934) On the proportions of organic derivatives in sea water and their relation to the composition of plankton. *James Johnstone Memorial Volume*, pp. 177–192. Liverpool: Liverpool University Press.

Redfield AC (1958) The biological control of chemical factors in the environment. *American Journal of Science* 46: 205–221.

Rue EL and Bruland KW (1997) The role of organic complexation on ambient iron chemistry in the equatorial Pacific Ocean and the response of a mesoscale iron addition experiment. *Limnology and Oceanography* 42: 901–910.

Smith SV (1984) Phosphorus versus nitrogen limitation in the marine environment. *Limnology and Oceanography* 29: 1149–1160.

Stanton TP, Law CS, and Watson AJ (1998) Physical evolutation of the IronEx I open ocean tracer patch. *Deep-Sea Research Part II* 45: 947–975.

Takeda S and Obata H (1995) Response of equatorial phytoplankton to subnanomolar Fe enrichment. *Marine Chemistry* 50: 219–227.

Trick CG and Wilhelm SW (1995) Physiological changes in coastal marine cyanobacterium *Synechococcus* sp. PCC 7002 exposed to low ferric ion levels. *Marine Chemistry* 50: 207–217.

Turner SM, Nightingale PD, Spokes LJ, Liddicoat MI, and Liss PS (1996) Increased dimethyl sulfide concentrations in seawater from *in situ* iron enrichment. *Nature* 383: 513–517.

Upstill-Goddard RC, Watson AJ, Wood J, and Liddicoat MI (1991) Sulfur hexafloride and helium-3 as sea-water tracers: deployment techniques and continuous underway analysis for sulphur hexafloride. *Analytica Chimica Acta* 249: 555–562.

Van den Berg CMG (1995) Evidence for organic complesation of iron in seawater. *Marine Chemistry* 50: 139–157.

Watson AJ, Liss PS, and Duce R (1991) Design of a small-scale *in situ* iron fertilization experiment. *Limnology and Oceanography* 36: 1960–1965.

APPENDICES

APPENDIX 1. SI UNITS AND SOME EQUIVALENCES

Wherever possible the units used are those of the International System of Units (SI). Other "conventional" units (such as the liter or calorie) are frequently used, especially in reporting data from earlier work. Recommendations on standardized scientific terminology and units are published periodically by international committees, but adherence to these remains poor in practice. Conversion between units often requires great care.

The base SI units

Quantity	Unit	Symbol
Length	meter	m
Mass	kilogram	kg
Time	second	s
Electric current	ampere	A
Thermodynamic temperature	kelvin	K
Amount of substance	mole	mol
Luminous intensity	candela	cd

Some SI derived and supplementary units

Quantity	Unit	Symbol	Unit expressed in base or other derived units
Frequency	hertz	Hz	s^{-1}
Force	newton	N	$kg\,m\,s^{-2}$
Pressure, stress	pascal	Pa	$N\,m^{-2}$
Energy, work, quantity of heat	joule	J	$N\,m$
Power	watt	W	$J\,s^{-1}$
Electric charge, quantity of electricity	coulomb	C	$A\,s$
Electric potential, potential difference, electromotive force	volt	V	$J\,C^{-1}$
Electric capacitance	farad	F	$C\,V^{-1}$
Electric resistance	ohm	ohm (Ω)	$V\,A^{-1}$
Electric conductance	Siemens	S	Ω^{-1}
Magnetic flux	weber	Wb	$V\,s$
Magnetic flux density	tesla	T	$Wb\,m^{-2}$
Inductance	henry	H	$Wb\,A^{-1}$
Luminous flux	lumen	lm	$cd\,sr$
Illuminance	lux	lx	$lm\,m^{-2}$
Activity (of a radionuclide)	becquerel	Bq	s^{-1}
Absorbed dose, specific energy	gray	Gy	$J\,kg^{-1}$
Dose equivalent	sievert	Sv*	$J\,kg^{-1}$
Plane angle	radian	rad	
Solid angle	steradian	sr	

*Not to be confused with Sverdrup conventionally used in oceanography: see SI Equivalences of Other Units.

SI base units and derived units may be used with multiplying prefixes (with the exception of kg, though prefixes may be applied to gram $= 10^{-3}$kg; for example, 1 Mg $= 10^6$ g $= 10^6$kg)

Prefixes used with SI units

Prefix	Symbol	Factor
yotta	Y	10^{24}
zetta	Z	10^{21}
exa	E	10^{18}
peta	P	10^{15}
tera	T	10^{12}
giga	G	10^{9}
mega	M	10^{6}
kilo	k	10^{3}
hecto	h	10^{2}
deca	da	10
deci	d	10^{-1}
centi	c	10^{-2}
milli	m	10^{-3}
micro	μ	10^{-6}
nano	n	10^{-9}
pico	p	10^{-12}
femto	f	10^{-15}
atto	a	10^{-18}
zepto	z	10^{-21}
yocto	y	10^{-24}

SI Equivalences of Other Units

Physical quantity	Unit	Equivalent	Reciprocal
Length	nautical mile (nm)	1.85318 km	km = 0.5396 nm
Mass	tonne (t)	10^3 kg = 1 Mg	
Time	min	60 s	
	h	3600 s	
	day or d	86 400 s	s = 1.1574×10^{-5} day
	y	3.1558×10^7 s	s = 3.1688×10^{-8} y
Temperature	°C	°C = K − 273.15	
Velocity	knot (1 nm h^{-1})	0.51477 m s^{-1}	m s^{-1} = 1.9426 knot
		44.5 km d^{-1}	
		16 234 km y^{-1}	
Density	gm cm^{-3}	tonne m^{-3} = 10^3 kg m^{-3}	
Force	dyn	10^{-5} N	
Pressure	dyn cm^{-2}	10^{-1} N m^{-2} = 10^{-1} Pa	
	bar	10^5 N m^{-2} = 10^5 Pa	
	atm (standard atmosphere)	101 325 N m^{-2} = 101.325 kPa	
Energy	erg	10^{-7} J	
	cal (I.T.)	4.1868 J	
	cal (15°C)	4.1855 J	
	cal (thermochemical)	4.184 J	J = 0.239 cal

(*Note*: The last value is the one used for subsequent conversions involving calories.)

Energy flux	langley (ly) min^{-1} = $cal\,cm^{-2}\,min^{-1}$	$697\,W\,m^{-2}$	$W\,m^{-2} = 1.434 \times 10^{-3}$ $ly\,min^{-1}$
	$ly\,h^{-1}$	$11.6\,W\,m^{-2}$	$W\,m^{-2} = 0.0860\,ly\,h^{-1}$
	$ly\,d^{-1}$	$0.484\,W\,m^{-2}$	$W\,m^{-2} = 2.065\,ly\,d^{-1}$
	$kcal\,cm^{-2}\,y^{-1}$	$1.326\,W\,m^{-2}$	$W\,m^{-2} = 0.754\,kly\,y^{-1}$
Volume flux	Sverdrup	$10^6\,m^3\,s^{-1}$ $3.6\,km^3\,h^{-1}$	
Latent heat	$cal\,g^{-1}$	$4184\,J\,kg^{-1}$	$J\,kg^{-1} = 2.39 \times 10^{-4}$ $cal\,g^{-1}$
Irradiance	Einstein $m^{-2}\,s^{-1}$ (mol photons $m^{-2}\,s^{-1}$)		

*Most values are taken from or derived from *The Royal Society Conference of Editors Metrication in Scientific Journals*, 1968, The Royal Society, London.

The SI units for pressure is the pascal ($1\,Pa = 1\,N\,m^{-2}$). Although the bar ($1\,bar = 10^5\,Pa$) is also retained for the time being, it does not belong to the SI system. Various texts and scientific papers still refer to gas pressure in units of the torr (symbol: Torr), the bar, the conventional millimetre of mercury (symbol: mmHg), atmospheres (symbol: atm), and pounds per square inch (symbol: psi) – although these units will gradually disappear (see Conversions between Pressure Units).

Irradiance is also measured in $W\,m^{-2}$. Note: 1 mol photons $= 6.02 \times 10^{23}$ photons.

The SI unit used for the amount of substance is the mole (symbol: mol), and for volume the SI unit is the cubic metre (symbol: m^3). It is technically correct, therefore, to refer to concentration in units of $mol\,m^3$. However, because of the volumetric change that sea water experiences with depth, marine chemists prefer to express sea water concentrations in molal units, $mol\,kg^{-1}$.

Conversions between Pressure Units

	Pa	kPa	bar	atm	Torr	psi
1 Pa =	1	10^{-3}	10^{-5}	$9.869\,23 \times 10^{-6}$	$7.500\,62 \times 10^{-3}$	$1.450\,38 \times 10^{-4}$
1 kPa =	10^3	1	10^{-2}	$9.869\,23 \times 10^{-3}$	$7.500\,62$	$0.145\,038$
1 bar =	10^5	10^2	1	$0.986\,923$	750.062	145.038
1 atm =	$101\,325$	101.325	$1.013\,25$	1	760	14.6959
1 Torr =	133.322	$0.133\,322$	$1.333\,22 \times 10^{-3}$	$1.315\,79 \times 10^{-3}$	1	$1.933\,67 \times 10^{-2}$
1 psi	6894.76	$6.894\,76$	$6.894\,76 \times 10^{-2}$	$6.804\,60 \times 10^{-2}$	$51.715\,07$	1

psi = pounds force per square inch.
$1\,mmHg = 1\,Torr$ to better than $2 \times 10^{-7}\,Torr$.

APPENDIX 2. USEFUL VALUES

Molecular mass of dry air, $m_a = 28.966$

Molecular mass of water, $m_w = 18.016$

Universal gas constant, $R = 8.31436\,\mathrm{J\,mol^{-1}K^{-1}}$

Gas constant for dry air, $R_a = R/m_a = 287.04\,\mathrm{J\,kg^{-1}K^{-1}}$

Gas constant for water vapor, $R_v = R/m_w = 461.50\,\mathrm{J\,kg^{-1}K^{-1}}$

Molecular weight ratio $\varepsilon \equiv m_w/m_a = R_a/R_v = 0.62197$

Stefan's constant $\sigma = 5.67 \times 10^{-8}\,\mathrm{W\,m^{-2}K^{-4}}$

Acceleration due to gravity, $g\,(\mathrm{m\,s^{-2}})$ as a function of latitude φ and height $z\,(\mathrm{m})$

$$g = (9.78032 + 0.005172\sin^2\varphi - 0.00006\sin^2 2\varphi)(1 + z/a)^{-2}$$

Mean surface value, $\bar{g} = \int_0^{\pi/2} g\cos\varphi\,d\varphi = 9.7976$

Radius of sphere having the same volume as the Earth, $a = 6371\,\mathrm{km}$ (equatorial radius $= 6378\,\mathrm{km}$, polar radius $= 6357\,\mathrm{km}$)

Rotation rate of earth, $\Omega = 7.292 \times 10^{-5}\,\mathrm{s^{-1}}$

Mass of earth $= 5.977 \times 10^{24}\,\mathrm{kg}$

Mass of atmosphere $= 5.3 \times 10^{18}\,\mathrm{kg}$

Mass of ocean $= 1400 \times 10^{18}\,\mathrm{kg}$

Mass of ground water $= 15.3 \times 10^{18}\,\mathrm{kg}$

Mass of ice caps and glaciers $= 43.4 \times 10^{18}\,\mathrm{kg}$

Mass of water in lakes and rivers $= 0.1267 \times 10^{18}\,\mathrm{kg}$

Mass of water vapor in atmosphere $= 0.0155 \times 10^{18}\,\mathrm{kg}$

Area of earth $= 5.10 \times 10^{14}\,\mathrm{m^2}$

Area of ocean $= 3.61 \times 10^{14}\,\mathrm{m^2}$

Area of land $= 1.49 \times 10^{14}\,\mathrm{m^2}$

Area of ice sheets and glaciers $= 1.62 \times 10^{13}\,\mathrm{m^2}$

Area of sea ice $= 1.9 \times 10^{13}\,\mathrm{m^2}$ in March and $2.9 \times 10^{13}\,\mathrm{m^2}$ in September (averaged between 1979 and 1987)

APPENDIX 8. ABBREVIATIONS

AABW	Antarctic Bottom Water
AAIW	Antarctic Intermediate Surface Water
AASW	Antarctic Surface Water
AATSR	Advanced ATSR
ABE	Autonomous Benthic Explorer
ABF	Angola–Benguela Front
ABW	Arctic Bottom Water
ACC	Antarctic Circumpolar Current
ACD	aragonite compensation depth
ACOUS	Arctic Climate Observations Using Underwater Sound (project)
ADAS	Airborne Diode Array Spectrometer
ADCP	acoustic Doppler current profiler
ADEOS	Advanced Earth Observing Satellite
ADL	aerobic diving limits
ADV	Adventure Bank Vortex
ADW	Adriatic Deep Water
AEE	anomalously enriched elements
AFC	Automatic Flow Cytometry
AFGP	antifreeze glycopeptides
AGCM	atmospheric general circulation model(s)
AGDW	Aegean Deep Water
AIRSAR	Airborne SAR
AIS	Atlantic Ionian Stream
AISI	Airborne Imaging Spectrometer
AIW	Atlantic Intermediate Water
ALACE	Autonomous Lagrangian Circulation Explorer
ALOS	Advanced Land Observing Satellite
AMC	axial magma chamber
AMIP	Atmospheric Model Intercomparison Project
AMS	accelerator mass spectrometry
AMSR	Advanced Microwave Scanning Radiometer
AMT	Atlantic Meridional Transect
AOCI	Airborne Ocean Color Instrument
AOGCM	atmosphere–ocean general circulation models
AOL	Airborne Oceanographic LIDAR
AOP	apparent optical property
APC	Advanced Piston Corer
APE	available potential energy
APFZ	Antarctic Polar Frontal Zone
APTS	astronomical polarity timescale
ARIES	Autosampling and Recording Instrumental Environmental Sampler
ASDIC	Antisubmarine Detection Investigation Committee
ASMR	Advanced Scanning Microwave Radiometer
ASP	amnesic shellfish poisoning
AST	axial summit trough
ASUW	Atlantic Subarctic Upper Water
ASW	Arabian Sea Water; *or* antisubmarine warfare
ATM	Airborne Topographic Mapper
ATOC	Acoustic Thermometry of Ocean Climate
ATP	adenosine triphosphate
ATSR	Along-Track Scanning Radiometer

AUV	autonomous underwater vehicle
AVHRR	Advanced Very High Resolution Radiometer
AVIRIS	Airborne Visible/Infrared Imaging Spectrometer
AVP	axial volcanic ridge(s)
AVPPO	Autonomous Vertically Profiling Plankton Observatory
AW	Atlantic Water
AXBT	air-launched XBT
AXSV	air-launched XSV
AZP	azaspirazid shellfish poisoning
BB	broadband
BBL	benthic boundary layer
BBW	Bengal Bay Water
BCD	bacterial carbon demand
BGE	bacterial growth efficiency
BGHS	base of gas hydrate stability
BI	baroclinic instability
BIO	Bedford Institute of Oceanography (Canada)
BIOMASS	Biological Investigations of Marine Antarctic Systems and Stocks
BIONESS	Bedford Institute of Oceanography Net and Environmental Sensing System
BMC	Brazil/Malvinas Confluence
BOD	biological oxygen demand
BP	bacterial production
BR	bacterial respiration
BSR	bottom-simulating reflector
BTM	Bermuda Testbed Mooring
BWT	bottom water temperature
CASI	Compact Airborne Spectrographic Imager
CBDW	Canadian Basin Deep Water
CCAMLR	Commission for the Conservation of Antarctic Marine Living Resources
CCD	calcite compensation depth
CCrD	carbonate critical depth
CDOM	colored dissolved organic matter
CDW	Circumpolar Deep Water; *or* Cretan Deep Water
CFA	carbonate fluoroapatite; *or* continuous flow analyser
CFC	chlorofluorocarbon
CFP	ciguatera fish poisoning
CFT	controlled flux technique
CHAMP	Challenging Minisatellite Payload
Chl-a	chlorophyll-*a*
CIW	Californian Intermediate Water; *or* Cretan Intermediate Water
CLE/CSV	competitive ligand equilibration with cathode stripping voltammetry
CLIVAR	Climate Variability and Predictability Program
CMA	Chemical Manufacturers Association
CNES	Centre Nationale d'Etudes Spatiales
COIS	Coastal Ocean Imaging Spectrometer
COS	carbonyl sulfide
COT	cost of transport
CPR	Continuous Plankton Recorder
CRM	chemical remanent magnetization
CS_2	carbon disulfide
CSA	Canadian Space Agency
CSO	combined sewer overflow
CSSF	Canadian Scientific Submersible Facility
CTD	conductivity, temperature, and depth; *or* conductivity–temperature–depth (profiler)
CUFES	continuous underway fish egg sampler

CZCS	Coastal Zone Color Scanner
D/O	Dansgaard–Oeschger
DCC	direct current condenser
DCM	deep chlorophyll maximum
DCMU	3-(3,4-dichlorophenyl)-1,1-dimethylurea
DDT	dichlorodiphenyltrichloroethane
DIC	dissolved inorganic carbon
DIN	dissolved inorganic nitrogen
DIP	dissolved inorganic phosphorus
DMGe	dimethylgermanic acid
DMHg	dimethylmercury
DMS	dimethyl sulfide
DMSb	dimethylantimonate
DMSP	Defense Meteorological Satellite Program; *or* 3-(dimethylsulfonium) propionate
DNA	deoxyribonucleic acid
DOC	dissolved organic carbon
DOM	dissolved organic matter
DON	dissolved organic nitrogen
DOP	dissolved organic phosphorus
DPASV	differential pulse anodic stripping voltammetry
dpm	disintegrations per minute
DRM	detrital remanent magnetization/depositional detrital remanent magnetization
DSDP	Deep Sea Drilling Project
DSL	deep scattering layer
DSP	diarrhetic shellfish poisoning
DVM	diel vertical migration
DW	Deep Arctic Water
DWBC	deep western boundary current
DWT	deadweight tonnage
EAC	East Australian Current
EACC	East Africa Coastal Current
EAIS	East Antarctic ice sheet
EASIW	Eastern Atlantic Subarctic Intermediate Water
EBDW	Eurasian Basin Deep Water
ECMWF	European Centre for Medium-Range Weather Forecasts
EEM	excitation–emission matrix (spectroscopy)
EEZ	exclusive economic zone
EF	enrichment factor
EGC	East Greenland Current
EIC	Equatorial Intermediate Current
EKE	eddy kinetic energy
ELM	external limiting membrane
EM	electromagnetic
EMDW	Eastern Mediterranean Deep Water
EMT	Eastern Mediterranean Transient
ENACW	Eastern North Atlantic Central Water
ENPCW	Eastern North Pacific Central Water
ENPTW	Eastern North Pacific Transition Water
ENSO	El Niño Southern Oscillation
Envisat	Environmental Satellite
EOS	Earth Observing System
EPI	epifluorescence microscopy
EPR	East Pacific Rise
EPS	extracellular polysaccharides
ERS	Earth Resources Satellite

ESA	European Space Agency
ESD	equivalent spherical diameter
ESMR	Electrically Scanning Microwave Radiometer
ESPCW	Eastern South Pacific Central Water
ESPIW	Eastern South Pacific Intermediate Water
ESPTW	Eastern South Pacific Transition Water
ESS	evolutionarily stable strategy
ESSE	error subspace statistical estimation
ESTAR	Electrically Scanning Thinned Array Radiometer
ETR	electron transport rate
EUC	Equatorial Undercurrent
FAO	(UN) Food and Agriculture Organization
FBI	fresh water–brackish water interface
FCLS	ferrochrome lignosulfate
FDOM	fluorescent (dissolved) organic matter
FFP	fast field program
FIA	flow injection analyzer
FID	flame ionization detector/detection
FLIP	Floating Instrument Platform
FMI	Formation Micro-Image
FMP	Fishery Management Plan
FMS	formation scanner
FOM	figure of merit
FPA	fixed-potential amperometry
FRAM	Fine Resolution Antarctic Model
FRR	fast repetition rate
FRRF	fast repetition rate fluorometry
FSU	Florida State University
FTIR	Fourier transform infrared spectrometry/spectrometer
FY	first-year (ice)
GAC	global area coverage
GBRUC	Great Barrier Reef Undercurrent
GC	gas chromatography
GCM	general circulation model
GCOS	Global Climate Observing System
GEF	Global Environmental Facility
GEOHAB	Global Ecology and Oceanography HABs (program)
GEOSECS	Geochemical Ocean Sections Study
GLI	Global Imager
GLOBEC	Global Ocean Ecosystem Dynamics
GLORIA	geological long-range ASDIC
GMOs	genetically modified organism(s)
GOCE	Gravity Field and Steady-state Ocean Circulation Explorer
GODAE	Global Ocean Data Assimilation Experiment
GOM	Gulf of Mexico
GOOS	Global Ocean Observing System
GPS	Global Positioning System
GPTS	geomagnetic polarity timescale
GRACE	Gravity Recovery and Climate Experiment
GSNW	Gulf Stream North Wall (index)
GSSP	Global Boundary Stratotype Section and Point
GT	gross tonnage
HAB	harmful algal bloom
HCB	hexachlorobenzene
HCH	hexachlorocyclohexane

HE	Halmahera Eddy
HEXMAX	HEXOS Main Experiment
HEXOS	Humidity Exchange Over the Sea (experiment)
HIRIS	High-Resolution Imaging Spectometer
HMW	high-molecular weight
HNDA	high natural dispersing areas
HNLC	high-nutrients, low-chlorophyll
HOPLASA	Horizontal Plankton Sampler
HOPS	Harvard Ocean Prediction System
HOT	Hawaiian Ocean Time Series
HPLC	high-performance liquid chromatography
HRGB	hard rock guide base
HRPT	high-resolution picture transmission
HRS	high-resolution sampler
HSSW	high-salinity shelf water
HST	high stand system tract
IABO	International Association for Biological Oceanography
IAPSO	International Association for the Physical Sciences of the Ocean
IAS	Intra-Americas Sea
IBM	individual-based model
IC	integrated circuit
ICCAT	International Commission for the Conservation of Atlantic Tuna
ICES	International Council for the Exploration of the Sea
ICJ	International Court of Justice
ICNAF	International Commission for the Northwest Atlantic Fisheries
ICP	International Conferences on Paleo-oceanography
ICP-MS	inductively coupled plasma mass spectrometry
ICRP	International Commission on Radiological Protection
ICSU	International Council for Science (formerly International Council of Scientific University)
ICZM	Integrated Coastal Zone Management
IEW	Indian Equatorial Water
IFQ	Individual Fishery Quota
IFREMER	l'Institut Francais de Recherche pour l'Exploitation de la Mer
IGBP	International Geosphere–Biosphere Program
IGW	internal gravity waves
IIOE	International Indian Ocean Experiment
IIP	(US Coast Guard) International Ice Patrol
IIW	Indonesian Intermediate Water
IKMT	Isaacs–Kidd midwater trawl
IMO	International Maritime Organization
INDEX	Indian Ocean Experiment
INPFC	International North Pacific Fisheries Commission
IOC	International Oceanographic Commission
IOCCG	International Ocean Color Coordinating Group
IODP	Integrated Ocean Drilling Program
IOP	inherent optical properties
IPCC	Intergovernmental Panel on Climate Change
IPI	interpulse interval
IPNV	infectious pancreatic necrosis virus
IPSFC	International Pacific Salmon Fisheries Commission
IR	infrared
IRD	iceberg rafted detrital
IRONEX	Iron Enrichment Experiment
IronEx II	Iron Fertilization Experiment II
ISAV	infectious salmon anemia virus

ISE	ion-selective electrode
ISV	Ionian Shelfbreak Vortex
ISW	ice shelf water
ITCZ	Intertropical Convergence Zone
ITQ	individual transferable quota
IUCN	International Union for the Conservation of Nature
IUW	Indonesian Upper Water
IVF	*in vivo* fertilization
JAMSTEC	Japan Marine Science and Technology Center
JERS	Japan Environmental Resources Satellite
JGOFS	Joint Global Ocean Flux Study
JLS	join, leave, or stay
JOI	Joint Oceanographic Institutions Incorporated
JOIDES	Joint Oceanographic Institutions for Deep Earth Sampling
JPL	Jet Propulsion Laboratory
K/T	Cretaceous/Tertiary (boundary)
KE	kinetic energy
KHI	Kelvin–Helmholtz Instability
KYM	krill yield model
LAA	large amorphous aggregates
LAC	local area coverage
LADCP	Lowered Acoustic Doppler Current Profiler
LADS	Laser Airborne Depth Sounder
LARS	launch and recovery system
LBMP	land-based marine pollution
LCDW	Lower Circumpolar Deep Water
LDEO	Lamont–Doherty Earth Observatory
LDW	Levantine Deep Water
LHPR	Longhurst–Hardy plankton recorder
LIA	Little Ice Age
LIDAR	light detection and ranging
LIP	large igneous province
LIW	Levantine Intermediate Water
LLD	liquid line of defense
LME	large marine ecosystems
LMW	low-molecular weight
LNG	liquefied natural gas
LNHC	low-nitrate, high-chlorophyll
LNLC	low-nitrate, low-chlorophyll
LOAEL	lowest observed adverse effect level
LOCHNESS	Large Opening/Closing High Speed Net and Environmental Sampling System
LOICZ	Land Ocean Interaction in the Coastal Zone (program)
LOOPS	Littoral Ocean Observing and Prediction System
LPG	liquefied petroleum gas
LSFC	laser-stimulated fluorescence of chlorophyll
LST	low stand systems tract
MAB	Mid-Atlantic Bight
MAD	Magnetic Airborne Detector
MAR	Mid-Atlantic Ridge
MARPOL	Marine Pollution (treaty)
MASZP	moored, automated, serial zooplankton pump
MAW	Modified Atlantic Water
mbsf	meters below seafloor
MBT	mechanical bathythermograph
MC	Mindanao Current; *or* Mozambique Current

MCC	Maltese Channel Crest
MCP	Medieval Cold Period
MCSST	multichannel SST
ME	Mindanao Eddy
MFP	matched field processing
MIJ	Mid-Ionian Jet
MIZ	marginal ice zone
MIZEX	marginal ice zone experiments
MMAs	monomethylarsenate
MMD	mass median diameter
MMGe	monomethylegermanic acid
MMHg	monomethylmercury
MMJ	Mid-Mediterranean Jet
MMSb	monomethylantimonate
MOBY	Marine Optical Buoy
MOC	Meridional Overturning Circulation
MOCNESS	Multiple Opening/Closing Net and Environmental Sensing System
MOCS	Multichannel Ocean Color Sensor
MODE	Mid-Ocean Dynamics Experiment
MODIS	Moderate Resolution Imaging Spectroradiometer
MOM	modular ocean model
MOR	mid-ocean ridge
MORB	mid-ocean ridge basalt
MPA	marine protected area
MPS	multiple plankton sampler
MSFCMA	Magnuson–Stevens Fishery Conservation Management Act
MSL	mean sea level
MST	Mediterranean Salt Tongue; *or* multisensor track
MSVPA	multispecies virtual population analysis
MSY	maximum sustainable yield
MUC	Mindanao Undercurrent
MVDF	maximum variance distortion filter
MW	Mediterranean Water; *or* molecular weight
MWDW	modified warm deep water
MWP	Medieval Warm Period
MY	multiyear (ice)
NABE	North Atlantic Bloom Experiment
NADPH	reduced form of nicotinamide–adenine dinucleotide phosphate
NADW	North Atlantic Deep Water
NAFO	Northwest Atlantic Fishery Organization
NAO	North Atlantic Oscillation
NASCO	North Atlantic Salmon Conservation Organization
NASDA	Japanese National Space Development Agency
NASF	North Atlantic Salmon Fund
NCAR	National Center for Atmospheric Research
NCC	Norwegian Coastal Current
NCEP	National Centers for Environmental Prediction
NDBC	National Data Buoy Center
NDSF	National Deep Submergence Facility
NEΔT	noise equivalent temperature difference
NEADS	North East Atlantic Dynamics Study
NEAFC	North-east Atlantic Fisheries Commission
NEC	North Equatorial Current
NECC	North Equatorial Countercurrent
NEE	nonenriched elements

NEMO	Naval Earth Map Observer
NEW	North-east Water
NGCC	New Guinea Coastal Current
NGCUC	New Guinea Coastal Undercurrent
NLSST	nonlinear SST
NM	normal mode
NMC	North-east Monsoon Current
NMHC	nonmethane hydrocarbons
NMR	nuclear magnetic resonance
NOAA	National Oceanographic and Atmospheric Administration
NOAEL	no observed adverse effect level
NOSAMS	National Ocean Sciences Accelerator Mass Spectrometry facility
NOSS	National Oceanographic Satellite System
NPAFC	North Pacific Anadromous Fisheries
NPOESS	National Polar-Orbiting Satellite System
NPP	NPOESS Preparatory Program
NRM	natural remanent magnetization
NROSS	Navy Remote Ocean Observing Satellite
NSP	neurotoxic shellfish poisoning
NSSC	Northern Subsurface Countercurrent
NTU	nephelometric turbidity unit
NWP	numerical weather prediction
OBS	ocean bottom seismograph
OCMIP	Ocean Carbon Model Intercomparison Project
OCTS	Ocean Color and Temperature Sensor
ODAS	Ocean Data Acquisition System
ODE	ordinary differential equation(s)
ODP	Ocean Drilling Program
OGCM	ocean general circulation model
OI	optimal interpolation
OML	ocean mixed layer
OMZ	oxygen minimum zone
OOP	object-oriented programming
OPA	Oil Pollution Act
OPC	optical plankton counter
OSC	overlapping spreading centers
OTEC	Ocean Thermal Energy Conversion
OTIP	Optimal Thermal Interpolation Scheme
PAH	polycyclic aromatic hydrocarbons
PALACE	Profiling Autonomous Lagrangian Circulation Explorer
PALK	potential alkalinity
PAM	pulse amplitude modulation
PAR	photosynthetic available radiation
PBB	passive broadband
PCBs	polychlorobiphenyls/polychlorinated biphenyls
PCG	Panama–Columbia Gyre
PCGC	preparatory capillary gas chromatography
PCR	polymerase chain reaction
PCS	pressure core sampler
PCUC	Peru–Chile Undercurrent
PDM	particulate detrital matter
pDRM	postdepositional detrital remanent magnetization
PDW	Pacific Deep Water
PE	parabolic equation
PEAS	possible estuary-associated syndrome

PEW	Pacific Equatorial Water
PF	Polar Front
PFSST	Pathfinder SST
PGE	platinum-group elements
PHILLS	Portable Hyper-spectral Imager for Low-Light Spectroscopy
PICES	North Pacific Marine Science Organization
PML	Polar Mixed Layer
PMM	photomultiplier module
PMT	photomultiplier tube
PN	particulate nitrogen
PNB	passive narrowband
PNM	primary NO_2 maximum
POC	particulate organic carbon
POCM	Parallel Ocean Climate Model
POEM	Physical Oceanography of the Eastern Mediterranean
POM	particulate organic matter
PON	particulate organic nitrogen
POP	persistent organic pollution
PP	primary production
ppbv	parts per billion by volume
PPF	Pump and Probe Fluorometer
ppmv	parts per million by volume
pptv	parts per trillion by volume
PRF	pulse repetition frequency
PRN	pseudo random noise
PS II	Photosystem II
PSC	Pacific Salmon Commission
PSIW	Pacific Subarctic Intermediate Water
PSMSL	permanent service for mean sea level
PSP	paralytic shellfish poisoning
PSSF	Passive Solar Stimulated Fluorescence
PSU	practical salinity unit
PSUW	Pacific Subarctic Upper Water
PSW	Polar Surface Water
PTCS	pressure and temperature core sampler
PV	potential vorticity
PW	Polar Water; *or* Pacific Water
QSU	quinine sulfate unit
RAD	ridge axis discontinuity
RAR	real aperture radar
RCB	rotary core barrel
REE	rare-earth elements
REMUS	remote environmental measuring units
rf	radiofrequency
RMT	rectangular mouth opening trawl
RNA	ribonucleic acid
RO	reverse osmosis
ROFI	region of fresh water influence
ROS	reactive oxygen species
ROV	remotely operated vehicle
ROWS	Radar Ocean Wave Spectrometer
RPE	retinal pigment epithelium
RR	refracted-refracted (rays)
rRNA	ribosomal RNA
RS	remote sensing

RSPGIW	Red Sea–Persian Gulf Intermediate Water
RSR	refracted-surface-reflected (rays)
RTE	radiative transfer equation
RTR	relative tide range
RV	research vessel
SACW	South Atlantic Central Water
SAH	South Atlantic High
SAHFOS	Sir Alister Hardy Foundation for Ocean Science
SAMW	Subantarctic Mode Water
SAR	synthetic aperture radar
SAS	SeaWiFS Aircraft Simulator
SASW	Subantarctic Surface Water
SAV	submerged aquatic vegetation
SAVE	South Atlantic Ventilation Experiment
SC	Somali Current
SCAR	Scientific Committee for Antarctic Research
SCOR	Science Commission on Oceanic Research
SCUBA	self-contained underwater breathing apparatus
SCV	sub-mesoscale coherent vortex
SeaWiFS	Sea-viewing Wide Field-of-view Sensor
SEC	South Equatorial Current
SECC	South Equatorial Countercurrent
SEM	scanning electron microscopy
SF_6	sulfur hexafluoride
SHOALS	Scanning Hydrographic Operational Airborne LIDAR Survey
SICW	South Indian Central Water
SIMBIOS	Sensor Intercomparison and Merger for Biological and Interdisciplinary Oceanic Studies
SINODE	Surface Indian Ocean Dynamic Experiment
SIO	Scripps Institution of Oceanography
SIPPER	Shadowed Image Particle Profiling and Evaluation Recorder
SIR	Shuttle Imaging Radar
SLFMR	Scanning Low Frequency Microwave Radiometer
SMC	South-west Monsoon Current
SMMR	Scanning Multichannel Microwave Radiometer
SMOS	Soil Moisture and Ocean Salinity
SMOW	standard mean ocean water
SMR	Scanning Microwave Radiometer
SNP	soluble nonreactive phosphorus
SNR	signal-to-noise ratio
SOC	Southampton Oceanography Centre (UK)
SOFAR	sound fixing and ranging
SOI	Southern Oscillation Index
SOIREE	Southern Ocean Iron Enrichment Experiment
SP	short period (instrumentation)
SPCZ	South Pacific Convergence Zone
SPE	solid-phase extraction
SPM	suspended particulate matter
SPMW	subpolar mode water
SRA	Scanning Radar Altimeter
SSB	spawning stock biomass
SSH	sea surface height
SSL	sound-scattering layer
SSM/I	Special Sensor Microwave/Imager
SSSC	South Subsurface Countercurrent
SST	sea surface temperature

SSU rRNA	small subunit rRNA
SSXBT	submarine-launched XBT
STCC	Subtropical Countercurrent
STD	salinity, temperature, and depth (measuring instrument)
STMW	Subtropical Mode Water
SVD	singular-value decomposition
SWG	Science Working Group(s)
T/P	Topex/Poseidon (system)
TAC	total allowable catch
TAP	Transarctic Arctic Propagation (experiment)
TBT	tributyltin
TDN	total dissolved nitrogen
TEM	transmission electron microscopy
TEP	transparent exopolymer particles
TEU	twenty-foot equivalent unit
THC	thermohaline circulation
TIMS	thermal ionization mass spectrometry
TIROS-N	Television and Infrared Observing Satellite-version N
TKE	turbulent kinetic energy
TLC	total lung capacity
TLE	total lipid extract
TMI	TRMM Microwave Imager
TMS	tether management system
TMW	Transitional Mediterranean Water
TMZ	turbidity maximum zone
TOBI	Towed Ocean Bottom Instrument
TOC	total organic carbon
TOGA	Tropical Ocean–Global Atmosphere (program)
TOMS	Total Ozone Mapping Spectrometer
TON	total oxidized nitrate
TOVS	Total Ozone Vertical Sounder
TRM	thermoremanent magnetization
TRMM	Tropical Rainfall Measuring Mission
TS curve	temperature–salinity curve
TST	transgressive stand systems tract
TTO	Tropical Tracers in the Ocean (experiment)
TTONAS	TTO North Atlantic Study
TTOTAS	TTO Tropical Atlantic Study
TU	tritium unit
TW	tropical waters
UBL	under-ice boundary layer
UCDW	upper Circumpolar Deep Water
ULES	upward-looking echo sounders
UML	upper mixed layer
UN(O)	United Nations (Organization)
UNCED	United Nations Conference on Environment and Development
UNCLOS	United Nations Convention on the Law of the Sea
UNEP	United Nations Environmental Program
UNESCO	United Nations Educational, Scientific, and Cultural Organization
UOR	undulating oceanographic recorder
uPDW	Upper Polar Deep Water
UTC	Coordinated Universal Time
UUV	unmanned underwater vehicle
UVP	underwater video profiler
VACM	vector averaging current meter

VAD	vertical advection diffusion
VHSV	viral hemorrhagic septicemia virus
VIIRS	Visible and Infrared Imaging Radiometer Suite
VISR	Visible and Infrared Scanning Radiometer
VLCC	very large crude carrier
VMCM	vector measuring current meter
VMM	volume-weighted mean concentration
VMS	vehicle monitoring system
VOC	vapor-phase organic carbon or volatile organic compounds
VPR	video plankton recorder
VRM	viscous remanent magnetization
VSF	volume scattering function
WAIS	West Antarctic Ice Sheet
WASIW	Western Atlantic Subarctic Intermediate Water
WCP	World Climate Program
WCR	warm core ring
WHO	World Health Organization
WHOI	Woods Hole Oceanographic Institution
WIW	Winter Intermediate Water
WMDW	Western Mediterranean Deep Water
WMO	World Meteorological Organization
WNACW	West North Atlantic Central Water
WNPCW	Western North Pacific Central Water
WOCE	World Ocean Circulation Experiment
WSC	West Spitsbergen Current
WSDW	Weddell Sea Deep Water
WSPCW	Western South Pacific Central Water
WWGS	Winter Weddell Gyre Study
WWSP	Winter Weddell Sea Project
XBT	expendable bathythermograph
XCP	expendable current profiler
XCTD	expendable conductivity–temperature–depth probe
XKT	expendable optical irradiance probe
XSV	expendable sound velocity probe

INDEX

Notes

Cross-reference terms in italics are general cross-references, or refer to subentry terms within the main entry (the main entry is not repeated to save space). Readers are also advised to refer to the end of each article for additional cross-references - not all of these cross-references have been included in the index cross-references.

The index is arranged in set-out style with a maximum of three levels of heading. Major discussion of a subject is indicated by bold page numbers. Page numbers suffixed by T and F refer to Tables and Figures respectively. vs. indicates a comparison.

This index is in letter-by-letter order, whereby hyphens and spaces within index headings are ignored in the alphabetization. For example, 'oceanography' is alphabetized before 'ocean optics.' Prefixes and terms in parentheses are excluded from the initial alphabetization.

Where index subentries and sub-subentries pertaining to a subject have the same page number, they have been listed to indicate the comprehensiveness of the text.

Abbreviations used in subentries

AUV - autonomous underwater vehicle
$\delta^{18}O$ - oxygen isotope ratio
ENSO - El Niño Southern Oscillation
NADW - North Atlantic Deep Water
ROV - remotely operated vehicle
SST - sea surface temperature

Additional abbreviations are to be found within the index.

A

AAAS (American Association for the Advancement of Science), 29
AABW *see* Antarctic Bottom Water (AABW)
Abalone (*Haliotis* spp.)
  stock enhancement/ocean ranching programs, 227T, 231–232, 232F
'Acqua alta,' Italy, 459
Actinomycetes
  definition, 312
  marine-derived, 309
Adaptations
  noise masking, 149
Adductors, 191
Adée penguin (*Pygoscelis adeliae*),
  response to climate change, 533, 533F, 534F
  prehistoric, 529–530
Adjustable proportion fluid mixture (APFM), 293
Administrative costs, marine protected areas, 126, 126F

Admiralty law, 17
  Law of the Sea distinguished from, 17
  see also National Control and Admiralty Law
Adriatic Sea
  hypoxia, historical data, 346
Advective feedback, 497
A-E index (*Ammonia parkinsoniana* over *Elphidium* spp.), 349
  hypoxia, 349
Aerial traps, 70
Aerosols
  particles, atmospheric contaminant deposition, 384–385
Africa
  anthropogenic reactive nitrogen, 390–391, 391T
African Humid Period, 481
AFS (American Fisheries Society), 119
Aggregates, marine
  definition, 265
  gravel extraction *see* Offshore gravel mining
  sand extraction *see* Offshore sand mining

supply and demand outlook, 272
Agriculture, ocean thermal energy conversion, 296
Ahermatypic, definition, 400
Air
  pollution
    coral impact, 397
    see also Pollutants; Pollution
Airborne marine pollutants,
    environmental protection and Law of the Sea, 25–26
Air conditioning
  ocean thermal energy conversion, 296
Airlift mining system, 278, 279F
Air pollution
  coral impact, 397
  see also Pollutants; Pollution
Air–sea gas exchange
  carbon dioxide cycle, 487
  see also Carbon dioxide (CO$_2$)
Alaska
  earthquake (1964), 448F
  tsunami (1964), 444
    seafloor displacement, 447–448

Printed and bound by CPI Group (UK) Ltd, Croydon, CR0 4YY

03/10/2024

01040311-0012